All-Russian Conference with the School for Young Scientists Thermophysics and Physical Hydrodynamics 2016

AA006143

Journal of Physics: Conference Series Volume 754

Crimea
19-25 September 2016

Editors:

D. M. Markovich **S. V. Golovin**

ISBN: 978-1-5108-3178-0
ISSN: 1742-6588

Printed from e-media with permission by:

Curran Associates, Inc.
57 Morehouse Lane
Red Hook, NY 12571

Some format issues inherent in the e-media version may also appear in this print version.

Printed by Curran Associates, Inc. (2016)

For permission requests, please contact the Institute of Physics
at the address below.

Institute of Physics
Dirac House, Temple Back
Bristol BS1 6BE UK

Phone: 44 1 17 929 7481
Fax: 44 1 17 920 0979

techtracking@iop.org

Additional copies of this publication are available from:

Curran Associates, Inc.
57 Morehouse Lane
Red Hook, NY 12571 USA
Phone: 845-758-0400
Fax: 845-758-2633
Email: curran@proceedings.com
Web: www.proceedings.com

TABLE OF CONTENTS

PREFACE

HEAT TRANSFER AND HYDRODYNAMICS IN SINGLE PHASE FLOWS

HYDRODYNAMICS AND HEAT MASS TRANSFER IN MULTIPHASE FLOWS

PHASE TRANSITIONS

REACTING FLOW DYNAMICS, DETONATION PROCESSES

NUMERICAL METHODS IN THERMOPHYSICS AND PHYSICAL HYDRODYNAMICS

ELECTROPHYSICAL PHENOMENA IN GASEOUS AND LIQUID MEDIA

HEAT TRANSFER AND HYDRODYNAMICS IN INDUSTRIAL PROCESSES AND ENVIRONMENT PROTECTION

15 September 2016

Preface for the TPH 2016 – Thermophysics and physical hydrodynamics conference

PREFACE

Proceedings of the TPH 2016 – Thermophysics and physical hydrodynamics conference

D. M. Markovich, S.V. Golovin

The origins of the Conference start from 1970 in the Soviet Union, Novosibirsk. It was organized by Kutateladze Institute of Thermophysics SB RAS. The name of the conference was "Actual problems of thermophysics and physical hydrodynamics". The conference has been organized under this name up to 2015. The conference chairs were Academician V.E. Nakoryakov, Prof. D.M. Marckovich and Prof. S.V. Alekseenko. Peer reviewed proceedings of the conference have been published in the format of printed books. In 2016 the conference is reorganized in a new format with a shorter name: "Thermophysics and physical hydrodynamics". The conference took place in Yalta, a beautiful city in Crimea on the bank of the Black Sea. Lavrentev Institute of Hydrodynamics and the National committee on Heat and Mass Transfer are among other conference organizers besides Kutateladze Institute of Thermophysics.

The present Conference covers the following topics: heat transfer and hydrodynamics in single phase and multiphase flows, phase transitions, reacting flows, detonation processes, experimental and numerical techniques in thermophysics and physical hydrodynamics, heat transfer and hydrodynamics on micro- and nanoscale and in industrial processes. The proceedings contain 91 papers grouped by topic.

The scientific committee appreciates the enormous work of the editorial board and reviewers in the preparation of this volume. We would like to express our sincere thanks to all authors for their research contributions, and also to organizers of the conference for their valuable spadework.

Journal of Physics: Conference Series **754** (2016) 011002 doi:10.1088/1742-6596/754/1/011002

Peer review statement

All papers published in this volume of *Journal of Physics: Conference Series* have been peer reviewed through processes administered by the proceedings Editors. Reviews were conducted by expert referees to the professional and scientific standards expected of a proceedings journal published by IOP Publishing.

Microconvection in vertical channel at given heat flux

V B Bekezhanova[1,2] and I A Shefer[2]

[1] Institute of Computational Modeling SB RAS, Akademgorodok, 50/44, Krasnoyarsk, 660036, RUSSIA

[2] Institute of Mathematics and Computer Sciences, Siberian Federal University, Svobodny, 79, Krasnoyarsk, 660041, RUSSIA

E-mail: bekezhanova@mail.ru, ilya.shefer@gmail.com

Abstract. A problem on stability of the viscous heat-conducting liquid flow in the vertical channel at given heat flux on the permeable solid walls is studied. The two-dimensional flow is described by an exact invariant solution of the microconvection equations. The investigation of the exact solution allows one to find out the extent of influence of the thermal load, gravity and the system geometry on the flow structure. Stability of the solution is investigated in the framework of the linear theory. The spectrum of the spatial characteristic perturbations is analyzed in the space of problem parameters. Typical forms of the hydrodynamic and thermal disturbances are presented and dependence of characteristics of the arising structures on the thermal load and gravity is established. Convective cells, hydrothermal rolls and polygonal structures can appear in the channel. By weak gravity the hydrothermal rolls are not formed. Changing heat flux and disturbance wave length lead to deformation of the cells and complication of the spatial form of the structures.

1. Introduction

A complex of problems related with modeling of flows arises by prediction of the dynamics of different technological nonisothermal processes in which liquids are used as working media (e.g. crystal growing, thermal coating, fluidic cooling). One of the problems is the search of possibilities to control the flow through the external effects or choice of liquids with different thermophysical properties. By mathematical modeling we can obtain a priori evaluations of all characteristics. Fulfillment of the estimates allows one to realize stable regimes both in laboratory experiments and in industrial plants. For this purpose the problem on stability of convective regimes is studied. Mechanisms and character of possible crisis phenomena are investigated.

Studies in space materials science are motivated by opportunity to use long-term weightlessness in work processes. Under microgravity there is no the Archimedes force, which causes the density stratification in the field of gravity, natural convection is weak and does not lead to the mixing of liquid layers with various temperature. It opens up crucially new possibilities for obtaining qualitatively new (advantaged) materials, improvement of the existing ones and modification of the fluidic technologies.

The mathematical model option has the most important significance in the theoretical investigations. As a result of elaborations of the Oberbeck – Boussinesq approximation of the Navier – Stokes equations the models of microconvection [1, 2] were obtained. In [3, 4] the results of investigations of gravitational and non-gravitational convection were generalized.

Mathematical models of convection in microgravity and microscales and the results of the liquid motions analysis in the framework of these models were presented in [5, 6].

Dissimilar internal and external factors influence the origin and the characteristics of the convective flows. Interaction of the factors can lead to the growth of liquid motion intensity and the motion can become unstable. The most important condition of correct work of the experimental or industrial equipment using liquid system is the stable state of the working fluid. Thus, it is necessary to ensure the stability of the basic state (equilibrium or motion) of the working liquid medium. For example, in the processes of the crystal growing from molten mass the appearing cellular perturbations worsen quality and properties of the obtained pattern. Therefore, efficiency of the technological process can be increased by means of stabilization of convective flow of the liquid molten mass [7, 8, 9, 10].

By mathematical modeling a special attention is paid to construction and investigation of the exact solutions describing the convective flows. The importance of exact solutions is explained by opportunities of a rapid, accurate and effective analysis of the physical factors, defining the main flow structure. In the case of a multiparameter problem we can examine influence of the separate physical factors on the flow characteristics and criteria of its stability.

In the present paper the exact invariant solution of the microconvection equations describing the non-isothermal flow in a vertical layer is investigated. The properties of the spatial normal perturbations of the main flow and the influence of the gravity and external thermal load on their structure are studied.

2. Microconvection equations and exact invariant solution

We consider non-isothermal flow of viscous heat-conducting liquid in the vertical channel $\Omega = \{|\zeta| < 1, -\infty < \xi < \infty, -\infty < \vartheta < \infty\}$ with immovable solid walls $\zeta = \pm 1$.

2.1. Governing equations and basic parameters

For description of the flow the micronvection equations of the isothermally incompressible liquid [1] are used. In the non-dimensional form the equations are [6]

$$\boldsymbol{w}_\tau + \boldsymbol{w}\nabla\boldsymbol{w} + \varepsilon\,\mathrm{rot}\boldsymbol{w} \times \nabla\theta + \varepsilon^2\mathrm{div}(\nabla\theta\otimes\nabla\theta - |\nabla\theta|^2 I) =$$

$$= (1+\varepsilon\theta)(-\nabla q + \Delta\boldsymbol{w})\mathrm{Pr} - \varepsilon\boldsymbol{\eta}(\tau)\mathrm{Pr}\,\theta, \tag{1}$$

$$\mathrm{div}\boldsymbol{w} = 0, \tag{2}$$

$$\theta_\tau + \boldsymbol{w}\cdot\nabla\theta + \varepsilon|\nabla\theta|^2 = (1+\varepsilon\theta)\Delta\theta, \tag{3}$$

where $\boldsymbol{w}$, q are the modified velocity and the analogue of modified pressure, θ is the temperature, $\varepsilon = \beta\theta_*$ is the Boussinesq parameter (β, θ_* are the heat expansion coefficient and temperature drop, respectively), $\mathrm{Pr} = \nu/\chi$ is the Prandtl number (ν, χ are the viscosity and thermal diffusivity coefficients, respectively), $\boldsymbol{\eta} = a^3\boldsymbol{g}(t)/(\nu\chi)$ is the vector parameter of microconvection (a, $\boldsymbol{g}$ are the layer thickness and the gravitation vector, respectively), $\otimes$ denotes the tensor product. There is multiplier $\varepsilon\boldsymbol{\eta}(\tau)\mathrm{Pr}$ in equation (1) which defines vector of the Grashof numbers $\mathbf{Gr} = \varepsilon\boldsymbol{\eta}(\tau)\mathrm{Pr} = \beta\theta_* a^3\boldsymbol{g}(t)/\chi^2$. If $|\boldsymbol{g}| = g$ and g is the gravity acceleration then $\eta = a^3 g/(\nu\chi)$ and $\mathrm{Gr} = \beta\theta_* g a^3/\chi^2$. The microconvection parameter η characterizes the relation of the velocities orders originated by the volumetric expansion and buoyancy forces.

No-slip condition and heat flux are imposed at solid walls ($\zeta = 1$)

$$\boldsymbol{w} + \varepsilon\nabla\theta = 0, \qquad \partial\theta/\partial n = D(\boldsymbol{\zeta},\tau), \quad \boldsymbol{\zeta} = (\zeta,\xi,\vartheta), \tag{4}$$

where D is the non-dimensional heat flux.

To close the problem statement the condition of zero mass flow of the liquid through any cross section of the layer $\xi = \mathrm{const}$ is imposed for this solution.

2.2. Stationary solution

System (1) – (3) in plane stationary case admits invariant solution [11]

$$\boldsymbol{w} = (u(\zeta), v(\zeta), 0), \quad \theta = \theta(\zeta), \quad q = \varphi\xi + r(\zeta), \tag{5}$$

where $\varphi = \text{const}$, $r(\zeta)$ are to be defined.

Substitution of (5) into the governing equations, taking into account (4) and condition of zero mass flow, leads to the following relations

$$\varphi = 1 - \frac{p_1}{2p_2\mathrm{Pe}}, \quad r(\zeta) = \text{const}, \quad \theta(\zeta) = 1 - \mathrm{Pe}\,\zeta, \quad u(\zeta) = \text{const} = u_0,$$

$$v(\zeta) = \frac{p_1}{4p_2\mathrm{Pe}}\zeta^2 - \frac{p_3}{2\mathrm{Pe}^2}\zeta + \frac{p_1}{4p_2\mathrm{Pe}} - \frac{p_4}{2\mathrm{Pe}^2} + \frac{1+\varepsilon-\mathrm{Pe}\,\zeta}{\mathrm{Pe}^2}\left[\ln(1+\varepsilon-\mathrm{Pe}\,\zeta) - 1\right],$$

where $\mathrm{Pe} = u_0 a/\chi$ is the Peclet number and p_j constants ($j = 1, .., 4$) can be written as

$$p_1 = (\ln f_2^* - \ln f_1^*)\Big[f_1^*\ln f_1^* + f_2^*\ln f_2^* + \frac{1+\varepsilon}{\mathrm{Pe}}(f_1^*\ln f_1^* - f_2^*\ln f_2^*)\Big] + 4\mathrm{Pe},$$

$$p_2 = 1 + \varepsilon + (\ln f_2^* - \ln f_1^*)\Big(\frac{\mathrm{Pe}}{2} - \frac{(1+\varepsilon)^2}{2\mathrm{Pe}}\Big),$$

$$p_3 = f_1^*\ln f_1^* - f_2^*\ln f_2^* + 2\mathrm{Pe}, \quad p_4 = f_1^*\ln f_1^* + f_2^*\ln f_2^* - 2(1+\varepsilon),$$

$$f_1^* = 1 + \varepsilon - \mathrm{Pe}, \quad f_2^* = 1 + \varepsilon + \mathrm{Pe}.$$

Function q is defined up to a constant, we can assume $r \equiv 0$ without loss of generality. Furthermore, form of solution (4) restricts values of heat fluxes at the channel walls $D(-1) = D(1) = D$ and then $u_0 = -\varepsilon D/\eta$.

Thus, exact solution (5) describes a composite flow which is the superposition free-convective flow with a transverse uniform flux caused by seepage through permeable boundaries of the vertical channel. Flow consists of two counter fluxes. Convective circulation is in channel. Liquid rises near the hot wall and drop down near the cool one. Temperature is changed linearly. Equality of heat fluxes at walls is provided.

3. Stability of the microconvective flow

If the microconvection parameter $\eta < 1$, then we have a microconvective flow. Small values η can be achieved both due to the smallness of linear thickness of the channel a (convection in microscales) and due to the smallness of the gravity acceleration g (convection in microgravity). Stability of the microconvective flow in the vertical channel with permeable walls is investigated in the framework of the linear theory. Influence of the thermal load at boundaries channel and gravity on the typical form of the perturbations is studied.

3.1. Problem for the small disturbances

Let us suppose $\tilde{w}(\zeta,\tau) = \boldsymbol{w}(\zeta,\tau) + \boldsymbol{W}(\zeta,\tau)$, $\tilde{q}(\zeta,\tau) = q(\zeta,\tau) + Q(\zeta,\tau)$, $\tilde{\theta} = \theta(\zeta,\tau) + T(\zeta,\tau)$, where $\boldsymbol{W} = (U, V, W)$, Q, T are the small non-stationary disturbances, $\boldsymbol{w}(\zeta,\tau)$, $q(\zeta,\tau)$, $\theta(\zeta,\tau)$ are basic solutions (5). Now functions $\tilde{w}, \tilde{q}, \tilde{\theta}$ are the solutions of problem (1) – (3), (4).

We consider the normal spatial perturbations

$$(\mathbf{W}, Q, T) = (\mathbf{W}(\zeta), Q(\zeta), T(\zeta)) \cdot \exp\left[i(\alpha_1\xi + \alpha_2\vartheta - \mathrm{C}\tau)\right] \tag{6}$$

with wavenumbers α_1, α_2 along the ξ and ϑ axes, respectively, and complex decrement $\mathrm{C} = C_r + iC_i$, determining the perturbations development with time. Linearized problem for the amplitudes of disturbances (6) has the following form

$$-1<\zeta<1: \quad i(\alpha_1 v - \mathrm{C})U + \mathrm{Pe}\, U' - [i\alpha_1\, \varepsilon\, v_\zeta + \varepsilon^2\theta_\zeta(\alpha_1^2+\alpha_2^2)]T =$$

$$= (1+\varepsilon\theta)[-Q' + U'' - (\alpha_1^2+\alpha_2^2)U]\mathrm{Pr}, \tag{7}$$

$$i(\alpha_1 v - \mathrm{C})V + (\mathrm{Pe}+\varepsilon\theta_\zeta)V' + (v_\zeta - i\alpha_1\,\varepsilon\theta_\zeta)U + (\varepsilon v_\zeta - i\alpha_1\varepsilon^2\theta_\zeta)T' +$$

$$+ i\alpha_1\varepsilon^2\theta_{\zeta\zeta}T = (1+\varepsilon\theta)[-i\alpha_1 Q + V'' - (\alpha_1^2+\alpha_2^2)V]\mathrm{Pr} + \frac{\mathrm{Gr}\, T}{1+\varepsilon\theta}, \tag{8}$$

$$i(\alpha_1 v - \mathrm{C})W + (\mathrm{Pe}+\varepsilon\theta_\zeta)W' - i\alpha_2\varepsilon\theta_\zeta U - i\alpha_2\varepsilon^2\theta_\zeta T' =$$

$$= (1+\varepsilon\theta)[-i\alpha_2 Q + W'' - (\alpha_1^2+\alpha_2^2)W]\mathrm{Pr}, \tag{9}$$

$$U' + i\alpha_1 V + i\alpha_2 W = 0, \tag{10}$$

$$i(\alpha_1 v - \mathrm{C})T + (\mathrm{Pe}+2\varepsilon\theta_\zeta)T' + \theta_\zeta U = (1+\varepsilon\theta)[T'' - (\alpha_1^2+\alpha_2^2)T] + \varepsilon\theta_{\zeta\zeta}T, \tag{11}$$

$$|\zeta| = 1: \quad U + \varepsilon T' = 0, \quad V = 0, \quad W = 0, \quad T = 0. \tag{12}$$

where prime denotes the differentiation with respect to ζ. The obtained problem (7) – (12) is the spectral problem with respect to complex decrement C and defines spatial perturbations depending on α_1, α_2, ε, Pe, Pr and Gr parameters.

3.2. Analogue of the Squire transformation

The transformations reducing problem (7) – (12) to the equivalent 2D problem with additional parameter characterizing the spatial perturbations are found. Actually, if (8) and (9) are multiplied by $i\alpha_1$ and $i\alpha_2$, respectively, and the obtained products are summed then the result is

$$i(\alpha_1 v - \mathrm{C})\widetilde{Z} + (\mathrm{Pe}+\varepsilon\theta_\zeta)\widetilde{Z}' + (i\alpha_1 v_\zeta + \alpha^2\,\varepsilon\theta_\zeta)U + (i\alpha_1\varepsilon v_\zeta + \alpha^2\varepsilon^2\theta_\zeta)T' -$$

$$-\varepsilon^2\alpha^2\theta_{\zeta\zeta}T = (1+\varepsilon\theta)[\alpha^2 Q + \widetilde{Z}'' - \alpha^2\widetilde{Z}]\mathrm{Pr} + i\alpha_1\frac{\mathrm{Gr}\, T}{1+\varepsilon\theta}, \tag{13}$$

where

$$\widetilde{Z} = i\alpha_1 V + i\alpha_2 W, \quad \alpha^2 = \alpha_1^2 + \alpha_2^2. \tag{14}$$

The parameter $\lambda = \alpha_1/\alpha$ is introduced to characterize the spatial perturbations and $\lambda \in [0,1]$. Value $\lambda = 0$ corresponds to perturbations with $\alpha_1 = 0$, which are periodical with respect to ϑ and do not depend on longitudinal coordinate ξ. The disturbances are not plane one because $V \neq 0$ in this case. In the stationary regime liquid particles move along helix line whose axes are parallel to the direction of the basic motion (so called "spiral" perturbations). Other limiting case $\lambda = 1$ corresponds to plane perturbations with $\alpha_2 = 0$.

The following transformations are supplemented

$$\widetilde{U} = i\alpha U, \quad \widetilde{Q} = i\alpha Q, \quad \widetilde{T} = i\alpha T, \tag{15}$$

then (13) takes form

$$i(\alpha\lambda v - \mathrm{C})\widetilde{Z} + (\mathrm{Pe}+\varepsilon\theta_\zeta)\widetilde{Z}' + (\lambda v_\zeta - i\alpha\,\varepsilon\theta_\zeta)\widetilde{U} + (\varepsilon\lambda v_\zeta - i\alpha\varepsilon^2\theta_\zeta)\widetilde{T}' +$$

$$+ i\alpha\varepsilon^2\theta_{\zeta\zeta}\widetilde{T} = (1+\varepsilon\theta)\left[-i\alpha\widetilde{Q} + \widetilde{Z}'' - \alpha^2\widetilde{Z}\right]\mathrm{Pr} + \frac{\lambda\mathrm{Gr}\,\widetilde{T}}{1+\varepsilon\theta}. \tag{16}$$

Equations (7), (10), (11) are rewritten

$$i(\alpha\lambda v - \mathrm{C})\widetilde{U} + \mathrm{Pe}\,\widetilde{U}' - [i\alpha\,\varepsilon\,\lambda v_\zeta + \varepsilon^2\theta_\zeta\alpha^2]\widetilde{T} = (1+\varepsilon\theta)[-\widetilde{Q}' + \widetilde{U}'' - \alpha^2\widetilde{U}]\mathrm{Pr}, \tag{17}$$

$$\widetilde{U}' + i\alpha\widetilde{Z} = 0, \tag{18}$$

$$i(\alpha\lambda v - \mathrm{C})\widetilde{T} + (\mathrm{Pe} + 2\varepsilon\theta_\zeta)\widetilde{T}' + \theta_\zeta\widetilde{U} = (1+\varepsilon\theta)[\widetilde{T}'' - \alpha^2\widetilde{T}] + \varepsilon\theta_{\zeta\zeta}\widetilde{T}. \tag{19}$$

It is obviously that structure of the equations for new functions $\widetilde{U}$, $\widetilde{Z}$, $\widetilde{Q}$ $\widetilde{T}$ is the same that one for U, V, Q T in the case of plane problem. In addition $\widetilde{U}, \widetilde{Z}, \widetilde{T}$ functions defined by (14) and (15) satisfy the boundary conditions which are equivalent to (12):

$$|\zeta| = 1: \qquad \widetilde{U} + \varepsilon\widetilde{T}' = 0, \quad \widetilde{Z} = 0, \quad \widetilde{T} = 0. \tag{20}$$

Thus, obtained transformation (14), (15) is the analogue of the Squire transformation [12] reducing the problem on the stability about three-dimensional perturbations to the corresponding two-dimensional problem. The parameter λ is to be fixed one for calculation of the characteristics of the spatial perturbations through solutions of the plane problem.

3.3. Stability of the shortwave "spiral" perturbations

Stability of the basic flow as concerning shortwave "spiral" ($\lambda = 0$) perturbations is proven analytically. Actually, let us consider amplitude equations (16) – (19) under $\lambda = 0$ and $\alpha \to \infty$. Terms containing C, the highest derivative and the highest powers of α are remained in the equations, then for $\widetilde{Z}$ function we have the following boundary value problem

$$\widetilde{Z}'' = \left(\alpha^2 - \frac{i\mathrm{C}}{\mathrm{Pr}(1+\varepsilon\theta)}\right)\widetilde{Z} = 0, \quad \widetilde{Z}(-1) = \widetilde{Z}(1) = 0. \tag{21}$$

Equation (21) is multiplied by complex conjugate value $\widetilde{Z}^*$ and product is integrated over the segment $[-1, 1]$. Taking into account the boundary conditions we obtain

$$\int_{-1}^{1} |\widetilde{Z}'|^2\, d\zeta = \frac{i\mathrm{C}}{\mathrm{Pr}} \int_{-1}^{1} \frac{|\widetilde{Z}|^2}{1+\varepsilon\theta}\, d\zeta - \alpha^2 \int_{-1}^{1} |\widetilde{Z}|^2\, d\zeta.$$

The latter is rewritten in the form

$$\frac{i\mathrm{C}}{\mathrm{Pr}} \int_{-1}^{1} \frac{|\widetilde{Z}|^2}{1+\varepsilon\theta}\, d\zeta = \int_{-1}^{1} |\widetilde{Z}'|^2\, d\zeta + \alpha^2 \int_{-1}^{1} |\widetilde{Z}|^2\, d\zeta. \tag{22}$$

Decrement C is the pure imaginary number with the negative imaginary part from (22). The result is right in virtue of the positivity of the all subintegral terms and does not depend on temperature function θ for the basic flow. Thus, all shortwave "spiral" perturbations fade out monotonically with time.

4. Typical forms of perturbations

Calculations of the spectrum of perturbations for the melt of silicon dioxide was performed by the method of stepwise integration with orthogonalization [13]. In fact, for the specific model medium the variations of λ and α correspond to the changes of configuration of perturbations waves, Pe connects with the transversal velocity or/and channel thickness, Gr relates to the gravity action or/and channel thickness. Taking into account dependence of the transversal velocity u_0 on heat flux value D (see subsection 2.1) the contribution of the given heat flux into the change of the typical form of the perturbations is also characterized by the Peclet number Pe.

We assume the silicon dioxide melt fills the vertical channel. Walls of the channel are permeable and constant temperature drop θ_* between them are kept, on the walls the constant heat flux are held. The following values of physical parameter for chosen model medium are used: $\nu = 2.65 \cdot 10^{-3}$ cm^2/s, $\chi = 0.49$ cm^2/s, $\beta = 0.75 \cdot 10^{-5}\,{}^o$K^{-1}, $k = 0.042$ W/(cm·K). Calculations were performed under $\mathrm{Pr} = 5.41 \cdot 10^{-3}$ and $\varepsilon = 7.5 \cdot 10^{-5}$ in the all cases.

In Fig. 1 – 6 the typical hydrodynamic and thermal perturbations under different values of heat flux on the channel boundaries, gravity acceleration, wave length and λ parameter are shown. Values of the Peclet number in captions to the figures correspond to different values of the heat flux: $\mathrm{Pe}_1 = 8.93 \cdot 10^{-5}$ for 5 W/cm^2, $\mathrm{Pe}_2 = 35.71 \cdot 10^{-5}$ for 20 W/cm^2, $\mathrm{Pe}_3 = 178.57 \cdot 10^{-5}$ for 100 W/cm^2.

Analysis of the eigenfunctions reveals the perturbations are the convective cells, hydrothermal rolls or deformed structures of the complex spatial form. Type of patterns depends on the problem parameters. Spatial sizes depend on wave length of disturbances ($\lambda_w = 2\pi/\alpha$). The formation of the oscillatory regimes is possible in the considered system. Arising structures oscillate periodically ($C_r \neq 0$) and vanish with time ($C_i < 0$). Existence of the oscillatory regimes is possible in virtue of properties of the spectral problem (7) – (12) and related with its non-self-conjugation.

Fig. 1, 2 present the disturbances which are formed in the channel with thickness $a = 0.1$ cm under gravity acceleration $g = 2g_0 \cdot 10^{-3}$, $g_0 = 981$ cm/s^2. The hydrothermal rolls arise due to the interaction of the perturbations with basic flow in the channel under weak thermal load (Fig. 1a, 2a). Liquid particles rise up along the roll axis to the hot wall direction and go down to the cold wall direction. With increasing of the value of heat flux (and, consequently, the transversal velocity) the spatial form of the structures is complicated (Fig. 1b, 2b). Under essential thermal load the transition to the polygonal cells is observed. In the case the form of disturbances is entirely defined by growing transversal velocity (Fig. 1c, 2c).

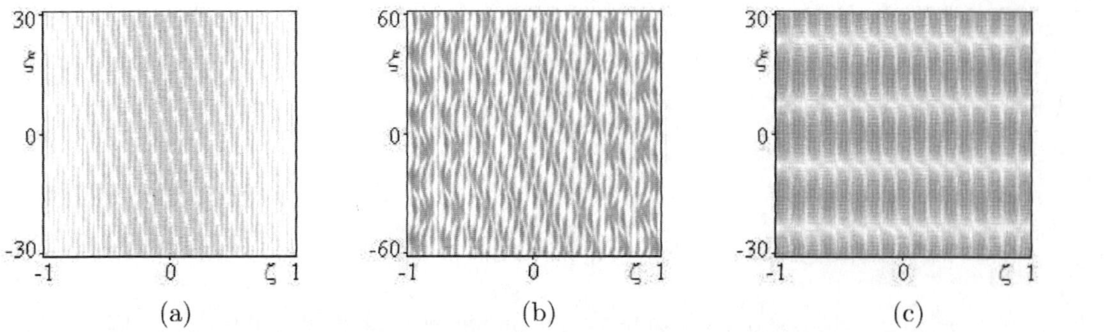

Figure 1. Distribution of the hydrodynamic perturbations in the vertical layer under $\eta = 1.51$, $\mathrm{Gr} = 61.27 \cdot 10^{-8}$: (a) — Pe_1, $\lambda = 1/\sqrt{2}$, $\alpha_1 = 0.2$; (b) — Pe_2, $\lambda = 0.2$, $\alpha_1 = 0.2$; (c) — Pe_3, $\lambda = 1/\sqrt{2}$, $\alpha_1 = 0.2$.

The hydrodynamic and thermal perturbations presented in Fig. 3, 4 were obtained for $a = 0.1$ cm, $g = g_0 \cdot 10^{-4}$ cm/s^2. With attenuation of the gravitational action the hydrothermal rolls is changed by large-scale convective cells (Fig. 3a), which break down by growing the thermal load. The large-scale cells are changed by patterns of the complex spatial form (Fig. 3b). Further increasing the heat flux at boundaries (and, consequently, the transversal velocity) leads

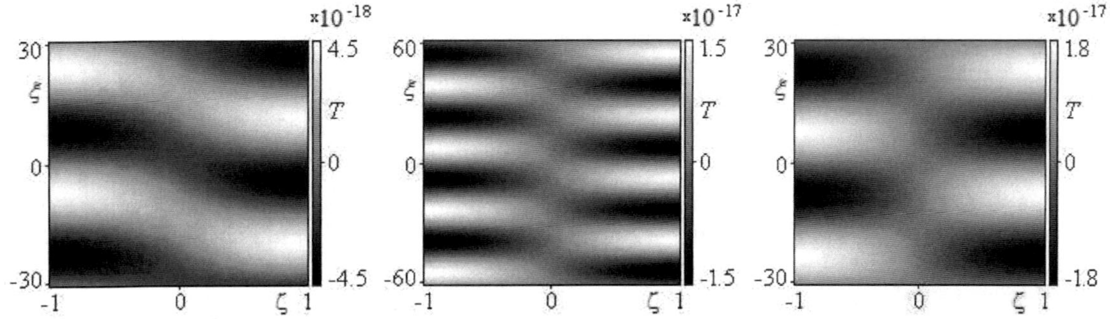

(a) (b) (c)

Figure 2. Distribution of the thermal perturbations in the vertical layer under $\eta = 1.51$, $\mathrm{Gr} = 61.27 \cdot 10^{-8}$: (a) — Pe_1, $\lambda = 1/\sqrt{2}$, $\alpha_1 = 0.2$; (b) — Pe_2, $\lambda = 0.2$, $\alpha_1 = 0.2$; (c) — Pe_3, $\lambda = 1/\sqrt{2}$, $\alpha_1 = 0.2$.

to formation of the convective cells (Fig. 3c). Temperature perturbations result in appearance of the thermal rolls (Fig. 4a-c).

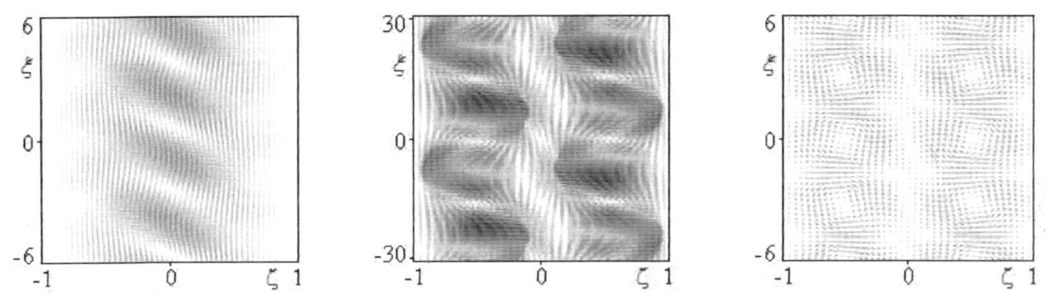

(a) (b) (c)

Figure 3. Distribution of the hydrodynamic perturbations in the vertical layer under $\eta = 0.0755$, $\mathrm{Gr} = 3.06 \cdot 10^{-8}$: (a) — Pe_1, $\lambda = 1$, $\alpha_1 = 1$; (b) — Pe_2, $\lambda = 0.2$, $\alpha_1 = 0.2$; (c) — Pe_3, $\lambda = 1$, $\alpha_1 = 1$.

The hydrodynamic and thermal perturbations presented in Fig. 5a, 6a are obtained for $a = 0.1$ cm, $g = g_0 \cdot 10^{-4}$, in Fig. 5b-c, 6b-c, for $a = 1$ cm, $g = g_0 \cdot 10^{-6}$. Decrease of the wave numbers causes the size reduction of the patterns. At the weak gravity the thermal rolls are changed by thermal "spots" (Fig. 6). The presence of the significant transversal velocity prevents the forming hydrothermal rolls.

It should be noted that the hydrothermal rolls are also destructed at weaking gravity force. Changing of thermal load and configuration of the perturbations wave can lead to deformation of the convective cells and more complex spatial form of the arising structures. Defining typical form of the perturbations and crisis mechanisms will allow one to solve the problem on suppression of the disturbances in the melt under obtaining ultra-pure high-quality crystals.

Acknowledgments
This work was supported by the Russian Foundation for Basic Research (project No. 14-01-00067). The first author also acknowledges the Government of the Russian Federation, for

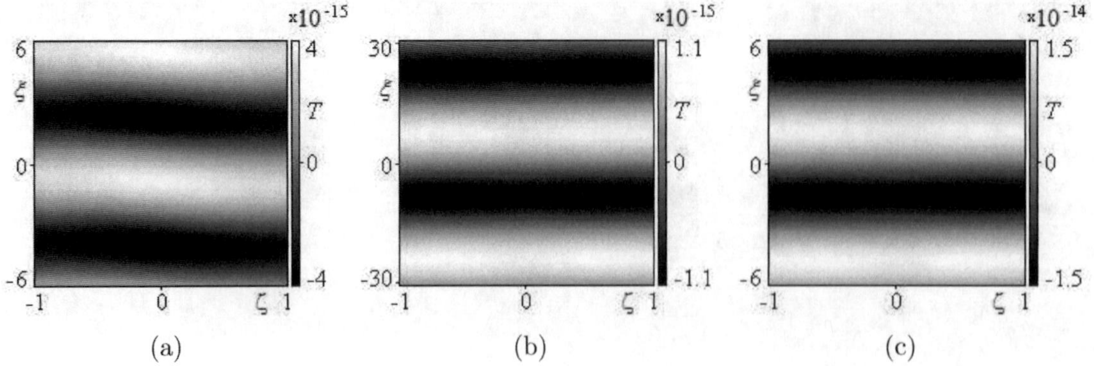

Figure 4. Distribution of the thermal perturbations in the vertical layer under $\eta = 0.0755$, $\mathrm{Gr} = 3.06 \cdot 10^{-8}$: (a) — Pe_1, $\lambda = 1$, $\alpha_1 = 1$; (b) — Pe_2, $\lambda = 0.2$, $\alpha_1 = 0.2$; (c) — Pe_3, $\lambda = 1$, $\alpha_1 = 1$.

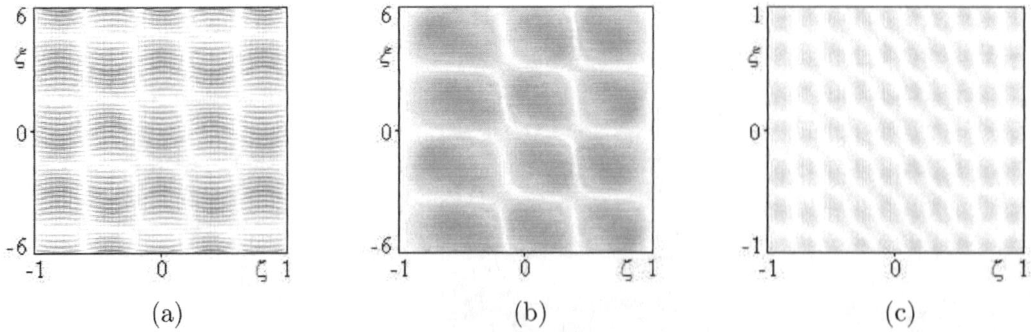

Figure 5. Distribution of the hydrodynamic perturbations in the vertical layer under $\mathrm{Pe} = 178.57 \cdot 10^{-5}$: (a) — $\eta = 0.0755$, $\mathrm{Gr} = 3.06 \cdot 10^{-8}$, $\lambda = 0.2$, $\alpha_1 = 1$; (b) — $\eta = 0.755$, $\mathrm{Gr} = 30.63 \cdot 10^{-8}$, $\lambda = 0.2$, $\alpha_1 = 1$; (c) — $\eta = 0.755$, $\mathrm{Gr} = 30.63 \cdot 10^{-8}$, $\lambda = 0.2$, $\alpha_1 = 10$.

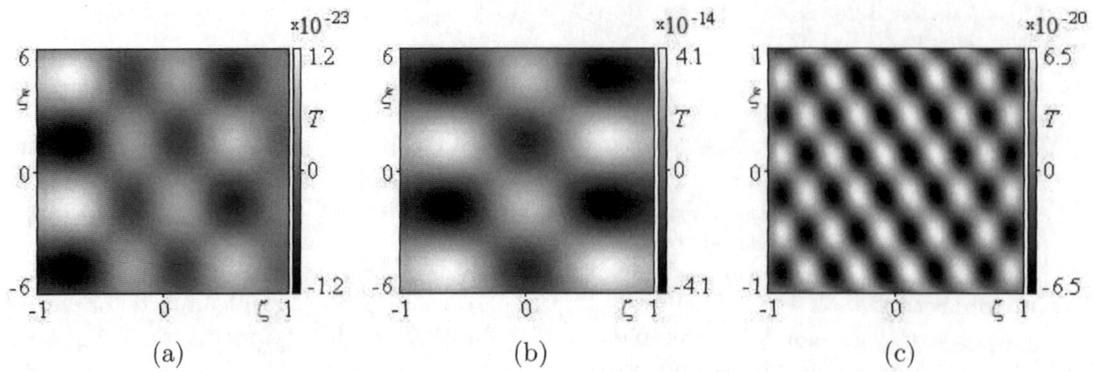

Figure 6. Distribution of the thermal perturbations in the vertical layer under $\mathrm{Pe} = 178.57 \cdot 10^{-5}$: (a) — $\eta = 0.0755$, $\mathrm{Gr} = 3.06 \cdot 10^{-8}$, $\lambda = 0.2$, $\alpha_1 = 1$; (b) — $\eta = 0.755$, $\mathrm{Gr} = 30.63 \cdot 10^{-8}$, $\lambda = 0.2$, $\alpha_1 = 1$; (c) — $\eta = 0.755$, $\mathrm{Gr} = 30.63 \cdot 10^{-8}$, $\lambda = 0.2$, $\alpha_1 = 10$.

partial support, grant for studying under the supervision of leading scientists in the Siberian Federal University (project No. 14.Y26.31.0006).

References

[1] Pukhnachev V V 1992 *Modeling in Mechanics* **6(23)** 47-56 (in Russian)
[2] Perera P S, Sekerka R F 1997 *Phys. Fluids* **9(2)** 376-391
[3] Polezhaev V I, Bello M S, Verezub N A, Dubovik K G, Lebedev A P, Nikitin S A, Pavlovskii D S, Fedyushkin A I 1991 *Convective Processes in Migrogravity* (Moscow: Nauka)
[4] Polezhaev V I 2006 *Fluid Dynamics* **41(5)** 736-754
[5] Monti R 2001 *Physics of Fluids in Microgravity* (London: Taylor & Francis)
[6] Andreev V K, Gaponenko Yu A, Goncharova O N, Pukhnachev V V 2012 *Mathematical Models of Convection* (Berlin/Boston: Walter de Gruyter)
[7] Müller G 1988 *Convection and Inhomogeneities in Crystal Growth from the Melt* (Berlin: Springer-Verlag)
[8] Lebedev A P, Polezhaev V I 1990 *Uspekhi Mekhaniki* **1** 3-51 (in Russian)
[9] Bessonov O A, Brailovskaya V A, Polezhaev V I 1997 *Fluid Dynamics* **32(3)** 379-386
[10] Nikitin N V, Nikitin S A, Polezhaev V I 2003 *Uspekhi Mekhaniki* **4** 63-105 (in Russian)
[11] Andreev V K, Kaptsov O V, Pukhnachev V V, Rodionov A A 1998 *Applications of Group-Theoretical Methods in Hydrodynamics* (Netherlands: Kluwer Academic Publishers)
[12] Squire H B, 1933 *Proc. Roy. Soc.* **A 142** 621-628
[13] Godunov S K, 1961 *Uspekhi Math. Nauk* **16(3)** 171-174 (in Russian)

Mathematical simulation of nonisothermal filling of plane channel with non-Newtonian fluid

E Borzenko[1], K Ryltseva[2], O Frolov[1] and G Shrager[3]
[1] Associate professor, Tomsk State University, Tomsk, Russia
[2] Student, Tomsk State University, Tomsk, Russia
[3] Professor, Tomsk State University, Tomsk, Russia

E-mail: borzenko@ftf.tsu.ru

Abstract. In this paper, the fountain flow of a non-Newtonian fluid during the filling of a plane vertical channel with due account of dissipative heating is investigated. The rheological features of the medium are defined by Ostwald de Waele power-law with exponential temperature dependence of viscosity. The numerical solution of the problem is obtained using a finite-difference method, based on the SIMPLE algorithm, and the method of invariants for compliance with the natural boundary conditions on free surface. It was shown that the flow separates into a two-dimensional flow zone in the vicinity of the free surface and a one-dimensional flow zone away from it. The parametrical investigations of kinematic and thermophysical properties of the flow and the dependence of the free surface behavior on the basic criteria and rheological parameters are implemented.

1. Introduction

During the processing of polymeric materials the injection molding, based on the mold filling with a liquid medium, is a widespread technology. The flow of processed composition is characterized by a complicated rheological behavior and, generally, under nonisothermal conditions. The forecasting of quantity of the products requires the prior investigations of the technological process, which can be implemented by the physical and mathematical simulation methods. The mathematical simulation method is preferable as it is more efficient and informative than the physical simulation, which is used for verification of the theoretical results on the last investigation stage. The process of filling is accompanied by an existence of the free surface, nonisothermicity, which is connected with dissipative heating, chemical conversion, heat-exchange boundary conditions, and dependence of the rheological characteristics of the medium on the temperature. The results of researches of nonisothermal complicated rheological media flow in the channels without considering the free surface are introduced in [1, 2]. There are articles [3-6], which contain the description of computational technologies and the results of filling process investigations with taking into account the free surface. The aim of this work is to investigate the influence of the medium rheological parameters on the shape of free surface, and the kinematic of a fountain flow during the filling of the plane channel on the base of rheological Ostwald-de Waele equation.

2. Formulation of the problem

The nonisothermal filling of a planar vertical channel with a fluid of the complicated rheology in a gravity field is considered. The fluid flow direction is opposite to the gravity force. The flow area is shownin Fig. 1. The mathematical basis of the flow is formed by motion, heat-transfer, and continuity equations written in dimensionless variables as follows:

$$\mathrm{Re}\frac{d\mathbf{V}}{dt} = -\nabla p + \nabla\bullet(2B\mathbf{E}) + \mathbf{W},\ \mathrm{Pe}\frac{dT}{dt} = \Delta T + \mathrm{Br}BA^2,\ \nabla\bullet\mathbf{V} = 0.$$

Here, **V** is the dimensionless velocity vector; p is the dimensionless pressure; T is the dimensionless temperature; t is the dimensionless time; **W** = {W, 0} is a dimensionless vector; A is the dimensionless intensity of a strain-rate tensor **E**. The system of equations is completed by a rheological power-law, which defines the apparent viscosity as follows [6]:

$$B = \exp\left[-CT\right]A^{m-1},$$

where m is the nonlinearity degree.

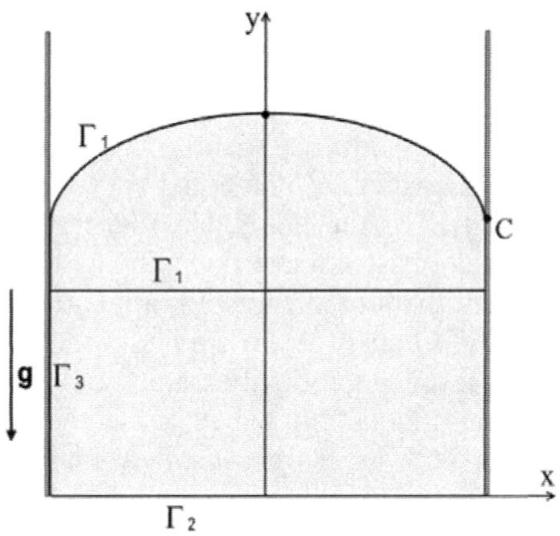

Figure 1. Calculation domain

The dimensionless similarity criteria are given by formulas

$$\mathrm{Re} = \frac{\rho L^m U_0^{2-m}}{\mu_0},\ \mathrm{Pe} = \frac{c\rho UL}{\lambda},\ \mathrm{W} = \frac{\rho L^{m+1} g}{\mu_0 U_0^m},\ \mathrm{Br} = \frac{\mu_0 U^{m+1} L^{1-m}}{\lambda T_0},\ \mathrm{C} = kT_0.$$

Here, ρ is the fluid density; L is the half-width of the channel; U_0 is the average velocity in the inlet section; μ_0 is the power-law rheological parameter; k is the parameter of nonisothermicity; c is the heat capacity; λ is the heat conductivity coefficient; g is the gravity acceleration; T_0 is the dimensional wall temperature.

At the inlet boundary Γ_2 the velocity and the temperature profiles, which correspond to a one-dimensional nonisothermal fluid flow with a specified constant flow rate in the infinite channel, are assigned. The typical longitudinal velocity and temperature distributions at the inlet boundary are illustrated in Fig. 2. On the rigid walls Γ_3 the no-slip boundary conditions are realized, and the dimensionless temperature is set to zero. On the free surface Γ_1 the boundary conditions are the equality of the shear stress to zero, the equality of normal stress to the external pressure, and the absence of the heat flux. At the initial time moment, the free surface is plane and it is located in a distance from the inlet boundary, and the temperature of fluid is equal to the temperature of the wall in the area of initial filling. Moreover, the free boundary moves in compliance with the kinematic condition. The surface tension forces are not considered. The problem formulation is dimensionless. The values of dimensionless criteria are assigned so that the stationary solution for the nonisothermal flow in the plane infinite channel could exist [7].

Consequently, the boundary conditions are written as follows:

$$\Gamma_1: \frac{\partial u_n}{\partial s} + \frac{\partial u_s}{\partial n} = 0, \quad p = 2B\frac{\partial u_n}{\partial n}, \quad \frac{\partial T}{\partial n} = 0; \tag{1}$$

$$\Gamma_2: u = 0, \quad v = f_1(x), \quad T = f_2(x);$$

$$\Gamma_3: u = 0, \quad v = 0, \quad T = 0;$$

$$\Gamma_4: u = 0, \quad \frac{\partial v}{\partial x} = 0, \quad \frac{\partial T}{\partial x} = 0.$$

The conditions (1) are written in a local Cartesian coordinate system (n, s), which is normally related with the free surface. The free boundary Γ_1 moves in compliance with the kinematic condition written in the Lagrangian representation as follows:

$$\frac{dx}{dt} = u, \quad \frac{dy}{dt} = v. \tag{2}$$

The analysis of the classical mathematical model of fluid dynamics using the motion equations, natural boundary conditions on the free surface, and no-slip conditions on the moving contact line for the values of the dynamic contact angle, which is not equal to 0 and π, has shown that in the determination of the dynamic characteristics of the flows, singularities arise that lead to an infinite increase in the flow parameters as the three-phase contact line (TPCL) is approached [8, 9].Therefore, the problem formulation takes into account that at the contact point C the condition of the shear stress absence is realized (1), the normal velocity is equal to zero, and in the small vicinity of the contact line the tangential velocity on the wall decreases asymptotically from the value at the point C to zero. It is supposed, that the flow regime and the interaction behavior of the phases on the contact line correspond to the edge angle equal to π with an insignificant capillary effects in comparison with inertial, viscous, and gravitational forces in the flow. Using such a boundary conditions eliminates the speciality of the traditional formulation of the problem and at the same time has a little effect on the kinematic characteristics of the flow is a small vicinity of the contact point [10].

3. Method of solution

The formulated problem is solved using a finite-difference method. A staggered difference grid covers the solution domain. The SIMPLE algorithm [11] is applied to calculate the unknown variables at the internal nods. Near the free surface the irregular nodes appear. In these nodes, the values of unknown variables are defined using linear interpolation of data from the free surface and regular nodes. The free boundary is represented as a set of uniformly distributed marker particles: the first marker is located on the line of symmetry and the last marker is on the TPCL. The velocity components are calculated using the method of invariants [12], which is based on the assumption of no shear stresses and the continuity equations for each marker that designate the free boundary at a given time moment. The markers of the free surface, except for the last one, move in accordance with the discrete analog of the kinematic condition (2). The motion of the contact point is implemented in accordance with the slip condition and the value of the dynamical edge angle is equal to π.

The calculation method was tested on the problem of power-law fluid flow in a planar channel at a specified flow rate with consideration for dissipative heating and exponential temperature dependence of viscosity. The parabolic velocity profile and zero temperature were specified at the inlet of a channel, and the soft boundary conditions were specified at the outlet of a channel. On the rigid walls, the no-slip boundary conditions were realized and the temperature was set to zero. In this case, a length of the channel was selected that it was sufficient to set steady state flow in the cross section of the outlet.

The results of calculations were compared with the solution of a one-dimensional problem that described the nonisothermal fluid flow in the infinite channel at a specified flow rate. The mathematical statement of that problem included the following system of equations:

$$\frac{d}{dx}\left(e^{-CT}\left|\frac{dv}{dx}\right|^{m}\right) = -\delta p, \quad \frac{d^2T}{dx^2} + \mathrm{Br}e^{-CT}\left|\frac{dv}{dx}\right|^{m+1} = 0, \tag{3}$$

and the boundary conditions on the rigid wall were similar to the conditions of two-dimensional mathematical statement. Here, δp is the pressure drop along the channel. The equation for the temperature calculation [7] is carried out from the differential equation system (3)

$$\frac{d^2T}{dx^2} + \mathrm{Br}\exp(CT/m)\left(\delta p\, x\right)^{1+1/m} = 0 .$$

The approximate solution of this equation with required accuracy is obtained using the sweep method. For calculation of the velocity the numerical integration of the first equation in system (3) is implemented. The comparison of the calculated data with a solution of the one-dimensional problem is shown in Fig. 2. Table 1 contains a quantitative confirmation of approximation convergence based on the values of the relative errors $e_1$ande_2, which are defined using the formulas

$$e_1 = \frac{v_{axis}^{numerical} - v_{axis}^{approximate}}{v_{axis}^{approximate}} \cdot 100\%,\ e_2 = \frac{T_{axis}^{numerical} - T_{axis}^{approximate}}{T_{axis}^{approximate}} \cdot 100\%,$$

where $v_{axis}^{numerical}$, $T_{axis}^{numerical}$ are the values on the centerline at the outlet boundary; $v_{axis}^{approximate}$, $T_{axis}^{approximate}$ are the results of a one-dimensional problem solution.

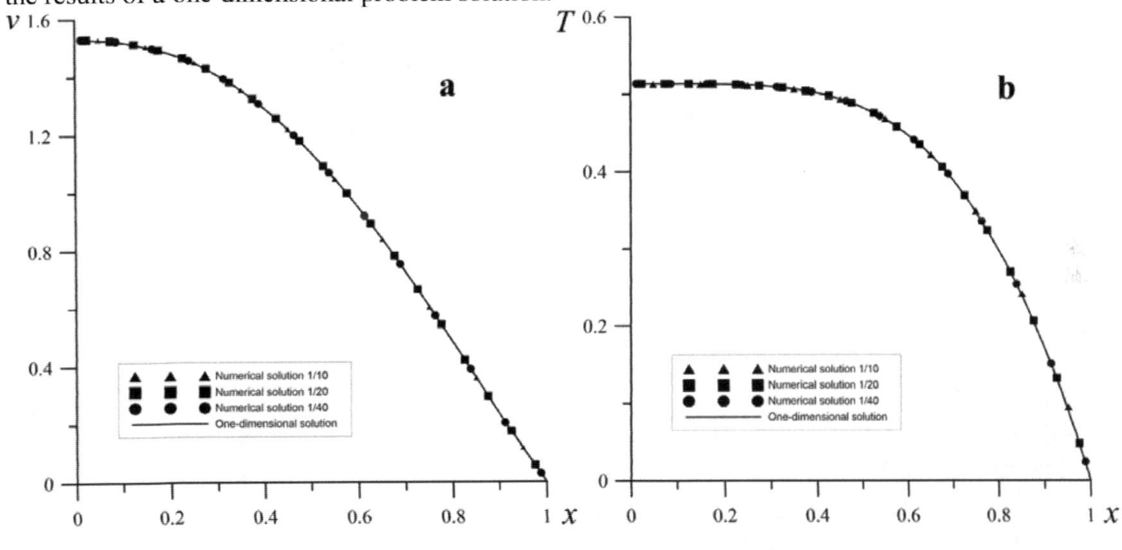

Figure 2. Distributions of the velocity (a) and temperature (b) at the outlet boundary (Re=0.1, W=0, Pe=100, Br=1, C=1)

Table 1. Values of the relative errors $e(h)$, where h is a step of the square computational mesh

h	$e_1(h)$	$e_2(h)$
1/10	0.102	0.186
1/20	0.083	0.046
1/40	0.075	0.027

4. Results and discussion

According to the flow behavior, the stream can be separated into a three-dimensional flow zone near the free surface and a one-dimensional flow zone far away from it. A parabolic longitudinal velocity profile typical for a steady state flow in the channel is realized in a one-dimensional flow zone. The deceleration of liquid particles that approach the flow front leads to the appearance of transverse velocity component, and the velocity vector turns towards the rigid wall. In course of time, the stable state distribution of the characteristics is developed in the fountain flow zone (Fig. 3).

Journal of Physics: Conference Series **754** (2016) 022002 doi:10.1088/1742-6596/754/2/022002

The effect of the Br parameter and nonlinearity degree *m* on the steady state temperature distribution near the free surface is shown in Fig. 4. It is evident that the temperature maximum is located in a fountain flow zone and its value increases with an increase of Br.

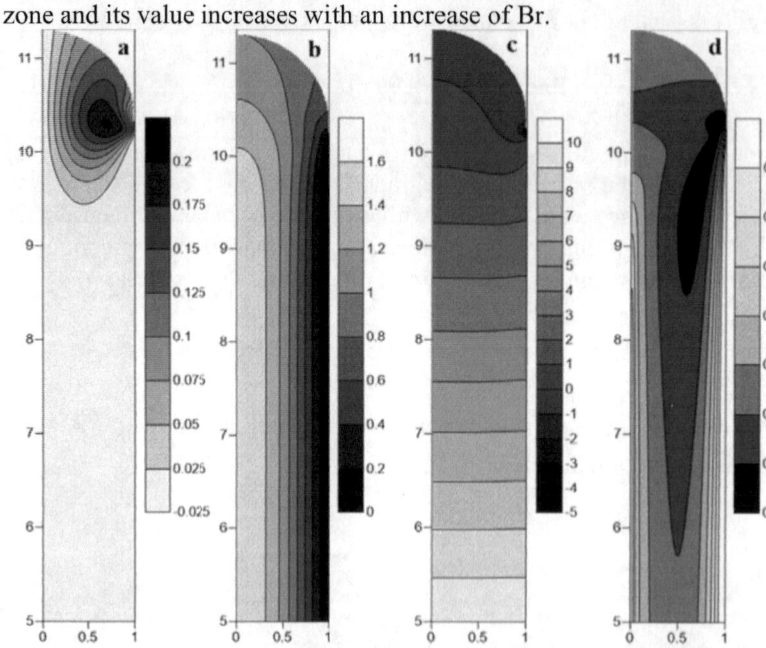

Figure 3. Distribution of the transverse velocity (a), longitudinal velocity (**b**), pressure (c), apparent viscosity (d)
(Re=0.1, Pe=100, *m*=0.9: C=1, Br=1, W=5)

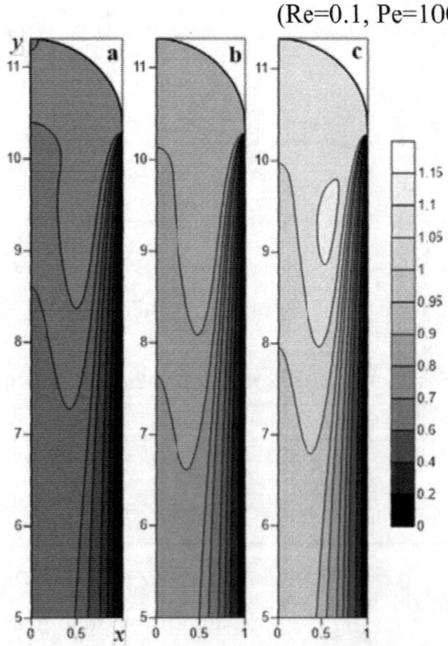

Figure 4. Temperature distribution as a function of Br parameter
(Re=0.1, Pe=100, *m*=0.9, W=5: C=1, a – Br=1, b – Br=1.5, c – Br=2)

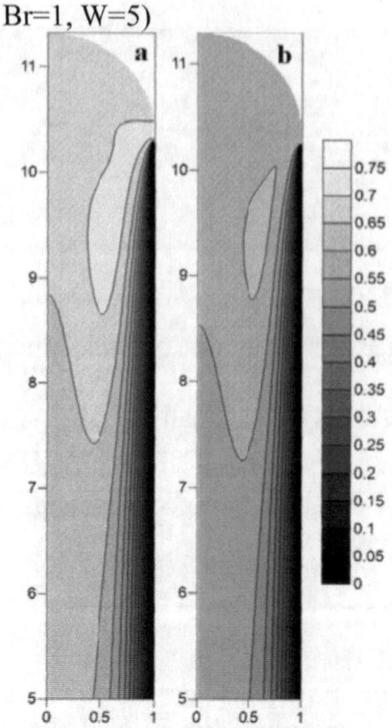

Figure 5. Temperature distribution as a function of nonlinearity degree
(Re=0.1, Pe=100, Br=1, W=5: C=1, a – *m*=1, b – *m*=0.8)

5. Conclusions

The numerical simulation has shown a separation of the flow into a two-dimensional flow zone in the vicinity of the free surface and a one-dimensional flow zone away from it, where the temperature maximum is located in two-dimensional flow zone. The effect of a viscous dissipation and rheology on the temperature distribution was demonstrated. With an increase of Br from 1 to 2 the temperature maximum increases from 0.67 to 1.11, and with a decrease of nonlinearity degree from 1 to 0.8 the temperature maximum decreases from 0.72 to 0.61.

6. Acknowledgments

This work was supported by the Russian Foundation for Basic Research (project no.15-08-03935a).

References

[1] Baranov A V 2010 *Compos. Mech. Des.* **16** 384–399
[2] Hassan H, Regnier N, Pujos C and Defaye G 2008 *Polym. Eng. Sci.* **48** 1199–1206
[3] Nguyen-Chung T, Mennig G 2001 *Rheol. Acta* **40** 67–73
[4] Otmani R E, Zinet M, Boutaous M and Benhadid H 2011 *J. Appl. Polym. Sci.* **121** 1579–1592
[5] Wang W, Li X and Han X 2012 *Polym. Eng. Sci.* **52** 42–51
[6] Borzenko E I, Frolov O Y and Shrager G R 2014 *Fluid Dyn.* **49** 37–45.
[7] Kutepov A M, Polyanin A D, Zapryanov Z D et al 1996 *Chemical Hydrodynamics: Handbook* (Moscow: Byuro Kvantum)
[8] Huh C and Scriven L E 1971 *J. Colloid Interface Sci.* **35** 85–101
[9] Pukhnachev V V and Solonnikov V A 1982 *J. Appl. Math. Mech.* **46** 771–779
[10] Borzenko E I, Shrager G R 2015 *J. Appl. Mech. Tech. Ph+* **56** 167–176
[11] Patankar S *Numerical heat transfer and fluid flow. Series on computational methods in mechanics and thermal science* (New York: Taylor & Francis)
[12] Vasenin I M, Sidonsky O B and Shrager G R 1974 *Proc. USSR Acad. Sci.* **217** 295–298

An experimental investigation of the interaction between a pair of precessing vortices in a tangential vortex chamber

S V Dremov[1,2], S I Shtork [1,2], S G Skripkin [1], and I K Kabardin [1]

[1]Kutateladze Institute of Thermophysics, SB RAS, Novosibirsk 630090, Russia
[2]Novosibirsk State University, Novosibirsk, 630090,Russia

E-mail: supersonny@mail.ru

Abstract. This paper is devoted to experimental investigation of the interaction between the pair of precessing vortices in a tangential vortex chamber. The test section was the tangential vortex chamber with a cylindrical working area. The liquid was tangentially fed into the chamber through 12 rectangular nozzles. The swirl parameter varied in the range $0 \div 6.6$, and the Reynolds number varied within $6000 \div 52000$. On the basis of visualization materials the dependence of the precession frequency of the system of two vortices was obtained. For the quantitative investigation the optical methods of laser Doppler anemometry (LDA) and the Particle Image Velocimetry (PIV) were used. Information about vortices and recirculation zones was obtained from the optical measurements.

1. Introduction

The investigations of large-scale vortex structures are topical, because such structures regularly occur in nature and are frequently encountered in technical devices, such as combustion chambers [1, 2]. Special attention should be paid to the vortex structures which are generated in hydraulic turbines operating under off-design conditions. It has been identified that the double vortex structure may be formed behind the hydraulic turbine runner. Transition from single-helical to double-helical configuration even for a short time can lead to a considerable change in precession frequency beyond the design values [3, 4]. Despite the available investigations including the above mentioned parameters the transition from single to double vortex has not been fully defined. Thus it makes the study of the double vortex structure and defining its characteristics very important and useful. Besides its significance for practical applications, it is also of a great interest from scientific point of view [5].

The aim of this work is the experimental study of the interaction between the pair of precessing vortices in the tangential vortex chamber.

The experiments were carried out in a closed hydrodynamic loop for different values of input parameters. The fluid, sc. the distilled water, was supplied via the centrifugal pump from the tank to the working section (fig. 1). The flow rate was varied in the range from 3 to 25 m^3/h and was measured by an ultrasonic flow meter with an error less than 1.5%. The work area where the double vortex structure emerged was a cylindrical channel with diameter D = 190 mm and a length of 600 mm. The working section was made of transparent Plexiglas, which enabled the flow investigation using optical methods. Fluid was fed into the chamber through 12 tangential rectangular nozzles that were arranged in three tiers and grouped in corner blocks (fig. 1).

The main parameters were the geometrical swirl parameter S and the Reynolds number Re. The geometrical swirl parameter was determined as follows: $S = D^2 \sin(\gamma) / (N\sigma_n)$, where, N = 12 is the number of nozzles, σ_n = 14mm x 23mm = 322 mm^2 is the nozzle cross-section area, and γ is the nozzle turning angle relative to the channel center. The angle γ was varied from 0 to 45 degrees. The flow rate Q was maintained in the range from 3 to 25 m^3/h. The swirl parameter was changed in the range 0 ÷ 6.6. The Reynolds number was defined as Re = DU_0/ν, where D = 190 mm was the chamber diameter, $U_0 = Q / N\sigma_n$ was the mean velocity calculated by dividing the flow rate Q by the nozzle cross-section area $N\sigma_n$ and ν was the kinematic viscosity of water. The Reynolds number varied in the range of 6000 ÷ 52000.

2. Experimental Methods and Results

The first stage of the work was the flow visualization using the LED light source to illuminate the working area from the bottom (fig. 6). The flow pattern was captured by the camera Canon EOS 7D (fig. 2) which allowed taking still images of the flow structure and video recording with a frame rate of 60 Hz. The visualization has revealed the presence of two twisted precessing vortices (fig.2).

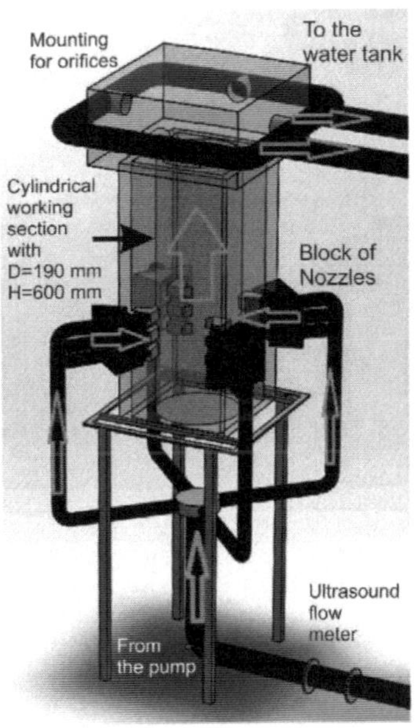

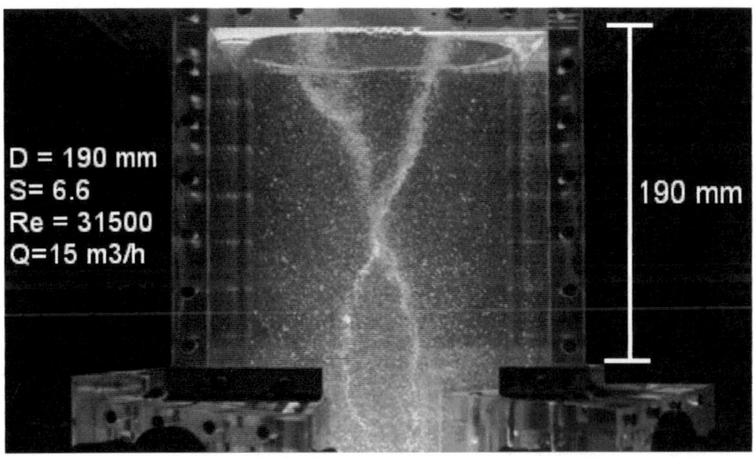

Figure1. The vortex chamber. **Figure2.** Visualization of the pair of interacting vortices.

Based on visualization materials the dependence of the precession frequency of the system of two vortices on the fluid flow rate has been built. The dependence is described by a linear function (1) within an error of 10 % (fig. 3).

$$f\,[Hz] = 0.97 \cdot Q\,[m^3/h] - 0.45 \qquad (1)$$

Figure 4 shows the regime map of helical vortex pair generation. The filled circles correspond to the existence of the vortex pair. The transparent circles show parameters where the vortex structure is not observed. When the swirl parameter takes values in the range from 0 to 1.2, the precessing vortex pair does not appear, but with an increase in the flow swirling the vortex structure becomes visible. At maximum swirl parameter S = 6.6 the double helix can be observed at a flow rate above 9.5 m^3/h. With the flow rate decrease below this value the helical vortex structure begins to vanish (fig. 4).

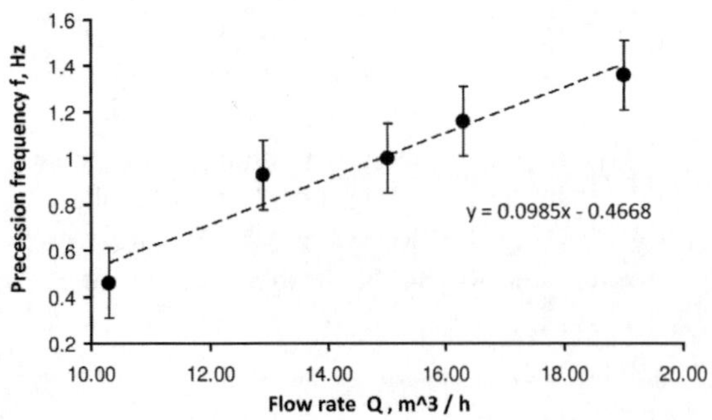

Figure 3. The dependence of the precession frequency of the pair of vortices on the flow rate.

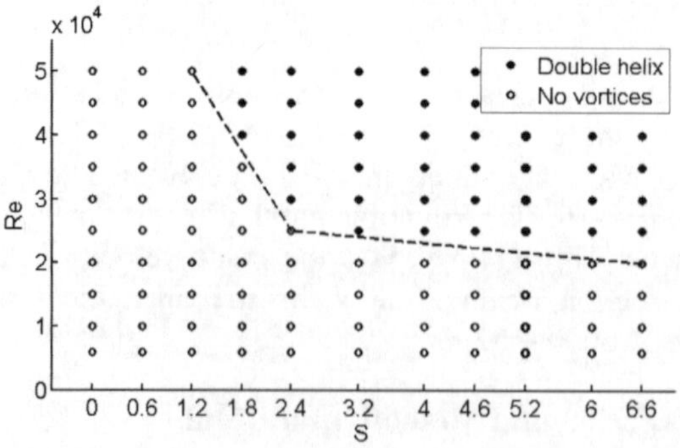

Figure 4. The regimes map of a double vortex system.

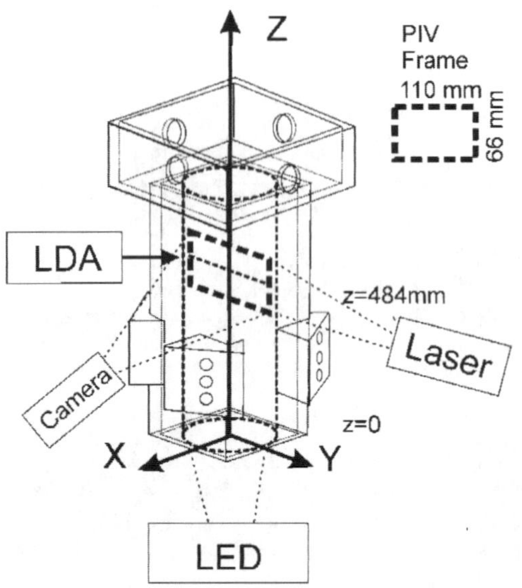

Figure 5. The scheme of velocity measurement by LDA and PIV methods in vertical cross-section.

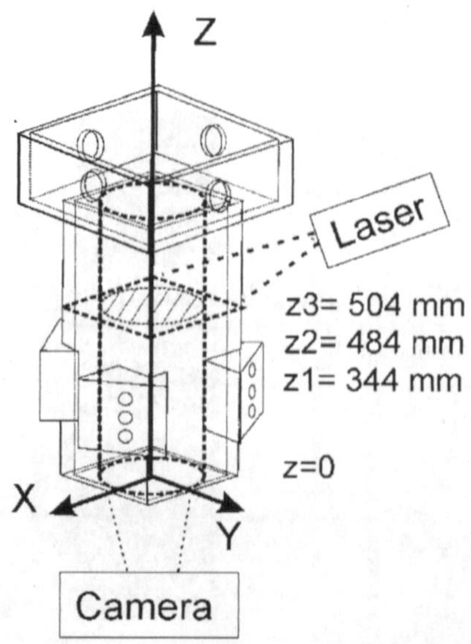

Figure 6. The scheme of visualization and velocity measurement by PIV method in horizontal cross-section.

The velocity profiles in the vortex chamber were measured using the laser Doppler anemometer (LDA) LAD-05 with an adaptive temporal selection of the velocity vector. The LDA was designed and manufactured at the Kutateladze Institute of Thermophysics SB RAS in Novosibirsk. The scheme of velocity measurement in the vortex chamber is shown in fig. 5. The measurements were carried out at height Z = 484 mm and Y = 0 in the range of the coordinate X = -0.5 D ÷ 0.5 D. The LDA served to measure two projections of the velocity vector Vy and Vz in the range 0.001- 30 m/s. The relative error did not exceed 0.1%. The size of the measuring volume was 0.1x0.1x0.5 mm. The positioning device allowed moving the control volume in the region of 250 x 250 x 250 mm with an accuracy of 0.1 mm. The number of LDA measurements at each point was equal to 5000, and the maximum acquisition time was 60 seconds. The reliability of determining the average rate was guaranteed at a level of 95%. As scattering particles for obtaining the Doppler signal the polystyrene particles with 20 microns diameter and 1.069 g/mm^3 density were used.

For instantaneous flow field diagnostics the 2D PIV instrumentation of POLIS Company was used. The PIV system was designed and manufactured at the Kutateladze Institute of

Thermophysics SB RAS. As an optical source for forming the light sheet we used Nd: YAG pulsed laser POLIS v3.2 with the following characteristics: wavelength of 532 nm, light sheet thickness of 1 mm, the energy pulse power of120 mJ, and the operation frequency of 2 Hz. Measurements were performed in a vertical cross-section with coordinates Z = 451 ÷ 517 mm and Y = -95 ÷ 95 mm. To study the vorticity field and the tangential velocity in horizontal sections the measurements were performed at three heights Z_1 = 344 mm, Z_2 = 484 mm and Z_3 = 504 mm (fig. 5) in the areas X = - 55 ÷ 55 mm and Y = -33 ÷ 33 mm and X = -55 ÷ 55 mm and Y = -55 ÷ 55 mm. Images were registered by POLIS camera v1.0 with lenses Nikon AF 28 mm f/2.8D Nikkor at a resolution of 1280x768. To calculate the two-dimensional velocity field the Actual flow software Version 1.16.7.0 was used.

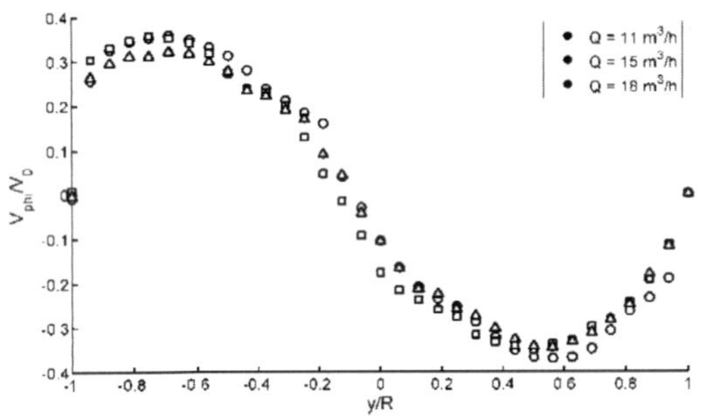

Figure 7. Profiles of tangential velocity at different fluid flow rates. The swirl parameter S= 6.6.

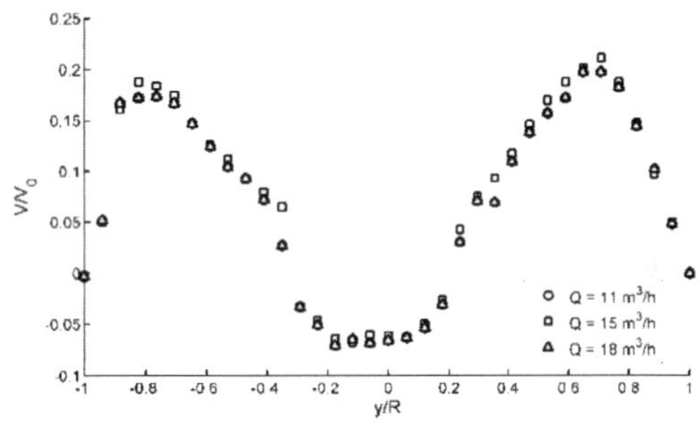

Figure 8. Profiles of axial velocity at different fluid flow rates. The swirl parameter S = 6.6.

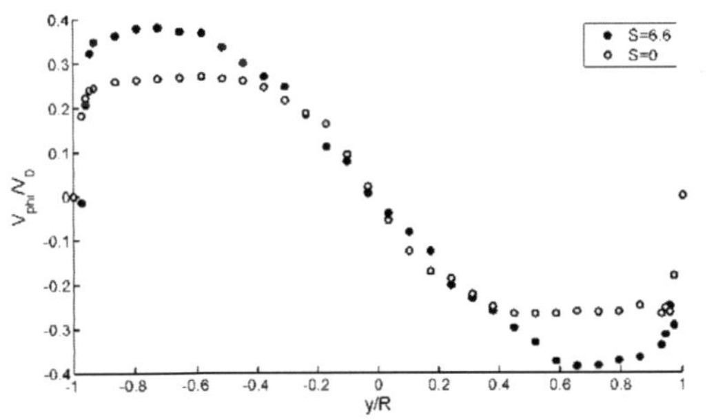

Figure 9. Profiles of tangential velocity at different values ofswirl parameter. Flow rate Q =11.6 m^3/h.

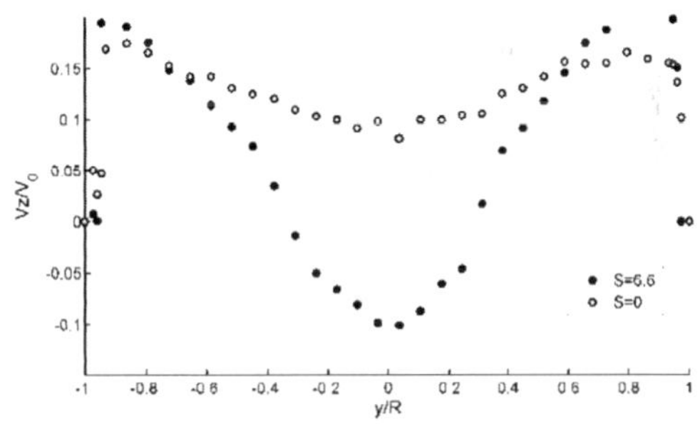

Figure 10. The dependence of the axial velocity on the swirl parameter. Flow rate Q = 11.6 m^3/h.

Figures 7, 8, 9, and 10 show the profiles of tangential and axial velocity components, obtained by the LDA technique for various values of Reynolds number and swirl parameter. Figures 7 and 8 illustrate the dependence of the tangential and axial velocity on the flow rate. The geometric swirl parameter remains constant and has a value of 6.6. According to graphs 7 and 8 the flow rate does not affect the flow pattern. The dimensionless velocity profiles remain unchanged. Figures 9 and 10 show the dependence of the tangential and axial velocity on the geometric swirl parameter. For these profiles the flow rate remains constant at 11.6 m^3/h. The axial velocity profiles indicate the presence of the reverse flow zone in the range X/R = -0.25 ÷ 0.25 (fig. 8). From the last two dependencies (figs. 9 and 10) it can be seen that velocity profiles change with different swirl parameters. Even if the value of the geometric swirl parameter is equal to zero, the flow swirl is present. This is due to the fact that the real swirl parameter is not zero, since it was impossible to ideally direct all incoming nozzle jets in the center of the vortex chamber. As expected this leads to flow swirling. On the other hand, the system is not fully isolated and flow swirling occurs under the action of external forces. Interesting effects can be seen in the fig 10. With reducing the swirl parameter the recirculation zone, which correlates with negative values of the dimensionless axial velocity, decreases until complete disappearance in spite of significant tangential velocity. Absence of the reverse flow zone at the parameter S close to zero explains why we did not see the double vortex precession at low swirl. It is usually supposed that namely the central recirculation zone is associated with the formation of a precessing vortex structure [2].

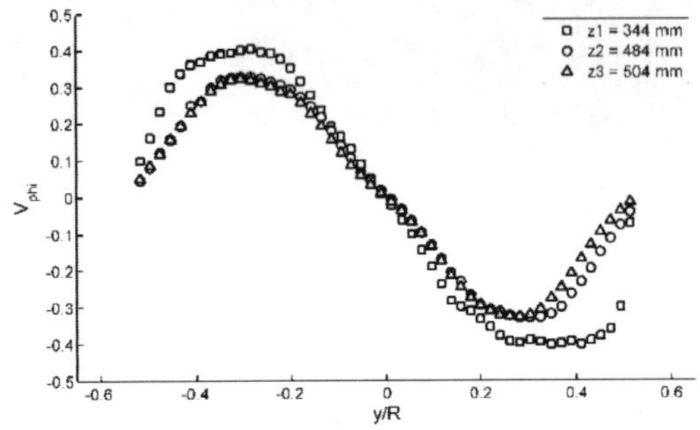

Figure 11. Tangential velocity profiles in different horizontal cross-section.

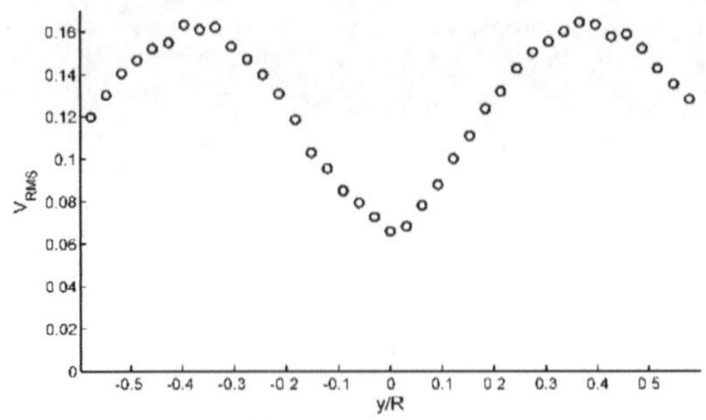

Figure 12. The velocity RMS.

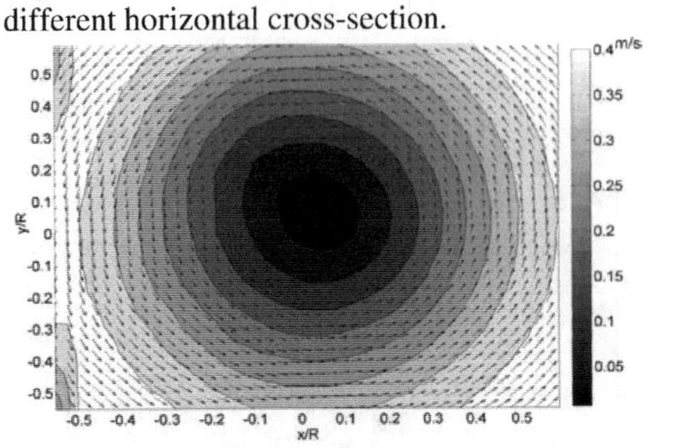

Figure 13. The field of mean velocity in horizontal cross-section with height z = 484 mm. The contour map presents the velocity vector length.

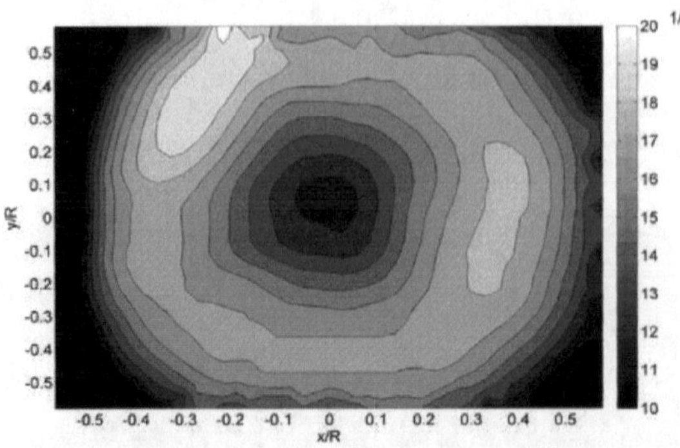

Figure 14. The average vorticity distribution in the horizontal cross-section with height z = 484 mm.

Figure 11 presents the tangential velocity profiles for three different cross-sections of the vortex chamber. The difference between the velocity values in sections at heights Z_3 =504 mm, Z_2 = 484 mm and Z_1 = 344 mm can be explained by the fact that the first two cross-section heights differ by only 20 mm and, therefore, have a similar profile. Section at height Z_1 = 344 mm is located much lower than Z_1 and Z_2 and slightly above the top nozzle of the vortex chamber. In this section, the velocity was measured at a strong influence of the flow from the nozzle. This mean flow velocity in the nozzle is larger than the mean flow velocity in the working section.

Figure 12 presents the standard deviation of velocity. Since the vortices induce velocities with different signs in the center of the vortex chamber the resulting velocity value at that point tends to zero and the pulsations tend to minimum. Two peaks correspond to the radius of the vortex core localization. In that area the pulsations tend to maximum.

Figure 13 shows the mean velocity field in the cross section at z = 484 mm, depicted by the vector and contour plots. The velocity reaches its maximum value at the edges of the frame and is minimal at the center. Figure 14 presents the mean vorticity field in the vortex chamber cross-section. The mean vorticity was calculated based on 100 samples. The vorticity field had a shape of a ring, which visually matched to the area of vortex precession.

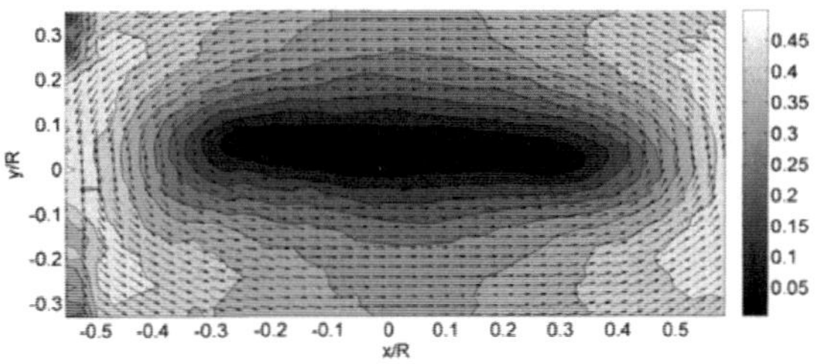

Figure 15. Phase averaged velocity field of interacting precessing vortex cores. The contour map shows the velocity vector length

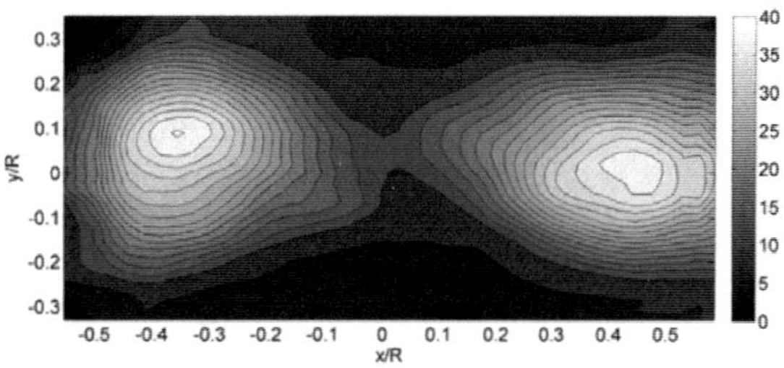

Figure 16. Phase averaged vorticity field.

Figure 15 illustrates the phase averaged velocity field of interacting precessing vortex cores. There is also acontour map showing the velocity. The central region of low velocity corresponds to the interaction of two rotating vortex cores. Figure 16 illustrates the instantaneous (phase averaged) vorticity field with clearly visible vortex cores. Averaging was performed over 50 images, taken in the same phase.

Conclusions

The paper presents an experimental study of the interaction between the pair of precessing vortices in the tangential vortex chamber. At the first stage the flow visualization was performed. It showed the presence of two twisted precessing vortices, unvaryingly reproduced in a broad range of operating parameters. The regime map was drawn on the basis of the visualization data. The dependence of the precession frequency of two vortices on the flow rate was obtained. Optical methods of Laser Doppler Anemometry (LDA) and Particle Image Velocimetry (PIV) were used for quantitative investigations of the vortex structure. Velocity profiles proved the presence of the reversal flow zone and of the two vortices inside the vortex chamber. The interaction of the vortices was shown to form the region with a zero instantaneous circumferential velocity in the center of the channel. The vortices position was determined in the horizontal cross-section by the vorticity field.

Acknowledgments

Research was supported by the Russian Science Foundation (Project № 14-29-00093).

References

[1] Alekseenko S V, Kuibin P A and Okulov V L 2007 Th*eory of concentrated vortices. An introduction* (Berlin , Heidelberg, New york: Springer)
[2] Gesheva E S, Litvinov I V, Shtork S I and AlekseenkoS V 2014 *Thermal Engineering* **61** (9) 649–657
[3] Skripkin S G, Tsoy M A, Shtork S I and Hanjalić K 2016 *Journal of Hydraulic Research,***54** (4) 450-460
[4] Dörfler P, Sick M and Coutu A 2014 *Flow-induced pulsation and vibration in hydroelectric machinery* (London: Springer)
[5] Naumov I V, Dvoinishnikov S V, Kabardin I K and Tsoy M A 2015 *Physics of Fluids* **27** 124103

Acoustic streaming with heat exchange

A A Gubaidullin[1,2,3] and A V Pyatkova[1,2,4]

[1] Tyumen Branch of the Khristianovich Institute of Theoretical and Applied Mechanics, Siberian Branch of Russian Academy of Sciences, 74 Taymirskaya Street, Tyumen, Russia
[2] Tyumen State University, 6 Volodarskogo Street, Tyumen, Russia
[3] Tyumen industrial University, 38 Volodarskogo Street, Tyumen, Russia
[4] Institute of Mechanics and Engineering, Kazan Science Center, Russian Academy of Sciences, 2/31 Lobachevskogo Street, Kazan, Russia

E-mail: annyakovenko@yandex.ru

Abstract. Acoustic streaming in a cylindrical cavity with heat exchange is numerically investigated. The cavity is filled with air. The boundaries of the cavity are maintained at constant temperature. The features of acoustic streaming manifesting with the decrease in the frequency of vibration in comparison with the resonant frequency are determined. The influence of the nonlinearity of process on acoustic streaming is shown. The nonlinearity is caused by the increase of the vibration amplitude.

1.Introduction

Acoustic streaming is a directed time average mass transfer by steady vortices formed in addition to the periodic motion of the medium in the sound field [1]. The first theoretical analysis of acoustic streaming in standing waves between parallel planes was performed by Rayleigh [2]. He analytically described stationary motion consisting of a series of vortices with a definite sense of rotation. Acoustic streaming inside the boundary layer is called the inner streaming or Schlichting streaming. Acoustic streaming outside the boundary layer is called outer streaming or Rayleigh streaming [3]. In [4, 5], for the example of vibration with the minimum resonance frequency, it was analytically demonstrated that in the case of a “narrow” tube, for which the ratio of the channel radius to the acoustic boundary layer thickness was below 5.7, only Schlichting streaming was present.

Due to the numerical methods and experimental investigations it became possible to describe the strongly nonlinear streaming, when the analytical solution is no longer valid. Nonlinear effects in acoustic streaming are widely investigated [6-11]. In [12-14] the nonlinear effects manifesting with the increase in vibration frequency at fixed vibration amplitude are described. In [15], the case of thermally insulated cavity walls is compared with the case of walls maintained at constant temperature for weakly nonlinear processes and vibration frequencies below resonance.

In the present study, acoustic streaming in a cylindrical cavity at below resonance vibration frequencies is investigated. The constant temperature walls are considered. For the five vibration frequencies the influence of heat exchange to the acoustic streaming is shown. Furthermore, for the closest to the resonant vibration frequency the influence of the nonlinearity on acoustic streaming is studied.

2.Problem formulation

A circular cylindrical cavity of length L and radius R_0 with impermeable ends is considered (figure 1). The cavity is filled with a perfect viscous gas (air). Let gas in the cavity be initially at rest at constant temperature T_0 and constant pressure p_0. The system is disturbed from equilibrium by vibration action $A\cos(\omega t)$ with constant amplitude A and frequency ω, so that this action causes harmonic vibration of the entire cavity along its axis. We consider the case of walls maintained at constant temperature. The thermal conductivity, heat capacity and viscosity are assumed constant.

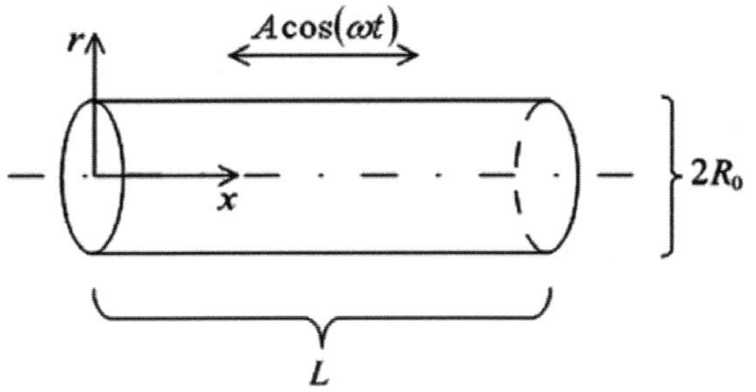

Figure 1. Schematic of the problem.

The system of equations describing gas motion with respect to the vibrating cavity in a cylindrical coordinate system in terms of dimensionless variables has the following form:

$$\frac{\partial \tilde{\rho}}{\partial \tau}+\frac{\partial \tilde{\rho} U}{\partial X}+\frac{1}{R}\frac{\partial R\tilde{\rho} V}{\partial R}=0,$$

$$\frac{\partial \tilde{\rho} U}{\partial \tau}+\frac{\partial \tilde{\rho} UU}{\partial X}+\frac{1}{R}\frac{\partial R\tilde{\rho} UV}{\partial R}=-\frac{\partial P}{\partial X}+\mathrm{N}\left(\frac{\partial^2 U}{\partial X^2}+\frac{1}{R}\frac{\partial}{\partial R}\left(R\frac{\partial U}{\partial R}\right)\right)+\frac{1}{3}\mathrm{N}\left(\frac{\partial^2 U}{\partial X^2}+\frac{1}{R}\frac{\partial}{\partial X}\left(\frac{\partial RV}{\partial R}\right)\right)+$$
$$\tilde{\rho}\tilde{A}\Omega^2\cos(\Omega\tau),$$

$$\frac{\partial \tilde{\rho} V}{\partial \tau}+\frac{\partial \tilde{\rho} UV}{\partial X}+\frac{1}{R}\frac{\partial R\tilde{\rho} VV}{\partial R}=-\frac{\partial P}{\partial R}+\mathrm{N}\left(\frac{\partial^2 V}{\partial X^2}+\frac{1}{R}\frac{\partial}{\partial R}\left(R\frac{\partial V}{\partial R}\right)\right)+\frac{1}{3}\mathrm{N}\left(\frac{1}{R}\frac{\partial}{\partial R}\left(R\frac{\partial V}{\partial R}\right)+\frac{\partial}{\partial R}\left(\frac{\partial U}{\partial X}\right)\right)-$$
$$\frac{4}{3}\mathrm{N}\frac{V}{R^2},$$

$$\frac{\partial \tilde{\rho}\Theta}{\partial \tau}+\frac{\partial \tilde{\rho} U\Theta}{\partial X}+\frac{1}{R}\frac{\partial R\tilde{\rho} V\Theta}{\partial R}=\gamma\Gamma\left(\frac{\partial^2 \Theta}{\partial X^2}+\frac{1}{R}\frac{\partial}{\partial R}\left(R\frac{\partial \Theta}{\partial R}\right)\right)-\gamma(\gamma-1)P\left(\frac{\partial U}{\partial X}+\frac{1}{R}\frac{\partial RV}{\partial R}\right)-\frac{2}{3}\gamma(\gamma-1)\mathrm{N}\cdot$$
$$\left(\frac{\partial U}{\partial X}+\frac{1}{R}\frac{\partial RV}{\partial R}\right)^2+\gamma(\gamma-1)\mathrm{N}\left(\frac{\partial U}{\partial R}+\frac{\partial V}{\partial X}\right)^2+2\gamma(\gamma-1)\mathrm{N}\left(\left(\frac{\partial U}{\partial X}\right)^2+\left(\frac{\partial V}{\partial R}\right)^2+\frac{V^2}{R^2}\right),$$

$$P=\frac{\tilde{\rho}(\Theta+1)}{\gamma}.$$

The initial and boundary conditions are as follows:

$\tau=0:\ U=0,\ V=0,\ \Theta=0,\ P=1/\gamma\approx 0.71,\ \tilde{\rho}=1,$

$X=0:\ U=0,\ V=0,\ \Theta=0,$

$X=1:\ U=0,\ V=0,\ \Theta=0,$

$R=\tilde{R}_0:\ U=0,\ V=0,\ \Theta=0.$

The following dimensionless variables and parameters are introduced:

$$X=\frac{x}{L},\ R=\frac{r}{L},\ \tau=\frac{tc_0}{L},\ P=\frac{p}{\gamma p_0},\ \widetilde{\rho}=\frac{\rho R_g T_0}{p_0},\ \Theta=\frac{T-T_0}{T_0},\ U=\frac{u}{c_0},\ V=\frac{v}{c_0},\ \mathrm{N}=\frac{\nu}{c_0 L},\ \Gamma=\frac{\chi}{c_0 L},$$

$$\Omega=\frac{\omega L}{c_0},\ \widetilde{A}=\frac{A}{L},\ \widetilde{R}_0=\frac{R_0}{L}.$$

Here, t is time, x and r are the spatial coordinates, u and v are the axial and radial particle velocity components, respectively, ρ is density, p is pressure, T is temperature, c_0 is the adiabatic speed of sound in the undisturbed medium, R_g is the gas constant, ν is the kinematic viscosity and χ is the thermal diffusivity.

To calculate the velocity components of acoustic streaming (time-averaged rate of mass transfer), we use the formulas [6]:

$$U_{st}=\frac{\langle\widetilde{\rho}U\rangle}{\langle\widetilde{\rho}\rangle},\ V_{st}=\frac{\langle\widetilde{\rho}V\rangle}{\langle\widetilde{\rho}\rangle},$$

where $\langle\ \rangle$ indicates averaging over the period $\widetilde{\mathrm{T}}=\frac{2\pi}{\Omega}$, U_{st} is the dimensionless axial velocity component of acoustic streaming and V_{st} is the dimensionless radial velocity component of acoustic streaming.

3.Numerical method and parameters of calculations

The calculations were performed using an implicit numerical scheme correct to first order in time and space. Discretization of equations was performed by the control volume method with approximation of the convective diffusion flow by a power law. The numerical method is described briefly in [15] and in more detail on the example of the one-dimensional case in [16]. The following dimensionless parameters were used in the computations described below: $\mathrm{N}=8.6\cdot10^{-6}$, $\Gamma=1.2\cdot10^{-5}$, $\gamma=1.4$, $\widetilde{R}_0=0.02$. The parameters of the gas corresponded to the thermophysical properties of air at a temperature of 300 K. The dimensional cavity length was 0.005 m. We considered the following vibration frequencies: Ω = 0.5, 1, 1.5, 2 and 2.5; with a dimensionless resonance vibration frequency of about π. The amplitude of vibration $\widetilde{\mathrm{A}}$ was varied from 0.01 to 0.5. The computation grid for the vibration frequencies Ω = 0.5–2 had 1002×22 nodes; for the frequency $\Omega=2.5$ – 1002×42 nodes. Thus, no less than 4 nodes were located within the acoustic boundary layer thickness, which is determined in the dimensionless form by the formula $\delta_\nu=\sqrt{2\mathrm{N}/\Omega}$.

4.Results and discussion

Let us consider acoustic streaming at the case of week nonlinearity at the vibration amplitude $\widetilde{A}=0.01$. At Ω = 2.5, the value of the ratio between the cavity radius $\widetilde{R}_0$ and acoustic boundary layer thickness $\widetilde{\delta}_\nu$ is 7.6. Thus, acoustic streaming contains both Schlichting and Rayleigh streaming [4,5]. This is illustrated in figure 2(a). With decrease in vibration frequency, the ratio $\widetilde{R}_0/\widetilde{\delta}_\nu$ decreases and Rayleigh streaming vortices must be reduced, as compared to Rayleigh streaming vortices, and be disappeared with a further decrease of this ratio. This follows from the theory of acoustic streaming in the absence of heat exchange and was illustrated in [15] for the similar problem formulation at adiabatic boundary conditions. However, in the case of acoustic streaming with heat exchange, Schlichting streaming vortices decrease in size and disappear (figures 2(b-e)) with decrease in vibration frequency in comparison with the resonance frequency.

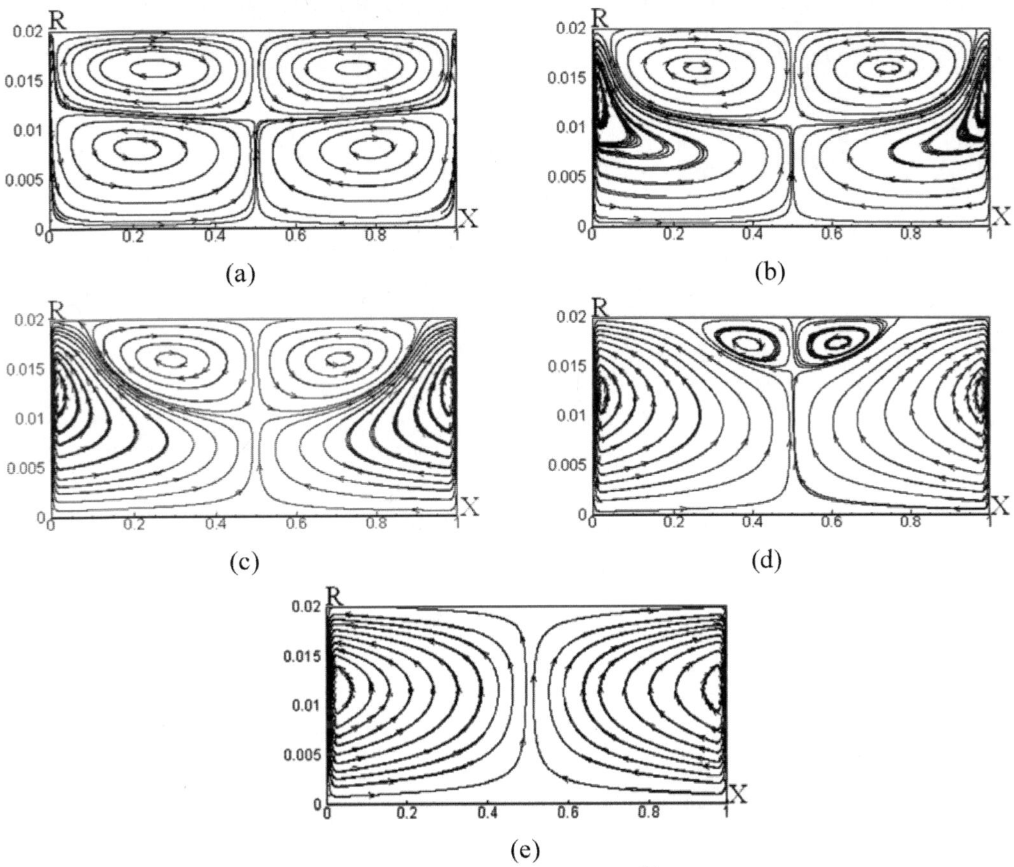

Figure 2. Streamlines of the streaming flow at $\widetilde{A} = 0.01$; (a) – $\Omega = 2.5$, (b) – $\Omega = 2$, (c) – $\Omega = 1.5$, (d) – $\Omega = 1$, (e) – $\Omega = 0.5$.

At the lowest vibration amplitude from the considered range $\widetilde{A} = 0.01$, the process intensity is low and the period average temperature, density and pressure coincide with their initial distributions. With the increase in vibration amplitude, the influence of the nonlinearity on the process increases and the distortion of acoustic streaming vortices is observed. Let us show this by the example of the frequency $\Omega = 2.5$. Figure 3 represents the period average distributions of temperature and streamlines of acoustic streaming with increase of the amplitude of vibration. It is seen that acoustic streaming is changed. The vortices, which were initially located near the lateral surface of the cylinder, are moved to the central part of the cavity. Also, additional small vortices are formed. It is seen that the period average temperature is negative (lower than the initial temperature) in the central part of the cavity. Note that the area of negative temperature decreases by increasing the frequency of vibration, but the minimum value of the period average temperature becomes closer to the initial temperature.

The period average density also becomes nonuniform with increasing of vibration amplitude (figure 4). In this case, there is also the area in the central part of the cavity in which the period average density is less than the initial value, but the size of this area changes not so quickly. The minimum value of the period average density decreases with the increase of the amplitude of vibration.

Journal of Physics: Conference Series **754** (2016) 022004 doi:10.1088/1742-6596/754/2/022004

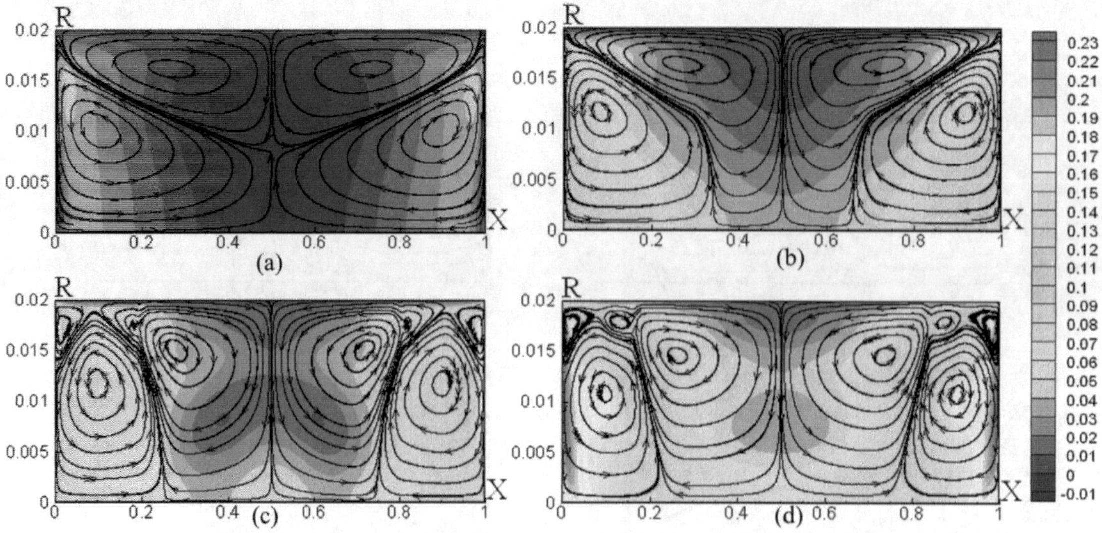

Figure 3. Period average distributions of temperature and streamlines at $\Omega = 2.5$;

(a) – $\widetilde{A} = 0.05$, (b) – $\widetilde{A} = 0.1$, (c) – $\widetilde{A} = 0.2$, (d) – $\widetilde{A} = 0.3$.

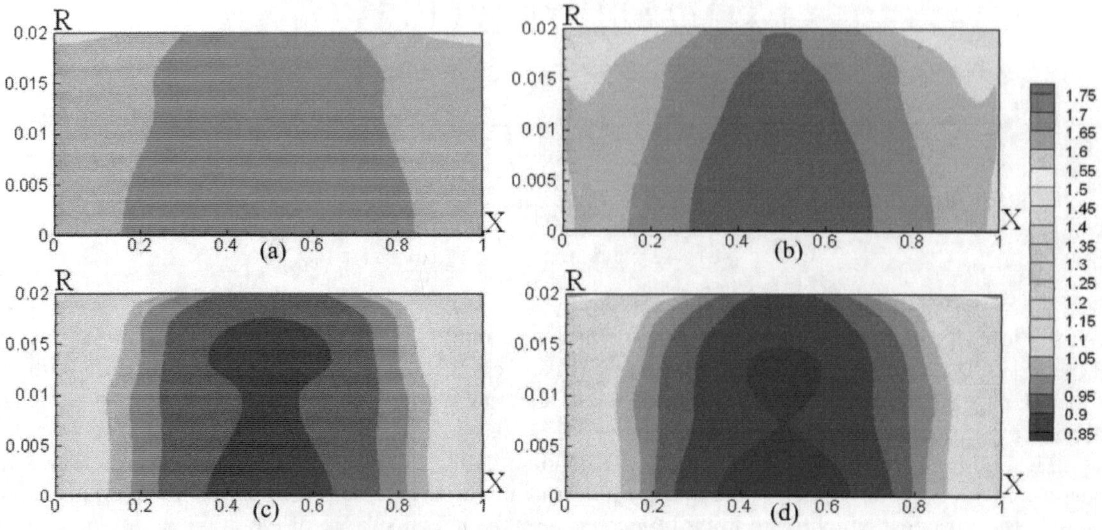

Figure 4. Period average distributions of density at $\Omega = 2.5$;

(a) – $\widetilde{A} = 0.05$, (b) – $\widetilde{A} = 0.1$, (c) – $\widetilde{A} = 0.2$, (d) – $\widetilde{A} = 0.3$

5.Conclusions

Acoustic streaming in a cylindrical cavity at the vibration frequencies less than the resonant one and isothermal boundary conditions was investigated. When the frequency decreases from the resonance, it is possible to observe a decrease in size of the vortices of Schlichting streaming, as compared to the vortices of Rayleigh streaming, and their subsequent disappearance. For the case of near resonance frequency, it was shown that the enhancement of nonlinearity of the process by increasing the amplitude of vibration leads to the shift of the streaming vortices and the formation of additional vortices. Also, the period average temperature and density became nonuniform.

Acknowledgments
The study was performed by a grant from the Russian Science Foundation (project No. 15-11-10016).

References

[1] Nyborg W L 1965 Acoustic streaming *Physical Acoustics* vol 2 ed W P Mason (New York: Academic Press) chapter 11 pp 265–331
[2] Rayleigh L 1884 *Philos. Trans. R. Soc. London* **175** 1–21
[3] Zarembo L K 1971 Acoustic streaming *High-Intensity Ultrasonic Fields* ed L D Rozenberg (New York: Plenum) part III pp 156–64
[4] Hamilton M F, Ilinskii Y A and Zabolotskaya E A 2003 *J. Acoust. Soc. Am.* **113** 153–60
[5] Hamilton M F, Ilinskii Y A and Zabolotskaya E A 2003 *J. Acoust. Soc. Am.* **114** 3092–101
[6] Aktas M K and Farouk B 2004 *J. Acoust. Soc. Am.* **116** 2822–31
[7] Nabavi M, Siddiqui K and Dargahi J 2009 *Wave Motion* **46** 312–22
[8] Daru V, Baltean-Carlès D, Weisman C, Debesse P and Gandikota G 2013 *Wave Motion* **50** 955–63
[9] Reyt I, Daru V, Bailliet H, Moreau S, Valière J-C, Baltean-Carlès D and Weisman C 2013 *J. Acoust. Soc. Am.* **134** 1791–1801
[10] Reyt I, Bailliet H and Valière J-C 2014 *J. Acoust. Soc. Am.* **135** 27–37
[11] Gubaidullin A A and Yakovenko A V 2015 *J. Acoust. Soc. Am.* **137** 3281–7
[12] Gubaidullin A A and Yakovenko A V 2014 *Thermophysics and Aeromechanics* **21** 589–99
[13] Gubaidullin A A and Yakovenko A V 2015 *High Temp.* **53** 73–9
[14] Gubaidullin A A and Yakovenko A V 2014 *High Temp.* **52** 264–70
[15] Gubaidullin A A and Pyatkova A V 2016 *Acoust. Phys.* **62** 300–5
[16] Gubaidullin A A and Yakovenko A V 2013 *Thermophysics and Aeromechanics* **20** 277–88

Linear and nonlinear development of controlled disturbances in the supersonic boundary layer on a swept wing at Mach 2.5

G L Kolosov[1,2] and A D Kosinov[1,2]

[1] Khristianovich Institute of Theoretical and Applied Mechanics, Siberian Branch of Russian Academy of Sciences, Institutskaya str. 4/1, Novosibirsk 630090, Russia
[2] Novosibirsk State University, Novosibirsk 630090, Russia

E-Mail: kolosov@itam.nsc.ru

Abstract. Experimental data on the linear and nonlinear wave train development in 3D supersonic boundary layer over a 45° swept-wing at Mach number 2.5 are presented. Travelling artificial disturbances were introduced in the boundary layer by periodical glow discharge at frequencies 10 and 20 kHz. The spatial-temporal and spectral-wave characteristics of the wave train of unstable disturbances in the linear region are obtained. It is shown that the additional peaks in β'-spectra arise for both subharmonic and fundamental frequencies. The experiments indicate the presence of subharmonic resonance mechanism in 3D boundary layer at Mach number 2.5.

1. Introduction

The problem of the transition to turbulence in swept-wing boundary layers has been the focus of considerable attention by researchers for more than 30 years [1, 2]. High practical relevance is directly related to airplane wings. However, a variety of factors involved in the process of laminar–turbulent transition pose many difficulties in its study. The fundamental property of a swept-wing boundary layer is the presence of crossflow, which leads to the appearance of new instabilities in comparison with a two-dimensional boundary layer. The modern theoretical concept of the transition mechanisms in a three-dimensional (3D) boundary layer gives four basic types of flow instabilities: (1) attachment line boundary layer instability on the swept leading edge; (2) stationary crossflow instability; (3) crossflow instability to the travelling disturbances; and (4) possible instability of Tollmien–Schlichting waves. Their relative role in the laminar-turbulent transition strongly depends on environmental conditions, such as level of free stream disturbances, surface quality and Mach number.

Most of the theoretical and experimental results on stability and transition of the 3D boundary layers are obtained for subsonic flow. In such flows there are two main scenarios for transition [1]. One such scenario is connected with the domination of stationary disturbances (crossflow vortices), and can occur if the free-stream turbulence level is small, i.e., the stationary disturbances are larger, relative to the initial amplitude of the non-stationary disturbances [3]. In this case, stationary vortices, created as a result of 3D roughness near the leading edge, are linear developed and eventually saturate to become subjected to high-frequency secondary instability with further laminar-flow breakdown [4, 5]. The other scenario for transition is the nonlinear interaction between travelling modes and stationary vortices, which can be observed in natural conditions when the level of unsteady freestream disturbance is high. Deyhle and Bippes [6] pointed out that, for turbulence levels $Tu > 0.2\%$ and smooth surfaces, travelling

Journal of Physics: Conference Series **754** (2016) 022005 doi:10.1088/1742-6596/754/2/022005

modes begin to dominate. Observations of travelling instability modes were also made under controlled disturbance conditions when wave packets were generated by a localized pulse [7]. We should note that this is in fact the only paper by Wasserman and Kloker [8] that scrutinizes the crossflow breakdown by travelling modes in detail by using direct numerical simulation (DNS).

Supersonic boundary layer is far more complex than incompressible boundary layer. Major differences between incompressible and compressible flows can also be expected both for the linear and nonlinear regimes. Several studies have been devoted to the evolution of travelling disturbances in 3D supersonic boundary layer and it was suggested that the transition is more likely to be dominated by them [9, 10]. To excite controlled travelling disturbances in a supersonic boundary layer, a glow discharge technique was developed at ITAM SB RAS [11]. Such technique allowed to scrutinize the linear development of a wave train on a swept wing with 45°-sweep angle at Mach number M = 2.0 and obtain the main stability characteristics for disturbances with frequencies 10 and 20 kHz [12, 13]. Moreover, at ITAM SB RAS the technique of pulsed excitation of a wave packet with a broad disturbance spectrum is developed [14]. We can expect that Yatskikh et. al. will also provide the detailed wave analysis for a wide range of frequencies in 3D supersonic boundary layer.

Later stages of the transition are characterized by nonlinearity of the disturbance flow field. The early nonlinear transition regime for the 2D boundary layer at moderate supersonic Mach numbers has been studied extensively by Kosinov and co-workers [15]. They have focused on an 'asymmetric subharmonic resonance' breakdown mechanism. The viability of this kind of resonance in 2D compressible boundary layer was confirmed theoretically [16]. Principal possibility of occurrence of resonant triplets in swept-wing boundary layers is known over a long period of time [17]. However, to date, very few experimental investigations of the subharmonic resonance in 3D supersonic boundary layer have been carried out. The goal of the present study is to perform the controlled disturbance experiments in the swept-wing supersonic boundary layer at M = 2.5 and to obtain the wave characteristics of the most unstable disturbances in the region of their linear and weakly nonlinear development.

2. Experimental setup and data processing

The experiments were conducted in the T-325 low noise supersonic wind tunnel of ITAM SB RAS at M = 2.5 and unit Reynolds number $Re_1 = 5\times10^6$ m^{-1}. A swept-wing model was used in the experiments. The model had a sweep angle of 45° and a slightly blunted leading edge of radius 0.2 mm. The swept wing had a 2.6% lens-shape airfoil at the upper side and flat surface at the bottom side with maximum thickness of 12 mm. The curvature radius of the upper surface of the model was approximately 4000 mm. A sketch of the experimental model and the coordinate systems are shown in figure 1; dimensions are given in mm. The average roughness of the upper surface of the model was no more than 1 micron. The model was fixed in the central plane of the test section at approximately a zero angle of attack with an alignment error of 0.06°. A source of localized artificial disturbances was built in the model, similar to that described by Kosinov et al. [11]. Controlled pulsations had been generated using high-frequency glow discharge in a chamber and were injected into the boundary layer through an aperture 0.4 mm in diameter in the upper surface of the model. The source aperture was located at $x_0 = 56.6 \pm 0.3$ mm from the leading edge on the center line of the model symmetry. A sine-wave generator and amplifier with transformer output were used to produce high voltage at a frequency of 10 kHz. The actuator worked at a frequency of 20 kHz, thereby enabling artificial perturbations to be excited at frequencies of 10 and 20 kHz in the boundary layer.

The wind tunnel had an automated measuring system that allowed for the determination of the flow parameters in "real time". The flow disturbances were measured by a constant-temperature anemometer (CTA). A tungsten hotwire of 10 μm in diameter and 1.6 mm in length was used. The axis of the hotwire was parallel to the z-axis. The probe was moved in the x-, y-, z-directions with the help of traversing gears with which the wind tunnel was equipped. The measurement accuracy was about 0.1 mm in the x- and z-directions and about 0.01 mm in the y-direction. Hotwire measurements were performed at a fluctuation maximum in the boundary layer and at y = const when the probe moved parallel to the leading edge. The fluctuation maximum was determined at $z' = 0$. Because the overheat ratio was taken to be about 0.8, the measured disturbances of up to 95% consisted of mass flow fluctuations.

The pulsation measurements were synchronized with a disturbance source and sine wave generator of a special device in a computer-automated measurement and control crate. The AC and DC signals from the CTA were written to the PC using a 12-bit analog-to-digital converter with sampling rate 750 kHz and by DC voltmeter, respectively. Four time traces of 65536 points in length were measured and recorded in each space position of the hotwire. This was sufficient to detect even very small amplitudes of the controlled pulsations after averaging. To analyze the evolution of natural fluctuations, the full time traces were considered.

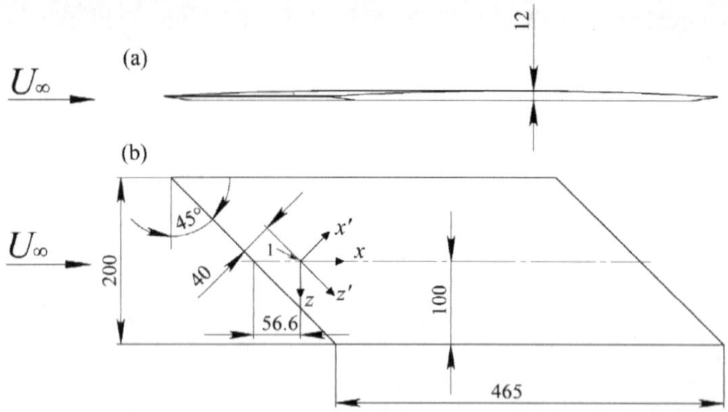

Figure 1. A sketch of the swept wing and coordinate systems used; (a) – side view; (b) – plan view; (1) marks the disturbance source.

The obtained fluctuation time traces normalized by the mean voltage CTA output at each spatial measurement point were transformed to the mass flow fluctuations according to the relation $m(x',z',t) = e'(x',z',t)/(Q{\cdot}E(x',z'))$. Here $e'(x',z',t)$ is the fluctuation signal, $E(x', z')$ the mean voltage from CTA output, and Q the probe sensitivity to the mass flow fluctuations. Note that the probe measures the x-component of the mass flow fluctuation. The frequency-wave disturbance spectra at a fixed x'-coordinate were determined using the discrete Fourier transform in the form:

$$m_{f\beta'} = \frac{\sqrt{2}}{T\delta_n} \sum_{j,k} m(z'_j, t_k) \exp[i(\beta' z'_j - \omega t_k)] \Delta t \Delta z'_j , \tag{1}$$

where T is the time-length of realization, β' the spanwise component of wave vector, and δ_n the boundary layer thickness scale for normalizing the spectra in β'; $\delta_n = 1$ mm. The Fourier transform in time was done using the Hanning window. After the Fourier transform, the amplitude and phase of the controlled disturbances were determined from

$$A_{f\beta'} = \sqrt{\mathrm{Re}^2(m_{f\beta'}) + \mathrm{Im}^2(m_{f\beta'})}, \tag{2}$$

$$\Phi_{f\beta'} = \arctan[-\mathrm{Im}(m_{f\beta'}) / \mathrm{Re}(m_{f\beta'})], \tag{3}$$

The streamwise wavenumber α_r is defined as

$$\alpha_r = \frac{\Delta\Phi_{f\beta'}}{\Delta x}, \tag{4}$$

where Δx is the distance between the measured sections in the x-direction.

3. Results

3.1. Natural development of disturbances

Before considering controlled disturbances, firstly we need to determine the regions of linear and nonlinear development of natural disturbances. Downstream distributions of the mean voltage E and

RMS mass flow pulsations <*m*> are presented in figure 2a. Note that the longitudinal coordinate x is measured from the disturbance source. The measurements were performed at constant mass flow that corresponds to the maximum in the disturbance amplitude profile. The region where a monotonic increase of <*m*> is started is usually associated with the finishing of the linear stage of the laminar-turbulent transition and the onset of the nonlinear evolution of disturbances. From this data, it can be seen that the transition onset occurs at $x \approx 50$ mm (i.e., at a distance of around 106 mm from the leading edge of the model or up to Reynolds number $Re_x = Re_1 \times (x + 56.6 \text{ mm}) \approx 0.53 \times 10^6$). Figure 2b shows the downstream evolution of the amplitude spectrum of the natural mass-flow fluctuations. The first three lines belong to the linear pulsation growth. Here the pulsation amplitude amplification occurs within the frequency range $f = 4$–30 kHz. In the region of nonlinear disturbance growth ($x > 50$ mm), both the low- and high-frequency disturbances with frequencies up to $f = 200$ kHz have rapid growth. At the last measured point ($x = 129$ mm), the spectrum is close to turbulent flow.

Hence, from this data, it can be concluded that the source of the controlled disturbances is located in the linear regime of the natural fluctuation development and, additionally, that the fundamental (20 kHz) and subharmonic (10 kHz) frequencies fall within the frequency range for most unstable natural disturbances.

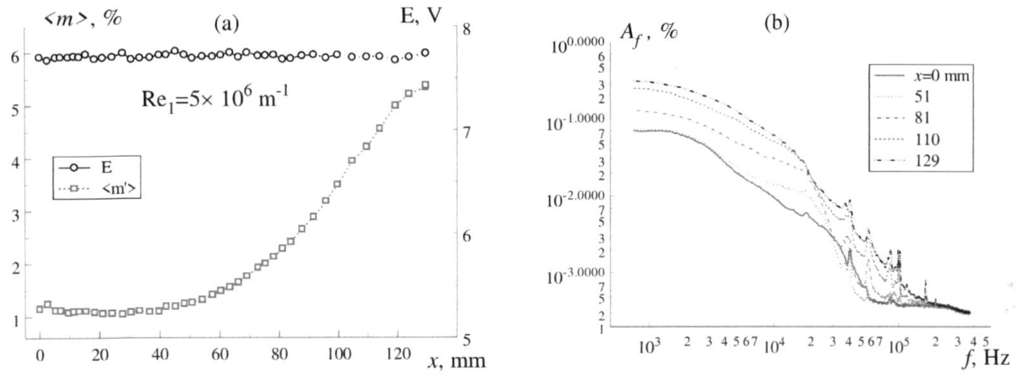

Figure 2. (a) Evolution of the mean voltage E and RMS mass flow fluctuations <*m*> downstream. (b) Amplitude spectrum of the natural disturbances.

3.2. Linear development of disturbances

In this section the controlled disturbance development was measured in three different sections at a fixed coordinate x' by means of displacing the anemometer probe along the z' coordinate, that is parallel to the leading edge of the model at $x = 40$, 50 and 60 mm. The measurements were made in the layer of maximum fluctuations, that corresponds to the condition $y/\delta \approx 0.6$, where δ is the boundary layer thickness.

The amplitude distributions of perturbations for both frequencies in the spanwise direction are shown in figure 3. One can see that in some places inside the wave packet the amplitude grows downstream, whereas the main amplitude maximum is slowly displaced in the positive direction of the z'-coordinate. Recall that the asymmetry of the amplitude distributions along the spanwise z'-coordinate, relative to the position of the discharge ($z' = 0$ mm), is associated with the presence of crossflow on the swept wing. The wave train is spread along the leading edge of the model in the direction opposite to the crossflow that is in agreement with previously obtained results for Mach number 2.0 [12, 13]. Note, that the amplitude for disturbances with frequency 10 kHz is even higher than that for disturbances with frequency 20 kHz.

Figure 4 shows the amplitude β'-spectra of the controlled disturbances at frequencies 10 and 20 kHz. As distinct from the flat plate case, the swept-wing β'-spectra are asymmetric. The amplitude growth is observed in the wide range of the spanwise wavenumbers, but the amplitude maximum is located at

$\beta' = 0.65$ rad/mm for 10 kHz and $\beta' = 0.9$ rad/mm for 20 kHz. These values correspond to the linear peaks in downstream wave train development. We should note that along with the linear peaks there is a number of additional peaks. Their existence is not explained by the linear theory and be discussed in the section 3.3.

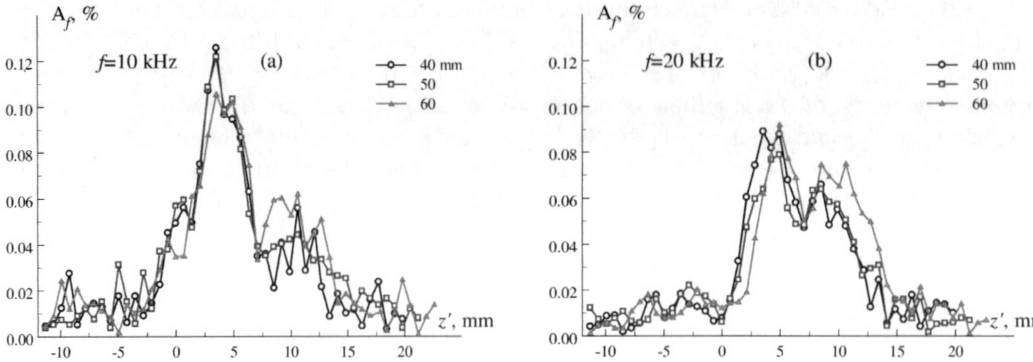

Figure 3. Amplitude distributions of disturbances in the spanwise direction. (a) $f = 10$ kHz and (b) $f = 20$ kHz.

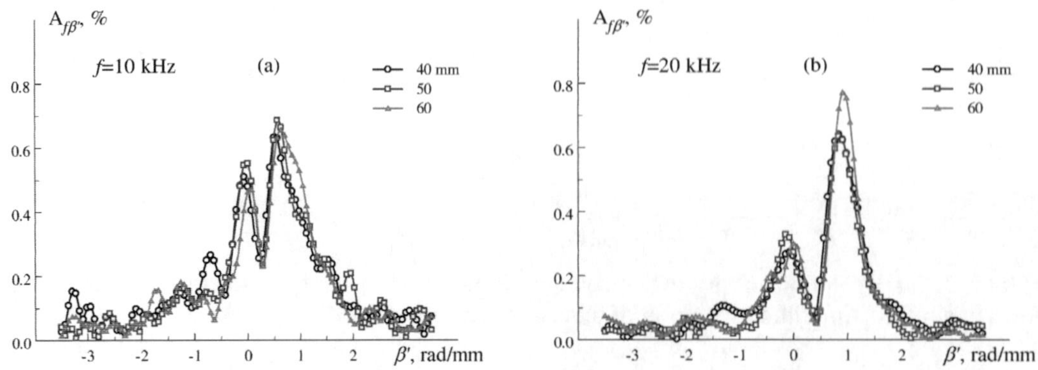

Figure 4. Amplitude β'-spectra of disturbances. (a) $f = 10$ kHz and (b) $f = 20$ kHz.

3.3. Nonlinear development of disturbances

Before trying to identify the nonlinear mechanisms in the experiment, the resonance conditions must be given. Generally, this condition for a three-wave triad is a phase synchronization of all three instability waves [18, 19]:

$$f_1 = f_2 + f_3, \quad \beta'_1 = \beta'_2 + \beta'_3, \quad \alpha'_{r1} = \alpha'_{r2} + \alpha'_{r3}, \tag{5}$$

For the subharmonic resonance mechanism the conditions (5) go with $f_1 = 20$ kHz and $f_2 = f_3 = 10$ kHz.

Figure 5 shows the amplitude β'-spectra of the controlled disturbances at frequencies 10 and 20 kHz for the sections $x = 60$, 70 and 80 mm. As previously obtained in 3.2, the greatest amplitude growth is observed at $\beta' = 0.75$ rad/mm for 10 kHz and $\beta' = 0.89$ rad/mm for 20 kHz that corresponds to the linear development of disturbances. In addition to the main linear peaks we can see that a second maximum forms near $\beta' \approx 0.13$ rad/mm for 10 kHz. Its appearance, as previously mentioned, can not be explained by the linear theory, since it is a result of nonlinear wave interactions.

To explain the appearance of additional peaks in amplitude β'-spectra at 10 kHz the estimation of dispersion relation for wave numbers $\alpha_r(\beta)$ (see formula (4)) and the verification of subharmonic resonance condition (5) were made. Figure 6 shows the streamwise wavenumber α_r as a function of the spanwise wavenumber β for both frequencies 10 and 20 kHz. Resonance conditions for wave numbers were chosen in accordance to maxima in amplitude β'-spectra for linear waves. In our case wave triplets can be find at $\beta' = 0.14$; 0.75 rad/mm (or $\beta = 0.03$; 0.89 rad/mm) at frequency 10 kHz and $\beta' = 0.89$ rad/mm (or $\beta = 0.9$ rad/mm) at 20 kHz. Graphically the subharmonic resonance is presented in figure 6.

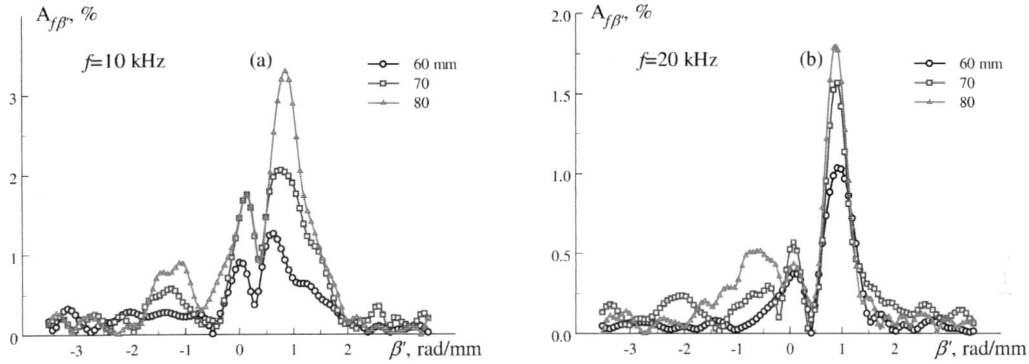

Figure 5. Amplitude β'-spectra of disturbances. (a) $f = 10$ kHz and (b) $f = 20$ kHz.

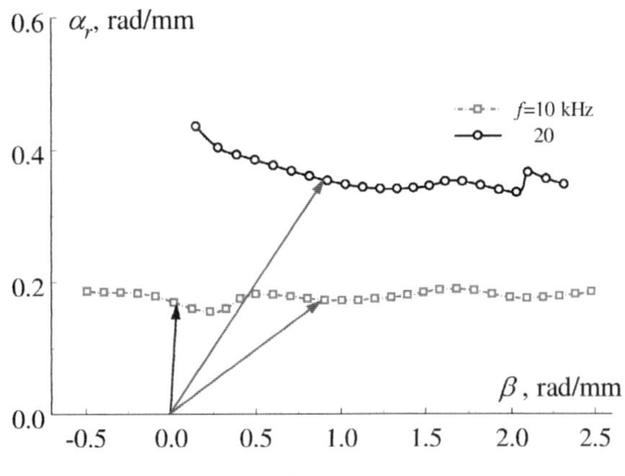

Figure 6. Dispersion relation and wave triplet.

Regarding other additional peaks in β'-spectra for both frequencies (at $\beta' \approx -1$ rad/mm for 10 kHz and $\beta' \approx -0.5$ rad/mm for 20 kHz), it can be caused by another nonlinear mechanism, for example the oblique breakdown mechanism [18] where the resonance triplet contains the steady wave. As due to a clean and polished surface of the model we can obtain the experimental data regarding only the travelling modes. The confirmation the existence of oblique breakdown mechanism in 3D supersonic boundary layer needs to introduce the controlled modulation of the base flow that is beyond this study.

4. Conclusion

An experimental study of nonlinear interactions of controlled disturbances in supersonic boundary layer on smooth swept wing at Mach 2.5 was carried out. The region of linear and nonlinear stage of natural disturbance evolution was defined. The amplitude distributions of the controlled disturbances are shown to be spatially asymmetric which is attributable to the presence of a crossflow in the three-dimensional boundary layer. It is found that for the linear development of disturbances the amplitude maxima are located at $\beta' = 0.65$ rad/mm for 10 kHz and $\beta' = 0.9$ rad/mm for 20 kHz. Downstream nonlinearity of the controlled disturbance evolution was detected and appearance of additional peaks in β'-spectra at subharmonic frequency can be explained by subharmonic resonance. It was experimentally demonstrated nonlinear amplification of stable in the linear sense disturbances in a wide range of β' wave numbers in the 3D supersonic boundary layer.

Acknowledgments

The work was supported by the Russian Foundation for Basic Research (Grant No. 16-31-00290).

References

[1] Bippes H 1999 *Prog. Aerospace Sci.* **35** 363
[2] Saric W S, Reed H L and White E B 2003 *Annu. Rev. Fluid Mech.* **35** 413
[3] Malik M R, Li F and Chang C L 1994 *J. Fluid Mech.* **268** 1
[4] Poll D I A 1985 *J. Fluid Mech.* **150** 329
[5] White E B and Saric W S 2005 *J. Fluid Mech.* **525** 275
[6] Deyhle H and Bippes H 1996 *J. Fluid Mech.* **316** 73
[7] Gaponenko V R, Ivanov A V and Kachanov Y S 1995 *Thermophys. Aeromechanics* **2** 287
[8] Wasserman P and Kloker M 2003 *J. Fluid Mech.* **483** 67
[9] Cattafesta III L N, Iyer V, Masad J A, King R A and Dagenhart J R 1995 *AIAA J.* **33** 2032
[10] Li F and Choudhari M 2011 *Theor. Comput. Fluid Dyn.* **25** 65
[11] Kosinov A D, Maslov A A and Shevelkov S G 1990 *J. Fluid Mech.* **219** 621
[12] Ermolaev Yu G, Kolosov G L, Kosinov A D and Semenov N V 2014 *Fluid Dynamics* **49** 188
[13] Kosinov A D, Kolosov G L, Semionov N V and Yermolaev Yu G 2016 *Phys. Fluids* **28** 064101
[14] Yatskikh A A, Ermolaev Yu G, Kosinov A D and Semionov N V 2015 *Thermophys. Aeromechanics* **22** 17
[15] Kosinov A D, Maslov A A and Semionov N V 1997 *J. Appl. Mech. Tech. Phys.* **38** 45
[16] Zelman M B and Maslennikova I I 1993 *J. Fluid Mech.* **252** 449
[17] Tumin A 1995 *Int. J. Non-Linear Mechanics* **30** 661
[18] Mayer C S J, Wernz S and Fasel H F 2011 *J. Fluid Mech.* **668** 113
[19] Kachanov Y S and Levchenko V Y 1984 *J. Fluid Mech.* **138** 209

A model of pyrolysis in a staged scheme of low-grade solid fuel gasification

A A Levin, V A Shamansky, A N Kozlov

Department of Thermal Power Systems, Melentiev Energy Systems Institute (ESI) SB RAS, 130, Lermontov str., Irkutsk, Russian Federation

Corresponding author's e-mail address: levin@isem.irk.ru

Abstract. The development and evolution of the theory of solid fuel combustion took place in the mid-20th century. The most studied research subject was sized fossil coal. At the same time the research on the fuels with a high yield of volatile matter (biomass) was limited due to insufficient data on kinetics of physical and chemical processes that occur at their heating. Obviously, the creation of a prospective low-tonnage technology for staged gasification of wooden biomass is possible only when a detailed mechanism of these processes is well understood and their kinetic parameters are known. This paper presents the initial results of the development of a model of wood pyrolysis in a screw reactor as the first stage of the multistage gasification process. One of the currently set goals in this research is to develop a mathematical model of heat and mass transfer processes to perform optimization calculations.

Introduction

Modern world trends to involve low-grade solid resources in the energy production are getting urgent for Russia and particularly for the regions that lack their own generating capacities [1]. The development and main evolution of the theory of solid fuel combustion occurred in the mid-20th century and the most studied research subject was sized fossil coal. At the same time the research into fuels with a high yield of volatile matter (biomass) was limited due to insufficient data on kinetics of physical and chemical processes that occur at their heating. It is obvious that a prospective low-tonnage technology for staged gasification of wooden biomass can only be created on the basis of comprehension of a detailed mechanism of these processes and knowledge of their kinetic parameters. At present the mechanisms of wood pyrolysis, gasification and combustion are dealt with in many studies [2-4]. This is explained by a growing interest of the world energy in renewable energy sources. Some technical difficulties, however, hinder the wide-scale adoption of the technology. These are:

- The number of operating parameters, that affect the operation of gas generator is great (exceeds 11). This circumstance complicates optimization of operating conditions of the plant.
- Low temperature of the third stage of the process prevents the complete coal residual combustion, thus leading to a decrease in the efficiency of gasification and formation of a great amount of coke and ash waste.
- Operation of modern gas consuming equipment (internal combustion engine, fuel cell, gas micro-turbine, etc.) requires additional gas cleaning from tar to a level of < 5- 100 mg/nm3. Fine gas cleaning makes the technology considerably more expensive. It is necessary to optimize stages II and III of the process to obtain virtually tar free gas.

This paper presents the initial results of the development of a model of wood pyrolysis in a screw reactor as the first stage of the multistage gasification process.

In terms of design the pyrolyzer represents a recuperative heat exchanger where a mix of stack and recirculation gases is considered as a heat carrier. To prevent sintering of particles the screw transportation of fuel is used.

To describe kinetics of pyrolysis and transportation of volatile products inside wood particles and their transfer to gas phase, the research was conducted using the simultaneous thermal analysis system: the weighting system of Netzsch STA 449 F1, quadruple mass-spectrometer QMS 409 C Aeolos and the system of pulse gas supply PulseTA (Unique research high-temperature plant). For the indicated plant we have developed the original techniques for interpretation of measurements, including the techniques for a technical analysis of fuels and determination of the kinetic coefficients and pyrolysis mechanism.

The model was verified by both the data of publications and the results of full-scale experiments using a laboratory plant for multistage gasification of low-grade solid fuel [5].

One of the goals set in this research is the development of a mathematical model of heat and mass transfer processes. The model will allow us to do optimization calculations. Thus, geometrical sizes of the pyrolyzer, thermal-physical properties of solid and gas phases, operating parameters of the plant can vary to perform the optimization calculations. Here the temperature profiles and rates should be calculated with a relatively low error within a short time period.

Model of pyrolyzer

Detailed numerical modeling of the pyrolyzer with an output of 30 kW was done in the Comsol Multiphysics Software that also enables us to optimize the design and operating parameters of the pyrolysis process in the screw reactor. The sizes applied in modeling the screw pyrolyzer are presented in Fig.2. 40 per cent of interior space is filled with solid mass, which should provide easy circulation of emerging gases around the screw. The screw represents a spiral with a wall 3 mm thick, which is wound on a hollow shaft. The interior space of the pyrolyzer with solid fuel and formed products is separated from heating gases by a wall 5 mm thick. The diameters of the interior space of the pyrolyzer and its external case are 149 mm and 200 mm, respectively.

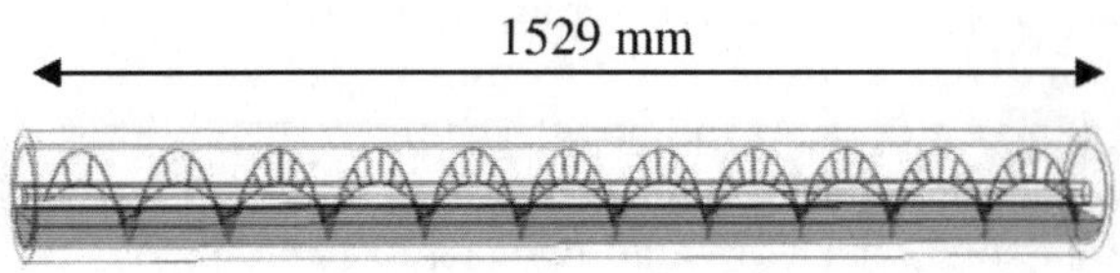

Figure 1. View of the screw pyrolyzer. The color highlights the area with porous solid fuel.

Modeling is performed for a homogeneous medium, considering physical properties (porosity, permeability etc.) of the solid phase, through which the pyrolysis products move:

$$\frac{\rho}{\varepsilon}\left((u_1\cdot\nabla)\frac{u_1}{\varepsilon}\right)=\nabla\cdot\left[-pI+\frac{\mu}{\varepsilon}\left(\nabla u_1+(\nabla u_1)^T\right)-\frac{2\mu}{3\varepsilon}(\nabla\cdot u_1)I\right]-\left(\frac{\mu}{k}+\beta_F|u_1|+\frac{Q_{br}}{\varepsilon^2}\right)u_1+F \quad (1)$$

$$\nabla\cdot(\rho u_1)=Q_{br} \quad (2)$$

The flow of stack gases was considered to be non-isothermal laminar:

$$\rho\left((u_2\cdot\nabla)u_2\right)=\nabla\cdot\left[-pI+\mu\left(\nabla u_2+(\nabla u_2)^T\right)-2/3\mu(\nabla\cdot u_2)I\right]+F \quad (3)$$

$$\nabla\cdot(\rho u_2)=0 \quad (4)$$

The effective heat transfer in pyrolyzer was considered according to the equation

$$\rho C_p u\cdot\nabla T+\nabla\cdot(-k_{eff}\nabla T)=Q \quad (5)$$

In the area of high temperature heads we took into account the radiation heat transfer:

$$Q=c_0(T_2^4-T_1^4) \quad (6)$$

The temperature profile included the areas of heating gases, separating steel wall, released pyrolysis products, porous medium and steel screw.

The boundary conditions selected to solve the problem are:

- Inlet fuel temperature – 20°C;
- Temperature of inlet heating gases – 600°C;
- Fuel flow rate in pyrolyzer – 1 mm/s;
- Rate of inlet heating gases varied in the range – 0,1÷1 m/s.

The thermal analysis was applied to study the pyrolysis of a fuel particle (pine). To this end the pine samples (chips, 1x1 cm in size, with an average mass of 145 mg) were heated in the flow of inert gas (argon) up to 600°C at various heating rates (1, 5, 10, 15, 30 K/min). The weight loss of the samples was recorded (Fig.2). Figure 2 shows that the sample is dried to the temperature of 120°C. The amount of moisture produced during heating at a minimum heating rate is higher than at the other heating rates. This can be explained by the fact that the fuel particle stays in the thermal analysis furnace for a much longer time, which fosters water release from the particle depth. In the temperature interval 200 – 400°C the high volatile yield occurs (about 65%). According to the shape of thermal analysis curve of weight loss we can suppose that this process runs in two stages in temperature intervals 200 – 300°C and 300 – 350°C, at a heating rate of 1 K/min. This is caused by the fact that in the beginning the volatiles leave the particle surface without destroying the structure, then at higher temperatures the particle is destroyed and volatiles are formed from its depth. In the temperature interval 400 – 600°C there is a deep decomposition of fuel particle with breaking the carbon skeleton bonds C-C and C-H. Weight loss at the heating rates other than the heating rate of 1 K/min shifts to the region of higher temperatures (by about 50°C). In this case the mechanism of fuel particle decomposition remains the same and ceases to depend on the heating rate.

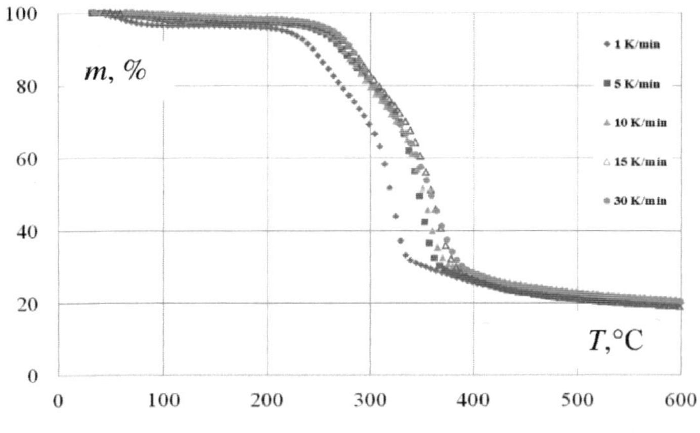

Figure 2. Relationship between the mass of fuel particle in the pyrolysis process and the temperature at stationary heating

Thus, the mechanism of fuel particle pyrolysis can be described by two successive stages

Biomass→Volatiles1+Char
Char→Volatiles2+ASH

Each stage can be described by the equation of the first order rate:

$$k = A \cdot \exp\left(-E/_{RT}\right), \tag{7}$$

where k – reaction rate constant, s^{-1}; A – pre-exponential factor, s^{-1}; E – activation energy, kJ/mole; R – gas constant, T – temperature, K.

With further processing of thermal gravimetric curves according to equation (7) we can construct the Arrhenius plot *ln(k)-1/T*. The plot makes it possible to determine the activation energy and pre-exponential factor (Fig.3).

The kinetic coefficient values calculated for stage 1 were E=31/3 kJ/,ole, lmA=3 s-1 and for stage 2 - E=23.5/mole, lnA=2 s-1. The obtained kinetic coefficients were applied in further modeling.

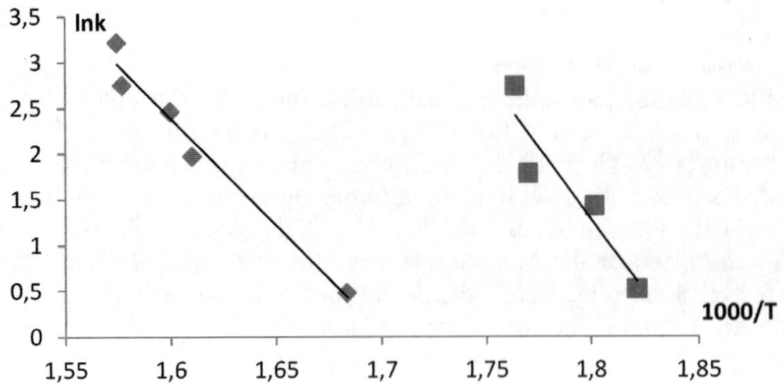

Figure 3. The Arrhenius plots obtained at pyrolysis of fuel particles at different heating rates.

Results and discussion

The constructed calculated mesh consisted of 604 thousand elements of three types (tetrahedrons, prisms and pyramids) and had the minimum size of 0.2 mm. The change in the value of calculated elements corresponded to the gradients of the main parameters to be determined and was carried out automatically by the Comsol Multiphysics Software.

One of the decisive factors of heat transfer and, as a result, the pyrolysis efficiency, is the volumetric rate of heating gases. In the course of numerical experiments the main attention was paid to the calculated model stability under changing boundary conditions.

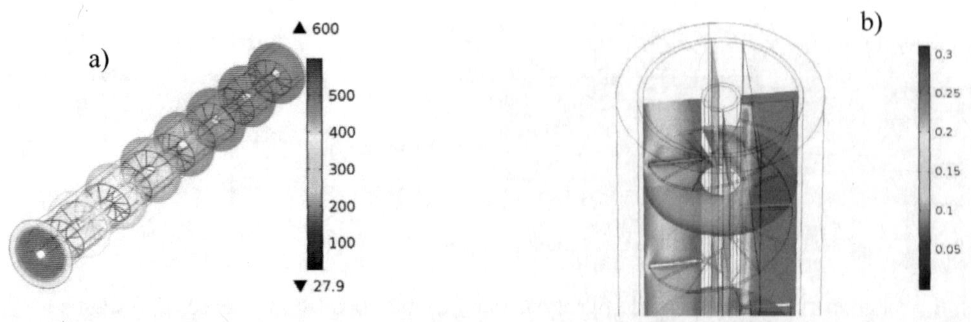

Figure 4. Example of calculated temperature values (a) and flow rate of pyrolysis products (b) at an average gas flow rate at the inlet section w = 0.1 m/s.

Figure 4 presents an example of temperature profiles and flow rate of gases formed in the process of pyrolysis that were obtained from the calculation. The emerging local areas of accelerated gas flow in the space between the screw shaft and porous medium are well seen. Consideration of the radiation heat transfer in the model led to a 40-fold increase in the time of calculation. Comparison of calculation results with the model (1)–(5) and (1)–(6) allows us to draw a conclusion about an insignificant contribution of the radiation heat transfer to the total thermal balance of the pyrolysis process in the modeled screw reactor. This made it possible to simplify the model and thus considerably reduce the computational effort, which is critical in the forthcoming optimization problem of working parameters and sizes of a staged gasification scheme to be developed.

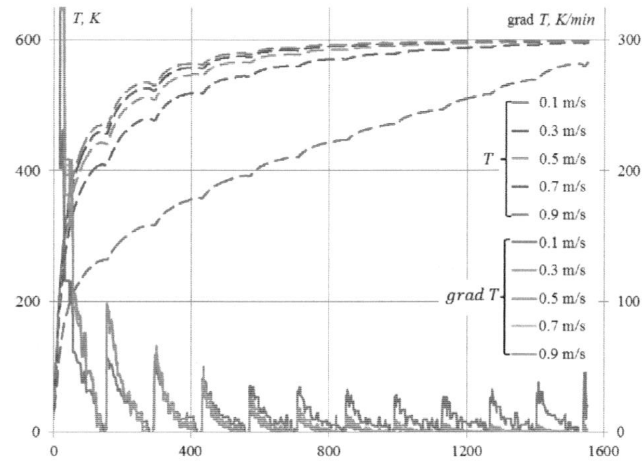

Figure 5. Calculated distribution of temperature values and its derivative versus various rates of heating gases.

Figure 5 presents changes in the calculated values of solid fuel temperatures and its heating rate determined by the temperature gradient, considering the condition of solid particle travel at a speed of 1mm/s. Non-uniform character of temperature rise is explained by the fact that the volume of fuel packed is crossed by the screw blade that has a higher temperature. Since the largest yield of volatile components occurs in the range of temperatures 200÷500 °C (Fig. 2), we can make a conclusion that the chemical kinetics equations determined at $dT/d\tau = 30$ K/min can be applied.
One of the preliminary conclusions of the conducted parametric research is the existence of an asymptote for the relationship between the solid fuel residence time in the temperature conditions required for pyrolysis and the amount of coming heating gases (Fig.4).

Conclusions

A model of heat exchange and aerodynamics was developed for pyrolysis in a staged scheme of solid fuel gasification. The model is intended for high speed optimization calculations. For the range of considered operating conditions we show the possibility of applying stationary experimental relationships ($dT/d\tau = 30$ K/min) to determine the rate of change in the density of solid fuel, volatile yield, etc. The results of the parametric research demonstrated the existence of an asymptote for an optimal amount of heating gases, that ensures the completeness of the first stage of pyrolysis.

[1] O.V.Marchenko, S.V.Solomin 2013 *Renewable energy sources for autonomous energy systems* (LAP Lambert Academic Publishing)
[2] Abhishek Sharma, Vishnu Pareek, Dongke Zhang 2015 Biomass pyrolysis. A review of modelling, process parameters and catalytic studies *Renewable and Sustainable Energy Reviews* Volume 50 pp 1081-1096
[3] Andrés Anca-Couce 2016 *Reaction mechanisms and multi-scale modelling of lignocellulosic biomass pyrolysis* Progress in Energy and Combustion Science Volume 53 pp 41-79.
[4] Xiangmei Meng, Wiebren de Jong, Fqatemeh Sadat Badri etc. 2012 Combustion study of partially gasified willow and DDGS chars using TG analysis and COMSOL modeling *Biomass and bioenergy* Volume 39 pp 356–369.
[5] Kozlov A, Svishchev D, Donskoy I, Shamansky V, Ryzhkov A. 2015 A technique proximate and ultimate analysis of solid fuels and coal tar *Journal of Thermal Analysis and Calorimetry* Volume 122 No. 3 pp 1213-1220.

Acknowledgments
The research was performed at Melentiev Energy Systems Institute SB RAS under the support of Russian Science Foundation (Grant № 16-19-10227).

Journal of Physics: Conference Series **754** (2016) 022007 doi:10.1088/1742-6596/754/2/022007

Simulation of helium–methane mixture separation on selectively permeable membranes

V S Naumkin

Kutateladze Institute of Thermophysics, Siberian Branch of the Russian Academy of Science, Ac. Lavrentiev ave., 1, Novosibirsk, Russia

E-mail: vsnaumkin@itp.nsc.ru

Abstract. In the article, the helium-methane mixture separation on the various types of membranes was considered. A flat membrane module was studied. It was made of two channels connected by a semipermeable membrane. It was shown that high membrane selectivity could not always provide a high degree of mixture separation.

1. Introduction

Currently, natural and petroleum gases are the raw material for production of helium. Helium content in the Russian oil fields in the Far East and Eastern Siberia is up to 0.6% (mole fraction). The method of low-temperature rectification is used to extract helium in the industrial scale [1]. Technical and economic evaluations show that it is promising to use the membrane technology for helium extraction [2]. The combined process is the most promising: enrichment of helium gas mixture using membranes and subsequent processing of helium concentrate on the low-temperature plants.

Wide-scale testing the plate membrane modules on natural gas containing about 5% helium was carried out by "Union Carbide Corporation" in 1970-72 years. The two-stage system with asymmetric membranes made of cellulose acetate allowed obtaining the helium concentrate with 82.5% of helium content at 62% recovery. Productivity was 18.7 m^3/h.

In Russia, now there is an experimental pilot plant for membrane helium extraction from natural gas [3]; it is possible to increase the helium content in helium concentrate from 0.5 to 30% by volume for two stages of separation. In [3], it is also pointed out that despite availability of highly selective membranes for gas separation to solve a particular task, it is necessary to take into account many factors determining the problem: pressure, concentration of the feed mixture components, and others.

The authors of [4] proposed to use the cenospheres to extract helium from a gas mixture. In [4], the cenospheres are placed in a container with gas mixture, helium diffuses into the cenospheres. Then, the pressure has been sharply decreased. By pressure decreasing, helium has been started to come out of cenospheres. The advantage of this method is high selectivity of helium extraction. The main disadvantage is the need to “recharge” cenospheres, i.e., impossibility of continuous operation.

In [5], it is proposed to use a xeolite membrane for separating mixtures containing methane, but it does not consider the possibility of helium-methane mixture separation. An advantage of xeolite membranes as compared to cenospheres is continuity of work. In [5], the xeolite membrane selectivity is considered in detail for pair gases: nitrogen/methane, carbon dioxide/methane. The helium flow through the xeolites was experimentally determined in [6].

Most studies of methane mixture separation consider the removal harmful impurities from natural gas: carbon dioxide, hydrogen sulfide, water vapor, etc. [7-10]. Harmful impurities can impede methane transport and contribute to corrosion of gas pipelines.

According to Australian Society membrane [11] and [12], polymer membranes can have high selectivity (about 4000) for the pair of gases helium/methane.

2. Problem formulation

In this paper, helium-methane mixture separation with the helium mass fraction of 0.1% on the membranes with different permeability coefficients is considered (Table 1).

Gas separation was simulated in a flat co-current flow membrane module, consisting of two channels, connected by a membrane with different permeability coefficients for helium and methane (Fig. 1). Module length is 1 m, the height of each channel is 2 mm. The influence of sidewalls was neglected. The approximation of narrow channel was used to simulate the flow. In this case, the gas dynamics equations are similar to the boundary layer equations. The flow of the gas mixture is stationary and laminar. The flows of gas mixture components through membrane were determined by difference of partial pressures above and below the membrane. The pressure in the lower channel was assumed to be P=40 atm., in the top channel, it was 1 atm. The properties of gas mixture components were determined by the polynomial dependence of [13]. It was believed that the thickness of the membrane selective layer is $\delta=10^{-6}$ m and $\delta=10^{-5}$ m. The influence of substrate membrane is neglected. The gas flow in the bottom channel is $G=10^{-2}$ kg/s, the Reynolds number is calculated by the hydraulic diameter and it is $\mathrm{Re}_d = 1770$.

Table 1. Helium and methan permeabilities and selectivity $\left[Cm^3_{STP} \cdot Cm / \left(Sec \cdot Cm^2 \cdot Cm.\ of\ \ Hg\right)\right]$ [12].

№	Membrane	He	CH_4	Selectivity He/CH_4
1.	Silicone rubber	2.3×10^{-8}	5.9×10^{-8}	0.39
2.	Phenylene silicone rubber	1.5×10^{-8}	2.0×10^{-8}	0.75
3.	Nitrile silicone rubber	0.79×10^{-8}	1.0×10^{-8}	0.79
4.	Polycarbonate	6.7×10^{-9}	0.36×10^{-9}	19
5.	Teflon FEP	6.2×10^{-9}	0.14×10^{-9}	44
6.	Polystyrene	3.5×10^{-9}	0.23×10^{-9}	15
7.	Trithene B	3.4×10^{-9}	0.0084×10^{-9}	400
8.	Ethyl cellulose	3.1×10^{-9}	0.64×10^{-9}	4.9
9.	Ethylene-vinyl acetate	2.1×10^{-9}	1.1×10^{-9}	1.9
10.	Viton A	1.7×10^{-9}	0.016×10^{-9}	110
11.	Polyvinyl chloride (plasticized)	1.4×10^{-9}	0.2×10^{-9}	7
12.	Polyvinyl fluoride	1.8×10^{-10}	0.0065×10^{-10}	280
13.	Mylar	1.0×10^{-10}	0.006×10^{-10}	170
14.	Saran	6.6×10^{-12}	0.025×10^{-12}	260

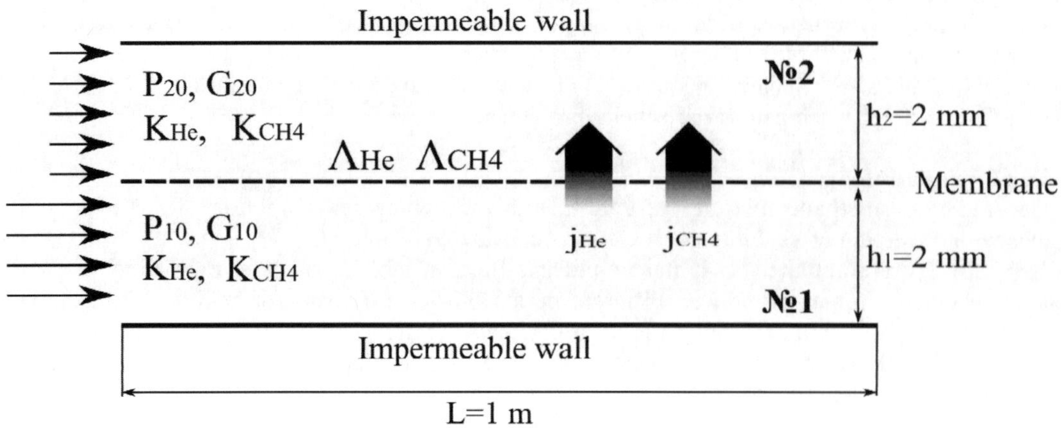

Figure 1. Gas flows in the membrane module.

3. Results

Distributions of helium and methane flows along the channel length for selective layer membrane thickness $\delta=10^{-6}$ m are shown in Fig. 2. It can be seen for all investigated membranes that the helium mass flow is much less than the methane flow. This is because the methane is the main component of the mixture. For this reason, the methane flow is constant all along the membrane length. The helium mass flow is reduced by the length of the membrane module.

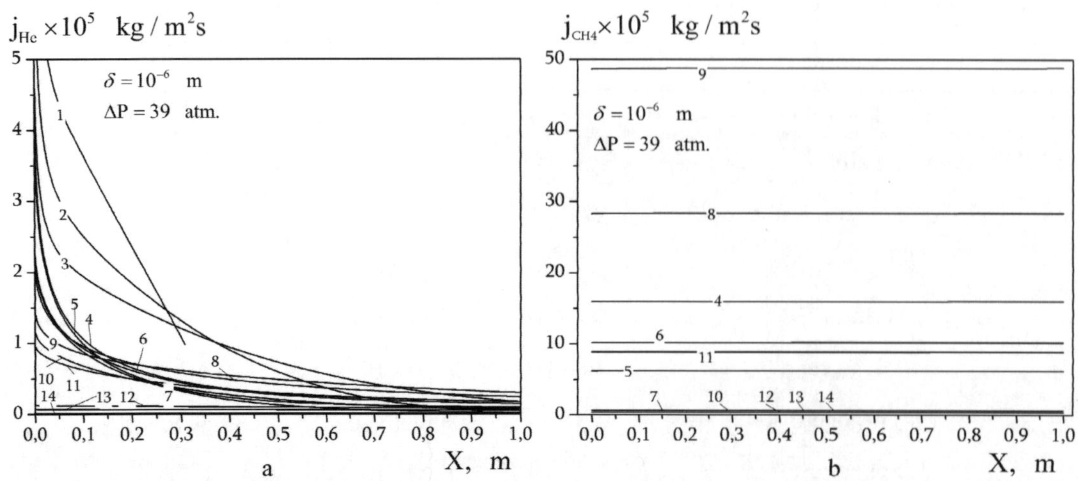

Figure 2. Longitudinal (a) helium and (b) methane flow distributions for various types of membrane: 1) Silicone rubber, 2) Phenylene silicone rubber, 3) Nitrile silicone rubber, 4) Polycarbonate, 5) Teflon FEP, 6) Polystyrene, 7) Trithene, 8) Ethyl cellulose, 9) Ethylene-vinyl acetate, 10) Viton A, 11) Polyvinyl chloride, 12) Polyvinyl fluoride, 13) Mylar, 14) Saran.

In both channels, there is a change in composition because the membrane is passed by the mixture components at different speeds. Fig. 3 shows a change in helium concentration along the channel, where the gas is blown. It is seen that most helium permeable membrane (No. 1, 2 and 3) as well as the least permeable (No. 12, 13, 14) have the worst separate properties. The first group combines the membranes, highly permeable to methane, the second group combines the membranes with a low permeability coefficient. Membrane No. 7 has the greatest separation factor among all examined membranes. This membrane allows obtaining the most enriched helium concentrate. Only membrane No.10 Viton A matches with membrane No. 7. The separation factor of membrane No.10 is three times less than this factor of membrane No. 7. Two above-mentioned membranes keep 70% of helium in the separated mixture.

However, increasing the selective layer thickness causes membrane No. 5 being more efficient as compared to No. 7 and 10, although it has the separation factor less than other membranes. This is because membrane No. 5 has the higher coefficient of helium permeability than membranes No. 7 and No. 10.

In the lower duct, there is a decrease in helium concentration (Fig. 4). Membranes with the highest permeability coefficient (No. 1, 2 and 3) keep the least amount of helium in the separated mixture but at a small thickness of the selective layer, almost all fed mixture permeates through the membrane. Membranes No. 12, 13, 14 due to low permeability coefficients keep up to 90% of helium in the mixture. Other membranes (No. 4 -11) keep, in average, 60% of helium in retante.

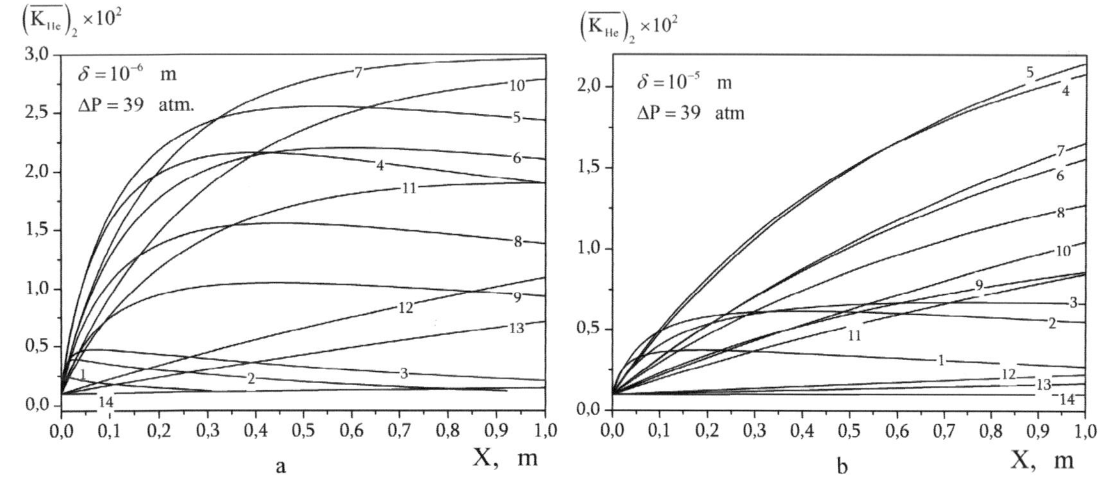

Figure 3. Mass concentration of helium in channel 2. Membrane thickness is $\delta=10^{-6}$m (a) and $\delta=10^{-5}$m (b).

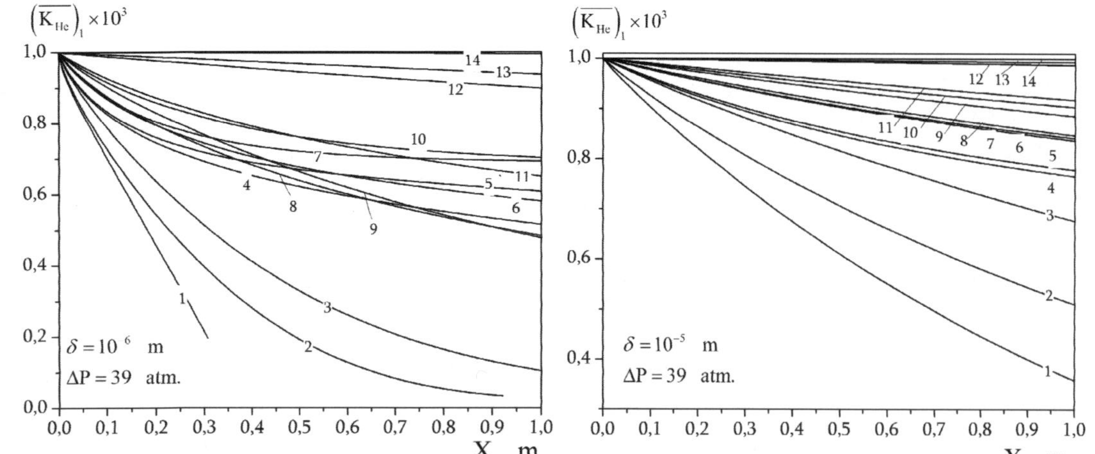

Figure 4. Mass concentration of helium in channel 1. Membrane thickness is $\delta=10^{-6}$m (a) and $\delta=10^{-5}$m (b).

4. Conclusion

The high selectivity of membrane is not always able to provide a high degree of mixture separation. Membrane No. 7 Trithene B with selectivity of helium/methane gas pair of 400, has the best gas separation properties as compared to other membranes only at a small thickness of the selective layer ($\delta=10^{-6}$ m). By increasing the selective layer thickness, membrane No. 7 demonstrates the worse separation degree than membrane No. 5 Teflon FEP with selectivity of 44 because permeability of membrane No. 5 for helium is higher than membrane permeability coefficient of membrane No. 7. These data are in good agreement with the data of [12].

Acknowledgments
This research was supported by Russian Science Foundation, project no. 14-19-00352.

References

[1] Molchanov S A 2011 *Zaschita okrushayuschey sredy v neftegasovom komplekse* **11** 36-42
[2] Laguntsov N I, Kurchatov I M, Karaseva M D, Solomahin V I 2014 Membrany i Membrannye Tekhnologii (International Academic Publishing Company Nauka/Interperiodica) **4** 272-279.
[3] Gas industry of Russia 2013 **11** 698
[4] Dolgushev S V, Fomin V M 2008 Computational continuum mechanics (Permian: Institute of Continuous Media Mechanics of the Ural Branch of RAS) **1** 66-75
[5] Ting Wu, Merritt C Diaz, Yihong Zheng, Rongfei Zhou, Hans H Funke, John L Falconer, Richard D Noble. 2015 Journal of Membrane Science (Elsevier Science Publishing Company, Inc.) 201–209
[6] Jonas Hedlund, Mattias Grahn, Danil Korelskiy, Mark Rayson, Sven Oberg, Patrick R. Briddon 2012 Journal of Membrane Science (Elsevier Science Publishing Company, Inc.) 271–77
[7] Mahdi Hedayat, Mohammad Soltanieh, Seyyed Abbas Mousavi. 2011 Journal of Membrane Science (Elsevier Science Publishing Company, Inc.) 191– 7
[8] Raja Swaidan, Xiaohua Ma, Eric Litwiller, Ingo Pinnau 2013 Journal of Membrane Science (Elsevier Science Publishing Company, Inc.) 387–94
[9] Raja Swaidan, Bader S Ghanem, Eric Litwiller, Ingo Pinnau Journal of Membrane Science (Elsevier Science Publishing Company, Inc.) 95–102
[10] Hafez Maghsoudi, Mohammad Soltanieh 2014 Journal of Membrane Science 159–65
[11] https://www.membrane-australasia.org/polymer-gas-separation-membranes
[12] Stern S A, Sinclair T F, Gareis P J, Vanldieck N P, Mohr P H 1965 Industrial and engineering chemistry **57**
[13] Gordon S, McBride B J 1994 NASA RP1311 (Washington) **1**

Turbulence and heat exchange inside the dome room of lidar station. Experiment and simulation

V V Nosov, V P Lukin, E V Nosov, A V Torgaev

V.E. Zuev Institute of Atmospheric Optics SB RAS, 1, Academician Zuev square, Tomsk, 634021, Russia

Nosov@iao.ru

Abstract. The structure of air turbulent motion inside the dome room – Primary Mirror (diameter 2.2 m) closed shaft at Siberian Lidar Station of V. E. Zuev Institute of Atmospheric Optics of Siberian Branch of the Russian Academy of Science has been experimentally and theoretically studied. Research is needed to forecast a laser radiation distortion. Results of experiments performed using the ultrasonic compact portable weather station are presented. The main heat exchange directions of air flows inside the shaft are determined. Theoretical results have been obtained by numerical solving of boundary value problem for Navier-Stokes equations. Solitary large vortices (coherent structures, topological solitons) are observed inside the shaft. Coherent decay of such vortices generates the coherent turbulence. It is determined that inside the primary mirror shaft one may expect the weakening of phase fluctuations of optical radiation, and thus improvement the lidar optical imaging. This increases the efficiency of Lidar Station.

1. Introduction

The coherent structures [1-5] are educed in experimental measurements of turbulence both in open atmosphere and indoors [6-11]. A hydrodynamic coherent structure – the compact formation, including the long-living spatial vortex structure (cell), resulting from the prolonged exposure of thermodynamic gradients, and the products of its discrete coherent cascade decay. In broad sense, the coherent structure – the solitary solitonic solution of hydrodynamics equations (topological 3D-soliton, solitary wave). It is either one-soliton solution or one soliton in multisoliton solution [6-11].

As stated in [6-11], the turbulence, resulting from decay of the major vortex in a single coherent structure, is coherent and deterministic. It contains both large-scale and small-scale decay products. Therewith, the turbulence one-dimensional spectrum (of velocity components and temperature) has the faster decrease in inertial interval (usually 8/3-power decrease, which becomes the stronger 12/3-decrease at high-frequency part of inertial interval) in comparison with Kolmogorov 5/3-power decrease [6-11, 16].

This paper presents the results of experimental and theoretical study of heat exchange and turbulence structure inside the dome room – closed shaft of the Primary Mirror (diameter 2.2 m) at Siberian Lidar Station of V. E. Zuev Institute of Atmospheric Optics of Siberian Branch of the Russian Academy of Science (IAO SB RAS) (figure 1). This research is needed to forecast a laser radiation distortion near the Primary Mirror and near the focus inside the dome room of lidar station. It allows to estimate (and forecast) the efficiency of the lidar station.

2. Experiment

The experimental studies of heat exchange characteristics and turbulence characteristics inside the dome room have been implemented in December 2015. Measurements were executed by the compact portable ultrasonic weather station at different points (figure 2) inside the shaft.

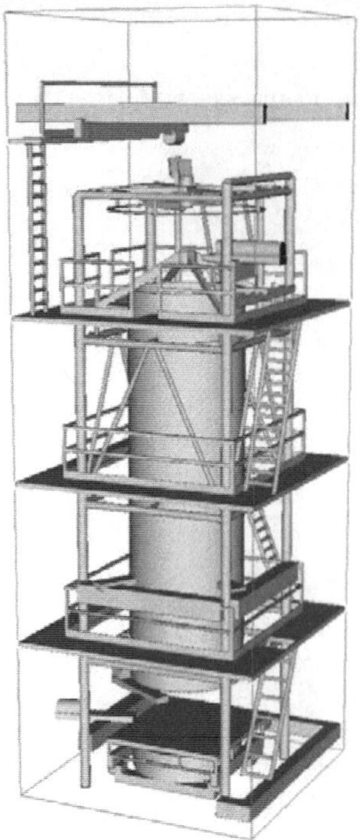

Figure 1. Model of the Primary Mirror shaft. Common South-East corner view

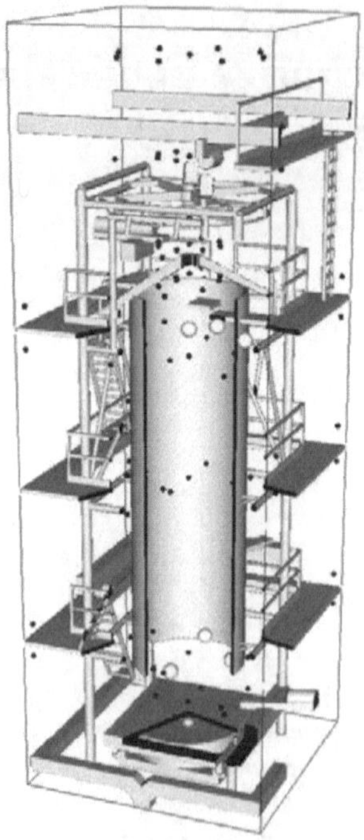

Figure 2. Location scheme of registration points: light spheres – experiment, dark dots - simulation. Near sector is truncated. North-West view

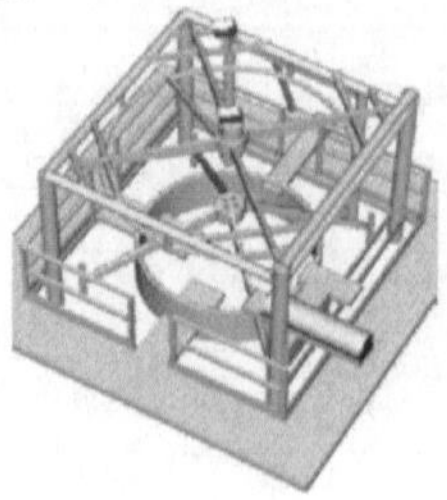

Figure 3. Model fragment – constructions and equipment at the focus. South-East upper view

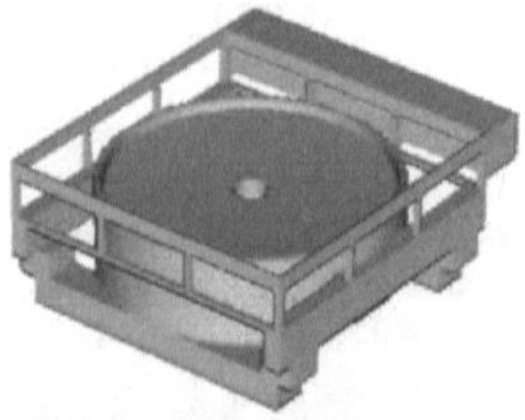

Figure 4. Model fragment – Primary Mirror (diameter 2.2 m). Shield is folded. South-East view

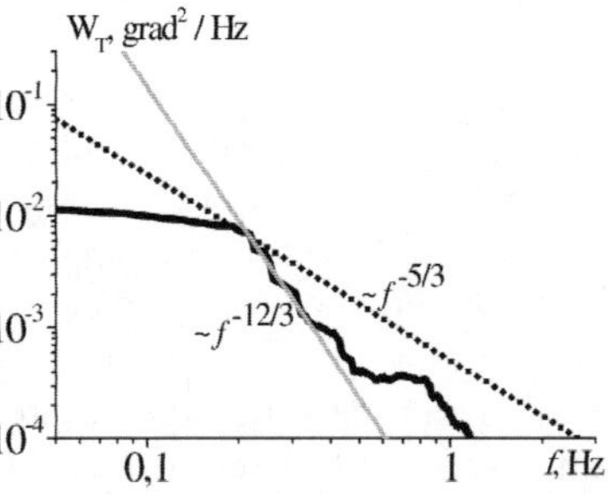

Figure 5. Experimental temporal frequency spectrum of temperature fluctuations at lower point

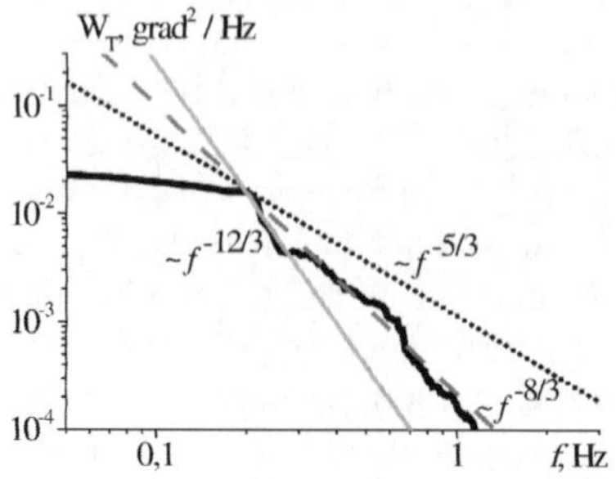

Figure 6. The same as figure 5, at upper point

Earlier, authors of this paper have performed the similar measurements inside the domes of several different telescopes, for example, astronomical telescopes AZT-33 and AZT-14 at Sayan Solar Observatory, Institute of Solar-Terrestrial Physics of RAS [16-19], and Large Altazimuth Telescope

(Bolshoi Teleskop Alt-azimutalnyi, BTA) at Special Astrophysical Observatory of RAS in the Caucasus Mountains [12, 13].

The theoretical results have been obtained by the solving numerically of boundary value problem for the equations of hydrodynamics (Navier-Stokes equations).

Figures 5, 6 show the experimental temporal frequency spectrum of temperature fluctuations W_T for some measuring points. In these cases the relatively extended inertial interval with 8/3- or 12/3-power decrease is observed in the spectra, it corresponds to the coherent turbulence [6-13, 19, 20].

3. Simulation

Our earlier numerical simulation results [12, 13] show that the solitary large vortices (coherent structures, topological solitons) are observed inside closed air volume over the homogeneously heated surface. The coherent decay of such vortices generates the coherent turbulence.

To explore the structure of air turbulent motion inside closed volume (without exchange of internal and external medium through the borders) the boundary value problem for hydrodynamics equations (Navier–Stokes) is solving. For numerical solving the hydrodynamics boundary value problem we have used The Gerris Flow Solver [14] – "a Free Software program for the solution of the partial differential equations describing fluid flow" (Navier–Stokes). The open-source code is available free of charge. The efficiency and the required accuracy of Software [14] had been tested and confirmed on the wide class of 100 typical test problems [15], which solutions give the satisfactory results.

The model of the shaft with constructions and equipment have been simulated (figures 1, 3, 4) on the basis of in-situ measuring the dimensions and the surfaces temperature, the assembly drawings have also been used.

4. Boundary Value Problem

For the numerical simulation we formulated the following boundary value problem for Navier-Stokes equations:

Dirichlet boundary conditions: temperature of shaft surfaces – corresponds to the experimentally measured temperature; boundaries velocity: *x*-, *y*-, *z*-component of velocity is zero (no-slip impermeable boundaries).

Initial conditions: temperature of medium (air) indoors – corresponds to the experimentally measured temperature; pressure P_0 = 1013.25 mbar.

Medium: the fluid like the air: kinematic viscosity $\nu = 17.3 \cdot 10^{-6}$ m^2/s; density $\rho = 1.2$ kg/m^3.

Constructions and equipment: indoors there are large optical elements and constructions of the lidar (figures 1, 3, 4), including Primary Mirror (elliptical paraboloid): diameter 2.2 m, focal length 10 m.

The shaft (figure 1) dimensions (w×h×d): 5.2×15.6×5.2 m. Location scheme of registration points inside the shaft is on figure 2.

As a result of solving the boundary value problem, the pattern of air motion inside the shaft was found in parameters of velocity vector field, temperature scalar field and pressure scalar field (figures 7, 8). We can see the solitary large vortices (coherent structures, topological solitons) indoors. We consider these vortices together as an analogue of Benard convection cell, formed in the closed rectangular volume with internal constructional contents of complex geometry.

5. Evolution of convective cells (theoretical and experimental data)

Figures 7, 8 illustrate the patterns of air flows inside the shaft model at different simulation timepoints. We note the prevailing influence of major inhomogeneities in the air motion. The geometry of constructions (the horizontal gangways and the vertical cylindrical blind) spatially causes the large vortex cells within each floor. The insignificant contribution of small vortices should appear as a more rapid decrease of the turbulence spectrum at high frequencies.

Experimental measurements have revealed the presence of air flows (up to 0.4 m/s) inside Primary Mirror shaft throughout its height (15.6 m), including the flows in technological gaps between the construction elements.

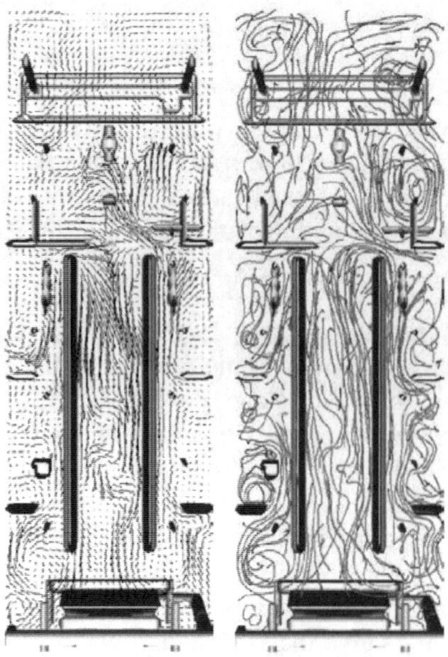

a b

Figure 7. Pattern of turbulent air motion inside Primary Mirror shaft on simulation timepoint t = 00:00:31, (a) velocity field at vertical cross-section layer North-South, (b) streamlines at the same layer. East side view

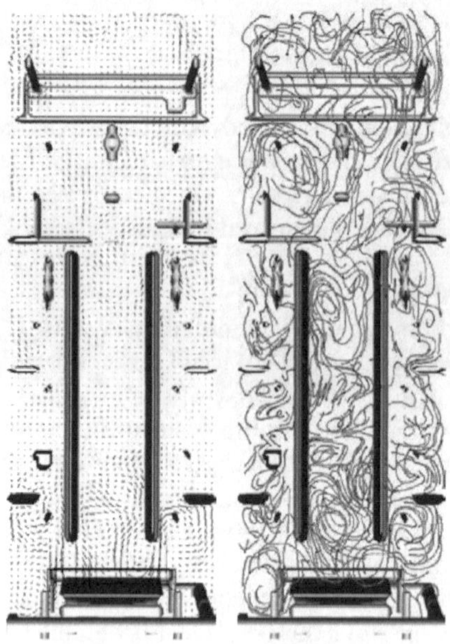

a b

Figure 8. The same as figure 6, on t = 01:48:16

The following heat-exchange directions of air motion inside the shaft were identified as a result of theoretical and experimental studies: 1) vertical downward direction inside the blind (made as a vertical tube of circular cross-section); 2) vertical upward – on external southern side of the blind (in gap between the blind and the horizontal gangways); 3) vertical downward one – on external northern side of the blind. The causes of these air flows, as it turned out, were the gradients of average temperature: 1) increasing the temperature at shaft lower part at places adjoining the shaft entrance doors from the adjacent warm room; 2) lowering the temperature at shaft upper part as a result of external cooling of the shaft sliding roof.

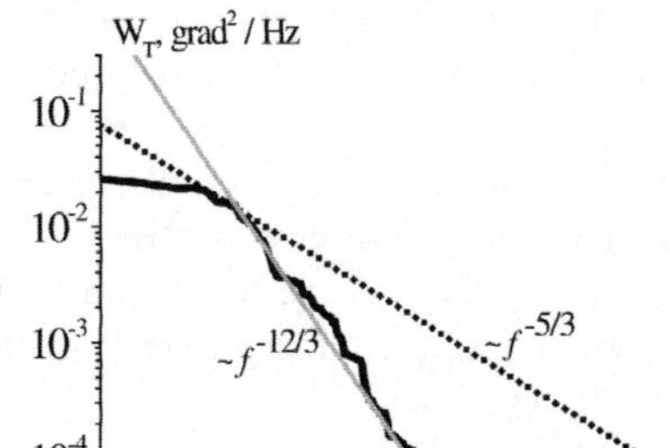

Figure 9. Temporal frequency spectrum of temperature fluctuations in simulation at lower point

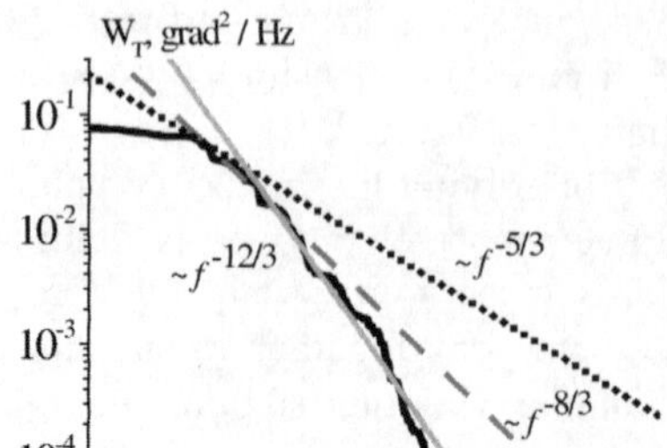

Figure 10. The same as figure 9, at upper point

6. Spectral analysis

Figures 9, 10 present the temporal frequency spectrum of temperature fluctuations in simulation. We see in the theoretical spectra like in the experimental ones (figures 5, 6) there is a relatively extended inertial interval with 8/3- or 12/3-power law. It corresponds to the coherent turbulence too.

The vertical directions of the air flows, similar to experimentally fixed ones, are observed in the simulation mainly at non-stationary mode. The air, chilled by contact with the shaft cold roof, falls down mostly inside the blind, which due to its geometry (vertical tube) collects it into a relatively narrow downdraft. The warm air from the rabbet ledge of shaft entrance door on ground floor (being displaced by the downdraft cold air) rises in the structural gaps between the gangways and the blind.

Such experimentally observed directions of air flows occur less frequently and become less expressed at steady thermal mode in the developed model. This follows from the fact that the shaft is not hermetically closed: the cold outdoor air flows into the shaft while the warm air flows out through the small gaps at the junction of shaft roof. The model developed here assumes a hermetically sealed room what leads to the pattern of air motion insignificantly different from the experimental one. At the same time, spectral analysis and comparison of the experimental and simulation data point to the fact that the developed model of shaft reasonably well reproduces the structure of turbulent air motion inside the shaft, confirming the presence of a coherent turbulence.

7. Conclusion

Thus, the scientific results of experimental and theoretical studies performed in the closed shaft at lidar station can be formulated as follows:

1. As a result of theoretical and experimental studies the heat-exchange directions of air motion inside the shaft were specified: 1) vertical downward direction inside the blind (made as a vertical tube of circular cross-section); 2) vertical upward – on external southern side of the blind (in gap between the blind and the horizontal gangways); 3) vertical downward one – on external northern side of the blind. The gradients of average temperature cause these air flows.

2. Both experiment measurements and numerical simulation indicate the presence of coherent turbulence inside the Primary Mirror shaft at lidar station. As is known, the coherent turbulence (compared to Kolmogorov turbulence) is depleted of small-scale inhomogeneities [6-11, 19, 20]. Therefore, on the basis of data [6-11], inside the closed shaft we expect the weakening of the phase fluctuations of the optical radiation, and therefore [6-11], we also expect the improvement of lidar optical images.

3. We consider the performed complex of actions for studying the structure of air turbulent motions inside the closed shaft at lidar station (the experimental measurement and numerical simulation without exchange of internal and external medium through the borders) as the first step to study the efficiency of Siberian Lidar Station of IAO SB RAS. The next step – the study of the structure of air turbulent motions inside the open shaft in conditions of free heat exchange and mass exchange between the shaft and open atmosphere.

Acknowledgements

The authors consider it their pleasant duty to express the sincere appreciation to Stéphane Popinet for his great selfless contribution to Science development in the form of free software Gerris created by him, which allowed to authors to focus their efforts in the study of hydrodynamics laws.

The authors are grateful to the Russian Foundation for Basic Research for financial support (grant no. 15-05-05404).

References

[1] Townsend A A 1956 *The structure of turbulent shear flow* 1st Ed. (Cambridge: Cambridge Univ. Press)

[2] Townsend A A 1959 *J. Fluid Mech.* Temperature fluctuations over a heated horizontal surface **5** 209-241

[3] Batchelor G K 1967 *An introduction to fluid dynamics* (Cambridge: Cambridge Univ. Press)

[4] Monin A S, Yaglom A M 1971 *Statistical Fluid Mechanics: Mechanics of Turbulence* Vol. 1 (Cambridge, Mass.: MIT Press)

[5] Monin A S, Yaglom A M 1975 *Statistical Fluid Mechanics: Mechanics of Turbulence* Vol. 2

(Cambridge, Mass.: MIT Press)

[6] Nosov V V, Grigorjev V M, Kovadlo P G, Lukin V P, Nosov E V, Torgaev AV 2007 *Atmospheric and oceanic optics* Astroclimate of specialized rooms of the Large solar vacuum telescope. Part 1 **20** 926-934

[7] Nosov V V, Grigorjev V M, Kovadlo P G, Lukin V P, Nosov E V, Torgaev AV 2008 *Atmospheric and oceanic optics* Astroclimate of specialized rooms at the Large Solar Vacuum Telescope. Part 2 **21** 180-190

[8] Nosov V V, Lukin V P, Nosov E V, Torgaev AV, Grigoriev V M, Kovadlo P G 2013 *Mathematical Models of Non-linear Phenomena, Processes and Systems* Coherent structures in the turbulent atmosphere, eds A B Nadycto et al. (N.Y.: Nova Science Publishers) chapter 20 pp 297-330

[9] Nosov V V, Kovadlo P G, Lukin V P, Torgaev AV 2013 *Atmospheric and Oceanic Optics* Atmospheric coherent turbulence **26** 201-206

[10] Nosov V V, Lukin V P, Nosov E V, Torgaev A V, Grigoriev V M, Kovadlo P G 2009 *Proc. SPIE* Coherent structures in turbulent atmosphere **7296** 729609-1-17

[11] Nosov V V, Lukin V P, Nosov E V, Torgaev A V 2016 *Applied Optics* Causes of non-Kolmogorov turbulence in the atmosphere **55** B163-B168

[12] Nosov V V, Lukin V P, Nosov E V, Torgaev AV 2014 *Proc. SPIE* Simulation of coherent structures (topological solitons) indoors by numerical solving of hydrodynamics equations **9292** 92920U-1-14

[13] Nosov V V, Lukin V P, Nosov E V, Torgaev AV 2015 *Optika Atmosfery i Okeana - Atmospheric and Oceanic Optics* Modelirovaniye kogerentnyh struktur (topologicheskih solitonov) v zakrytyh pomeshcheniyah putyom chislennogo resheniya uravneniy gidrodinamiki [Simulation of coherent structures (topological solitons) inside closed rooms by solving numerically hydrodynamic equations] **28** 120-133 (in Russian)

[14] Popinet S 2001-2015 *The Gerris Flow Solver* A free, open source, general-purpose fluid mechanics code. http://gfs.sourceforge.net (accessed 28.06.2016)

[15] Popinet S *100 Gerris Tests V.1.3.2.* http://gerris.dalembert.upmc.fr/gerris/tests/tests/index.html (accessed 28.06.2016)

[16] Nosov V V, Grigor'ev V M, Kovadlo P G, Lukin V P, Papushev P G, Torgaev AV 2006 *Solnechno-Zemnaya Fizika - Solar-Terrestrial Physics* Rezultaty izmereniy astroclimaticheskih harakteristik podkupolnogo prostranstva teleskopa AZT-33 Sayanskoy solnechnoy observatorii Instituta solnechno-zemnoy fiziki SO RAN [The measurements results of the astroclimatic characteristics in dome room of AZT-33 telescope at Sayan solar observatory of Institute of Solar-Terrestrial Physics SB RAS] **9** 101-103 (in Russian)

[17] Nosov V V, Grigoriev V M, Kovadlo P G, Lukin V P, Papushev P G, Torgaev AV 2008 *Proc. SPIE* Repeated testing of under dome astroclimate of AZT-33 telescope **7296** 729608-1-5

[18] Nosov V V, Grigoriev V M, Kovadlo P G, Lukin V P, Papushev P G, Torgaev AV 2008 *Proc. SPIE* Astroclimate inside the dome of AZT-14 telescope of Sayan Solar Observatory **6936** 69361R1-4

[19] Nosov V V, Grigor'ev V M, Kovadlo P G, Lukin V P, Nosov E V, Torgaev AV 2012 *Izvestia vuzov. Fizika* Kogerentnye struktury - elementarnye sostavlyayushchiye atmosfernoi turbulentnosti [Coherent structures - elementary components of the atmospheric turbulence] **55** 236-238 (in Russian)

[20] Nosov V V, Grigor'ev V M, Kovadlo P G, Lukin V P, Nosov E V, Torgaev AV 2013 The International Conference dedicated to the memory of academician A.M.Obukhov "Turbulence, atmosphere and climate dynamics" Abstracts Kogerentnye sostavlyayushchiye turbulentnosti [Coherent components of the turbulence] (Moscow: GEOS, IFA RAN) pp 43-47 (in Russian)

Heat and mass transfer with evaporation cooling of a porous plate

S N Makarova[1,2] and A A Shibaev[1,2]

[1] Kutateladze Institute of Thermophysics Siberian Branch of Russian Academy of Science, 1 Lavrentiev ave., Novosibirsk, Russia
[2] Nikolaev Institute of Inorganic Chemistry Siberian Branch of Russian Academy of Science, 3 Lavrentiev ave., Novosibirsk, Russia

alex-shib@yandex.ru

Abstract. In this paper the results of experimental and theoretical investigation of heat and mass transfer with adiabatic evaporation of bicomponent water/ethanol fluid to an air flow are presented. An innovative test section for the wind tunnel with an active thermal stabilization system, maintaining the cuvette temperature equal to the evaporation surface temperature, is used to provide the evaporation adiabatic conditions. The wall temperature obtained experimentally shows the presence of expressed quasi-stationary evaporation area, qualitatively similar to sublimation curves of volatile organometallic compounds. A theoretical model based on the similarity of heat and mass transfer processes for each of the evaporating solution component is suggested. This model allows to determine evaporation surface temperature (sublimation temperature) accounting for radiation effect.

1. Introduction

Transition-of-phase heat and mass transfer processes became a frequent practice in aeronautical and heat engineering. Suffice to point out evaporation and sublimation cooling processes, which are used as an effective thermal protection of gas-turbine engine blades. The important scientific and technical objective is rise in field reliability of cooled turbine blades.

Heat and mass transfer processes with evaporation of pure (single component) liquids at flat, cylindric and spherical surfaces are described in detail in the literature [1, 2]. At the same time, heat and mass transfer effects with bicomponent liquids evaporation are still poorly understood, despite their widespread use in technological processes. The problem becomes more complicated in case of binary liquid evaporation, as far as it is necessary to take into account the deviation from ideal mixture, azeotropism, compatibility of the liquid components, which are complex functions of temperature, pressure and composition. Apart from this, the concentration of easily volatile substance in the solution is decreasing more intensively in unsteady evaporation regime due to the difference of components fugacity. Evaporation of aqueous ethanol, methanol and acetone droplets results in temperature non-uniformity of a droplet surface with the passage of time [3]. It is necessary to take special actions to provide composition constancy of liquid mixture over time in steady evaporation regime. Steady evaporation of binary aqueous ethanol, methanol and acetone at a porous sphere surface in a continuous liquid phase feed regime is experimentally studied in [4]. There is suggested a correlation for dimensionless temperature of adiabatic binary liquid mixtures evaporation with temperatures of pure liquids adiabatic evaporation used as representative. The influence of composition of bicomponent liquid film flowing on the inner wall of a vertical cylinder on heat and mass transfer intensity is studied numerically in [5]. It is shown an addition of methanol in very small quantities to benzene, results in a significant increase of the liquid evaporation intensity. The influence of evaporation surface geometry on heat transfer coefficients during flow of freons binary mixture film is considered in experimental studies [6-8]. It is observed that changing the nature of film wave flow at Reynolds numbers Re>400 leads to the intensification of heat transfer during evaporation at transversely finned surface compared with smooth and diamond surfaces. The heat transfer coefficient for surface with net coating in evaporation regime and nucleate boiling, is significantly higher than for smooth and structured surfaces.

Control of heat influx to evaporation surface during experimental studies under laboratory conditions is one of the problems to solve. Spurious radiative and conductive heat influxes emitted by components of an apparatus are comparable to convective heat fluxes under investigation when evaporation surface temperature decreases sufficiently, heat and diffusion flows are small. The important objective is elimination or accurate control of addition heat fluxes. Ideal study conditions are known as an adiabatic evaporation. The adiabatic evaporation regime in practice realizes by means of heat-reflecting shielding [4, 9], small area of evaporation surface (drops, wetted fibers) and minimal size of thermometric probes (0.05...0.2 mm) [10]. An investigation into the influence of flow dynamics in a boundary layer on heat and mass transfer processes during evaporation requires a large test section area. Providing adiabatic conditions with evaporation surface rise becomes sophisticated problem since contact area with metallic elements of the apparatus, and area of radiative and conductive heat transfer rise.

The thermocouple measurements of evaporation surface temperature of aqueous ethanol at a flat porous plate under conditions close to adiabatic were carried out in this paper. The innovative thermal stabilization system of the test section containing Peltier coolers with feedback is used. This system allows to maintain the temperature of liquid inside the porous plate equal to the evaporation surface temperature. The obtained thermograms are used for verifying the theoretical model of binary liquid evaporation based on the assumption of temperature distribution similarity of a vapor-air flow and mixture components concentration over the boundary layer thickness.

2. Experimental setup and measurement

Heat and mass transfer during evaporation of aqueous ethanol at the porous plate with dimensions 100x150x10 mm into the boundary layer of air is studied. The innovative thermal stabilization system of the test section is used. The system makes use of Peltier elements with feedback and allows the temperature of liquid inside the plate to be equal to the temperature of evaporation surface. The porous plates are made of TZMK-10 (super-pure amorphous quartz fiber) with porosity of 90...95% and thermal conductivity of 0.05 W/m·deg. The test section design appears in figure 1.

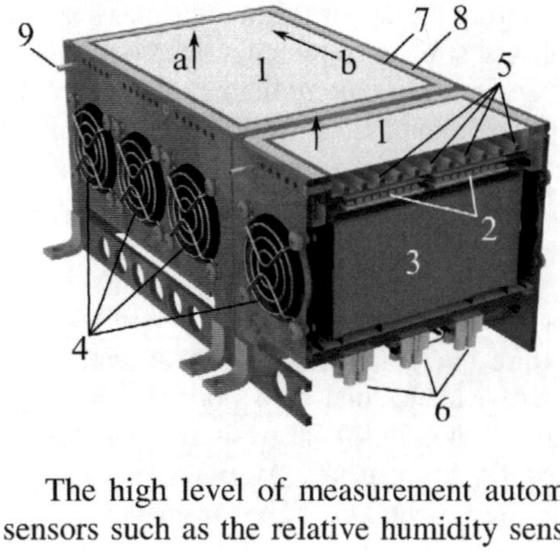

Figure 1. The test section design: 1 – porous plates; 2 – Peltier coolers; 3 – heatsinks; 4 – cooling fans; 5 – fluid inlet tube; 6 – commutator switch; 7 – cuvette; 8 – heat insulation; 9 – supply pipe; a – flow direction of evaporating fluid; b – main flow direction.

The high level of measurement automation achieved by using own software and modern set of sensors such as the relative humidity sensor SHT-15 (accuracy of at least 2%RH), absolute pressure sensor BMP180 (accuracy of at least 0.2%), multi-channel thermocouple measurement unit (TermoLab32 and ADC LCard 14-440. The Peltier coolers were controlled by remotely adjustable power supplies (four APS-7305L units). The location scheme of chromel-alumel microthermocouples and block diagram of the thermal stabilization system are displayed in figure 2. The average temperature of bottom cuvette wall determined by means of thermocouples ($p_1 \ldots p_6$) is maintained equal to the average temperature of evaporation surface obtained from thermocouples ($s_1 \ldots s_6$) when the thermal stabilization system is turned on.

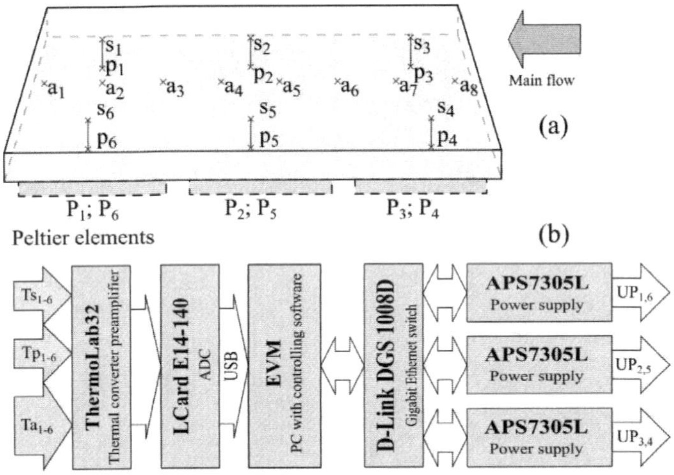

Figure 2. The location scheme of thermo-couples in porous plate (a): $P_1 \ldots P_6$ – set of Peltier elements with feedback by dint of thermocouples $s_1, p_1 \ldots s_6, p_6$; $a_1 \ldots a_6$ – measuring thermocouples; block diagram of thermal stabilization system (b).

The experiments were carried out using an open-circuit subsonic wind tunnel with cross-section at a nozzle exit with dimensions 108x108 mm. The flow velocity at the nozzle was maintained at 2 m/s. Pre-measured by LDA turbulence at the given flow velocity doesn't exceed 2%. The relative humidity of air didn't exceed 0.5%RH. Preliminary experiments have shown there is a significant error in measurements when the thermal stabilization system is turned off. The error in the determination of surface evaporation temperature in the conducted experiments is able to rise up to 50%.

3. Results and discussion

Series of measurements of aqueous ethanol evaporation at the high-porous plate interface in dry air was conducted. The surface evaporation temperature at different points of the porous plate depending on time is obtained. The data on temperature change during evaporation of a solution with mass ethanol concentration of 93.1 % are displayed in figure 3.

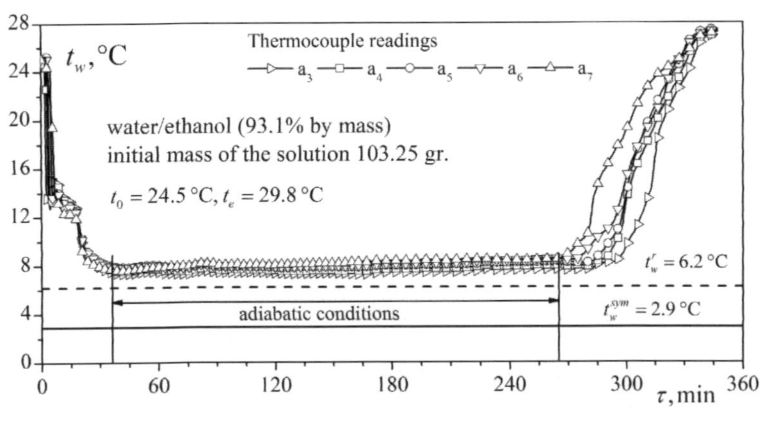

Figure 3. Temperature change at various surface points of the porous plate depending on time. The data referred to the evaporation process of water/ethanol solution with a mass concentration of ethanol of 93.1%.

There are three sections on the thermogram curve: the initial section corresponding to liquid cooling from room temperature to the temperature of steady evaporation regime; the section of steady adiabatic evaporation; the section corresponding to porous wall heating up to room temperature when evaporation process is over. The quasi-stationary evaporation regime occupies more than 3 hours, and the temperature exhibits a minimum at all thermocouple readings on this section. The equal temperature values fixed by thermocouples installed streamwise show that the composition of liquid mixture changes insignificantly and evaporation conditions are adiabatic on this section. The surface temperature obtained in experiment on this section is thus the adiabatic evaporation temperature of binary liquid with predetermined composition. The study results for mass ethanol concentration K_L in the solution of 0; 46.8; 75.5; 93.1% are listed in table 1 (the temperatures indicated in centigrade degrees, °C).

Table 1. The experimental conditions and study results for water/ethanol solutions.

K_L,%	t_0	t_e	t_w	t_w^{sym}	δ^r,°	t_w^r
0	21.0	26.4	8.8	5.5	968	9.0
46.8	21.4	25.4	3.0	-0.1	749	4.5
75.6	22.6	26.4	4.7	1.5	422	4.7
93.1	21.5	26.2	4.5	1.7	397	5.6
93.1	24.5	29.8	6.6	2.9	330	6.2

The upper limit of concentration is governed by the azeotropic solution point at standard pressure. Figure 3 and Table 1 show experimental values of surface evaporation temperature t_w distinguish from the temperature t_w^{sym} values, calculated using similarity of heat and mass transfer processes [11]. According to [12] the heat addition due to radiation sometimes significantly affects the surface evaporation temperature. The temperature of structural components near the test section was t_e during the experiments. The radiation influence in heat and mass transfer principles similarity [11] may be considered as:

$$t_w^r = t_0 - \left(K_{1w} Le_1^m / K_{3w} + K_{2w} Le_2^m / K_{3w}\right)\left(r/c_{p0} - \delta^r\right), \tag{1}$$

in this equation $\delta^r = \varepsilon_{eff}\sigma\left(T_e^4 - T_w^4\right)/j_w c_{p0}$ – the correction factor, depending on radiative heat transfer intensity and evaporation intensity, $\varepsilon_{eff} = 0.9$ – the reduced emissivity factor, calculated using emissivity factor of porous quartz, ethanol and surrounding surfaces, covered with organic dyes, $\sigma = 5.67 \cdot 10^{-8}$ W/m^{2}deg^4 – blackbody emission coefficient. To calculate the surface temperature t_w^r we derived the average evaporation intensity j_w from experimental data as a ratio of the initial mass of solution, filled in cuvette, to the evaporation surface area and to solution complete evaporation time. The complete evaporation time of solution was determined by the thermograms. The moment of time when surface temperature exceeds the main flow temperature and variation of the average surface temperature over the past 10 minutes doesn't exceed 0.5 deg. is considered the evaporation process end. The average evaporation intensity over plate area can be determined theoretically using mass transfer principles for components of the mixture under standard conditions. In laminar flow regime A=0.664; n=0.5; m=2/3:

$$\frac{j_w}{\rho_0 u_0} = \frac{A}{\mathrm{Re}_L^n}\left(\frac{K_{1w}}{K_{3w} Sc_1^m} + \frac{K_{2w}}{K_{3w} Sc_2^m}\right). \tag{2}$$

Figures 4 and 5 show the evaporation surface temperature depending on the water/ethanol fluid composition. Figure 4 shows the experimental points for adiabatic evaporation surface temperature distribution of water and aqueous ethanol with azeotropic composition over the length of plate. Note, the experimental data are almost identical with the results of heat and mass transfer numerical simulation computed considering the radiative heat fluxes according to [11]. In this case, wetted wall temperature rises over the length of plate. The evaporation surface temperature remains constant over the entire length in the absence of radiative heat transfer. In this case the calculated value of temperature is significantly below the experimental data.

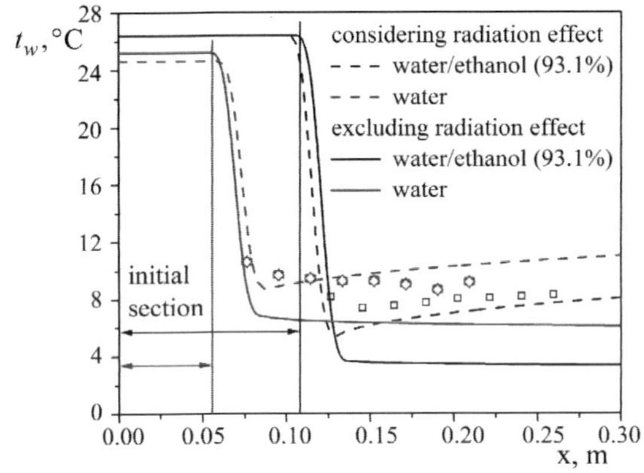

Figure 4. The temperature of porous plate surface depending on the length of plate in comparison with simulation data [11].

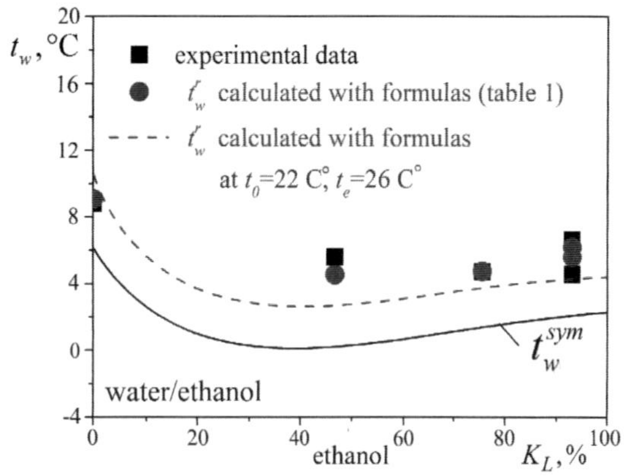

Figure 5. The temperature of porous plate surface plotted against binary liquid composition.

Figure 5 shows t_w^{sym} calculated by means of a formula (1) excluding the impact of δ^r. Here we can see lower values of the evaporation surface temperature. The data calculated with formulas (1) and (2) considering the radiation effect at the main air flow temperature of t_0 = 22 °C and room temperature of t_e = 26 °C (dash line t_w^r in figure 5) describe experimental data more accurately. The temperature reaches minimum value when mass content of ethanol in the solution is about 40%.

4. Conclusions

The conducted experiments showed an applicability of the developed equipment for study of heat and mass transfer during evaporation of binary liquid. The simple ratio for determining the temperature of wetted wall during evaporation of binary liquid was derived. The surface temperatures of aqueous ethanol evaporation calculated with account for radiation are close to ones obtained experimentally and by numerical simulation.

Acknowledgments

This work was supported by the Russian Science Foundation (grant No. 16-19-10325).

References

[1] Erbil H Y 2012 *Advances in Colloid and Interface Science* **170** 67–86

[2] Eames I W, Marr N J and Sabir H 1997 *Int. J. Heat Mass Transfer* **40** 2963–73

[3] Terekhov V I and Shishkin N E 2010 *Polzunov bulletin* **1** 55–9

[4] Terekhov V I and Shishkin N E 2009 *Thermophysics and Aeromechanics* **16** 239–45

[5] Baumann W W and Thiele F 1990 *Int. J. Heat Mass Transfer* **33** 267–73

[6] Pecherkin N I, Pavlenko A N and Volodin O A 2011 *Thermophysics and Aeromechanics* **18** 579–89

[7] Volodin O A, Pavlenko A N and Pecherkin N I 2013 *High Temperature* **51** 785–94

[8] Volodin O A, Pavlenko A N, Pecherkin N I and Serdyukov V S 2014 *Bulletin of Novosibirsk*

State University: Physics **9** 70–8
[9] Haji M and Chow L C 1988 *Journal of heat transfer* **4** 185–92
[10] Lukashov V V 2003 *Theor. Found. Chem. Eng.* **37** 325–9
[11] Makarov M S and Makarova S N 2016 *Thermophysics and Aeromechanics* **23** 23–32
[12] Boyarshinov B F, Volchkov E P and Terekhov V I 1985 *Bulletin of Siberian Branch of the Academy of Sciences of the USSR: Series of technical sciences* **16** 13–22

Turbulent convective flows in a cubic cavity at high Prandtl number

A Vasiliev, A Sukhanovskii, P Frick

Institute of Continuous Media Mechanics - Academ. Korolyov, 1, Perm, 614013, Russia

E-mail: `vasiliev.a@icmm.ru`

Abstract. Characteristics of turbulent convective flows in a cubic cell is studied experimentally for high values of Prandtl number. The first set was carriied out with propylene glycol (Pr = 64 and the second one with 25% water solution of propylene glycol (Pr = 24). It was found that increasing of Pr from 6.1 to 24 leads only to the slight change of intensity of the flow but during the next increasing of Pr from 24 to 64 the flow changes its structure.

1. Introduction

Turbulent free convection in closed cavities is a basic object for modeling convective processes in engineering applications but even in case of a simple geometry it is a complex problem. Convection in a cubic cavity with vertical temperature difference is characterized by number of regimes and was studied extensively for moderate values of Rayleigh number $Ra < 10^8$ [1]. Specific of convective flows under vertical temperature difference is their three-dimensional structure even at small supercriticalities. Experimental studies (these results were reviewed in [2]) showed that the case of strictly vertical temperature difference is characterized by a variety of large-scale flows.

Among the problems of large-scale dynamics of Rayleigh-Bénard convection we point out the studies of inversions of large-scale circulation in cavities of different geometry (mostly in cylinders) [3, 4, 5, 6, 7, 8]. Spontaneous inversions of large-scale circulation (LSC) for rectangular cavity with $\Gamma = 0.25$ were found in [9]. Regimes of LSC in rectangular tank of different aspect ratio was studied experimentally in [10], where it was shown that the LSC strongly depends on the small variation of the cavity thickness. Three-dimensional convective flows in a cubical cavity were experimentally and numerically studied in [11, 12], where the variety of dynamical regimes was observed.

This study is focused on the influence of physical properties of a fluid on the structure and characteristics of turbulent flows. Earlier, the influence of kinematic viscosity on turbulent characteristics was studied in [12]. Experiments were carried out for water at $T_0 = 25$ °C and 50 °C and values of Prandtl number Pr = 6.1 and Pr = 3.5 respectively. Experimental results of [12] did not reveal strong dependence of the structure and characteristics of convective flows on Prandtl number. Here we present new experiments where we considered fluids at fixed values Rayleigh number but for substantially higher values of Prandtl number (64 and 24).

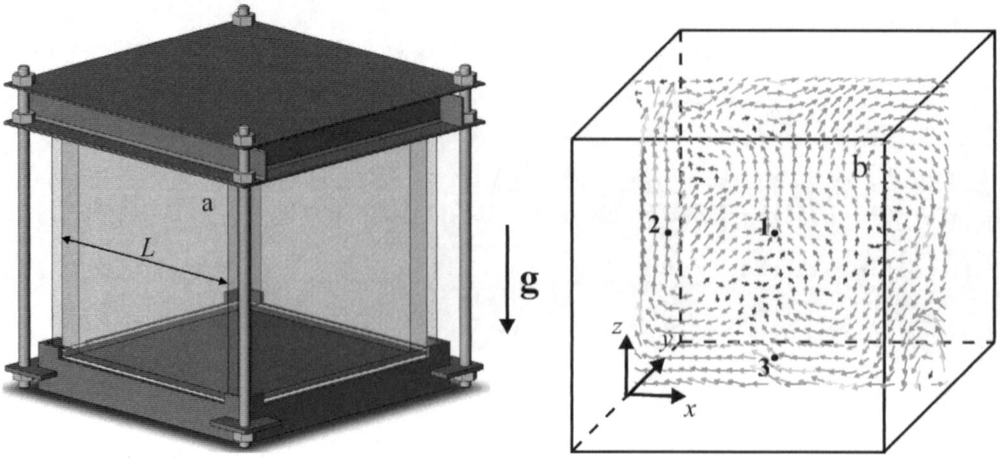

Figure 1. a – experimental cell, b – the middle vertical cross-section with an example of PIV velocity field.

2. Experimental setup

Experimental studies of turbulent convection were carried out in a cubic cell with the side length $L = 250$ mm (Fig. 1,a). Vertical sides had been made of plexiglas of 25 mm thickness. The top and bottom walls were used as heat exchangers. The temperature in copper heat exchangers was kept constant by recirculation of fluid from a thermostatic bath and controlled by embedded thermocouples with accuracy 0.1 °C. Propylene glycol and its 25% water solution were used as working fluids.

In order to reduce the heat losses the side walls were heat-insulated by cellular polystyrene plates of 40 mm thickness, with a plexiglas window for PIV measurements.

Governing parameters are Prandtl and Rayleigh numbers

$$Pr = \frac{\nu}{\chi}, \qquad Ra = \frac{g\beta L^3 \Delta T}{\nu\chi},$$

where g is the gravitational acceleration, β is the coefficient of thermal expansion, ν is the kinematic viscosity, χ is thermal diffusivity, and ΔT is the temperature difference between the hot and cold plates. Measurements of instantaneous velocity fields were carried out in the middle vertical cross-section (Fig. 1,b) by PIV technique using PIV system POLIS. Time shift between two consecutive PIV images were varied from 40 to 100 ms in order to get optimal tracers displacement for different values of Rayleigh number. Interrogation window was 32x32 pixels with 50% overlap. Spatial resolution for most of experiments was about 3 mm. The relative error of the velocity measurements was accurate to within 5%.

Temperature measurements were done by copper-constantan thermocouple array, located in the middle horizontal cross-section. Thermocouple module (NI 9213) was used to measure the voltage of the thermocouple at a sampling rate of 10 Hz.

Table 1 shows the characteristics of experiments that were realized.

3. Results

One of the main parameters which describes the structure and characteristics of convective flows is Prandtl number. At high Prandtl numbers heat transfer is provided by convection

Table 1. Characteristics of experiments

Experiment	ΔT, °C	Pr	Ra
I	5	64	$9.4 \cdot 10^8$
II	8.6	64	$1.6 \cdot 10^9$
III	25	64	$4.7 \cdot 10^9$
IV	5	24	$1.3 \cdot 10^9$
V	10	24	$2.7 \cdot 10^9$
VI	25	24	$5.2 \cdot 10^9$

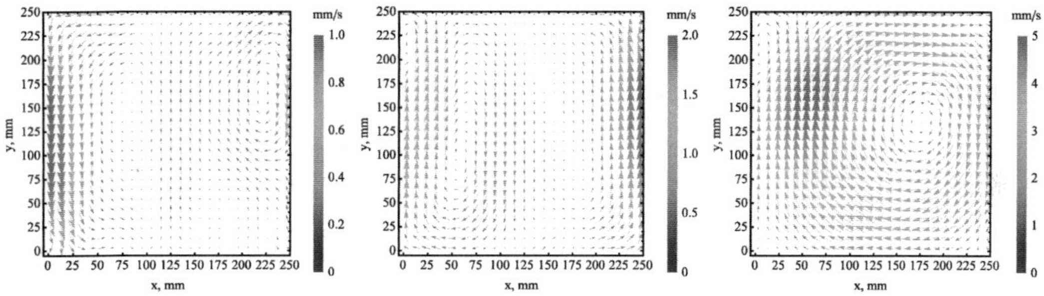

Figure 2. Time-averaged velocity fields in central vertical cross-section, experiments I-III (from left to right).

when at low Prandtl numbers thermal conductivity is more efficient (flor liquid metals). Earlier convection in a cubic cell for two values of Prandtl was studied in [12]. As a next step it is very interesting to study what would happen with a flow structure if we replace water with a fluid with substantially higher value of Prandtl number. Using propylene glycol (experiments I–III) and its 25% water solution (experiments IV–VI) as working fluids allowed us to carry out experiments for Rayleigh numbers comparable to the ones in [12] but for high Prandtl numbers (64 and 24). Time-averaged velocity fields in central vertical cross-section obtained for propylene glycol are shown in Fig. 2.

First remarkable difference from experiments with water is the large-scale flow structure. All experiments in [12] had large-scale circulation (LSC) oriented along one of the diagonals. Fig. 2 shows that for experiment I there is no LSC which occupies the whole cell. We can assume that for this experiment the upward convective flow is mainly located in the center and downward flows are near the side walls. Increasing of Rayleigh number leads to the change of the flow structure and in experiment III we can see LSC which occupies the whole cell. The structure of LSC in a central vertical cross-section is asymmetric.

The distribution of turbulent pulsations (Fig. 3) is also very different in comparison with water experiments ([12]). In water experiments the maximums of turbulent pulsations are located near corners of the cell. In experiments with propylene glycol the turbulent pulsations are located mostly in the central part of the cell with dominance of pulsations of vertical velocity. It means that turbulent flow of substantially more viscous fluid are strongly anisotropic. Even when LSC is formed (experiment III) the level of turbulent pulsations in a center of the cell is still high. We assume that thermal plumes is the source of these intensive turbulent pulsations.

For description of the temporal evolution of the large-scale circulation we decompose the

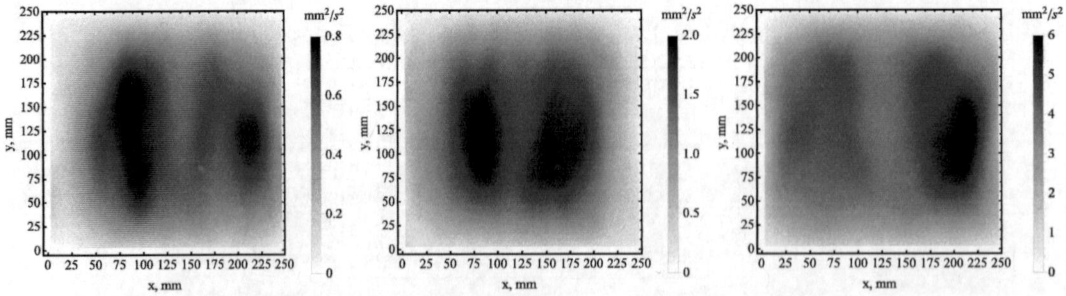

Figure 3. Experimental distributions of the mean turbulent kinetic energy in central vertical cross-section, experiments I-III (from left to right).

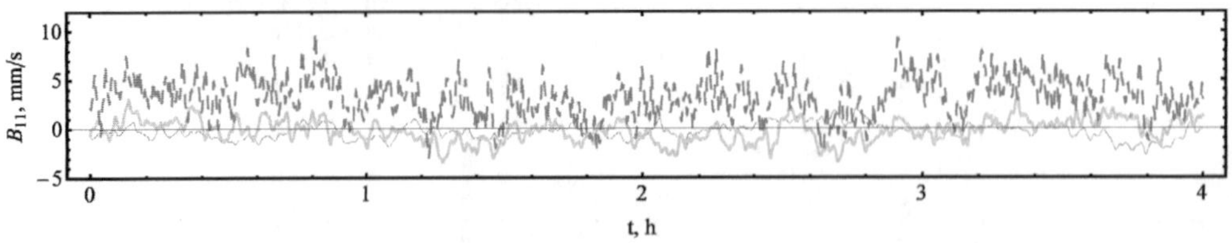

Figure 4. Variations of B_{11}: thin blue line – Experiment I, thick green line – Experiment II, dashed red line – Experiment III.

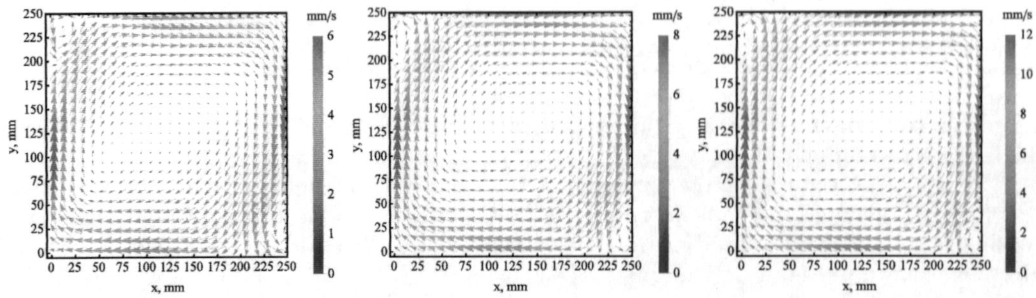

Figure 5. Time-averaged velocity fields in central vertical cross-section, experiments IV-VI (from left to right).

vertical component of the velocity into Fourier series

$$v_z(t,x,z) = \sum_n \sum_m B_{nm}(t) \cos(\frac{\pi n x}{L}) \sin(\frac{\pi m z}{L})$$

and we analyze the time variations of the amplitudes of lowest modes, assuming that the intensity of LSC is characterized by $B_{11}(t)$. Time variation of $B_{11}(t)$ is shown in Fig. 4. It is clear that LSC exist only in experiment III at relatively high Rayleigh number but time variation of LSC is non-periodic without reversals (change of LSC direction).

Second series of experiments (IV-VI) was carried out with 25% water solution of propylene glycol (Pr=24). The goal was to consider the case when Pr is still high in comparison with water but substantially less than The general structure of mean velocity fields (Fig. 5) are similar to the ones in water experiments [12]. Increasing of Pr from 6.1 to 24 leads only to the slight change of intensity of the flow but during the next increasing of Pr from 24 to 64 the flow changes its structure.

4. Results

The main goal of this study was revealing the influence of the variation of Prandtl number on characteristics of turbulent convective flows in a cubic cell. The first set of experiments was carried out with propylene glycol (Pr = 64) and the second one with 25% water solution of propylene glycol (Pr = 24). It was found that increasing of Pr from 6.1 to 24 leads only to the slight change of intensity of the flow but during the next increasing of Pr from 24 to 64 the flow strongly changes its structure. This result deserves special attention because it proves that variation of Prandtl number for fluids with relatively high values of Pr leads to the remarkable evolution of the flow. It can be important for description of thermal convection in different technological processes with highly viscous fluids.

The financial support of grant RFBR No.16-41-590406 is gratefully acknowledged.

References

[1] Valencia L, Pallares J, Cuesta I and Grau F X 2007 *International journal of heat and mass transfer* **50** 3203–3215
[2] Zimin V D and Frick P G 1988 *Turbulent Convection* (Nauka)
[3] Funfschilling D and Ahlers G 2004 *Physical Review Letters* **92** 194502
[4] Xi H D and Xia K Q 2007 *Physical Review E* **75** 066307–+
[5] Xi H D and Xia K Q 2008 *Physical Review E* **78** 036326–+
[6] Funfschilling D, Brown E and Ahlers G 2008 *Journal of Fluid Mechanics* **607** 119–139
[7] Brown E and Ahlers G 2009 *Journal of Fluid Mechanics* **638** 383–+
[8] Xi H D, Zhou S Q, Zhou Q, Chan T S and Xia K Q 2009 *Physical Review Letters* **102** 044503–+
[9] Sugiyama K, Ni R, Stevens R J A M, Chan T S, Zhou S Q, Xi H D, Sun C, Grossmann S, Xia K Q and Lohse D 2010 *Physical Review Letters* **105** 034503–+
[10] Vasiliev A Y and Frick P G 2011 *JETP Letters* **93** 330–334 ISSN 0021-3640
[11] Bolshukhin M A, Vasiliev A Y, Budnikov A V, Patrushev D N, Romanov R I, Sveshnikov D N, Sukhanovsky A N and Frick P G 2012 *Computational continuum mechanics* **5** 469–480
[12] Vasiliev A, Sukhanovskii A, Frick P, Budnikov A, Fomichev V, Bolshukhin M and Romanov R 2016 *International Journal of Heat and Mass Transfer* **102** 201 – 212 ISSN 0017-9310

Journal of Physics: Conference Series **754** (2016) 032001 doi:10.1088/1742-6596/754/3/032001

Linear waves on a surface of vertical rivulet

S P Aktershev[1,2], S V Alekseenko[1,2] and D G Arkhipov[1,2]

[1] Kutateladze Institute of Thermophysics, Lavrentiev Ave. 1, Novosibirsk 630090, Russia
[2] Novosibirsk State University, Pirogova St. 2, Novosibirsk 630090, Russia
E-mail: sergey-aktershev@mail.ru

Abstract. The type of film flow whereby the fluid flows in the form of many streamlets is typically called a rivulet flow. Whereas an individual streamlet bounded by two contact lines is called a rivulet. Special attention has been paid to rivulet flows because of their practical value for a variety of devices in power engineering and chemical technology, such as absorbers, distillation columns, evaporators, and heat exchangers for the liquefaction of natural gas. In the present paper the waves in vertical rivulet are investigated analytically. The Kapitza-Shkadov model is used to describe the wavy rivulet flow since it was well proven in the study of nonlinear waves in falling liquid films over a wide range of Reynolds numbers. The equations of the wavy rivulet flow are derived on the basis of the weighed residual method. These equations turn out to be the projections of the Shkadov's model equations on system of basis functions, constructed in special way. Linearizing these equations results in the dispersion relations for plane waves. The stability criterion for rivulet flows is deduced, and the analysis of dispersion relations depending on dimensionless parameters is carried out.

1. Introduction

Compared to uniform liquid film flows, rivulet dynamics exhibit several interesting features, such as the presence of movable contact lines and wetting angle hysteresis. Most theoretical studies, beginning with the pioneering study of Towell and Rothfeld [1], are devoted to stationary and smooth (without waves) rivulets flowing down an inclined plane as well as curved surfaces [2, 3]. Profiles of smooth rivulets and steady regimes of rivulet flows were calculated using both analytical methods in a lubrication approximation [2, 3] and numerical methods based on Navier-Stokes equations [4, 5]. In the simplest case of a thin rivulet flowing down a vertical flat surface, the rivulet profile was shown to represent an arc of a circle.

The stability of a moving rivulet has been studied in a limited number of works. A linear stability analysis was performed in [6-10] based on Navier-Stokes equations for the normal mode (plane waves). In these theoretical studies the stability calculations of rivulet are executed in lubrication approximation, i.e. at very small Reynolds numbers and for plane modes of perturbations only. Alekseenko, Markovich and Shtork [11] performed the first detailed experimental study of the wave motion of a rectilinear rivulet for the case of liquid flow along the lower outer wall of an inclined tube. With application of the imposed periodic pulsations of the flow rate the periodic non-linear waves have been studied, new wave modes are found, and comparison to known two-dimensional film flow is made. In the papers [12, 13] method of PIV was applied in the first time to measure instantaneous velocity field in wavy rivulet. Alekseenko *et al.* [14] were the only researchers to present the results of experimental investigation on periodic, nonlinear waves on a rectilinear rivulet flowing down a vertical wall. Waves in rivulet flow were studied in a wide range of a Reynolds number and forcing frequency for various values of angles of wetting. In [15] the results of numerical simulations of three-dimensional waves on the surface of a rivulet flowing down a vertical plate are presented. Various characteristics of linear and nonlinear regular waves in the rivulet are obtained through numerical calculations as a function of the forcing frequency at different Reynolds numbers and contact wetting angles. Calculations were executed only for those conditions which were implemented in experiments [14] and comparison with experimental data was made. The comparison shows that the applied model adequately describes the shape of the wave surface of a rivulet.

In the present paper the waves in vertical rivulet are investigated theoretically. On the basis of a weighed residual method the approach allowing simplify a problem is developed and dispersion relations for linear waves are deduced analytically at moderate Reynolds numbers.

2. Theoretical model

Let us consider a thin liquid layer flow over a vertical flat plate as a rivulet of constant width $2b$. We introduce a Cartesian coordinate system *Oxyz* with the *Ox*-axis directed along gravity and the *Oy*-axis directed along the normal to the plate (figure 1*b*). We assume that the layer thickness h is substantially smaller than the wavelength λ (long-wave approximation) and the rivulet semi-width b. All of these assumptions are in good agreement with experimental observations [14]. As a first approximation with respect to the small parameters $\varepsilon_\lambda = h/\lambda$ and $\varepsilon_b = h/b$, the waves in a liquid layer may be described by the integral boundary layer (IBL) model [15], which was developed for three-dimensional long waves in a falling liquid film. The corresponding momentum and mass balance equations in dimensionless variables x/H_0, z/H_0, h/H_0, q/q_m, m/q_m, t/t_m are as follows

$$\frac{\partial q}{\partial t}+\frac{6}{5}\left(\frac{\partial}{\partial x}\frac{q^2}{h}+\frac{\partial}{\partial z}\frac{qm}{h}\right)=\frac{3}{\mathrm{Re}_m}\left(h-\frac{q}{h^2}\right)+hWe\frac{\partial\Delta h}{\partial x},$$

$$\frac{\partial m}{\partial t}+\frac{6}{5}\left(\frac{\partial}{\partial z}\frac{m^2}{h}+\frac{\partial}{\partial x}\frac{qm}{h}\right)=hWe\frac{\partial\Delta h}{\partial z}-\frac{3m}{\mathrm{Re}_m h^2}, \qquad (1)$$

$$\frac{\partial h}{\partial t}+\frac{\partial q}{\partial x}+\frac{\partial m}{\partial z}=0.$$

Here, $q(x,z,t)=\int_0^h u\,dy$ and $m(x,z,t)=\int_0^h w\,dy$ are the liquid flow rates per unit width along the *Ox* and *Oz* axes, respectively, $\Delta h=\partial^2 h/\partial x^2+\partial^2 h/\partial z^2$. We choose the unperturbed film thickness H_0 at the symmetry axis of the rivulet as the scale and introduce the scales of velocity $gH_0^2/3\nu$, time $t_m=3\nu/gH_0$, and flow rate $q_m=gH_0^3/3\nu$. Here, $\mathrm{Re}_m=gH_0^3/3\nu^2$ is the Reynolds number,

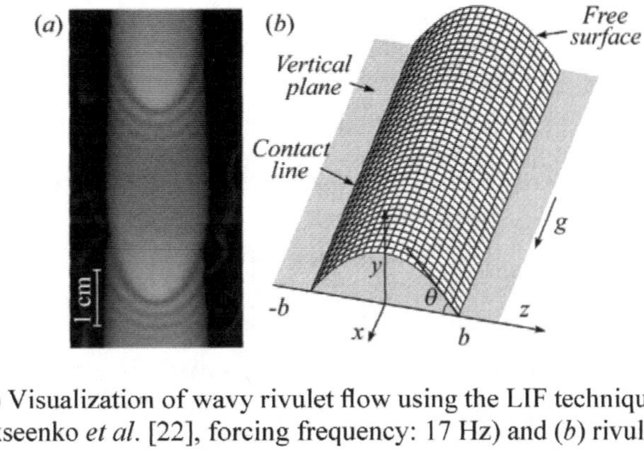

Figure 1. (*a*) Visualization of wavy rivulet flow using the LIF technique (experimental setup of Alekseenko *et al.* [22], forcing frequency: 17 Hz) and (*b*) rivulet flow diagram.

$We = \left(3Fi/\mathrm{Re}_m^5\right)^{1/3}$ is the Weber number, $Fi = \sigma^3/\rho^3 g\nu^4$ is Kapitza number, σ is the surface tension, ρ is the density, and ν is the kinematic viscosity of the fluid. With regard to the rivulet flow, these equations must be supplemented by the follows boundary conditions at the contact line (fixed contact line): $h|_{z=\pm b} = 0$, $q|_{z=\pm b} = 0$, $m|_{z=\pm b} = 0$. It is follows from (1), that h and q are even functions of coordinate z, and m is odd function of coordinate z. In case of a undisturbed flow (without waves) the partial derivatives on x and t in the equations (1) and also the flow rate m are equal to zero, thus a cross-section of rivulet is a parabola $h_0(z) = 1 - z^2/b^2$ and $q_0(z) = h_0^3(z)$.

Supposing $h = h_0\tilde{h}$, $q = q_0\tilde{q}$, $m = q_0\tilde{m}$ we introduce a new functions $\tilde{h}$, $\tilde{q}$, $\tilde{m}$. Instead of coordinate z we use the variable $\xi = z/b$. The function $\tilde{h}$, $\tilde{q}$, $\tilde{m}$ should have the finite values at a contact line, thus boundary conditions will be automatically fulfilled, since $h_0|_{\xi=\pm 1} = q_0|_{\xi=\pm 1} = 0$. Using $\tilde{h}$, $\tilde{q}$, $\tilde{m}$, we rewrite the equations (1) as follows

$$
\begin{aligned}
&h_0^3\frac{\partial\tilde{q}}{\partial t} + \frac{6}{5}\frac{\partial}{\partial x}\left(\frac{\tilde{q}^2}{\tilde{h}}h_0^5\right) + \frac{6}{5b}\frac{\partial}{\partial\xi}\left(\frac{\tilde{q}\tilde{m}}{\tilde{h}}h_0^5\right) = \frac{3h_0}{\mathrm{Re}_m}\left(\tilde{h} - \frac{\tilde{q}}{\tilde{h}^2}\right) + Weh_0\tilde{h}\frac{\partial L}{\partial x},\\
&h_0^3\frac{\partial\tilde{m}}{\partial t} + \frac{6}{5}\frac{\partial}{\partial x}\left(\frac{\tilde{q}\tilde{m}}{\tilde{h}}h_0^5\right) + \frac{6}{5b}\frac{\partial}{\partial\xi}\left(\frac{\tilde{m}^2}{\tilde{h}}h_0^5\right) = \frac{We}{b}h_0\tilde{h}\frac{\partial L}{\partial\xi} - \frac{3h_0}{\mathrm{Re}_m\tilde{h}^2}\tilde{m},\\
&h_0\frac{\partial\tilde{h}}{\partial t} + h_0^3\frac{\partial\tilde{q}}{\partial x} + \frac{1}{b}\frac{\partial}{\partial\xi}\left(\tilde{m}h_0^3\right) = 0.
\end{aligned}
\tag{2}
$$

Here, $h_0 = 1 - \xi^2$, $L \equiv \Delta\left(h_0\tilde{h}\right) = \left(1-\xi^2\right)\dfrac{\partial^2\tilde{h}}{\partial x^2} + \dfrac{1}{b^2}\left(\left(1-\xi^2\right)\dfrac{\partial^2\tilde{h}}{\partial\xi^2} - 4\xi\dfrac{\partial\tilde{h}}{\partial\xi} - 2\tilde{h}\right)$.

3. The application of a weighed residual method for wavy rivulet flow model

On interval $-1 < \xi < 1$ we define two different systems of orthogonal basis polynomials: even polynomials $\Phi_j(\xi)$ and odd polynomials $\Psi_j(\xi)$, $j = 1,2,3,\ldots$ Let's spread out even functions $\tilde{h}$ and $\tilde{q}$ abreast on $\Phi_j(\xi)$, and odd function $\tilde{m}$ abreast on $\Psi_j(\xi)$:

$$
\tilde{h} = \sum_{j=1}^{\infty} H_j(x,t)\Phi_j(\xi),\quad \tilde{q} = \sum_{j=1}^{\infty} Q_j(x,t)\Phi_j(\xi),\quad \tilde{m} = \sum_{j=1}^{\infty} M_j(x,t)\Psi_j(\xi).
\tag{3}
$$

As a whole, relations (3) do not satisfy (2) (substitution in the equations gives the residual). According to a weighed residual method, the projections of a residual to basis functions should be equated to zero, as a result we obtain the system of differential partial equations for H_j, Q_j, M_j. Coefficients of these equations are calculated through integrals from product of basis polynomials on function from ξ. in the equations (2).

We define a scalar product of polynomials $\Phi_k(\xi)$ and $\Phi_j(\xi)$ by means of weight function $g(\xi) = \left(1-\xi^2\right)^3$ as follows: $\langle\Phi_k|\Phi_j\rangle = \int_{-1}^{1} g(\xi)\Phi_j(\xi)\Phi_k(\xi)d\xi$.

Let's take the first even polynomial $\Phi_1 = 1$. The second even polynomial we take as $\Phi_2 = \Phi_1 + a_2\xi^2$, where the constant a_2 is calculated by means of a condition of orthogonality $\langle\Phi_1|\Phi_2\rangle = 0$. Thus, we obtain $\Phi_2 = 1 - 9\xi^2$. Each subsequent even polynomial Φ_{n+1} is defined by recurrence formula $\Phi_{n+1} = \Phi_1 + a_2\Phi_2 + a_3\Phi_3 + \ldots a_n\Phi_n + a_{n+1}\xi^{2n}$, where constants $a_2, a_3, \ldots a_{n+1}$ are calculated from conditions of orthogonality of Φ_{n+1} to previous polynomials $\Phi_1, \Phi_2, \ldots \Phi_n$. Similarly, for odd polynomials we take $\Psi_1 = \xi$, and each subsequent polynomial Ψ_{n+1} is defined as $\Psi_{n+1} = \Psi_1 + a_2\Psi_2 + a_3\Psi_3 + \ldots a_n\Psi_n + a_{n+1}\xi^{2n+1}$, where constants $a_2, a_3, \ldots a_{n+1}$ are calculated by means of conditions of orthogonality of Ψ_{n+1} to previous polynomials $\Psi_1, \Psi_2, \ldots \Psi_n$. Thus, we obtain the second odd polynomial $\Psi_2(\xi) = \xi - 11\xi^3/3$. Note the important property of the proposed basis, namely, near to a contact line the profiles of a thickness and flow rate are similar geometrically to corresponding profiles for a undisturbed flow, that is $h \sim (1-\xi^2)$ и $q \sim (1-\xi^2)^3$. Thus the contact angle varies according to a wave phase.

4. Plane waves

In the most simple case each basis consist of one element: $\Phi = 1$, $\Psi = \xi$, that is $\widetilde{h} = H(x,t)$, $\widetilde{q} = Q(x,t)$, $\widetilde{m} = \xi M(x,t)$. It corresponds to a case of plane waves for which profile of wavy rivulet remains a parabola in any section. Let's project the equations (2) on correspondent polynomials. For this purpose the first and third equations (2) multiply on $\Phi_1 = 1$, and the second equation (2) multiply on $\Psi_1 = \xi$. Then the equations are integrated on a variable ξ (integrals are calculated analytically), and we obtain

$$\begin{aligned}
&\frac{\partial Q}{\partial t} + \frac{32}{33}\cdot\frac{\partial}{\partial x}\left(\frac{Q^2}{H}\right) = \frac{35}{8\,\mathrm{Re}_m}\left(H - \frac{Q}{H^2}\right) + \frac{We}{6}H\left(7\frac{\partial^3 H}{\partial x^3} - \frac{35}{2b^2}\frac{\partial H}{\partial x}\right),\\
&\frac{\partial M}{\partial t} + \frac{96}{143}\cdot\frac{\partial}{\partial x}\left(\frac{QM}{H}\right) = \frac{96}{143}\cdot\frac{M^2}{bH} - \frac{63}{8\,\mathrm{Re}_m}\frac{M}{H^2} - \frac{21}{4b}WeH\frac{\partial^2 H}{\partial x^2},\\
&\frac{\partial H}{\partial t} + \frac{24}{35}\frac{\partial Q}{\partial x} = 0.
\end{aligned} \tag{4}$$

The first and third equations (4) are very similar to the equations of model [15] for two-dimensional waves in a film, but differ in coefficients. Unlike a film flow, there are terms containing additional parameter b in (4).

Let's linearize the equations (4) concerning small perturbations, supposing $H = 1 + \widetilde{H}$, $Q = 1 + \widetilde{Q}$, where $\widetilde{H} << 1$, $\widetilde{Q} << 1$, $M << 1$. As a result we obtain

$$\begin{aligned}
&\frac{\partial \widetilde{Q}}{\partial t} + \frac{32}{33}\left(2\frac{\partial \widetilde{Q}}{\partial x} - \frac{\partial \widetilde{H}}{\partial x}\right) = \frac{35}{8\,\mathrm{Re}_m}\left(3\widetilde{H} - \widetilde{Q}\right) + \frac{We}{6}\left(7\frac{\partial^3 \widetilde{H}}{\partial x^3} - \frac{35}{2b^2}\frac{\partial \widetilde{H}}{\partial x}\right),\\
&\frac{\partial M}{\partial t} + \frac{96}{143}\frac{\partial M}{\partial x} = -\frac{21\cdot We}{4b}\frac{\partial^2 \widetilde{H}}{\partial x^2} - \frac{63}{8\,\mathrm{Re}_m}M,\\
&\frac{\partial \widetilde{H}}{\partial t} + \frac{24}{35}\frac{\partial \widetilde{Q}}{\partial x} = 0.
\end{aligned} \tag{5}$$

4.1. Dispersion relations

Let's present perturbations in the form of a traveling wave: $\widetilde{H} = H_a \exp(ik(x - Ct) + \beta t)$, $\widetilde{Q} = Q_a \exp(ik(x - Ct) + \beta t)$, $M = M_a \exp(ik(x - Ct) + \beta t)$. Here, H_a, Q_a, M_a. are amplitudes, k is a wavenumber, C is a wave phase velocity, β is a time increment of the wave. Substituting it in the equations (5), after some transformations yields system of two dispersion relations for β and C:

$$\left(\beta Re_m+\frac{35}{16}\right)^2=(k\,Re_m)^2\left(\left(C-\frac{32}{33}\right)^2-E\right)+\left(\frac{35}{16}\right)^2,$$
$$\left(\beta Re_m+\frac{35}{16}\right)\left(C-\frac{32}{33}\right)=\frac{157}{66}. \qquad (6)$$

Here, $E=\frac{41}{35}\left(\frac{16}{33}\right)^2+\frac{2We}{b^2}+\frac{4We}{5}k^2$. Distinction between dispersion relations for waves in a rivulet and in a film consists not only in coefficients of (6) but also available additional parameter b. The system (6) the same as in case of a film flow, is easily reduced to one quadratic equation with respect to $Y=\left(\beta \mathrm{Re}_m+\frac{35}{16}\right)^2$. Having the value of Y we obtain $\beta=\frac{1}{\mathrm{Re}_m}\left(-\frac{35}{16}\pm\sqrt{Y}\right)$ $C=\frac{1}{66}\left(64\pm\frac{157}{\sqrt{Y}}\right)$. Here, sign "plus" corresponds to unstable mode, and the sign "minus" corresponds to stable mode (in the same way, as for waves in a film flow). For unstable mode $C>0$ and the perturbations are growing ($\beta>0$) in some range of a wavenumber. For stable mode $C<0$, and the perturbations are damping at any wavenumber ($\beta<0$).

4.2. Stability criterion of rivulet flow

In an asymptotics $\beta \mathrm{Re}_m \ll 1$ (at $k\to 0$) from the second equation (6) we obtain $C=72/35$. Then from the first equation (6) we find $\beta=\frac{8k^2\,\mathrm{Re}_m}{35}\left(A-\frac{4We}{5}k^2\right)$, where $A=\frac{64\cdot 79}{121\cdot 35}-\frac{2We}{b^2}$. From here it follows that long-wave disturbances can be unstable if $A>0$. Thus a criterion of instability can be written as a condition for the rivulet half-width $b>\sqrt{\frac{35\cdot 121\cdot We}{32\cdot 79}}\approx\sqrt{1.6752\cdot We}$. This condition been applied to dimensional rivulet half-width is $b_{\dim}>\left(\frac{3\nu^2}{g}\right)^{1/3}\sqrt{\frac{1.6752\cdot(3Fi)^{1/3}}{\mathrm{Re}_m}}$. Thus, at $b<b_{\min}$ the rivulet flow is stable with respect to 2D disturbances of any wavelength.

Calculated dispersion dependences are shown in figure 2 for $\mathrm{Re}_m = 25$ and various values of the dimensionless parameter b. From figure it is clear that reducing of parameter b leads to decreasing of the increment, and increasing of the phase velocity. Figure 3 shows dependence of increment on the wave number for various values of Reynolds number. From figure it is clear, that the range of instability on a wave number extends with increase of Re_m. The maximum of increment also increases with growth of Re_m. Figure 4 shows the dependence of increment on a wave number for $\mathrm{Re}_m = 10$ and

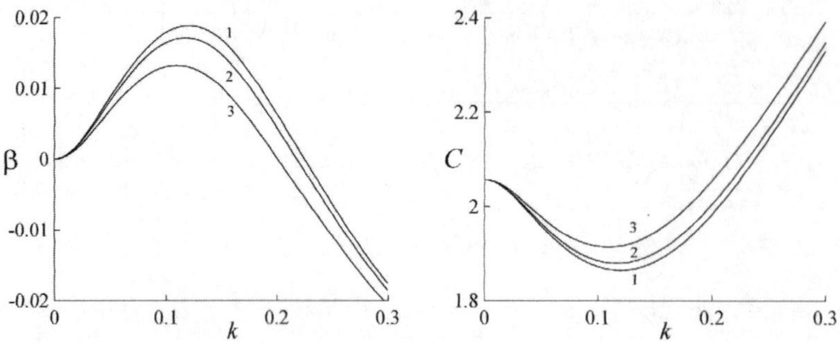

Figure 2. Increment β and phase velocity C of plane waves at $\mathrm{Re}_m = 25$ for various values of dimensionless parameter b: 1− 50, 2− 25, 3− 15 (water, $Fi^{1/3}$=3270).

various values of Kapitza number. From figure it is clear that area of instability and the maximum of increment decrease with parameter $Fi^{1/3}$ growth.

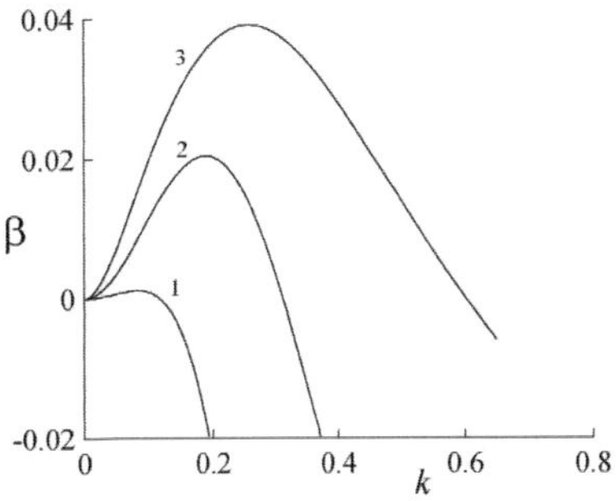

Figure 3. Dependence of increment of plane waves on a wavenumber for b = 10, $Fi^{1/3}$=300 and various values of Reynolds number Re_m: 1– 5, 2– 10, 3– 20.

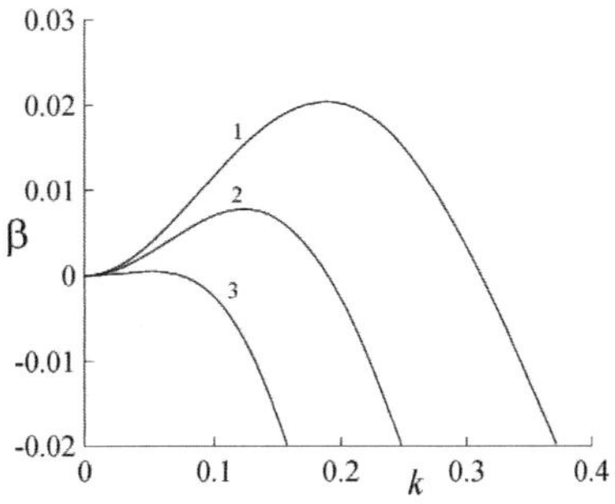

Figure 4. Dependence of increment of plane waves on a wavenumber for Re_m = 10, b = 10 and various values of parameter $Fi^{1/3}$: 1– 300, 2– 600, 3– 1200.

5. Conclusions

The 3D wavy rivulet flow has been studied theoretically. The approach reducing a problem to 2D one has been proposed on the basis of weighted residual method. The equations of the 2D rivulet flow have been deduced using projections of initial equations on the system of orthogonal polynomials. The system of equations for plane waves in the vertical rivulet has been deduced, and dispersion relations have been obtained. The stability criterion of rivulet flow has been deduced, and the analysis of influence of dimensionless parameters on dispersion dependences has been carried out.

Acknowledgments
The work was supported by the Russian Science Foundation (grant No. 14-29-00093).

References

[1] Towell G D and Rothfeld L B 1966 *AIChE J.* **12** 972-980
[2] Wilson S K and Duffy B R 1998 *Phys. Fluids* **10** 13-22
[3] Paterson C, Wilson S K and Duffy B R 2013 *Eur. J. Mech.* B **41** 94-108
[4] Perazzo C A and Gratton J 2004 *J. Fluid Mech.* **507** 367-379
[5] Tanasijczuk A J, Perazzo C A and Gratton J 2010 *Eur. J. Mech.* B **29** 465-471
[6] Davis S H 1980 *J. Fluid Mech.* **98** 225-242.
[7] Weiland R H and Davis S H 1981 *J. Fluid Mech.* **107** 261-280
[8] Young G W and Davis S H 1987 *J. Fluid Mech.* **176** 1-31
[9] Diez J, Gonzalez A and Kondic L 2012 *Phys. Fluids* **24** 032104
[10] Benilov E S 2009 *J. Fluid Mech.* **636** 455-474
[11] Alekseenko S V, Markovich D M and Shtork S I 1996 *Phys. Fluids* **8** 3288-3299
[12] Alekseenko S V, Bobylev A V and Markovich D M 2008 *J. Engng. Thermophys.* **17** 259-272.
[13] Alekseenko S V, Antipin V A, Bobylev A V and Markovich D M 2007 *Exp. Fluids* **43**, 197-207
[14] Alekseenko S V, Bobylev A V, Guzanov V V, Markovich D M and Kharlamov S M 2010 *Thermophys. Aeromech.* **17** 371-384
[15] Alekseenko S V, Aktershev S P, Bobylev A V, Kharlamov S M and Markovich D M 2015 *J. Fluid Mech.* **770** 350-373
[16] Demekhin E A and Shkadov V Ya 1984 *Fluid Dynamics* **19** 689-695

Thermocapillary deformation in a locally heated layer of silicone oil

E V Barakhovskaia[1,2], I V Marchuk[1,2] and A A Fedorets[1,3]

[1] Kutateladze Institute of Thermophysics, Russian Academy of Sciences, Prosp. Lavrentyev 1, Novosibirsk, 630090, Russia
[2] Novosibirsk State University, Pirogova Str. 2, Novosibirsk, 630090, Russia
[3] Tyumen State University, Volodarskogo Str. 6, Tyumen, 625003, Russia
E-mail: ella94@bk.ru

Abstract. The processes of heat and mass transfer in systems with liquid-gas interface are of interest to a wide range of problems. Thermocapillary flows have an important role in such systems. Thermocapillary deformation of silicone oil layer was investigated using laser scanning confocal microscope Zeiss LSM 510 Meta. The numerical solution of the problem was obtained in the lubrication approximation theory for two-dimensional axisymmetric thermocapillary flow. The model takes into account the surface tension, viscosity, gravity and heat transfer in the substrate. Evaporation is neglected. The numerical algorithm for the joint solution of the energy equation and the evolution equation for the liquid layer thickness has been developed. Stationary solutions have been obtained by the establishment method. The dependences of the depth of thermocapillary deformation on the layer thickness were obtained for silicone oils of different viscosities. It was found that the value of the relative deformation decreases nonlinearly with increasing the initial layer thickness. There is a good qualitative agreement of numerical results and experimental data.

1. Introduction

The processes of heat and mass transfer in systems with liquid-gas interfacial surface are of interest to a wide range of fundamental and applied problems. Thermocapillary flows have an important role in such systems. Thermocapillary flows can be caused by even minor irregularities of interfacial surface temperature [1-3]. To a large extent thermocapillary effect appears in thin liquid layers with local heating. It is perspective to use thermocapillary effect in the locally heated thin liquid layers to determine the properties of the liquid and the layer thickness [4-5]. A method of measuring the liquid layer thickness using the laser-induced thermocapillary effect is proposed in [4]. A method for identifying liquids is described in [5]. Most of theoretical studies is considered that processes in thin liquid layers are modeled using an evolution equation for the layer thickness, resulting in the lubrication approximation (long-wave approximation) [6]. Velocity, temperature, pressure of the liquid, etc. are defined as a function of film thickness (the solution of this equation). This approach eliminates the complexity of the problem, caused by the presence of the free surface. However, the evolution equation is nonlinear and has a high order, and its solution is generally searched numerically. Steady state thermocapillary flows and deformations in locally heated horizontal liquid layer were numerically modelled in [7] for the axisymmetric case, the heat transfer in the substrate was not considered. In [8] unsteady solutions, with a local sharp changing in temperature of the surface layer, were obtained numerically for axisymmetric statement. In all these papers heat transfer in the substrate is not taken into account for calculating, so it is the reason that did not allow performing a quantitative comparison of the calculation results with the experimental data. The aim of this work is the experimental and theoretical study of thermocapillary deformations in the locally heated horizontal liquid layers with different properties using a confocal microscopy and numerical modeling based on the thin layer approximation.

2. Description of the experiment

Thermocapillary deformation of dimethylpolysiloxane layer (further PMS or silicone oil) was investigated using laser scanning confocal microscope Zeiss LSM 510 Meta. The main technical characteristics of the device: the length of the scanning beam wavelength - 405, 458, 477, 488, 514, 543, 633 nm (sources are 4 different types of lasers), scanning speed is up to 5 frames/s (at a resolution of 512 x 512 pixels), the maximum resolution frame is 2048 x 2048 pixels, the scan area is 1,4 x 1,4 mm. According to the results of mathematical processing of layered scans, the device creates a three-dimensional image of the sample with the elements of volume reconstruction, animation and quantitative analysis. This microscope allowed obtaining highly accurate data on thermocapillary deformation at the stage of steady flow. Measurements were conducted for five types of silicone oil: PMS-5, PMS-50, PMS-100, PMS-200 and PMS-400. The taken types of silicone oil cover the range of viscosity from 5 to 365 cP, other thermal properties are slightly varied. OPHIR NOVA meter with detector model 2A-SH (operating range from 60 mW to 2 W, precision ± 3%) was used to control the power of the laser beam, inducing thermocapillary flow.

The experimental scheme is shown in figure 1a, where 1 microscope lens, 2 scanning beam microscope (wavelength $\lambda = 488$ nm), 3 glass cuvette at the bottom of which a layer of (the cuvette radius $R_c = 18$ mm), 4 colored lacquer (figure 1b) based on nitrocellulose, 5 layer of silicone oil, 6 ray semiconductor laser (wavelength 650 nm, power $P = 16.5 \pm 0.5$ mW). The lacquer coating was transparent to the beam of scanning microscope, thus, absorbed about 99% of the semiconductor laser. In this way the bottom layer was induced by the heat source (heater), which was characterized by the parameter $P_H = P / S_H$, where S_H - heating area. In the experiment, radius of the spot that is inducing laser on the substrate was R_h =90 μm (figure 1c), hence $P_H = 0.62$ W/mm^2.

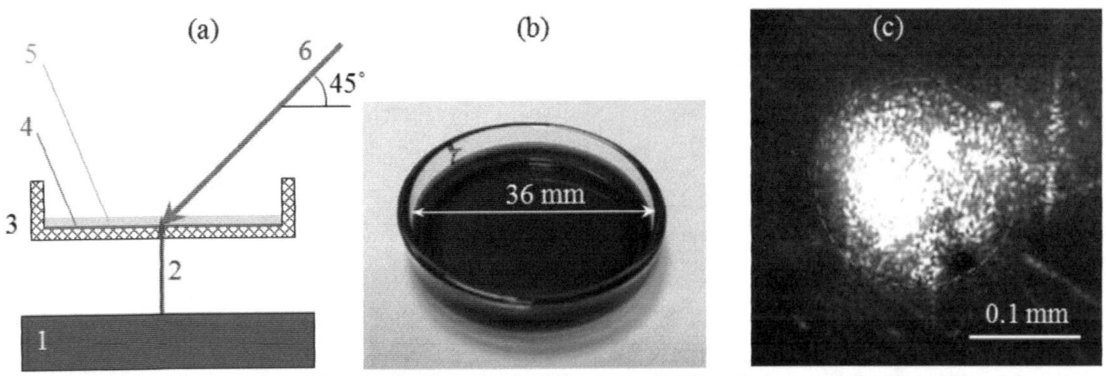

Figure 1. (a) The experimental scheme; (b) Photograph of a glass cuvette with a blue colored bottom; (c) The spot of laser which is inducing a localized heating of the silicone oil layer.

In the preparatory stage of the experiment it was formed a uniform layer of liquid in a cuvette, using a high-precision dispenser. The thickness of not deformed layer h_0 was controlled by the 3D image from the microscope and, if necessary, altered by the addition/subtraction liquid. Thermocapillary deformation depth $\Delta h = h_0 - h_C$ (hereinafter h_C is residual layer thickness in the center of the thermocapillary deformation) was determined by two complementary layer 3D images when the heater is off and on, figure 2. Moreover, from the moment when the laser heating is on to starting scanning, enduring a time of about 100 seconds required for the output thermocapillary flow on the stationary regime. If the duration of the scanning cycle is about 2 minutes, then because of microvibrations, interfacial surface of the liquid-gas gets a few blurred, but with a thin dark line through the middle. This ensured the reproducibility of the thickness measurement with pixel precision 3D image, i.e. about 3.5μm.

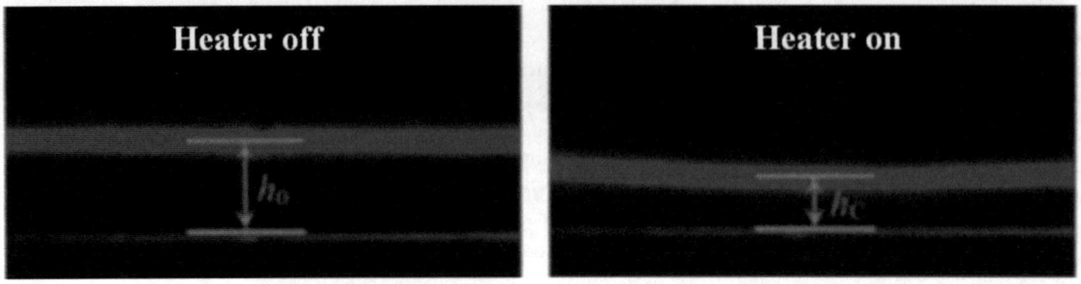

Figure 2. Thermocapillary deformation of PMS-5(non-deformed layer thickness h_0 = 228 μm)

3. Numerical modelling

The problem of thermocapillary deformation of the locally heated horizontal layer of silicone oil has been solved using the lubrication approximation theory for two-dimensional axisymmetric statement. The model takes into account important parameters such as capillary pressure, viscosity and gravity. Evaporation is neglected. Initially the liquid layer has flat surface and uniform temperature. The substrate is locally heated. Deformations of the liquid surface are determined by the properties of the liquid, substrate and heater. Stationary solutions have been obtained by the establishment method.

Dynamics of thin films is well described by the evolution equation, which has been obtained using the lubrication approximation theory [6,8,9]. The evolution equation in cylindrical coordinates for the axial symmetrical case has the following form:

$$h_t + \frac{1}{r}\frac{\partial}{\partial r}\left(r\left[\frac{h^3}{3\mu}\frac{\partial}{\partial r}\left(\rho g h + \sigma H\right) + \frac{h^2}{2\mu}\sigma_T \frac{\partial T}{\partial r} \right]\right) = 0 \ . \tag{1}$$

where h is the thickness of the liquid layer, μ is dynamic viscosity, ρ is density, g is gravitational acceleration, σ is surface tension, σ_T is linear coefficient, which determines the dependence of the surface tension on temperature, T is temperature, $H = h_{rr}/\left(1+h_r^2\right)^{3/2} + h_r/r\left(1+h_r^2\right)^{1/2}$ is double mean curvature of the liquid surface. Equation (1) is a nonlinear differential equation of the first order in time and the fourth order in spatial variables relative to unknown function *h(t,r)*.

Boundary conditions for equation (2) have a clear physical meaning:

$h_r(t,0) = 0$ the condition of the axial symmetry is given in the center of the cuvette ,
$h_r(t,R_c) = 0$ the contact angle $\pi/2$ is given on the border of the cuvette,
wetting angle value does not matter in this problem (2)
$q(t,0) = 0$ flow rate is equal to 0 in the center,
$q(t,R_c) = 0$ the condition of impermeability of liquid through the walls.

Here $q = \frac{h^3}{3\mu}\frac{\partial}{\partial r}\left(\rho g h + \sigma H\right) + \frac{h^2}{2\mu}\sigma_T\frac{\partial T}{\partial r}$ is the liquid flow rate along the substrate.

The temperature of the liquid layer and cuvette is determined by the energy equation in cylindrical coordinates:

$$\rho C_p\left(\frac{\partial T}{\partial t}+u\frac{1}{r}\frac{\partial(rT)}{\partial r}+v\frac{\partial T}{\partial z}\right)=\frac{1}{r}\frac{\partial}{\partial r}\left(\lambda\left(r\frac{\partial T}{\partial r}\right)\right)+\frac{\partial^2 T}{\partial z^2}+Q\ , \tag{3}$$

where, ρ is density, λ is coefficient of thermal conductivity; C_p is specific heat of the medium (solid or liquid). $Q=Q(t,r,z)=\begin{cases}0, \text{ outside the heating area}\\ const>0, \text{ in the heating area}\end{cases}$ is bulk density of the heat sources. Heating area is a thin round layer in the centre of the cuvette bottom with radius R_h, where the laser beam emission is absorbed. Boundary conditions are defined as follows:

$$\left.\frac{\partial T}{\partial r}\right|_{r=0}=0,\ \left.\frac{\partial T}{\partial r}\right|_{r=R}=0,\ \lambda\left.\frac{\partial T}{\partial n}\right|_{W}=\alpha_w\left(T_W-T_a\right),\ \lambda\left.\frac{\partial T}{\partial n}\right|_{S}=\alpha\left(T_S-T_a\right)\ , \tag{4}$$

axial symmetry condition, adiabatic side wall, the convective heat transfer coefficient is specified on the bottom wall of the cuvette and on the liquid surface. Here index W indicates to condition on the bottom wall, index S indicates to the liquid surface, T_a is a temperature of the ambient.

Initially the liquid layer has flat surface, temperature of the liquid surface and cuvette is uniform:

$$h|_{t=0}(r,z)=h_0,\quad T|_{t=0}(r,z)=T_0=const \tag{5}$$

4. Calculating method

Splitting into physical processes, such as thermal conduction and flow of the liquid, is used for calculations. The grid in the space variables in the liquid and solid phases is uniform: $r_i=i\cdot dr$, $i=0,..,N$. The time step τ is also uniform. The value of the liquid layer thickness in the time t_k and i node: $h_i^k=h(r_i,t_k)$.

The evolution equation of the liquid layer thickness (1) is approximated at grid's nodes with finite volume method [10]. A discrete analogue of eq. (1) is written for each volume:

$$\frac{h_i^{k+1}-h_i^k}{\tau}+\frac{1}{r_i}\frac{q_{i+1/2}r_{i+1/2}-q_{i-1/2}r_{i-1/2}}{dr}=0 \tag{6}$$

Newton's method is used to solve system of nonlinear algebraic equations (6). Numerical linearization is used to calculate the Jacobians [9]. The scheme has the second-order approximation for the spatial coordinates and the first-order in time.

When the temperature in the liquid is calculated, deformations of the surface are not taken into account. Since the laser beam locally heats the bottom of the cuvette, it is considered that the source of the heat is concentrated only in one top layer of nodes of the cuvette. The method of fractional steps is used to solve the difference analog of the energy equation (3), [11]:

$$\begin{cases}\dfrac{T_{ij}^{n+1/2}-T_{ij}^{n}}{dt}=\dfrac{\lambda}{2\rho Cp}\left(\dfrac{\left(1+\dfrac{dr}{2r}\right)T_{i+1,j}^{n+1/2}-2T_{ij}^{n+1/2}+\left(1-\dfrac{dr}{2r}\right)T_{i-1,j}^{n+1/2}}{dr^2}+\dfrac{T_{i,j+1}^{n}-2T_{ij}^{n}+T_{i,j-1}^{n}}{dz^2}\right)+\dfrac{Q_{ij}}{2\rho Cp}\\[2ex] \dfrac{T_{ij}^{n+1}-T_{ij}^{n+1/2}}{dt}=\dfrac{\lambda}{2\rho Cp}\left(\dfrac{\left(1+\dfrac{dr}{2r}\right)T_{i+1,j}^{n+1/2}-2T_{ij}^{n+1/2}+\left(1-\dfrac{dr}{2r}\right)T_{i-1,j}^{n+1/2}}{dr^2}+\dfrac{T_{i,j+1}^{n+1}-2T_{ij}^{n+1}+T_{i,j-1}^{n+1}}{dz^2}\right)+\dfrac{Q_{ij}}{2\rho Cp}\end{cases} \tag{7}$$

Here $Q_{ij}=P/\pi R_h^2 dz$ at $i<N\,R_h/R_c$, $j=\mathrm{M}/2+1$ and $Q_{ij}=0$ for others i, j.

Boundary and initial conditions are set in the finite differences equations:

$$j = 1,..,M,\ T(j,0) = T(j,2) \text{ symmetry, } T(j,N+1) = T(j,N-1) \text{ adiabatic right wall.} \quad (8)$$

$$i = 1,..,N,\ T(0,i) = T(2,i) - \frac{\alpha}{\lambda}(T(1,i) - T_a)2dz \text{, heat transfer coefficient on the liquid surface}$$
$$T(M+1,i) = T(M-1,i) - \frac{\alpha_w}{\lambda_w}(T(M,i) - T_a)2dz, \text{ heat transfer coefficient on bottom wall} \quad (9)$$

Initial conditions: $T^0_{i,j} = 0,\ i = 0,..,N,\ j = 1,..,M-1$.

The resulting system of linear algebraic equations is solved at each step with the tridiagonal matrix algorithm. Calculations are performed sequentially. The time step for the evolution equation is done after the time step for the energy equation.

The mathematical model accounted such defining parameters as the geometry of the problem, parameters of the liquid, properties of the substrate and the heater materials, heating methods. The calculation results have shown that all these parameters have a significant impact on the distribution of the heat and deformations of the liquid surface.

5. Analysis of the experimental results and calculations

Dependences on the silicone oil thickness of the thermocapillary deformation's depth have been measured experimentally and calculated by the model (figure 3). Numerical calculations were made for two marks of silicone oil: PMS-5 and PMS-50. The thinner the layer, the greater the difference Δh, when other things being equal. In thin layers thermocapillary deformation depth reaches 30% or more of the initial thickness, but quickly decreases with increasing h_0 (see figure 3). When the layer thickness is $h > 400 \mu m$ values of deformations are practically the same, because viscosity variations for thick layers have less effect on the value of deformation. There is a good qualitative agreement between the calculating results and experimental data. The differences in the calculating results for PMS-5 and PMS-50 are explained by differences in the coefficients of surface tension and coefficients of dependence of the surface tension on temperature for these liquids.

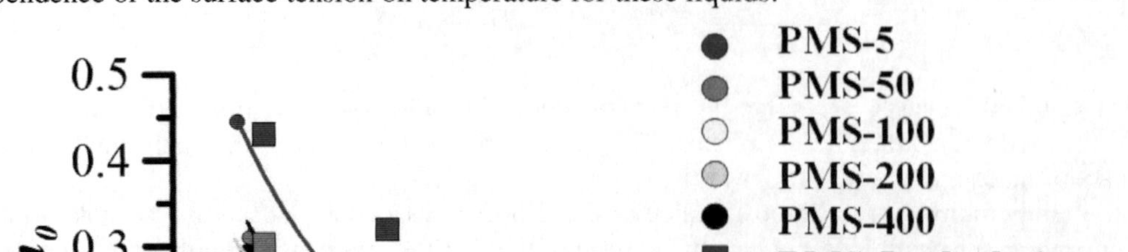

Figure 3. Thermocapillary deformation depth dependence on the thickness of the layer of silicone oil. h_0 – thickness of non-deformed layer, $\Delta h = h_0 - h_C$ - thermocapillary deformation depth, h_C - layer thickness over the center of the heater at the stage of steady thermocapillary flow.

6. Conclusion

There have been measured and numerically calculated deformations in locally-heated horizontal layers of silicone oils of different types and thickness. It has been found that the value of the relative deformation of the layer decreases nonlinearly with increasing the layer thickness, when other conditions being equal. The results of modeling using the thin layer approximation are predicting well the main features of the experimental dependences of the thermocapillary deformations on the layer thickness for silicone oils of different marks. So the calculation results are in good qualitative and quantitative agreement with experimental data. The experimental results can be used to test a wide class of computer programs that simulate the processes of heat and mass transfer in multiphase systems with liquid-gas interfaces.

Acknowledgments The study was financially supported by the Russian Science Foundation (Project 15-19-20049).

References

[1] Kabov O A 2010 *International J. of Emerging Multidisciplinary Fluid Sciences* Interfacial thermal fluid phenomena in thin liquid films **2 (2-3)** 87-121

[2] Wedershoven H M J M 2014 *Appl. Phys. Lett.* Infrared laser induced rupture of thin liquid films on stationary substrates **104** 054101-1–054101-4

[3] Kupershtokh A L, Ermanyuk E V and Gavrilov N V 2015 *Communications in Computational Physics* The Rupture of Thin Liquid Films Placed on Solid and Liquid Substrates in Gravity Body Forces **17(05)** 1301-19

[4] Bezuglyĭ B A and Fedorets A A 2001 *Tec. Phys. Let.* Measuring the thickness of thin liquid films on solid surfaces using the laser-induced thermocapillary response **27 (5)** 359-361

[5] Bezuglyj B A, Fedorets A A, Tarasov O A and Ivanova N A 2004 *Poverkhnost Rentgenovskie Sinkhronnye i Nejtronnye Issledovaniya* Photo-induced capillary effects: New application fields **6** 80-84

[6] Oron A, Davis S H and Bankoff S G 1997 *Rev. of Modern Phys.* Long-scale evolution of thin liquid films **69(3)** 931-980

[7] Marchuk I V 2009 *JET* Thermocapillary deformation of a thin locally heated horizontal liquid layer **18(3)** 227-237 DOI: 10.1134/S1810232809030047

[8] Marchuk I V 2015 *JET* Thermocapillary deformation of a horizontal liquid layer under flash local surface heating **24(4)** 381-385 DOI: 10.1134/S181023281504013X

[9] Marchuk I V and Kabov O A 2016 *Doklady Physics* Model of filmwise vapor condensation on curved surfaces **61(1)** 19-23. DOI: 10.1134/S102833581601002X

[10] Samarskij A A and Nikolaev E S 1978 *Methods of Solving Grid Equations* (Moscow: Nauka)

[11] Yanenko N N 1971 *The Method of Fractional Steps* (New York: Springer-Verlag, Berlin-Heidelberg)

Characteristics of film flow during transition to three-dimensional wave regimes

A Bobylev[1], V Guzanov[1], A Kvon[1] and S Kharlamov[1]

[1]Institute of Thermophysics, Siberian Branch of RAS, Novosibirsk, Russia

E-mail: guzanov@itp.nsc.ru

Abstract. Results of experimental study of transition processes from two-dimensional to three-dimensional wave regimes for the case of isothermal falling liquid film are presented. Experiments were carried out using liquids with various physical properties in the range of Reynolds numbers $5 < Re < 100$. It is shown that the process of transition to the three-dimensional wave motion is accompanied by formation of rivulets and that the downflow evolution of the rivulets usually has non-monotonic character. The statistical analysis of the 3D wave's characteristics allowed us to characterize in detail different scenarios of downflow evolution of 3D waves and to determine the regimes of fully developed steady-state 3D wave flow. Noticeable differences between natural and forced wave evolution have been observed for the relatively small Reynolds numbers whereas for the higher Reynolds numbers fast evolution of all investigated statistical characteristics with achievement of the steady-state 3D wave regimes of film flow occurs.

1. Introduction

Three-dimensional wave regimes are considered to be the last stage of film flow evolution at moderate Reynolds numbers. In such regimes the film surface is covered by the numerous 3D waves chaotically interacting with each other. Usually 3D waves arise due to transverse instability of nonlinear 2D waves which in turn can be formed during the process of natural wave evolution or generated by artificial modulation of liquid flow rate. The case of transition to three-dimensional wave regimes for excited 2D waves is more investigated by now. Recent experimental ([1], [2], [3]) and theoretical ([4], [5], [6]) investigations show good agreement in characteristic dimensions of 3D waves despite some differences in their shape. Using modern experimental techniques, in the first place Laser Induced Fluorescence (LIF) technique, allow us to reveal sufficient redistribution of liquid in transverse direction in the processes of 2D – 3D transition ([2], [3]) for the case of isothermal film flow, which leads to the formation of well pronounced rivulets on the time-averaged fields of film thickness. Such behavior is unexpected for isothermal film flow and previously was not observed in experiments basically because of using local probes or shadowgraph technique adjusted for registration of the waves' parts which have high curvature of free boundary. Main aim of this work is to investigate experimentally general dependencies of process of transition from two-dimensional to three-dimensional wave motion and associated formation of rivulets on the vertically falling liquid film surface in the range of $5 < Re < 100$ for the downstream distances up to 140 cm. The main consideration was given to the analysis of statistical characteristics of the waves at different distances from the film inlet and make up accurate classification of wave regimes in the specified range of flow rates.

2. Experimental techniques and operating conditions

The liquid film was formed on the vertical transparent glass plate with width of 50 cm and length of 140 cm by adjustable slot distributor. The film freely flows down onto one side of the plate by the action of gravity. Three types of working fluids were used in experiments: water ($\nu = 0.99*10^{-6}$ m^2/s, $\rho = 998$ kg/m^3, $\sigma = 0.072$ kg/s^2), water-glycerol solution WGS1 ($\nu = 1.6*10^{-6}$ m^2/s, $\rho = 1050$ kg/m^3, $\sigma = 0.072$ kg/s^2) , and water-glycerol solution WGS2 ($\nu = 2.2*10^{-6}$ m^2/s, $\rho = 1070$ kg/m^3, $\sigma = 0.072$ kg/s^2) where ν – kinematic viscosity, ρ – density and σ – surface tension of liquid. That allowed us to cover range of Kapitsa numbers $\gamma = \sigma\rho^{-1}\nu^{-4/3}g^{-1/3}$, where g is gravity acceleration, from $\gamma \approx 1100$ for more viscous WGS2 up to $\gamma \approx 3700$ for water. The range of investigated Reynolds numbers $5 < Re = q/\nu < 100$, where q – volumetric flow rate per unit width. The 2D – 3D transition processes in the case of the natural wave evolution as well as for the case of initially regular excited 2D waves were investigated. For the last case the regular two-dimensional waves were forced by the periodic flow rate modulation with prescribed frequency F. With help of regular 2D waves the uniformity of local flow rate distribution over the plate's surface was controlled. In the case of uniform distribution of liquid the wave's front in the upper part of the flow is a straight line perpendicular to the flow direction. The procedure of control and adjustment of uniformity of liquid distribution was performed obligatory before each experiment.

The shadowgraph imaging was used for registration of the wave patterns over whole area of the plate. The white matt screen placed behind the glass plate on the variable distance was used for the technique realization. The halogen light source and registering camera were placed from the film side. Varying the distance between the screen and the plate allows obtaining shadow images for parts of liquid surface with different curvature. When the screen is placed close to the glass plate, shadow images represent the areas with high curvature and are similar to those obtained in other works (e.g. [1]). If the screen is placed at the distances of a few centimeters from the plate, the film areas with much lower curvature form evident shadows on the screen. The distance between the screen and the glass plate was selected before each experiment in order to register capillary ripples having high curvature as well as the rivulets having lower curvature of film surface (figure 1).

The Laser-Induced Fluorescence (LIF) technique was used for measurement of instant film thickness distribution with high time and spatial resolutions. High-speed camera registers LIF images with frequency of 1 kHz on the measurement areas with sizes from 13x13 cm^2 up to 19x19 cm^2 dependently on flow conditions; this leads to spatial resolution from 0.13 mm/pix to 0.19mm/pix accordingly. Record length of each set of LIF images was 2 seconds. For each investigated regime the records were obtained independently at the different distances from film inlet. Various statistical characteristics were obtained for each record. As it turned out, the most informative for evaluating the downstream wave evolution were such statistical characteristics as the film thickness probability density function (PDF), dispersion D and spectral power density Y(f) of film thickness temporal variation.

3. Experimental results

For all investigated liquids three typical scenarios of wave evolution depending on Reynolds number can be singled out in the range of $5 < Re < 100$. Major difference between liquids consists in different values of boundary Reynolds numbers which exact definition is the complex problem and lies beyond the scope of this work.

First scenario of wave evolution is observed for low $Re < Re_1$ when transition to three-dimensional wave regimes does not occur even in the bottom part of the test section. The waves have a shape close to two-dimensional up to the end of test section and time-average film thickness distributions over the plate are flat. Re_1 is different for the different liquids and lies in the range from 5 to 15. Over this value ($Re > Re_1$) transition from two-dimensional to three-dimensional wave regimes occurs which is characterized by fast degradation of transverse size of the waves. In the upper part of the flow initially 2D waves, no matter natural or excited, are breaking down into numerous 3D waves with characteristic dimension of 1 – 2 cm in transverse and longitudinal directions.

For $Re_1 < Re < Re_2$ (where $40 < Re_2 < 60$) the second scenario of wave evolution with well pronounced rivulets is observed. In this case evolution of natural and excited waves is different and rivulets are observed up to the end of the test section (figure 1). Forming 3D waves move mainly upon the rivulets and form chains. At the same time wave motion between rivulets looks depressed. Despite some differences in the regimes of natural and excited wave evolution the shapes of 3D waves in bottom part of the test section have similar characteristics.

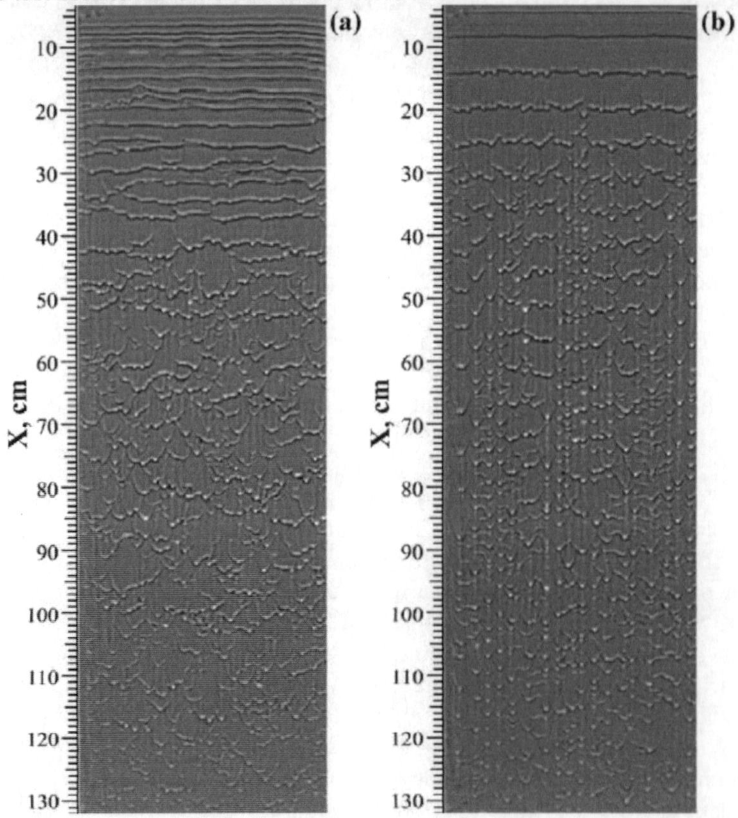

Figure 1. Shadowgraph images of water film at Re = 25. a) natural wave evolution, b) evolution of excited waves, F = 8Hz. X – downstream distance from the film inlet.

Because the formation of rivulets in the film flow is usually connected with non-isothermal effects, direct measurements of film surface temperature were performed with help of Titanium ATR-570M infrared scanner. Measured with total error not exceeded 0.05K temperature fields of the film surface have shown that the film is isothermal on the whole flow area.

In third scenario of wave evolution which is observed for Reynolds number $Re > Re_2$ the 2D-3D transition occurs in the upper part of the flow over a few wavelengths of 2D waves. Rivulets formation in such regimes is not monotonic. Till 40 – 50 cm amplitude of rivulets can increase and further downstream they begin to decay up to full disappearance at the lower part of the test section. Difference between cases of natural and forced wave evolution is observed only in the upper part of the plate at the distances up to a few wavelengths from the film inlet, after that the wave patterns for both cases become similar and appeared to be steady state 3D wave flow regime.

Detailed analysis of time-averaged film thickness distributions in dependence on different time-averaging intervals allows us to reveal formation of short-living rivulets shifting at random in transverse direction. Chaotic behavior of such rivulets leads to their disappearance or sufficient decrease of their amplitude on time-averaged film thickness distribution for large values of averaging time. As it turned out such short-living rivulet is the chain consisting of at least 5 – 8 3D waves. And such chains look like the main structures of regimes with high $Re > Re_2$.

Differences between scenarios and the direction of wave evolution can also be observed through analysis of statistical characteristics. As an example, probability density functions (PDF) of the liquid film thickness (h) at different distances from the film inlet are shown in figure 2. Distributions have characteristic asymmetric right-tailed shape. For the high Reynolds numbers (the case of $Re > Re_2$) the film thickness PDFs almost don't change with downstream distance (figure 2(a)) and are identical for the cases of natural and forced wave evolution. However transform of the PDFs with shifting of the most probable film thickness value with downstream distance is observed for the lower Re (case of $Re_1 < Re < Re_2$) (figure 2(b)). In this case the difference between the cases of natural and forced wave evolution can be observed up to the end of test section (lines 4 and 5 in figure 2(b)).

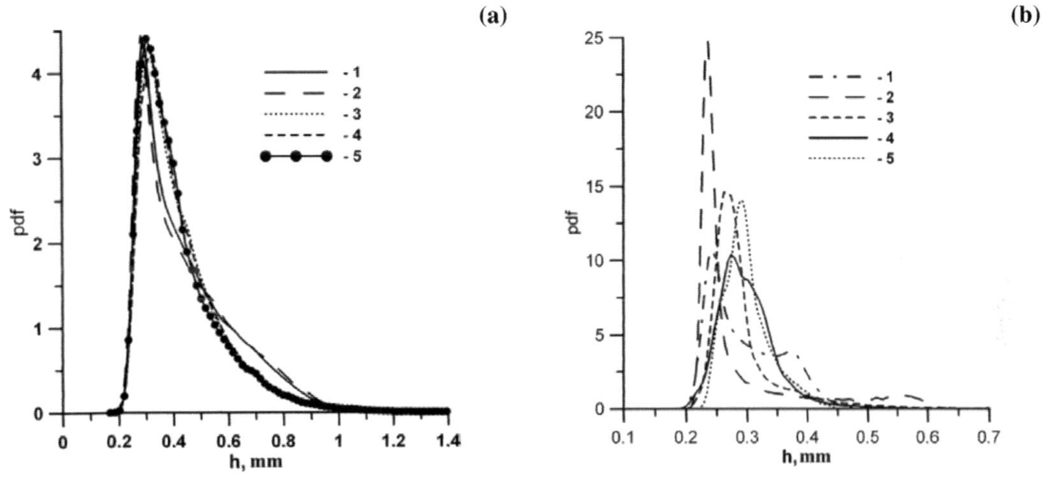

Figure 2. Film thickness PDFs, working liquid - WGS2. a) Re = 54, distance X from the film inlet: 1 – X = 26 см; 2 – X = 26 cm, F = 7 Hz; 3 – X = 92 cm; 4 – X = 133 cm, F = 7 Hz; 5 – X = 133 cm. b) Re = 19, 1 – X = 26 cm; 2 – X = 26 cm, F = 7 Hz; 3 – X = 92 cm; 4 – X = 133 cm; 5 – X = 133 cm, F = 7 Hz.

Values of the dispersion D were obtained by averaging of the calculated local D values over the each transverse section of the measurement area. Some examples of characteristic downstream evolution of D are shown in figure 3. Initial growth of D with downstream distance in the upper part of flow is attributed to the initial developing and growth of 2D waves in this region. Farther downstream, in the region of flow where 2D waves are decayed into 3D waves significant diminishing of D is observed for all investigated regimes. But if for the high Reynolds number (case of $Re > Re_2$) after fast evolution in the upper part of flow D reaches steady-state value for the both, natural and forced 2D waves (lines 3 and 4 in figure 3), in the case of lower Reynolds number (case of $Re_1 < Re < Re_2$) downstream evolution of D is different: the gradual decreasing of D is observed up to the end of the plate (lines 1 and 2 in figure 3). It should be underlined that in the flow regimes for the case of $Re_1 < Re < Re_2$ transition from 2D to 3D waves is accompanied with formation of well pronounced rivulets observed not only on the time-averaged but also on instant film thickness distribution.

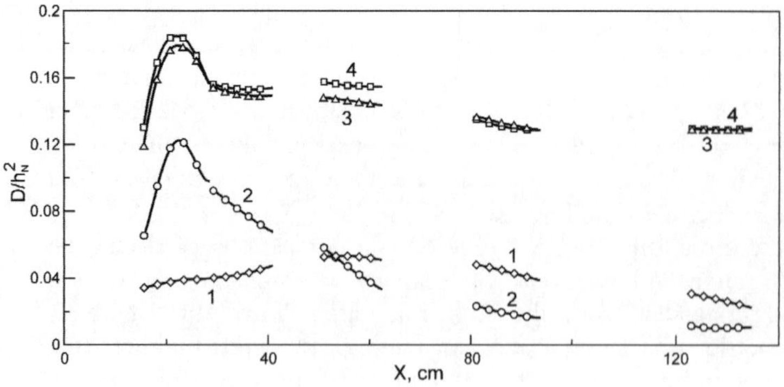

Figure 3. Film thickness dispersion D normalized to the squared undisturbed film thickness (h_N^2). Working liquid – WGS2. 1 – Re = 19, natural waves; 2 – Re = 19, forced waves F = 7 Hz; 3 – Re = 54, forced waves F = 7 Hz; 4 – Re = 54, natural waves.

Fast forming of the steady-state three-dimensional wave motion for the case of large Re > Re_2 is also observed while analyzing spectral power density (figure 4(a)). Also the difference in spectral characteristics between the cases of natural and excited waves is revealed only on short distances in the upper part of the flow and farther downstream the spectral characteristic for the both cases become identical. It is interestingly that for the steady-state flow regimes in the area of the high frequencies the spectral power decays by law of Y ~ f^{α}, with index of power α = -2.8. This index is close to that obtained in the work [7] and as it turned out it doesn't depend on the liquid properties.

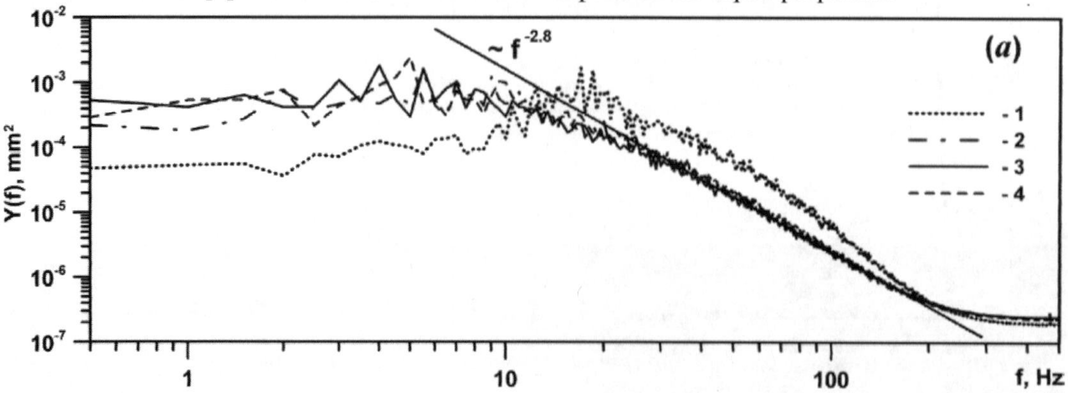

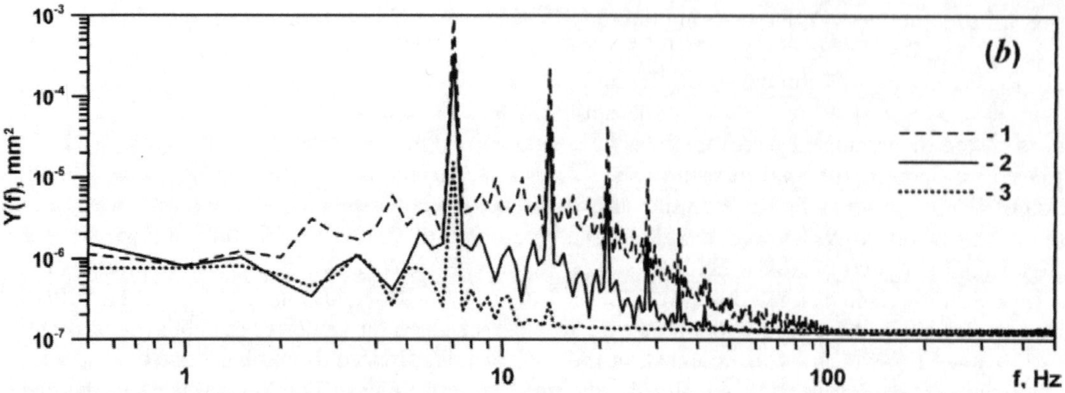

Figure 4. Spectral power density Y(f) at various distances from the distributor. Working liquid – WGS2 a) Re = 54, natural waves, 1 – X = 21 cm, 2 –X = 87 cm, 3 – X = 128 cm; 4 – forced F = 7 Hz, X = 128cm. b) Re = 13.5, forced waves F = 7 Hz. 1 – X = 21 cm, 2 – X = 56 cm, 3 – X = 128 cm.

For the cases of $Re_1 < Re < Re_2$ spectral power density shows significant difference between the regimes of forced and natural waves on all distances downstream the film inlet. Also spectra revealed downstream decay of wave motion (figure 4(b)) with gradual downstream decrease of spectral power especially in the range of high frequencies.

4. Conclusions

Performed experimental study of wave evolution on isothermal film flowing down a big plate reveals three typical scenarios of wave evolution depending on the Reynolds number in the range of $5 < Re < 100$.

For low $Re < Re_1$, where Re_1 is different for the different working liquids and lies in the range from 5 to 15, no transition to three-dimensional wave is observed and the waves have a shape close to two-dimensional up to the end of test section. In this case time-averaged film thickness distributions over the plate are flat.

For $Re_1 < Re < Re_2$, where $40 < Re_2 < 60$, the second scenario of wave evolution with well pronounced rivulets forming in the process of 2D - 3D wave transition is observed. For this range of Reynolds numbers no steady-state 3D wave regimes of film flow have been observed. All considered statistical characteristics of the waves show gradual downstream decaying of 3D waves at all distances from the film inlet.

For $Re > Re_2$ fast evolution of the film flow toward steady state 3D wave regimes is observed. Rivulets formation in such regimes is not monotonic. The amplitudes of rivulets are increased in the upper part of the flow and further downstream the rivulets begin to decay up to full disappearance at the lower part of the test section. Stabilizing of the dispersion value as well as of the PDFs and spectral power densities of film thickness fluctuations for such regimes apparently indicates that at these regimes stabilization of 3D wave motion occurs.

Acknowledgments
This work was supported by the Russian Foundation for Basic Research (project no. 15-01-06702).

References

[1] Park C D, Nosoko T 2003 Three-dimensional wave dynamics on a falling film and associated mass transfer *AIChE J* **49(11)** 2715–27

[2] Alekseenko S V, Guzanov V V, Markovich D M and Kharlamov S M 2012 Specific features of a transition from the regular two-dimensional to three-dimensional waves on falling liquid films *Tech. Phys. Lett.* **38(8)** 739–42

[3] Kharlamov S M, Guzanov V V, Bobylev A V, Alekseenko S V and Markovich D M 2015 The transition from two-dimensional to three-dimensional waves in falling liquid films: Wave patterns and transverse redistribution of local flow rates *Phys. of Fluids* **27** 114106

[4] Kalliadasis S, Ruyer-Quil C, Scheid B and Velarde M G 2012 *Falling liquid films* (London: Springer-Verlag)

[5] Scheid B, Ruyer-Quil C, Manneville P 2006 Wave patterns in film flows: modelling and three-dimensional waves *J. Fluid Mech* **562** 183-222

[6] Dietze G, Kneer R 2011 Flow separation in falling liquid films *Frontiers in Heat and Mass Transfer* **2** 033001

[7] Chu K J, Dukler A E 1975 Statistical characteristics of thin, wavy films III. Structure of the large waves and their resistance to gas flow *AIChE J* **21(3)** 583–93

Thermocapillary instability of a liquid layer on interior surface of a rotating cylinder

O A Burmistrova

Lavrentyev Institute of Hydrodynamics SB RAS, Lavrentyev prosp. 15, 630090, Novosibirsk, Russia
E-mail: oksanabur@mail.ru

Abstract. Stability problem of a liquid, situated on the inner surface of a cylinder, which rotates about its axis with a constant angular velocity, is considered. The flow is assumed to be non-isothermal, and thermocapillary instability is investigated. The temperature at the free boundary satisfies the condition of the third kind with a given Biot number. It is supposed that the free boundary is undeformable. The exact solution of Navier–Stokes and heat conduction equations is obtained. Neutral disturbances of this solution are considered, where Marangoni number is taken as the spectral number. Asymptotics of critical values of Marangoni number for long and short waves are found analytically. Neutral curves are constructed numerically for various values of independent dimensionless parameters. The dependences of critical Marangoni number on Reynolds number, Biot number and the aspect ratio are investigated.

1. Introduction

The isothermal problem of a liquid layer on the surface of the rotating cylinder was studied in [1]–[9].

In [1, 3, 4, 6] the liquid layer on the outside of the cylinder was considered. In [1] principle of change of stability in weightless conditions was investigated. In [3] the existence of the stationary solution of the problem with free boundaries was proven in the exact statement; the evolution equation of the film dynamics was obtained in a thin layer approximation. This equation was studied in [6] later. In [4] a plane problem was considered by the techniques of lubrication theory, the behaviour of liquid rings around the cylinder was described.

The motion on the interior surface of a cylinder was not sufficiently studied. In [2] the bifurcation of rotationally symmetric motions was investigated; a branch condition in terms of critical Weber number was obtained. In [5] the plane problem was considered; steady-state liquid-film profiles were found. In [7] the effect of surface tension on the stability of the film was examined in the case of the plane problem. Three-dimensional coating and rimming flow was studied in [8]. In [9] inertial instability of flows on the inside or outside of a rotating horizontal cylinder was investigated.

In the present work the flow is assumed to be non-isothermal. Reasoning by analogy with [1], axisymmetric disturbances are considered as more dangerous. Thermocapillary instability of a liquid layer on the interior surface of a rotating cylinder is studied.

2. Mathematical model

Let the viscous incompressible thermo-conducting liquid partially filling the space between two neighboring cylindrical surfaces of radii r_1 and $r_2 < r_1$. Surface $r = r_2$ is the free one. The outer surface is rigid and rotates about its axis with a constant angular velocity ω. We will discuss the motion in a cylindrical coordinate system r, θ, z. It is assumed that gravity force is absent. Let us suppose that liquid density ρ, kinematic coefficient of viscosity ν and thermodiffusion coefficient χ are constant while the surface tension coefficient σ is a linear function of temperature T:

$$\sigma = \sigma_0 - \kappa(T - T_0),$$

where σ_0, κ and T_0 are positive constants. The temperature at the free boundary satisfies the condition

$$\frac{\partial T}{\partial \boldsymbol{n}} = q(T_\Gamma - T),$$

where q is the interfacial heat transfer coefficient, T_Γ is the given temperature of the environment, $\boldsymbol{n}$ is the unit normal vector to the free boundary. The temperature at the solid wall satisfies the condition

$$T = T_S,$$

where $T_S \neq T_\Gamma$ is the given positive constant.

The basic solution to the Navier–Stokes and heat conduction equations has the form

$$V = (0, \omega r, 0),\ \ P = \frac{\rho\omega^2 r^2}{2} + \widetilde{p}_0,\ \ \widetilde{T} = \frac{\mathrm{Bi}(T_S - T_\Gamma)}{1 + \mathrm{Bi}\ln a}\ln\frac{r}{r_2} + \frac{T_S + T_\Gamma \mathrm{Bi}\ln a}{1 + \mathrm{Bi}\ln a},$$

where $\widetilde{p}_0$ is the constant, $a = \frac{r_1}{r_2} > 1$ is the aspect ratio, $\mathrm{Bi} = qr_2$ is Biot number.

We introduce dimensionless parameters, choosing as scales of length, velocity, pressure and temperature the quantities r_2, ωr_2, $\rho\omega^2 r_2^2$ and $\delta T = \frac{\mathrm{Bi}(T_S - T_\Gamma)}{1 + \mathrm{Bi}\ln a}$. The basic solution in terms of dimensionless variables has the form

$$V = (0, r, 0),\ \ P = \frac{r^2}{2} + p_0,\ \ \widetilde{T} = \ln r + c, \tag{1}$$

where p_0 and $c = \frac{T_s + T_\Gamma \mathrm{Bi}\ln a}{\mathrm{Bi}(T_s - T_\Gamma)}$ are constants.

Then we impose small axisymmetric disturbances on the basic solution. We seek velocity, pressure and temperature fields of the form

$$V^1 = (u, \upsilon + r, w),\ \ p^1 = P + \frac{p}{\mathrm{Re}},\ \ T^1 = \widetilde{T} + \mathrm{Pr}\,T,$$

where $\mathrm{Re} = \frac{\omega r_2^2}{\nu}$ is Reynolds number, $\mathrm{Pr} = \frac{\nu}{\chi}$ is Prandtl number. Navier–Stokes and heat conduction equations and boundary conditions are linearized. Reasoning by analogy with [10], we assume that the stability loss of the basic solution leads to an appearance of a new stationary solution. In this case Marangoni number $\mathrm{Ma} = \frac{\kappa\delta T r_2}{\rho\nu\chi}$ is taken as the spectral parameter. It is assumed that the free boundary is undeformable. The boundary value problem with a fixed domain is obtained:

$$\Delta u - \frac{u}{r^2} - p_r = -2\,\mathrm{Re}\,\upsilon,\ \ \Delta\upsilon - \frac{\upsilon}{r^2} = 2\,\mathrm{Re}\,u,\ \ \Delta w - p_z = 0,\ \ \frac{1}{r}(ru)_r + w_z = 0,\ \ \Delta T = \mathrm{Re}\frac{u}{r}, \tag{2}$$

$$r = a:\ u = \upsilon = w = T = 0, \tag{3}$$

$$r = 1:\ u = \upsilon_r - \upsilon = -T_r + \mathrm{Bi}T = \mathrm{Re}(w_r + u_z) - \mathrm{Ma}T_z = 0, \tag{4}$$

where $\Delta = \frac{1}{r}\frac{\partial}{\partial r}(r\frac{\partial}{\partial r}) + \frac{\partial^2}{\partial z^2}$.

Independent dimensionless parameters of the problem are Re, Bi and a. It should be noted that in the full statement (a deformable free surface), $\mathrm{Cr}=\frac{\kappa\delta T}{\sigma_0}$ (crispation number) is also the key dimensionless parameter, which characterizes the degree of deformation of the free surface by thermocapillary forces. Cr is small and in analogy with [10, 11] it is neglected.

We seek the solution of the problem (2) – (4) in the form

$$u = A(r)\cos(kz),\ \ \upsilon = B(r)\cos(kz),\ \ w = C(r)\sin(kz),\ \ p = \xi(r)\cos(kz),\ \ T = \zeta(r)\cos(kz),$$

and obtain the spectral problem

$$\frac{1}{r}(rA')' - (k^2 + \frac{1}{r^2})A + 2\,\mathrm{Re}\,B - \xi' = 0,\ \ \frac{1}{r}(rB')' - (k^2 + \frac{1}{r^2})B - 2\,\mathrm{Re}\,A = 0,$$

$$\frac{1}{r}(rC')' - k^2C + \xi k = 0,\ \ \frac{1}{r}(rA)' + kC = 0,\ \ \frac{1}{r}(r\zeta')' - k^2\zeta - \mathrm{Re}\frac{A}{r} = 0,$$

$$r = a:\ A = B = C = \zeta = 0,$$

$$r = 1:\ A = B' - B = -\zeta' + \mathrm{Bi}\,\zeta = \mathrm{Re}\,C' + \mathrm{Ma}\,\zeta k = 0.$$

After elimination of B, C and ξ, the system and boundary conditions take the form

$$(L^3 - 4\,\mathrm{Re}\,k^2)A = 0, \tag{5}$$

$$\frac{1}{r}(r\zeta')' - k^2\zeta - \mathrm{Re}\frac{A}{r} = 0, \tag{6}$$

$$A(a) = A'(a) = L^2A(a) = A(1) = ((\frac{\partial}{\partial r} - 1)L^2)A(1) = 0, \tag{7}$$

$$\zeta(a) = -\zeta'(1) + \mathrm{Bi}\,\zeta(1) = 0, \tag{8}$$

$$\mathrm{Re}(A''(1) + A'(1)) = \mathrm{Ma}\,\zeta(1)k^2, \tag{9}$$

where $L = \frac{\partial^2}{\partial r^2} + \frac{1}{r}\frac{\partial}{\partial r} - (\frac{1}{r^2} + k^2)$.

The spectral problem (5) – (9) for the parameter Ma is degenerate. Ma can be found uniquely for a given set of determinative parameters. The solution of the problem is determined up to an arbitrary multiplicative constant. So it can be supposed that

$$A''(1) + A'(1) = 1, \tag{10}$$

and A can be found from the boundary value problem (5), (7), (10). Then ζ is found from the boundary value problem (6), (8). It can be shown using variation of parameters that $\zeta(1) \neq 0$. So Ma is defined uniquely from (9).

3. Asymptotics and numerical results.

Asymptotics of neutral curves for the problem (5) – (9) are found analytically.

Asymptotics for $k \to 0$ (long waves) have the form

$$A_0(r) = \frac{2(a^4 - r^2)(r^2 - 1)\ln a - (a^2 - 1)((a^2 - r^2)(r^2 - 1) + 2(a^2 - 1)r^2 \ln r)}{4r(a^4 - 1 - 4\ln a)},$$

$$\zeta_0(r) = \frac{\mathrm{Re}}{64(a^4 - 1 - 4\ln a)(1 + \mathrm{Bi}\ln a)}(-16a^4 \ln^3 a(1 + \mathrm{Bi}\ln r) + (a^2 - 1)(5a^4 - r^2(r^2 - 12) -$$
$$4a^2(r^2 + 3) + (\mathrm{Bi}(11 - 16a^2 + 5a^4) - 12 + 8r^2(a^2 - 1))\ln r - 8a^2\ln^2 r) - 2\ln^2 a(4 + 4a^2 - 3\mathrm{Bi} +$$
$$4\mathrm{Bi}\,r^2 - \mathrm{Bi}\,r^4 + 4a^4(1 + \mathrm{Bi}(r^2 - 1)) - 4a^2(a^2 - 1)\mathrm{Bi}\ln r - 8a^4\mathrm{Bi}\ln^2 r) + \ln a(11\mathrm{Bi} - 12 - 8r^2 -$$
$$12\mathrm{Bi}\,r^2 + 2r^4 + \mathrm{Bi}\,r^4 + a^4(14 - 8r^2 - 4\mathrm{Bi}(r^2 - 1)) + a^2(12 - \mathrm{Bi}(15 - 16r^2 + r^4)) + 2(4 - 3\mathrm{Bi} +$$
$$4\mathrm{Bi}\,r^2 - 8a^2\mathrm{Bi}r^2 + a^4(8 + 3\mathrm{Bi} + 4\mathrm{Bi}\,r^2))\ln r + 8a^2\ln^2 r(\mathrm{Bi} - a^2(\mathrm{Bi} - 2)))),$$

$$\mathrm{Ma} = \frac{64\left(3 - 4a^2 + a^4 + 4\ln a\right)\left(1 + \mathrm{Bi}\ln a\right)}{k^2\left((a^2 - 1)^2(5a^2 - 11) + 2\ln a(3(a^4 + 2a^2 - 3) - 4\ln a(2a^4\ln a + a^4 + a^2 + 1))\right)} + O(1), k \to 0.$$

Asymptotics for $k \to \infty$ (short waves) have the form

$$A_\infty(r) = -\frac{1}{2}k(r-1)e^{-k(r-1)}, \ \zeta_\infty(r) = -\frac{\mathrm{Re}}{8k^2}(1 + k(r-1) + k^2(r-1))e^{-k(r-1)},$$

$$\mathrm{Ma} = 8k^2 + O(1), k \to \infty.$$

It should be noted that asymptotics for critical Ma values coincide with Pearson's short-wave asymptotics for Ma [10].

Critical values of Marangoni number Ma^* and wave number k^* are obtained numerically using Runge–Kutta method of the fourth order for various values of independent dimensionless parameters; the neutral curves on k, Ma plane are constructed.

The figure 1 shows the neutral curves for different values of Bi when a = 1.1, Re = 1. Critical values are $k^* = 19.94$, $\mathrm{Ma}^* = 8793.2$ for $\mathrm{Bi} = 1$; $k^* = 22.58$, $\mathrm{Ma}^* = 12107.78$ for $\mathrm{Bi} = 10$; $k^* = 23.74$, $\mathrm{Ma}^* = 15653.48$ for $\mathrm{Bi} = 20$. Thus, with increasing Bi, Ma^* and k^* also increase.

It should be noted that for water, when $\mathrm{Ma}^* = 8793.2$ and $r_2 = 3 \cdot 10^{-2} m$, then $\mathrm{Cr} = 5 \cdot 10^{-4}$. Thus, the neglect of Cr is justified.

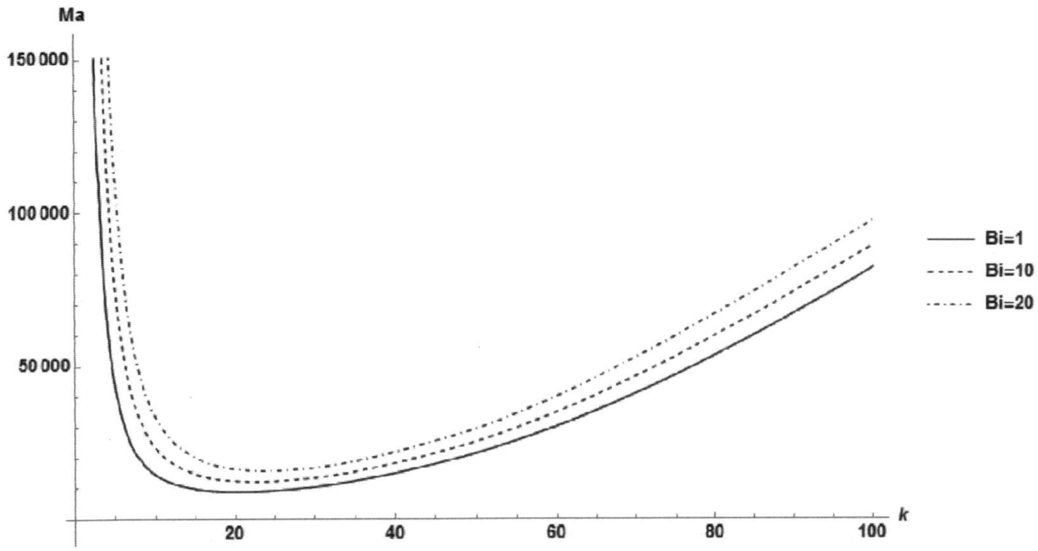

Figure 1. The neutral curves for a = 1.1, Re = 1 and various values of Bi.

The figure 2 shows the neutral curves for different values of a when Bi = 1, Re = 1. Critical values are $k^* = 19.94$ $\text{Ma}^* = 8793.2$ for $a = 1.1$; $k^* = 13.72$, $\text{Ma}^* = 4107.37$ for $a = 1.15$; $k^* = 10.57$, $\text{Ma}^* = 2406.07$ for $a = 1.2$ Thus, with increasing a, Ma^* and k^* decrease.

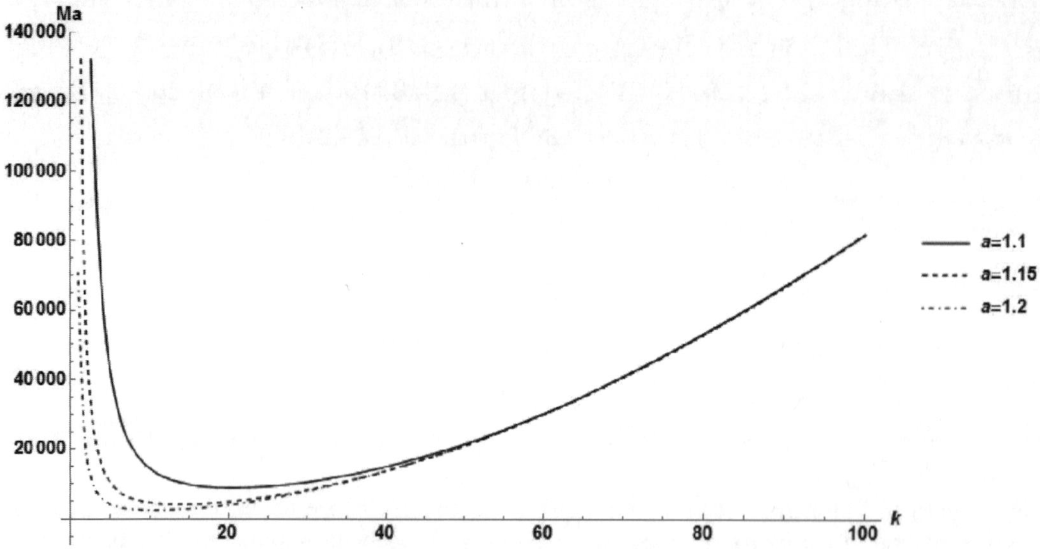

Figure 2. The neutral curves for Bi = 1, Re = 1 and various values of a.

The figure 3 shows the neutral curves for different values of Re when Bi = 1, a = 1.2. Critical values are $k^* = 10.57$, $\text{Ma}^* = 2406.07$ for $\text{Re} = 1$; $k^* = 12.88$, $\text{Ma}^* = 6441.86$ for Re=500. Thus, with increasing Re, Ma^* and k^* also increase.

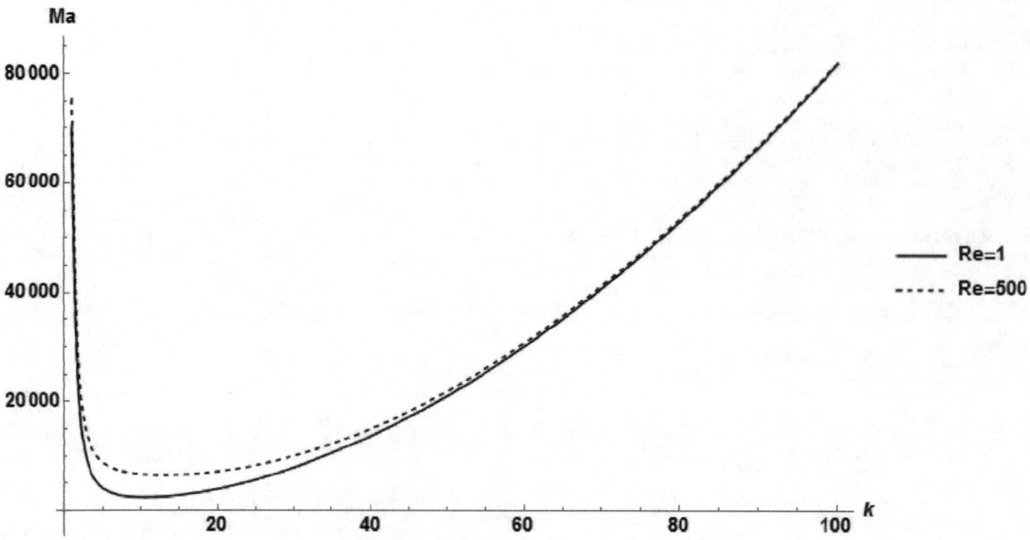

Figure 3. The neutral curves for Bi = 1, a = 1.2 and various values of Re.

4. Conclusion.

For non-isothermal liquid, situated on the inner surface of the rotating cylinder, the exact solution of Navier–Stokes and heat conduction equations is obtained. The basic solution has only the azimuthal component of the velocity as non-zero, which is proportional to radial coordinate *r*. The pressure is the quadratic function of *r* and the temperature is the logarithmic function of *r*. Small disturbances are considered, where Marangoni number is chosen as the spectral parameter.

Asymptotics of neutral curves for long and short waves are found analytically. The main term of asymptotics has the order of $O(k^{-2})$ when $k \to 0$ and $O(k^2)$ when $k \to \infty$. Critical values of Marangoni number and wave number are obtained numerically for various values of independent dimensionless parameters. The dependences of critical Marangoni number on Reynolds number, Biot number and the aspect ratio are investigated. It is obtained that the critical Marangoni number and the critical wave number increase with increasing Re and Bi, and decrease with increasing *a*.

It should be noted that critical values of Rayleigh number Ra^* in the terms of the temperature gradient is proportional to $(a-1)^4$ (see [12]) when Ma is not proportional to any degree of $a-1$. Thereby, it could be concluded that for thin layers ($a-1 \ll 1$) thermocapillary mechanism of instability dominates and the neglect of convection induced by buoyancy force is justified.

Acknowledgments
This work was supported by a grant of President of Russia for support of Russian Leading Scientific Schools NSh-8146.2016.1. The author is grateful to V.V. Pukhnachev for helpful discussions.

References

[1] Yih C S 1960 *Proc. Roy. Soc. A* **258** 63–86
[2] Pukhnachev V V 1973 *J. Appl. Mech. Tech. Phys.* **14** 253–8
[3] Pukhnachev V V 1977 *J. Appl. Mech. Tech. Phys.* **18** 344–51
[4] Moffatt H K 1977 *J. de Mécanique.* **16** 651–73
[5] Johnson R E 1988 *J. Fluid Mech.* **190** 321–42
[6] Hinch E J and Kelmanson M A 2003 *Proc. R. Soc. Lond. A* **459** 1193–213
[7] Benilov E S, Kopteva N and O'Brien S B G 2005 *Q. J. Mech. Appl. Math.* **58** 158–200
[8] Leslie G A, Wilson S K and Duffy B R 2013 *J. Fluid Mech.* **716** 51–82
[9] Benilov E S and Lapin V N 2013 *J. Fluid Mech.* **736** 107–29
[10] Pearson J R A 1958 *J. Fluid Mech.* **4** 489–500
[11] Andreev V K, Zakhvataev V E and Ryabitskii E A 2000 *Thermocapillary instability* (Novosibirsk: Nauka)
[12] Shaidurov G F, Shliomis M I and Yastrebov G V 1969 *Fluid Dynamics* **4** 55–8

Numerical modeling of gravity-driven bubble flows with account of polydispersion

A S Chernyshev and A A Schmidt

Computational Physics Laboratory, Ioffe Institute, Politekhnicheskaya 26, Saint Petersburg, 194021, Russia

E-mail: alexander.tchernyshev@mail.ioffe.ru

Abstract. The present study is focused on the investigation of motion of bubble-liquid medium inside the bubble columns or vertical pipes with account of polydisperse phenomena by the means of numerical simulation. The underlying mathematical model is based on the Euler-Euler approach with interphase interaction described by the momentum and mass transfer between phases, along with the k-ω-SST turbulence model which includes turbulence generation by the bubble motion and bubble path dispersion. Polydispersion is taken into account by the multi-class model with piecewise-constant distribution of bubble sizes per cell. Simulation of downward flow inside the straight vertical pipe resulted in maximum of the bubble void fraction close to the pipe center which is in good correlation with the experimental data. Simulation of multiphase flow inside rectangular bubble column with off-center sparger resulted in vertical bubble-liquid jet which is biased towards nearby wall with the correct prediction of attachment point location.

1. Introduction

Investigations of bubbly flows are of great importance since such flows can be frequently observed in natural phenomena and technological processes. Such investigations involved, for example, the study of the suppression of the bubble formation process, which is important in such facilities as heat exchangers or fuel pipelines and closely related to the cavitation phenomena [1, 2], or dedicated to the optimization of chemical reactors [3]. Direct measurements of the entire flow pattern are not always possible, thus numerical modeling is employed.

Different approaches are used to describe bubble-liquid medium (see, for example, [4]). It is found that two most widely used approaches, the Euler-Euler and the Euler-Lagrange, are suitable for real-scale problems with significant number of bubbles, while with the increase of bubble quantity the latter one becomes more resource demanding.

Treatment of the variation of bubble sizes in the framework of the Euler-Euler approach can be done in a number of ways. The simplest one is to assume that locally all bubbles have same size equals to the Sauter mean diameter [5], which can be used in some problems but fails when bubbles have wide variation in sizes and local size constrain can not be matched. Polydispersion can be implemented in that cases by introducing the population balance equation [6]. Such model provides description of the bubble size variation in each point of space, but due to the integral-differential nature of the equation the solution procedure became complex. Another approach is a multi-class or multi-size-group (MUSIG) [7, 8] model. This method consists in grouping all bubbles according to their sizes in a number of classes from minimal to maximal size, the number of classes is the same for the entire domain. Mass and momentum conservation equations are formulated and solved for each class of bubbles, with the conservation equations for liquid phase. This model provides unified description of all phases and classes of bubbles with the possibility of simple incorporation inter-phase and class-to-class mass and momentum exchange processes.

The presented work is focused on the improvement of the previously developed Euler-Euler model by incorporation of the polydisperse multi-class approach and conducting simulations for justification of the proposed model and numerical algorithm.

2. Mathematical model

The mathematical model is an extension of the previously developed Euler-Euler model with single-size bubbles [3, 9]. Since N classes are introduced and mass and momentum conservation equations for dispersed and liquid phases will have the following form:

$$
\begin{gathered}
\frac{\partial \alpha_{ib}\rho_b^0}{\partial t}+\operatorname{div}\left(\alpha_{ib}\rho_b^0\vec{V}_{ib}\right)=0,\\
\frac{\partial \alpha_{ib}\rho_b^0\vec{V}_{ib}}{\partial t}+\operatorname{div}\left(\alpha_{ib}\rho_b^0\vec{V}_{ib}\vec{V}_{ib}+p\bar{E}\right)=S_{ib},\\
\frac{\partial\left(1-\sum_{i=1}^{N}\alpha_{ib}\right)\rho_l^0}{\partial t}+\operatorname{div}\left(\left(1-\sum_{i=1}^{N}\alpha_{ib}\right)\rho_l^0\vec{V}_l\right)=0,\\
\frac{\partial\left(1-\sum_{i=1}^{N}\alpha_{ib}\right)\rho_l^0\vec{V}_l}{\partial t}+\operatorname{div}\left(\left(1-\sum_{i=1}^{N}\alpha_{ib}\right)\rho_l^0\vec{V}_l\vec{V}_l+p\bar{E}-\bar{\tau}_l\right)=S_l.
\end{gathered}
\tag{1}
$$

Subscript i for the first two equations of the dispersed phase refers to the i-th class of bubbles, i=1,...,N. This system should be accomplished by conservation equations of number densities per each class:

$$
\frac{\partial N_{ib}}{\partial t}+\operatorname{div}\left(N_{ib}\vec{V}_{ib}\right)=0. \tag{2}
$$

The initial distribution of bubbles by their sizes in the bulk liquid, or distribution of bubbles at inlet boundaries can be described by Log-Normal or the Rosin-Rammler distributions:

$$
\begin{gathered}
\mathrm{LN}:\frac{1}{R\sigma\sqrt{2\pi}}\cdot\exp\left(-\left(\ln(R)-\mu\right)^2\cdot 0.5\sigma^{-2}\right),\\
\mathrm{RR}:\frac{k}{\lambda}\left(\frac{R}{\lambda}\right)^{k-1}\cdot\exp\left(-\left(R/\lambda\right)^k\right), R\geq 0.
\end{gathered}
\tag{3}
$$

Mathematical expectations in that distributions are taken equal to most probable bubble size, which can be estimated from the problem set-up parameters or taken from the experimental data. The number of classes N is usually in the range from 5 to 10, and selected on the criteria of converged solution and invariance of the solution from N.

2.1. *Turbulence*

The *k*-*ω*-SST model is used to describe turbulence:

$$\begin{gathered}\frac{\partial(\rho_l k)}{\partial t}+\mathrm{div}(\rho_l \vec{V}_l k)\\ =\widetilde{P}-\beta^*\rho_l\omega k+\mathrm{div}\left[(\mu_{lam}+\sigma_k\mu_{turb})\,\mathrm{grad}\,k\right]+\sum_{i=1}^{N}S_{ibk},\\ \frac{\partial(\rho_l\omega)}{\partial t}+\mathrm{div}(\rho_l\vec{V}_l\omega)\\ =\alpha_{k\omega}\frac{\rho_l}{\mu_{turb}}\widetilde{P}-\beta\rho_l\omega^2+\mathrm{div}\left[(\mu_{lam}+\sigma_\omega\mu_{turb})\,\mathrm{grad}\,\omega\right]+2(1-F_1)\rho_l\sigma_{\omega 2}\frac{1}{\omega}\frac{\partial k}{\partial x_i}\frac{\partial\omega}{\partial x_i}+\sum_{i=1}^{N}S_{ib\omega},\end{gathered} \tag{4}$$

with modifications according to the bubble turbulence generation and dissipation terms:

$$S_{ibk}=\frac{3}{8}\frac{C_{iD}}{R_{ib}}\alpha_{ib}\rho_l\left|\vec{V}_{irel}\right|^3,\quad S_{ib\omega}=0.8\frac{k}{\mu_{turb}}S_{ibk}. \tag{5}$$

2.2. *Interphase momentum transfer*

The Stokesian drag force F_{iD}, lift force F_{iL} and turbulent dispersion force F_{iTD} are included into the momentum transfer equation per each phase:

$$S_{ib}=\alpha_{ib}\overset{0}{\rho_b}\vec{g}+\vec{F}_{iD}+\vec{F}_{iL}+\vec{F}_{iTD},\quad S_l=\left(1-\sum_{i=1}^{N}\alpha_{ib}\right)\overset{0}{\rho_l}\vec{g}-\sum_{i=1}^{N}\left(\vec{F}_{iD}+\vec{F}_{iL}+\vec{F}_{iTD}\right), \tag{6}$$

$$\vec{F}_{iD}=\frac{3\rho_l}{8R_{ib}}C_{iD}\vec{V}_{irel}\left|\vec{V}_{irel}\right|,\vec{V}_{irel}=\vec{V}_l-\vec{V}_{ib}, \tag{7}$$

$$\vec{F}_{iL}=C_{iL}\alpha_{ib}\rho_l\vec{V}_{irel}\times\mathrm{rot}\,\vec{V}_l, \tag{8}$$

$$\vec{F}_{iTD}=-1.2C_{iD}\frac{3}{8}\frac{\left|\vec{V}_{irel}\right|}{R_{ib}}\mu^{eff}\nabla\alpha_{ib}, \tag{9}$$

expressions for C_{iD} and C_{iL} were taken from [10] and [11], respectively.

3. Numerical method

The numerical method is based on the finite-volume approach and unstructured grids with co-located storage of flow variables. Pressure and velocity are found in separate fashion by modified phase coupled SIMPLE (PC-SIMPLE) algorithm which is based on the original SIMPLE proposed by Patankar [12]. The modified PC-SIMPLE includes only carrier fluid fluxes for pressure correction, since fluxes of dispersed phase are negligibly small due to low density of gas phase. Such assumption makes additional constraint on the bubble volume fraction which should not exceed 20% and is satisfied in the following simulations. A second order accuracy over spatial coordinates provided by TVD schemes and iterative process is realized through the implicit first-order pseudotime marching scheme. Elimination of the effect of mesh skewness is done by the orthogonal correction in calculation of viscous fluxes and curvilinear gradient calculation scheme.

4. Results

4.1. *Downward flow in a straight pipe*

Simulations of downward bubbly flow inside a vertical straight pipe were carried out according to the conditions mentioned in [13]. Water with immersed air bubbles is supplied to the top end of the straight vertical pipe with the diameter D=25.4 mm and length L=150D. At the inlet, the superficial liquid velocity V_L=1.25 m/s, superficial gas velocity V_G=0.087 m/s and gas void fraction α=0.067. The most probable bubble size is 2 mm diameter, all bubbles distributed uniformly across inlet area. Experiments were carried out at normal conditions with temperature T=297 K and pressure P=101350 Pa.

Numerical simulations conducted on axisymmetrical geometry with 50 cells across radial coordinate with denser mesh near pipe walls, which provides required resolution of flow gradients. Due to significant difference in density, the effect of gravity was included into the simulation setup which resulted in the buoyant (Archimedean) force. Data was extracted from the radial cross-section which is located at the position z=133D from the inlet location.

The simulation results and experimental data points are presented in the figure 1 and figure 2 – void fraction and velocity distributions, respectively. It can be seen that lift force, which acts along the direction perpendicular to the slip velocity between bubble and liquid, at large distances can significantly alter the profile of bubble concentration, dragging the bubbles toward center of the pipe. Such behavior is a common for downward flows, for upward flows opposite behavior will be observed. Such flow changes can be favored for heat transfer pipelines/heat exchangers, since reduction of bubbles near walls intensified wall-liquid heat transfer. The velocity distribution reaches maximum value close to the center of the pipe, with local maximum near the wall. Such behavior can be explained by taking into the consideration several competitive processes: the Stokesian drag and wall friction forces slow down the flow, while the gravity force accelerates the flow. Due to the drop of the bubble concentration near the wall, impact of the drag force reduces and liquid velocity increases with the formation of the localized velocity peak. All results obtained in simulations are in good agreement with the experimental data.

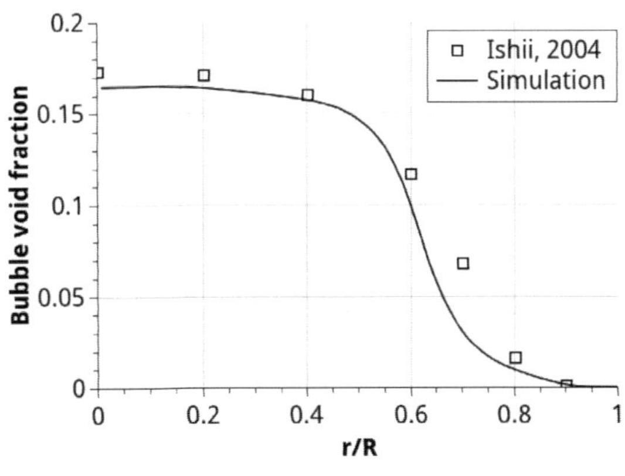

Figure 1. Bubble void fraction distribution at z=133D. Solid line is the simulation, squares are the experimental data [13].

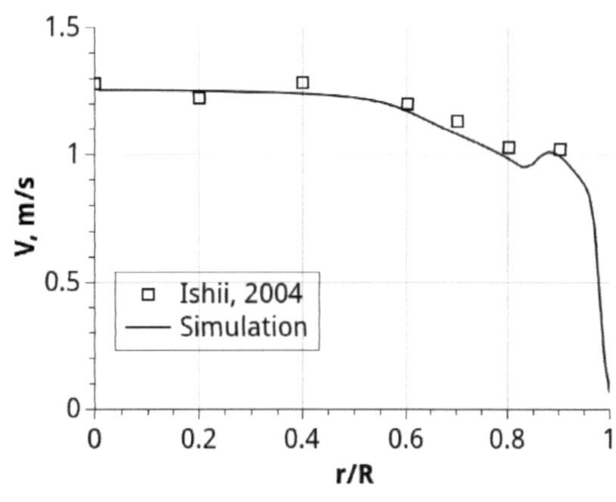

Figure 2. Velocity distribution at z=133D. Solid line is the simulation, squares are the experimental data [13].

4.2. Bubble column reactor

The flow pattern inside a full-scale laboratory bubble column reactor was investigated in [14]. The reactor has parallelepiped shape with dimensions of 500 mm width, 80 mm depth and 1500 mm hight. A bubble injector (sparger) has depth of 80 mm and mounted in the bottom wall at the distance of 160 mm from the left wall. The entire volume of the column is filled by pure water and air bubbles are supplied into the reactor through the sparger with the superficial velocity V_G=0.66 mm/s. The mean diameter of the initial bubble was taken equal to 2 mm. The experiments were carried out at normal conditions with room temperature T=297 K and pressure P=101350 Pa.

The hexahedral orthogonal mesh was constructed with dense mesh towards walls to provide adequate resolution of boundary layers. The top boundary condition was assumed as free surface allowing air bubbles freely leave the domain while slip boundary condition was applied for the liquid phase. The buoyant force in the current setup is the main driver of the flow, thus gravity is also included.

Simulation results of bubble void fraction, velocity and experimental velocity vector field are presented in figures 3-5. It can be seen that due to the lift and turbulence dispersion forces bubble column is smeared with reduction of maximum void fraction towards free surface. Bubbles near the sparger due to their high concentration effectively accelerate surrounding liquid, which leads to the turbulence intensification and formation of strong velocity gradients in the liquid phase. Both of that effects intensifies lateral forces, which, in turn, leads to significant smearing of the bubble column. Turbulence intensification can lead to the more intense diffusion of dissolved gas and, therefore, intensification of chemical processes.

Comparing the simulated (figure 4) and experimental (figure 5) velocity distributions, one can see that slight difference in gradient prediction exists, which is more pronounced in simulations. That can be dedicated to some uncertainties in experiment measurements since during LDA measurements nearby bubbles can alter measured data and due to the fact that experimental image was obtained from the averaging procedure over large period of time. Some discrepancies can also occur because of impact of the top boundary condition which didn't represent actual shape of water surface. However, the overall picture of the flow predicted correctly, with top recirculation zone close to the left wall and jet attachment point which lies 300 mm from the bottom wall.

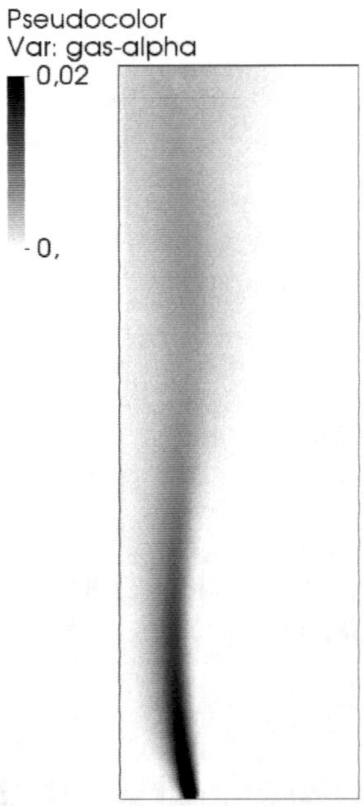

Figure 3. Bubble void fraction distribution in mid-plane obtained in simulation.

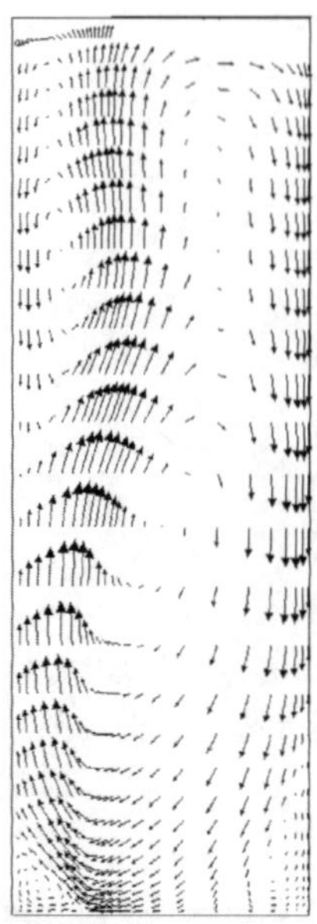

Figure 4. Liquid velocity vector field in mid-plane obtained in simulation, max value 0.24 m/s.

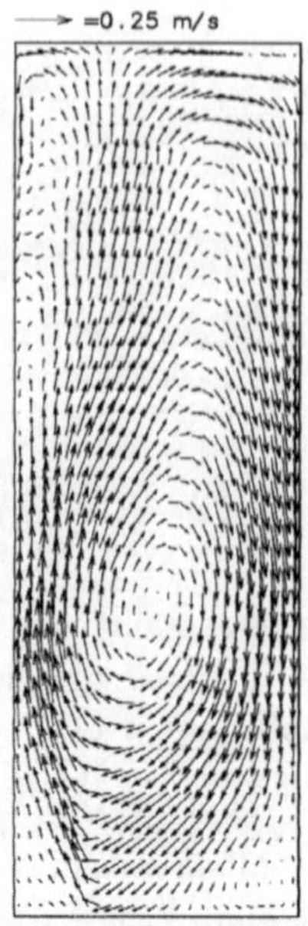

Figure 5. Experimental snapshot of the liquid velocity distribution in mid-plane.

5. Conclusion

The mathematical model based on the Euler-Euler approach and capable of description of polydisperse flows by the means of multi-class model was developed. The numerical algorithm was accordingly modified and numerical simulations were carried out for two distinctive cases of bubbly flows. It was proved that the proposed model and algorithm are able to correctly predict the flow structure under the real flow conditions, with qualitative and quantitative agreement with the available experimental data. Both downward flow inside straight pipe and upward bubbly flow inside rectangular bubble column were investigated, with comparison of velocity and bubble void fraction distributions.

References

[1] Chernyshev A, Kitanin E, Kumzerova E and Schmidt A 2007 *Proc. Int. Conf. on Multiph. Flow (Leipzig Germany)* Paper S7_Tue_D_29
[2] Iben U, Ivanov N, Isaenko I and Schmidt A 2015 *Tech. Phys. Lett.* **41**(12) 1159–62
[3] Chernyshev A and Schmidt A 2013 *AIP Conf. Proc.* **1558** 1071–4
[4] Loth E 2000 *Prog. Energy Comb. Sci.* **26** 161–223
[5] Yao W and Morel C 2004 *Int. J. Heat Mass Transfer* **47** 307–28
[6] Wang T, Wang J and Jin Y 2005 *Chem. Eng. Sci.* **60** 6199–209
[7] Bannari R, Kerdouss F, Selma B, Bannari A and Proulx P 2008 *Comp. and Chem. Eng.* **32** 3224–37
[8] Morel C, Ruyer P, Seiler N and Laviéville J M 2010 *Int. J. Multiph. Flow* **36** 25–39
[9] Chernyshev A S and Schmidt A A 2013 *Tech. Phys. Lett.* **39**(6) 548–51
[10] Lain S, Sommerfeld M, Broder D and Goz M 2002 *Int. J. Multiph. Flow* **28** 1381–407
[11] Tomiyama A, Tamai H, Zun I and Hosokawa S 2002 *Chem. Eng. Sci.* **57** 1849–58
[12] Patankar S V 1980 *Numerical Heat Transfer and Fluid Flow* (New York: McGraw-Hill)
[13] Ishii M, Paranjape S, Kim S and Sun X 2004 *Int. J. Multiph. Flow* **30** 779–801
[14] Sokolichin A and Eigenberger G 1999 *Chem. Eng. Sci.* **54** 2273–84

Stability of a dome-shape liquid film for conditions when it flows around a plate

R A Dekhtyar[1,2], E Yu Slesareva[1], and V V Ovchinnikov[1]

[1]SB RAS, Kutateladze Institute of Thermophysics, 630090 Novosibirsk, Russia

Corresponding author: dekhtyar@itp.nsc.ru

Abstract. The experimental research of hydrodynamic stability of a dome-shape film of liquid for the conditions, when it flows around a thin plate has been carried out. Experiments have been executed for a case of Savart "water bell". An axisymmetrical film has been formed by the impact of a liquid jet with the width of 10 mm on a solid disc with the diameter of 14.5 mm. The width of the thin plate, streamlined by a dome-shape water film, was varied from 0.05 to 3.5 mm. The wide side of the plate was located at two angles relative to the direction of the flow velocity vector in a film: along and across, at various distances from the disc. Stability of the "water bell" surface has been considered and general stability criterion showing its sensitivity to the angle of plate orientation has been derived. The critical Weber number, above which the discontinuity of bell surface appears, and threshold value of the Weber number, below which it coalescences, have been defined. It was shown, that the critical and threshold Weber numbers depend on the Reynolds number.

1. Introduction

The use of a liquid film in heat exchangers (condensers, absorbers, distillation columns, etc.) is important for studying such flow. The structure of the thin liquid film flow influences significantly the local and average characteristics of heat and mass transfer. It is known that intensification of heat and mass transfer in the case of developed wave surface occurs both because of an increase in the surface of heat and mass transfer, and due to a change in the velocity profile of the liquid film [1, 2]. One of the forms of the liquid film is the dome-shaped film. Such films are formed, when the liquid flow impinges a solid disk. The first investigation of the dome-shaped liquid films, the so-called «water bells», has been published in 1833 by Felix Savart [3]. In [4, 5], the models of a «water bell» are submitted. The liquid film velocity U depending on distance Y from a disk surface is presented by dependence [4, 5]

$$U = \sqrt[2]{U_0^2 + 2 \cdot g \cdot Y}\ , \tag{1}$$

where g is gravity acceleration, U_0 is velocity on the brink of a disk. The liquid film thickness δ at section Y is calculated in following form

$$\delta = \frac{Q}{2\pi \cdot R \cdot U}\ , \tag{2}$$

here Q is the volumetric rate of liquid flow, R is radius of a «water bell» profile in the cross-section at altitude Y.

In [6], the flow film is checked experimentally by puncturing the bell. In the period immediately following the puncture, when the pressurized air has not yet escaped from the cavity through the growing hole, the bell shape remains globally unchanged. One then sees that, a) the border of the hole, disturbed upstream, is entrained by the flow as expected for usual supercritical liquid; (b) the border of the hole, disturbed downstream, recedes against the flow, as it corresponds to the subcritical flow. To generate the developed wave surface, the aggregated thin rods are used. At interaction of the dome-shaped film flow with a rod, destruction of the liquid film surface can occur. It is demonstrated in figure 1, the flow regime changes, when rod assemblage is shifted up by 20 mm for constant water flow rate.

The rod diameter and water flow rate are insufficient for determination of the flow regime. It is required to define the additional criteria, allowing us to predict the flow regime considering the rod position relative to the «water bell» surface. The problem can be simplified if we observe the conditions of stability for a thin plate with width ζ. The purpose of this work is obtaining and analysis of data on hydrodynamic stability of a dome-shaped liquid film in the flow around a thin plate.

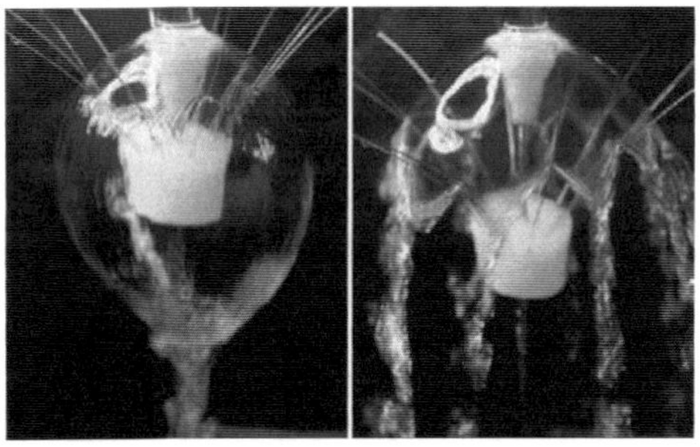

Figure 1. Influence of position of 12 aggregated thin rods relative to the «water bell» surface on the flow regime. On the left, there is a continuous film flow, on the right, it is broken. Diameter of rods is 0.35 mm, flow rate is Q = 55 ml/s, U_0 = 0.74 m/s.

2. A measurement procedure

Experiments with a «water bell», formed at a leak-in of a water jet on a disk with the diameter of 14.5 mm were carried out. The jet was formed at the exit of a brass tube with the inner diameter of 10 mm at the distances from 5 to 60 mm between a tube edge and disk surface. The flow rate was measured by the pressure drop on the calibrated glass tube of 400-mm length and 9.5-mm diameter. Flow rate was calibrated by measuring the rate of the weight change in the filled volume. Calibrating accuracy was 0.3 ml/s. The flow patterns were recorded by a digital camera with the speed of 300 fps. The study the effect of the plate width on the liquid film flow, wedge-shaped plate from stainless steel of 0.1-mm thickness was used, the plate width was changed from 0.05 to 3.5 mm at the length of 34.5 mm. Size ζ was determined in the area of plate surface intersection with the liquid film. In experiments, ζ was determined by length $\mathcal{L}$, obtained at video frame processing (figure 2). In experiments, the plate wide side was mounted at two angles relative to the flow velocity vector; in the film, it was located along and across at a various distance from the disc. By means of a microscrew, the plate was moved along the normal to the film surface and this allowed precise determination of size ζ at the moment of destructing or rebuilding the liquid film surface to the dome shape.

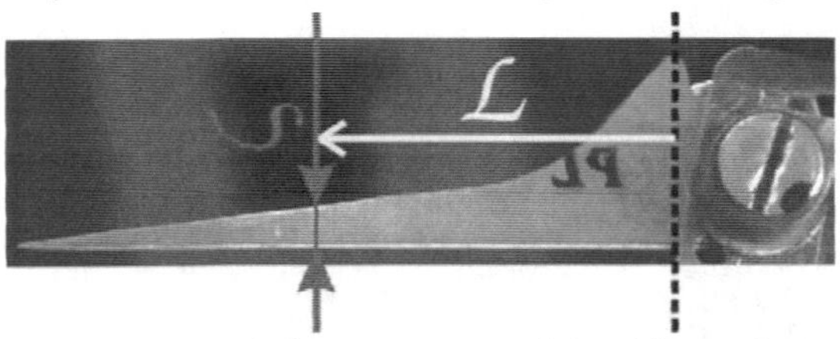

Figure 2. The scheme of determination of the value of the plate wide side ζ by measuring length $\mathcal{L}$. $\zeta = f(\mathcal{L})$.

Values U and δ are computed by 'dependences (1) and (2)', accordingly. Processing the video frames of a «water bells» has shown that dependence of radius profile R on location Y can be presented by smooth functions in the following form:

$$R = A \cdot Y^a \cdot \exp\left[\left(\frac{Y-C}{B}\right)^b\right] \quad (3)$$

Examples of comparisons of the «water bell» shapes with calculation by 'dependence (3)' are presented in figure 3 for various flow rates Q. As is follows from figure 3, the dependence describes the shape film well, if it does not consider the ripples on the surface. Thus, using 'equations (1-3)', for any point of the dome-shape film, it is possible to calculate liquid velocity U and film thickness δ. These values are necessary for computation of the dimensionless parameters, which define the stability conditions of the dome-shape film.

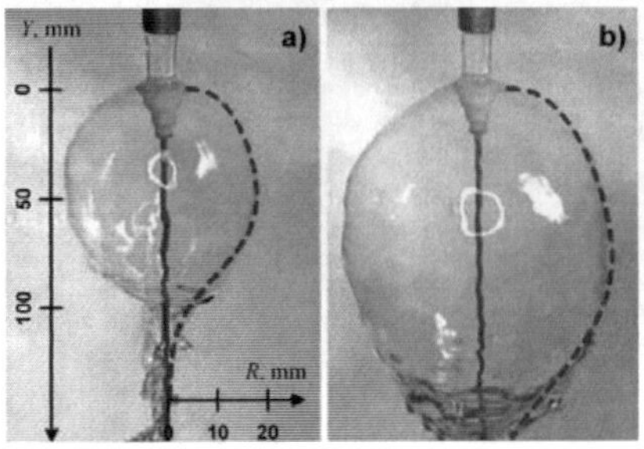

Figure 3. Video frames of the flow of dome-shaped liquid film for different flow rates Q, **a)** 52 ml/s; **b)** 73 ml/s. Dashed line is calculation R (Y) by 'dependence (3)'.
For sizes in "mm", the value of parameters A, a, B, b in 'dependence (3)' is:
a) A = 3.9; a = 0.5; B = 132; b = 12; C = -50.
b) A = 5.9; a = 0.5; B = 164; b = 8; C = -34.

3. Discussion of results

Hydrodynamic stability of a dome-shape liquid film was studied experimentally for the conditions, when it flows around a thin rod and thin plate, and when a water droplet falls on the film surface. The experiment has been executed for the case of Savart "water bell".

When a water droplet falls on the film surface, the wavetrain from the centre at the droplet fall point is formed. This centre is displaced downwards along the flow. If the wave amplitude reaches the magnitude commensurable with the film thickness, a hole is formed the wavetrain centre on the surface. This hole expands with approximately constant velocity and displaces simultaneously downstream. However, the complete destruction of the bell does not happen, or the hole is carried away from a water bell, or the upper edge of the hole moves in the direction of the disk, where the boundary line stops, and reverse motion of the hole boundary line occurs, and this initiates the rebuilding of the film surface. The video frames of the wave and hole on the surface after water droplet fall on the dome-shape film are presented in figure 4. In this case, the droplet falls on the upper part of the bell, the hole on the surface appears through 73 ms after collision in a medial part of the bell. The average velocity of hole expansion is 2 m/s.

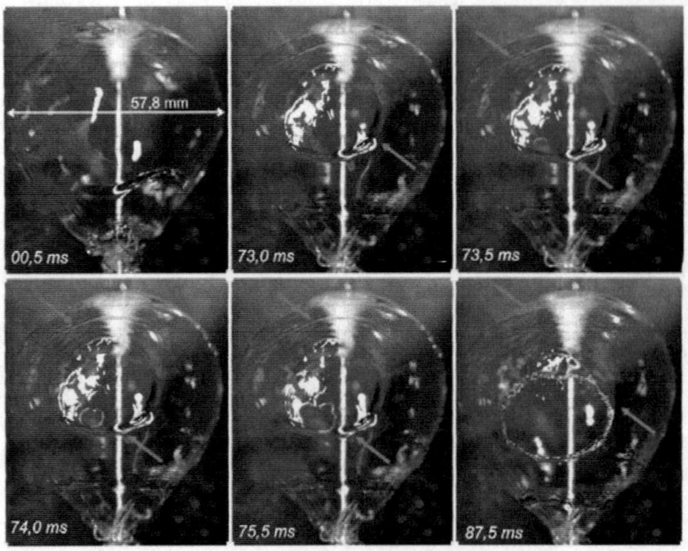

Figure 4. Video frames of wave (blue arrow) and hole (red sign) development on a surface after water droplet fall (red arrow, 0.5 ms) on a dome-shape film. Flow rate is $Q = 48$ ml/s; $U_0 = 0.73$ m/s. Video speed is 5000 fps.

The experimental research showed stability of the "water bell" surface and derived the general stability criterion showing its sensitivity to the angle of location on the plate wide side. Plate width ζ

has critical value ζ_C above which the surface of a dome-shape liquid film is broken below the plate and threshold value ζ_T below which it coalesces.

Three samples of video frames are shown in Figure 5 at different plate widths, total water flow rate of 73 ml/s and initial velocity of 1.2 m/s. For this case, the plate wide side is disposed along the flow velocity vector. It is possible to see the different conditions on the liquid surface, while the dome-shape liquid film flows around a plate. On the left of the picture, the wavetrain (wave regime) is generated in front and behind the plate for case, when $\zeta < \zeta_C$. Discontinuity of the liquid film (discontinuity regime) occurs below the plate in the picture centre for the case, when $\zeta = \zeta_C$. Coalescence of the liquid film (coalescence regime) is observed on the right of the picture for case, when $\zeta < \zeta_T$.

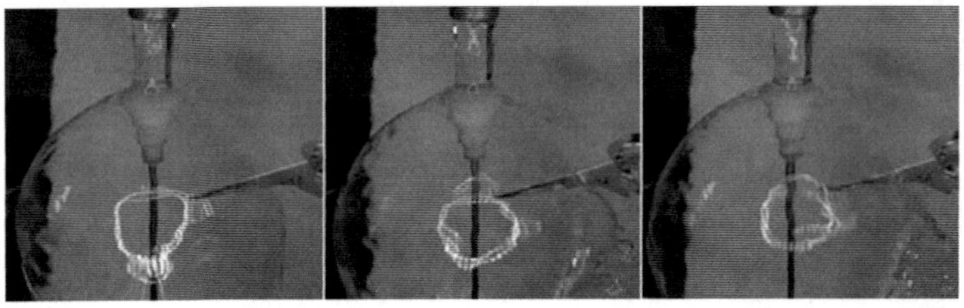

Figure 5. Patterns of the dome-shaped film, when the plate wide side is located along the flow velocity vector. On the left - wave regime, in the centre - discontinuity, on the right - coalescence. Flow rate is Q = 73 ml/s; U_0 = 1.2 m/s.

Using the obtained data on velocity U, film thickness δ, plate width ζ, dimensionless criteria have been counted at the moment of destructing or rebuilding of the liquid film on the «water bell» surface. These are Reynolds number $\mathrm{Re}_\delta = \dfrac{U \cdot \delta}{\nu}$ and Weber number $\mathrm{We}_\zeta = \dfrac{\rho \cdot U^2 \cdot \zeta}{\sigma}$, where ν is kinematic viscosity of liquid, ρ is density of liquid, σ is surface tension. Two types of conditions on the liquid surface for dimensionless parameters We_ζ vs. Re_δ are presented in figure 6. The first type is when $\zeta = \zeta_C$, the second type is when $\zeta = \zeta_T$. It is possible to see that for the case, when the plate wide side is disposed along the flow velocity vector, coalescence of the film surface occurs at the widths less, than that corresponding to discontinuity.

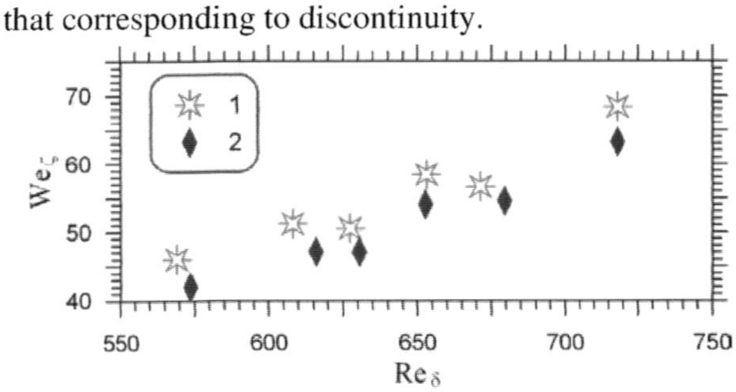

Figure 6. The flow pattern map as function We_ζ of Re_δ for the case when the plate wide side is located along the flow velocity vector. 1 - discontinuity of a liquid film below a plate; 2 – coalescence of a liquid film below a plate.

Like in figure 5, three samples of the video frames are shown in Figure 7 at different plate widths, under the total water flow rate of 52 ml/s and initial velocity of 0.94 m/s. For this case, the wide side of the plate is disposed across the flow velocity vector. The wave regime, when $\zeta < \zeta_C$, is shown on the left picture; discontinuity regime, when $\zeta = \zeta_C$, is in the picture centre; and coalescence regime, when $\zeta < \zeta_T$, is on right of the picture. The flow pattern map as We_ζ function of Re_δ for the case, when the plate wide side is disposed across the flow velocity vector, is presented in figure 8. It is possible to see that for the case, when the plate wide side is disposed across the flow velocity vector, film surface discontinuity occurs if the width is much smaller, than that at coalescence. Line 3 indicates that in this case, film surface coalescence occurs, if plate width ζ is close to film thickness δ.

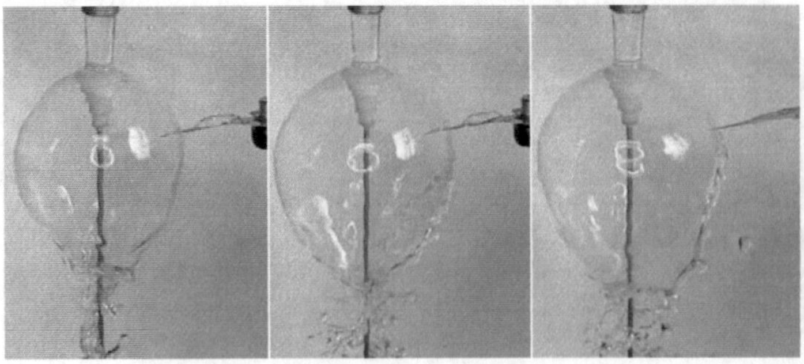

Figure 7. Patterns of the dome-shaped film, when the plate wide side is located across the flow velocity vector. On the left - wave regime, in the centre - discontinuity, on the right - coalescence. Flow rate is Q = 52 ml/s; U_0 = 0.94 m/s.

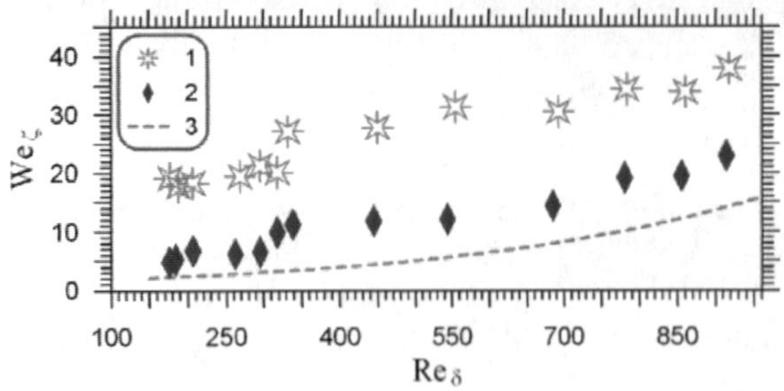

Figure 8. The flow pattern map as function We_ζ of Re_δ for the case, when the plate wide side is located across the flow velocity vector. 1 - discontinuity of a liquid film below a plate; 2 - coalescence of a liquid film below a plate; 3 - Weber number We_ζ at $\zeta = \delta$.

Thus, the experimental research shows that the critical and threshold Weber numbers depend on the Reynolds number at the inter-section point. If rod diameter d is less than local film thickness δ, then the flow without separation is quite possible under the conditions, when the dome-shape film flows around the rod.

4. Conclusion

The experimental research shows that rod diameter d, plate size ζ and water flow rate Q are insufficient for determination of the flow regime. Stability of the dome-shape film surface has been considered and general stability criterion showing its sensitivity to the angle of plate orientation has been derived. If the plate wide side is oriented across the flow velocity vector, discontinuity of the film surface occurs, when the plate width is much smaller than that at coalescence. If the plate wide side is oriented along the flow velocity vector, coalescence of the film surface occurs, when the plate width is slightly less than that at discontinuity. We have determined the critical Weber number above which the discontinuity of bell surface appears and the threshold Weber number below which it coalescences. It is shown that the threshold and critical local Weber numbers depend on the local Reynolds number at the inter-section line of the bell with the plate.

This work was supported by the Russian Science Foundation (project no. 15-19-10025).

References

[1] Alekseenko S V, Nakoryakov V E, Pocusaev B G 1994 *Wave Flow of Liquid Films* (New York: Begell house)
[2] Nakoryakov V E, Misyura S Y and Elistratov S L 2013 *J. Engin. Therm.*, **22(1)**, 1
[3] Savart F 1833 *Ann. De Chim.*, **54** 56
[4] Taylor G 1959 *Proc. R. Soc. London A*, **253(13)** 289
[5] Parlange J Y 1967 *J. Fluid Mech.* **29(2)** 361
[6] Lhuissier H and Villermaux E 2012 *J. Fluid Mech.* **693** 508

Anomalous waves in gas-liquid mixtures near gas critical point in Gardner equation approximation

V G Gasenko

Institute of Thermophysics, Lavrentiev avenu 1, Novosibirsk, Russian Federation
E-mail: gasenko@itp.nsc.ru

Abstract. The waves in a bubbled incompressible liquid with Van der Waals gas in a bubbles being near critical points is considered in a frame of Gardner equation. It is shown that both coefficients on quadratic and cubic nonlinear terms in Gardner equation change the sign near gas critical point and it results the anomalous waves: negative and limited solitons, kinks, antikinks and breathers. The dynamics and interactions of these waves was studied numerically by high accuracy Fourier methods with periodically boundary conditions. In particular it is revealed that limited solitons always arise from initial distribution with a few identical soliton's pair and stand stable in their form after numerous interactions.

1. Introduction

The existence of rarefaction shock waves in Van der Waals gas was firstly shown by Zeldovich [1] under necessary condition of negative derivation $\partial^2 V / \partial P^2 < 0$. Zones of negative derivation bounded by curves 3,5 and 4,5 under adiabatic and isothermal conditions respectively was calculated in [1] and are shown on figure 1. Adiabatic anomalous zone appears only if adiabatic index $\gamma < 1.049$ is unreal.

The situations radically changes if gas behavior is near isothermal. This condition can be achieved in a small gas bubbles in a liquid under intense heat exchanging during the wave's pass. Since the second derivation near critical point can be vanished, the next term $\partial^3 V / \partial P^3$ of series was considered. In this paper the strict yielding of Gardner equation for nonlinear waves in homogeneous bubbled liquid and its numerical calculations are presented. It appears that anomalous zones of positive third adiabatical and isothermal derivation being shown between curves 2,5 and 4,5 on figure 1 respectively in a frame of Gardner equation are much greater then for ordinary rarefaction shock waves in [1].

The Gardner equation was firstly yielded for inner waves in two layer liquid [2] where anomalous zones of changes the signs of quadratic and cubic terms is too narrow to observe it experimentally, the waves in bubbled liquid near critical point for some gases in bubbles as freons is easy to achieve and thus we predict the future experiments in bubbled liquid near critical point and observing all kind of exotic waves which exist in a frame of Gardner equation with changing signs of nonlinear terms.

2. Derivation of Gardner equation

The homogeneous model of bubbled liquid in acoustic approximation $P_{xx} = \rho_{tt}$ [3] with mixture density definition $\rho = \rho_1(1-\varphi) + \rho_2\varphi$, where $\varphi = 4/3\pi R^3 N$ is gas void fractions, N is bubble numbers in mixture unit volume in incompressible liquid results following wave equation for dimensionless pressure $p = P/P_0$ and dimensionless bubble's volume $v = R^3/R_0^3$

$$p_{xx} = -\gamma v_{tt} / c_0^2 . \quad (1)$$

Here $c_0^2 = \gamma P_0 / \rho_0 \varphi_0$ is sound velocity in a mixture, indexes 1,2 and 0 refer to liquid, gas and initial values respectively. For bubble's volume we use dissipationless Reley equation

$$v^{-1/3} v_{tt} - \frac{1}{6} v^{-4/3} v_t^2 = \frac{w_0^2}{\gamma}(p_2 - p), \tag{2}$$

where $w_0^2 = 3\gamma P_0 / \rho_1 R_0^2$ is bubble's resonant frequency, $p_2 = P_2 / P_0$ is dimensionless gas pressure, R_0 is initial bubble's radii. In adiabatic equation of state for Van der Waals gas

$$\left(P_2 + a/V^2\right)(V-b)^\gamma = \left(P_0 + a/V_0^2\right)(V_0 - b)^\gamma,$$

two parameters being expressed through critical values $a = 3P_k V_k^2$, $b = V_k/3$ results its dimensionless equation of state $p_2(v)$

$$p_2(v) = (1+a_0)(1-b_0)^\gamma (v-b_0)^{-\gamma} - \frac{a_0}{v^2}, \tag{3}$$

where $a_0 = 3/\tilde{V}^2 \tilde{P}$, $b_0 = 1/3\tilde{V}$, $\tilde{V} = V_0/V_k$, $\tilde{P} = P_0/P_k$. For weak but finite amplitudes waves nonlinear system (1), (2), (3) can be reduce to Gardner equation if we expand (3) to Taylor's series

$$\begin{aligned} p_2(v) &= 1 - A(v-1) + B(v-1)^2 - C(v-1)^3, \\ A &= -\frac{dp_2}{dv}\bigg|_{v=1} = \gamma\frac{1+a_0}{1-b_0} - 2a_0, \\ B &= \frac{1}{2}\frac{d^2 p_2}{dv^2}\bigg|_{v=1} = \frac{\gamma(\gamma+1)}{2}\frac{1+a_0}{(1-b_0)^2} - 3a_0, \\ C &= -\frac{1}{6}\frac{d^3 p_2}{dv^3}\bigg|_{v=1} = \frac{\gamma(\gamma+1)(\gamma+2)}{6}\frac{1+a_0}{(1-b_0)^3} - 4a_0. \end{aligned} \tag{4}$$

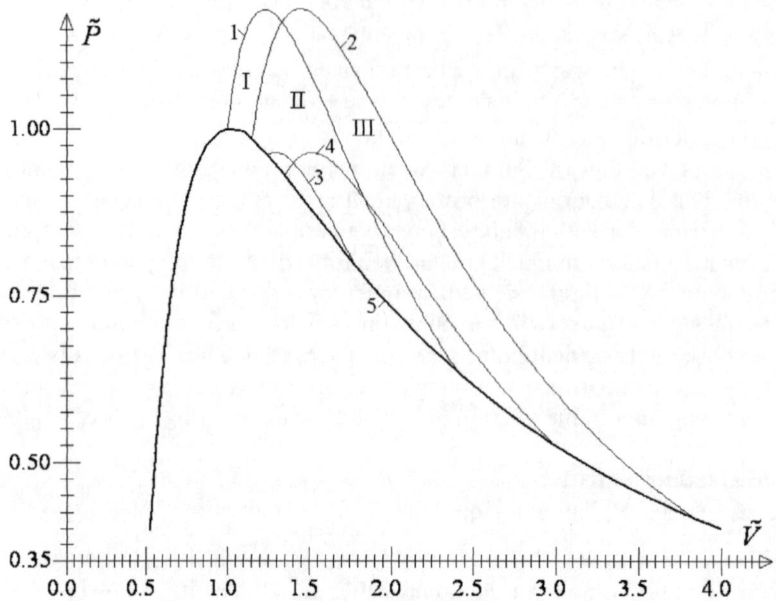

Figure 1. The borders of anomalous zones : 1 – $\alpha_T = 0$, 2 – $\beta_T = 0$, 3 – $\alpha_S = 0$, 4 – $\beta_S = 0$, 5 – binodal.

Substitute (4) into linearized Reley equation, twice differentiate it by x and take into account (1) we come to Boussinesq equation for volume perturbation $\tilde{v} = v - 1$

$$\frac{A}{\gamma}\tilde{v}_{xx} - \frac{1}{c_0^2}\tilde{v}_{tt} - \frac{B}{\gamma}\tilde{v}^2_{xx} + \frac{C}{\gamma}\tilde{v}^3_{xx} + \frac{1}{w_0^2}\tilde{v}_{xxtt} = 0. \quad (5)$$

Transition from (5) to Boussinesq equation for dimensionless pressure perturbation $\tilde{p} = p - 1$ can be done with linear link $\tilde{v} = -A^{-1}\tilde{p}$ in (4)

$$\tilde{p}_{xx} - \frac{1}{c^2}\tilde{p}_{tt} - \frac{B}{\gamma A^2}\tilde{p}^2_{xx} + \frac{C}{\gamma A^3}\tilde{p}^3_{xx} + \frac{1}{w^2}\tilde{p}_{xxtt} = 0, \quad (6)$$

where $c^2 = c_0^2 A/\gamma$, $w^2 = w_0^2 A$. For the waves propagating in one direction (6) reduce to Gardner equation, which preferable to write in dimensionless variables $x' = xk$, $t' = ckt$, $k = \sqrt{2}w/c$, $u = \tilde{p}(\gamma+1)/12\gamma$ in coordinate system moving with c velocity (all primes further are omitted)

$$u_t + 6\alpha u u_x + 6\beta u^2 u_x + u_{xxx} = 0, \quad (7)$$

where equation coefficients

$$\alpha = \frac{(\mathrm{d}^2 p_2/\mathrm{d}v^2)_S}{12(\mathrm{d}p_2/\mathrm{d}v)_S^2}, \quad \beta = -\frac{(\mathrm{d}^3 p_2/\mathrm{d}v^3)_S}{24(\mathrm{d}p_2/\mathrm{d}v)_S^3} \quad (8)$$

can be expressed through coefficients A, B, C from (4). The equations of curves $\alpha = 0$ and $\beta = 0$ result to follow equations on dimensionless $\tilde{P}\tilde{V}$ diagram

$$\alpha = 0: \quad \tilde{P} = \frac{2(3\tilde{V}-1)^2}{\gamma(\gamma+1)\tilde{V}^4} - \frac{3}{\tilde{V}^3}, \quad \beta = 0: \quad \tilde{P} = \frac{8(3\tilde{V}-1)^3}{3\gamma(\gamma+1)(\gamma+2)\tilde{V}^5} - \frac{3}{\tilde{V}^3}., \quad (9)$$

The curves of zero derivations (9) are shown in figure 1 as α_T, β_T under isothermal conditions ($\gamma = 1$) and as α_S, β_S under adiabatic conditions ($\gamma = 1.049$). Curves $\alpha_T = 0$, $\beta_T = 0$ and bimodal curve divide areas on $\tilde{P}\tilde{V}$ diagram near critical point onto three isothermal anomalous zones: zone I ($\alpha_T, \beta_T < 0$), zone II ($\alpha_T < 0,\ \beta_T > 0$) and zone III ($\alpha_T, \beta_T > 0$). Analogous adiabatic zones linked with curves α_S, β_S are much less and so do not marked on figure 1. Beyond these three zones the cubic nonlinearity looses his sense and instead (7) we must take ordinary KdV equation ($\alpha = 1,\ \beta = 0$).

3. Analytical and numerical solutions of Gardner equation

The Gardner equation (7) which was firstly derived for inner waves in two layer liquid [2] is now well studied since it is can be transferred to modified Korteveg-de-Vries equation (mKdV)

$$Q_t - 3\alpha^2/2\beta \cdot Q_x + 6\beta Q^2 Q_x + Q_{xxx} = 0, \quad (10)$$

by Gardner substitution $u = Q - \alpha/2\beta$ and thus becomes fully integrable by Inverse Scattering Transform Method [4]. In turn the KdV equation

$$Q_t + 6QQ_x + Q_{xxx} = 0,$$

can be transform to Gardner equation (7) by generalized Miura transform $Q = u/\alpha + u^2\beta/\alpha^3 + u_x(-\beta)^{1/2}/\alpha^2$, but reverse transform from Gardner to KdV is possible only if $\beta < 0$ when differential relation between u and Q becomes real.

Equation (7) has the stationary solution in a form of positive u_+ and negative u_- solitons

$$u_{\pm}(\xi)=\frac{U}{\alpha \pm \sqrt{\alpha^{2}+\beta U}\,\mathrm{ch}(\xi\sqrt{U})}, \tag{11}$$

where $\xi = x - Ut$, which transform to limited soliton with amplitude $u_{max} = \alpha / \beta$ if $\beta < 0$ and U coming to $-\alpha^2 / \beta$. Solitons of both polarities $u_{\pm}$, which exist in zones II, III are shown in figure 2. Transition of negative solitons to limited soliton in zone I if $U \Rightarrow -\alpha^2 / \beta$ are shown on figure 3.

Limited soliton presents itself superposition of kink ($u_m > 0$) and antikink ($u_m < 0$) step waves

$$u(\xi) = u_m \mathrm{th}\left(u_m \sqrt{-\beta}\xi\right) - \frac{\alpha}{2\beta}, \quad U = 2u_m^2 \beta - \frac{3\alpha^2}{2\beta}. \tag{12}$$

with shifted phase and amplitudes $u_m = \pm\sqrt{\alpha} / 2\beta$. Another valuable amplitude of kink and antikink (12) is $u_m = \pm\sqrt{3}\alpha / 2\beta$ when their propagation velocities U are zero.

More information about stationary waves gives the analyses of phase plane (u, u_ξ), searching out separatrix, limit cycles and singularity points, where du_ξ / du does not determined. According to first integral of (7) $u_\xi(u) = \pm u(U - 2\alpha u - \beta u^2 / 2)^{1/2}$ there are always three singular points: $u_\xi = 0$, $u_1 = 0$, $u_{2,3} = [-2\alpha \pm (4\alpha^2 + 2\beta U)^{1/2}] / \beta$. First point is always a saddle. Second and third are centre points if $\beta > 0$. Separatrix which refer to solitons begin and end at saddle, passing around the centers where there are numerous closed curves referring to conoidal waves. If $\beta < 0$ first and second points are the same, but third one is saddle and thus exists only one polarity solitons and conoidal waves. In special case $U = -\alpha / \beta$ two separatrix referring to kink and antikink connect both saddles. Phase plane for the case $\alpha = \beta = -1$, $U = 0.9$ are shown on figure 4. Bold lines are the separatrix.

Second class of analytical solutions of Gardner equation are breathers. They exist only for positive $\beta > 0$ and represent itself the envelop waves on substrate. The breather of mKdV at $\beta = 1$

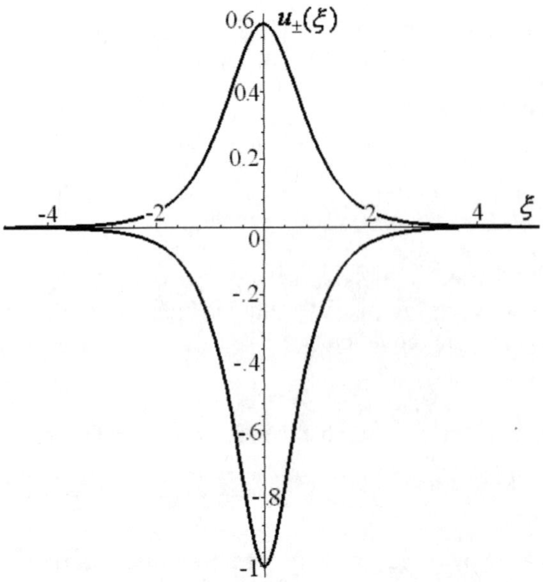

Figure 2. Positive u_+ and negative u_- solitons (11) at zones II, III at $\alpha = 1, \beta = 5$, $V = 3$.

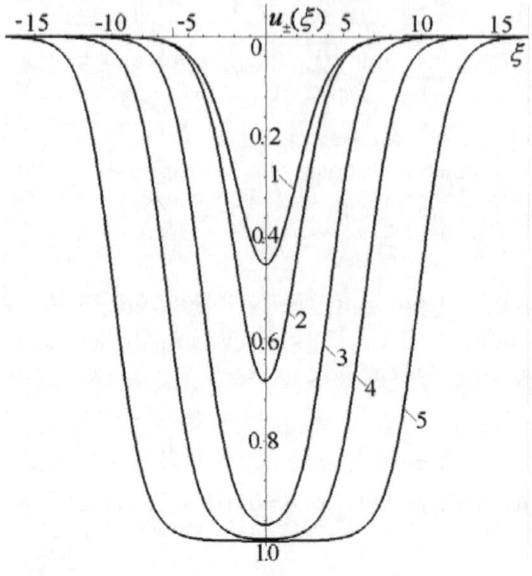

Figure 3. Normal and limited solitons (11) at zone I: $\alpha = 1, \beta = -5$, 1 – $U = 0.7$, 2 – $U = 0.9$, 3 – $U = 0.999$, 4 – $U = 1 - 10^{-5}$, 5 – $U = 1 - 10^{-7}$.

$$u(x,t) = -\frac{2k_1}{\mathrm{ch}\,\vartheta_1} \cdot \frac{\sin\vartheta_2 + k_1/k_2 \cos\vartheta_2 \,\mathrm{th}\,\vartheta_1}{1 + k_1^2/k_2^2 \cos^2\vartheta_2 / \mathrm{ch}^2\vartheta_1}, \tag{13}$$

is also a breather solution of (7) after Gardner substitution. Here $\vartheta_1 = k_1 x - (k_1^3 - 3k_1k_2^2)t + \ln(k_2/k_1)$, $\vartheta_2 = k_2 x - (3k_2k_1^2 - k_2^3)t$, $k_{1,2}$ are wave numbers.

Numerical solutions of Gardner equation was found by high accuracy Fourier method with periodical boundary conditions. Due to connection the end and beginning of x integrating interval in Fourier method, wave dynamics and it's interactions can be observed during unlimited time.

The breakup of initial negative Gaussian wave onto four pairs of ordinary soliton and one limited soliton are shown in figure 5. It is well seen that after fifteen round race of limited soliton and more then hundred interactions soliton pairs are stayed stable. Soliton's pair dynamics is observed firstly.

More complicate interactions of three breathers (13) with common $k_1 = 0.5$ and $k_2 = 1,\ 2.5,\ 5$ are shown in figure 6. The interactions arise because of different and negative propagating velocities which are proportional to k_2, so the first and the fastest breather with $k_2 = 5$ run down the second and third ones at $t = 0.523$ and $t = 1.866$ respectively. After a long time $t = 17.17$ breathers running in circled integrating interval and nineteen breather's elastic interactions their forms do not changed.

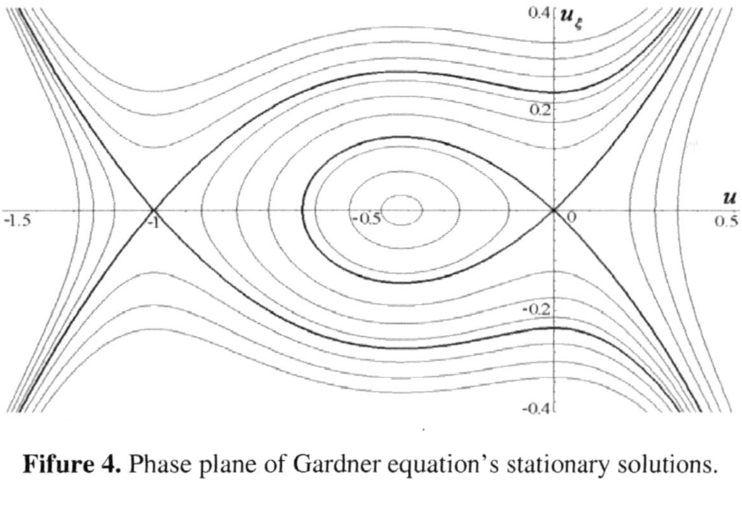

Fifure 4. Phase plane of Gardner equation's stationary solutions.

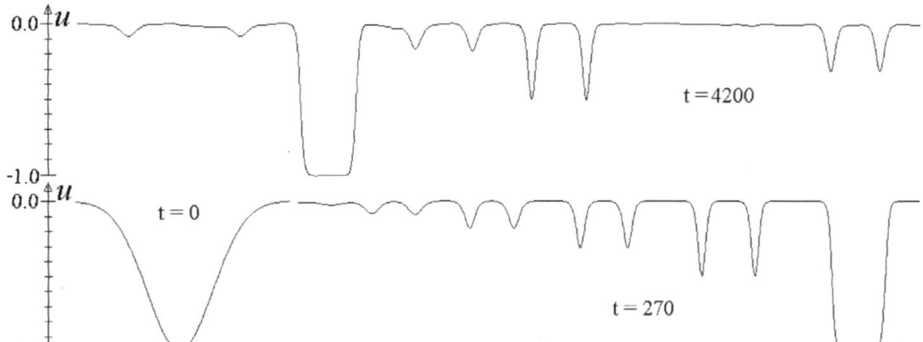

Figure 5. Long time evolution and interactions of soliton pair and limited soliton precipitated from initial Gaussian wave at $\alpha = \beta = -1$.

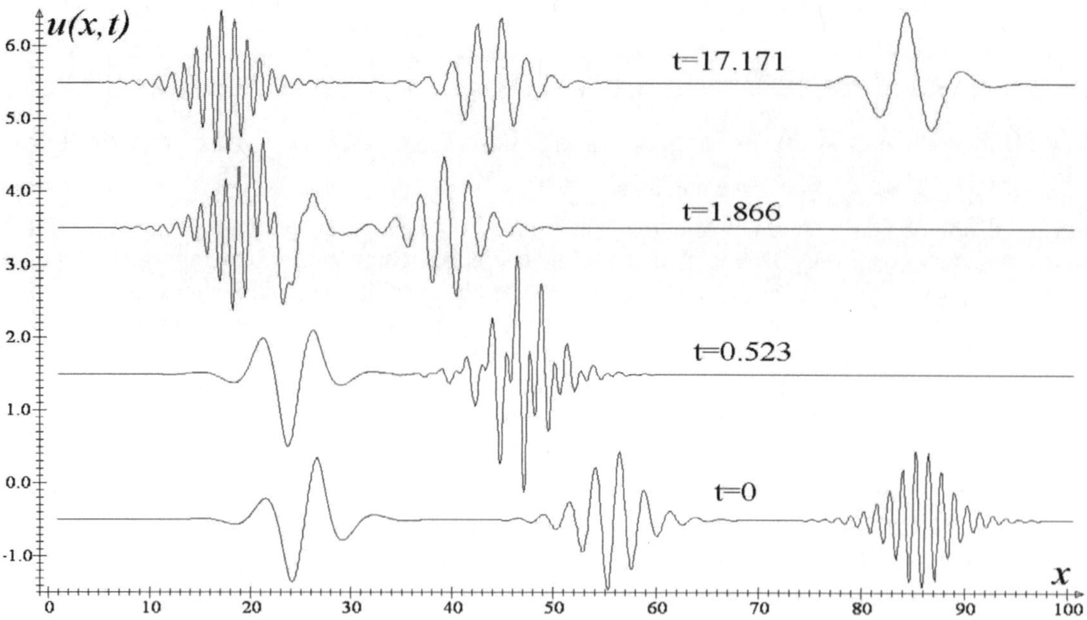

Figure 6. Interactions of three breathers (13) at $\alpha = \beta = 1$.

4. Conclusions

It is shown that in bubbled incompressible liquid with Wan der Waals gas in bubbles pressure waves can be correctly considered in a frame of Gardner equation, with coefficients of quadratic and cubic nonlinear terms being proportional to $\partial^2 V / \partial P^2$ and $\partial^3 V / \partial P^3$ respectively.

The analyze of nonlinear coefficients on P,V diagram showed three ample anomalous zones where their signs are differed from those ones in ideal gas if gas behavior is near isothermal.

Since near isothermal behavior of gas bubbles in a liquid is easy to achieve, all known solutions of Gardner equation like negative and limited solitons, kinks, antikinks and breathers in bubbled liquid near critical point are the anomalous waves and can be experimentally observed instead of rarefaction shock waves predicted by Zeldovich in pure nonideal gas being always adiabatic in shock waves.

As illustrations the dynamics and interactions of these anomalous waves was studied numerically by high accuracy Fourier methods with periodically boundary conditions. In particular it is revealed that limited solitons always arise from negative initial distribution with a few identical soliton's pair and stand stable in their form after numerous interactions.

5. Acknowledgments

This paper was supported by Russian Science Foundation grant № 15-19-10025.

References

[1] Zeldovich Ya B 1946 On possibility of rarefaction shock waves existence *Zh. Eksp. Teor. Fiz.* **16** 363
[2] Kakutani T, Yamasaki N 1978 Solitary waves in two-layer fluid *J. Phys. Soc. Jap.* **45** 674
[3] Van Wingaarden L. 1972 One-dimensional flow of liquids containing small gas bubbles *Ann. Rev. Fluid Mech.* **4** 369
[4] Wiles J W 1981 On internal solitary waves *Tellus* **38** 394
[5] Slunyaev A V, Pelinovsky E N 1999 Soliton dynamics of large amplitude *Zh. Eksp. Teor. Fiz.* **116** 318

Investigation of the two-layer fluid flows with evaporation at interface on the basis of the exact solutions of the 3D problems of convection

O N Goncharova[1,2] and O A Kabov[1,3]

[1] Institute of Thermophysics SB RAS, Ac. Lavrentieva ave 1, Novosibirsk, Russia
[2] Altai State University, pr Lenina 61, Barnaul, Russia
[3] National Research Tomsk Polytechnic University, Lenin ave 30, Tomsk, Russia

E-mail: go_n@mail.ru

Abstract. New physical experiments in the Institute of Thermophysics SB RAS allow one to investigate structure of the flows of liquid layers being under action of the co-current gas flux. The flow topology is determined by four main mechanisms: natural and thermocapillary convection, tangential stresses induced by the flow of gas and mass transfer due to evaporation at the interface. Mathematical modeling of the fluid flows in an infinite channel of the rectangular cross section is carried out on the basis of a solution of special type of the convection equations. The effects of thermodiffusion and diffusive thermal conductivity in the gas phase and evaporation at the thermocapillary interface are taken into consideration. Numerical investigations are performed for the liquid-gas (ethanol-nitrogen) system under normal and low gravity.

1. Introduction

The present consideration is devoted to modeling of the convective fluid flows with evaporation using the new exact solutions of the convection equations. Obtaining the exact solutions of the governing equations has become more relevant because of the effective opportunity to model the real fluid flows [1, 2]. The solution of special type of the Boussinesq approximation of the Navier-Stokes equations can be called a generalization of the well-known Ostroumov-Birikh solutions [3, 4] for three dimensional case with respect to evaporation at interface. The group nature of the Birikh solutions has been analyzed in [5]. In [6] the three dimensional exact solution has been constructed to describe the two-layer fluid flows in an infinite channel in the non-axis-symmetrical case.

The fluid flows with an interface being under action of an adjacent gas flux are the subject of many experimental and theoretical investigations in the last decade [1, 2]. A closer interest in such problems is explained by need of study of the phenomena in the liquid media and features of the flow structure caused by the gas flow and gas flow related evaporation effects. The physical experiments are an important motivation for improvement of theoretical understanding of evaporative convection. One of the first examples of the solutions of the problems of two-layer flows with mass transfer at the "liquid-liquid" interface has been obtained in [7]. In [8, 9] the exact solutions, described the two-layer flows with evaporation, have been constructed in the case of a given specific gas flow rate [8] and under closed flow conditions in each phase [9].

Stability of the convective flows with evaporation and the effects of the intensity of the thermal load on the channel walls on the stability characteristics have been investigated in the problem statement without taking into account the Soret effect [10].

2. New exact solutions of the two-layer fluid flows with evaporation at interface

The stationary solution of the Boussinesq approximation of the Navier-Stokes equations

$$(\mathbf{v}_i \cdot \nabla)\mathbf{v}_i = -\frac{1}{\rho_i}\nabla p_i + \nu_i \Delta \mathbf{v}_i - \mathbf{g}(\beta_i T_i + \underline{\gamma C}), \quad (1)$$

$$div\ \mathbf{v}_i = 0, \quad (2)$$

$$\mathbf{v}_i \cdot \nabla T_i = \chi_i(\Delta T_i + \underline{\delta \Delta C}), \quad (3)$$

$$\mathbf{v}_2 \cdot \nabla C = D(\Delta C + \alpha \Delta T_2) \quad (4)$$

are constructed. The equations (1)-(3) without marked terms are used to model the flows of the liquid in the lower layer (figure 1). The equations (1)-(4) are considered to model the flows in the upper layer so that the Dufour and Soret effects or the effects of diffusive thermal conductivity and thermodiffusion [11-14] in the gas phase are taken into account. Two layers are separated by the thermocapillary interface. By construction of the exact solution of the equations (1)-(4) the interface will be given by the relation $x = 0$. The Cartesian coordinate system is chosen so that the gravity acceleration vector $\mathbf{g}$ ($g = |\mathbf{g}|$) is directed opposite to the Ox axis (figure 1).

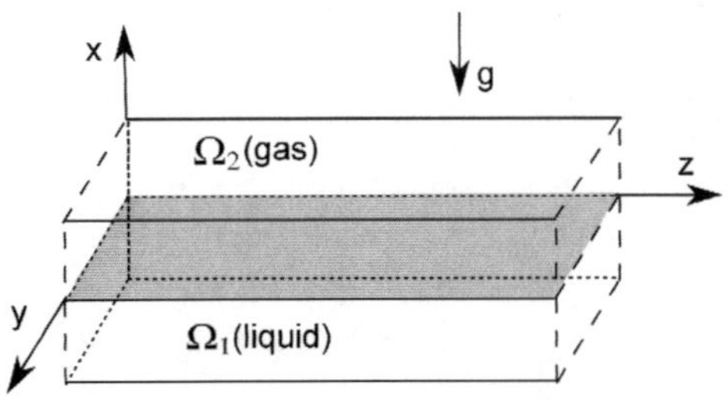

Figure 1. Flow domain.

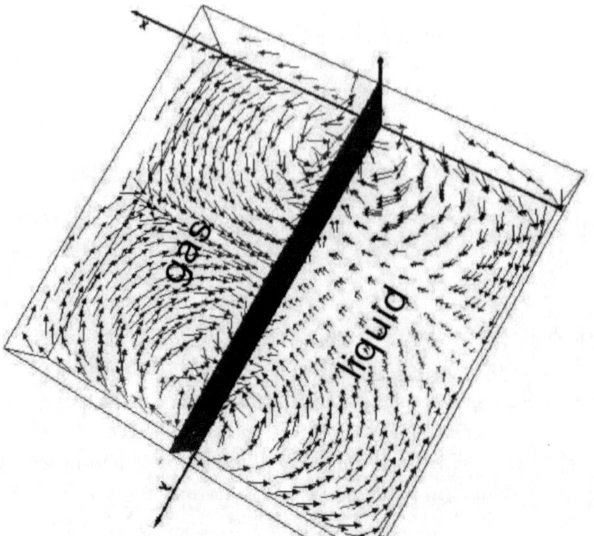

Figure 2. Flow picture: velocity field; $Gr = 470$, $\tilde{T} = 0.5$.

The fluids (the liquid and the gas-vapor mixture) fulfill the infinite horizontal layers Ω_1 and Ω_2:

$$\Omega_1 = \{(x, y, z) : -x_0 < x < 0,\ 0 < y < y^0,\ -\infty < z < \infty\},$$

$$\Omega_2 = (x, y, z) : 0 < x < x^0,\ 0 < y < y^0,\ -\infty < z < \infty\}.$$

We assume that the boundaries of the domains Ω_1, Ω_2 defined by $x = -x_0$, $x = x^0$, $y = 0$ $y = y^0$ are the fixed impermeable boundaries.

The exact solution is characterized by dependence of the components of the liquid $\mathbf{v}_1 = (u_1, v_1, w_1)$ and gas velocity $\mathbf{v}_2 = (u_2, v_2, w_2)$ on the transverse coordinates (x, y). The functions of temperature T_1, T_2, pressure (deviation of pressure from the hydrostatic one) p_1, p_2 and vapor concentration C have the terms Θ_1, Θ_2, q_1, q_2, Φ similarly depending on the transverse coordinates (x, y):

$$u_i = u_i(x, y), v_i = v_i(x, y), w_i = w_i(x, y), \tag{5}$$

$$p_1 = -A\rho_1\beta_1 gxz + q_1(x, y),\ p_2 = -A\rho_2\beta_2 gxz + B\rho_2\gamma gxz + q_2(x, y), \tag{6}$$

$$T_1 = -Az + \Theta_1(x, y),\ T_2 = -Az + \Theta_2(x, y), \tag{7}$$

$$C = Bz + \Phi(x, y). \tag{8}$$

The following notations are introduced: ρ_i is the density, ν_i is the kinematic viscosity coefficient, χ_i is the thermal diffusivity coefficient, β_i is the thermal expansion coefficient, γ is the concentration coefficient of density, the coefficients α and δ characterize the Soret and Dufour effects in the gas-vapor layer. The coefficients A and B determine the longitudinal temperature and concentration gradients along the interface. Subscript i is responsible for belonging to the liquid Ω_1 $(i = 1)$ or gas-vapor phase Ω_2 $(i = 2)$ (see figure 1).

The kinematic

$$u_1 = u_2 = 0 \tag{9}$$

and dynamic conditions (projection on the vectors tangent to the interface) should be fulfilled on the interface $x = 0$:

$$\rho_1\nu_1(u_{1y} + v_{1x}) - \rho_2\nu_2(u_{2y} + v_{2x}) = -\sigma_T\Theta_{1y}, \tag{10}$$

$$\rho_1\nu_1 w_{3x} - \rho_2\nu_2 w_{2x} = A\sigma_T. \tag{11}$$

Here $\sigma = \sigma_0 - \sigma_T(T - T_0)$ is the surface tension ($\sigma_T = const > 0$) . The dynamic boundary condition in projection on the normal vector $\mathbf{n}$ that coincides with the unit vector of the Ox axis is [6]:

$$-q_1 + q_2 + 2(\rho_1\nu_1 u_{1x} - \rho_2\nu_2 u_{2x}) = 0. \tag{12}$$

Conditions of continuity of tangential velocities and temperature are assumed to be fulfilled on the thermocapillary interface $x = 0$:

$$v_1 = v_2,\ w_1 = w_2,\ T_1 = T_2. \tag{13}$$

At $x = 0$ the heat transfer condition with respect to the diffusive mass flux due to evaporation and the mass balance equation are formulated with respect to the Dufour and Soret effects:

$$\kappa_1 T_{1x} - \kappa_2 T_{2x} - \delta\kappa_2 C_x = -\lambda M,\ M = -D\rho_2(C_x + \alpha T_{2x}). \tag{14}$$

The linearized form of an equation for saturated vapor concentration at interface is introduced [8-10]:

$$C|_{x=0} = C_*(1 + \varepsilon(T_2 - T_0)). \tag{15}$$

Here C_* is the saturated vapor concentration at $T_2 = T_0$ (T_0 is equal to 20° in [7], [9]), $\varepsilon = \lambda\mu/(R^* T_0^2)$, λ is the latent heat of evaporation, μ is the molar mass of the evaporating liquid, R^* is the universal gas constant. The relation (15) is a consequence of the Clapeyron – Clausius equation and the Mendeleev – Clapeyron equation for an ideal gas.

On the fixed impermeable walls of the channel the no-slip conditions for velocity fields

$$x = x_0,\ x = x^0,\ y = 0,\ y = y^0 :\ \mathbf{v}_i = 0 \tag{16}$$

and the conditions of thermal insulating of the lateral walls

$$x = -x_0 : T_{1x} = 0;\ x = x^0 : T_{2x} = 0;\ y = 0 : T_{1y} = 0, T_{2y} = 0;\ y = y^0 : T_{1y} = 0, T_{2y} = 0 \quad (17)$$

are prescribed. The case of absence of vapor flux on the upper and lateral rigid boundaries is studied in the present statement:

$$x = x^0 : C_x = 0;\ y = 0,\ y = y^0 : C_y = 0. \quad (18)$$

The analytical technique to model three-dimensional convection of the immiscible fluids in the stationary case with evaporation at the interface is complemented by the numerical investigations so that construction of the three-dimensional flows in a channel of the rectangular cross-section with the interface will be completed. Constructing of the stationary solutions of type (5) - (8) for the two-layer convective fluid flows is reduced to sequential solving of the corresponding two-dimensional problems for the functions $w_1(x, y)$ and $w_2(x, y)$, $\Theta_1(x, y)$ and $\Theta_2(x, y)$, and $\Phi(x, y)$ and for the stream functions $\psi_1(x, y)$, $\psi_2(x, y)$ and vorticity and $\omega_1(x, y)$, $\omega_2(x, y)$ instead of the transverse components u_i, v_i of the velocity vectors ($u_i = (\psi_i)_y$, $v_i = -(\psi_i)_x$, $i = 1, 2$). The two-dimensional problems are computed in the domains $\tilde{\Omega}_i$ ($\tilde{\Omega}_1 = \{(x, y) : -x_0 < x < 0,\ 0 < y < y^0\}$, $\tilde{\Omega}_2 = \{(x, y) : 0 < x < x^0,\ 0 < y < y^0\}$).

2.1. General scheme of solution of the coupled problem

(i) We assume that the velocity components u_i, v_i are found and will proceed from a given state.

(ii) With given u_i, v_i we solve numerically the corresponding problems to find the third components w_i ($i = 1, 2$) of the velocity vectors.

(iii) To find the unknown terms Θ_i ($i = 1, 2$) in the temperature functions (7) and Φ in the concentration function (8) we solve numerically the equations followed from the equations (3), (4) with the boundary conditions which are the consequence of the relations (13)-(18) on interface and fixed boundaries.

(iv) With found Θ_i and Φ we solve numerically the systems of the equations for the stream functions and vorticity ψ_i, ω_i ($i = 1, 2$) with boundary conditions followed from the relations (9)-(13) and (16) on interface and fixed boundaries.

(v) Transition to the step (i) should be carried out. The components u_i, v_i ($i = 1, 2$) of the velocity vectors will be recalculated. Iteration process is organized similarly to [6] with use of the convergence criteria.

The numerical procedure is based on the longitudinal transverse finite difference scheme known as the method of alternating directions [6, 15]. Note that the conditions of continuity of temperature (13) and of the heat transfer at interface (14) are used to find the interface temperature and, consequently, the temperature distributions in both phases. The form of the exact solution (7), (8) dictated by the convection equations (1)-(4) will lead to the dependence between A and B due to the interface conditions (15). Continuity of the tangential velocities (see (13)) and dynamic condition (11) help to find the values of w_1 and w_2 at interface and the third velocity components in both phases. The kinematic condition (9), the consequences of the no-slip conditions (16) in terms of ψ_1, ψ_2 are used as the boundary conditions to find the stream functions. The dynamic conditions (10), (12) written in terms of the stream function and vorticity allows us to compute the interface values of vorticity ω_1 and ω_2 and, consequently, the vorticity functions in both domains $\tilde{\Omega}_1$ and $\tilde{\Omega}_2$.

3. Numerical results

Numerical investigations are carried out in order to describe the possible flow topology for the liquid-gas system like "ethanol-nitrogen" with following physicochemical properties [16]: $\rho_1 = 0.79 \cdot 10^3$ (kg/m^3), $\rho_2 = 1.2$ (kg/m^3) , $\nu_1 = 0.15 \cdot 10^{-5}$ (m^2/s), $\nu_2 = 0.15 \cdot 10^{-4}$ (m^2/s), $\beta_1 = 0.108 \cdot 10^{-2}$ (K^{-1}), $\beta_2 = 0.367 \cdot 10^{-2}$ (K^{-1}), $\gamma = -0.62$, $\kappa_1 = 0.1672$ (W/(m K)), $\kappa_2 = 0.2717 \cdot 10^{-1}$ (W/(m K)), $\sigma_T = 0.8 \cdot 10^{-4}$ (N/(m K)), $D = 0.135 \cdot 10^{-4}$ (m^2/s), $\varepsilon = 0.01$ (K^{-1}), $C_* = 0.1$.

We investigate the flow topology computed with following values of the Grashoff number $Gr = \{47000, 470\}$, which correspond to the conditions of normal ($g = g_0 = 9.81$m/s^2) and low gravity ($10^{-2}g_0$ m/s^2) and to the characteristic temperature drop $T_* = 10K$. Here $Gr = \beta_1 T_* g l^3/\nu_1^2$, l is the characteristic length ($l = 10^{-2}$ m). The non-dimensional longitudinal temperature gradient $\tilde{T}$ corresponds to the dimensional one A: $\tilde{T} = Al/T_*$. The non-dimensional Soret $\overline{\alpha}$ and Dufour $\overline{\delta}$ parameters are determined as follows: $\overline{\alpha} = \alpha T_*$, $\overline{\delta} = \delta/T_*$.

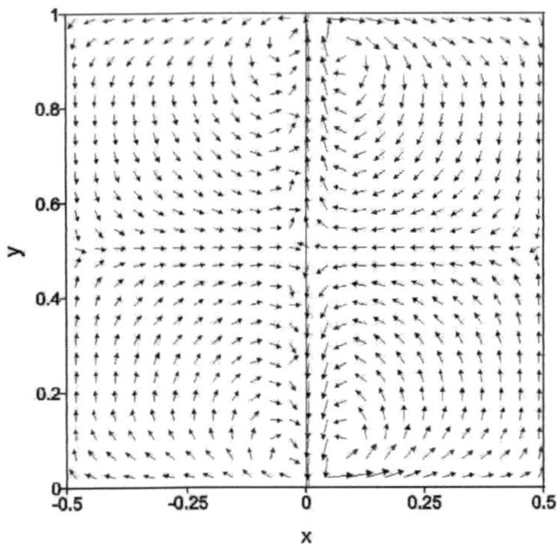

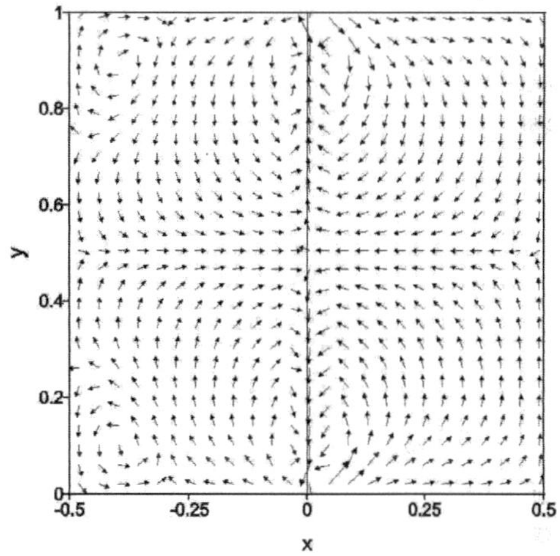

Figure 3. 2D-projection of velocity field (2D cross-section picture): $Gr = 470$, $\tilde{T} = 0.5$ (left); $Gr = 47000$, $\tilde{T} = 0.1$ (right).

Figure 2 presents the possible three dimensional flow pictures under conditions of low gravity with following values of the parameters: $Gr = 470$, $\tilde{T} = 0.5$, $\overline{\alpha} = 0.001$, $\overline{\delta} = 0.00001$ (we look from "above" from the direction of z- axis on the flow structure). The translational character of the liquid flow can be characterized as a dominant one in the case of flow structure presented in figure 2. In order to visualize a rotational character of the flows we multiply the first and second velocity components of the liquid by factor 10. Flow of the gas-vapor mixture in the upper layer is characterized by clearly marked symmetric two-vortex structure with the centers located close to the interface in the "corners". Two slight rotating vortexes transfer the liquid particles in the z- direction (upward motion). To describe the flow topology more correctly we present also the flow projection in the cross-sections. The 2D cross-section picture (see figure 3, left picture) confirms a rotational nature of the flow. The flow of the liquid in figure 2 can be characterized as a rotational motion of a smaller intensity than in the gas layer.

The intensity of the liquid flow depends on intensity of the gravitation field or interface temperature regime. More intensive rotational flow has been observed by normal gravity ($Gr = 47000$). Figure 3 (right picture) demonstrates more complicated character of the fluid flow. Two "corner" vortex structures appear by weaker action of the longitudinal temperature gradient along the interface (here $\tilde{T} = 0.1$; the same values of the Soret and Dufour parameters have been used: $\overline{\alpha} = 0.001$, $\overline{\delta} = 0.00001$). It has been found that in this case the flows have a combined translational and rotational character. The motion becomes more complicated and we can observe the reverse flows near the solid "bottom" wall.

4. Conclusions

In this paper a stationary coupled problem of the gravitational and thermocapillary convection with respect to the diffusive mass transfer at the interface due to evaporation is studied. In the problem of convective fluid flows a new exact solution is constructed in the three-dimensional case. This solution is the analogue of the Ostroumov–Birikh solution of the convection equations. In gas-vapor phase the Dufour and Soret effects are taken into account additionally. It describes a joint flow of an evaporating viscous heat-conducting liquid and gas-vapor mixture in an infinite horizontal channel of the rectangular cross section. Reduction of the common problem to several two-dimensional problems can be carried out so that the real three-dimensional flows are described by the two-dimensional statements. The flows of both fluids (of a liquid and gas-vapor mixture) are computed. They are characterized as translational motion and progressively rotational flow and can be realized in the various forms. The numerical investigations allow one to analyse the possible flow structure with respect to the intensity of the gravitation field and longitudinal temperature gradient created on the interface.

Acknowledgments

The authors gratefully acknowledge support of this work by the Russian Science Foundation (project RSF 15-19-20049).

References

[1] Lyulin Y and Kabov O 2014 Evaporative convection in a horizontal liquid layer under shear-stress gas flow *Int. J. Heat Mass Transfer* **70** 599–609
[2] Goncharova O N, Rezanova E V, Lyulin Yu V and Kabov O A 2015 Modeling of two-layer liquid-gas flow with account for evaporation, *Thermophysics and Aeromechanics* **22(5)** 655–661
[3] Ostroumov G A 1952 *Free convection under the conditions of an internal problem* (Moscow-Leningrad: Gostekhizdat Press)
[4] Birikh R V 1969 Thermocapillary convection in a horizontal layer of liquid *Journal of Applied Mechanics and Technical Physics* **3** 43–45
[5] Pukhnachov V V 2000 Group-theoretical nature of the Birikh's solution and its generalizations (Krasnoyarsk: Book of Proc. Symmetry and differential equations) 180–183
[6] Goncharova O N, Kabov O A and Pukhnachov V V Solutions of special type describing the three dimensional thermocapillary flows with an interface *Int. J. Heat Mass Transfer* **55** 715–725
[7] Shliomis M I and Yakushin V I 1972 Convection in a two-layers binary system with an evaporation, *Collected papers: Uchenye zapiski Permskogo Gosuniversiteta, seriya Gidrodinamika* **4** 129–140
[8] Goncharova O N, Hennenberg M, Rezanova E V and Kabov O A 2013 Modeling of the convective fluid flows with evaporation in the two-layer systems *Interfacial phenomena and heat transfer* **1(3)** 317–338
[9] Goncharova O N and Rezanova E V 2014 Example of an exact solution of the stationary problem of two-layer flows with evaporation at the interface *Journal of Applied Mechanics and Technical Physics* **55(2)** 247–257
[10] Bekezhanova V B and Goncharova O N 2016 Stability of the exact solutions describing the two-layer flows with evaporation at interface (accepted for publication) *Fluid Dynamics Research Journal*
[11] De Groot S R 1951 *Thermodynamics of irreversible processes* (Amsterdam: North-Holland publishing Co.; New York: 1nterscience Puhlishers Inc)
[12] De Groot S R and Mazur P 1984 *Thermodynamics of irreversible processes* (London: Dover 1984)
[13] Prigogine I and Dufay R 1954 *Chemical thermodynamics* (London, New York: Longmans, Green)
[14] Gebhart B, Jaluria Y, Mahajan R L and Sammakia B 1988 *Bouyancy-induced flows and transport* (Berlin, Heidelberg, New York, London, Paris, Tokyo: Springer Verlag)
[15] Roache P J 1976 *Computational Fluid Dynamics* (Albuquerque: Hermosa Publishers)
[16] Weast C R C 1979 *Handbook of Chemistry and Physics* (Boca Raton, Florida: CRC Press Inc)

Investigation of the properties of different surfaces used in evaporation systems

E O Kirichenko[1,2] and E Ya Gatapova[1]

[1]Kutateladze Institute of Thermophysics, Siberian Branch of Russian Scademy of Sciences, Novosibirsk,
[2]Novosibirsk State University

E-mail: kirichenko_ekaterina@outlook.com

Abstract. This paper presents a study of the properties of surfaces with nanocoatingsof two types:NOA 81 photoresist with silanizationand NOA 65 photoresist. These surfaces are resistant to thermal and mechanical loads. The structure of the surfaces was investigated by scanning electron microscopy. The contact angle hysteresis for these surfaces wasdeterminedby the standard DSA-100 KRUSSprocedure, in which theadvancing and receding contact angles are measured, and also by a second method – measuring the contact angle under isothermal droplet evaporation. The contact angle hysteresis values measured by the two different methods are in satisfactory agreement. The contact angle hysteresis was about 20°for the NOA 81 substrateand about 50°for NOA 65.

1. Introduction

The rapid development of electronic technology and the miniaturization of various devices have necessitated a substantial increase in the performance of heat-exchange devices and the use of new materials. Nano- and micro-structured surfaces are promising for use in microfluidics and microsystems. One of the major problems today is the cooling of microelectronics, where the change in the surfacechemical composition, morphology, and properties can play a key role. For example, at present, there is a transition in the production of processors from 14 nm topology to7 nm technology. This transition was made possible bythe use of a silicon-germanium alloy in the transistor design instead of pure silicon, the traditional material for microchip manufacture. The new technology has led to adecrease in the size of the microprocessor building block, an increase in transistor switching speed, and a reduction in power consumption. Next-generation devices may have higher heat fluxes and pulsed loads [1]. Knowledge of the morphology and contact angle hysteresis of these surfaces is required to determine their possible applications [2].

The objectives of this work were to study the morphology and determine the contact angle hysteresis (CAH) forsurfaces with different coatings. Coatings were chosen so as to be resistant to thermal and mechanical loads and hold promise for use in various evaporator cooling systems. We study glass surfaces with two different types of nanocoating: surfaces obtained by coating with NOA 81 photoresist and silanization and surfaces obtained by coating with NOA 65 photoresist. The type of applied coating affects the properties of the surface.

2. Types of coatings and contact angle hysteresis

NOA 81 (Norland Optical Adhesive 81) is a single-component liquid adhesive which, when exposed to ultraviolet light, hardens in a few seconds to form a hard solid polymer. NOA 81 forms a solid film without becoming brittle. The filmhas low elasticity, providingvibration or thermal relaxation and long-term performance of the adhesive bond. The solidified film withstands temperatures from -150 °C to 125 °C.

Silanization is the treatment of a glass or quartz surface with silanizing agents such as trimethylchlorosilane, dimethyldichlorosilane, disilazane, etc. The silanol groups (-Si-OH) present on the glass surface react with the silanizing agent to form a chemically sewn hydrophobic siloxane film.

These surfaces were examined with a HitachiS3400N scanning electron microscope. A photograph of a glass substrate with a NOA 81 photoresist coating and silanizationis shown in Fig. 1. Inclusions are uniformly distributed across the surface, which is typical of the absorption of trichloro-octadecylsilaneon silica or glass surfaces. The typical size of the inclusions is found to be 0.5–1 μm, as seen in Figure 1b.

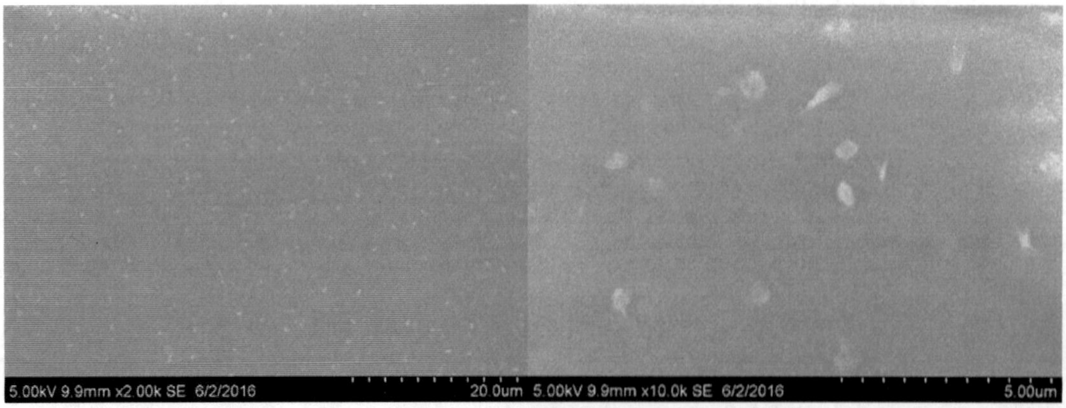

a б

Figure 1. SEM on HitachiS3400N,NOA 81 with silane. a - scale 20 μm : 10, b - scale 5 μm : 10.

NOA 65is a clear, colorless liquid photopolymer, which also hardens when exposed to ultraviolet light. It has good elasticity, temperature range from -15 to 60 °C.

On the test surfaces, the contact angle and the contact anglehysteresis were measured. The difference between the advancing and receding contact angles is the contact angle hysteresis [3]. The advancing contact angle ismeasured when the solid/liquid contact area increases, and the receding contact angle isobserved when this area decreases. The interval may be 10° or more if the surfaces are not specially treated. Thehysteresis value is affected by the main three factors: surface roughness [4, 5], chemical impurities or inhomogeneities in the structure of the solid surface, and substances (surfactants, polymers, etc.) dissolved in the liquid [6].

3. Experimentalsetup and methods

The DSA-100E (KRUSSGmbH, Hamburg)drop shape analyzercan be used to measure the contact angle by the sessile drop method. The contact angle is measured in the range 1-180° with a measurement error of ± 0.1°. The DSA-100E includes an automatic carousel-type liquid supply system (step 0.1 μl), an optical system (7xzoom, fieldofview3.7x2.7 - 23.2x17.2 mm), a digital camera (up to 311 frames/s with a maximum resolution of 780 x 580 /780 x 60 pixels), and a computer-controlled table with motorized axes for samples.

Contact angle hysteresis was determined using two different methods. In the first method, the advancing and recedingcontact angles were measured (the standard KRUSS procedure, Fig. 2). A 10 μl drop is dosed onto the surface at a constant rate of 0.05 μl/min, and the advancing contact angle is measured. After the maximum volume is reached, the liquid is pumped, and the recedingcontact angle is measured. The difference between the advancing and recedingcontact angles is the contact angle hysteresis. As the second method we studiedisothermal droplet evaporation. The evaporation of a 5 μl water droplet into the atmosphere is considered, andthechange in the contact angle throughout the evaporation of the droplet is determined. In both cases, the hysteresis is determined in the vicinity of the zero contact-line speed.

а б

Figure 2. Advancing (a) and receding (b) contact angles on a substrate coated with NOA 65.

4. Results

On the NOA 81 surface, the contact angle was 110°,andon the NOA 65 surface, it was 82°. Figure 3 shows the dependence of the contact angle on the contact line speed on a substrate with a NOA81 photoresist coating and silanization. Hysteresis is determined in the vicinity of the zero contact line speed, and on this surface, it is 20° (Fig. 3).

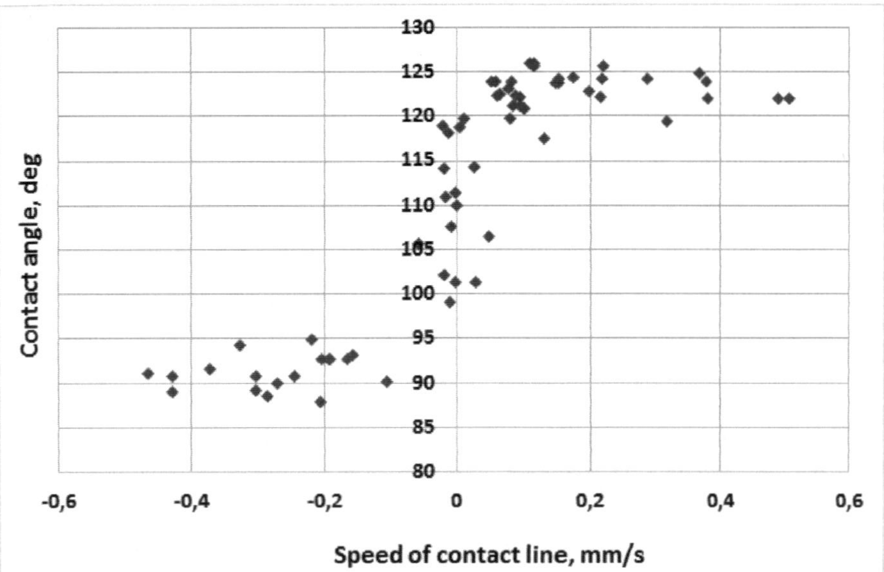

Figure3. Contact line speed versus contact angle on a substrate coated with NOA 81 and silane.

Figure 4 shows the dependence of the contact angle on the contact line speed on a substrate coated with NOA65 photoresist. For this surface, the hysteresis is found to be about 50°.

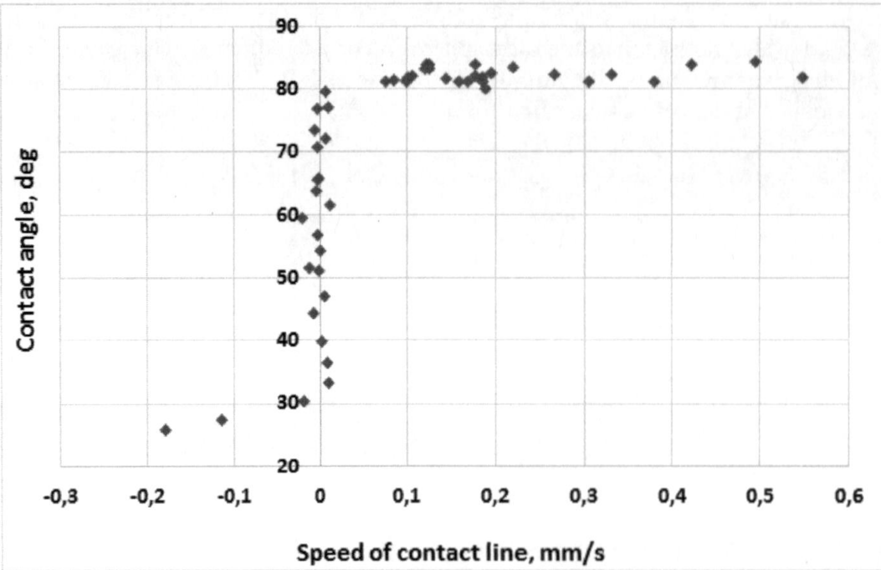

Figure 4. Contact line speed versus contact angle on a substrate coated with NOA 65.

It is shown that the contact angle hysteresis depends on the type of coating applied. The contact angle hysteresis values obtained by the method of isothermal droplet evaporation into the atmosphereare in agreement with the results of the first method and are about 20°and 50°for NOA81 substrates with silanization and NOA65 substrates, respectively. Thetest surfaces can be used to produce controlled flows. Surfaces obtained by applying NOA65 photoresist can provide evaporation under pinning conditions (CAH=50°). Surfaces obtained by applying NOA81 photoresist and silanizationcan be used to produce flows with a constant contact angle.

5. Conclusions

Glass surfaces with nanocoatings of two different types — a NOA 65 coating and a NOA 81 coating with silanization — were studied. These coatings are sufficiently resistant to thermal and mechanical loads. The characteristic sizes of the inclusions present in the NOA 81 coatingwith silanization were measured with a scanning electron microscope to be 0.5–1 μm. The contact angle hysteresis was studied by two methods on a KRUSS DSA-100 apparatus.The contact angle hysteresis is about 20° for NOA 81 substrates with silanization and about 50° for NOA 65. The results obtained by the standard KRUSS procedures are in good agreement with the results obtained by isothermal droplet evaporation.

The authors gratefully acknowledge the support of this work by the Ministry ofEducation and Science of the Russian Federation (Agreement 14.616.21.0016, project identifier RFMEFI61614X0016).

References

[1] KrausA., Bar-CohenA.A. 2015*Mechanical EngineersHandbook.* IV:1:12:1-47

[2] Gatapova E.Ya., Semenov A.A., Zaitsev D.V., Kabov O.A. 2014. *Colloids and Surfaces A: Physicochemical and EngineeringAspects* V. 441.P. 776 - 785.

[3] DeGennesP.G.1985*Reviews of Modern Physics*. V. 57. P. 827—863.
[4] TrillatJ., Fritz K. 1937 *J. de Chim. Phys.* T. 35. P. 45.
[5] Mason S.G 1978. *Ed. By Padday. – New York: Avademic Press* – P. 321
[6] Chappuis J. 1984. *Multiphase Science and Technology* Eds G.F. Hewitt, J. Delhaye, N. Zuber. New York: H – P. 387.

The superfluid helium flow in the channel with porous insert at the presence of longitudinal heat flux

P V Korolyov, A P Kryukov, Yu Yu Puzina

Low Temperature Department, Moscow Power Engineering Institute,
14 Krasnokazarmennaya, Moscow, 111250 Russia

E-mail: korolyov2007@yandex.ru

Abstract The processes of heat and mass transfer during the flow of helium-II (He-II) in a channel with porous backfilling placed in a particular section of its length are studied. Heat flux is directed along the axis of the channel in such a way that on one side of the backfilling vapor plug is formed. Calculation of steady-state transport processes at vapor–He-II interfaces is carried out using methods of molecular-kinetic theory. The normal fluid flow in the pores in laminar and turbulent regimes is described by equations taking into account features of heat and mass transfer in superfluid helium. The relationships between the length of the porous insert and the velocities of fluids for different flow regimes are formulated. The results of the calculations are analyzed by comparison with previous data for the flow of He-II in the individual capillary.

1. Introduction

Development of cryogenic systems for special purposes associated with a high level of heat fluxes at helium temperature level, including space applications, makes it necessary to investigate the dynamics of helium-II within the porous body. The processes of heat transfer at the superfluid helium (He-II) flow in a porous body are studied in [1] with respect to a new type of superconducting cable insulation. These studies related to the improvement of the Large Hadron Collider, the increase in the power of magnets and a corresponding increase in heat capacity. Authors investigated the problem of heat transfer through a porous medium filled with superfluid helium assuming that the two-fluid model of Landau describes the behavior of He-II at low heat flux or small velocity in a porous medium. The method of volume averaging has been applied to the pore-scale equations. The permeability was calculated on an array of simple unit cells in the temperature range from 1.4K to 2.1K and was found to be equal to the intrinsic permeability (Stokes' flow), independently of the heat flux and temperature. Heat flow through superfluid helium contained in porous media is examined in [2]. In particular, heat transfer experiments were performed on He-II contained in a bed of polyethylene spheres of uniform size arranged in random packs. Measured results include the steady-state temperature drops across the three random packs of spheres (35, 49, and 98 μm diameter) and the associated steady heat inputs. Bath temperatures range from 1.7 to 2.1 K to help grasp the superfluid effects. Two pure flow regimes (laminar and turbulent) are decipherable from the heat flux dependence of the temperature gradient. Laminar permeability and shape factor results are compared to past studies of He-II in porous media and in channel flows.

The main aim of this paper is the investigation of heat and mass transfer processes for superfluid helium dynamics in porous media. Thus, the new possibilities for interpreting the effects in a quantum liquid and development of new equipment for various scientific applications will be opening up.

2. Problem

The problem of the helium-II flow in the vapor-filled capillary at the presence of the longitudinal heat flux was considered in paper [3]. It was found that the helium-II column moves to the heater if the length of this liquid helium plug in the capillary more than certain "critical" length, not away from heater, unlike ordinary liquids [4]. If the helium-II column length is "critical" length than the helium-II flow velocity is zero. This critical length does not depend on the prescribed heat flux for the laminar regime of the normal fluid and superfluid flows. For other fluids flows regimes [5] the helium-II column movement to the heater is possible if certain conditions related to the length of the helium-II column and the heat flux density (which dependent on the capillary diameter).

The experimental data confirming the theoretical model were obtained early in paper [6]. In the present investigation a single capillary is replaced by a system of a capillary channels formed by the monodisperse backfilling of metallic balls in a pipe (porous insert). In this case the hydraulic diameter of the capillary channels within the porous media is definitely less than the diameter of the capillary considered in [3, 6]. The change of the heat and mass transfer processes characteristics in a quantum liquid may be expected in this case.

A porous insert (monodisperse backfilling of metallic spheres of diameter of d) is placed in the channel (pipe) of diameter of d_c. The porous insert length is L_2. The channel is filled with helium, i.e. the superfluid helium column of known length, L_0, is placed in this channel (see figure 1). A heat flux, whose density, q_w, is assumed to be constant thereafter, is delivered to the left-hand interface. The side surface of a channel is adiabatically insulated. The vapor pressure, P_b, is maintained constant in immediate proximity (in macroscopic sense) from the right-hand interface (and the vapor–He-II interface temperature, T_b, corresponding saturation pressure is constant). It is required to define the relationship between the velocity of liquid helium flow in the capillary and structural characteristics of the monodisperse backfilling (porosity of m, metallic spheres diameter of d, length of L_2). The authors considerate only steady-state flow because they believe that the unsteady-state flow problem requires a special analysis. Generally speaking, the steady-state flow is impossible in this case as length of a liquid helium plug decreases. However, quasi-stationary flow is possible at small intensity of evaporation. For such flow regime the same method of description as for steady-state flow is applicable as a first approximation. The length of the liquid helium plug, L_0, is accepted constant. The problem is axisymmetric, gravity is neglected.

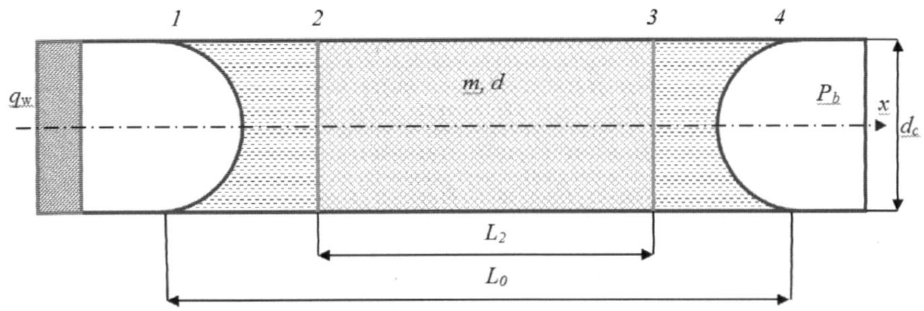

Figure 1. The channel with porous backfilling

It is known that in accordance with two-liquid model of Landau the heat is transferred in He-II by normal fluid flow. Equations describing flow of superfluid helium are derived in the beginning and then required velocities are obtained from the system of these equations.

3. Equations system

Let's write equation which is the relationship between the pressure, P'', in vapor volume near the heater and the heat flux density, q_w. The solution of steady-state Boltzmann kinetic equation for problems of vaporization-condensation in linearized statement, i.e. at $q_w\left(P_S(T_1)\sqrt{2RT_1}\right)^{-1} << 1$ gives the relationship (from paper [7]):

$$P'' = P_S(T_1) + \frac{\sqrt{\pi}}{4}\frac{q_w}{\sqrt{2RT_1}}, \tag{1}$$

where R is helium gas constant, T_1 is temperature of interface surface, $P_S(T_1)$ is saturation pressure at the temperature T_1. It is necessary to note that the relation (1) is valid only for problems where the mass flux through an interface is equal to zero during steady stage (in a macroscopic time scale). But the vapor pressure differs from the pressure value in a similar problem where interface is unpenetrable for a mass flow. This fact is confirmed by the numerical solutions [8].

On the right-hand interface the solution of steady-state Boltzmann kinetic equation in linearized statement gives the following result:

$$P_b = P_S(T_4) - 0.6 \cdot \frac{q_w\sqrt{2\pi RT_4}}{\Lambda}, \tag{2}$$

where $P_S(T_4)$ is saturation pressure at the temperature T_4, $q_w/\Lambda=j$ is mass flux density, Λ is specific evaporation heat, ρ_b'' is vapor density along a saturation curve. It is necessary to note that the equations (1) and (2) are written for a case when a condensation coefficient is equal to 1.

The pressure in a liquid near left-hand interface, P_1, connects with the pressure in a vapor, P'', by the conservation equation for impulse which has following form: $P_1 + 4\sigma/d_c = P''$, where σ is surface-tension coefficient at vapor-liquid interface. In the same time the pressure in a liquid near right-hand interface, P_4, connects with the pressure in a vapor, P_b, by the following impulse conservation equation: $P_4 + 4\sigma/d_c = P_b + j^2/\rho_b''$ (if vapor density is constant), where ρ_b'' is vapor density near interface. As the terms $4\sigma/d_c$ in the impulse conservation equations writing above practically compensate each other and term j^2/ρ_b'' is negligible at small evaporation intensity then it is possible to accept that pressures in a liquid near the interfaces are equal to vapor pressures: $P_1=P''$ and $P_4=P_b$.

Using formulae (1), (2) and also Clapeyron-Clausius equation the pressure difference $P_1 - P_4$ can be written as

$$P_1 - P_4 = \frac{q_w\sqrt{2\pi RT}}{\Lambda}\left(0.6 + \frac{\Lambda}{8RT}\right) + \frac{\Lambda\rho'\rho''(T_1 - T_4)}{T(\rho' - \rho'')}, \tag{3}$$

where Λ is specific evaporation heat. In formula (3) and further it is necessary that T is the average liquid temperature, at the same time it is accepted that $T_1/T_4 \approx 1$. The main hydraulic resistance of the channel consists of resistances of a several sections along the channel length (a local resistances is neglected):

$$P_1 - P_4 = \xi\frac{\rho' V_n^2}{2}\frac{L_0 - L_2}{d_k} + \frac{\eta' V_n}{k_p}L_2 + b_n\rho_n m^2 V_n^2 L_2 + b_s\rho_s m^2 V_s^2 L_2, \tag{4}$$

where η' is a normal component viscosity, ρ' is liquid density, ρ_n is a normal component density, V_n is the normal fluid velocity in the channel, ξ is hydraulic resistance coefficient for the free sections of the channel length, k_p is permeability coefficient of a porous insert. The equation members $b_n\rho_n m^2 V_n^2$ and $b_s\rho_s m^2 V_s^2$ are entered by S.W. Van-Sciver in [9] for the description of a pressure gradient in the porous medium at turbulent normal fluid flow and vortex superfluid flow, where b_n и b_s are the corresponding empirical coefficients.

In the superfluid helium the heat flux connects with the normal fluid velocity. This relationship follows from the equations of two-fluid Landau hydrodynamics. In according with these equations the expression for a heat flux in the He-II for the considered problem has following form (see [10]):

$$q_w = \rho' ST(V_n - V), \tag{5}$$

where S is specific entropy of helium-II, V is the velocity of the liquid helium plug as whole.

The equation describing relationship between a temperature difference in helium-II and a pressure difference is derived from the two-fluid hydrodynamics equations and formulae of Gorter-Mellink theory:

$$gradP = \rho' S \cdot gradT + \rho' S \cdot \bar{f}_{GM}(T) \cdot \boldsymbol{q}^3, \tag{6}$$

where $\bar{f}_{GM}(T)$ is Gorter-Mellink function.

4. The flows regimes

For the cases of the laminar normal fluid and nonvortex superfluid flows and also the laminar superfluid and turbulent normal fluids flows the critical length values are found below. For other combinations of the normal fluid and superfluid flows regimes the authors plan to find the critical length values in future.

From the equation (5) for nonvortex superfluid flow regime the following relationship between a temperature difference and a pressure difference is obtained:

$$P_1 - P_4 = \rho' S \cdot (T_1 - T_4). \tag{7}$$

Temperature difference $T_1 - T_4$ from formula (7) is substituted in equation (3) and then pressure difference P_1–P_4 is derived. Further this pressure difference is substituted in equation (4). Taking into consideration the fact that normal fluid flow regime in pores of monodisperse backfilling is laminar we can write:

$$\frac{q_w\sqrt{2\pi RT}}{\Lambda\left(1-\dfrac{\Lambda\rho''}{TS(\rho'-\rho'')}\right)}\left(0.6+\frac{\Lambda}{8RT}\right)=\frac{32\eta' V_n}{d_c^{\,2}}(L_0-L_2)+\frac{\eta' V_n}{k_p}L_2. \tag{8}$$

The liquid velocity is defined from the equation (4):

$$V = V_n - \frac{q_w}{\rho' ST}. \tag{9}$$

If the porous insert is absent (L_2=0) then critical length, L_0, deduced in (8) becomes the solution obtained early in [3]. At $L_0 > L_{0cr}$ the liquid velocity is the negative, i.e. liquid helium plug moves to the heater when the heat flux is a constant. This result confirms the main conclusion of the works [3] where the helium-II flow in individual capillary is considered. Obviously, at V=0 and normal fluid flow velocity, V_n, determined from eq. (8) at L_2=0 the equation (9) gives the following expression for critical length, L_{0cr}:

$$L_{0cr} = \frac{\rho' ST\sqrt{2\pi RT}}{\Lambda\left(1-\dfrac{\Lambda\rho''}{TS(\rho'-\rho'')}\right)}\left(0.6+\frac{\Lambda}{8RT}\right)\frac{d_c^{\,2}}{32\eta'}. \tag{10}$$

The formula (10) does not differ from the solution presented in [3]. Having estimated values of the members of equation in a right part of the equation (8), we can see that the main hydraulic resistance of the channel is resistance of a porous insert. This result is explained by mainly strong influence of the permeability, k_p. The values, k_p, for various porous mediums are equal to 10^{-9}–10^{-12} m^2. Therefore the equation (8) is reduced to

$$\frac{q_w\sqrt{2\pi RT}}{\Lambda\left(1-\frac{\Lambda\rho''}{TS(\rho'-\rho'')}\right)}\left(0.6+\frac{\Lambda}{8RT}\right)=\frac{\eta' V_n}{k_p}L_2 \quad (11)$$

Critical length is not the length of liquid helium plug but only a porous insert length (i.e. L_2) as we see from eq. (11). Taking into account that V=0 in accordance with (9) we obtain $V_n = q_w/\rho' ST$ and the value, L_{2cr}, is expressed as

$$L_{2cr}=\frac{\rho' ST\sqrt{2\pi RT}}{\Lambda\left(1-\frac{\Lambda\rho''}{TS(\rho'-\rho'')}\right)}\left(0.6+\frac{\Lambda}{8RT}\right)\frac{k_p}{\eta'} \quad (12)$$

It is seen that the critical length of a porous insert does not depend on channel diameter d_c. The liquid column velocity is easily derived from eq. (9) with use eq. (8) and eq. (12):

$$V=\frac{q_w}{\rho' ST}\left(\frac{L_{2cr}}{L_2}-1\right) \quad (13)$$

One can see in formula (14) that the liquid column velocity is negative, i.e. helium column moves to the heater if a porous insert length, L_2, is more than, L_{2cr} (of course, only in the presence of a vapor cavity near the heater). The permeability coefficient, k_p, for a monodisperse backfilling is defined by the sphere diameter of d and porosity of m (depending on a spheres packing way) [11]: $k_p = 5.97\times 10^{-3}\frac{m^3}{(1-m)^2}d^2$.

The critical length of the liquid helium plug, L_{0cr}, for diameter of 100 microns is 0.6 m (at a temperature of liquid of 2K). For monodisperse backfilling if spheres diameter is 100 micron and porosity is 0.4 the porous insert critical length L_{2cr} is equal to 21 cm. In details the dependences of critical length of L_{2cr} on diameter of the particles (spheres) and the critical length, L_{0cr}, on the individual capillary diameter is shown in figures 2 and 3 respectively.

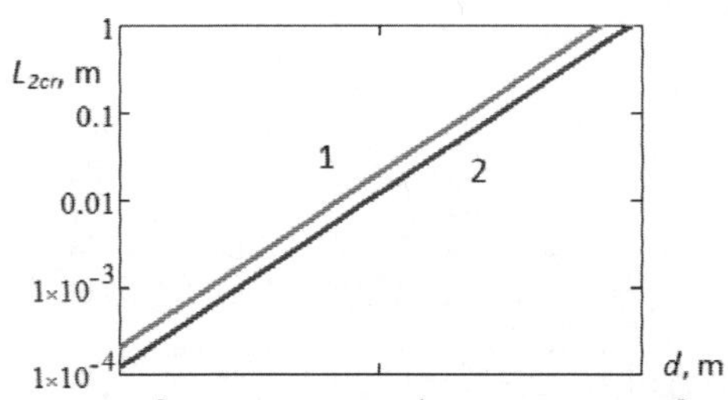

Figure 2. The dependence of critical length calculated from (12) on particles diameter. 1 – T_b=2K, 2 – T_b=1.8K.

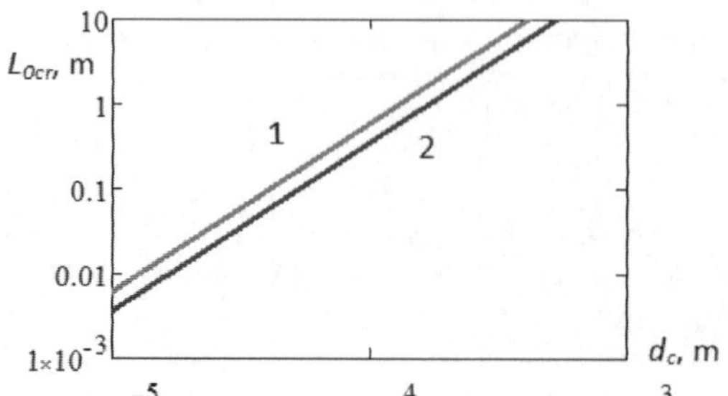

Figure 3. The dependence of critical length calculated from (10) on channel diameter. 1 – T_b=2K, 2 – T_b=1.8K.

In paper [9] it is noted that the pressure gradient in counterflow He-II followed to the form which implies the superfluid contribution to the turbulent friction factor is ‘‘negligible, if not non-existent’’. At L_2=L_{2cr} the liquid helium velocity V=0 and there is helium-II counterflow in the capillary channels of the porous medium. Therefore we believe that b_s=0. Then for nonvortex superfluid flow and turbulent normal fluid flow the porous insert critical length may be found:

$$L_{2cr} = \frac{\rho' ST\sqrt{2\pi RT}\left(0.6 + \frac{\Lambda}{8RT}\right)}{\Lambda\left(1 - \frac{\Lambda\rho''}{ST(\rho' - \rho'')}\right)\left(\frac{q_w b_n \rho_n m^2}{\rho' ST} + \frac{\eta'}{k_p}\right)} \quad (14)$$

In difference of a case of the laminar normal fluid and nonvortex superfluid flows in this case the critical length of L_{2cr} depends on the heat flux density. The coefficient b_n is defined as $b_n = 1.75\frac{1-m}{m \cdot d}$. This coefficient is presented in paper [9] and b_n is a dimensional value.

The dependence of the critical length, L_{2cr}, on a heat flux in the considered case is shown in figure 4.

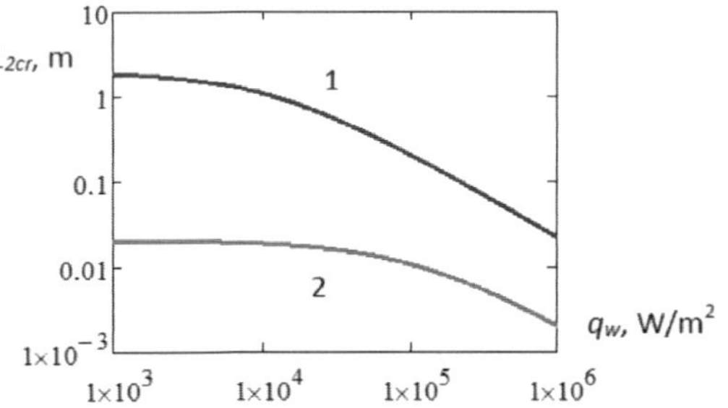

Figure 4. The dependence of critical length on heat flux calculated from (17): 1 – d=10^{-3} m, 2 – d=10^{-4} m.

5. Conclusion

The analysis of the different helium-II flow regimes in the channel filled with a vapor at the presence of the porous insert is shown that the liquid helium plug moves to the heater when the monodisperse backfilling length is more than determinate "critical" length. Thus, non-ideal thermomechanical effect is observed in considered system: in porous media the helium-II moves to the heat souse although the heat flux is supplied to the He-II through a vapor plug and the average pores diameter more than 1 micrometer in differ from the widely known experiments (see [12, 13]).

The critical length values of the porous insert for the different normal fluid and superfluid flows regimes are obtained. It is shown that the critical length depends on the structural characteristics of the monodisperse backfilling and also the heat flux density (but not for case when normal fluid and superfluid flows are laminar both). The dependences of the critical length on the spheres diameter of monodisperse backfilling and the heat flux are presented.

Acknowledgments

This work was supported by Russian Foundation of Basic Research (project No. 15-08-00980).

References

[1] Allain H, Baudouy B, Quintard M, Prat M 2010 *Int. J. of Heat and Mass Transfer* ***53*** 4852
[2] Vanderlaan M H and Van Sciver S W 2013 *Cryogenics* ***57*** 166
[3] Korolyov P V and Kryukov A P 2002 *Vestnik MEI* ***1*** 43 (in Russian)
[4] Kryukov A P 2000 *High Temperature* ***38*** 909
[5] Golovanov I N, Korolyov P V, Puzina Yu Yu 2010 *Voprosy elektromechaniki* ***119*** 6 49 (in Russian)
[6] Korolyov P V, Kryukov A P, Mednikov A F 2006 *Vestnik MEI* ***4*** 27 (in Russian)
[7] Muratova T M and Labuntsov D A 1969 *High Temperature* ***7*** 959
[8] Yastrebov A K and Kryukov A P 2006 *High temperature* ***44*** 556
[9] Vanderlaan M H and Van Sciver S W 2014 *Cryogenics* ***63*** 37
[10] Khurtin P V and Kryukov A P 2000 *Journal of Low Temperature Physics* ***119*** 413
[11] Vitkov G A, Holpanov L P, Sherstnev S N 1994 *Hydraulic resistance and heatmass transfer* (Moscow – Nauka) (in Russian)
[12] Putterman S J *Superfluid hydrodynamics* 1974 (Amsterdam – North-Holland Pub. Co)
[13] Frederking T H K, Yuan S W K, Carandang R M 1986 *Cryogenics* ***26*** 93

Effect of temperature on the convection flow within the liquid evaporation into the gas flow

A Kreta[1,2], Y Lyulin[2,3] and O Kabov[2,3]

[1]NovosibirskState University, Novosibirsk, Russia
[2]Kutateladze Institute of Thermophysics SB RAS, Novosibirsk, Russia
[3]Tomsk Polytechnic University, Tomsk, Russia

E-mail: lyulin@itp.nsc.ru

Abstract. Study of convection in a horizontal liquid layer (ethanol) evaporating into the gas flow (air) has been performed.The two-dimensional velocity field in the liquid layer is obtained using the PIVmethod. It is shown that the temperature rise induces significant changes in the convective flow structure. The main factor determining structure of convective flows within the liquid is the thermocapillary effect.

1. Introduction

Study of heat and mass transfer through the gas-liquid interface is one of the most important problems in the present time. The intensive evaporation from the liquid surface into the gas flow induces various convective flows within the liquid. The structure of the convection can have complicated form and causes a direct effect on the evaporation rate. The considered processes are commonly used in mini- and microchannels of the apparatus such as heat pipes, film evaporators, two-phase cooling systems, etc. The interest to the problem is provided by experiments under normal gravity and microgravity conditions as well as by the preparation of experiments on the International Space Station in the frame of the scientific project ''Convection and Interfacial Mass Exchange" (CIMEX) of the European Space Agency and Roscosmos [1, 2]. These experiments are aimed to study the features of the convective flows of a fluid under conditions of a gas flow and evaporation.

Evaporation of a liquid into air or inert gas implies a limitation of the evaporation rate by diffusion of vapor into the gas. However, the presence of an inert gas also strongly induces surface-tension-driven (thermal Marangoni) instabilities in the liquid,[3].These instabilities can enhance the heat transfer through the gas-liquid interface and increase evaporation form the liquid surface. The dynamics of thermal ripples at the interface of a volatile pure liquid (ethanol) has been studied experimentally and numerically in [4]. 2D and 3D numerical simulations has been performed in the work [5] to study of relation between convective thermal patterns and heat flux through an evaporating surface.The exact solution of the stationary problems of a convective fluid flow in a horizontal layer with a thermocapillary interface being under action of a gas flow has been constructed in [6, 7]. The work [6] demonstrated reverse flow of the fluid to the direction of the gas flow near the gas-liquid interface and possibility to control the convection in the fluid by means of the gas flow. It was shown experimentally in [8], that the action of thermocapillary effect resulting from temperature gradient at the gas–liquid interface of the falling liquid film may induce some structural changes in the flow and may lead to a reverse flow in the liquid film. Theoretical study of instabilities in a horizontal liquid layer in co-current gas flow with an evaporating interface is presented in [9].

The aim of this work is to experimentally study the structure of convective flows and to measure two-dimensional velocity field distribution in the liquid layer under the action of the gas flow using PIV method.

2. Experimental setup

The experimental investigations were conducted on the setup shown in Fig. 1. The setup consists of the following components: test section, gas and liquid loops, data acquisition system, thermal stabilization system, and optical techniques. Pure gas enters to the test section from the compressor. The mass flow controller sets flow rate of the gas at the inlet of the test section. The liquid is injected into the test section by a precision syringe pump. The working liquid evaporates under the gas flow and the vapor-gas mixture flows to the outlet of the gas channel in test section. The vapor-gas mixture flow at the outlet is measured by a mass flow meter. The liquid temperature in the test section is determined by the temperature of the substrate base, which controlled with the help of thermoelectric module. The position of the interface level is controlled with an accuracy of 10 micrometers using a Schlieren technique and the syringe pump [2].

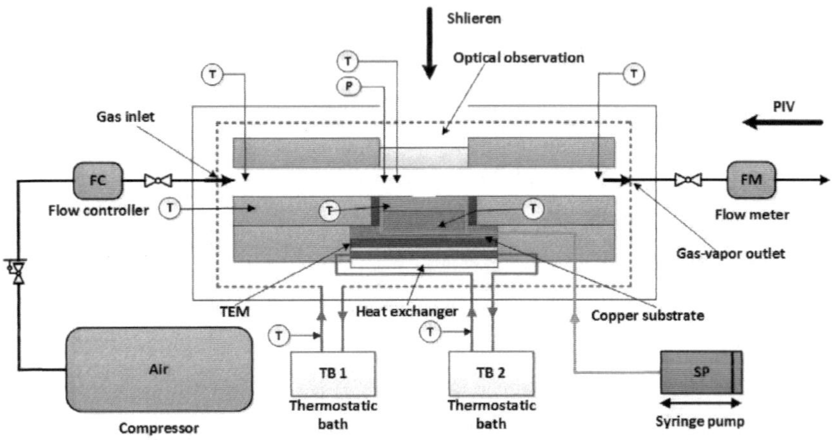

Figure 1. Scheme of the experimental setup.

Recording of the convective structures in the liquid layer under action of the gas flow is carried out using the Particle Image Velocimetry (PIV) method. The schematic diagram of the method is shown in Fig. 2. The method allows to determine the shape, velocity of the vortex structures in the liquid layer depending on various system parameters such as gas flow velocity, liquid temperature, the height of the liquid layer, etc. Detailed description of PIV method is given in [10].

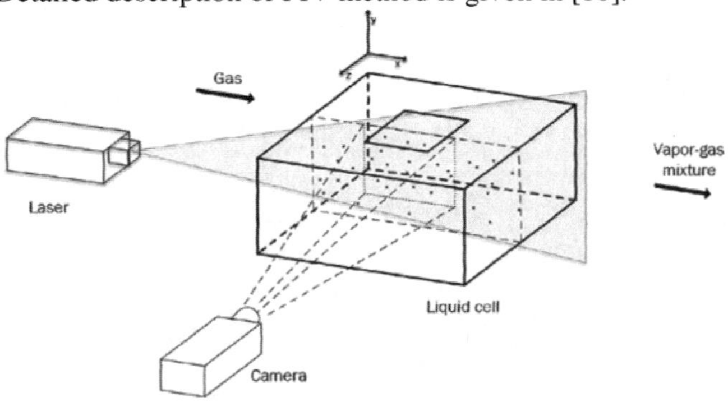

Figure 2. Schematic diagram of the PIV system.

Instantaneous two-component velocity field is measured in the central cross section of the liquid layer directed towards the gas flow. In the beginning of the experiment, the particles (tracers) with a size of 5 microns are added to the liquid. The size and density of the tracers are chosen in such a way as to avoid the influence of the particles on the liquid flow pattern. A pulsed laser sheet optics creates a thin (300 μm) light knife and illuminates small particles. The particles positions at the time of two consecutive laser flashes are recorded on two frames of a digital camera. The optical axis of the digital camera was normal to the laser sheet. The flow velocity is determined by calculating a displacement of the particles during the time between the laser flashes. The determination of the particles displacement is based on application of correlation methods to the images of a tracer particle using a regular decomposition into elementary regions. The data processing is carried out using the "ActualFlow" software program. The program implements two main methods for image processing: the standard and the iterative cross-correlation methods [11]. The standard cross-correlation method is based on the employment of fast Fourier transform using the correlation theorem. However, for a more accurate calculation of the velocity vector is used an iterative cross-correlation method for calculation of two-component velocity field. The basic idea of the method consists in multiple processing of one and the same area using the result of processing at the previous iteration to estimate the processing parameters for the next iteration. As a consequence of this approach, it is possible to obtain a significant increase in the accuracy of the obtained data [11]. The cross-correlation method is used as the primary method to calculate two-component velocity field.The size of the measuring area corresponds to 1290x272 pixels and 14,2x3 mm, respectively. The size of interrogation area is 32x32 pixels. The time delay between successive frames is equal to 100 mks.

3. Results and discussion

The experiments were conducted at atmospheric pressure in a test section, and a stationary 3 mm thick liquid layer. Ethanol was used as the working liquid. The evaporation surface area was 100 mm2 with the corresponding 10x10 mm cut-out in the plate. The gas flow rate was 1000 ml/min with that corresponds to gas average velocity to 0.138 m/s (Re=28), respectively. The temperature of the "liquid-gas" system ranged from 20°C to 50°C . The difference between the temperature of the gas and liquid does not exceed 0.1 °C.

Fig. 3 represents a two-component velocity field for gas flow rate of 1000 ml/min and for four "liquid-gas" temperature T= 20, 30, 40, 50, °C, correspondently . The calculation of the average velocity of convective structures is conducted as follows. First, the measuring area with a liquid vortex is determined visually. The measuring areas of the vortexes are indicated by red rectangles in Fig. 3. Next, a rectangular area with this vortex is selected in the "ActualFlow" program. The program displays a table with the coordinates and velocity projections along the axes x and y for a group of particles within the small interrogation area inside the measurement area. Extremely low velocity vectors are removed from the table. Removed vectors are located near borders of the measuring area and outside of the liquid convective vortex.The velocity of particles in the interrogation area is determined by the formula:

$$V=\sqrt{((V_x)^2+(V_y)^2)}$$

where, V_x and V_y – are x and y components of the velocity vector. The velocity of the convective vortex is calculated as the average value of the all velocitiesin the measuring area.

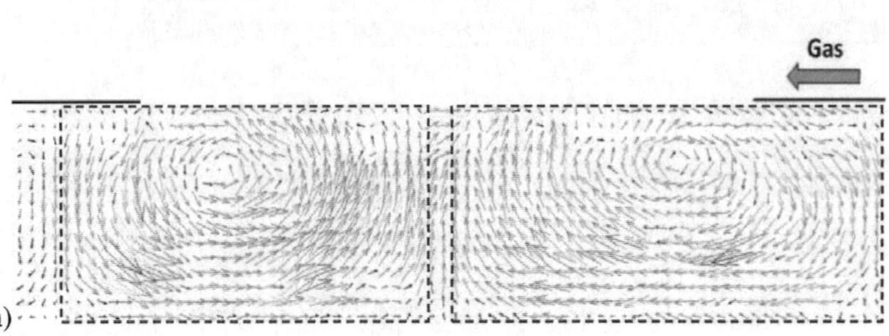

a)

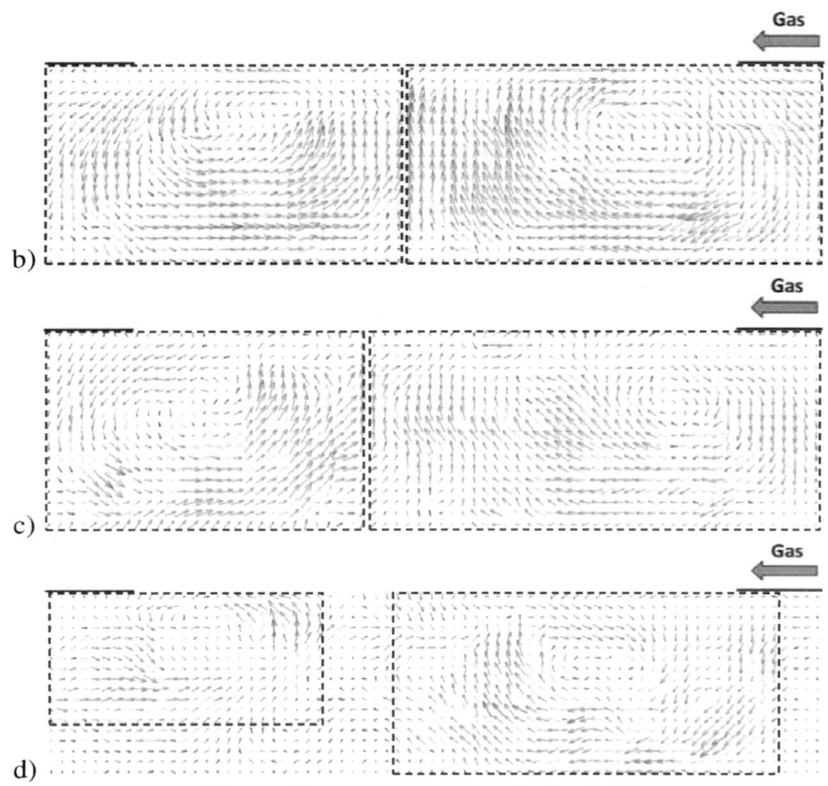

Figure 3. Two-component velocity field.
Liquid - ethanol.The layer thickness - 3 mm.
Gas flow rate - 1000 ml/min (Vgas=0.138 m/s, Re=28).
Temperature:a) 20 °C; b) 30 °C; c) 40 °C; d) 50 °C.

A two-component velocity field at a gas flow rate of 1000 ml/min and gas-liquid temperature 20 °C is shown in Fig. 3 a). It is found the formation of two symmetrical convective vortexes within the liquid layer. The first vortex at the initial area of the gas-liquid contact is moving in counter-current direction of the gas flow. A strong evaporation at the initial section of the gas-liquid interface leads to an interfacial temperature gradient. The interfacial temperature gradient due to thermocapillary effect induces a counter-current to the gas flow motion of the liquid at the interface. The second vortex moves in the same direction as the gas flow. This liquid movement occurs due to the action of the gas shear stress and the effect of the first vortex structure. The velocities of the two vortexes are equal and correspond to 0.241 m/s and the second one corresponds to 0.28 m/s.

When the temperature is increased up to 30 °C and 40 °C(Fig. 4 b, c) two convective liquidvortexes which move in opposite directions relative to each other are alsoobserved. However, there is agrowth of the vortex velocities for both temperaturewith reference to of 20 °C. For temperature of 30 °C the velocities of the first and second vortex structures are equal to 0.285 and 0.283 m/s, respectively. For the temperature of 40 °C there is a slight decrease of the velocities relatively temperature of 30 °C. The velocities of the first and second vortexes for temperature of 40 °C are equal to 0.272 m/s and 0.269 m/s, correspondently. It can be noted at the temperature rise of the "gas-liquid" system the size of the first convective vortex that moves in opposite direction to the gas flow becomes larger.The size of the second vortex isappropriatelyreduced.Velocity of the first vortex becomes also lager in comparison to the second one. Future growth of the "gas-liquid" temperature up to 50 °C (Fig. 3, d) leads to increase of size of the first convective vortex and the second liquid vortex is almost disappeared. Velocity of the first vortex of liquid is reduced to 0.228 m/c.

Most probably, such behaviour of the convective structure within the liquid is caused by increased intensity of the evaporation at the interface due to the temperature rise of "gas-liquid" system. More intensive evaporation can lead to increase of interfacial temperature gradient. The growth of temperature gradient increases the thermocapillary forces on the gas-liquid interface.It can be concluded that, one of governing factor determining structure of convective flows within the liquid is thethermocapillaryeffect.

4. Conclusion

Convection study by PIV method in a horizontal liquid layer evaporating into the gas flow has been performed. Effect of "gas-liquid" temperature on the convective flows in the liquid layer has been considered. It is shown that the temperature rise induces significant changes in the convective flow structure. The growth of temperature from 20 °Cto 50 °C leads to the expansion of the first convective vortex of the liquid that moves in opposite direction of the gas flow. For the maximum temperature (50 °C) of the "gas-liquid" system the second convective vortex whichcirculates in the same direction with the gas flow is almost disappeared.The main factor determining structure of convective flows within the liquid is the thermocapillary effect.

The work was financially supported by the Russian Ministry of Education and Science (Project identifier RFMEFI61314X0011).

5. References

[1] Balázs T.2012 *Microgravity Sci. Technol.***24** 189
[2] Lyulin Yu., Kabov O. 2014 *Int. J. Heat and Mass Transfer***70** 599-609
[3] Haut B., Colinet P., 2005*Journal of Colloid and Interface Science***285** 296
[4] Scheid B. et al. 2012 *Experiments in Fluids***52** 1107
[5] Machrafi H., Iorio C., Dauby P. 2014 *Interfacial Phenomena and Heat Transfer***2(3)**199
[6] Goncharova O., Kabov J. 2010*Int. Journal of Heat and Mass Transfer*, pp. 2795
[7] Goncharova O.N. et al. 2013 *Interfacial phenomena and heat transfer***1(3)** 317
[8] Kabov O. 2010 *International Journal of Emerging Multidisciplinary Fluid Sciences* 87-121
[9] Liu, R. and Kabov O.A. 2012 *Physical Review E***85**066305
[10] Raffel M., Willert C. 1988*Particle Image Velocimtry*
[11] Scarano F., Riethmuller M.L. 1999 *Exp. Fluids***26** 513–523

Two-phase microfluidics: thermophysical fundamentals and engineering concepts

V V Kuznetsov[1,2]

[1] Institute of Thermophysics Siberian Branch of Russian Academy of Sciences
1, Academician Lavrentyev Av., Novosibirsk, 630090, Russian Federation
[2] Novosibirsk State University, 2, Pirogova Str., Novosibirsk, 630090, Russian Federation

E-mail: vladkuz@itp.nsc.ru

Abstract. Thermophysical fundamentals and engineering concepts of the two-phase microfluidic devises based on controlled liquid decay are discussed in this paper. The results of an experimental study of the explosive evaporation at a thin film heater of the MEMS devise in application to thermal inkjet printing are presented. The peculiarities of homogeneous nucleation and bubble growth in the liquid subjected to pulse heating are discussed. Using experimental data a simple equation suitable for predicting the growth rate of a vapor bubble in a non-uniformly superheated liquid was obtained and used to complete a mathematical model of the self-consistent nucleation and vapor bubbles growth in the induced pressure field. The results of numerical calculations according to the proposed model showed good agreement with the experimental data on a time of nucleation and duration of the initial stage of an explosive evaporation of water.

1. Introduction

The revolutionary development of novel micro and nanotechnologies in different fields of the engineering occurs at the past twenty years [1]. Microfluidic devices based on these technologies get a very wide range of the applications due to significant advantages over their conventional counterparts, including fast response time and high intensity of the heat and mass transfer. Using advances in two-phase microfluidics the promising devices were developed including microchannel heat sinks and MEMS devices based on the controlled decay of metastable liquid. Difficulties encountered in development of the two-phase microfluidics is caused by qualitative changes in the regularities of fluid and heat flows with phase change due to the reduction of channels size and are poorly understood [2].

A promising direction in the development of two-phase microfluidics is the MEMS devices based on controlled explosive decay of the liquid micro volumes, including thermal inkjet printheads [3]. In these printheads, the ink droplet is ejected from a microfluidic chamber as the result of liquid pulsed heating and explosive evaporation arising, see figure 1. It is well-known that the condition for transition to the explosive boiling is achievement of the temperature of limit overheating $T_{\lim}/T_c = 0.905 + 0.095\cdot(T_s/T_c)^8$, where T_s is saturation temperature and T_c is critical temperature [4]. The problem of this temperature achievement (313.14 ^{0}C for water at atmospheric pressure) is the exponential growth of density fluctuations near spinodal line and a sharp decrease in the nucleation waiting time. It determines the possible time for explosive evaporation occurrence less then 2–3 microseconds. The regularities of explosive evaporation on a flat microheater were considered in

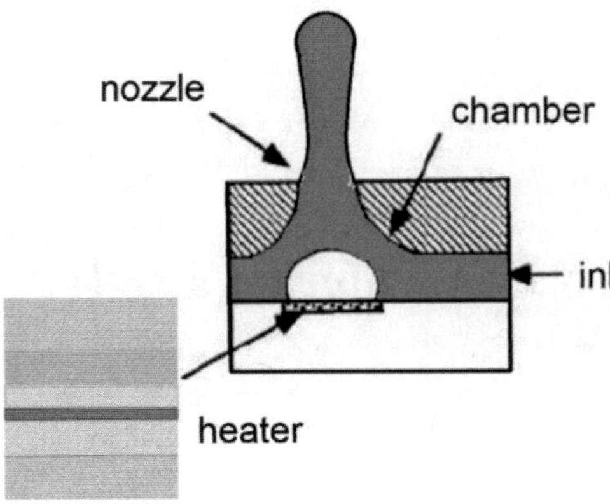

Figure 1. Scheme of the droplet formation in thermal inkjet printer.

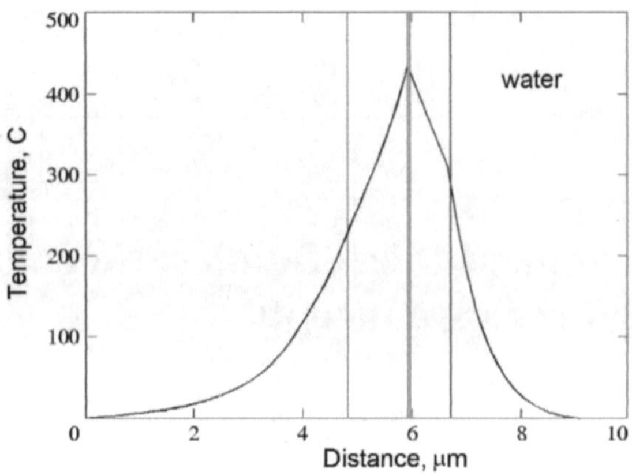

Figure 2. Temperature profile across microheater at time 3.3 μs from heating start.

[5–7], but the initial stage of an explosive evaporation, which determines the efficiency of droplet ejection, is poorly understood because of its short duration.

In present paper the thermophysical fundamentals of two-phase microfluidic devices based on the controlled decay of metastable liquid are considered. The results of an experimental study of the explosive evaporation at a thin film heater of the MEMS devise in application to thermal inkjet printing are presented. The theoretical model of explosive evaporation at a multilayer thin film heater which can be used for design of the two-phase microfluidic device based on the controlled decay of metastable liquid is proposed and discussed.

2. Experimental study of the explosive evaporation in two-phase microfluidics

A necessary condition for the stable realization of an explosive boiling in two-phase microfluidics is rapid removal of the heat from a heat release zone immediately after the onset of nucleation. Therefore design of the microheater, which provides fast heating of the liquid as far as rapid heat dissipation after heat release shutdown, is of paramount importance to ensure the stable operation of the MEMS devises. For design of thin film microheater (resistor), the numerical calculations were carried out using 3–D heat equation for multilayer resistor immersed in a liquid. It was found that the multilayer thin film resistor with optimized material and thickness of the layers has the necessary properties. In this case, the perspective is the use of thin film resistor made from tantalum-aluminum alloy as heat release layer and additional layers of another material to provide a thermal interface with the substrate of quartz glass and electrical insulation of the resistive layer. These layers were made by sequential PECVD process of layering deposition of the silicon dioxide, TaAl alloy (resistive layer), silicon nitride and silicon carbide on the glass substrate according to [8]. During microheater manufacturing the hard silicon carbide was used as the outer layer. The heater area equals to $1.1 \cdot 10^{-8}$ m^2. Numerical calculations defined the layer thicknesses, which permit multiple high growth rate of liquid temperature without microheater destruction. The example of temperature profile calculations for multi-layer microheater at time 3.3 μs from a heating start is shown in figure 2. As can be seen, the maximum temperature in the microheater corresponds to the resistive layer, and the choice of layer thicknesses allows us to achieve the necessary compliance of heat fluxes to the liquid and heat regulating layer below the resistive layer.

Key features of the experimental test section and measuring procedure are summarized here, and most details are available in a prior publication [7]. Studies of the initial stage of explosive evaporation of the water and alcohols are conducted using the thin film heater immersed in the cuvette with working liquid. Single rectangular pulses of a current are supplied to the resistive layer and effective heat flux reaches 2.2 GW/m^2 at the temperature growth rate of 700 MK/s. Effective heat flux

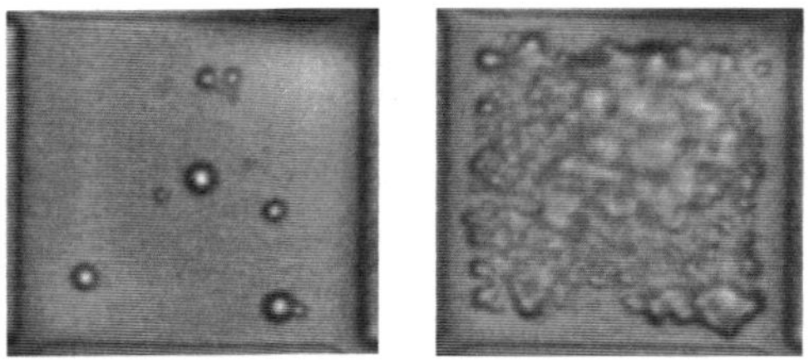

Figure 3. Successive stages of water evaporation with time frame of 200 ns.

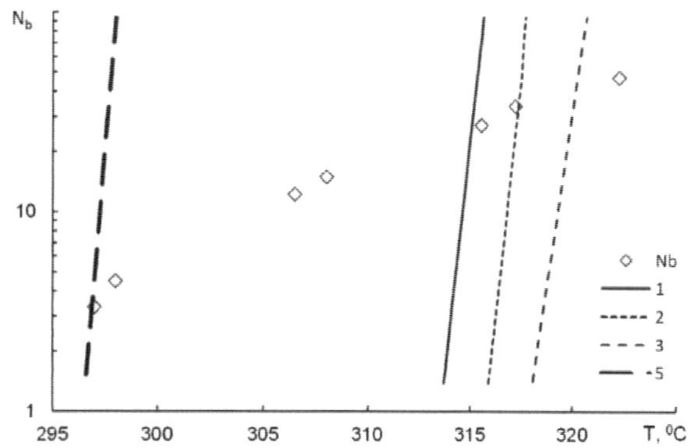

Figure 4. The dependence of the number of bubbles on surface temperature for water.

is defined as the electric power divided on the microheater area. In order to register a nucleation the optical technique based on measuring the intensity of the laser beam reflected from the mirror surface of the heater, pulsed laser and digital video camera are used. The inverted and normalized optical signal from a photodiode corresponds to the specific surface area of the heater occupied by vapor bubbles. Heat flux to the liquid, liquid temperature and its growth rate were determined by numerical solution of the 3–D heat equation for multilayer microheater.

The experiments show that before initiation of an explosive evaporation the surface temperature of heater significantly exceeds the saturation temperature and the liquid is in metastable state. At initial stage of an explosive boiling, the bubbles are uniformly distributed on the surface and their number is small, see figure 3. The observed duration of an explosive evaporation is less than 350 ns, and as a result of the spontaneous nucleation in superheated layer the microbubbles form a vapor film on a heater surface. Pressure in the vapor film exceeds significantly the pressure in a surrounding fluid, which leads to ejection of the ink droplet in ink–jet printhead.

3. Modeling of the explosive evaporation in two-phase microfluidics

The reason for the bubbles incipience during explosive evaporation is the homogeneous nucleation in the superheated liquid layer. The model of homogeneous nucleation is well known [9]. This model defines the critical work of the nucleation as follows:

$$W_* = 16\pi\sigma^3 / 3(p_s - p_l)^2(1 - \rho_v / \rho_l) \tag{1}$$

where σ is the surface tension, p_l is the liquid pressure, p_s, ρ_v, and ρ_l are the saturation pressure, vapor density and liquid density accordingly; as far as nucleation frequency:

$$J = N_1 B exp(-G) \tag{2}$$

The nucleation frequency is governed by Gibbs number derived from the nucleation critical work $G = \Psi W_*/kT$. Here N_l is the molecules number by volume unit, B is the kinetics coefficient, k is the Boltsman constant and Ψ is the parameter which shows the reduction of critical work of a nucleation due to poor surface wettability. To determine kinetics coefficient B the Kogan's approach [10] was used. Then total number of the bubbles generated near the heated surface during the time τ equals to:

$$N_b(\tau) = \int_0^\tau \left\{ \int_V J[T_l(t,x,y,z), p_l] dV \right\} dt \ . \tag{3}$$

For high heat fluxes the overheated layer thickness in a liquid is much less than the longitudinal microheater size. Therefore one can assume that the temperature varies mostly along the coordinate z

with normal orientation to the heater surface and the number of bubbles can be presented as follows:

$$N_b(\tau) = S_0 \int_0^{\tau} \left\{ \int_0^{\infty} J[T_l(z,t), p_l] dz \right\} (1 - \tilde{S}_v(t)) dt \,, \tag{4}$$

where S_o is the heater area and $\tilde{S}_v(t)$ is specific surface area of the heater occupied by vapor bubbles.

Measured dependence of the number of bubbles N_b on heater surface on the surface temperature is shown in figure 4 for the bubbles size larger than one micron. Points in figure 4 are the number of bubbles as function of surface temperature. Figure 4 shows also the calculated number of the bubbles at different liquid pressures for Ψ=1 as the lines denoted as 1–4. The calculations were done using equation (4) for liquid temperature profile showed in figure 2 while neglecting the influence of the specific surface area of the heater occupied by vapor bubbles. The line 5 in figure 4 shows the calculation of the number of bubbles for $\Psi = 0.25$. One can see that the bubbles arise at lower liquid superheat than it predicted by the theory of homogeneous nucleation with Ψ=1. It is connected with the reduction of the critical work of a nucleation due to poor surface wettability for water. In case of ethanol, the measured temperature of nucleation is in a good agreement with the prediction according to homogeneous nucleation theory for Ψ=1.

The important feature of a nucleation at the microheater is pressure growth near its surface during the bubbles incipience and growth. It follows from figure 4 where the bubbles number grows with the temperature not as quick as homogenous nucleation theory predicts at initial pressure 1 bar (line 1). For instance, at the temperature of 320 ^{0}C the observed bubbles number corresponds to the calculated value in the pressure range from 20 up to 50 bar. The reason of this trend is nucleation suppression due to bubbles growth and pressure increasing. To account this phenomenon when calculating explosive boiling the two-zone model of self-consistent nucleation and vapor bubbles growth in the pressure field induced by growing bubbles was proposed in [7]. In this model arising bubbles increase the pressure near the heater and the bubble nucleation takes place under magnified pressure p_l, which is defined by the second deviation from total volume of vapor generated during the time τ as well as by inertia of the attached liquid mass as follows:

$$p_l = p_{l\infty} + 8\rho_l l / S_0 3\pi^{3/2} \left(d^2 V_v / d\tau^2 \right), \tag{5}$$

were l is the heater transverse size and $p_{l\infty}$ is the pressure far from the heater.

The total vapor volume generated at the microheater surface during the time τ is determined by number of bubbles and their volumes. The volume of semi-spherical bubble at the time τ, if it was generated at the moment t, equals to $V_b(\tau - t) = 2\pi R_b^3(\tau - t)/3$, where $R_b(\tau - t)$ is the bubble radius. Considering that the bubbles are generated through homogeneous mechanisms, the vapor volume generated near heater surface and specific surface area of the heater occupied by vapor bubbles during time τ are as follows:

$$V_v(\tau) = S_0 \int_0^{\tau} \left\{ \int_0^{\infty} J[T_l(z,t), p_l] dz \right\} V_b(\tau - t)(1 - \tilde{S}_v(t)) dt \,, \tag{6}$$

$$\tilde{S}_v(\tau) = S_v(\tau)/S_0 = \int_0^{\tau} \left\{ \int_0^{\infty} J[T_l(z,t), p_l] dz \right\} \pi R_b^2(\tau - t)(1 - \tilde{S}_v(t)) dt \,. \tag{7}$$

The nucleation occurs in short run that is much less that the liquid heating time. Thus, in the first approximation, the task of liquid temperature calculation for determination of the J function is independent on the nucleation problem solution. Hence, the temperature of the microheater surface and distribution of the liquid temperature can be determined from the solution of heat equation, see figure 2.

To solve the equations (1), (2), (5) – (6) we need in the model of bubble growth near the spinodal

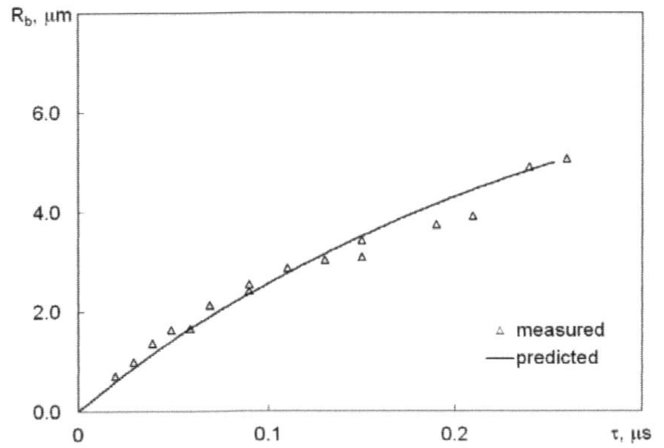

Figure 5. The dependence of the bubble radius on time during the nucleation in water.

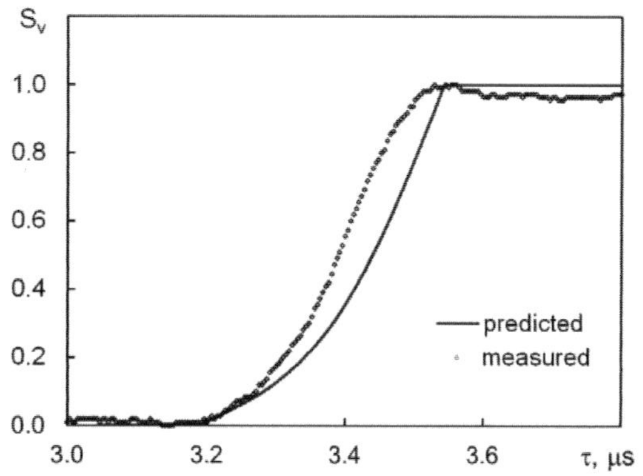

Figure 6. The dependences of the specific surface area of a heater occupied by vapor bubbles on time.

line. The bubble growth rate in overheated liquid is yielded by Rayleigh's equation as follows:

$$R_b \frac{d^2 R_b}{d\tau^2} + 1.5\left(\frac{dR_b}{d\tau}\right)^2 = \frac{p_l(R_b) - p_l(\infty)}{\rho_l}, \tag{8}$$

where $p_l(R_b) = p_v - 4\mu_l/R_b \cdot dR_b/dt - 2\sigma/R_b$ is the pressure at the bubble wall, $p_l(\infty)$ is the pressure in near heater area and far from the bubble, p_v is the vapor pressure and μ_l is the liquid viscosity. Right after nucleation, the vapor pressure equals to saturation pressure at surrounding liquid temperature and is compensated by the capillary pressure. During bubble growth the vapor pressure is reduced as a result of bubble wall cooling due to intensive evaporation. Therefore the bubble growth rate near a spinodal line is essentially smaller than that for the inertia controlled stage of bubble growth:

$$\frac{dR_{bs}}{dt} = \sqrt{\frac{2(p_s - p_l(\infty))}{3\rho_l}}. \tag{9}$$

Figure 5 shows the experimental data on dependence of the bubble radius on time during the nucleation in water at reduced heat pulse energy. For this case the number of bubbles is small and we can neglect the neighboring bubbles interaction. The analysis of this data shows that the rate of bubble growth can be approximated by modified equation (9) as follows:

$$\frac{dR_b(\tau - t)}{dt} = \frac{dR_{b,s}}{dt} g(\tau - t), \tag{10}$$

were the function $g(\tau - t)$ can be fitted as exponential function $g(\tau - t) = g_0 \cdot e^{-(\tau - t)/\tau_T}$. Here τ_T is the characteristic time depending on physical properties of the liquid and vapor. For water one can use the reasonable approximation g_0 = 0.45, τ^{-1} = 3.9 μs^{-1} for calculation of the rate of bubble growth, see figure 5.

The proposed model of homogeneous nucleation and bubbles growth near spinodal line at the flat microheater consists of the equations (1), (2), (5)–(6) and heat equation for multilayer microheater immersed in the liquid. These equations were solved numerically to predict the regularities of an explosive evaporation in two-phase microfluidic devise. Figure 6 shows the dependences of the specific surface area of the heater occupied by vapor bubbles on time which were obtained for water in the calculation (line) and in the experiment (points). One can see that account of the pressure growth near a heater surface and experimentally determined equation for the bubble growth rate near a spinodal line allow to predict reasonably the time of nucleation and the duration of an initial stage of water explosive evaporation in the MEMS devise with multilayer microheater.

4. Conclusions

Presented results show high efficiency of the two-phase microfluidic devise based on controlled liquid decay. The design of multi-layer microheater for two-phase microfluidics was proposed using numerical simulations. It allows receiving the abnormally high growth rate of the temperature for the liquid micro volume. The regularities of initial stage of the explosive evaporation of water on the heater with an external silicon carbide layer when exposed to high-intensity heat pulses were established. Based on experimental data the equation for the bubbles growth rate near a spinodal line was proposed. It allows us to complete the mathematical model of self-consistent nucleation and growth of the vapor bubbles in induced pressure field. The results of numerical calculations of the water explosive evaporation according to the proposed model show good agreement with the experimental data on the time of nucleation and duration of the initial stage of explosive evaporation in the MEMS devise with multilayer microheater. Presented results reveal the basic regularities of the explosive evaporation in two-phase microfluidics with pulsed heating. It allows us to create more efficient two-phase microfluidic devices based on the controlled liquid decay, including thermal ink jet prinheads.

Acknowledgments

This work performed in the Kutateladze Institute of Thermophysics SB RAS by a grant from the Russian Science Foundation (project No 16-19-10519).

References

[1] Abramson A R and Tien C L 1999 Recent Developments in Microscale Thermophysical Engineering *J. Microscale Thermophys. Eng.* **3** 229–44
[2] Kuznetsov V V 2010 Heat and Mass Transfer with Phase Change and Chemical Reactions in Microscale *Proc. Int. Heat Transfer Conf. IHTC14 (Washington, DC)* vol **8** pp 209–28
[3] Hue P Le 1998 Progress and Trends in Ink-jet Printing Technology *J. Imaging Science and Technology* **42** 49 – 62
[4] Lienhard J H 1976 Correlation for the limiting liquid superheat *Chem. Engr. Sci.* **31** 847–9
[5] Zhao Z, Glod S and Poulikakos D 2000 Pressure and Power Generation during Explosive Vaporization on a Thin-Film Microheater *Int. J. Heat Mass Transfer* **43** 281–96
[6] Hong Y, Ashgriz N and Andrews J 2004 Experimental study of bubble dynamics on a micro heater induced by pulse heating *J. Heat Trans.* **126** 259 –71
[7] Kuznetsov V V, Vasserman E S 2004 Explosive vaporization dynamics on a flat microheater *Proc. 3rd Int. Symp. on Two-Phase Flow Modeling and Experimentation* (*Pisa*) vol 1 (Piza: Edizioni ETS) p II-1279
[8] Bhaskar E V and Aden J S 1985 Development of the thin-film structure for the ThinkJet printhead *Hewlett-Packard Journal* **36** 27 – 33
[9] Skripov V P 1974 *Metastable liquids* (New York: John Wiley) p 272
[10] Kagan Y 1960 The kinetics of boiling of a pure liquid *Russ. J. Phys. Chem.* **34** 42 – 8

Hydrodynamic structure of a bubbly flow in an annular channel: experimental study by means of PIV/PFBI/PTV

P D Lobanov[1] and M V Timoshevskiy[1,2]

[1] Kutateladze Institute of Thermophysics, Siberian Branch of the Russian Academy of Sciences, 1, Lavrentyev Ave., Novosibirsk, 630090, Russia
[2] Department of Physics, Novosibirsk State University, 2, Pirogov Str., Novosibirsk, 630090, Russia

lobanov@itp.nsc.ru

Abstract. In the paper, an experimental investigation of the local structure of a bubbly upward flow in an annular channel was carried out by means of a combination of PIV/PFBI/PTV techniques. PIV was applied to measure velocity distributions and turbulent characteristics in the continuous phase, PFBI approach was applied to visualize bubbles in the flow and evaluate their positions and sizes and the simplest PTV method was employed to determine the bubble velocities. The flow was studied at the Reynolds number of 12,500 and different void fractions $\beta = 0$, 1 and 2%. The mean air bubble diameter was estimated to be about 0.8 mm for all β. Bubble concentration was observed to increase near the channel walls. Rising velocity of the gas bubbles was measured in various locations across the annulus duct and it was found that it is substantially higher for the bubbles moving in the central part of the channel. An increase of the void fraction resulted in redistribution of the mean velocity in the flow cross-section and intensification of the turbulent fluctuations up to three times for $\beta = 2\%$.

1. Introduction

Two-phase bubbly flows are used in various hydrotechnical and hydropower engineering facilities, which makes their investigation vitally important and highly urgent. In most of the studies on the heat transfer and hydrodynamics in bubbly media available in literature at present, high-Reynolds-number dispersed flows in vertical pipes are mainly considered (e.g. [1–5]). Depending on the gas concentration, size distribution of bubbles, Reynolds number, characteristic turbulence scales and other conditions, various effects can be observed. For instance, injection of a gas into a moving liquid is well known [6, 7] to cause a noticeable modification of the flow hydrodynamic structure (even at low gas fractions) due to redistribution of gas bubbles across a duct and, thereby, alteration of the mean flow and intensification or suppression of turbulence, e.g. this phenomenon is determinative for heat transfer processes in the near-wall region.

In spite of the fact that the literature is abundant in researches on bubbly pipe flows, it lacks for experimental studies on bubbly flows in complex geometries, such as annular channels, vertical rod assemblies, tubes with partially obscured cross-section and others. Moreover, results obtained for a pipe flow can be hardly translated to a complicated flow which seem, however, quite similar at first glance, e.g. annular flow where liquid moves through a gap between two coaxial tubes. Annular channel is a simplified test object for modeling the flow in fuel assemblies of nuclear power units. Presently, there are only a small number of papers dealing with this topic in the literature. As an example, Ozar et al. [8] measured dispersed phase characteristics (local gas fraction, specific interfacial area and phase slip velocity) but the mean and turbulent characteristics of the continuous phase left undetermined. All these imply the investigation of bubbly flows in annular channels is of a high importance.

Probably, the most crucial problem that one faces when studying a bubbly flow is the choice of measurement techniques. Among a large number of different approaches for two-phase media diagnostics, two main categories are distinguished: intrusive and contactless. All contact methods allow only point measurements of various flow characteristics and most of them have a high temporal resolution. In an overwhelming majority of the researches, right intrusive techniques are conventionally applied. Non-intrusive methods are, as a rule, optical and are conversely employed to visualize or measure instantaneous spatial distributions of certain quantities within a flow region of interest but their temporal resolution is relatively low, except for LDA and PDA. The principal limitation of all optical approaches is that they can be applied only when the bubble density is sufficiently low (commonly at void factions of about 1%). Nevertheless, during the last two decades the optical methods have been developed considerably and are now utilized more and more frequently, with their temporal resolution increasing progressively. This makes their use nowadays more preferable and profitable.

In the paper, a bubbly upward flow in an annular channel is studied experimentally at the same Re = 12,500 for several gas volume fractions β = 0, 1, 2% and the same bubble size by means of a combination of high-speed particle image velocimetry and planar fluorescence for bubbles imaging – PFBI (see [9, 10]) approaches. We report here on a visual analysis of high-speed images registered by a shadow photography (SP) and PFBI techniques to determine the mean bubble size, their velocities and distributions across the annular flow, as well as on PIV measurements of spatial distributions of instantaneous velocity inside the annulus, with a special emphasis put on the effect of dispersed phase on the flow mean velocity and turbulent fluctuations profiles.

2. Experiment

The bubbly annular flow was reproduced in the hydrodynamic rig in Kutateladze Institute of Thermophysics SB RAS. The experimental setup is of closed type and continuous operation. Its test channel is formed by two concentric tubes of roughly 2.7 m height made of transparent organic glass. The internal tube is constituted from three aligned sections with the outer diameter of D_{INT} = 20 mm, while the external one is composed of five in-line parts with the inner diameter of D_{EXT} = 42.2 mm. So, the hydraulic diameter of the annular channel is equal to D_H = 22.2 mm. The inlet of the test section equipped with an immersion unit filled with quiescent distilled water to reduce optical distortions is located at the distance of $70D_H$ from the annulus inlet for hydrodynamic stabilization of the flow (see Fig. 1). The distance from the test section outlet to the annulus outlet is $35D_H$ to minimize the influence of downstream transformation of the pipeline, which can cause flow separation, acceleration, bending, vortex generation and so on.

The operating fluid was also distilled water which was driven by a centrifugal pump equipped with a control module to vary the rotor speed. The flow rate was measured by an ultrasonic flowmeter. The water temperature was maintained constant at 30±0.1 °C in the test section by means of a thermostatic regulator filled with tap water as a heat-transfer agent and containing a tubular electrical heater, cooling circuit and aquarium pump to permanently blend the heat carrier. Cooling and heating in the thermostat was implemented by a PID-control system activating an electromagnetic valve to start/stop the coolant supply to the cooling circuit and a switching relay to loop/break the electric line to the heater. The liquid temperature was measured by heat-variable resistors placed in the test section, heat exchanger and settling tank. The Reynolds number based on the superficial liquid velocity $V_0 = 4\cdot Q_W/[\pi\cdot(D_{EXT}^2-D_{INT}^2)]$ = 0.45 m/s, where Q_W is the water volume rate (measurement error is 2%), and D_H equaled approximately to 12,500.

In order to saturate the flow with bubbles, an air-water mixer of a special design was used to provide a quasi-monodisperse size distribution of the bubbles. The air was supplied to the mixer by a compressor through a couple of filters of rough (5 μm) and fine (0.01 μm) cleaning. The air overpressure at the mixer inlet was specified in the range of 9.8–19.6 kPa depending on the air volume flow rate Q_A by a high-precision reducer. The air temperature, pressure and flow rate were measured together by a thermal mass flowmeter (TSI model 4140) with the measurement uncertainties 0.1 °C, 0.1 kPa and 2%, respectively. Different volume gas fractions $\beta = Q_A/(Q_W+Q_A)$ in the annular flow were achieved by changing Q_A using a needle valve. In experiments, β possessed the following values: 0, 1 and 2%. The mean diameter of the bubbles D_B was approximately 0.8 mm for all β.

Figure 1. A photograph of the test section of the annular channel with the immersion unit filled with distilled water. Fragments of the laser head and camera lens are also visible in the picture, which displays the arrangement of the measurement system.

For the implementation of PFBI approach (which details are given in [9, 10]) to visualize bubbles in the flow, the water was merged with Rhodamine 6G as a fluorescent dye, the concentration of which was relatively low (about 86 μg/l). So, the water properties, especially viscosity and surface tension, can be considered unchanged. As bubbles illuminated by a laser light produce bright glares in images and, thereby, contaminate raw data, a laser-induced fluorescence (LIF) approach was applied to avoid their undesirable effect on PIV measurements. For this, fluorescent seeding particles made of polymethyl methacrylate filled with Rhodamine B of MicroParticles GmbH production (hydrophobic, size distribution 1–20 μm, wavelength range 550–700 nm) were added into the operating liquid to perform PIV measurements.

In order to illuminate and register the bubbles and tracers suspended in the flow, a PIV-system with a high temporal resolution consisting of a pulsed Nd:YAG Photonics Industries DM-532-50 laser (wavelength 532 nm, repetition rate 15 kHz, pulse duration 10 ns, pulse energy 15 mJ), Photron FASTCAM SA5 CMOS-camera (digit capacity 12 bits, resolution 1,024x1,024 pix., acquisition rate 7 kHz) equipped with a Sigma DG Macro 105 mm f/2.8D lens, a system of extension rings of 116 mm cumulative length and a low-pass optical filter (transmission edge at 570 nm) and Berkeley Nucleonics Corporation pulse/delay generator (model 575) for external synchronization was employed (Fig. 1). The distance between the camera and the measurement plane was roughly 395 mm. The magnification of the optical system was 1.8, which allowed us to reduce the plane dimensions of the measurement region approximately to 11.5x11.5 mm. The thickness of a laser light sheet was about 1 mm in the measurement section that passed through the annulus axis. In the experiment, the recording rate was 5 kHz.

The raw data (a series of 10,000 PFBI snapshots for each regime) gathered continuously during 2 seconds were processed using a PC with "ActualFlow" software [11]. At first, two procedures, namely subtraction of the mean two-frame intensity field averaged over the whole sample range and masking, were successively applied to enhance the quality of the registered images and to remove the areas corresponding to shadows from the subsequent calculations. Velocity fields were calculated using the iterative cross-correlation algorithm with a continuous window shift and deformation and 75% overlap of the interrogation windows. In addition, at this step of processing the local particle concentration was accounted for. In order to have a relatively large dynamic range, the initial size of the interrogation window was chosen to be 64x64 pixels but it was subsequently reduced so that the final interrogation window was 8x8 pixels, which provided a high enough spatial resolution. The obtained instantaneous velocity vector fields were then validated with the following procedures: peak validation with the threshold of 2.0, adaptive median 7x7 filter and cluster validation with the coefficient of 50 [12].

3. Results

Results of the bubbly flow visualization by both SP and PFBI are shown together for different void fractions in Fig. 2 to facilitate a direct comparison between the methods. As seen, an analysis of the images is considerably complicated due to several reasons. First, both types of the images are noticeably distorted from the outer side (at the left) because of the annulus external tube curvature. Second, PFBI patterns of the bubbles located close to the channel walls are partial (Fig. 2-b) because, in the experiment, these bubbles are not illuminated by the fluorescent dye emission from the side of both internal and external tubes, so automated identification of such bubbles is extremely difficult. Further, out-of-focus bubbles corresponding to dark blurred regions in the PFBI images (Fig. 2-b) and spatially situated in front of the measurement section often overlap in-focus bubbles with sharp bright borders. This can result in an underestimation of the local gas fraction and an increase of its measurement uncertainty, especially in the near-wall region at higher β. Finally, the bubble and tracer patterns are substantially merged (Fig. 2-b), which makes it difficult to distinguish the signals from the dispersed and continuous phases and, thereby, correctly calculate the mean velocity and turbulent characteristics in the flow.

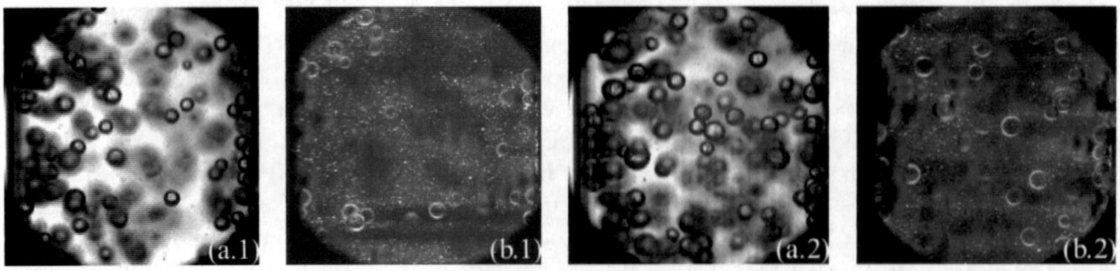

Figure 2. Examples of instantaneous images of the bubbly annular flow captured by (a) shadow photography and (b) PFBI approach for (1) $\beta = 1\%$ and (2) $\beta = 2\%$. The external wall of the annular channel is at the left-hand side of the images, while the inner one is at the right. In case of PFBI, bright spots correspond to seeding particles used for PIV measurements.

Figure 2 also demonstrates that the bubble shape is spherical. The mean bubble size was estimated to be roughly $D_B = 0.8$ mm. Most of the bubbles were found to move close to the annular channel walls rather than in its central part, which is connected with the balance of forces acting on the bubbles in the flow. This implies that the profile of the local gas concentration across the annulus is saddle-shaped. According to [1-3], such a trend is typical for upward pipe flows containing small bubbles. A saddle-shaped profile can cause a redistribution of local characteristics in a two-phase flow in comparison with those for the single-phase flow and, thereby, change heat- and mass-transfer parameters of the flow [2]. In order to recognize bubbles in the PFBI images, a correlation-based algorithm [9] was used. This processing procedure divides a bubble pattern into several sectors and calculates the correlation coefficient between a part of raw PFBI image and a corresponding one of a 2D Gaussian mask independently for each of the sectors and, thus, allows an identification of overlapped and partially illuminated bubble patterns. Its output is positions of bubble centers and their radii. Analyzing the shift between bubble positions in successive images, it is possible to evaluate its instantaneous velocity (the simplest PTV approach).

Time variation of the measured bubble velocities in the annular upward flow is presented in Fig. 3. For this, all the bubbles were divided into three categories depending on their transverse coordinate: located in the central part of the channel and near the internal and external walls, using the criterion of $0.2 < x/L < 0.8$ for the central region, where $L = (D_{EXT}-D_{INT})/2 = 11.1$ mm. The bubble velocity is observed to change in time by maximum 40% (Fig. 3). For approximation, the mean velocity of the bubbles moving in the central part was also calculated with the following formula [13]:

$$V_B = C \cdot V_0 + V_T, \quad (1)$$

where V_B is the measured bubble velocity in the two-phase annular flow, C is a proportionality coefficient equaled to 1.2 [13], V_T = 0.09 m/s is the terminal velocity of an air bubble rising in a quiescent water. As seen (Fig. 3), this approximation is in a good agreement with the measurements. It is worth noting that such a trend is common for all bubbles moving in the central part of the annulus. Thus, this value of C = 1.2 can be recommended for the drift-flux model under conditions similar to the present ones.

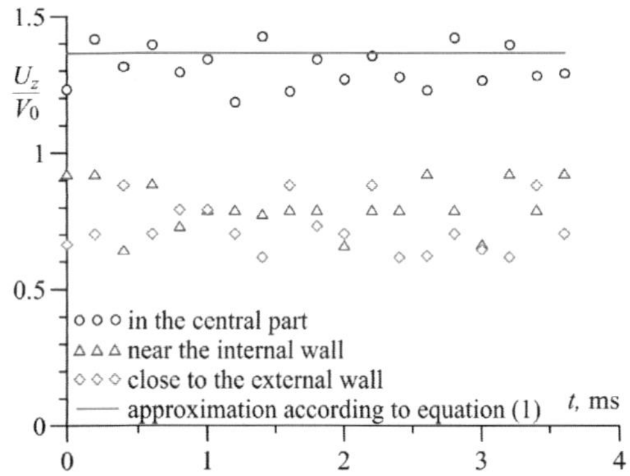

Figure 3. Time dependence of the rising velocity of air bubbles in different locations across the annulus (β = 1%). The estimation of the bubble velocity in each location was based on consideration of only one bubble.

In the central part of the annulus, the mean liquid velocity is equal to about 1.2 for all β (see Fig. 4-a), which is characteristic for turbulent fluid motion in channels. Unlike the case of fluid motion in axisymmetric pipes, the velocity maximum of the single-phase annular flow is not at the duct center but somewhat (x/L = 0.461) shifted to its inner wall. Turbulent fluctuations are in the range of 0.06–0.08V_0 in the central part of the annular channel and gradually grow up to 0.18V_0 when approaching the walls (Fig. 4-b). An increase of the volume gas fraction leads to a further displacement of the mean velocity maximum to the internal wall (Fig. 4-a), so that it is located at x/L = 0.425 and 0.387 for β = 1 and 2%, respectively. Turbulence intensity rises across the whole annulus duct when the volume gas fraction increases. For β = 2%, it reaches 0.12V_0 in the central flow region and $\tilde{v}_z/V_0$ = 0.28–0.3 near the walls (Fig. 4-b). Moreover, the transverse dimension of the near-wall region where turbulent fluctuations are high is several times larger for the external wall (0.1L) compared to the internal one (about 0.02L).

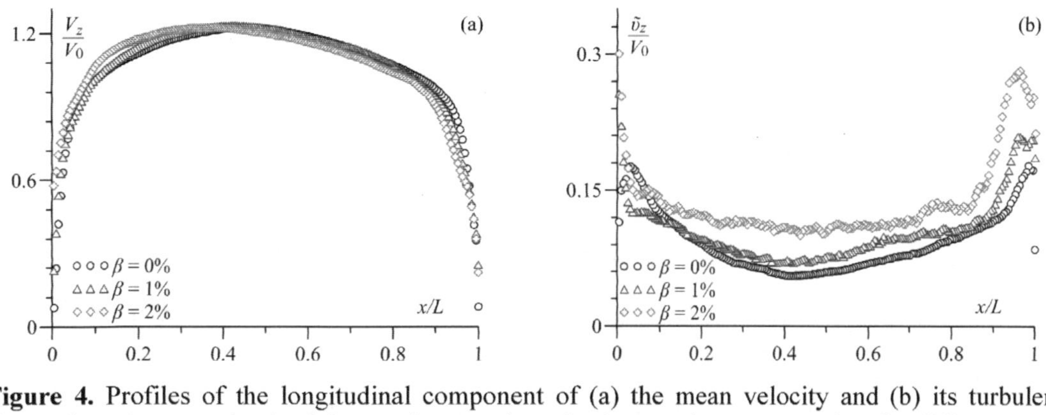

Figure 4. Profiles of the longitudinal component of (a) the mean velocity and (b) its turbulent fluctuations (r.m.s. values) of the continuous phase in the annulus cross-section for different void fractions. z- and x-axes are directed downstream and crosswise. The internal and external walls of the annulus are located at x/L = 0 and 1, respectively. Only every fifth measured point is shown.

4. Summary

The bubbly annular upward flow with void fractions up to $\beta = 2\%$ and the mean bubble size equal to 0.8 mm was for the first time studied by a combination of PIV/PFBI/PTV approaches at the Reynolds number Re = 12,500. Under such conditions, the air bubbles were found to tend to migrate to the near-wall regions. Bubble velocities are higher in the central part of the annulus compared to those in the near-wall regions. An increase of the volume gas fraction results in redistribution of the mean velocity across the channel so that its maximum shifts to the inner wall and intensification of turbulent fluctuations over the whole channel cross-section up to two or three times for $\beta = 2\%$.

Acknowledgments

The experimental study was financially supported by a grant from the Russian Foundation for Basic Research (Project No. 15-38-21040-mol_a_ved supervised by PhD P. Lobanov).

References

[1] Serizawa A, Kataoka I and Michiyoshi I 1975 Turbulence structure of air-water bubbly flow – II. Local properties *Int. J. Multiphase Flow* **2** 235–46

[2] Nakoryakov V E, Kashinsky O N, Burdukov A P and Odnoral V P 1981 Local characteristics of upward gas-liquid flows *Int. J. Multiphase Flow* **7** 63–81

[3] Wang S K, Lee S J, Jones Jr O C and Lahey Jr R T 1987 3-D turbulence structure and phase distribution measurements in bubbly two-phase flows *Int. J. Multiphase Flow* **13** 327–43

[4] Kashinsky O N and Randin V V 1999 Downward bubbly gas-liquid flow in a vertical pipe *Int. J. Multiphase Flow* **25** 109–38

[5] Hibiki T, Ishii M and Xiao Z 2001 Axial interfacial area transport of vertical bubbly flows *Int. J. Heat Mass Transfer* **44** 1869–88

[6] Kashinsky O N, Lobanov P D, Pakhomov M A, Randin V V and Terekhov V I 2006 Experimental and numerical study of downward bubbly flow in a pipe *Int. J. Heat Mass Transfer* **49** 3717–27

[7] Lu J and Tryggvason G 2007 Effect of bubble size in turbulent bubbly downflow in a vertical channel *Chem. Eng. Sci.* **62** 3008–18

[8] Ozar B, Jeong J J, Dixit A, Juliá J E, Hibiki T and Ishii M 2008 Flow structure of gas-liquid two-phase flow in an annulus *Chem. Eng. Sci.* **63** 3998–4011

[9] Akhmetbekov Ye K, Alekseenko S V, Dulin V M, Markovich D M and Pervunin K S 2010 Planar fluorescence for round bubble imaging and its application for the study of an axisymmetric two-phase jet *Exp. Fluids* **48(4)** 615–29

[10] Dulin V M, Markovich D M and Pervunin K S 2012 The optical principles of PFBI approach *AIP Conf. Proc.* **1428** 217–24

[11] Akhmetbekov Ye K, Bilsky A V, Lozhkin Yu A, Markovich D M, Tokarev M P and Tyuryushkin A N 2006 Software for experiment management and processing of data obtained by digital flow visualization techniques (ActualFlow) *Numer. Methods Program.* **7** 79–85 (in Russian)

[12] Kravtsova A Yu, Markovich D M, Pervunin K S, Timoshevskiy M V and Hanjalić K 2014 High-speed visualization and PIV measurements of cavitating flows around a semi-circular leading-edge flat plate and NACA0015 hydrofoil *Int. J. Multiphase Flow* **60** 119–134

[13] Nicklin D J 1962 Two-phase bubble flow *Chem. Eng. Sci.* **17** 693–702

Critical heat flux and dynamics of boiling in nanofluids at stepwise heat release

M I Moiseev[1] **and D V Kuznetsov**

Kutateladze Institute of Thermophysics SB RAS, Novosibirsk, Russia

[1]moiseevmikhail@gmail.com

Abstract. In this paper results of an experimental study on critical heat flux and dynamics of boiling crisis onset in nanofluids at stepwise heat generation are presented. Freon R21 with three types of nanoparticles - SiO_2, Cu and Al_2O_3 was used as test fluid. Critical heat fluxes and temperatures of boiling initiation were obtained. It was shown that the addition of nanoparticles increased CHF at stepwise heat generation by up to 21%. Under conditions of the experiment transition to film boiling occurred via evaporation fronts. Data on propagation velocity and structure of evaporation fronts were obtained; the spectral analysis of fluctuations of the evaporation front interface was carried out. The characteristic frequencies and amplitudes of interface fluctuations were determined depending on the velocity of evaporation front propagation. It was shown that the addition of nano-sized particles significantly affects development of interface instability and increases the front velocity.

1. Inroduction

Boiling is one of the most effective ways to remove heat from a surface, thus it is widely used in various technical applications. But increasing requirements for miniaturization and performance of heat equipment stimulate researchers to look for new methods to improve boiling heat transfer and increase critical heat flux. During the last two decades one of the promising trends in boiling heat transfer enhancement is usage of nanofluids. The addition of nanoparticles can significantly increase both the heat transfer coefficient and CHF [1]. However, the obtained results are often contradictory and depend on many factors, so the practical use of nanofluids needs further systematic research. In addition, there is a necessity in the experiments to be carried out in nanofluids under non-stationary heat release which simulate different emergency processes (depressurization of devices, removal of rods in nuclear power plants, etc.). Under such conditions interphase boundary of a vapor bubble can lose hydrodynamic stability [2, 3]. At presence of overheated wall it leads to initiation of self-sustaining evaporation front in the layer of metastable liquid near the heater and transition to film boiling at much lower heat fluxes than critical heat flux at quasi-stationary heating. There are papers dedicated to experimental and theoretical investigation of evaporation fronts in pure fluids [4-8], but at present only few studies were made with nanofluids [9-10]. The available works demonstrate an increase in non-stationary CHF when adding nanoparticles.

In present work experimental study of boiling crisis in nanofluids at stepwise heat generation was performed. Freon R21 was used as base fluids, and three types of nanoparticles, SiO_2, Al_2O_3 and Cu were used to prepare nanofluids. Critical heat fluxes and temperatures of boiling initiation were measured, and dynamics of evaporation front propagation was studied.

2. Experimental setup and procedure

Experiments were carried out in Freon R21 at reduced pressures P/Pc= 0.037 (0.193 MPa). For experimental studies of heat transfer and transient dynamics at non-stationary heat release under free convection the setup, shown schematically in figure 1, was developed.

On bed 1, the frame 2 with refill tank 3 and working chamber 4, housing the working section 10, is hinged. The working chamber is a sealed cylindrical stainless steel vessel. The inner diameter of the chamber is 250 mm, and the height of the workspace is 250 mm. The working chamber is equipped with four windows 5. The bellows 6 with an adjusting screw 7 allows creating the required pressure (up to 0.4 MPa) in the working chamber at the closed valve 8. To set the desired temperature of liquid in the bottom of the chamber, heat exchanger 9 was located.

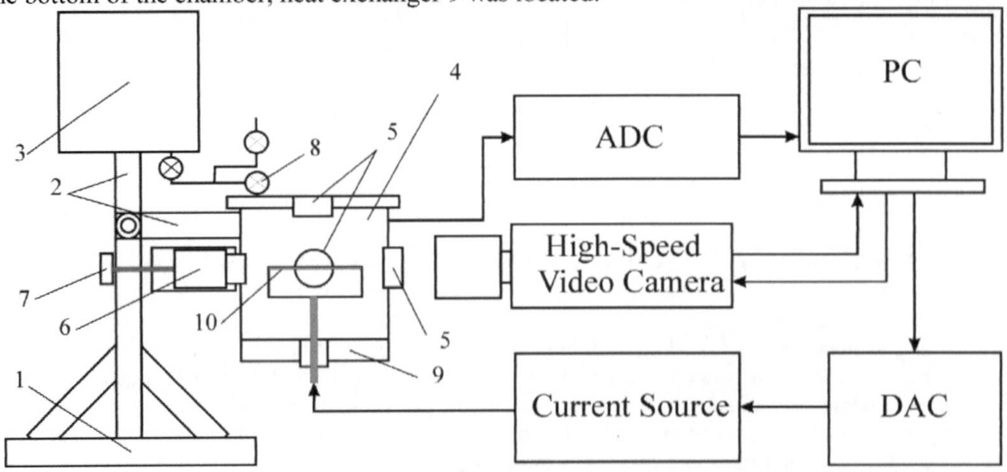

Figure 1. Scheme of experimental setup.

The working section was made of the stainless steel tube with outer diameter of 3 mm, wall thickness of 0.5 mm and length of 50 mm. A thin-film platinum thermometer was installed inside the tube to measure the temperature in the quasi-stationary regimes. The tube ends were hermetically sealed. The tube was heated by electric current from controllable current source. In experiments with non-stationary heat release the rectangular current pulse of 420 A was supplied to the working section. Since thermal inertia in the section was rather high, at the times of the evaporation front development (about 50 ms) wall temperature was growing with almost a constant rate of 2180 K/s. Temperature of the heater was determined by the current pulse duration. After current was turned off, the wall temperature remained almost constant at the times of the evaporation front propagation. Numerical calculation using thermal conductivity equations has shown that for the time of front passage (much lesser than the time of convection development) the wall temperature before the front dropped no more than by 0.2 K. Thus in each experiment, the propagation of self-sustained evaporation front was investigated at the pre-set constant wall temperature.

The working section temperature until the moment of vapor phase emergence was measured using temperature dependence of the heater resistance. For this purpose, two conductors of 0.05-mm diameter were welded in the middle of the tube to measure the voltage drop in the region of 30-mm length. In each measurement, calibration was performed by the temperature of undisturbed fluid. This technique allowed measuring the average temperature of the working section with an error less than ±1.5 K. After the vapor phase formation, the wall temperature was determined numerically based on the equations of non-stationary thermal conductivity. This approach is valid under conditions of our experiments since the times of the evaporation front propagation (about 30 ms) were much less than the time of convection development (about 200 ms). Comparison of the temperature measured using heater resistance and obtained from numerical solution showed excellent agreement of the experiment and calculations.

Visual observations of vapor phase formation and propagation on the heating surface were performed using a high-speed digital video camera Phantom v7.0. The shooting rate was 25 000 frames per second with the exposure of 26 microseconds. When analyzing the results of high-speed digital video-shooting, the line of the interfacial boundary was determined on each frame using specially developed software. The obtained data was used to study dynamic properties of evaporation front boundary.

Three types of nano-additives were used. The first type is Tarkosil-T20 SiO_2 particles with average diameter of 20 nm, specific surface area 140 m^2/g, 0.003% vol, produced by electron beam evaporation method. The second type is Al_2O_3 particles with diameter of 50 nm, specific surface area 30 m^2/g, 0.001% vol. The third type is passivated Cu particles with diameter of 80 nm, specific surface area 14 m^2/g, 0.0013% vol. Al_2O_3 and Cu nanoparticles were produced using electric explosion of wire. Nanofluids preparation procedure was as follows. The powder was loaded into the working volume, and then vacuum exhaust, filling with fluid and mixing were performed.

3. Experimental results

3.1. Critical heat flux

In experiments to determine critical heat flux, rectangular current impulses with magnitude of 420 A were supplied, starting with the short impulse duration and, thus, low wall overheat. After each impulse, we waited for the system to return in the initial state. If vapor phase was not registered for 5 attempts, we supposed that evaporation front initiation doesn't occur at the adjusted wall overheat, and the impulse duration was increased by 1 ms, which corresponds to wall overheat increase by 2 K.

Figure 2 shows the dependence of the heat flux density q, transmitted by the wall to the fluid, on the temperature of the wall overheating ΔT. The value of q was obtained from the numerical solution of the heat equation. Vertical lines mark the minimum wall overheats, at which the evaporation front propagation was recorded. In pure Freon R21, the minimal overheat is 41 K and corresponding heat flux density is 0.159 MW/m^2. It is 3 times less than CHF in Freon R21 at quasi-stationary heat release. In R21+Al_2O_3 nanofluid no difference from pure fluid was observed. In R21 + Cu nanofluid minimal overheat of front initiation was increased by 7 K, and corresponding heat flux was 0.172 MW/m^2. In R21 + SiO_2 nanofluid, minimal wall overheat was higher by 18 K and heat flux was 0.192 MW/m^2.

Thus, the addition of nanoparticles has led up to 18 K increase in minimal wall overheat required to evaporation front initiation, which is equivalent to a 21% increase in non-stationary CHF.

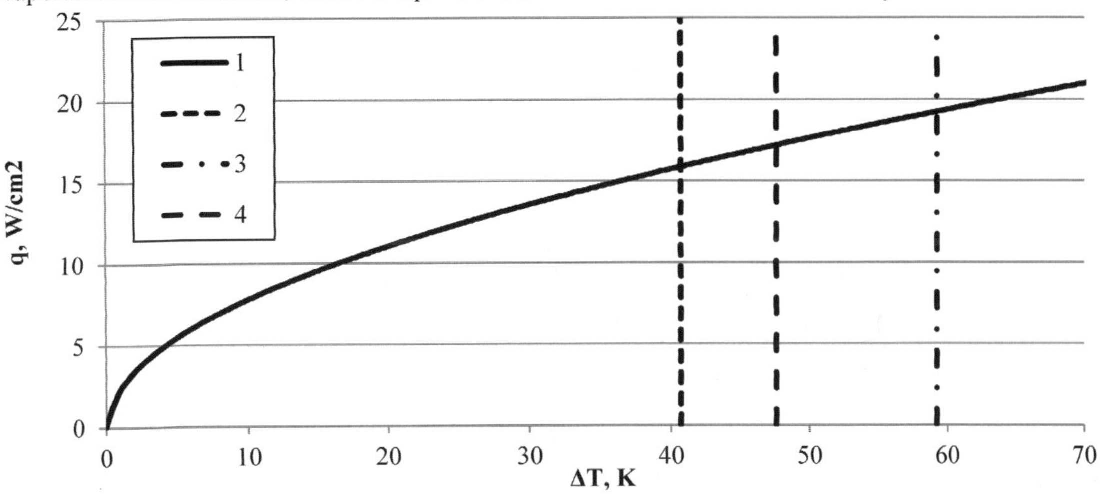

Figure 2. 1 – heat flux; 2 –front initiation in pure R21 and R21+Al_2O_3; 3 – front initiation in R21+SiO_2; 4 – front initiation in R21+Cu.

3.2. Evaporation front

As was mentioned before, at nonstationary heat release emergence of a single vapor bubble leads to initiation of self-sustaining evaporation front in the layer of metastable liquid near the heater wall (Figure 3). Interphase boundary has perturbations on two scales: large-scaled waves with wavelength 1.5-2 mm, which is close to capillary length in Freon R21 (about1.7 mm), and small-scaled perturbations with size 100-200 μm. In previous works it was shown that these perturbations are caused by development of hydrodynamic instability, and they can increase heat transfer through the interphase boundary, which leads to increase in evaporation rate and front velocity [2, 3].

Figure 3. Development of the evaporation front in pure Freon R21. Time from the beginning of heating: (a) – 28 ms; (b) – 29.4 ms; (c) – 32.8 ms.

Figure 4 shows dependence of the front velocity on the wall overheat. At low overheats ($\Delta T < 60$ K) the slope of the experimental data is relatively small, and there is no difference in velocity for pure fluid and nanofluids. At higher overheats ($\Delta T > 65$ K), corresponding to the development of the hydrodynamic instability, slope of the experimental data increases. Front velocity in nanonfluids is higher compare to the base fluid by 25% on an average, but there is no clear distinction between fluid with different nanoparticles.

To investigate large-scaled pulsations of the front interface, we used the dependence of x coordinate on time over one of the heater generatrices, i.e. for the fixed value of y-coordinate. The linear trend, complying with the distance run at constant speed V_{fr}, was deducted from the resulting dependence. The fast Fourier transformation was applied to the remained coordinate deviations from the path at a constant velocity. In all the spectra, the pronounced peak of main harmonic was observed. Using the amplitude and frequency of the main harmonic, we determined the characteristic velocity pulsations and acceleration of the interface.

Dependences of frequency and amplitude of the main harmonic on average velocity are shown in figures 5 and 6. In pure liquid, the frequency of the main harmonic increases with a rise of the average velocity. The amplitude doesn't show any clearly distinguishable trend in studied range of velocities. In SiO_2 and Al_2O_3 nanofluids the similar tendency is observed, but fluctuation frequencies decrease in comparison with pure liquid. The Cu nanofluid was too opaque to perform spectral analysis, thus corresponding data is not presented.

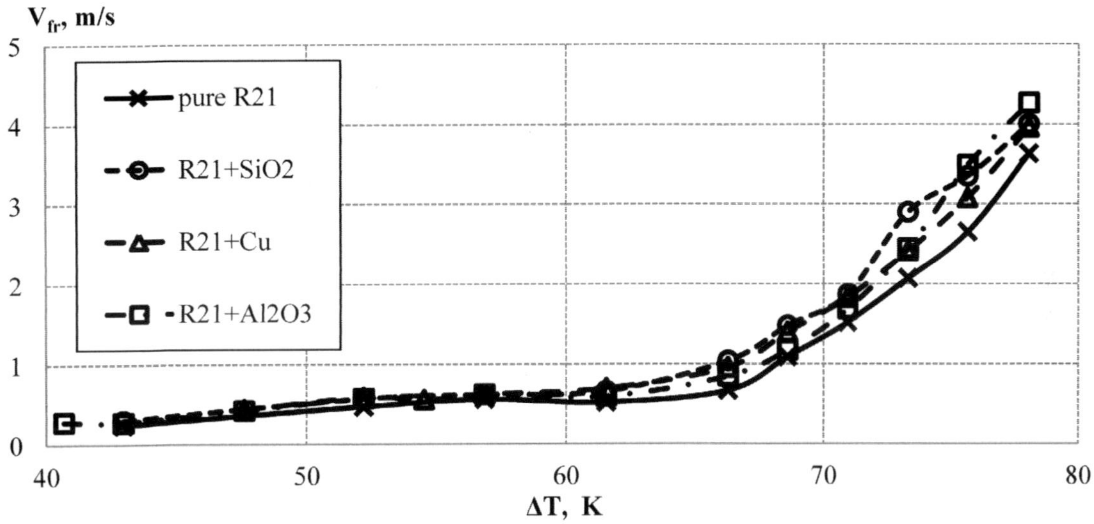

Figure 4. Evaporation front velocity vs wall overheat.

The amplitude value of interface acceleration, defined as the product of main harmonic amplitude on the square of angular frequency, can be estimated from the obtained data. These values, normalized to gravity acceleration, are shown in figure 7. In pure liquid, acceleration can reach values up to 160 g at high wall overheats. In nanofluids, interface acceleration was several times lower than in the pure liquid.

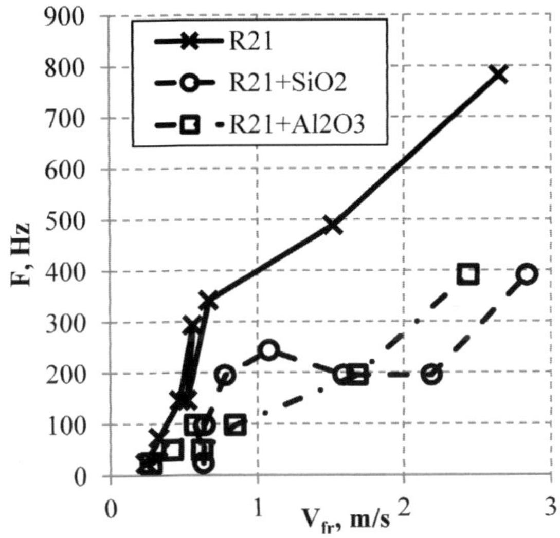

Figure 5. Interface oscillation frequencies.

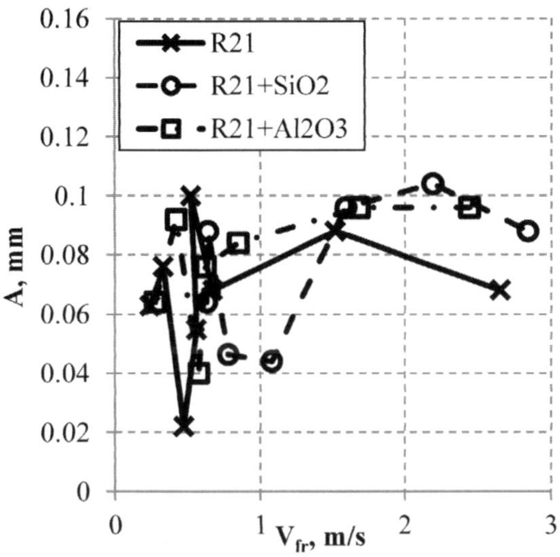

Figure 6. Interface oscillation amplitudes.

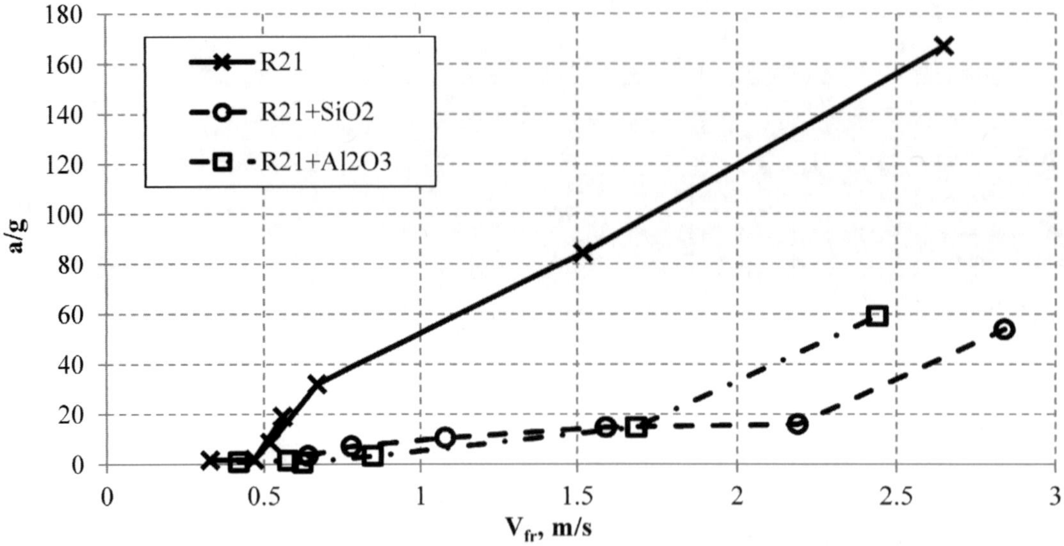

Figure 7. Interface oscillation accelerations.

4. Conclusions

Addition of nanoparticles provided up to 21% increase in CHF under non-stationary heat release. Wall overheat corresponding to boiling initiation also increased by up to 18 K. Temperature of heterogeneous nucleation depends on the surface roughness, and all types of nanofluids caused visible modification of the heater surface. But effect on the temperature of boiling onset at stepwise heat release significantly differs for different nanoparticle types. Detailed analysis of the surface is needed to explain such results.

Addition of nanoparticles also increases the velocity of evaporation front by 25% at high wall overheat. Experiments show that nanoparticles changed parameters of interface pulsations. It is shown in [4] that interface acceleration affects significantly the small-scaled instability development and intensification of heat transfer through the interface. A decrease in acceleration at presence of nanoparticles gives an opportunity for instability development in wider range of front velocities and interface perturbation scales, which can explain increased front velocity.

Acknowledgements
This research has been performed at IT SB RAS with the support of the RFBR (project № 13-08-00178) and Program of Fundamental Research SB RAS (project No F3 III. 18.2.3)

References

[1] Ciloglua D and Bolukbasib A 2015 *App. Therm. Eng.* **84** 45–63
[2] Syromyatnikov S and Pavlov P 1998 *High Temperature* **36** 282–6
[3] Zudin Y. 2015 *Thermophysics and Aeromechanics* **22** 285–96
[4] Pavlenko A, Tairov A, Zhukov V, Levin A and Moiseev M 2014 *JET* **23** 173–93
[5] Pavlenko A and Lel V 1999 *Thermophysics and Aeromechanics* **6** 105–17
[6] Kuznetsov D, Moiseev M and Zhukov V 2015 *MATEC Web of conferences* **23** 115–20
[7] Stutz B and Simoes-Moreira J 2013 *Int. J. Heat and Mass transfer* **56** 683–93
[8] Pavlov P and Vinogradov V 2010 *High Temperature* **48** 683–90
[9] Misyura S 2016 *Exp. Therm. Fluid Sci.* **75** 43–53
[10] Sharma V, Buongiorno J, McKrell T and Hu L. 2013 *Int. J. Heat and Mass Transfer* **63** 425–31

Theoretical research of the carbon dioxide injection process into the rock saturated with ice

N G Musakaev[1,2] and M K Khasanov[3]

[1]Tyumen Branch of Khristianovich Institute of Theoretical and Applied Mechanics of SB RAS, Taymirskaya Str. 74, Tyumen 625026, Russia
[2]Tyumen State University, Volodarskogo Str. 6, Tyumen 625003, Russia
[3]Sterlitamak Branch of Bashkir State University, Lenina Str. 37, Sterlitamak 453103, Russia
E-mail: musakaev@ikz.ru

Abstract. The results of the theoretical research of the carbon dioxide injection process into the reservoir initially saturated with methane and ice are presented. In the rectilinear-parallel approach the analytical solutions which describe the distributions of temperature and pressure in the reservoir were built. It is shown, that modes with different quantity of the interphase boundaries of the carbon dioxide hydrate formation are possible.

1. Introduction

Currently, the greenhouse effect is one of the global environmental problems facing mankind. The main reason for this effect is industrial gases entering the atmosphere, primarily carbon dioxide. This raises the problem of reducing the carbon dioxide concentration in the atmosphere, for example, due to its disposal in underground reservoirs [1-3]. However, at long-term storage of CO_2 in gaseous or liquid state, there is a danger of emission into the atmosphere [4]. Subsurface CO_2 storage through clathrate hydrate formation is a novel option for the reduction of atmospheric carbon content and permanent underground CO_2 disposal over geological periods. A number of researchers proposed the carbon dioxide injection into permafrost, because the upper boundary of the carbon dioxide hydrate stability in the permafrost begins from the depths of 100-150 meters and the bottom boundary – 500-800 meters [4]. In addition, the CO_2 hydrate formation in a porous medium can be used to increase the capacity of the underground reservoirs for the gas store. The fact of the matter is that in a hydrate state the same gas mass can be stored under considerably lower pressure and in the less volume [5]. Technological ideas must be confirmed by the valid theoretical researches. Therefore the construction of the mathematical model of the dioxide carbon injection process into a reservoir partially saturated with ice is the actual problem.

2. Mathematical model

Assume that a reservoir, initially filled with methane and ice, fills half-space $x > 0$. Before the injection of gas, pressure p_0 and temperature T_0 in the reservoir correspond to thermodynamic conditions of the existence of methane and ice in a free state:

$$t=0, x>0: \quad T=T_0, \; p=p_0, \quad S_i=S_{i0}. \tag{1}$$

Assume that through the left boundary of the reservoir ($x = 0$) the carbon dioxide is being injected. The pressure p_w and temperature T_w of the injected gas correspond to the existence conditions of the mixture of CO_2 and its hydrate, and are supported at this boundary at constant level:

$$t>0, x=0: \quad T=T_w, \; p=p_w. \tag{2}$$

Conducted estimations show that at the CO_2 injection into a reservoir saturated with methane and ice the diffusion mixing can be neglected. This is because the mass transfer rate due to filtration is much greater than the mass transfer rate due to diffusion. Also, in the considered range of pressures and temperatures the viscosity of CO_2 is about one and a half times higher than the viscosity of CH_4. Therefore in this work we consider that the gas mixing process has no effect to the stability of the displacement front of methane with carbon dioxide [6]. In this case we can assume that at the injection of carbon dioxide into a reservoir saturated with methane and ice two characteristic zones are formed. The pores in the zone $[0, x_{(n)}]$ are filled with CO_2 and its hydrate, and the porous medium in the zone $[x_{(n)}, \infty)$ is saturated with methane and ice (fig. 1). Thereby, $x_{(n)}$ is the coordinate of the movable boundary of phase transitions.

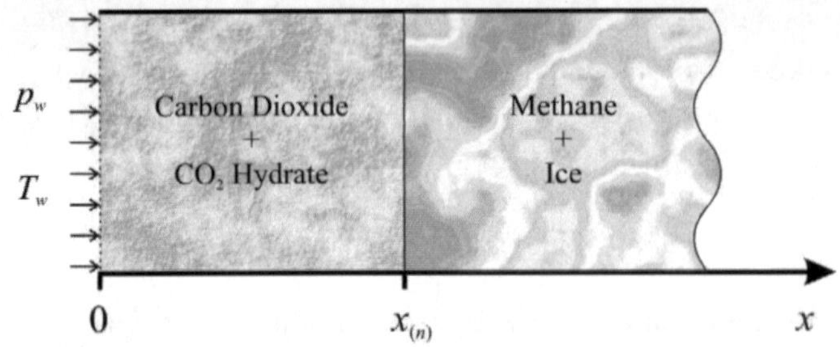

Figure 1. Scheme of the problem of the carbon dioxide injection into a porous medium taking into account the CO_2 hydrate formation.

We adopt several assumptions enumerated here. CO_2 hydrate is a two-component system with the mass concentration of gas – G. We will neglect the relationship of the mass concentration of the gas in a hydrate with pressure and temperature. Suppose that porosity m is constant; rock skeleton, the hydrates of carbon dioxide and ice are all incompressible and immovable. Methane and carbon dioxide are the calorically perfect gases. In this paper we research the processes, the duration of which is considerably longer than the typical time of temperature equilibrium $\tau \approx d^2/\chi$ (d is the typical pore dimensions, $\chi = \lambda/\rho c$ is the thermal diffusivity of a porous medium). Therefore, we assume that the temperatures of the porous medium and of the saturant (carbon dioxide and its hydrate or methane and ice) are similar (one temperature process).

With consideration of the accepted assumptions the system of main equations can be written in the following form [5, 7, 8]:

$$\frac{\partial}{\partial t}\left(m\rho_g S_g\right) + \operatorname{div}\left(m\rho_g S_g \vec{\upsilon}_g\right) = 0, \tag{3}$$

$$mS_g\vec{\upsilon}_g = -\frac{k_g}{\mu_g}\operatorname{grad} p, \tag{4}$$

$$\frac{\partial}{\partial t}\left(\rho c T\right) + \rho_g c_g m S_g \vec{\upsilon}_g \operatorname{grad} T = \operatorname{div}\left(\lambda \operatorname{grad} T\right), \tag{5}$$

$$p = \rho_g R_g T, \qquad \sum_{j=g,i,h} S_j = 1, \qquad k_g = k_0 S_g^3,$$

$$\rho c = (1-m)\rho_{sk}c_{sk} + m\sum_{j=g,i,h}\rho_j S_j c_j, \qquad \lambda = (1-m)\lambda_{sk} + m\sum_{j=g,i,h} S_j\lambda_j .$$

Hereafter, subscripts sk, h, i and g are related to the parameters of the rock skeleton, hydrate CO_2, ice and gas (methane or carbon dioxide), respectively; ρ_j ($j = sk, h, i$ и g) is the true density of the j-th phase, S_j ($j = g, i, h$) is the j-th phase saturation, k_0 is the absolute reservoir permeability, υ_g and μ_g are

Journal of Physics: Conference Series **754** (2016) 032015 doi:10.1088/1742-6596/754/3/032015

the velocity and dynamic viscosity of the gas phase, p is the pressure, T is the temperature, c_j ($j = g, i, h$) is the specific heat capacity of the j-th phase, λ_j ($j = sk, g, i, h$) is the thermal conductivity coefficient of the j-th phase, ρc and λ are the specific volumetric heat capacity and the thermal conductivity coefficient of the system «reservoir – fluid».

The temperature and pressure values at the hydrate formation surface are related by the condition of phase equilibrium [9]:

$$T_s(p) = T_0 + T_* \ln(p/p_{s0}), \tag{6}$$

where p_{s0} is the equilibrium pressure corresponding to the initial temperature T_0, T_* is the empirical parameter.

In addition, at the boundary $x_{(n)}$ the relations, which follow from conditions of mass and heat balance, are realized [5, 8]:

$$\left[m\left(S_g \rho_g \left(\upsilon_g - \dot{x}_{(n)} \right) - S_h \rho_h G \dot{x}_{(n)} \right) \right] = 0, \tag{7}$$

$$\left[m\left(S_h \rho_h (1-G) + S_i \rho_i \right) \dot{x}_{(n)} \right] = 0, \quad \left[\lambda \operatorname{grad} T \right] = \left[m L_h \rho_h S_h \dot{x}_{(n)} \right].$$

were $[\psi]$ is the jump of the parameter ψ at the boundary $x_{(n)}$, L_h is the heat of the carbon dioxide hydrate formation, S_h is the CO_2 hydrate saturation. The dot over $x_{(n)}$ denotes the movement velocity of this boundary. Temperature and pressure at the boundary $x_{(n)}$ are considered to be continuous.

3. Calculation results

The system of equations (1)-(7) can be solved analytically. To do this we introduce the self-similar variable: $\xi = x/\chi^{1/2}t$. For this variable the system of equations (3)-(5), with application of Leibenzon linearization, can be presented in the form of the system of ordinary differential equations of the 2nd order:

$$-\frac{\xi}{2}\frac{dp_{(j)}^2}{d\xi} = \eta_{(j)} \frac{d}{d\xi}\left(\frac{dp_{(j)}^2}{d\xi} \right), \quad -\frac{\xi}{2}\frac{dT_{(j)}}{d\xi} = \frac{Pe_{(j)}}{2p_0^2}\frac{dp_{(j)}^2}{d\xi}\frac{dT_{(j)}}{d\xi} + \frac{d}{d\xi}\left(\frac{dT_{(j)}}{d\xi} \right) \quad (j = 1, 2), \tag{8}$$

$$\eta_{(j)} = \frac{\kappa_{(j)}}{\chi}, \quad \kappa_{(j)} = \frac{k_{g(j)} p_{(j)}}{m S_{g(j)} \mu_{g(j)}}, \quad Pe_{(j)} = \frac{\rho_{g0} p_0 c_{g(j)} k_{g(j)}}{\lambda \mu_{g(j)}}, \quad k_{(j)} = k_0 S_{g(j)}^3 \quad (j = 1, 2).$$

Here, the subscript 1 in brackets refers to the parameters of the zone $[0, \xi_{(n)}]$, 2 – $[\xi_{(n)}, \infty)$.

The piezoconductivity equation is nonlinear, because of pressure conductivity factor $\kappa_{(j)}$ include the unknown function $p_{(j)}$. To obtain approximate analytical solution we used the Leibenson linearization. For this purpose the variable pressure $p_{(j)}$ in the parameter $\kappa_{(j)}$ assumed constant and equal to the initial reservoir pressure p_0. Note that this will not make a significant error in the solution, because in the work we consider the cases of small pressure gradients ($\Delta p << p_0$).

After the piezoconductivity and temperature conductivity equations (8) integration the solutions for the pressure and temperature distributions in the first and second zones can be obtained.

$$p_{(1)}^2 = p_{(n)}^2 + \left(p_w^2 - p_{(n)}^2 \right) \int_{\xi}^{\xi_{(n)}} \exp\left(-\frac{\xi^2}{4\eta_{(1)}} \right) d\xi \Big/ \int_{0}^{\xi_{(n)}} \exp\left(-\frac{\xi^2}{4\eta_{(1)}} \right) d\xi \qquad \left(\xi \in [0, \xi_{(n)}] \right),$$

$$T_{(1)} = T_{(n)} + \left(T_w - T_{(n)} \right) \int_{\xi}^{\xi_{(n)}} \exp\left(-\frac{\xi^2}{4} - \frac{Pe_{(1)}}{2p_0^2} p_{(1)}^2 \right) d\xi \Big/ \int_{0}^{\xi_{(n)}} \exp\left(-\frac{\xi^2}{4} - \frac{Pe_{(1)}}{2p_0^2} p_{(1)}^2 \right) d\xi,$$

$$p_{(2)}^2 = p_0^2 + \left(p_{(n)}^2 - p_0^2\right)\int_{\xi}^{\infty}\exp\left(-\frac{\xi^2}{4\eta_{(2)}}\right)d\xi \Bigg/ \int_{\xi_{(n)}}^{\infty}\exp\left(-\frac{\xi^2}{4\eta_{(2)}}\right)d\xi \qquad \left(\xi \in [\xi_{(n)}, \infty]\right),$$

$$T_{(2)} = T_0 + \left(T_{(n)} - T_0\right)\int_{\xi}^{\infty}\exp\left(-\frac{\xi^2}{4} - \frac{Pe_{(2)}}{2p_0^2}p_{(2)}^2\right)d\xi \Bigg/ \int_{\xi_{(n)}}^{\infty}\exp\left(-\frac{\xi^2}{4} - \frac{Pe_{(2)}}{2p_0^2}p_{(2)}^2\right)d\xi .$$

On the basis of the conditions (7) and using the obtained solutions we can write the equations for determination of the coordinate of the phase transition boundary $\xi_{(n)}$ and the values of the parameters on this boundary.

When constructing the mathematical model it was assumed that the first zone saturated with CO_2 hydrate and in the second zone saturated with ice. Therefore, the constructed model adequately describes the process if the local temperature in the zone $[0, \xi_{(n)}]$ is lower than the local temperature of the carbon dioxide hydrate decomposition, and the local temperature in the zone $[\xi_{(n)}, \infty)$ is lower than 273 K (ice point). Let us call this the condition of thermodynamic consistency. The local temperature of the CO_2 hydrate dissociation is determined from the formula (6) by using the obtained pressure distribution.

Fig. 2 shows the distributions of temperature and pressure for the carbon dioxide injection pressure p_w = 1.2 MPa.

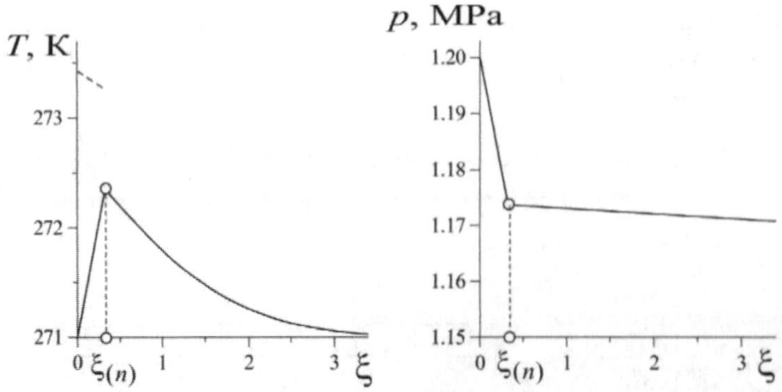

Figure 2. The pressure and temperature depending on the self-similar coordinate ξ. p_w = 1.2 MPa.

Here and on the following figures dashed lines correspond to the equilibrium temperature of CO_2 hydrate formation. The following parameters were used for calculation: $m = 0.25$, $S_{i0} = 0.6$, p_0 = 1.15 MPa, T_0 = 271 K, T_w = 271 K, $k_0 = 2\cdot10^{-15}$ m^2, $\rho c = 2.5\cdot10^6$ J/(K·m^3), λ = 2 W/(m·K), G = 0.28, specific gas constant for carbon dioxide R_{gc} = 189 J/(K·kg), specific gas constant for methane R_{gm} = 520 J/(K·kg), ρ_h = 1100 kg/m^3, ρ_i = 900 kg/m^3, specific heat of carbon dioxide c_{gc} = 800 J/(K·kg), specific heat of methane c_{gm} = 1560 J/(K·kg), coefficient of dynamic viscosity of carbon dioxide $\mu_{gc} = 1.3\cdot10^{-5}$ Pa·s, coefficient of dynamic viscosity of methane $\mu_{gm} = 10^{-5}$ Pa·s, $L_h = 1.6\cdot10^5$ J/kg, T_{*gih} = 19 K, T_{*glh} = 7.6 K, p_{s0} = 1.02 MPa [6, 8, 10].

Fig. 2 shows that at the relatively low pressure of the carbon dioxide injection the reservoir temperature in the zone $[0, \xi_{(n)}]$ is below the equilibrium CO_2 hydrate decomposition temperature, and in the zone $[\xi_{(n)}, \infty)$ is below the ice point. Therefore, in this case, the solution with the frontal boundary of phase transitions provides adequate mathematical description of the process.

Fig. 3 shows that a further increase of p_w leads to increasing the reservoir temperature, on a certain stretch after the phase transition boundary (in the zone $[\xi_{(n)}, \infty)$), above 273 K, and ice is overheating in this zone. Consequently, in this case, it is necessary to introduce a second phase-transition boundary of the ice melting. And the third (intermediate) zone is appearing in the reservoir, which is saturated with methane and water.

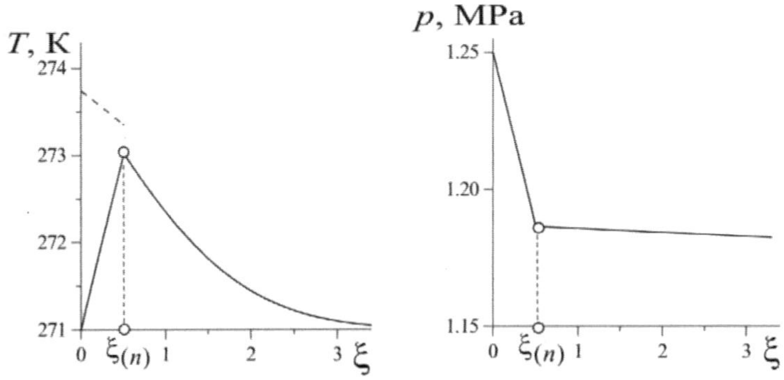

Figure 3. The pressure and temperature depending on the self-similar coordinate ξ. p_w = 1.25 MPa.

Fig. 4 shows that at the higher value of the injection pressure of carbon dioxide the reservoir temperature on a certain stretch of the zone [0, $\xi_{(n)}$] becomes higher the equilibrium temperature of the CO_2 hydrate decomposition. This corresponds to the overheating of the mixture of carbon dioxide and its hydrate at this region. Consequently, in this case, it is necessary to introduce the third boundary of phase transitions, because the zone, which is saturated with carbon dioxide and water, is appearing in the reservoir. This zone is located between zones "CO_2 + CO_2 hydrate" and "CH_4 + water".

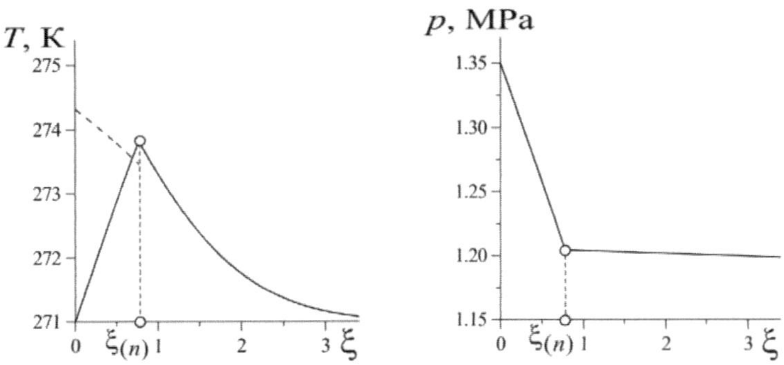

Figure 4. The pressure and temperature depending on the self-similar coordinate ξ. p_w = 1.35 MPa.

4. Conclusions

The mathematical model of the carbon dioxide injection process into the reservoir, which is initially saturated with methane and ice, was presented in this work. This model takes into account the CO_2 hydrate formation. The self-similar solutions of the one-dimensional problem were built. These solutions describe the distributions of main parameters in each of the allocated zones of the reservoir. On the basis of obtained solutions it was shown that different situations can exist at the carbon dioxide hydrate formation, notably with the one, two or three interphase boundaries.

Acknowledgments

This work was supported by Grant of the President of the Russian Federation for State Support of Leading Scientific Schools of the Russian Federation (NSh-6987.2016.1).

References

[1] Jadhawar P, Mohammadi A, Yang J and Tohidi B 2006 *Advances in the Geological Storage of Carbon Dioxide* vol 65 ed S Lombardi, L K Altunina and S E Beaubien (Dordrecht: Springer) 111–126

[2] Oldenburg C M, Pruess R and Benson S M 2001 *Energy and Fuel* **15** 293–298

[3] Duchkov A D, Sokolova L S, Ayunov D E and Permyakov M E 2009 *Earth's Cryosphere*, **13 (4)** 62–68

[4] Chuvilin E M and Guryeva O M 2009 *Earth's Cryosphere*, **13 (3)** 70–79
[5] Shagapov V Sh, Musakaev N G and Khasanov M K 2015 *Int. J. of Heat and Mass Transfer* **84** 1030–39
[6] Tsypkin G G 2014 *Fluid Dynamics* **49** 789–795
[7] Nigmatulin R I 1991 *Dynamics of Multiphase Media* (New York: Hemisphere Publ. Corp.)
[8] Khasanov M K, Musakaev N G and Gimaltdinov I K 2015 *J. of Engineering Physics and Thermophysics* **88** 1052–61
[9] Byk S Sh, Makogon Y F and Fomina V I 1980 *Gas Hydrates* (Moscow: Khimiya)
[10] Musakaev N G and Borodin S L 2016 *Heat and Mass Transfer* **52** 981–991

Bubbly free and impinging jets: experimental study by means of PIV and PFBI

K S Pervunin[1,2] and M V Timoshevskiy[1,2]

[1] Kutateladze Institute of Thermophysics, Siberian Branch of the Russian Academy of Sciences, 1, Lavrentyev Ave., Novosibirsk, 630090, Russia
[2] Department of Physics, Novosibirsk State University, 2, Pirogov Str., Novosibirsk, 630090, Russia

pervunin@itp.nsc.ru

Abstract. In the paper, an experimental investigation of the turbulent structure of bubbly free and impinging jets was carried out by means of PIV and PFBI techniques. PIV was applied to measure velocity distributions and turbulent characteristics in the continuous phase, while the PFBI approach was applied to visualize bubbles in the flow and evaluate their sizes. The flow was studied at the Reynolds number of 12,500 and three void fractions $\beta = 0$, 1 and 2%. The mean air bubble diameter was estimated to be roughly 0.8 mm for all β. It was revealed that, in the free jet, the air bubbles reduce the jet core size by more than 30% at the distance of a half of the nozzle diameter. In the two-phase jet, the radial velocity fluctuations rise faster in the mixing layer in the initial jet region between the nozzle edge and the downstream position of 40% of the nozzle diameter but further downstream they are suppressed about two times by the bubbles. In the impinging jet, the bubbles produce a high peak of the radial component of turbulent fluctuations near the wall at the distance of 70% of the nozzle diameter from the jet axis.

1. Introduction

Two-phase bubbly flows are of a great interest to study because of their wide spreading in nature and intense using in various technical applications. In practice, liquids are often saturated with gas bubbles intentionally to intensify mass transfer between the phases due to a large interfacial area (even at relatively low void fractions) and to increase mixing rate by vortex generation in the bubble wakes. At present, it is generally recognized that continuous and dispersed phases in two-phase media influence each other reciprocally. Depending on the size distribution of bubbles and characteristic scales of turbulence, different effects can be distinguished. For instance, bubble trajectories are affected by the anisotropy of turbulent fluctuations in liquids (turbulent dispersion), while the motion of individual bubbles and their clusters modifies distributions and intensities of turbulent characteristics (turbulence modulation) through a number of physical mechanisms.

Nowadays, mathematical modeling of turbulent flows is substantially complicated, particularly when they are two-phase, as the Navier–Stokes equations governing a liquid flow are nonlinear, which makes it necessary to average them to simplify numerical calculations. Nevertheless, the averaged equations remain unclosed and, therefore, the use of empirical considerations is required to solve the closure problem. The further development and verification of modern closure models are impossible without thorough and detailed information on distributions of various characteristics of both continuous and dispersed phases measured simultaneously, time dynamics of the dispersed phase associated with the downstream evolution of large-scale vortices, interaction of small-scale turbulence with individual dispersed inclusions and so on that can be initially gathered only in systematic and accurate experimental studies using advanced whole-field measurement techniques with a high spatial and temporal resolution.

Presumably, the most convenient and promising tool to investigate bubbly flows is presently Planar Fluorescence for Bubbles Imaging – PFBI [1, 2] since it allows local visualization of round bubbles in the same section illuminated by a laser light sheet as that employed for PIV measurements. Thus, using a combination of PIV, PFBI and PTV methods, it is possible to estimate simultaneously instantaneous planar distributions of velocities and its fluctuations of continuous and dispersed phases as well as local void fraction within a flow region of interest, including joint correlations of fluctuations of different quantities for the both phases up to the third-order statistical moments. According to [1], the PFBI technique can be hardly applied to study bubbly flows with the volume gas fractions higher than 2%. Nevertheless, this limitation is less rigorous compared to the one for other optical methods for bubbly flow diagnostics (typically about 1%) and does not seem so critical for the present research.

In the paper, bubbly free and impinging round jets are studied experimentally at the Reynolds number Re = 12,500 for several gas volume fractions $\beta = 0, 1, 2\%$ and the same mean bubble size of roughly $D_B = 0.8$ mm by means of a combination of high-speed particle image velocimetry and planar fluorescence for bubbles imaging approaches. We report here on a visual analysis of high-speed images registered by a shadow photography technique to determine the mean bubble size, bubble spatial distributions as well as on PIV measurements of spatial distributions of instantaneous flow velocity, with a special emphasis put on the effect of the dispersed phase on the mean flow field and its turbulent fluctuations.

2. Experimental conditions and measurement techniques

The bubbly free and impinging submerged round jets were reproduced in the hydrodynamic rig in Kutateladze Institute of Thermophysics SB RAS. The experimental setup is of closed type and continuous operation. Its test section is made of organic glass with the following dimensions (HxWxL): 425x300x300 mm (see Fig. 1). The operating fluid was distilled water which was driven by a centrifugal pump equipped with a control module to vary the rotor speed. The liquid volume rate was measured by an ultrasonic flowmeter. The water temperature was maintained constant at 30±0.1 °C in the test section by means of a thermostatic regulator filled with tap water as a heat-transfer agent and containing a tubular electrical heater, cooling circuit and aquarium pump to permanently blend the heat carrier. Cooling and heating in the thermostat was implemented by a PID-control system activating an electromagnetic valve to start/stop the coolant supply to the cooling circuit and a switching relay to loop/break the electric line to the heater. The liquid temperature was measured by heat-variable resistors placed in the test section, heat exchanger and settling tank.

The bubbly upward jets were formed by a round converging nozzle of 110 mm height with an inner diameter at its outlet edge $D_N = 15$ mm. The ratio of the inlet cross-section area of the nozzle to the outlet one was 16. In case of the impinging jet, the distance between the nozzle outlet and the impingement surface was $H = 3D_N$. The Reynolds number based on the superficial liquid velocity $V_0 = 4 \cdot Q_W/(\pi \cdot D_N^2) = 0.663$ m/s, where Q_W is the water volume rate (measurement error is 2%), and D_N equaled to Re = 12,500. In order to saturate the flow with bubbles, an air-water mixer of a special design was used to provide a quasi-monodisperse size distribution of air bubbles. The air was supplied to the mixer by a compressor through a couple of filters of rough (5 μm) and fine (0.01 μm) cleaning. The air overpressure at the mixer inlet was specified by a high-precision reducer in the range of 9.8–19.6 kPa depending on the air volume rate Q_A. The air temperature, pressure and volume rate were measured together by a thermal mass flowmeter (TSI model 4140) with the measurement uncertainties 0.1 °C, 0.1 kPa and 2%, respectively. The length of a stainless steel pipe with 20 mm inner diameter between the nozzle and mixer was 0.56 m. Different values of the ratio of air volume rate to the full flow rate $\beta = Q_A/(Q_W+Q_A)$ in the jet flows were achieved by changing Q_A using a needle valve. In experiments, β possessed three values: 0, 1 and 2%. The mean diameter of the air bubbles D_B was approximately 0.8 mm for all β.

Figure 1. A photograph of the test section of the hydrodynamic rig in IT SB RAS. The nozzle in configuration of the impinging jet is visible inside the test section. The flow is illuminated by the laser light.

In order to implement the shadow photography technique, a 100 W white LED-lamp with a diffusive screen for uniform illumination placed at the distance of 330 mm to the measurement section was utilized as a continuous backlight source. The bubble patterns were recorded by a Photron FASTCAM SA5 CMOS-camera (digit capacity 12 bits, resolution 1,024x1,024 pix., acquisition rate 7 kHz) equipped with a Nikon AF Micro-Nikkor 60 mm f/2.8D lens. For planar fluorescence imaging (PFBI approach), the water was merged with Rhodamine 6G as a fluorescent dye, the concentration of which was very low (about 20 μg/l). So, the water properties, especially its viscosity and surface tension, can be considered unchanged. In order to illuminate and register the bubbles suspended in the flow, a PIV system with a high temporal resolution consisting of a pulsed Nd:YAG Photonics Industries DM-532-50 laser (wavelength 532 nm, repetition rate 15 kHz, pulse duration 10 ns, pulse energy 15 mJ), Photron FASTCAM SA5 camera with the same lens as in the SP method and a low-pass optical filter (transmission edge at 570 nm) and Berkeley Nucleonics Corporation pulse/delay generator (model 575) for external synchronization was employed. The distance between the camera and the measurement section that passed through the jet axis was roughly 360 mm. The thickness of a laser light sheet was about 0.8 mm. The plane dimension of the measurement region was approximately 63.5x63.5 mm. In the experiment, the recording rate was 2 kHz. As bubbles illuminated by a laser light produce bright glares in images and, thereby, contaminate raw PIV data, a laser-induced fluorescence (LIF) approach was applied to avoid their undesirable effect on the measurements. For this, fluorescent seeding particles made of polymethyl methacrylate filled with Rhodamine B of MicroParticles GmbH production (hydrophobic, size distribution 1–20 μm, wavelength range 550–700 nm) were added into the operating liquid for PIV measurements.

The raw data (a series of 10,000 PFBI snapshots for each regime) gathered continuously during 5 seconds were processed using a PC with "ActualFlow" software [3]. At first, two procedures, namely subtraction of the mean two-frame intensity field averaged over the whole sample range and masking, were successively applied to enhance the quality of the registered images and to remove the areas corresponding to the nozzle and shadows from the subsequent calculations. Velocity fields were calculated using the iterative cross-correlation algorithm with a continuous window shift and deformation and 75% overlap between the interrogation windows. In addition, at this step of processing the local particle concentration was accounted for. In order to have a relatively large dynamic range, the initial size of the interrogation window was chosen to be 64x64 pixels but it was subsequently reduced so that the final interrogation window was 8x8 pixels, which provided a high enough spatial resolution. The obtained instantaneous velocity vector fields were then validated with the following procedures: peak validation with the threshold of 2.0, adaptive median 7x7 filter and cluster validation with the coefficient of 50 (the details of PIV data processing can be found in [4]).

3. Results and discussion

In this section, some selected results of the experimental study of submerged round free and impinging bubbly jets are presented. We give a brief discussion of instantaneous images from the high-speed visualization by the PFBI approach in comparison with those for shadow photography (SP) technique followed by downstream evolutions of the mean velocity and turbulent fluctuations for the continuous phase extracted from the PIV measurements.

3.1. Bubbly free jet

Figure 2 shows jointly examples of the PFBI and SP visualization of the bubbly free jet for the same flow regime to facilitate a comparison between the methods. The bubble patterns in the SP images are highly overlapped even when the volume gas concentration is relatively low ($\beta = 1\%$, not shown). In case of $\beta = 2\%$ (Fig. 2-a), it becomes extremely difficult to correctly distinguish individual bubbles in some flow regions. This occurs because, in SP approach, all bubbles situated in the flow region of interest and coming into the view of the camera (i.e. located along its line of sight) are registered. An application of the PFBI method allows overcoming of this problem. In PFBI images (Fig. 2-b), the bubble patterns are also overlapped but they look indeed differently. The bubbles intersecting the measurement section produce bright rings with sharp borders (in-focus), while the intensity of those patterns formed by the bubbles somewhat shifted from the measurement section is noticeably reduced and their edges become blurred (out-of-focus) (see Fig. 2-b). Taking this into consideration, one can recognize in-focus overlapping bubbles by applying sophisticated algorithmic approaches for their identification, e.g. the correlation procedure [1].

Figure 2. Instantaneous photos of the initial region of the free bubbly jet captured by the (a) shadow photography and (b) PFBI techniques for the volume gas fraction $\beta = 2\%$. The nozzle outlet is at the bottom of images. The flow direction is upward. Small bright spots in the PFBI image are fluorescent tracer particles employed for the PIV measurements.

The measured mean velocity distributions of the carrier phase are presented in Fig. 3 at four cross-sections $z/D_N = 0.5$, 1.5, 2.5 and 3.5. The jet core was found to break up at approximately $z/D_N = 3$. In the single-phase flow, the mean liquid velocity profile, being flat at the nozzle outlet, gradually slopes and expands laterally downstream (Fig. 3a), so that initially stagnant water becomes disturbed at the distance of $r/D_N = 1$ from the jet axis in the cross-section $z/D_N = 3.5$. The mean velocity V_z/V_0 along the jet axis diminishes slowly downstream and equals to 1.05 at $z/D_N = 0.5$ and 1.03 at $z/D_N = 3.5$. In case of the two-phase flow, the mean velocity profiles are similar to the ones for $\beta = 0\%$ but, in the initial flow region, the jet core is substantially reduced: its width does not exceed $0.6D_N$ at $z/D_N = 0.5$, while it is equal to $0.9D_N$ in the single-phase jet at the same cross-section (see Fig. 3a). The mean velocity on the contrary increases in the jet core with the void fraction by 4% and 6% for $\beta = 1\%$ and 2%, respectively. The discrepancy in the V_z profiles decreases downstream and, at $z/D_N = 2.5$, the distributions become quite similar for all β so that the velocities differ less than by 1%. However, further downstream the maximum velocity drops faster in the bubbly flow rather than in the single-phase one. When $\beta = 2\%$, the maximum of V_z/V_0 at $z/D_N = 3.5$ is below by 5% as compared to that for $\beta = 0\%$ (Fig. 3a).

In the single-phase case, the crosswise turbulent fluctuations at the nozzle edge ($z/D_N \approx 0$) are sufficiently suppressed ($\langle \upsilon_r^2 \rangle / V_0^2 \sim 0.001$, Fig. 3b) whereas the amplitude of the streamwise ones is relatively high $\langle \upsilon_z^2 \rangle / V_0^2 \approx 0.055$ (Fig. 3c). This is because the longitudinal velocity fluctuations develop inside the supply duct but the transverse ones are conversely reduced when the flow is contracted by the nozzle. Downstream, $\langle \upsilon_r^2 \rangle$ begins to grow rapidly in the jet mixing layer starting from $z/D_N \approx 0.25$ and reaches its maximum of $0.03V_0^2$ at $z/D_N = 0.8$ but, further downstream, it decays down to $0.015V_0^2$ at $z/D_N \approx 3$ (where the jet core breaks up) and then remains constant (Fig. 3b). $\langle \upsilon_z^2 \rangle$ oppositely declines gradually with an increase of z/D_N until it takes the constant value of $0.035V_0^2$ at $z/D_N \approx 1.8$ (Fig. 3c). The gas injection modifies this behavior of the turbulent characteristics noticeably. In the two-phase jet with both $\beta = 1$ and 2%, $\langle \upsilon_r^2 \rangle$ rises quickly downstream from the very nozzle edge up to $0.015V_0^2$ for $\beta = 1\%$ and $0.012V_0^2$ for $\beta = 2\%$ at $z/D_N \approx 0.55$ because the bubbles force the flow turbulization in their wakes. However, further downstream they conversely suppress the turbulence so the amplitude of the transverse fluctuations does not change. The longitudinal fluctuations are also affected by the bubbles in a similar way: they are higher at the nozzle edge in the bubbly jet ($\langle \upsilon_z^2 \rangle / V_0^2 \approx 0.085$) than the ones for the single-phase flow but drop very fast downstream to $\langle \upsilon_z^2 \rangle / V_0^2 \approx 0.025$ at $z/D_N \approx 0.45$ (Fig. 3c).

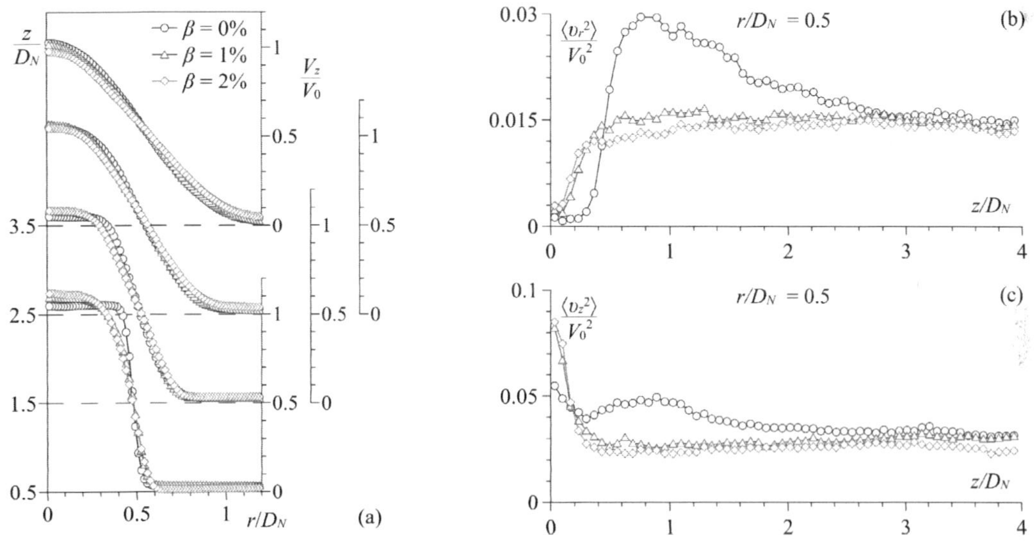

Figure 3. Downstream evolution of velocity characteristics of the continuous phase in the free bubbly jet at different void fractions: (a) the streamwise component of the mean velocity at four cross-sections and time-averaged squares of velocity fluctuations in the (b) transverse and (c) longitudinal directions along the jet mixing layer ($r/D_N = 0.5$). z and r are axial and radial coordinates, respectively. The reference point ($z/D_N = 0$, $r/D_N = 0$) is at the centre of the nozzle outlet.

3.2. Bubbly impinging jet

The mean liquid velocity decreases rapidly along the jet axis ($r/D_N = 0$) when the flow approaches the stagnant point at the wall (e.g., at $z/D_N = 2.9$, it equals only to $0.2V_0$, see Fig. 4a). The gas injection does not almost affect it in this longitudinal section though the buoyancy force acts on the bubbles and, thereby, the bubbles might accelerate the carrier phase (but this is not the case). Along the impingement surface ($z/D_N = 2.9$), the mean velocity magnitude **V** initially increases progressively with r/D_N up to $0.78V_0$ at $r/D_N = 0.85$ as a result of the flow deflection due to the wall influence but then declines monotonously (by the law of $1/r$) because of the flow expansion. The bubbles in general diminish **V**. This effect is not notably pronounced for $\beta = 1\%$ but clearly seen for $\beta = 2\%$ (Fig. 4a) so that the maximum of $\mathbf{V} = 0.61V_0$. The radial component of turbulent fluctuations gradually grows with

r/D_N in the single-phase jet. It becomes by an order of magnitude greater ($(\langle v_r^2\rangle/V_0^2 = 0.03)$ at $r/D_N = 2$ as compared to $\langle v_r^2\rangle/V_0^2 = 0.003$ at $r/D_N = 0$. The dispersed phase does not substantially change the intensity of the radial fluctuations at the jet axis (they reach only $\langle v_r^2\rangle/V_0^2 \approx 0.005$ for both $\beta = 1\%$ and 2%, Fig. 4b) but increases the amplitude of the axial ones (not shown) from $\langle v_z^2\rangle/V_0^2 = 0.004$ ($\beta = 0\%$) to 0.008 ($\beta = 2\%$). The main difference between the single- and two-phase impinging jets is that the bubbles cause the appearance of a maximum in the radial velocity fluctuations at $r/D_N = 0.7$ ($\langle v_r^2\rangle/V_0^2 = 0.035$ and 0.05 for $\beta = 1\%$ and 2%, Fig. 4b). Downstream, the fluctuations decay so that behind $r/D_N = 1.2$ no considerable discrepancy between the one- and two-phase jets can be observed.

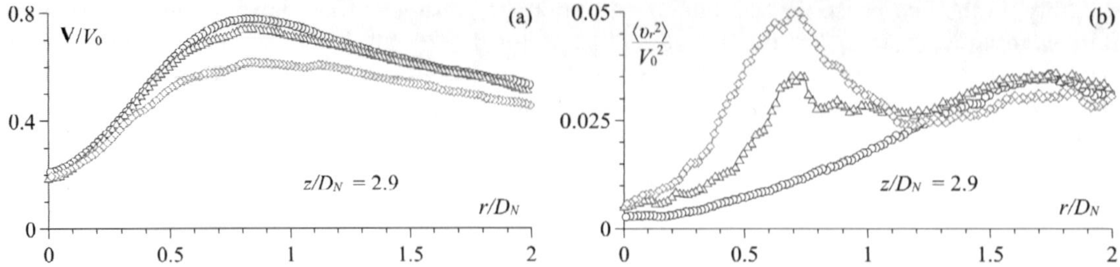

Figure 4. Downstream evolution of velocity characteristics of the continuous phase in the impinging bubbly jet close and parallel to the wall ($z/D_N = 2.9$) at different void fractions: (a) the modulus of the mean velocity and (b) time-averaged square of velocity fluctuations in the radial direction. The coordinate system is the same as in Fig. 3.

4. Summary

The bubbly free and impinging jets with void fractions up to $\beta = 2\%$ and the mean bubble size of 0.8 mm were studied by the PIV and PFBI methods at the Reynolds number Re = 12,500. Under such conditions, in case of the free jet, the air bubbles were found to reduce substantially the transverse dimension of the jet core (more than by 30% at $z/D_N = 0.5$) and cause a faster growth of the radial component of turbulent fluctuations in the mixing layer from the nozzle edge up to the distance of $z/D_N \approx 0.55$ but suppress them about two times further downstream as compared to the single-phase flow. In the impinging jet flow, the bubbles produce a maximum of the radial liquid velocity fluctuations near the wall at $r/D_N = 0.7$ which is absent in the single-phase jet.

Acknowledgments

The research was funded by a grant from the Russian Science Foundation (Project No. 14-19-01685 supervised by Prof. D.M. Markovich in IT SB RAS).

References

[1] Akhmetbekov Ye K., Alekseenko S V, Dulin V M, Markovich D M and Pervunin K S 2010 Planar fluorescence for round bubble imaging and its application for the study of an axisymmetric two-phase jet *Exp. Fluids* **48** 615–29

[2] Dulin V M, Markovich D M and Pervunin K S 2012 The optical principles of PFBI approach *AIP Conf. Proc.* **1428** pp. 217–24.

[3] Akhmetbekov Ye K, Bilsky A V, Lozhkin Yu A, Markovich D M, Tokarev M P and Tyuryushkin A N 2006 Software for experiment management and processing of data obtained by digital flow visualization techniques (ActualFlow) *Numer. Methods Program.* **7** 79–85 (in Russian)

[4] Kravtsova A Yu, Markovich D M, Pervunin K S, Timoshevskiy M V and Hanjalić K 2014 High-speed visualization and PIV measurements of cavitating flows around a semi-circular leading-edge flat plate and NACA0015 hydrofoil *Int. J. Multiphase Flow* **60** 119–134

Thickness measurement in a horizontal liquid layer when heated from a localized hot-spot

S E Spesivtsev[1,2], Yu V Lyulin[1,3], I V Marchuk[1] and O A Kabov[1,3]

[1] Institute of Thermophysics, 1, Lavrentiev Ave., 630090, Novosibirsk, Russia
[2] Novosibirsk State University, 2, Pirogov Ave., 630090, Novosibirsk, Russia
[3] Tomsk Polytechnic University, 30, Lenina Ave, Tomsk, 634050, Russia

Abstract. Thermocapillary breakdown of thin (0.3 – 0.7 mm) horizontal layers of liquid (ethanol) when heated from a localized hot spot was investigated experimentally. The effect of layer thickness on the breakdown dynamics was studied. Visualization and control of the liquid layer were carried out using schlieren and confocal techniques. Main steps of the breakdown process were determined with the help of both systems. Evolution of the layer thickness in the center of substrate was observed and the critical thickness of the layer was measured using confocal sensor.

1. Introduction

Investigation of heat transfer from a local heat source becomes one of the most important problems in thermophysics. The problem is closely connected to the cooling of microelectronic equipment [1]. Continuous development and complication of microchip structure in microelectronic devices produce nonuniform heat flux distribution on the chip surface and inevitably lead to degradation of its performance and reliability. Heat flux density in some regions is much higher than the chip average [2], of the order of 1 kW/cm². These specific regions called "hot spots" could have size from several hundred microns to 1-2 millimeters. However, using special localized cooling it is possible to produce large performance gains in microprocessors. Nowadays, there are several effective techniques for cooling of local hot spots such as spray cooling, boiling in microchannels, thermoelectric coolers. One of the promising methods for removing such high heat fluxes from a spotted heat source is technology based on evaporation of a thin liquid layer [3]. Dynamics of evaporation essentially depends on the conditions in the layer [4]. In particular, the breakdown of liquid layer leads to dramatic decreasing of heat transfer from a spotted heat source [5]. Processes of liquid layer rupture are actively studied experimentally [5-7] and theoretically in the present time [8-10]. The goal of the present work is to study the breakdown dynamics and measure the thickness of a horizontal evaporating liquid layer when heated from a localized hot spot.

2. Experimental setup

Experiments were conducted on the setup shown in Fig. 1. The working fluid is supplied to the test cell *1* with the help of the syringe pump *2* and a horizontal liquid layer *3* is formed. The thickness of liquid layer is varied from 300 to 700 μm. The layer of liquid is opened to the atmosphere and maintained on the surface of the working area using sharp edge on the substrate perimeter. Spot heating of the horizontal liquid layer takes place in the center of the substrate. The test cell consists of caprolon base *4*, metallic substrate *5* and the heating element *6*. The caprolon base has a special cut on the upper side for installation of the substrate and a central through hole with a diameter of 1.6 mm. The substrate is made of stainless steel and has a diameter of 50 mm and a thickness of 1 mm. In the center of the substrate is a closed hole with a diameter of 1.6 mm and a height of 0.8 mm. The heating element is made of brass and has a round tip with a diameter of 1.6 mm and a height of 3 mm. It is tightly inserted into the closed hole of the substrate through the caprolon base. Thermal paste is used for better thermal contact between the heater tip and the substrate. The distance between the tip and the upper side of substrate is 0.2 mm. The power of heating element is controlled by the power supply *7*. Insulating material *8* is located on the underside of the heater to minimize heat losses. Temperature in the test cell is measured by thermocouples *9* (type K) connected to the measuring system *10* with an accuracy of 0.1°C. Location of the thermocouples is shown in Fig. 1. Relative humidity and atmosphere temperature are measured using the thermohygrometer Testo 645 *11* with an accuracy of 2% and 0.1°C, respectively.

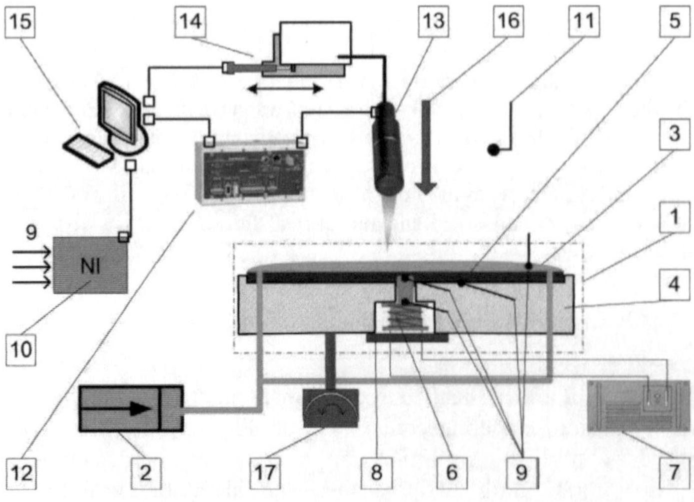

Figure 1. Scheme of the experimental setup

The heat flux density is determined by measuring the temperature difference between two different sections along the heater tip. The height of the horizontal liquid layer is maintained in constant position during the all experiment. Confocal system Micro-epsilon is used for measuring the layer thickness over the heating element. The system consists of the controller *12* and the sensor *13*. Sensors have the spatial resolution from 10 to 60 nm, the accuracy 0.35 - 0.7 μm, the spot diameter 6 - 16 μm and the measuring range 0.3 - 10 mm. The maximum temporal resolution is 100 μs. The sensor is fixed on the three-dimensional positioning system with high-speed linear actuator *14* on one of the horizontal axes. Linear actuator is controlled from personal computer *15* and by special software. The maximum speed the axis moves at is 104 mm/s. The sensor moves in range of 50 mm by steps of 1 μm. Also it moves in two other axes with the help of two hand-operated linear stages in range of 50 mm. For visualizing the deformation of the surface and recording the layer breakdown dynamics schlieren technique *16* with Photron FASTCAM 675K-M3 high-speed camera was used (at 5000 fps, 640x640 pixels, 25 μm/pix). The test cell was installed in a horizontal position with the help of two-axis goniometer *17*. Equilibrium contact angle of the substrate surface is defined by Young-Laplace equation [11] at room temperature of 25 ± 2 °C and is equal to $\theta = 8 \pm 1$ °. Profilometer "Micro Measure 3D station" was used for measuring the roughness of substrates. Average value of the substrate roughness is equal to $R_a = 0.037$ μm.

3. Results and discussion

3.1. Experimental results

Experiments were conducted at atmospheric pressure, temperature and relative humidity of 28±2°C and 25±3%, respectively. Layer thickness was ranged from 300 to 700 μm. Ethanol (95%) was used as a working fluid. Heater temperature range was 20 - 105°C. The temperature values were measured at the moment of layer breakdown. Heat flux density was varied from 0 to 95.3 W/cm². Injection liquid flow rate was up to 200 μl/min. During the experiment heat flux is increased up to a critical value at which the liquid layer ruptures. At this moment heating process is stopped to prevent the failure of heating element.

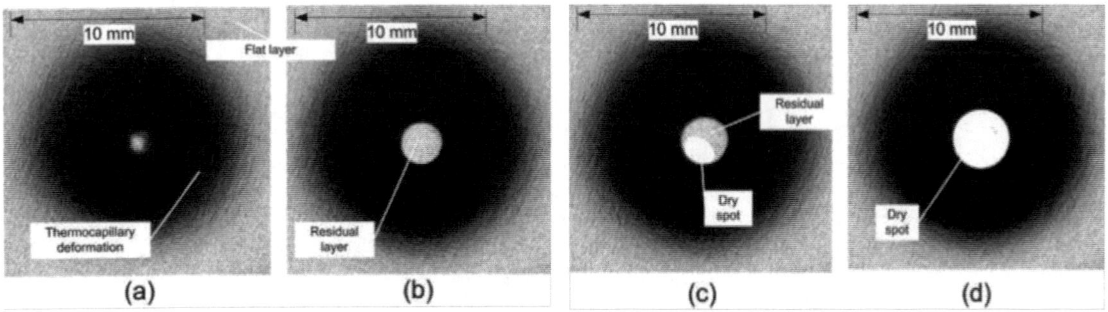

Figure 2. Visualization of breakdown dynamics and formation of the dry spots

It was found that breakdown process consists of several steps. At the beginning, thinning of the liquid layer over the heating area (Fig. 2a) due to the effect of thermocapillary forces [12, 13] and evaporation is observed. Further thinning leads to the formation of residual liquid layer in the area of the local heating, Fig. 2b. Then the residual layer evaporates until its thickness reaches the critical and breakdown of the liquid layer occurs, Fig. 2c, [5, 6]. After the breakdown the whole area of the local heating rapidly dries and quite symmetrical circular dry spot is formed, Fig. 2d.

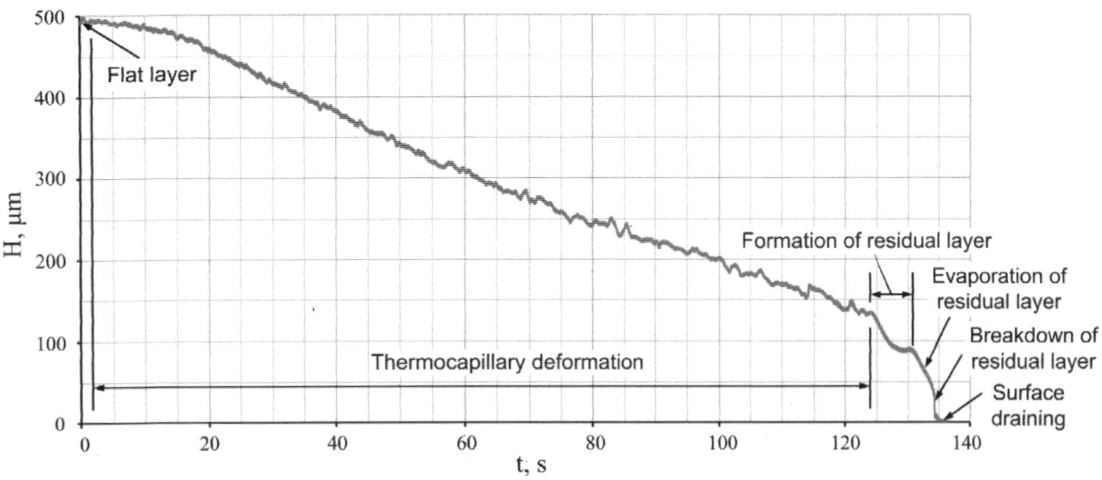

Figure 3. Dependence of the layer thickness over the heating area on time (initial layer depth 500 μm)

Evolution of the layer thickness in the center of substrate is observed with the help of the confocal system. All detected steps of the breakdown process observed by the schlieren technique were also confirmed using the confocal method. Dependence of the layer thickness over the heating area on time is shown in Fig. 3.

Evolution of the layer thickness before the breakdown was measured for different initial layer depth in range from 300 to 700 μm (Fig. 4). Pulsations of the layer thickness before formation of the residual layer have been observed. It is found that critical thickness of residual layer and thickness of residual film have no dependence on the initial layer depth and its values are 35 ± 2 μm and 10 ± 1 μm, respectively. It is assumed that critical thickness of the residual layer for present substrate mostly depends on the properties of working liquid. The time of dry spot formation is decreased when the layer thickness and, accordingly, the local heating intensity is increased. The time varies from 6 to 0.67 seconds. Temperature in the center of substrate increases and more intense evaporation of residual layer takes place. Thus, one of the main factors influencing the residual layer breakdown and the heating region draining is evaporation.

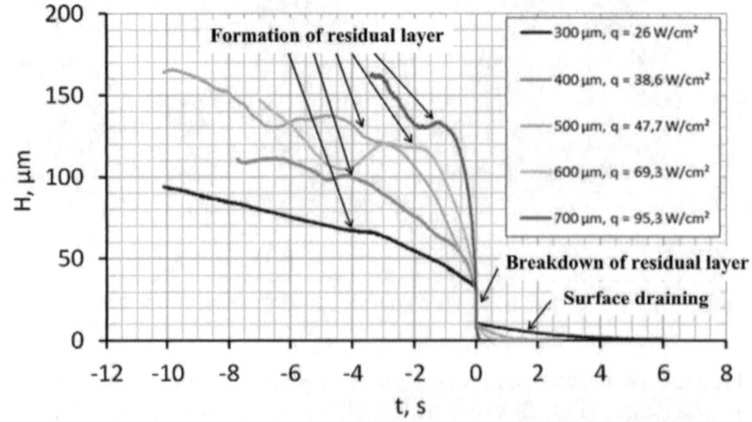

Figure 4. Dependence of the layer thickness over the heating area on time for different initial layer depth

3.2. Numerical modelling

The problem of thermocapillary deformations and breakdown of the locally heated horizontal liquid layer has been solved numerically in axisymmetric statement. The thin layer approximation model is used [12, 13]. Capillary pressure, viscosity and gravity are taken into account. Evaporating rate is supposed to be proportional to the temperature difference between temperatures of the liquid surface and ambient. Heat transfer in the substrate is also simulated. The deformation of the free surface has been calculated for the corresponding experimental conditions. The model predicts the formation of the thin residual layer of the liquid and film breakdown at sufficiently intensive heating.

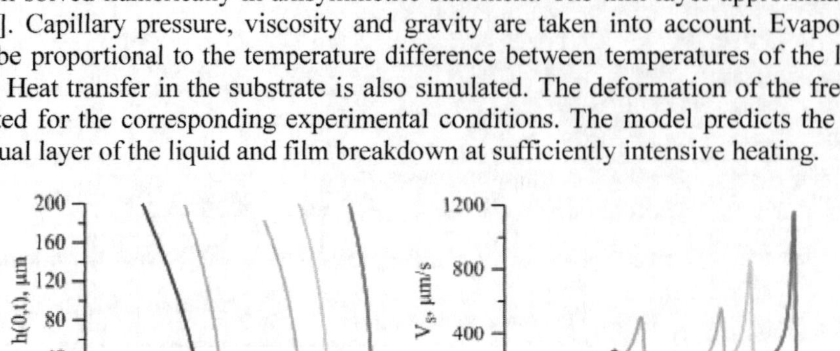

Figure 5. Calculated film thickness and surface velocity over the center of the heater (See parameters in Fig. 4)

Calculated film thickness dynamics over the heater center is shown in Fig. 5. After a period of rapid decrease in the thickness of the film a time of a slow decrease in the thickness of the film follows down to zero, Fig. 5a. The main mechanism of the outflow of the liquid from the heating zone in the first case is the thermocapillary shear stress, and in the second case this is the evaporation. One can see that points on curves in Fig 5a with maximal curvature divide the curves into two quasilinear parts. The film thickness for these points has quite the same value about 10 μm. One may conclude that at thickness less than 10 μm the main drainage mechanism is evaporation. At small liquid thickness the viscosity suppresses thermocapillary flows. Liquid surface velocity over the heater center is shown in Fig. 5b. Maximum in the curves correspond to the start of the decrease in the velocity. The film thickness for these points has quite the same value about 40 μm. Maximum of the slowdown acceleration of the liquid surface takes place at values of the film thickness about 20 μm.

4. Conclusions

Influence of the layer thickness on the breakdown dynamics of liquid layer was studied. The existence of residual liquid layer over the heating area before the breakdown has been proved. Pulsations of layer thickness over the heating area before the formation of residual layer have been found. Critical thickness of residual layer and thickness of residual film have no dependence on the initial layer depth. The main drainage mechanism for the residual layer is evaporation.

The study was financially supported by the Russian Science Foundation (Project 14-19-01755).

References

[1] Bar-Cohen A, Wang P 2012 *J. Heat Transfer* **134** 5
[2] Mahajan R, Chiu C, Chrysler G 2006 *Proceedings of the IEEE* **94** 8
[3] Gatapova E Ya, Kabov O A, Marchuk I V 2004 *Technical Physics Letters* **30** 5
[4] Fedorets A A, Marchuk I V, Kabov O A 2013 *Interfacial Phenomena and Heat Transfer* **1** 1
[5] Lyulin Yu V, Spesivtsev S E, Marchuk I V, Kabov O A 2015 *Technical Physics Letters* **41** 11
[6] Zaitsev D V, Rodionov D A, Kabov O A 2007 *Microgravity Science and Technology* **19** pp 100-103
[7] Zaitsev D V, Kabov O A 2007 *Microgravity Science and Technology* **19** 174-177
[8] Ajaev V S 2013 *Interfacial Phenomena and Heat Transfer* **1** 1
[9] Williams M B, Davis S H 1982 *J. Colloid Interface Sci.* **90** pp 220-228
[10] Burelbach J P, Bankoff S G and Davis S H 1990 *Phys. Fluids A* **2** pp 322–333
[11] Marchuk I V, Cheverda V V, Strizhak P A and Kabov O A 2015 *Thermophysics and Aeromechanics* **22** 3
[12] Marchuk I V 2009 *Journal Engineering of Thermophysics* **18** 3
[13] Marchuk I V 2015 *Journal Engineering of Thermophysics* **24** 4

The surface temperature of free evaporating drops

V Y Borodulin, V N Letushko, M I Nizovtsev and A N Sterlyagov*

Kutateladze Institute of Thermophysics, 1 Ac. Lavrentiev ave., Novosibirsk, Russia

*Corresponding author: sterlyagov@itp.nsc.ru

Abstract. Complex experimental and theoretical investigation of heat and mass transfer processes was performed at evaporation of free liquid drops. For theoretical calculation the emission-diffusion model was proposed. This allowed taking into account the characteristics of evaporation of small droplets, for which heat and mass transfer processes are not described in the conventional diffusion model. The calculation results of evaporation of droplets of different sizes were compared using two models: the conventional diffusion and emission-diffusion models. To verify the proposed physical model, the evaporation of droplets suspended on a polypropylene fiber was experimentally investigated. The form of droplets in the evaporation process was determined using microphotographing. The temperature was measured on the surfaces of evaporating drops using infrared thermography. The experimental results have showed good agreement with the numerical data for the time of evaporation and the temperature of evaporating drops.

Introduction

Interest in the study of evaporation of liquid droplets is associated with a variety of practical applications, ranging from fuel burning and intensification of heat and mass transfer processes to formation of nanostructures [1-3]. The study of the evaporation of liquid droplets is a classical fundamental problem, and it is the subject of a large number of works [4-7]. However in the processes associated with droplet evaporation, there are still many fundamental issues that need deeper study. At evaporation the droplet size continuously decreases, which leads to the necessity of solving this problem in non-stationary conditions. The droplet size is a significant parameter that determines the influence of certain factors (surface tension, gravity, the free path of vapor molecules, etc.) on the process of drop evaporation. A physical model of drop evaporation should be chosen depending on the most significant factors.

There are a number of physical models describing the behavior of evaporating liquid droplets that have been proposed and are currently used [10-13]. It should be noted that the majority of studies use the diffusion approach of Maxwell for calculations of droplets evaporation. The diffusion model describes well the evaporation of large drops, but has limitations in the description of the droplets of small size. In addition to the diffusive approach, some researchers propose gas-kinetic models to describe evaporation [8, 14]. Thus, in [8] it is demonstrated that consideration of the kinetic approach to small drops evaporation leads to an increase in the droplets evaporation time in comparison with purely diffusion mechanism of evaporation.

Experimental investigations of droplets evaporation widely apply a technique of drops suspended on a thin fiber, which allows obtaining a good approximation to the conditions of free drop evaporation [5].

The experiments usually pay main attention to determining the varying geometry of the drops using photo and video equipment [1, 12, 13]. The temperature of drops in these studies is measured using contact methods.

Drops suspended on a thermocouple have become a frequent practice allowing controlling the temperature change in the droplets in the evaporation process. For example, Watanabe et al. [12] investigated the evaporation of drops, suspended on a thin thermocouple, to study combustion of isolated droplets of liquid fuel. Gan et al. [1] conducted experiments on the evaporation of drops (suspended on thin K-type thermocouple) of different fuel mixtures with the addition of nanoparticles under natural and forced convection. A significant problem in such studies is heat supply to the droplet through the thermocouple due to high thermal conductivity of the thermocouple material, which leads to a significant error of the experiment.

Yang and Wong [13] studied the evaporation of drops, suspending them on the end of a thin quartz fiber to minimize the heat input to the drop. Similar experiments at elevated pressures and temperatures were carried out by Ghassemi et al. [15]. They found that at low ambient temperatures, the thermal conductivity of the fiber mostly influenced the early stage of evaporation, and at high temperatures the influence extended to the whole process. K. Han et al. [13] considered the effect of heat conduction of the thermocouple and the quartz fiber on the evaporation of suspended single droplets under different ambient temperatures. The results showed that the influence of quartz fiber on the evaporation of drops was substantially less than the one of the thermocouple. Nevertheless, there was an increase in evaporation rates, associated with the additional supply of heat. Thus, measuring the drop temperature with thermocouples implies significant errors due to the supply of heat through the thermocouple. In addition, such measurements are local in nature and do not provide information about the temperature distribution on the drop surface.

In recent years, temperature measurements on evaporating drop surfaces have been widely done using the infrared thermography technique [17-19]. The advantages of this method are contactlessness, high sensitivity and high-speed performance.

The aim of this work is a comparative analysis of the calculation results of free drops evaporation according to the diffusion model and the emission-diffusion model for different diameters of spherical water drops. To verify the proposed physical model the calculated changes in temperatures of suspended droplets and times of their evaporation were compared with the experimental results, obtained using the contactless infrared thermography and microphotographing.

1. Description of the computational model

For the calculations the emission-diffusion model of liquid droplets evaporation was used [8, 20]. The model is based on a dynamic equilibrium between the vapor flow, obtained in the framework of the gas-kinetic approach, and the diffusion flow. This approach, in contrast to the purely diffusional one, allows more accurately taking into account the characteristics of evaporation of small and large drops in the final stages of evaporation.

A conventional Maxwell diffusion model of droplets evaporation suggests that the concentration of vapor molecules in the vicinity of the drop surface is always equal to saturated vapor concentration at the temperature of the ambient gaseous medium. Obviously, this assumption is highly simplifying and is not always realizable. First, the surface emission of vapor permanently prevails over the diffusion flow, which entrains vapor into the ambient medium. Second, the assumption does not take into account the drop temperature and its dynamics. The drop temperature at the beginning of evaporation decreases rapidly and then for a certain time period keeps an approximately constant value that is lower than in the medium. In this case, the concentration of vapor molecules near the surface is determined not only by the ambient temperature.

The mentioned contradiction in a conventional diffusion model of drops evaporation can be surmounted involving the gas-kinetic description of the vapor flow from the droplet surface. At that, the flow on the surface must be equal to the diffusion vapor flow at a large distance from the droplet. The actual diffusion flow of vapor starts forming at a distance of several lengths of free path of vapor molecules from the surface. So the equality of the flows can be achieved by introducing some intermediate vapor layer. The emission of vapor molecules can be defined in the framework of Hertz–Knudsen approach. In this case, the mass flow of vapor J_{HK} emitted by the droplet surface to the medium may be written as follows [8, 21]:

$$J_{HK} = \frac{\gamma S}{\sqrt{2\pi R_V}} \cdot \frac{P_S(T_d)}{\sqrt{T_d}} \cdot (1-\phi), \tag{1}$$

where S is the surface area of the droplet, T_d is the temperature of the droplet, γ is the coefficient of accommodation, R_V is the gas constant for vapor, P_S is the saturated vapor pressure, and φ is the relative humidity of the intermediate vapor layer near the drop surface. For calculations l we used the traditionally accepted value of accommodation coefficient for water vapor γ=0.04 [8,22].
The diffusion flow of the vapor mass J_D at a distance from a spherical droplet taking into account the Stefan flow [5] may be expressed:

$$J_D = -2\pi D c_\mu \mu R \cdot Sh \cdot \ln\left(\frac{1-\phi\dfrac{P_S(T_d)}{c_\mu RT_d}}{1-\phi_\infty\dfrac{P_S(T_a)}{c_\mu RT_a}}\right), \tag{2}$$

where D is the molecular diffusion coefficient of vapor in the medium, μ is the molar mass of vapor, c_μ is the molar concentration of the vapor-gas mixture, R is the universal gas constant, φ_∞ is the relative humidity of the medium at a great distance from drops, T_a is the ambient air temperature. The equality between the vapor flows J_{HK} and J_D allows determining φ for the intermediate layer. To calculate the temperature drop let us write the equation of change in internal drop energy, at that supposing that the inhomogeneity of the temperature field inside the drop is small, and the drop has a temperature T_d. In this case, the change in the internal drop energy at its temperature change may be expressed:

$$dE = mc \cdot dT_d, \tag{3}$$

where m is the mass of the drop, and c is the specific heat capacity of fluid. The internal energy of the drop changes due to heat exchange with the medium and its evaporation. The total heat flux during heat exchange with the medium is denoted by Q_1, and the heat absorption at evaporation, by Q_2. Heat flux Q_1 includes conductive, free-convective, and radiant heat fluxes:

$$Q_1 = \alpha S \cdot (T_a - T_d) + \varepsilon\sigma S \cdot (T_a^4 - T_d^4), \tag{4}$$

where α is the heat transfer coefficient taking into account conduction and free convection, ε is the emissivity factor of the drop surface, and σ is the Stefan-Boltzmann constant.
The heat transfer coefficient is determined as follows:

$$\alpha = \frac{Nu \cdot 2R}{\lambda_a}, \tag{5}$$

where λ_a is the heat conductivity of the ambient medium.
Nu and Sh are numbers taking into account conductive and convective heat and mass transfer [16]:

$$Nu = 2 + 0.60 \Pr^{1/3} Gr^{1/4}, \; Sh = 2 + 0.60 \Pr^{1/3} Gr^{1/4} \tag{6}$$

where Gr is the Grashof number, determined by the droplet diameter d:

$$Gr = \frac{gd^3(T_a - T_d)}{v^2 T_a} \tag{7}$$

Pr is the Prandtl number, taken to be Pr = 0.7.
The value Q_2 can be determined taking into account the enthalpy of the vapor, heated to a temperature of the medium:

$$Q_2 = L\cdot(1+B)\frac{dm}{dt}, \tag{8}$$

where L is the specific heat of vaporization, taken at the ambient temperature T_a , the number B is

$$B = \frac{c\cdot(T_a - T_d)}{L}, \tag{9}$$

In the end, the equation of change in internal drop energy is:

$$mc\frac{dT_d}{dt} = 2\pi\lambda_a R\cdot Nu\cdot(T_a - T_d) + L\cdot(1+B)\cdot\frac{dm}{dt} + 4\pi R^2\varepsilon\sigma\cdot(T_a^4 - T_d^4), \tag{10}$$

In addition to the equation of change in internal energy (10) the calculated system of equations includes the equation of change in the drop mass (11) and the equations of equality of the mass vapor flow of Hertz-Knudsen and the diffusion flow (12).

$$\frac{dm}{dt} = -\frac{\gamma S}{\sqrt{2\pi R_V}}\cdot\frac{P_S(T_d)}{\sqrt{T_d}}\cdot(1-\phi), \tag{11}$$

$$\frac{\gamma S}{\sqrt{2\pi R_V}}\cdot\frac{P_S(T_d)}{\sqrt{T_d}}\cdot(1-\phi) = -2\pi Dc_\mu\mu R\cdot Sh\cdot\ln\left(\frac{1-\phi\dfrac{P_S(T_d)}{c_\mu RT_d}}{1-\phi_\infty\dfrac{P_S(T_a)}{c_\mu RT_a}}\right), \tag{12}$$

The differential equations (10), (11) were solved numerically by the method of Runge-Kutta-Merson of the fifth order with correction of the calculation step in each iteration. The value of φ was determined from the nonlinear equation (12) using the Newton method. Based on the developed algorithm for solving the system of equations we implemented a computer program for calculating the evaporation of free liquid drops with the use of emission–diffusion model.

2. Analysis of calculation results

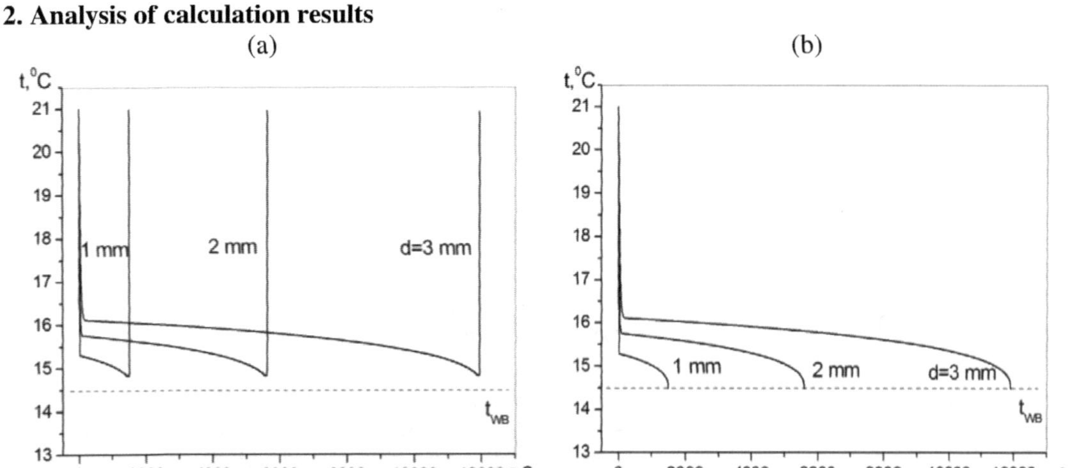

Figure 1. The temperature change of the large droplets at evaporation: (a) according to the emission-diffusion model, (b) according to the diffusion model.

Comparative calculations of evaporation of free spherical water drops of different diameters were performed with the use of both emission-diffusion model (EDM) and the conventional diffusion model (DM). All calculations were carried out at air temperature t_a = 21°C, relative humidity φ = 50% and initial droplet temperature t = 21°C. The calculation results of temperature change in large droplets with an initial diameter of 1mm, 2mm and 3 mm are shown in figure 1.

In the article, the conventional value, relative to which the drops are divided into small and large, is the initial diameter of about 25 microns. The choice of this value is due to the following. Diffusion model well describes the evaporation of drops at small Knudsen numbers, when the free path length of water molecules in the air is by 2-3 orders of magnitude smaller than the initial diameter of the drop. With decreasing diameter, this condition is violated and the diffusion model ceases to meet the necessary requirements. The present investigation and [8] show that in the studied range of temperatures and relative humidity it is at this diameter when the deviation of the drop evaporation time, determined by diffusion and emission-diffusion models, starts.

According to the results of calculations using the EDM (figure 1a) for the large droplets at the initial stage there was a sharp temperature decrease to that exceeding the wet bulb temperature t_{WB} by 1° - 1.5°C. After the initial cooling there was a long time period of relatively stable drop temperature. At the final stage of droplets evaporation there was a sharp increase of their temperature to ambient one.

Calculations of the evaporation time of large drops on DM (figure 1b) gave the close results to the calculations according to EDM. However, in contrast to them, at the final stage of evaporation the drop temperature did not increase to t_a. In the process of evaporation there was a gradual decrease of the drop temperature to the wet bulb temperature.

In contrast to the evaporation of large drops there were significant differences between the results of calculations according to EDM and DM for small drops. Figure 2 shows the results of calculations of temperature change of evaporating drops with an initial diameter of 5 μm, 10 μm and 20 μm.

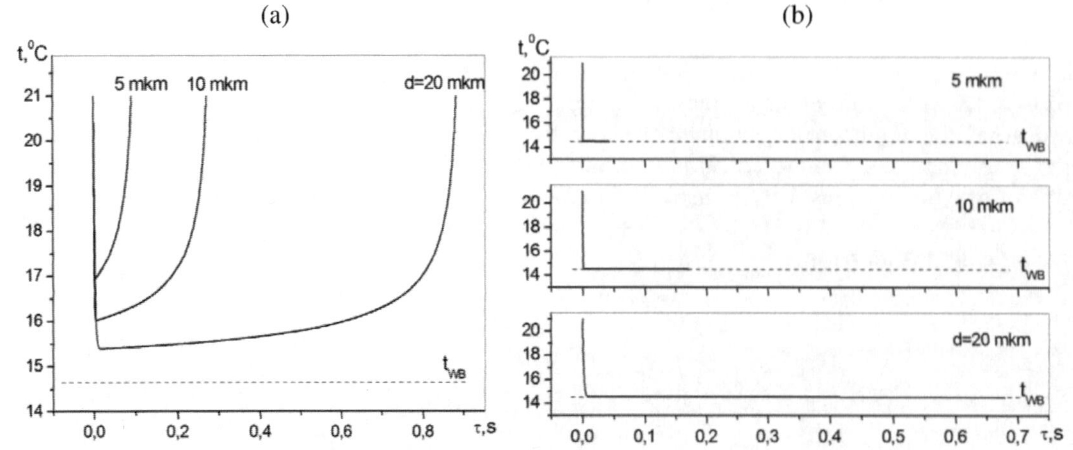

Figure 2. Temperature changes in small drops at evaporation: (a) according to emission-diffusion model, (b) according to diffusion model.

The calculation of the evaporation of drops was carried out using the two compared models under the same conditions. In all cases it was assumed that the drops have the same initial temperature. The drops were taken to be thermally uniform. The temperature field uniformity was assessed by Biot number Bi. For the conditions Bi = 0.09 – 0.13, when Bi<<1, the drops could be considered thermally uniform. Fig.2a shows the dependences of changes in the temperature of drops with micron dimensions in the evaporation process. It follows from the figure that the minimum temperature of drops cooling was 1-2 ^{0}C higher than the wet bulb temperature t_{WB}. After rapid cooling at the initial stage of evaporation, the temperature of drops gradually increased to ambient temperature. The minimum temperature of the drop cooling decreased with the increase in the initial diameter and approached the wet bulb temperature. The results of calculations in Fig.2b were obtained under the same conditions as for 2a, but using the diffusion model. It follows from the figure that at initial stage of evaporation the drops are cooled to the wet bulb temperature and keep it until their complete evaporation. Comparing Fig.2.a and Fig.2b it may be inferred that the emission-diffusion model gives a longer evaporation time than the diffusion model. The difference for the drop with d_0 =5 μm was more than 2 times.

3. Testing the computational model

For testing the proposed computational model a series of experimental studies of evaporation of water droplets of different diameters was performed. The experiments were conducted to measure the time of droplets evaporation and their temperature. Evaporation of free drops was simulated experimentally using suspended drops. Mono-fibers of different diameters, made of polypropylene with low thermal conductivity, were used as the suspension. Such experimental statement allowed minimizing the effect of the fiber and obtaining a good approximation to the conditions of free drops evaporation. The experiments were conducted at a constant temperature t_a = 27.5 ^{0}C and humidity φ = 27%. Absolute error of measurement of the ambient relative humidity did not exceed ± 3%, and of ambient temperature : ± 0.2 ^{0}C. Absolute measurement error for drop evaporation time did not exceed 2 seconds. Initial droplet diameters were measured by digital microphotography method with an accuracy of ± 10 μm. In the evaporation process, the temperature distribution on the drop surface was determined using infrared thermography [23, 24]. Temperature measurement was conducted with 5 seconds interval. In the experiments, we used the television camera NEC TH 7102WV, which operates in the spectral range 8÷14 μm and has a relative temperature accuracy 0.2 ^{0}C. The measurements were carried out with the use of micro lens TH 71-377 . As a result, the average temperatures of the surfaces of the evaporating water droplets with different diameters were obtained. Figure 3 presents the comparison of the obtained experimental data with the calculations results on the EDM.

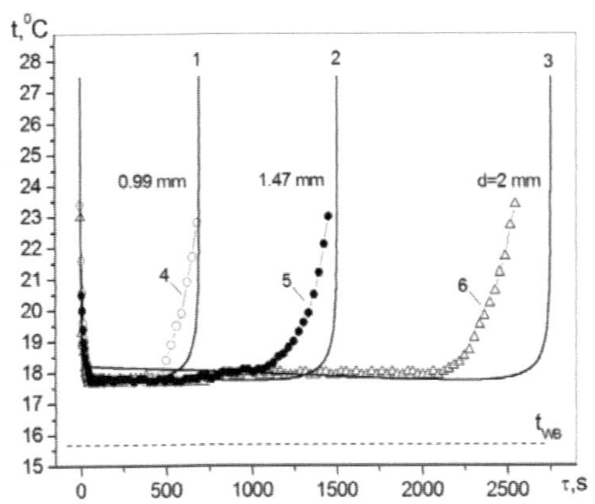

Figure 3. Comparison of the results of calculations and experiments on measuring temperature of the evaporating suspended droplets; calculations: 1- 0.99 mm, 2- 1.47 mm, 3- 2.0 mm; experiments: 4- 0.99 mm, 5- 1.47 mm, 6- 2.0 mm.

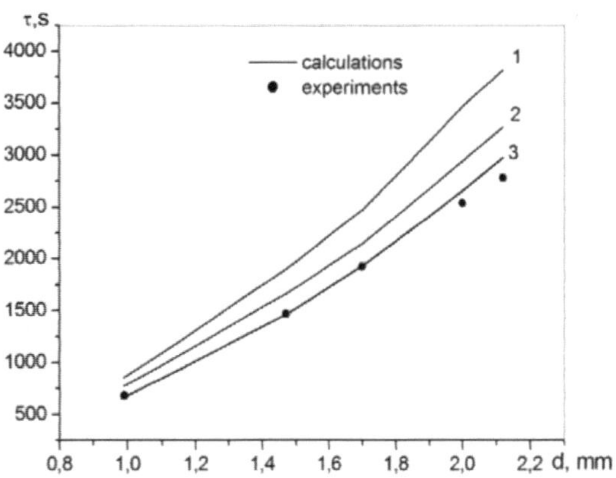

Figure.4. Comparison of results of calculations and experiments at the time of evaporation of the suspended drops. 1–without convection and supply of heat through the fiber; 2– with convection and without supply of heat through the fiber; 3– with convection and supply of heat through the fiber.

From figure 3 it follows that the calculation results are in good agreement with the experimental data at the initial stage of drop evaporation with a sharp decrease of temperature and at the stage with a slightly changed temperature of drops, exceeding t_{WB} by about 1.5°C. At the final stage of drop evaporation both the calculated and experimental data showed an increase in the drop temperature. However, the calculation shows a more dramatic increase in the temperature at the final stage and a longer total time of drop evaporation compared to the experiment. This was probably due to edge effects and deviation from the spherical shape of a drop at the end of evaporation.

In the process of drop evaporation microphotography was performed, which allowed recording the change of the drop shape and determining its geometrical dimensions. Using the proposed EDM the evaporation time of droplets of different diameter was calculated. Figure 4 shows the comparison of calculation results with experimental data.

The results of calculations demonstrate good agreement with experimental data taking into account free-convective heat exchange of the droplet with ambient air and additional supply of heat flux to the droplet along the polypropylene fiber.

Conclusions

The authors of the work have proposed and tested (in calculations of free spherical water droplets) the emission-diffusion model of evaporation, based on a dynamic equilibrium between the vapor flow, obtained in the kinetic approach of Hertz-Knudsen, and the diffusion flow of Maxwell. The proposed model allows calculating evaporation of droplets of various sizes.

The results for the time of evaporation of large water drops calculated using the emission-diffusion model are in good agreement with the results of calculations on the conventional diffusion model. For small water droplets, the results of calculations on the emission-diffusion model have given significantly longer evaporation times and higher minimum temperatures of droplets at evaporation than those obtained on the diffusion model.

Experiments with evaporation of the suspended drops have showed that the given free-convective heat transfer of drops and additional heat flux to the drops over the suspended fibers, the calculation results for the droplet evaporation times performed on the emission-diffusion model and their temperature during evaporation well coincided with experimental results.

Acknowledgments

The work was financially supported by the Russian Foundation for Basic Research (grant No. 15-08-00687) and by the grant of the RF President (grant No. NSh-8780.2016.8).

References

[1] Gan Y and Qiao L 2011 Evaporation characteristics of fuel droplets with the addition of nanoparticles under natural and forced convections *Int. J. Heat Mass Transf.* **54** 4913–22

[2] Nazarov A D, Serov A F and Terekhov V I 2014 The influence of gas coflow in a pulse aerosol on evaporation cooling process *High Temperature* **52** № 4 576-579

[3] Lebedev-Stepanov P V, Molchanov S P, Alfimov M V, Kadushnikov R M, Rubin N I and Shturkin N A 2011 Simulation of self-assembly of micro- and nanoparticles in an evaporating microdrop of solution *Nanotechnologies in Russia* **6** №1-2 79-87

[4] Spalding D B 1956 *Some Fundamentals of Combustion* (London: Pergamon) p 148

[5] Fuchs N A 1959 *Evaporation and Droplet Growth in Gaseous Media* (London: Pergamon)

[6] Seinfeld J H and Pandis S N 2006 *Atmospheric Chemistry and Physics: From Air Pollution to Climate* (New York: J. Wiley)

[7] Pruppacher H R and Klett J D 2010 *Microphysics of Clouds and Precipitation* (New York: Springer)

[8] Kozyrev A V and Sitnikov A G 2001 Evaporation of a spherical droplet in a moderate-pressure gas" *Phys. Usp.* **44** 725–733

[9] Kristyadi T, Depridurand V, Castanet G, Lemoine F, Sazhin S S, Elwardany A, Sazhina E M and Heikal M R 2010 Monodisperse monocomponent fuel droplet heating and evaporation, *Fuel* **89** 3995–4001

[10] Zhang H Q and Law C K 2008 Effects of temporally varying liquid-phase mass diffusivity in multicomponent droplet gasification *Combust. Flame* **153** 593–602
[11] Barash L Yu, Bigioni T P, Vinokur V M and Shchur L N 2009 Evaporation and fluid dynamics of a sessile drop of capillary size *Phys. Rev.* E **79** 046301
[12] Watanabe H, Suzuki Y, Harada T, Matsushita Y, Aoki H and Miura T 2010 An experimental investigation of the breakup characteristics of secondary atomization of emulsified fuel droplet *Energy* **35** 806–813
[13] Yang J R and Wong S C 2002 An experimental and theoretical study of the effects of heat conduction through the support fiber on the evaporation of a droplet in a weakly convective flow *Int. J. Heat Mass Transf.* **45** 4589–98
[14] Labuntsov D A 2000 the Physical basis of energy. Selected works on heat transfer, fluid dynamics, thermodynamics (Moscow: Moscow State University Press) p 388
[15] Ghassemi H, Baek S W and Khan Q S 2006 Experimental study on binary droplet evaporation at elevated pressures and temperatures *Combust.Sci. Technol.* **178** 1031–53
[16] Han K, Song G, Ma X and Yang B 2016 An experimental and theoretical study of the effect of suspended thermocouple on the single droplet evaporation *Appl. Therm. Eng.*, **101** 568–575
[17] Brutin D,·Zhu Z-Q,·Rahli O,· Xie J C,· Liu Q-S and·Lounes Tadrist 2010 Evaporation of Ethanol Drops on a Heated Substrate Under Microgravity Conditions *Microgravity Sci.Technol.* **22** 387–395
[18] Terekhov V I, Terekhov V V, Shishkin N E and Bi K Ch 2010 Heat and mass transfer in disperse and porous media experimental and numerical investigations of nonstationary evaporation of liquid droplets *J. of Engineering Phys. and Thermophys.* **83** № 5 883-890
[19] Yakhno T A, Sanina O A, Sanin A G, Yakhno V G and Volovik M G 2012 Thermographic investigation of the temperature field dynamics at the liquid-air interface in drops of water solutions drying on a glass substrate *Techn. Phys. The Rus. J. of Appl. Phys.* **57** № 7 915-22
[20] Nizovtsev M I, Sterlyagov A N and Shlupikov M Yu 2015 *Modern science: Collection of research papers* **1(15)** 101-105
[21] Xisheng Luo, Yu Fan, Fenghua Qin, Huaqiao Gui, Jianguo Liu 2016 A new model for the processes of droplet condensation and evaporation on solid surface *International Journal of Heat and Mass Transfer.* **100** 208–214.
[22] Knake O and Stranskiy I N 1956 Evaporation mechanism *Progress in Metal Physics* **6** 181–235
[23] Gatapova E Ya, Semenov A A, Zaitsev D V, Kabov O A 2014 Evaporation of a sessile water drop on a heated surface with controlled wettability *Colloids and Surfaces A: Physicochem.Eng.Aspects* **441** 776–785
[24] Kevin Gleason, Harish Voota, Shawn A Putnam 2016 Steady-state droplet evaporation: Contact angle influence on the evaporation efficiency *International Journal of Heat and Mass Transfer.* **101** 418–426.

Critical heat flux in locally heated liquid film moving under the action of gas flow in a mini-channel

E M Tkachenko[1,2], D V Zaitsev[1], E V Orlik[1] and O A Kabov[1,3]

[1]Kutateladze Institute of Thermophysics SB RAS, 1 Lavrentyev Ave., Novosibirsk 630090, Russia
[2]Novosibirsk State University, 2 Pirogova Str., Novosibirsk 630090, Russia
[3]Tomsk Polytechnic University, 30 Lenin Ave., Tomsk 634050, Russia

Corresponding author: egor_tkachenko@mail.ru

Abstract. Thin and ultra thin liquid films driven by a forced gas/vapor flow (stratified or annular flows), i.e. shear-driven liquid films in a narrow channel, is one of the promising candidate for the thermal management of advanced semiconductor devices with high local heat release. In experiments performed in this paper with locally heated shear-driven liquid films of water the effect of various conditions, such as flow rates of liquid and gas and channel height, on critical heat flux (CHF) was investigated. In experiments the record value of CHF as high as 540 W/cm^2 has been achieved. The heat spreading into the substrate and the heat loses into the atmosphere in total don't exceed 30% at heat fluxes higher than 200 W/cm^2. Comparison of shear-driven liquid films and gravity-driven liquid films showed that CHF in shear-driven films up to 10 times higher than in gravity-driven liquid films. Thus, prospect of using shear-driven films of water in modern cooling systems of semiconductor devices was confirmed.

1. Introduction

The development of modern microelectronic equipment requires efficient cooling systems. Increasing performance demands have resulted in greater non-uniformity of on-chip power dissipation, creating localized, "hot spots," often exceeding 1 kW/cm^2 in heat flux, which can degrade the processor performance and reliability [1]. Ultra thin liquid films driven by a forced gas/vapor flow in the mini-channel are promising for the use in thermal control systems of modern semiconductor devices [2]. Works [3-5] are devoted to analytical and numerical study of hydrodynamics and heat and mass transfer during joint motion of intensively evaporating liquid film and gas flow in minichannels. Authors of experimental works [6-8] established the basic laws of the flow and crisis phenomena in the liquid film moving under the action of the gas flow in the horizontal channel. First experiments with locally heated shear-driven liquid films have revealed that contrary to gravity-driven liquid films, shear-driven films are less likely to rupture. This provides a way to prevent and control hot dry patch formation by the shear stress induced by the gas flow. Disruption of the liquid film and formation of dry patches is important for the understanding of crisis phenomena in liquid films [9-11]. The main goal of the present work is to study rupture and CHF in locally heated shear-driven liquid films at heat fluxes higher than 500 W/cm^2. Also in this paper the data on heat loses into the atmosphere and heat spreading into the substrate are presented.

2. Experimental setup

Schematic of the test section is presented in figure 1. The main part of the test section is a thin stainless steel plate with a flush-mounted copper rod. At the working surface the rod has a 1×1 cm square head emulating surface of a computer chip. The rod is heated by a heating spiral coiled around its lower part. Such construction of the heater provides the condition at the heater surface T = const (which is confirmed by thermocouples measurements). The test section is covered with a transparent cover made of optical glass so that a flat channel is formed. We used six different channel heights, namely: 0.25, 0.5, 1, 1.2, 1.5 and 2.0 mm. Gas, pumped by a compressor, flows through the channel and passes to the atmosphere at the end of the test section. The liquid is supplied from a thermostat, gets into the channel through a liquid nozzle and flows under the friction of the gas along the stainless steel plate as a film. The liquid is accumulating at the bottom part of the test section and is returned into the thermostat. Distance from the gas inlet to the liquid nozzle is 57 mm and distance from the liquid nozzle to the heater is 32 mm. This provides steady flows of gas and liquid at the moment they reach the heater. Channel width is 40 mm. Several thermocouples are embedded in the stainless steel plate and in the copper rod, allowing determination of the working surface temperature. The temperature of the heater surface is calculated taking into account the depth at which the thermocouples are embedded (2 mm). All the thermocouples are individually calibrated to an accuracy of 0.1°C. The test section is oriented horizontally. When investigating liquid films falling under the action of gravity the test section is set vertically while the glass cover is removed.

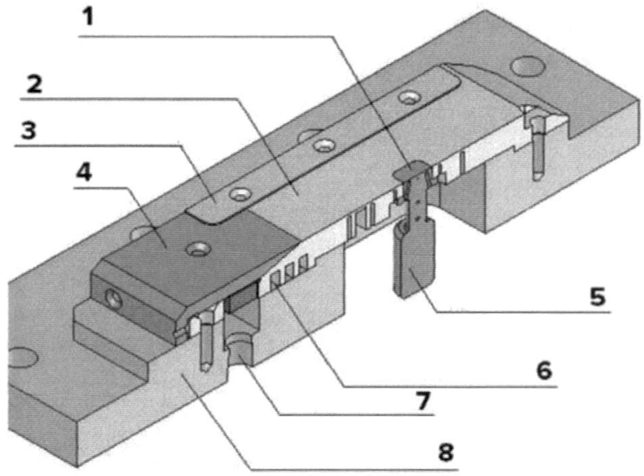

Figure 1. Design of test section: 1- source of the local heating; 2- stainless steel plate; 3- channel height control; 4- knife; 5- copper rod; 6- thermostabilizer; 7- liquid inlet; 8- textollite substrate.

Distilled water with initial temperature of 24°C is used as the working liquid. Air with temperature of 24-27 °C and relative humidity of 15-30% is used as the working gas. The experiments are carried out under stationary conditions. Superficial gas velocity (volumetric gas flow rate divided by cross-sectional area of the channel) U_{Sg} varies from 0.1 to 40 m/s.

The heat flux is determined by the electrical power dissipated on the heating spiral. Figure 2 presents data on the heat loses into the atmosphere and heat spreading into the steel plate. Thermal conductivity of copper is 400 W/mK which is almost 30 times higher than that of stainless steel (15 W/mK). This provides moderate heat spreading from the heater to the stainless steel plate: about 15% at q>200 W/cm^2, about 20% at q=100 W/cm^2 and up to 30% for smaller heat fluxes (according to estimation using measurements of thermocouples embedded into the stainless steel plate) (figure 2). Heat spreading into the atmosphere has been calculated as the ratio of the heat flux between two thermocouples embedded into the copper rod (estimated by measurements of thermocouples) and the total heat flux determined by the electrical power. In order to minimize heat spreading from the heating spiral into the atmosphere, the heating spiral is wrapped with a layer of heat insulation material (mineral wool). Heat loses into the atmosphere don't exceed 15-20% (figure 2). Thus, the heat loses into the atmosphere and heat spreading into the steel plate in total don't exceed 30% at heat fluxes higher than 200 W/cm^2. All data below are presented without taking into account the heat losses and heat spreading.

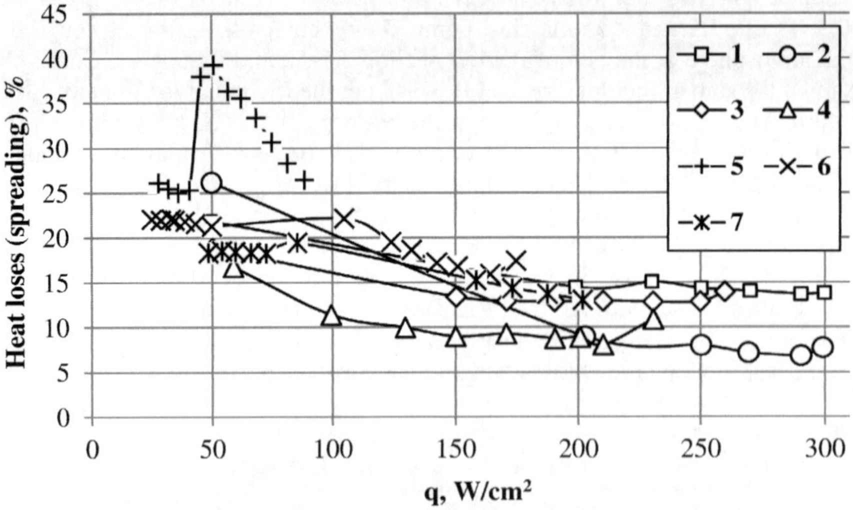

Figure 2. The heat loses into the atmosphere at $Re_l = 19$, H = 1.5 mm: 1 – $U_{Sg} = 31$ m/s; 2 – $U_{Sg} = 28$ m/s; 3 – $U_{Sg} = 14$ m/s; 4 – $U_{Sg} = 6$ m/s. The heat spreading into the steel plate at $U_{Sg} = 11$ m/s, H = 2 mm: 5 – $Re_l = 15$; 6 – $Re_l = 22$; 7 – $Re_l = 32$.

Figure 3 shows the distributions of the working surface temperature at the different heat fluxes using the measurements of thermocouples embedded into the test section. From the graph, it is seen that the construction of the test section provides a local heating.

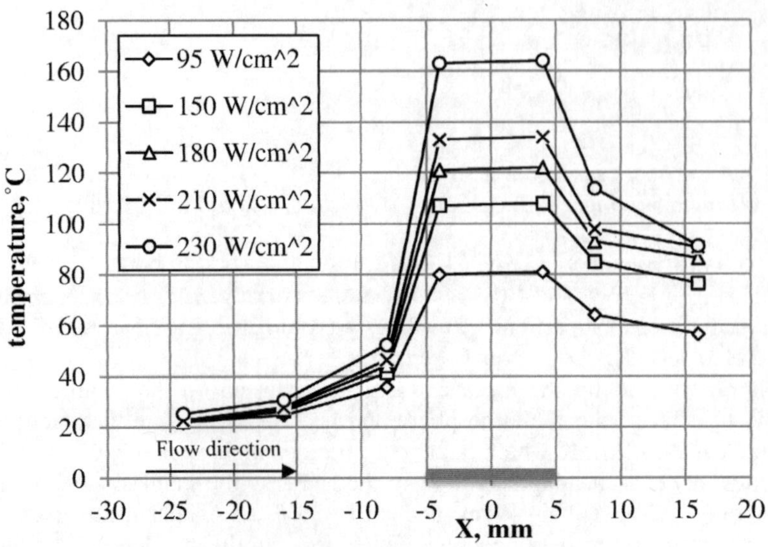

Figure 3. The distribution of the working surface temperature for different heat fluxes at $U_{Sg} = 5$ m/s and $Re_l = 18$, H = 0,25 mm. Red line shows the position of the heater.

3. Film rupture and CHF under local heating

Behaviour of the liquid film under local heating was investigated at H = 0.25, 0.5, 1, 1.2, 1.5 and 2 mm, in the ranges of q = 0-540 W/cm^2, Re_l = 8.5-50, Re_g = 350-2600, U_{Sg} = 3.8-40 m/s. Figure 4 illustrates the process of film disruption with increase of the heat flux at H = 0.5 mm, Re_l = 18, Re_g = 299 (U_{Sg} = 9.3 m/s). At a certain threshold heat flux q_{idp} the film ruptures with formation of a dry patch on the substrate. As usual, first dry patches initiate downstream of the heater along its lateral edges, as depicted in figure 4a. We deal with thermocapillary film rupture. Thermocapillary tangential stresses at the film surface, induced by the dependence of the liquid surface tension on the temperature, and tangential stresses, induced by the flowing gas, are the main competitive forces [12]. The film is thinnest along the lateral edges of the heater where temperature gradient at the film surface attains its maximum values. Film rupture is most likely to occur in this area. At a slightly higher heat flux $q_{idp,H}$ dry patches reach the heater (for smaller Re_g and smaller Re_l $q_{idp,H} = q_{idp}$). With further increase of the heat flux the heater is covered with metastable, intensively evaporating thin liquid film, with quickly emerging and disappearing small dry patches, figure 4b. And finally at the critical heat flux q_{cr}, the heater is suddenly dried out and its temperature starts to rise rapidly, figure 4c.

a) q = 100 W/cm^2, T_w = 83°C b) q = 250 W/cm^2, T_w = 140°C c) q = q_{cr} = 310 W/cm^2, crisis

Figure 4. Film rupture and crisis in shear-driven water film, heater 1x1 cm, H=0.5 mm, Re_l = 18, Re_g = 299 (U_{Sg} = 9.3 m/s), flow directed from top to bottom.

Figure 5 shows the effect of superficial gas velocity on the threshold heat flux for film rupture and on CHF, at Re_l = 14, and also the data for falling liquid film are presented (at Re_l = 14 $q_{idp} = q_{idp,H} = q_{cr}$). It is seen that at relatively small U_{Sg}, $q_{idp,H}$ and q_{cr} for shear-driven liquid film are close to those for falling liquid film. However at higher gas velocities, $q_{idp,H}$ for shear-driven liquid film is up to 3 times higher, while q_{cr}, almost linearly growing with U_{Sg}, at U_{Sg}=34 m/s is more than 10 times higher and reaches 250 W/cm^2. For shear-driven liquid film at increase of gas velocity, inertial force starts to dominate acting against thermocapillary forces, which provides more uniform liquid distribution over the heater as compared to gravity-driven liquid film. As a result, q_{cr} can be several times higher than q_{idp} and $q_{idp,H}$.

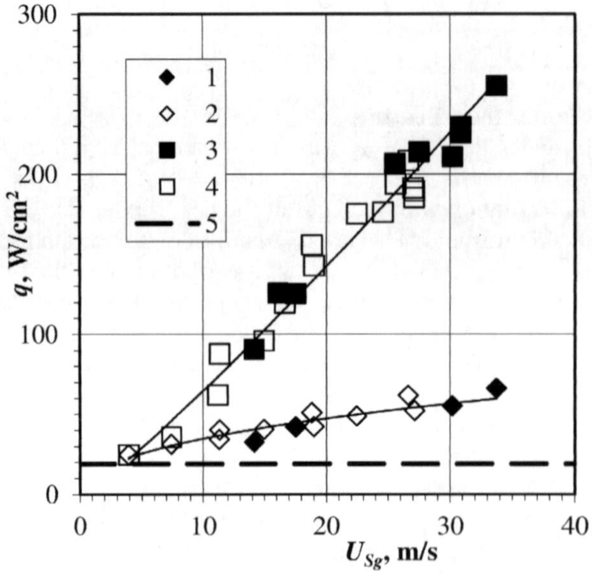

Figure 5. Effect of superficial gas velocity on film rupture and CHF, water-air flow, $Re_l = 14$. 1 – $q_{idp,H}$ at H = 1.2 mm; 2 – $q_{idp,H}$ at H = 1.5 mm; 3 – q_{cr} at H = 1.2 mm; 4 – q_{cr} at H = 1.5 mm; 5 – $q_{idp} = q_{idp,H} = q_{cr}$ for a falling liquid film. Curves, generalizing data for q_{idp}, $q_{idp,H}$ and q_{cr} are plotted.

Figure 6 shows the effect of gas velocity on CHF for a shear-driven liquid film for different Re_l and different H. It is seen that CHF grows with increase of both U_{Sg} and Re_l. In experiments the record value of CHF = 540 W/cm² was reached (point $Re_l = 50$, H=1 mm, U_{Sg} = 37 m/s). Unfortunately, the construction of the heater did not allow to achieve higher heat flux.

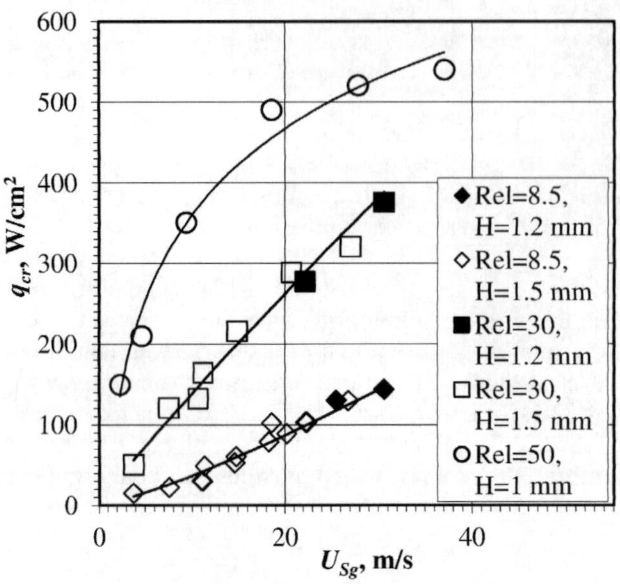

Figure 6. Effect of gas velocity on CHF for different liquid flow rates and different channel heights, water-air flow, horizontal channel, heater 1x1 cm. Curves, generalizing data are plotted for each Re_l.

4. Conclusions

From figures 5 and 6 it is seen that the channel height has no or little effect on CHF (within the range H=1.2-2.0 mm). For shear-driven water films the critical heat flux was found to be up to 10 times higher than that for a falling liquid film, and reaches up to 540 W/cm^2, which makes shear driven films (stratified and annular two-phase flow) very promising for high heat flux cooling applications. The heat loses into the atmosphere and heat spreading into the substrate in total don't exceed 30% at heat fluxes higher than 200 W/cm^2.

Acknowledgments

This work was supported by the Ministry of Education and Science of Russia (Agreement No 14.604.21.0053, project identifier RFMEFI60414X0053).

References

[1] Bar-Cohen A and Wang P 2012 Journal of Heat Transfer 134 (5) 051017-28
[2] Kabov O A, Lyulin Yu V, Marchuk I V and Zaitsev D V 2007 International Journal of Heat and Fluid Flow 28 103-112
[3] Gatapova E Ya and Kabov O A 2008 International Journal of Heat and Mass Transfer 51 (19) 4797-4810
[4] Goncharova O N and Kabov O A 2010 International Journal of Heat and Mass Transfer 53 (13-14) 2795-2807
[5] Kabova Y, Kabov O, Gambaryan-Roisman T, Stephan P and Kuznetsov V V 2014 International Journal of Heat and Mass Transfer 68 527-541
[6] Zaitsev D V, Rodionov D A and Kabov O A 2009 Technical Physics Letters 35 (7) 680-682
[7] Kabov O A and Zaitsev D V 2009 Multiphase Science and Technology 21 (3) 249-266
[8] Kabov O A, Chinnov E A and V Cheverda 2007 Microgravity Science and Technology 19 (3-4) 44-47
[9] Ajaev V S 2013 Interfacial Phenomena and Heat Transfer, 1 (1) 81-92
[10] Peles Y 2012 Contemporary Perspectives on Flow Boiling Instabilities in Microchannels and Minichannels (New York: Begell House) p 152
[11] Kadoura M and Chandra S 2013 Experiments in Fluids 54 (2) 1-11
[12] Chinnov E A and Kabov O A 2007 Microgravity Science and Technology. 19 (3-4) 18-22

Journal of Physics: Conference Series **754** (2016) 032020 doi:10.1088/1742-6596/754/3/032020

Simulating nonlinear steady-state traveling waves on the falling liquid film entrained by a gas flow

O Yu Tsvelodub[1, 2]

[1] Kutateladze Institute of Thermophysics SB RAS, Lavrentyev prosp. 1, Novosibirsk, 630090, Russia
[2] Novosibirsk State University, Pirogova Str. 2, Novosibirsk, 630090, Russia
E-mail: tsvel@itp.nsc.ru

Abstract. The article is devoted to the simulation of nonlinear waves on a liquid film flowing under gravity in the known stress field at the interface. In the case of small Reynolds numbers the problem is reduced to the consideration of solutions of the nonlinear integral-differential equation for film thickness deviation from the undisturbed level. Weakly nonlinear steady-state traveling solutions of the equation with wave numbers in a vicinity of neutral wave numbers are constructed analytically. The nature of the wave branching from the undisturbed solution is investigated. Steady-state traveling solutions, whose wave numbers within the instability area are far from neutral wave numbers, are found numerically.

1. Introduction and problem statement

Simultaneous liquid and gas flow is a classical problem of hydrodynamics. The solution to this problem in a full statement is associated with considerable computational difficulties. So, two stages of simulation are often distinguished: determining gas stresses on the film surface and further calculating the evolution of waves in the liquid. The liquid velocity is much smaller than the characteristic gas velocity, so the boundary surface is supposed to be rigid and stationary. In addition, due to the smallness of film thickness, the influence of interfacial perturbations on gas velocity may be considered as linear. As a result, the problem of computing the normal and shear stresses of gas on the surface is reduced to considering the impact of individual spatial harmonics. At the second stage of simultaneous flow study, the dynamics of nonlinear waves on the liquid film surface is examined.

This paper deals with the second stage of studies, namely simulating the dynamics of nonlinear waves on a liquid film, flowing under the action of gravity in the known stress field at the interface. The gas flow is turbulent and occurs in a vertical channel. Full statement of the problem for liquid includes Navier–Stokes and continuity equations with respective kinematic and dynamic boundary conditions.

If function $y = h(x,t)$ determining the position of the area boundary is single-valued there is a continuously differentiable coordinate transformation that transforms the area of the fluid flow into the band of constant thickness.

$$x = x, \qquad \eta = y / h(x,t), \qquad t = t. \tag{1.1}$$

In the paper [1] the system of hydrodynamic equations was presented in tensor form invariant under coordinate systems. Its use enabled obtaining a model system of equations for the considered flow to describe the evolution of long-wave perturbations of the interface at moderate Reynolds numbers.

In the case of low flow rates (Re ~ 1), due to the film thickness smallness compared to the wavelength, the solution of the system is represented as a series on small parameter. As a result, restricting the expansion to the first two terms, one evolutionary equation for the film thickness h was obtained:

$$h_t + \frac{\mathrm{Re}}{\mathrm{Fr}} h^2 h_x + \mathrm{Re}\,\tau_0 h h_x$$

$$+\varepsilon\frac{\partial}{\partial x}\left(\frac{1}{3}\varepsilon \mathrm{ReW}h^3h_{xxx}+\frac{2}{15}\frac{\mathrm{Re}^3}{\mathrm{Fr}^2}h^5h_x(h+\tau_0\mathrm{Fr})+\frac{1}{2}\mathrm{Re}h^2\tau_0\int\hat{h}_k k\tau(k)e^{ikx}\mathrm{d}k\right)=0. \qquad (1.2)$$

Here dimensionless parameters are introduced: Reynolds number $\mathrm{Re}=\rho h_0u_0/\mu$, Froude number $\mathrm{Fr}=u_0^2/gh_0$, parameter $\mathrm{W}=\sigma/\rho l_0u_0^2$, and $\varepsilon=h_0/l_0$ is relation of specific film thickness h_0 to specific wave length l_0. Direction of coordinate x coincides with the direction of gravity vector. In addition, in the equation (1.2) and in the dimensionless complexes we use the characteristic scales of velocity u_0 and time l_0/u_0. Here p is the pressure, σ is the surface tension, ρ is the density, μ is the dynamic viscosity of liquid, g is the acceleration of gravity, τ_0 is undisturbed component of gas shear stress on the film surface, $\tau(k)=\tau_r(k)+i\tau_{im}(k)$ are the Fourier components of tangential stresses of gas due to the boundary curvilinearity, and $\hat{h}(k,t)$ are the Fourier components of the surface form expansion:

$$\hat{h}(k,t)=\frac{1}{2\pi}\int_{-\infty}^{+\infty}h(x,t)e^{-ikx}\mathrm{d}x.$$

We should emphasize that the approximation of small Reynolds numbers (Re ~ 1) is used to derive the equation (1.2), and it is assumed that the Weber number is large: $\mathrm{W}\varepsilon\sim 1$. In the case of spatial periodic solutions of equation (1.2), the integral term is replaced by the respective Fourier series.

Restricting ourselves to perturbations of small but finite amplitude, introducing slow and fast times into consideration and using transformation $h=1+\varepsilon h_1$, $t_0=t$, $t_1=\varepsilon t$, from equation (1.2) we obtain:

$$\frac{\partial h_1}{\partial t_0}+\frac{\mathrm{Re}}{\mathrm{Fr}}\left(1+\mathrm{Fr}\tau_0\right)\frac{\partial h_1}{\partial x}=0, \qquad (1.3)$$

$$\begin{aligned}&\frac{\partial h_1}{\partial t_1}+\frac{\mathrm{Re}}{\mathrm{Fr}}\left(2+\mathrm{Fr}\tau_0\right)h_1\frac{\partial h_1}{\partial x}+\frac{\mathrm{W\,Re}\,\varepsilon}{3}\frac{\partial^4 h_1}{\partial x^4}\\&+\frac{2}{15}\frac{\mathrm{Re}^3}{\mathrm{Fr}^2}\frac{\partial^2 h_1}{\partial x^2}(1+\tau_0\mathrm{Fr})+\frac{1}{2}\mathrm{Re}\tau_0\int i\hat{h}_{1k}k^2\tau(k)e^{ikx}\mathrm{d}k=0.\end{aligned} \qquad (1.4)$$

Equation (1.3) implies that in the first approximation (fast times), perturbations of small but finite amplitude propagate with a characteristic constant velocity $c_0=\mathrm{Re}/\mathrm{Fr}\left(1+\mathrm{Fr}\tau_0\right)$. In this approximation, the motion occurs without changing the initial form of perturbations.

Equation (1.4) describes the nonlinear evolution of perturbations on large (slow) times. The characteristic longitudinal scale l_0 is defined so that the coefficients for the second and fourth derivatives in equation (1.4) are the same. This implies that the ratio for ε will take the form: $\varepsilon=2\mathrm{Re}^2/5\mathrm{W\,Fr}^2(1+\mathrm{Fr}\tau_0)$. Considering this choice after replacement $t=bt_1$, $h_1=AH$, $b=\mathrm{WRe}\varepsilon/3$, $A=2\mathrm{Fr}b/\left(\mathrm{Re}(2-\mathrm{Fr}\tau_0)\right)$, the equation (1.4) will be rewritten in the form:

$$\frac{\partial H}{\partial t}+2H\frac{\partial H}{\partial x}+\frac{\partial^2 H}{\partial x^2}+\frac{\partial^4 H}{\partial x^4}+B\int_{-\infty}^{\infty}ik^2\,\tau(k)\,\hat{H}(k,t)e^{ikx}\mathrm{d}k=0. \qquad (1.5)$$

Here $B=\dfrac{\mathrm{Re}\,\tau_0}{2b}\equiv\dfrac{3\tau_0}{2\mathrm{W}\,\varepsilon}$.

Thus, in the case of small Reynolds numbers, the problem of studying perturbations on the surface of the liquid film, flowing under gravity in the known stress field at the interface, is reduced to the analysis of one nonlinear integral-differential equation.

Equation (1.5) is an interesting example of model equations arising in the study of the evolution of perturbations in active-dissipative media. The instability of linear perturbations is determined by its terms with the second derivative and the term containing the integral (conditioned by stress perturbation at the interface of film–gas). Dissipation is provided by the fourth derivative, simulating capillary effects. Indeed, neglecting a nonlinear term in (1.5) and presenting its solution as $H \sim \exp(ik(x-ct))$, we obtain the following relation for dispersion:

$$c \equiv c_r + ic_i = i(k-k^3) + Bk\tau(k). \qquad (1.6)$$

Disturbances will be unstable if the imaginary part of the phase velocity c_i exceeds zero.

As the second term in the right part of (1.6), responsible for the perturbation stability, decreases faster than, for example, the first one, then the unstable are the long-wave perturbations. Their wave numbers are smaller than the neutral wave number k_n satisfying the equation:

$$1-k_n^2 + B\tau_{im}(k_n) = 0. \qquad (1.7)$$

As it can be seen from (1.7), the above described choice of the characteristic longitudinal scale l_0 implies that the neutral wave number $k_n = 1$ in the case of the freely falling film ($B = 0$). We will choose the parameters of the undisturbed flow, so that the neutral wave number k_n different from the unit. At the same time demand that this value k_n corresponds to a definite value τ_{im}. The value B corresponding to this situation is determined from (1.7):

$$B = \frac{k_n^2 - 1}{\tau_{im}(k_n)}.$$

2. Results of nonlinear wave calculation on the model equation

For finding solutions of equation (1.5) periodic on x the function H is presented as a spatial Fourier series:

$$H(x,t) = \sum_n H_n(t)\exp(iknx). \qquad (2.1)$$

Since H is real function, then $\bar{H}_{-n} = H_n$. The bar indicates the operation of complex conjugation. Steady-state traveling solutions, whose wave numbers within the instability area are far from neutral wave numbers, are found numerically.

Having substituted (2.1) into equation (1.5) we obtain an infinite system of ordinary differential equations for Fourier harmonics $H_n(t)$. Supposing that all $H_n(t)$ with indices $|n| \geq N$ are equal to zero, obtain its finite-dimensional counterpart.

At the points with neutral wave numbers k_n the steady-state traveling linear solutions bifurcate from the trivial solution $H = 0$. As it is seen from (1.6), they have the following phase velocity and frequency, respectively:

$$c_0 = Bk_n\tau_r(k_n), \qquad \omega_0 \equiv k_n c_0 = Bk_n^2\tau_r(k_n).$$

In the vicinity of the neutral wave number k_n, the solutions are represented in the form of a series (2.1). At the same time, we believe that

$$H_n = \delta^{|n|}\tilde{H}_n,$$
$$k = k_n + \delta^2 k_2 + \cdots \qquad (2.2)$$

and introduce the set of different times:

$$t_n = \delta^n t, \qquad n = 0, 1, 2, \cdots. \qquad (2.3)$$

Here δ is a small parameter.

Using the representations (2.2), (2.3) and substituting the series (2.1) into the equation (1.5), we collect the coefficients of the same degrees of parameter δ and equate them to zero. As a result, after a rather simple but cumbersome calculations we obtain:

$$H = A\exp[i(kx - \omega t)] + A^2 A_{H2} \exp[2i(kx - \omega t)] + C.C. \qquad (2.4)$$

Here $A = \delta|\tilde{H}_1|$, $k = k_n + A_k A^2$, $\omega = \omega_0 + A_\omega A^2$, $C.C.$ is the complex-conjugate expression.

From (2.4) it is clear that in the solution it is possible to take δ equal to 1, and to use the amplitude of the first harmonic A as the small parameter. Coefficients A_{H2}, A_k, A_ω depend only on the values $\tau(k_n)$, $d\tau(k)/dk|_{k_n}$, $\tau(2k_n)$. Due to the bulkiness their explicit forms are not presented here. The phase velocity with an accuracy of A^2 is:

$$c \equiv \frac{\omega}{k} = \frac{\omega_0 + A_\omega A^2}{k_n + A_k A^2} = c_0 + \frac{A^2}{k_n}(A_\omega - c_0 A_k).$$

Calculating the below presented results, we used data on stress pulsations obtained by the model of Benjamin in the work [1].

Figures 1and 2 show the dependences of the coefficients A_k, A_ω on the parameter B for $k_n = 1.1$ and $k_n = 1.5$, respectively. As can be seen from these figures, the type of branching here is soft, i.e. the correction to the wave number is negative. In other words, the wave numbers of the steady-state traveling regimes of small but finite amplitude are in the range of linear instability.

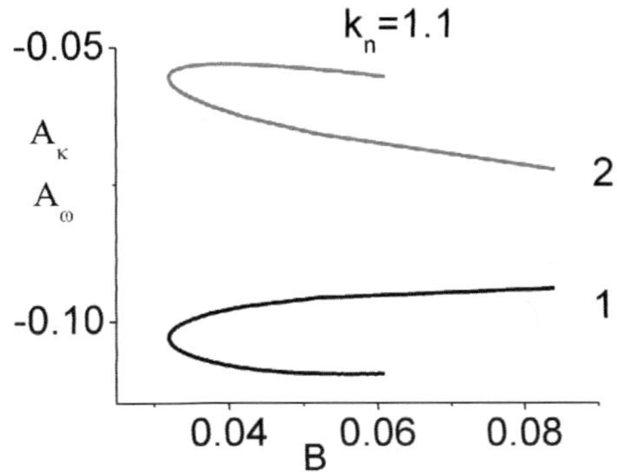

Figure 1. Dependencies of the coefficients A_k (curve 1) and A_ω (curve 2) on the parameter B. Neutral wave number $k_n = 1.1$.

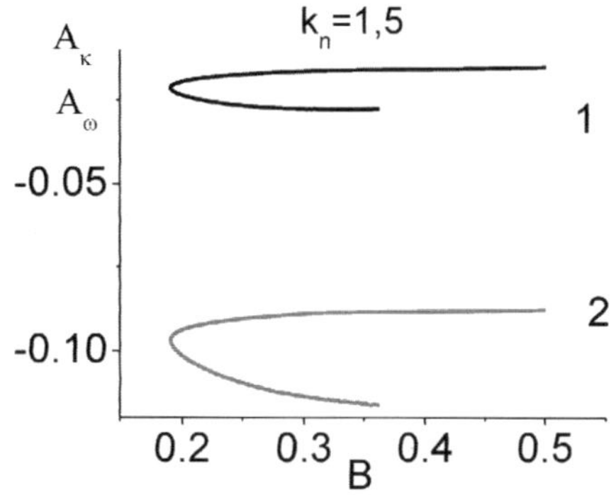

Figure 2. Dependencies of the coefficients A_k (curve 1) and A_ω (curve 2) on the parameter B. Neutral wave number $k_n = 1.5$.

To obtain the steady-state traveling solutions of this family (the first family) with wave numbers lying beyond the vicinity of the neutral wave number k_n, the problem was solved numerically. At truncating series (2.1) for values of N from 8 to 15. The evolution of the periodic perturbations was considered. The calculations show that in the region of wave numbers, where there is only the first family of steady-state traveling solutions, all initial perturbations evolve to the solution of this family.

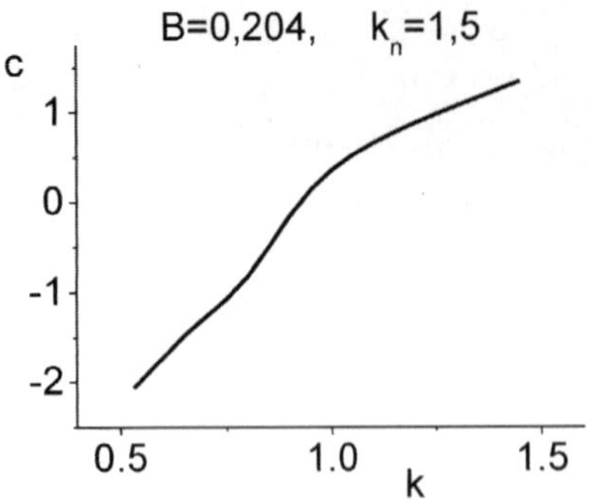

Figure 3. Dependence of the velocity of steady-state traveling solutions of the first family on the wave numbers *k*.

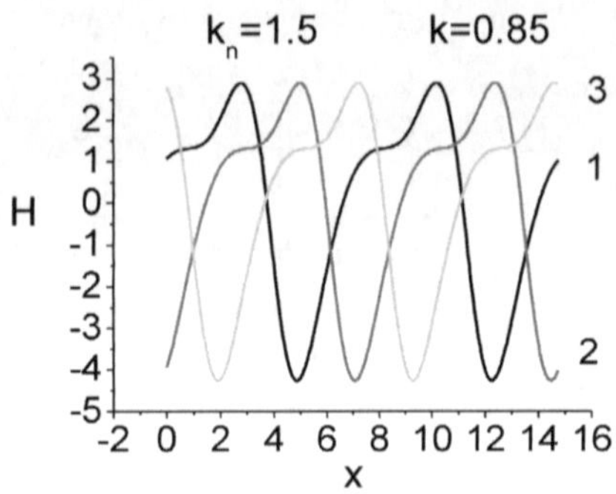

Figure 4. Surface profiles for three different time moments: t = 100 (*1*), 200 (2), 300 (3).

Figure 3 shows the dependence of the velocity of steady-state traveling solutions of the first family obtained at the parameter value $B = 0.204$. In this case, the neutral wave number $k_n = 1.5$. As an example, figure 4 shows the solution of this family with wave number $k = 0.85$. Here wave profiles are shown for three different time points. Two wave lengths $2\lambda\,(\lambda = 2\pi / k)$ are plotted along the abscissa axis. The wave velocity c is equal to -0.495.

3. Conclusion

Thus, nonlinear waves on the film of liquid flowing under gravity in the known stress field at the interface have been considered. In the case of small Reynolds numbers the problem was reduced to considering solutions of the nonlinear integral-differential equation for film thickness deviation from the undisturbed level. In this paper the nature of the wave branching from the undisturbed solution was investigated. The weakly nonlinear steady-state traveling solutions of this equation with wave numbers in the vicinity of neutral wave numbers were constructed analytically The steady-state traveling solutions of the first family, whose wave numbers within the instability area are far from neutral wave numbers, were found numerically. The examples of some periodic solutions to this equation were presented.

Acknowledgements

The work was performed with financial support of the Russian Science Foundation (grant No 16-19-10449).

References

[1] Vozhakov I S, Arkhipov D G and Tsvelodub O Yu 2015 *Thermophysics and Aeromechanics* **22** 191-202

Journal of Physics: Conference Series **754** (2016) 032021 doi:10.1088/1742-6596/754/3/032021

Modeling wave processes at the outflowing of a water coolant with supercritical initial parameters

I S Vozhakov [1,2], M V Alekseev[1], S I Lezhnin[1,2], N A Pribaturin[1]

[1] Institute of Thermophysics SB RAS, Ave. Ac. Lavrenteva 1, Novosibirsk 630090, Russia
[2] Novosibirsk State University, Pirogova St. 2, Novosibirsk 630090, Russia

E-mail: vozhakov@gmail.com

Abstract. Numerical simulation of the outflow of a coolant with supercritical initial parameters at a butt-break of high-pressure pipeline is carried out. The results of calculation of the outflow dynamics on a PV-diagram, as well as the pressure evolution are presented. It is shown that the flow rate weakly depends on temperature at its low values (up to $0,9\ T_c$). In the second region (from $0,9\ T_c$ to T_c), the coolant boiling occurs inside the channel, which leads to a sharp drop in the flow rate with increasing temperature. And the third area (above T_c) is typical for the gas coolant outflow, in which the density strongly depends on pressure and temperature.

1. Introduction

The most important task in the design of power plants is the creation of units that are resistant to malfunctioning, including different types of emergencies. The use of water coolant under high pressure in power plants in the accidents associated with pipeline breaks causes outflowing of the superheated coolant in the atmosphere, followed by rapid (explosive) fluid boiling. At that a non-stationary compression wave [1, 2, 3] runs to external medium and damages other elements of the unit. The danger of this emergency scenario increases significantly in the modern power plants due to growing power density, where working pressure and temperature approach or exceed the critical values [4]. In this regard it is extremely important to be able to predict the scenarios of the process development in a sudden depressurization of such a working region. In addition, the study of shock-wave processes in the vicinity of the termodynamic critical point is interesting not only from the point of view of practical value, but raises questions of a fundamental character.

To date a large number of experimental data on adiabatic boiling flows has been obtained [5, 6]. However, the research area for the majority of substances lies in a relatively narrow range of values of pressure and temperature. However, it is interesting to observe the change in the flow rate through the channel at supercritical initial pressure and at a consistent temperature rise across the critical point. An extreme change of physical properties in the area of supercritical single-phase states may have a significant impact on the coolant outflow, boiling and generation of shock wave phenomena. In the present work, the outflow of water coolant with supercritical parameters is numerically studied at sudden depressurization of the pipeline.

2. The calculation method

The study considers the axisymmetric problem of the boiling coolant outflow from the pipe. The problem is described by the Euler equations in combination with the continuity equation and the energy equation for the two-phase homogeneous mixture:

$$\frac{\partial}{\partial t}(\rho r) + \frac{\partial}{\partial z}(\rho r u) + \frac{\partial}{\partial r}(\rho r v) = 0 \tag{1}$$

$$\frac{\partial}{\partial t}(\rho r u) + \frac{\partial}{\partial z}\left(\rho r u^2\right) + \frac{\partial}{\partial r}(\rho r u v) = -r\frac{\partial p}{\partial z} \tag{2}$$

$$\frac{\partial}{\partial t}(\rho r v) + \frac{\partial}{\partial z}(\rho r u v) + \frac{\partial}{\partial r}\left(\rho r v^2\right) = -r\frac{\partial p}{\partial r} \tag{3}$$

$$\frac{\partial}{\partial t}(E r) + \frac{\partial}{\partial z}(u r (E + p)) + \frac{\partial}{\partial r}(v r (E + p)) = 0 \tag{4}$$

Here ρ – mixture density, u – axial velocity, v – radial velocity, p – pressure, E – energy.

To close the system of equations the equation of mixture state is used in the form that binds pressure, density and specific internal energy. The calculation of vapor and liquid properties is performed using the software package TTSE [7]. The solution to the original conservative system of equations is obtained by using the software package [8], applying the finite volume method coupled with the method of flow correction, FCT. The solution to the model equations by this method is equivalent to the application of some subgrid turbulence model [9]. The detailed description of the model is presented in works [10, 11, 12].

Simulation of liquid outflow from a vessel requires a simultaneous calculation of the flow in the vessel, in the pipeline and in the outside area. Conjugating the flow in the vessel and in the pipeline we apply a simplified "input area model" where an incompressible medium motion is described by one-dimensional Cauchy-Lagrange integral and which allows calculating pressure and velocity of fluid at a known pressure in the vessel. When simulating the external medium it is assumed that the vapor-liquid mixture outflows in the space, filled at atmospheric pressure by wet saturated vapor with density equal to that of air atmosphere at a temperature of 20 ^0C.

The process of non-equilibrium superheated boiling of the coolant is described using relaxation model, determining phase transition velocity on two parameters (temperature and pressure). It is assumed that the growth rate of the local mass vapor fraction X (equation of vapor generation) is proportional to its deviation from the equilibrium value X_s, i.e.

$$\frac{dX}{dt} = -\frac{X - X_s}{\theta} \tag{5}$$

where θ is the time of relaxation to equilibrium.

At that, it is assumed that parameters of the vapor phase (due to higher temperature diffusivity of vapor) correspond to equilibrium, and these of the liquid phase agree with the metastable state. The relaxation time θ is determined in [13] from experimental data as a function of two parameters: the volumetric steam fraction ε and the relative differential pressure $\psi = (P_s(T) - P)/P_s(T)$. On the basis of experimental data (pressures up to 1 MPa) processing the following is obtained:

$$\theta = 6.51 \times 10^{-4}\ \varepsilon^{-0{,}257}\psi^{-0{,}224} \tag{6}$$

For large pressures in the processing a different expression of the relative differential pressure is used: $\phi = (P_s(T) - P)/(P_c(T) - P_s(T))$, where P_c is the pressure in critical thermodynamical point. The corresponding formula for the relaxation time has the following form:

$$\theta = 3.84 \times 10^{-7}\,\varepsilon^{-0,54}\phi^{-1,76}. \tag{7}$$

3. The calculation results

In the vicinity of the critical point ($P_c = 21.8$ MPa, $T_c = 374\ ^0C(647\text{ K})$) the derivative of the enthalpy by temperature (Fig. 1) increases dramatically since thermophysical parameters also experience changes. One can see that the dependence of specific enthalpy of water on temperature at different pressures has a jump when pressure approaches the critical point. Therefore, the proximity of initial parameters of the coolant to the critical point significantly affects the coolant outflow. Let us consider the model problem on the outflow of water coolant with supercritical initial parameters in the atmosphere.

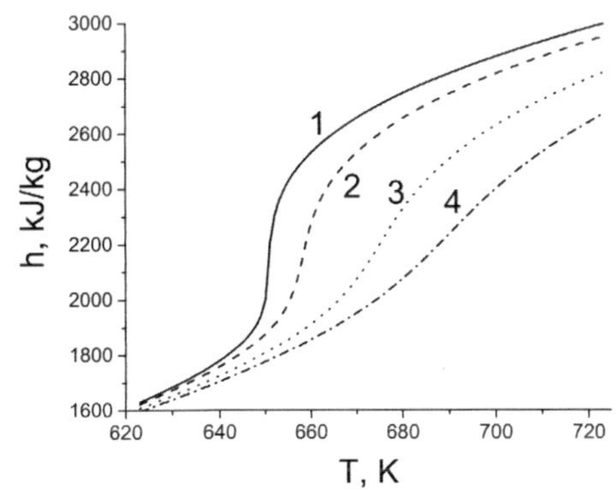

Figure 1. The dependence of specific enthalpy on temperature in the vicinity of the critical point (1 – $P_0 = 23$ MPa, 2 – $P_0 = 25$ MPa, 3 – $P_0 = 30$ MPa, 4 – $P_0 = 35$ MPa).

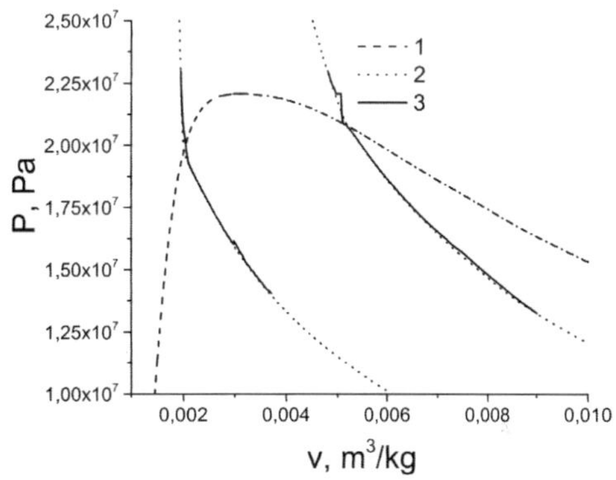

Figure 2. Comparison of theoretical isentropes and calculated curves of supercritical water outflow on PV-diagram (1 – boundary of two-phase area. 2 – theoretical isentropes, 3 – calculated curves).

Consider the PV-diagram with the calculated and theoretical curves for isentropic process (Fig. 2). The initial conditions of information output are chosen to take the calculated pressure at a distance of one caliber inwards from the pipe end on the pipe axis. The graph shows that in both cases the calculated curves with good accuracy coincide with the isentropic lines. Thus, in our case the liquid outflow is well described by the adiabatic approximation. At initial data $P_0 = 23$MPa and $T_0 = 370\ ^0C$ the adiabatic line on PV diagram runs to the left from the critical point and crosses the line of boiling, i.e. enters the two-phase area with the void fraction equal to null. When the initial data take values $P_0 = 23$ MPa and $T_0 = 380\ ^0C$, the adiabatic line passes to the right from the critical point and crosses the line of condensation, i.e. enters the two-phase region with the void fraction equal to one. Thus an insignificant difference in the initial data leads to a fundamentally different result of the coolant outflow.

Figure 3 shows calculated values of pressure depending on time. The process with initial temperature of $T_0 = 380\ ^0C$ is characterized by a sharp pressure drop after entering the two-phase area whereas a sharp pressure drop is not observed for the process with initial temperature of $T_0 = 370\ ^0C$. This is primarily due the fact that in the first case (curve 1) at the entrance to the two-phase area begins condensation, which considerably accelerates the pressure drop. In

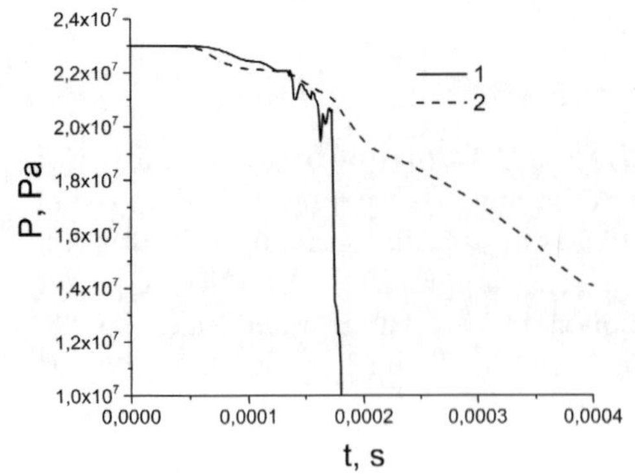

Figure 3. The evolution of pressure at supercritical fluid outflow (1 – P_0=23 MPa and T_0=370 ^0C, 2 - P_0=23 MPa and T_0=380 ^0C).

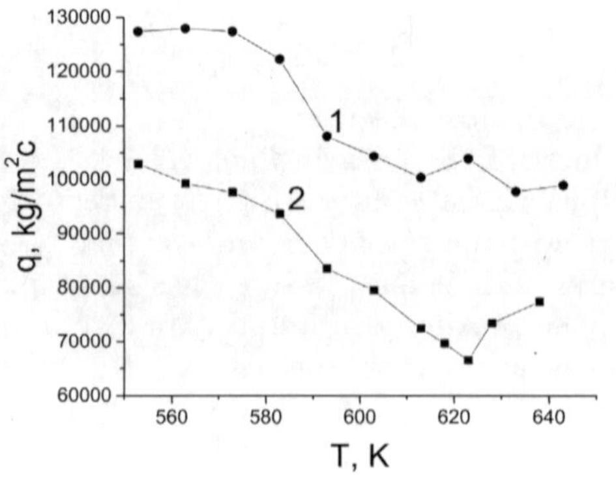

Figure 4. The dependence of the flow rate through the nozzle on the temperature in the vicinity of the critical point (1 – P_0=35 MPa, 2 – P_0=25 MPa).

the second case (curve 2) the coolant boiling occurs that compensates the pressure drop, caused by pipeline depressurization.

Figure 4 shows the dependence of the mass flow rate on temperature in the vicinity of the critical point at pressures above the critical. The flow rate weakly depends on temperature up to values 0.9 T_c. Further, there is a rapid reduction in the flow rate, and at $T = T_{min}(P)$ the flow rate curves have a minimum. With increasing pressure, the value $T_{min}(P)$ is shifted towards higher temperature values. This behavior may be explained by different outflow modes. At low values of temperatures there is a hydraulic outflow at which boiling occurs only beyond the nozzle. With increasing temperature the boiling front is displaced to the nozzle area, which leads to a sharp drop in the flow rate. The results have not been obtained for the case when the temperature is much higher that the T_c, but it may be assumed that they should match the flow rate characteristics for gas. That is in this area it is necessary to consider a strong density dependence on pressure and temperature.

4. Conclusions

Simulating the outflow of the coolant with supercritical parameters requires the enhanced prediction accuracy, as small differences in initial data can lead to fundamentally different results. The results of simulating the outflow of coolant with supercritical initial parameters are qualitatively consistent with the results of [14], which experimentally studied the discharge characteristics of short cylindrical channels at the outflow of nitrogen with supercritical parameters into the atmosphere. The calculations have proved the presence of three areas in the description of discharge characteristics of the coolant with supercritical parameters at its outflow from the channel. So the first area corresponds to the hydraulic outflow and is characterized by a weak dependence on the coolant temperature. In the second region, the coolant boiling occurs inside the channel, which leads to a sharp drop in the flow rate with increasing temperature. The third area is typical for the gas coolant outflow, in which the density strongly depends on pressure and temperature [14].

Acknowledgments

The authors express their gratitude to A.L. Sorokin for active participation in the development and numerical realization of the models.

The work was financially supported by the Russian Science Foundation (grant No14-29-00093).

References

[1] Edwards A R and O'brien T P 1970 *J. Br. Nucl. Energy Soc.* **9** 125–135

[2] Nigmatulin B I and Soplenkov K I 1980 *High Temperature* **18**(1) 118 – 131

[3] Goffman G V, Kroshilin A E and Nigmatulin B I 1981 *High Temperature* **19**(6) 897 – 905

[4] Schulenberg T, Starflinger J, Marsault P, Bittermann D, Maráczy C, Laurien E, à Nijeholt J L, Anglart H, Andreani M, Ruzickova M *et al.* 2011 *Nuclear Engineering and Design* **241** 3505–3513

[5] Pribaturin N A, Bezrukov Y A, Bykov M A, Krasnov S N, Onshyn V M, Lezhnin S I and Sorokin A L 2006 *Materials of RNKT* **5** 284–287 (in Russian)

[6] Reshetnikov A V, Mazheyko N A and Skripov V P 2000 *Journal of Applied Mechanics and Technical Physics* **41** 491–497

[7] 2003 *Guideline on the Tabular Taylor Series Expansion (TTSE) Method for Calculation of Thermodynamic Properties of Water and Steam Applied to IAPWS-95 as an Example. The International Association for the Properties of Water and Steam* Vejle, Denmark

[8] Boris J P, Landsberg A M, Oran E S and Garder J H LCPFCT - Flux-Corrected Transport Algorithm for Solving Generalized Continuity Equations. NRL/MR/6410-93-7192

[9] Belotserkovskii O M 1996 *U.S.S.R. Comput. Math. Math. Phys.* **22**(6) 166–183

[10] Alekseev M V, Lezhnin S I, Pribaturin N A and Sorokin A L 2014 *Thermophysics and Aeromechanics* **21**(6) 763–766

[11] Alekseev M V, Lezhnin S I and Pribaturin N A 2015 *Vestnik of TyumSU. Physical and Mathematical Modeling.* **1**(2) 75 – 84 (in Russian)

[12] Alekseev M V and Vozhakov I S 2015 *Vestnik of TyumSU. Physical and Mathematical Modeling.* **1**(4) 6 – 14 (in Russian)

[13] Downar-Zapolski P, Bilicky Z, Bolle L and Franco J 1996 *Int. J. Multiphase Flow* **22**(3) 473 – 483

[14] Maltsev S A, Baydakov V G and Skripov V P 1986 *High Temperature* **24**(4) 820 – 822

Simulating the nonlinear wave mode evolution in the problem of the falling thin layer of a viscous liquid

I S Vozhakov, D G Arkhipov and O Yu Tsvelodub

Kutateladze Institute of Termophysics SB RAS, Lavrentev Ave. 1, 630090, Novosibirsk, Russia
Novosibirsk State University, Pirogov Str. 2, 630090, Novosibirsk, Russia

E-mail: vozhakov@gmail.com

Abstract. The work is devoted to numerical simulation of the dynamics of nonlinear periodic waves on the surface of freely flowing liquid film realized with the use of the previously discovered symmetry of model equations. The calculation results have been presented for the dynamics of amplitudes of significant harmonics and for the wave surface evolution. They prove to be in good agreement with previously calculated steady-state traveling solutions obtained on the full and the reduced bases. In addition, the work has identified intervals of wavenumbers in which the evolution of perturbations with small initial amplitude leads to a wave signal periodically changing in time rather than to the steady-state traveling solution.

1. Introduction

The flow of a thin layer of viscous liquid (film) over an inclined solid plane has been studied for over a hundred years [1]. Until the middle of the last century, the interest in the falling films was almost exclusively utilitarian. In 1941, Friedman and Miller [2] noted: "In the general field of diffusional processes, including absorption, extraction, heat transfer, humidification, and distillation, the flow of thin liquid films is often encountered". In accordance with practical tasks, research works were mainly experimental in nature and were focused on a turbulent flow regime. Precision measurements of the fluid velocity on the surface of very thin laminar liquid layers freely flowing over a vertical tube [2] have revealed its significant exceedence over the velocity calculated with the trivial Nusselt solution for a smooth film. A qualitative explanation of this consisted in the necessity to take into account the wave formation on the liquid surface.

The impetus to a theoretical study of wave flows of liquid films was given in a series of brilliant works of Kapitza [3]. Notable photos of the film surface showed the existence of many regular wave flow regimes that were much more stable than the plane-parallel Nusselt flow. The sustained efforts of subsequent researchers to describe such regimes have been rewarded by the profound results that reveal the general nature of the hydrodynamic instability and have direct application in various fields of physics of continuous media. It is suffice to mention the widely known equation of Kuramoto-Sivashinskii, derived for the first time to simulate the very wave evolution on the liquid film surface [4]. The interaction of effects of long-wave pumping, short-wave dissipation and perturbation nonlinearity (each of them is expressed by only one term of the K-S equation) leads to an unusually rich picture of steady-state traveling wave modes, formed due to the instability of the trivial solution, as well as to the extremely complex dynamics of perturbations, demonstrating main features of the spatial-temporal deterministic chaos.

The full formulation of the problem of the wavy film flow includes the Navier – Stokes and the continuity equations with respective kinematic and dynamic boundary conditions. The major problem is the uncertainty of the moving boundary position, which is determined in the course of solving. If the effects of droplet entrainment and solid surface drying are excluded from consideration, the fluid flow area turns out to be simply connected. The presence of surface tension ensures no sharp edges on the film surface. Under these conditions, the function determining the surface position is often unambiguous. Then, there exists a continuously differentiable coordinate transformation, mapping the area of the fluid flow on the band of constant thickness:

$$x = x, \quad \eta = y/h(x,t), \quad t = t. \tag{1}$$

New variables (1) are not orthogonal, so the conventional formulation of the motion equations in vector form is not applicable. The transformation (1) is realized with the use of new variables in the equations, written in tensor form invariant to the coordinate systems. With the use of such approach for the film, freely falling over a vertical plane, in [5] the following system was obtained in the long-wave approximation:

$$\begin{gathered}\frac{\partial(hu)}{\partial t} + \frac{\partial(hu^2)}{\partial x} + \frac{\partial(huv)}{\partial \eta} = \frac{\sigma}{\rho} h \frac{\partial^3 h}{\partial x^3} + \frac{\mu}{\rho h}\frac{\partial^2 u}{\partial \eta^2} + gh, \\ \frac{\partial h}{\partial t} + \frac{\partial(hu)}{\partial x} + \frac{\partial(hv)}{\partial \eta} = 0.\end{gathered} \tag{2}$$

Here, h is the film thickness, u and v are the contravariant components of the longitudinal and transversal velocities, respectively, ρ is the density, μ is the dynamic viscosity, and σ is the surface tension of the liquid.

The boundary conditions of liquid no-slip and impermeability on the surface $\eta = 0$ are:

$$u(x,0,t) = 0, \quad v(x,0,t) = 0. \tag{3}$$

On the surface $\eta = 1$, the conditions of impermeability and shear stress absence take the form:

$$v(x,1,t) = 0, \quad \frac{\partial u}{\partial \eta}(x,1,t) = 0. \tag{4}$$

Let's perform a shifting transformation along the transverse coordinate in the considered problem:

$$\eta' = \eta - 1. \tag{5}$$

Now the flow region takes the band $\eta' \in [-1,0]$. It is easy to notice that the equations (2) are invariant under the transformation:

$$\begin{gathered}\eta' \to -\eta', \\ u(x,\eta',t) \to u(x,-\eta',t), \\ v(x,\eta',t) \to -v(x,-\eta',t).\end{gathered} \tag{6}$$

2. Calculation method

For a numerical study of the problem, the equations (2) and the boundary conditions are written in the following dimensionless form [6]:

$$\begin{gathered}\frac{\partial Q}{\partial t}+\frac{\partial}{\partial x}\left(\frac{Q^2}{h}\right)+\frac{\partial}{\partial \eta'}\left(\frac{QV}{h}\right)=\frac{1}{\varepsilon \mathrm{Re} h^2}\frac{\partial^2 Q}{\partial \eta'^2}+\frac{3h}{\varepsilon \mathrm{Re}}+\frac{18}{5}\varepsilon \mathrm{Re} h\frac{\partial^3 h}{\partial x^3},\\ \frac{\partial h}{\partial t}+\frac{\partial Q}{\partial x}+\frac{\partial V}{\partial \eta'}=0,\\ Q=V=0, \quad \text{at} \quad \eta'=-1,\\ \frac{\partial Q}{\partial \eta'}=0, \quad V=0, \quad \text{at} \quad \eta'=0.\end{gathered} \tag{7}$$

In the system (7), there are new functions $Q = hu$, $V = hv$ relative to which the second equation is linear. Here $\mathrm{Re} = \mathrm{g}h_0^3/3\nu^2$ is the Reynolds number, and $\varepsilon = h_0/l_0$ is the wavelength parameter. As characteristic scales the following are selected: the average thickness h_0 and the superficial velocity of a waveless Nusselt flow $u_0 = gh_0^2/3\nu$. The scales of the length l_0 and time l_0/u_0 are determined from the relationship:

$$\frac{h_0}{l_0}=\varepsilon=\sqrt{\frac{18}{5}\frac{\mathrm{Re}^{5/3}}{3^{1/3}\mathrm{Fi}^{1/3}}},$$

where $\mathrm{Fi} = \sigma^3/\rho^3 \mathrm{g}\nu^4$ is the film number, characterizing liquid properties. As it is shown in [6], this choice provides the value of neutral wave number $\alpha_n \approx 1$.

The problem was solved by the pseudospectral (collocation) method. Functions depending on transverse coordinate η' were expanded in a series on Chebyshev polynomials T_i. The work [6] showed that at moderate values of the Reynolds numbers the steady-state traveling solutions of the system (7) in the area extended over the transverse coordinate $\eta' \in [-1, 1]$ are characterized by the following symmetry, connected with transformation (6):

$$u(x,\eta',t)=u(x,-\eta',t), \ v(x,\eta',t)=-v(x,-\eta',t). \tag{8}$$

This allows choosing only even Chebyshev polynomials for the function Q, and only odd ones for the function V:

$$\begin{gathered}Q(x,\eta,t)=\sum_{i=1}^{M}Q_{2i}\left(x,t\right)\left(T_{2i}\left(\eta'\right)-1\right),\\ V(x,\eta,t)=\sum_{i=1}^{M}V_{2i+1}\left(x,t\right)\left(T_{2i+1}\left(\eta'\right)-\eta'\right).\end{gathered} \tag{9}$$

Here, M is a number of Chebyshev nodes.

This choice provides an automatic satisfaction of boundary conditions on the surface and on a solid wall. So, in a number of Chebyshev nodes:

$$\eta_j'=-\cos\left(\frac{j\pi}{2(M+1)}\right), \ j=1,...,M,$$

we obtain the set of equations:

$$\sum_{i=1}^{M} \frac{\partial Q_{2i}}{\partial t}\left(T_{2i}(\eta'_j)-1\right) = \frac{1}{\varepsilon \mathrm{Re} h^2} \sum_{i=1}^{M} Q_{2i} \frac{d^2 T_{2i}}{d\eta'^2}(\eta'_j) + \frac{18}{5}\varepsilon \mathrm{Re} h h_{xxx}$$
$$+\frac{3}{\varepsilon \mathrm{Re}} h - \frac{\partial}{\partial x}\left(\frac{\sum_{i=1}^{M} Q_{2i}\left(T_{2i}(\eta'_j)-1\right)\sum_{i=1}^{M} Q_{2i}\left(T_{2i}(\eta'_j)-1\right)}{h}\right) \tag{10}$$
$$-\frac{1}{h}\frac{\partial}{\partial \eta'}\left(\sum_{i=1}^{M} Q_{2i}\left(T_{2i}(\eta'_j)-1\right)\sum_{i=1}^{M} V_{2i+1}\left(T_{2i+1}(\eta'_j)-\eta'_j\right)\right),$$

$$\frac{\partial h}{\partial t} = -\frac{\partial q}{\partial x}.$$

Here $q(x,t) = \int_{-1}^{0} Q(x,\eta,t)d\eta'$ is a flow rate.

Functions that depend on the longitudinal coordinate were expanded into spatial Fourier series:

$$Q_{2i}(x,t) = \sum_{k=-\infty}^{\infty} Q_{2i}^{k}(t)\, e^{I\alpha k x},$$
$$V_{2i+1}(x,t) = \sum_{k=-\infty}^{\infty} V_{2i+1}^{k}(t)\, e^{I\alpha k x}, \tag{11}$$
$$h(x,t) = \sum_{k=-\infty}^{\infty} H^{k}(t)\, e^{I\alpha k x}.$$

Here α is the wave number of the periodic solution, I is the imaginary unit. The functions are real-valued, so $Q_{2i}^{-k} = \overline{Q}_{2i}^{k}$, $V_{2i+1}^{-k} = \overline{V}_{2i+1}^{k}$, $H^{-k} = \overline{H}^{k}$. The overline denotes the operation of complex conjugation.

Substituting representations (11) in (10) and limiting to finite numbers of harmonics we obtain the system of differential equations which was solved by the Runge-Kutta method of the 4-th order.

3. The calculation results

The evolution of periodic waves on freely falling thin layer of water ($\mathrm{Fi} = 4 \cdot 10^{10}$) for the Reynolds number $\mathrm{Re} = 20$ has been simulated for various values of the dimensionless wave number α. Plane-parallel Nusselt flow with harmonic perturbation of free surface $|H^1| = 0.45$ was selected as initial condition. Typical scenarios of the evolution of perturbations with small initial amplitude are presented in figures from 1 to 4. Figure 1 presents the evolution of modulus of amplitudes of the first three harmonics at the value $\alpha = 0.51$. It is seen that after the transient stage the amplitude of harmonics tends to some stationary value. At that forms the steady-state travelling wave, which profile is shown in figure 2. The wave profile and velocity fully agree with the calculations of steady-state travelling solutions in [6].

A different scenario is presented in figure 3 at the value $\alpha = 0.3$. It is seen that the modulus of harmonics' amplitudes changes periodically. Figure 4 shows the wave profile at time $t = 200$.

4. Conclusion

The numerical simulation of the freely-falling liquid film has been realized using the previously discovered symmetry of model equations. The calculation results have been presented for the

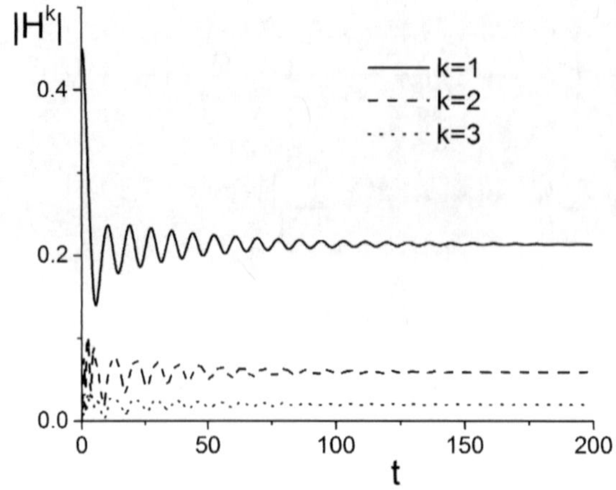

Figure 1. Evolution of the modulus of harmonics' amplitudes at $\alpha = 0.51$ (k is the number of harmonic).

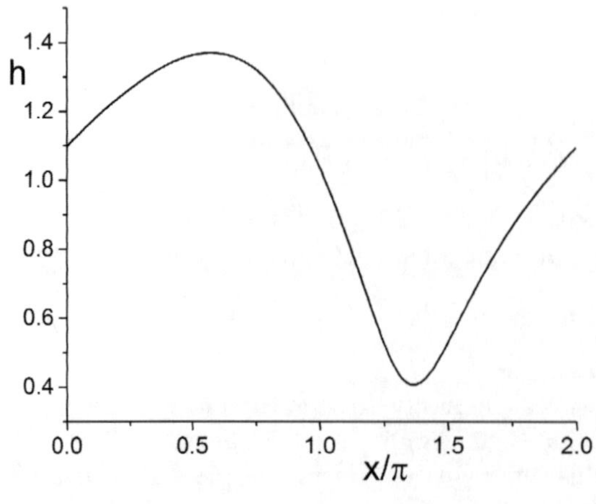

Figure 2. The steady state wave profile ($t = 200$, $\alpha = 0.51$).

dynamics of amplitudes of significant harmonics and for the wave surface evolution. They prove to be in good agreement with previously calculated steady-state traveling solutions obtained on the full and the reduced bases. In addition, the work has identified intervals of wavenumbers in which the evolution of perturbations with a small initial amplitude leads to a wave signal periodically changing in time rather than to the steady-state traveling solution.

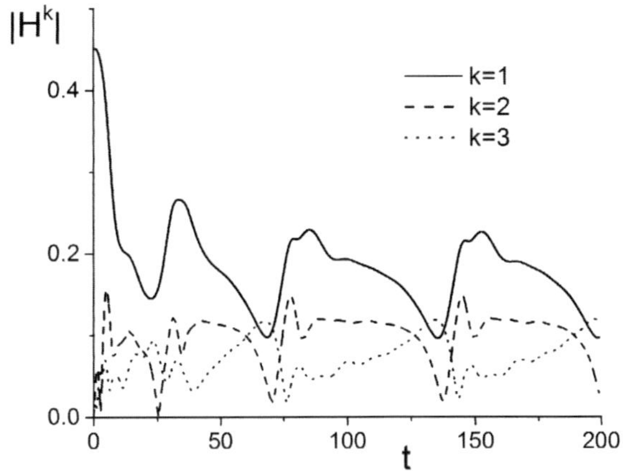

Figure 3. Evolution of the modulus of harmonics' amplitudes at $\alpha = 0.3$ (k is the number of harmonic).

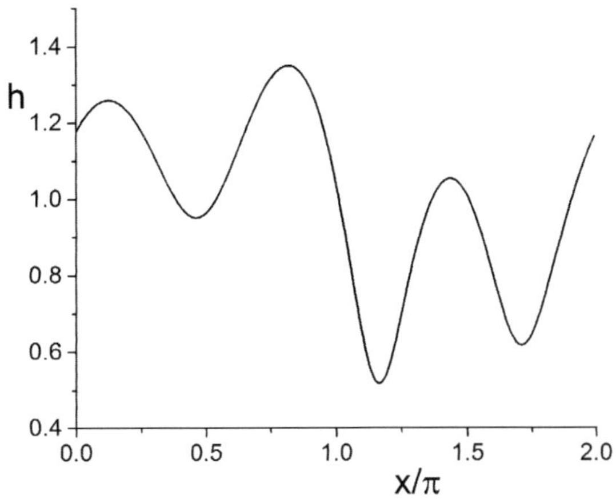

Figure 4. The wave profile ($t = 200$, $\alpha = 0.3$).

Acknowledgments
The work was financially supported by the Russian Science Foundation (grant No. 14-29-00093).

References

[1] Nusselt W 1916 *Zetrschr. Ver. Deutch. Ing.* **60** 541-6
[2] Friedman S J and Miller C O 1941 *Industrial Engineering Chemistry* **33** 885-91
[3] Kapitza P L 1948 *Zhurnal Eksperimentalnoi i Teoreticheskoi Fiziki* **18** 3-18
[4] Nepomnyashchii A A 1974 *Fluid Dynamics* **9** 354-9
[5] Alekseenko S V, Arkhipov D G and Tsvelodub O Yu 2011 *Doklady physics* **56** 22-5
[6] Arkhipov D, Vozhakov I, Markovich D and Tsvelodub O 2016 *European Journ. of Mechanics*-B/Fluids **59** 52-6

Features of shock-wave and vortex processes simulation at depressurization of circuits with superheated water coolant

M V Alekseev[1], M E Kuznetsova, S I Lezhnin, N A Pribaturin

Kutateladze Institute of Thermophysics SB RAS, Ave. Ak. Lavrenteva 1, 630090, Novosibirsk , Russia

E-mail: [1]`alekseev@itp.nsc.ru`

Abstract. Formation and evolution of pressure waves on the barrier were numerically simulated at non-staionary outflow of the superheated vapor from the vessel. The simulation conditions were fully consistent with the experimental conditions (1982) [1]. The calculated pressure fields were calculated both at the initial stage of non-stationary vapor outflow and at the quasistationary flow of superheated vapor on the barrier. The dynamics of interaction between shock waves and vortex structures on the barrier was studied at quasi-stationary outflow of the superheated vapor, and the experimental pressure profile on the target was compared with the calculated pressure profile. Comparison of the experimental data and the calculation results has showed acceptable agreement between the pressure profiles on the target up to a distance from the target to the nozzle approximately $X/D = 6$.

1. Introduction

The problem of depressurization of vessels or pipe lines with heated water coolant was formulated in the 60-80 years of the last century (see, e.g.[2, 3, 4]). This problem is related to prediction of emergency operation modes of power equipment in case of emergency depressurization. At that thermodynamic equilibrium and nonequilibrium models based on the hypothesis of heterogeneous nucleation were used to calculate the liquid water coolant boiling, caused by depressurization of high pressure vessel [5]. Accounting for possible bubble fragmentation due to the interfacial instability made the models too complicated [6]. Another approach to solving this problem is the use of non-equilibrium two-phase model, based on the experimental data for the relaxation time (transition) "non-equilibrium - equilibrium boiling" [7]. In [8] it was shown that the non-stationary relaxation model of boiling adequately describes the nonequilibrium boiling in subsonic and supersonic flows of boiling liquid in the channels. On the basis of the relaxation model and using the toolkit LCPFCT [9] the authors developed a software code for solving a wide class of problems [10, 11, 12] on depressurization of vessels or pipelines with water coolant. Averaging of the Euler equations [13] on the unit cell and the time interval is equivalent to the model of large-scale eddies with "subgrid turbulence model. The "subgrid turbulence model occurs implicitly, since the scale of large eddies is limited to the computational grid size. I.e., the scales of vortices for which direct modeling is impossible are calculated using a numerical viscosity.

The aim of this work is to test the program code for the problems of depressurizing vessels and channels with superheated vapor. As a test problem we selected the experimental work [1].

2. Calculation method

The calculation was performed using homogeneous model. At that the parameters of the vapor phase corresponded to saturation, and the liquid phase could be both metastable and equilibrium. The system of equations of the model includes the Euler equations in the axisymmetric approximation:

$$\frac{\partial}{\partial t}(\rho r)+\frac{\partial}{\partial z}(\rho r u)+\frac{\partial}{\partial r}(\rho r v)=0 \tag{1}$$

$$\frac{\partial}{\partial t}(\rho r u)+\frac{\partial}{\partial z}\left(\rho r u^2\right)+\frac{\partial}{\partial r}(\rho r u v)=-r\frac{\partial p}{\partial z} \tag{2}$$

$$\frac{\partial}{\partial t}(\rho r v)+\frac{\partial}{\partial z}(\rho r u v)+\frac{\partial}{\partial r}\left(\rho r v^2\right)=-r\frac{\partial p}{\partial r} \tag{3}$$

$$\frac{\partial}{\partial t}(E r)+\frac{\partial}{\partial z}(u r(E+p))+\frac{\partial}{\partial r}(v r(E+p))=0 \tag{4}$$

supplemented by the equation of the mass source associated with the phase transition:

$$\frac{\partial(\rho X_\Gamma)}{\partial t}+\frac{\partial(\rho u X_\Gamma)}{\partial z}+\frac{1}{r}\frac{\partial(\rho r v X_\Gamma)}{\partial r}=\Gamma \tag{5}$$

$$\Gamma=\frac{\rho(X_{SAT}-X_\Gamma)}{\theta} \tag{6}$$

Here z,r are the axial and radial coordinates, u,v are the axial and radial components of velocity, ρ is the density, P is the pressure, E is the specific internal energy, X_Γ is the mass, and X_{sat} is the equilibrium mass content of vapor (at given P and E), Γ is the rate of vapor generation, and θ is the relaxation time [7].

The density of homogenous two-phase mixture is expressed through the density of the phases and the mass vapor content:

$$\rho=\left(\frac{X_\Gamma}{\rho_{2,SAT}(P)}+\frac{1-X_\Gamma}{\rho_{1,M}(P,W_{1,M})}\right)^{-1} \tag{7}$$

and the specific enthalpy is determined by the following ratio:

$$W=X_\Gamma W_{2,SAT}(P)+(1-X_\Gamma)W_{1,M} \tag{8}$$

To describe of phase transition within the homogeneous two-temperature approximation of a vapor-liquid medium used phenomenological relaxation model (5)-(6), based on processing the experimental data for the time of relaxation (transition) "non-equilibrium - equilibrium boiling" [7]. Calculation of vapor and liquid properties (7)-(8) was carried out using the software package TTSE [14]. To solve the system of equations (1-4) the authors applied software package LCPFCT [9], which used a differential monotone conservative method of flow correction (Flux Corrected Transport). The goal of this study is to investigate the pressure field evolution during the transition from the non-stationary to quasi-stationary outflow of the coolant. In this case, it is required to calculate unsteady motion of ordered and large-scale turbulent structures. The scale of large eddies is limited to the computational grid size in the direct simulation. The scales of vortices for which direct simulation is impossible are modeled as a subgrid turbulence using the

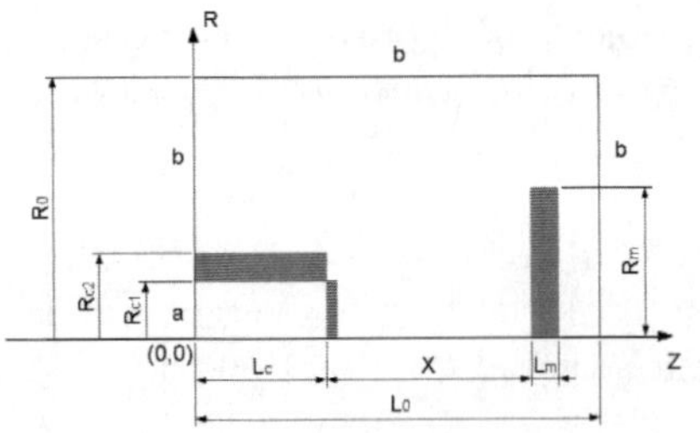

Figure 1. Calculation domain.

numerical viscosity. At finite difference approximation of Euler equations the viscosity terms of an order of $O(\Delta t, h^2)$ arise; at that the numerical viscosity $\varepsilon \approx |V|h$ allows modeling the subgrid turbulence viscosity.

The characteristic calculation domain is shown in figure 1. The calculation was carried out in the cylindrical coordinate system with a uniform grid. The radius of the computational domain was equal to R_0, and the length of the computational domain was L_0. The heated water coolant was filled in the pipe channel with an inner radius R_{c1} and the outer radius R_{c2}. The channel length was equal to L_{c1}. From the right end the pipe channel was closed by a diaphragm. From the left side a so-called "the Cauchy-Lagrange boundary condition" was set for the inflow in the channel [4]. The point of this "input section model" following. When the fluid flowing out from the vessel during the input section essentially not one-dimensional. The most significant feature is the inertia of the flow of fluid flowing into the tube, which affects the flow rate and pressure waves structure. It is proposed [4] on the approximation of the uniformly of the flow at the inlet and the potential flow to build analytic field and the velocity of the medium using the integral of the Cauchy - Lagrange, an equation for the fluid velocity in the inlet section to the outlet:

$$\frac{4D}{3\pi}\frac{\partial v}{\partial t} + \frac{v^2}{2} + \frac{p - p_0}{\rho_0},$$

where of D - the pipe diameter ; p_0, ρ_0 is the pressure and density of the medium in the vessel; p, v - pressure and velocity in the inlet section. As shown in [4] the proposed "input section model" should adequately describe the outflow of cooled liquid or vapor-liquid mixture at small pressure drops and high ones.

At the distance X the axisymmetric target in the form of disc with L_m thickness and R_m radius was set. On the boundaries of the computational domain the condition b of free inflow and outflow at pressure P_0 was used.

To select the optimal parameters of the external liquid-vapor medium at a temperature of 100^0C its density and sound velocity were compared with the density and sound velocity of standard air atmosphere at a temperature of 20^0C. The sound velocity in air was compared with the "frozen" sound velocity in liquid-vapor medium, which is more adequate for fast wave processes. The analysis showed [14, 11, 12], that the mass vapor content of the external vapor-liquid medium should be chosen equal to 0.5.

The initial conditions and the geometry of the numerical model corresponded to the experimental conditions [1]. Superheated vapor pressure at the entrance to the channel $P_0 = 4.02$ MPa, $T_0 = 258.24^O C$,the inner radius of the circular channel $R_{c1} = 5$ mm, the outer radius $R_{c2} = 9$ mm, and the channel length was equal to $L_{c1} = 100$ mm. Diameter meseni was $20D$.

3. Shock-wave and vortex effects in unsteady outflow of superheated vapor on the barrier

Figure 2 shows the dynamics of formation of shock-wave and vortex structures in non-stationary outflow of superheated vapor (field of normalized pressure gradient) when displacing the target to a distance of $X/D = 6.3$. It may be noticed that for the first 200 μs (figure 2(a-c)) of the process there is generation of compression waves that moves toward the target and is reflected from it. After the compression wave the structure of the Mach disk with barrel shocks begins to form. At the barrel shock due to the Kelvin — Helmholtz instability occurs the generation of vortex structures that interact with the primary compression wave and form the complex structure of the pressure fields ($t = 200 - 600$ μs) (figure 2(d-g)). After ($t = 600$ μs) the quasi-stationary jet flow occurs, and periodic vortex structures affecting the pressure field are generated.

Generation of periodic structures in the quasi-stationary outflow essentially depends on the distance between the target and the nozzle (figure 3). At a small distance $X/D = 0.5$ (figure 3.a) a narrow zone of supersonic flow is formed in the form of "disc" on the target and transits in a subsonic jet flow along the target. The pulsation of this boundary leads to the formation of periodic pressure waves along the target. When the distance $X/D = 3.3$ (figure 3.b) the formation of the Mach disk and the side barrel shocks occurs.

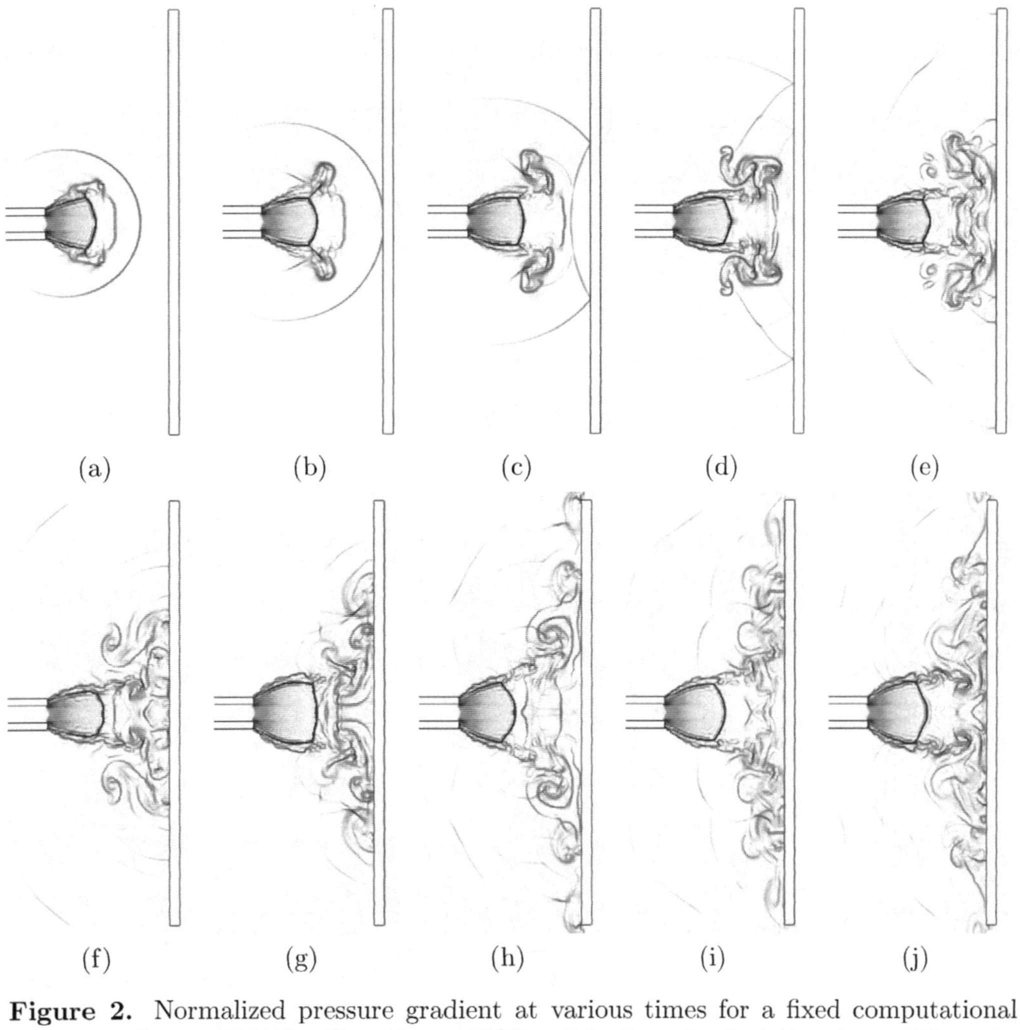

Figure 2. Normalized pressure gradient at various times for a fixed computational domain geometry, $P_0 = 4.02$ MPa; $D = 10$ mm;$X/D = 6.3$; time of calculation t: a - 100 μs, b - 140 μs, c - 180 μs, d - 250 μs, e - 350 μs, f - 450 μs, g - 600 μs, h - 800 μs, i - 1000 μs, j - 1130 μs

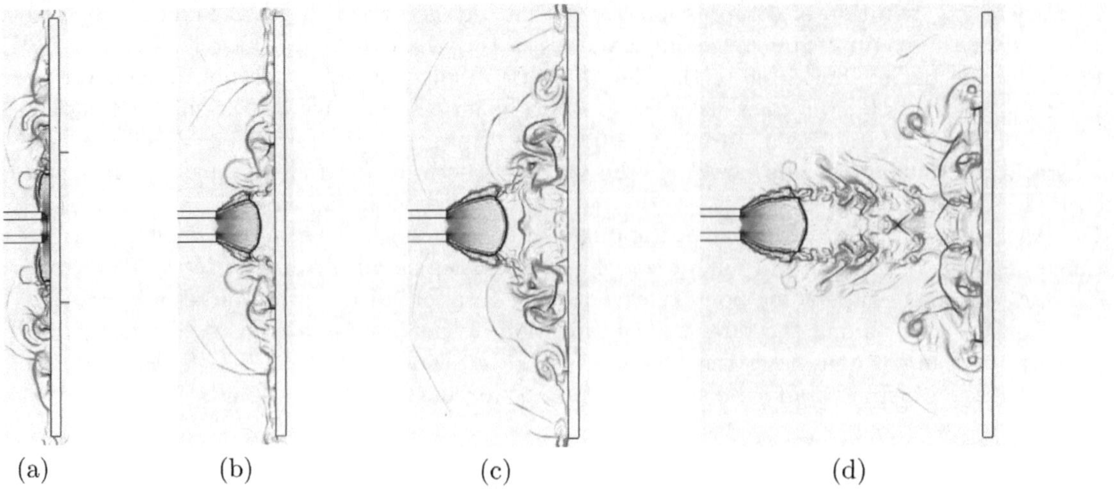

Figure 3. Normalized pressure gradient at the time of 850 μs for different computational domain geometry P_0 = 4.02 MPa; D = 10 mm; t = 850 μs; distance from the target to the nozzle: a - $X/D = 0.5$; b - $X/D = 3.3$; c - $X/D = 6.3$; d - $X/D = 13$.

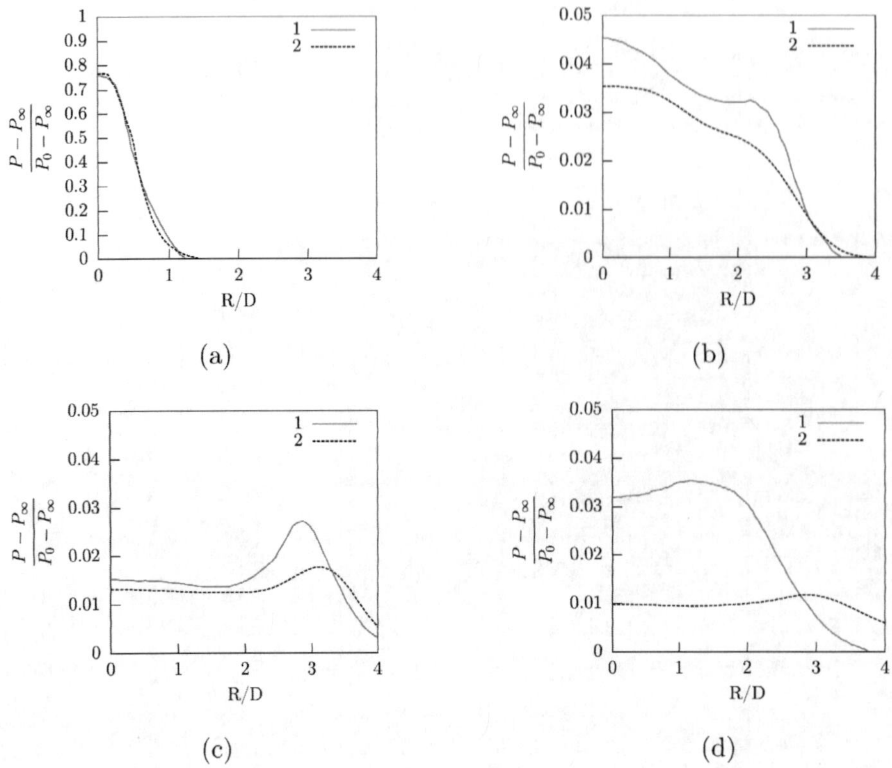

Figure 4. The profile of the dimensionless pressure on the target depending on the radius in calibers at the superheated vapor jet outflow: 1-experiment [1], 2-calculation; P_0 = 4.02 MPa; D = 10 mm; the distance of the target from the nozzle: a - X/D = 0.5; b - X/D = 3.3; c - X/D = 6.3; d - X/D = 13.

In addition to the formation of periodic pressure waves due to pulsation of the structure of the barrel shocks, the vortical structures are generated due to the Kelvin — Helmholtz instability on the side faces of the barrel shock structure. For a distance $X/D = 6.3$ (figure 3.c) the main contribution to the pressure pulsations is given by Kelvin — Helmholtz instability. When the distance $X/D = 13$ (figure 3.d) previous mechanisms of periodic structure generation are added with extra formation and collapse of off-design structures of pressure surges after the Mach disk and the target.

Figure 4 shows a comparison of the profiles of dimensionless pressure on the target depending on the radius in calibers, where 1 is the experiment [1], and 2 is the calculation. The calculation presents the pressure profile, obtained by averaging the non-stationary pressure profile over time of $100\mu s$, (the time of calculation of quasi-stationary process was $100ms$). The experimental pressure profile coincides with the calculated pressure profile if distance from the target is equal to $X/D = 0.5$ (figure 4.a). With increasing distance from the target to the nozzle to $X/D = 6.3$ the calculated pressure profiles on the target acceptably (but rather qualitative) repeat the profiles, obtained in the experiment (figure 4 (b,c)). When the distance between the calculated and experimental pressure profile on the target $X/D = 13$, the difference becomes very significant. This fact is connected with the methodological errors: a) the error resulting from the approximation of viscous effects in the numerical scheme; b) the error of modeling in "air atmosphere"; c) the absence of a boundary layer near the target. The strong divergence between the calculated and experimental data at a large distance from the nozzle due to the accumulation of errors mentioned above.

Acknowledgments
Authors express their gratitude to A.L. Sorokin for active participation in the development and numerical realization of the models.

The work was financially supported by the Russian Science Foundation (grant No14-29-00093).

References
[1] Masuda F and et al 1982 *Nuclear Engineering and Design* **67**(2) 273 – 286
[2] Edwards A R and O'brien T P 1970 *J. Br. Nucl. Energy Soc.* **9** 125–135
[3] Nigmatulin B I and Soplenkov K I 1980 *High Temperature* **18**(1) 118 – 131 (in Russian)
[4] Goffman G V, Kroshilin A E and Nigmatulin B I 1981 *High Temperature* **19**(6) 1240 – 1250
[5] Bolotnova R K and Buzina V A 2014 *Computational continuum mechanics* **7**(4) 343 – 352
[6] Ivashnev O E 2008 *Fluid Dynamics* **43**(3) 390 – 401
[7] Zapolski P D, Bilicky Z, Bolle L and Franco J 1996 *Int. J. Multiphase Flow* **22**(3) 473 – 483
[8] Artemov V, Minko K and Yan'kov G G 1996 *Thermal Engineering* **62**(12) 897 – 905
[9] Boris J P, Landsberg A M, Oran E S and Garder J H LCPFCT - Flux-Corrected Transport Algorithm for Solving Generalized Continuity Equations. NRL/MR/6410-93-7192
[10] Alekseev M V, Lezhnin S I, Pribaturin N A and Sorokin A L 2014 *Thermophysics and Aeromechanics* **21**(6) 763–766
[11] Alekseev M V, Lezhnin S I and Pribaturin N A 2015 *Vestnik of TyumSU. Physical and mathematical modeling.* **1**(2 (2)) 75– 84 (in Russian)
[12] Alekseev M V and Vozshakov I S 2015 *Vestnik of TyumSU. Physical and mathematical modeling.* **1**(4(4)) 6 – 14 (in Russian)
[13] Belotserkovskii O M 1996 *U.S.S.R. Comput. Math. Math. Phys.* **22**(6) 166–183
[14] Lezhnin S I and Pribaturin N A 1983 *Proceedings of Siberian branch of AS USSR. series "Technical sciences"* (2) 20 – 26 (in Russian)

Experimental study of heat transfer from the gas-vapor mixture to the surface in rectangular channel with the water vapor condensation

V V Bespalov[1], V I Bespalov[1] and D V Melnikov[1]

[1] Tomsk Polytechnic University, Tomsk, Russia

E-mail: bespalov_vv@mail.ru

Abstract. The laboratory setup was developed for studying the water vapor condensation process from the gas mixture on a vertical flat heat exchange surface in rectangular channel 10×2 cm. The experiments were carried out with normal pressure of steam-air mixture. The results allowed to determine the heat transfer coefficient from steam-gas mixture to the heat transfer surface. The dependences of the heat transfer coefficient on the initial moisture content and the mixture flow rate in the heat exchanger channel were obtained.

1. Introduction

The important task in sphere of energy saving is the flue gas heat recovery for energy and industrial enterprises [1]. The natural gas combustion products contain large amount of water vapor. Using the latent heat of water vapor condensation is called the deep flue gas heat recovery. Designing the surface-type heat exchangers to realize this task involves calculating the heat transfer surface area, which depends on the heat transfer coefficient. A known difficulty is the calculation of the heat transfer coefficient from the flue gases to the heat exchange surface where the water vapor condensation process takes place [2].

Existing methods are limited by the heat medium parameters and structural characteristics of the heat recovery units. Using the calculation method for designing of a new heat exchange unit requires a thorough analysis. Known techniques are based on experimental data on water vapor condensation on the surface of the tube bundle with smooth and finned tubes [1–4]. Criterial equations were obtained for certain ranges of speed and moisture values. Application of the tube bundle is caused by using water as the heated medium.

However, in some cases it is more appropriate to use another medium for heating. For example, the air is used as a heated medium during winter. That significantly increases the heat recovery depth and allows to condense the maximal moisture amount [5, 6]. Such heat exchangers are usually plate-type. The mathematical model of designed condensing plate-type heat exchanger for air heating contains a calculation of the heat transfer coefficient from the flue gases to the air flow, which strongly depends on the heat transfer on the air side as it has lower value. To test the mathematical model adequacy it is necessary to check the calculated heat transfer coefficient by experiment with vapor condensation. Due to lack of experimental data on the condensing surface heat exchangers the task was to carry out series of experiments on the water vapor condensation on a vertical heat exchange surface with the upward and downward gas-vapor mixture movement in a laboratory setting with close to practical speed and moisture values.

2. Experimental part

The experiment's goal is to determine the heat transfer coefficient from the gas-vapor mixture to a vertical heat exchange surfaces with water vapor condensation. The experimental setup is shown in Figure 1.

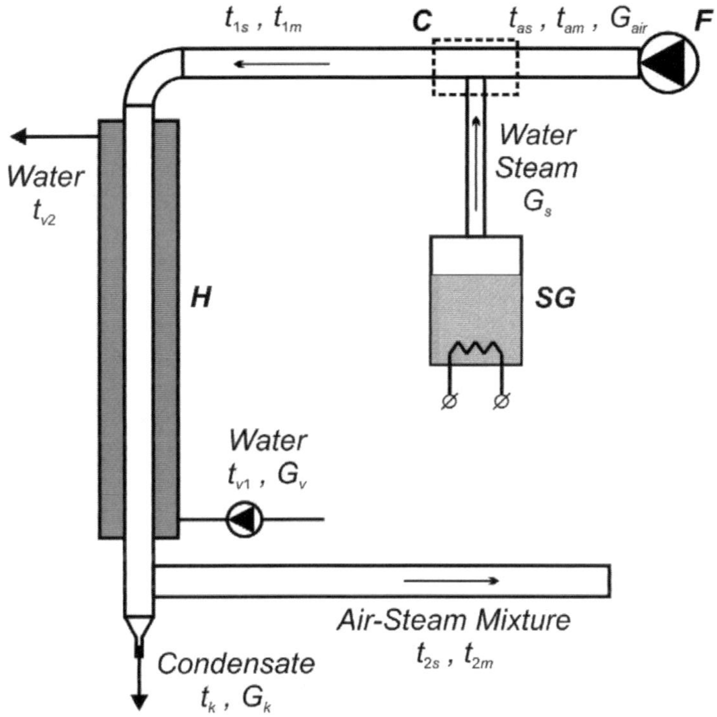

Figure 1. The experimental setup.

As a gas-vapor mixture the steam-air mixture was used. The air stream is fed by fan ***F*** to the mixing chamber ***C*** where it is humidified by the steam from the electric steam generator ***SG***. The resulting moisture content in mixture is calculated using the initial air moisture and amount of water evaporated in the steam generator. In addition, the moisture content in mixture flow is controlled by the testimony of dry and wet thermometers (t_{1s}, t_{1m}, t_{2s}, t_{2m}). The steam-air mixture is fed into a vertical heat exchanger ***H*** with the gas passage in form of rectangular cross section 10×2 *cm* which is surrounded by a cooling water channel. The water is used as a heated medium because the heat transfer coefficient from the wall to water is significantly higher than from the steam-air mixture to the wall. That allows to determine the last one with greater accuracy.

3. Results

During the experiments all the parameters of the steam-air mixture at the heat exchanger outlet (dry and wet bulb temperature) and condensate amount were measured. The cooling water parameters at exchanger inlet and outlet allow to calculate the heat balance. The overall heat transfer coefficient ***k*** is calculated by the heat transfer equation. The heat transfer coefficient α_1 from the gas-vapor mixture to the heat transfer surface is calculated from equation (1).

$$k = \frac{1}{\frac{1}{\alpha_1} + \frac{\delta}{\lambda} + \frac{1}{\alpha_2}}. \qquad (1)$$

Where α_2 – the heat transfer coefficient from the wall to water; δ/λ – the wall thermal resistance.

The first experiment was carried out without adding steam to the air flow. The founded heat transfer coefficient from the air to the wall was in good agreement with the calculations based on known criterial equation. When the vapor is added to the air stream the heat transfer coefficient increases by several times. The experiments allowed to obtain the heat transfer coefficient values with a margin error 7-10%. Some experimental results are given in the table 1 for mixture flow rate in the channel value about 1.8 *m/s*. The initial moisture content of the steam-air mixture was varied from 0.01 to 0.2 kilograms per kilogram of dry gas (*kg/kg.d.g*).

Table 1. The experimental results for mixture flow rate about 1.8 *m/s*.

Parameter	**Value**		
Initial moisture content of the steam-air mixture, *kg/kg.d.g*	0.01	0.1	0.21
Inlet temperature of mixture, °C	26	59	69
Outlet temperature of mixture, °C	17	42	49
Final moisture content of the steam-air mixture, *kg/kg.d.g*	0.01	0.06	0.05
Condensate flow, *g/s*	0	0.19	0.37
The overall heat transfer coefficient ***k***, $W/(m^2 \cdot K)$	11	44	100
The heat transfer coefficient from the wall to water α_2, $W/(m^2 \cdot K)$	135	204	236
The heat transfer coefficient from the steam-air mixture to wall α_1, $W/(m^2 \cdot K)$	12	58	174

Next, the effect of flow rate on the heat transfer coefficient value was studied. The flow rate was varied from 1.8 to 3.7 *m/s*. The results are shown in Figure 2.

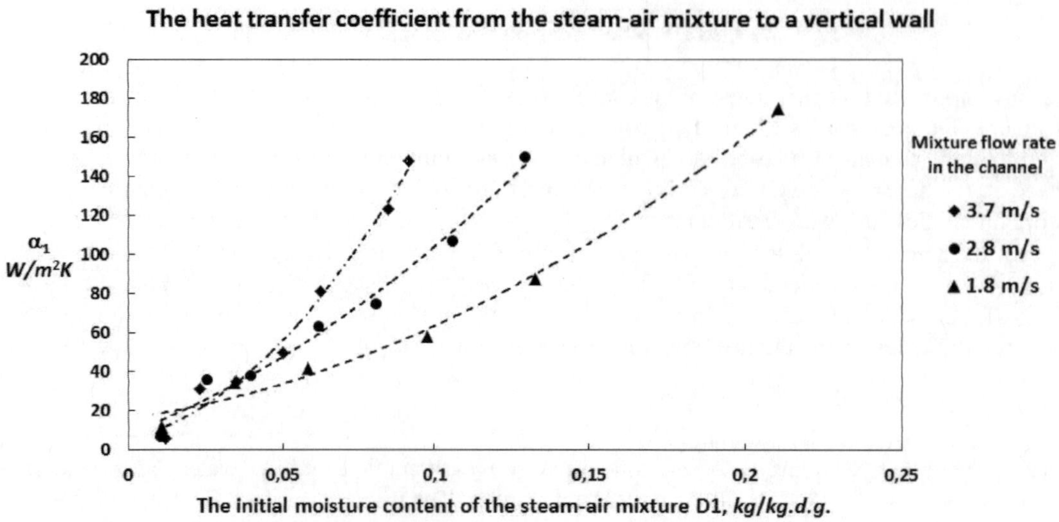

Figure 2. The dependence of heat transfer coefficient on the initial mixture moisture content at different flow rates.

4. Conclusion

The experimental data indicate that the heat transfer coefficient k is in the range from 25 to 100 $W/(m^2 \cdot K)$ and substantially less dependent on the flow rate than from initial moisture content. The water vapor condensation process from the air-steam mixture increases the heat transfer coefficient from mixture to the heat transfer surface by 1.5÷10 times. This means that the decline in the convective component of the heat transfer coefficient α_1 is compensated by mass transfer during the water vapor condensation. For getting numerical equation it is necessary to continue the experiments and analyze the results.

5. References

[1] Kudinov A and Kalmykov M 2002 *Thermal Engineering* **49** 685
[2] Kuzma-Kichta Y, Bukhonov D and Borisov Y 2007 Enhancement of heat transfer during condensation of water vapor from flue stack gases *Thermal Engineering* **54** 210
[3] Jeong K, Kessen M, Bilirgen H and Levy E 2010 Analytical modeling of water condensation in condensing heat exchanger *International Journal of Heat and Mass Transfer* **53** 2361
[4] Shi X, Che D, Agnew B and Gao J 2011 An investigation of the performance of compact heat exchanger for latent heat recovery from exhaust flue gases *International Journal of Heat and Mass Transfer* **54** 606
[5] Bespalov V V, Beljaev L A and Melnikov D V 2015 Using Air for Increasing the Depth of the Flue Gas Heat Recovery *MATEC Web of Conferences* **37** 01009
[6] Bespalov V V, Bespalov V I and Melnikov D V 2016 Evaluation of Heat Transfer Coefficients During the Water Vapor Condensation Contained in the Flue Gas *EPJ Web of Conferences* **110** 01007

Acknowledgments

The study was realized in "National research Tomsk polytechnic university" in framework of federal target program "Research and development in prior directions of scientific-technological complex development in Russia in 2014-2020 year", unique identifier of R&D project RFMEFI58114X0001.

On the mechanism of explosive eruption of mount erebus volcano: the dynamics of the rupture structure in a cavitating layer

E S Bol'shakova[1] and V K Kedrinskiy[1]

[1] Lavrentyev Institute of Hydrodynamics, Siberian Branch of Russian Academy of Sciences, Lavrentyev prosp. 15, Novosibirsk, 630090, Russia

E-mail: kedr@hydro.nsc.ru

Abstract. This paper presents the results of an experimental simulation of rupture development in heavily cavitating magma melt flow in volcanic conduits and its effect on the structure of explosive volcanic eruptions. The dynamics of the state of a layer of distilled water (similar in the density of cavitation nuclei to magma melt) under shock-wave loading was studied. The experiments were performed using electromagnetic hydrodynamic shock tubes (EM HST) with maximum capacitor bank energy of up to 100 J and 5 kJ. It was found that the topology of the rupture formed on the membrane surface did not change during its development. Empirical estimates were obtained for the proportion of the capacitor bank energy expended in the development of the rupture and the characteristic time of its existence. The study revealed a number of fundamentally new physical effects in the cavity dynamics in a cavitating medium: a cavitation "boundary layer" is formed on the surface of the quasi-empty rupture, which is transformed into a cluster of high energy density upon closure of the flow.

1. Introduction

Comparing the flow characteristics and structure in liquids containing a high density of micro-inhomogeneities (10^{12} m^{-3} in distilled water) under shock-wave (SW) loading with the natural processes occurring in volcanic conduits during explosive eruptions, it is easy to conclude that all of them are directly related to the hydrodynamics of high-velocity multiphase unsteady flows [1]. In many cases, the adequacy of the their representation by mathematical models and experimental designs in simulations of large-scale natural processes becomes obvious through the analysis of the pre-explosion state of Mount St. Helens volcano in May 1980 made by an international team [2]. The authors have reconstructed a model of the volcano architecture, which has proved an analog of the well-known hydrodynamic shock tube of I. Glass [3]. Studies of the mechanisms responsible for the explosive nature of volcanic eruptions often focus on two fundamental problems. One is to find mechanisms that determine the dynamics of the magma state behind the decompression wave front and its transition from the liquid melt state to the gas-particle state with the emission of giant ash clouds into the atmosphere [4-6]. The second problem is the discreteness of the eruption, which is observed in both closed (Mount St. Helens volcano) and open (Mount Erebus volcano) volcanic systems [6-12]. Obviously, the explosive nature of the decompression wave may well initiate ruptures in heavily cavitating magma flow in volcanic conduits even before the eruption. This hypothesis was made in [7], but the authors suggested that such processes could be characteristic of great depths and would not lead to an explosive eruption. Of interest is the architecture of Mount Erebus volcano – an open volcanic system of the explosive type with a lava lake on the crater floor, through which eruptions occur. The dynamics of tens of cyclic eruptions in the discharge zone has been studied using a radar system and the Doppler effect has been studied experimentally [2]. It has been shown that gas bubbles in the cavitating flow do not oscillate and thus cannot affect the cyclicity of eruptions of Mount Erebus volcano or similar systems. The aim of the present study was to experimentally simulate the development of a rupture in a cavitating liquid layer under SW loading in order to analyze the effect of the rupture dynamics on the cavitating flow structure.

2. Experimental setup

In the study we used an experimental setup based on SW loading of a liquid layer to produce rupture under conditions of rapidly developing bubble cavitation. The experiments were performed using electromagnetic hydrodynamic shock tubes (EM HST) with maximum capacitor bank energies of up to 100 J and 5 kJ, which makes it possible to analyze scale factors. A schematic diagram of the EM HST with the capacitor bank used and a general view of the formation dynamics of a characteristic eruption in a cavitating layer are presented in figure 1. The parameters of the electromagnetic unit provide an aperiodic discharge. Upon ignition of the discharge gap 3, the capacitor bank 1 (figure 1) is discharged into the flat spiral coil 2 placed under the conductive membrane 5, which serves as the bottom of the cell 4. The resulting current in the coil generates pulsed magnetic field B. The membrane 5 gains momentum as a result of the skin effect. The movement of the membrane generates a shock wave (SW) in the liquid layer 4 (distilled water). A high-speed "Photron FASTCAM SA 5" digital video camera (recording speed of up to 700 thousand frames per second) is synchronized with the ignition of the discharge gap 3.

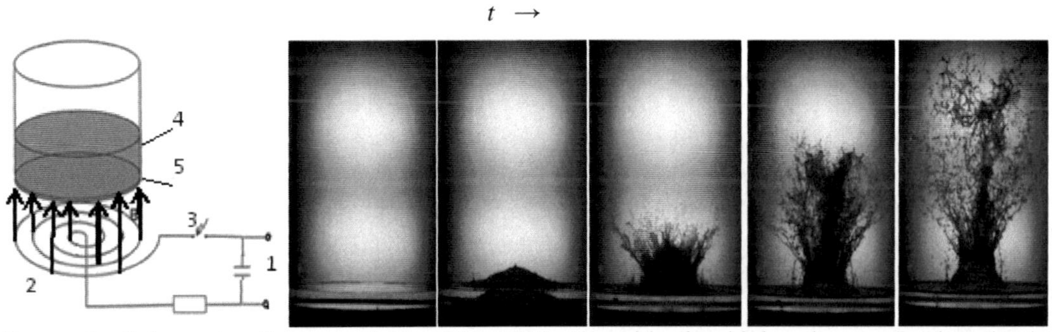

Figure 1. Schematic diagram of the EM HST (4) with the high-voltage capacitor bank (1); characteristic dynamics of eruption (experimental model).

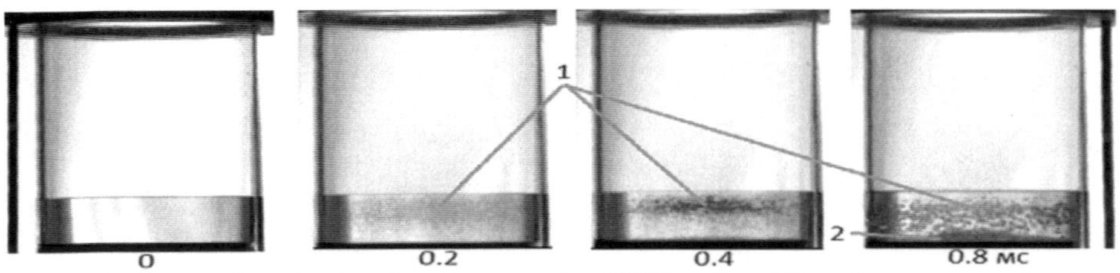

Figure 2. Development of cavitation in a layer under SW loading (1) and rupture nucleation (2) for a membrane diameter of 12 cm, a layer height of 4 cm, and a capacitor bank energy of 0.8 kJ.

3. Rupture dynamics in the cavitating liquid

The experiments showed that the shock wave reflected from the free surface of the layer initiated a rapid cavitation process in the layer (1) (figure 2). The initial position of the membrane is restored

Journal of Physics: Conference Series **754** (2016) 042003 doi:10.1088/1742-6596/754/4/042003

within a few microseconds for the small EM HST or within 10 μs for the large EM HST. The inertial motion of the impacted liquid leads to the formation of a rupture (2), whose dynamics and structure can be treated as a qualitative model of the initial stage of the Mount Erebus eruption process [13].

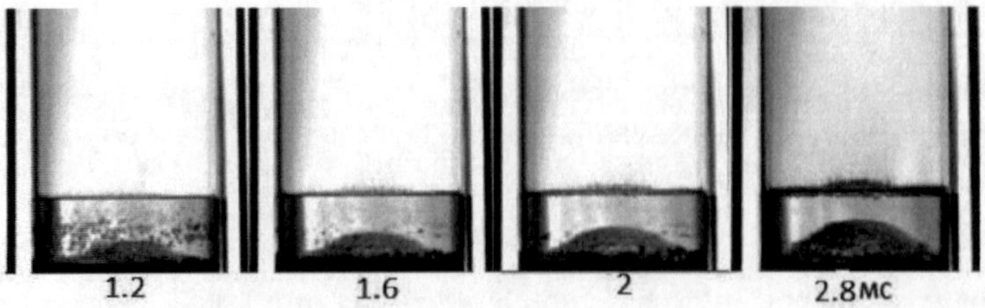

Figure 3. Development of rupture; the maximum volume is reached at $t = 2.8$ ms.

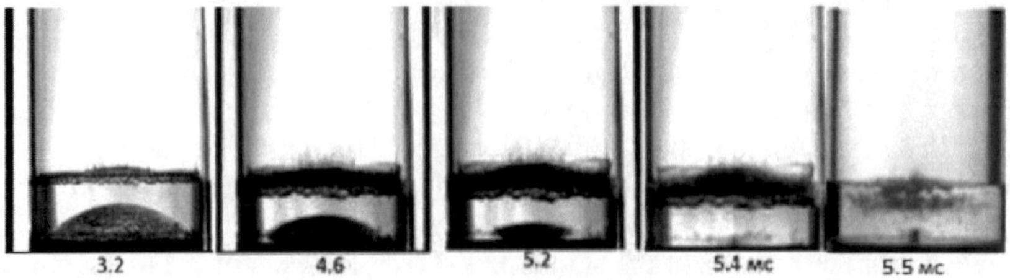

Figure 4. Closure of the rupture ($t = 3.2 - 5.2$ ms), the formation of a cluster ($t = 5.4$ ms), and the initial stage of its expansion ($t = 5.5$ ms).

A detailed analysis of the rupture structure (figures 3 and 4) shows that for different scales of SW loading of the layer, the rupture has the shape of a spherical segment and retains its topology throughout the period of existence. It is found that the collapse of the rupture on the membrane surface (solid wall) in the cavitating liquid does not produce a cumulative jet. The definition of the rupture as a spherical segment makes it possible to estimate its volume and potential energy $Q = VP_0$ at the time it reaches the maximum size. This result allows us to evaluate the proportion α of the capacitor bank energy E expended in the development of the rupture $Q = \alpha E$. It was found that $\alpha \approx 2\%$. It is assumed that the rupture is quasi-empty and the internal energy of the possible vapor-gas mixture in it is much less than the potential energy. A formula for the characteristic time of existence (formation–collapse) of the rupture was derived based on a combination of the determining parameters (the density ρ and the hydrostatic pressure P_0) within dimension theory:

$$T = \alpha^{1/3} \frac{E^{1/2} \rho^{1/2}}{P_0^{5/6}} \tag{1}$$

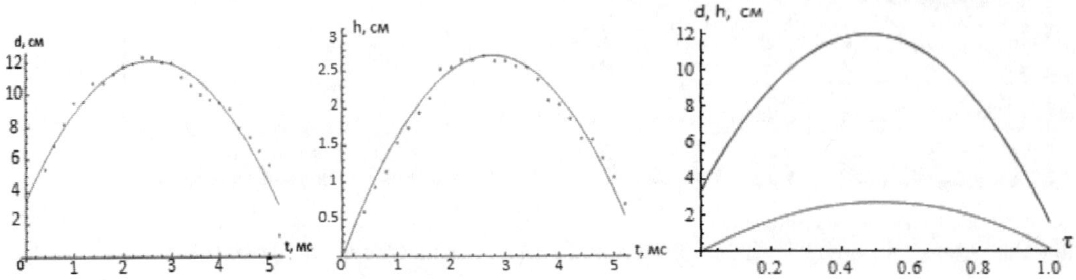

Figure 5. Dynamics of rupture parameters. In a dimension time – (a) width $d(t)$, (b) height $h(t)$ of rupture, at $E = 0.8$ кJ, and in a dimensionless time (c) – $d(\tau)$ (blue), $h(\tau)$ (red).

From an analysis of experimental data on the main geometrical characteristics of the rupture (diameter d and height h), we obtained semi-empirical relations of their dynamics:

$$d(t) = 3.45 + 6.71t - 1.3t^2, \qquad h(t) = 1.98t - 0.36t^2. \tag{2}$$

Introducing the dimensionless time ($\tau = t/T$), we obtain formulas in which d and h are in centimeters:

$$d(\tau) = 3.45 + 36.2\tau - 37.9\tau^2, \qquad h(\tau) = 10.7\tau - 10.5\tau^2, \tag{3}$$

which adequately describe the state of the rupture parameters throughout the range of its existence (figure 5).

Experiments with the small EM HST (membrane diameter of 4 cm, layer height ranges from 0.5 to 1.5 cm), with a capacitor bank energy less by more than an order of magnitude, show that the flows are practically similar and differ only in the characteristic linear and time parameters of rupture dynamics. It is found that the proportion of the energy expended in the development of rupture under these conditions is significantly lower $\alpha \approx 0.6\%$ due to the higher stiffness of the membrane.

One of the principal factors determining the characteristic rupture dynamics (unchanged topology, absence of cumulative flow) is the cavitating state of the surrounding liquid. In particular, the rupture development is accompanied by the formation of a kind of bubbly boundary layer on the surface (figure 6). The reason is obvious: the rupture is assumed to be quasi-empty and, hence, maintains low pressure throughout the period of development and closure up to the final phase. In this state, the rupture initiates the growth of microbubbles in the boundary layer.

Figure 6 shows an intermediate frames of the rupture initiated on a 4 cm membrane at a shock-loading energy below 100 J, proving that the scale factor does not affect the characteristics of the process studied.

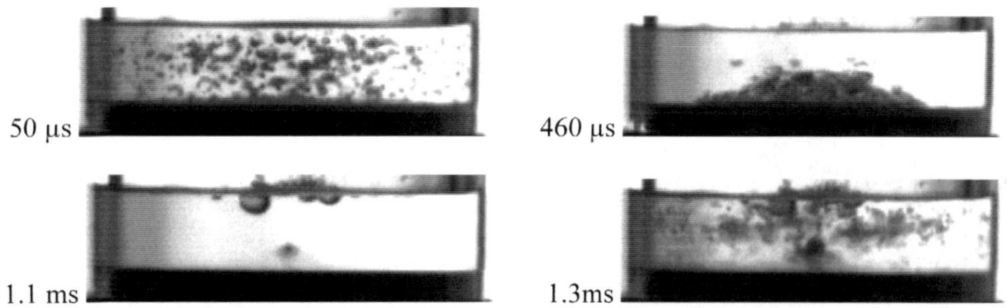

Figure 6. Structure of the rupture (membrane diameter of 4 cm, layer height of 1 cm): $t = 50$ and 460 μs (about V_{max}), t = 1.1 ms – at the collapse time (loading energy of 32 J); $t = 1.3$ ms – initial explosion.

Figure 7. Rupture structure ($t = 0.77$ ms) in the experiment using small EM HST for a cell diameter of 4 cm and a layer height of 1.5 cm. Right picture – the initial stage of cluster explosion (normal EM HST, $t = 5.5$ ms).

Analysis of the experimental data on the state dynamics of the rupture and boundary layer shows that the closure of the rupture in the cavitating liquid results in the formation of a highly compressed bubble cluster upon collapse of the boundary layer (figure 4, time $t = 5.4$ ms, the small dark area at the center of the membrane). The state of the bubble cluster at the time of closure of the rupture can be defined as a state with high internal energy. This is confirmed by experimental data on the generation of a fairly powerful secondary SW by the cluster at the initial stage of expansion. Radiation is detected from the formation of a secondary cavitation zone (figure 7, $t = 5.5$ ms) after the closure.

Acknowledgments
The study was financially supported by the Russian Foundation for Basic Researches (Grant No. 15-05-03336a).

References
[1] Kedrinskiy V K 2009 Hydrodynamic aspects of explosive eruptions of volcanoes: simulation problems *Shock Waves* **18**(6) 451–64
[2] Gerst A, Horst M, Kyle P R and Voege M 2006 The first second of a strombolian eruption: velocity observations at Erebus volcano, Antarctica *EOS Trans. Amer. Geophys. Union* **87**(52) Fall Meet. Suppl. Abstr. V31G-04
[3] Eichelberger J, Gordeev E and Koyaguchi T 2006 A Russian–Japan–US partnership to understand explosive volcanism http://www.uaf.edu/geology/ PIRE.pdf 1–4
[4] Glass I and Heuckroth L E 1963 Hydrodynamic shock tube *Phys. Fluids* **6**(4) 543–9
[5] Woods A W 1995 The dynamics of explosive volcanic eruptions *Rev. Geophys.* **33**(4) 495–530
[6] Dobran F 1992 Nonequilibrium flow in volcanic conduits and application to the eruptions of Mt. St. Helens on May 18, 1980, and Vesuvius in AD 79 *J. Volcanol. Geotherm. Res.* **49**(3–4) 285–311
[7] Gonnermann H M and Manga M 2007 The fluid mechanics inside a volcano *Annu. Rev. Fluid Mech.* **39** 321–56
[8] Denlinger R P and Hobbitt R P 1999 Cyclic eruptive behavior of silicic volcanoes *Geology* **27** 459–62
[9] Barmin A, Melnic O and Sparks S 2002 Periodic behavior in lava dome eruptions *Earth. Planet. Sci. Lett.* **199** 173–84
[10] Costa A, Melnik O, Sparks R S J and Voight B 2007 Control of magma flow in dykes on cyclic lava dome extrusion *Geophysical Research Lett.* **34** L02303 doi:10.1029/2006GL027466
[11] Kedrinskii V K 2014 Role of nuclei density as a "hidden" parameter in the formation of anomalous zones in a heavy cavitating magma *Journal of Applied Mechanics and Technical Physics* **55**(2) 276–81
[12] Kedrinskiy V K 2014 Structure of cavitation zones in a heavy magma under explosive character of its decompression *J. Acoust. Soc. Am.* **136**(4) Pt. 2 2290
[13] Kedrinskiy V K, Bol'shakova E S 2015 Empty cavity in a cavitating liquid: feature of flow structure *J. Acoust. Soc. Am.* **138**(3) Pt. 2 1829

Journal of Physics: Conference Series **754** (2016) 042004 doi:10.1088/1742-6596/754/4/042004

Kinetics of bulk crystallisation of supercooled melt

A.A. Chernov, A.A. Pil'nik

Kutateladze Institute of Thermophysics, Siberian Branch of the Russian Academy of Sciences, Novosibirsk 630090, Russia

E-mail: chernov@itp.nsc.ru

Abstract. The exact solution that fully describes the kinetics of the growth of a spherical crystal in supercooled melt is found. The kinetic model of nucleation-mediated crystallization is presented. It correctly takes into account the change in supercooling of the initial phase in the process of formation and evolution of a new phase.

1. Introduction

The process of crystallization occurs via fluctuation-induced nucleation and growth of centers of a new phase. Hence, the overall kinetics of such a process is determined by nucleation and growth rate of nuclei. These quantities are related to a set of various thermodynamic and kinetic parameters that are important under specific conditions and primarily to a degree of metastability of the parent phase. The main problem here is finding the time dependence of the fraction of the crystalline phase and the size distribution of the nuclei. The kinetic theory of fluctuation-induced nucleation and crystal growth is elaborated pretty well, whereas the thermophysical aspects of this problem are poorly investigated. In particular, the effect of heat release during phase transition on the rate of the process is insufficiently studied. Theoretical analysis of phase transition shows that for the most cases the process can be divided into two particular stages. The first one is the nucleation stage. The majority of new phase nuclei form during this particular stage. Nonlinear nature of nucleation expressed in strong dependence of nucleation rate from the degree of metastability. At the beginning of this stage the process is conditioned by the factors producing metastability, though the stage ends due to reduction in the degree of metastability. This reduction results from the transition of a portion of material from metastable phase to stable phase accompanied by release of latent heat of phase transition. Typically the portion of the new phase is relatively small at the moment of the end of nucleation stage. This is due to the fact that formation of even small portion of a new phase is enough for significant reduction in nucleation rate. The main part of initial phase transits to stable state during the subsequent growth of crystals already formed during nucleation stage. In studying the phase transition the nucleation stage attracts great interest, but it is a problem of great complexity. It involves thermodynamics of small systems, nucleation kinetics, knowing the growth mechanism in a wide range of parameters, correct accounting for effect that assembly of crystals has on metastability, etc. Combined with account for the factors producing metastability it all leads to challenging, nonlinear mathematical problem described by the system of integro-differential equations. At this time the complete solution of this problem is yet to be found.

2. The theory

2.1. Spherical crystal growth

Let a spherical crystal be nucleated and start to grow in a melt at the temperature $T_0 < T_{mel}$, where T_{mel} is the equilibrium phase-transition temperature. To find the growth rate, one has to solve the following boundary value problem [1, 2]. The heat transport in the crystal and melt isdescribed by the heat conduction equation

$$\rho_s c_s \frac{\partial T_s}{\partial t} = \frac{1}{r^2}\frac{\partial}{\partial r}\left(\lambda_s r^2 \frac{\partial T_s}{\partial r}\right); \quad \rho_l c_l \left(\frac{\partial T_l}{\partial t} + v_l \frac{\partial T_l}{\partial r}\right) = \frac{1}{r^2}\frac{\partial}{\partial r}\left(\lambda_l r^2 \frac{\partial T_l}{\partial r}\right), \tag{1}$$

where t is the time; r is the radial coordinate with the origin at the center of the crystal; T is the temperature; ρ, λ, c are the density, thermal conductivity, and specific heat, respectively (the subscripts "s" and "l" refer to the solid and liquid phases). It is noteworthy that heat conduction equation (1) includes the convection term describing the shrinkage of the substance under solidification, which leads to the emergence of a radial flow of the melt. The dependence of the melt velocity v_l on r is found from the continuity equation: $v_l = -k\dot{R}R^2/r^2$, where $k = \rho_s/\rho_l - 1$; R is the radius of the crystal, and $\dot{R}$ is the crystal growth rate.

The boundary conditions are as follows. Owing to the symmetry, there are no heat fluxes in the center of the crystal: $(\partial T_s/\partial r)_{r=0} = 0$. At the phase boundary, the temperatures of the solid and liquid phases are equal and the heat balance condition is fulfilled:

$$(T_s)_{r=R-0} = (T_l)_{r=R+0} = T_R(t); \tag{2}$$

$$\rho_s L\,\dot{R} = \lambda_s(\partial T_s/\partial r)_{r=R-0} - \lambda_l(\partial T_l/\partial r)_{r=R+0}, \tag{3}$$

where L is the specific heat of the phase transition and T_R is the temperature at the phase boundary, which is a sought function of time. Conditions (2) and (3) must be supplemented by the kinetic equation describing the dependence of the growth rate of the crystal on the supercooling of the melt at the crystallization front: $\dot{R} = \varphi(\Delta T)_{r=R}$, where $\Delta T = T_{mel} - T$ is the supercooling of the melt and φ is a function of the supercooling. In the case of a normal law of the crystal growth, this function is linear: $\varphi(\Delta T) = K\Delta T$, where K is the kinetic coefficient. This equation is often replaced by the condition that the melt temperature at the crystallization front coincides with the equilibrium phase transition temperature $T_R = T_{mel}$ (Stefan's approximation). However, this approximation holds only at the late stage of the process and only within a certain range of initial supercoolings.

For the growth of an isolated crystal, an additional boundary condition is the condition at infinity. It is the equality of the melt temperature and the initial temperature at a large distance from the crystal: $(T_l)_{r\to\infty} = T_0$. To describe the growth of the ensemble of nuclei, the cell models are used. According to these models, the volume of the crystallizing substance is divided into some regions (in the simplest case, spherical cells) with a single crystal growing in each region. Symmetry conditions are imposed at the cell boundaries.

The above problem can be simplified considerably under the assumption that the temperature gradients in the crystal are sufficiently low and the heat removal occurs entirely in the outermost region. In this case, only an external rather than conjugate problem should be solved. One more approximation is made under the assumption that a quasi-steady state is established quickly during the process; in this state, the temperature varies with time so slowly that one can set $\partial T/\partial t \simeq 0$. Then, the solution with the boundary conditions is [3]:

$$\theta_l(\tau,\bar{r}) = \mathbf{Ku}\,k^{-1}\frac{\exp(k\dot{\bar{R}}\bar{R}^2/\bar{r}) - 1}{\exp(k\dot{\bar{R}}\bar{R})}; \quad \dot{\bar{R}} = 1 - \mathbf{Ku}\,k^{-1}\{1 - \exp(-k\dot{\bar{R}}\bar{R})\}. \tag{4}$$

Here, $\theta = (T - T_0)/(T_{mel} - T_0)$; $\bar{r} = r/r_0$; $\bar{R} = R/r_0$; $\tau = t/t_0$; $\mathbf{Ku} = L/(c_l\Delta T_0)$ is the Kutateladze number; $r_0 = a_l/v_0$; $t_0 = a_l/v_0^2$; $v_0 = K\Delta T_0$; $a = \lambda/(\rho c)$.

In the case of zero shrinkage, the solution of (4) is simplified and the time dependence of the crystal radius can be found explicitly:

$$\theta_l(\tau,\bar{r}) = \mathbf{Ku}\,\dot{\bar{R}}\bar{R}^2/\bar{r}; \quad \bar{R}(\tau) = \mathbf{Ku}^{-1}(\sqrt{1+2\mathbf{Ku}\,\tau}-1). \tag{5}$$

As follows from (5), crystallization at $\tau \ll \mathbf{Ku}^{-1}$ proceeds isothermally, the supercooling of the melt at the crystallization front coincides with the initial one $\theta_R \approx 0$), and the growth rate of the crystal is determined only by the kinetics of the phase transition: $\bar{R}(\tau) \approx \tau$. At long times ($\tau \gg \mathbf{Ku}^{-1}$), the melt temperature at the crystallization front gradually approaches the equilibrium value ($\theta_R \to 1$) and the role of the thermal conductivity increases and becomes decisive. A quasi-steady-state temperature boundary layer is formed around the crystal and the time dependence of the crystal radius no longer includes the kinetic coefficient: $\bar{R}(\tau) \approx \sqrt{2\mathbf{Ku}^{-1}\tau}$.

It should be mentioned that the solution of (4) is correct only at low growth rates of the crystal. In the case of large deviations from equilibrium, another approach is needed [2, 4]. Change to the variables τ and $\chi = \bar{r}/\bar{R}$ reduces the boundary value problem to the problem with immobile boundaries:

$$\frac{\partial\theta_l}{\partial\tau} - \frac{\dot{\bar{R}}}{\bar{R}}\left(\frac{k}{\chi^2}+\chi\right)\frac{\partial\theta_l}{\partial\chi} = \frac{1}{\bar{R}^2}\frac{1}{\chi^2}\frac{\partial}{\partial\chi}\left(\chi^2\frac{\partial\theta_l}{\partial\chi}\right); \tag{6}$$

$$(\theta_l)_{\chi=1} = \theta_R; \quad \bar{R}\dot{\bar{R}} = -\{(k+1)\mathbf{Ku}\}^{-1}\left(\partial\theta_l/\partial\chi\right)_{\chi=1}; \quad \dot{\bar{R}} = 1-\theta_R; \tag{7}$$

$$(\theta_l)_{\chi\to\infty} = 0 \tag{8}$$

The quasi-steady-state solution of (6) with boundary conditions (7)–(8) is

$$\theta_l(\tau,\chi) = \theta_R(\tau)\left\{1 - \frac{\int_1^\chi \exp\left(-\left(\frac{\zeta^2}{2}-\frac{k}{\zeta}\right)\bar{R}\dot{\bar{R}}\right)\frac{d\zeta}{\zeta^2}}{\int_1^\infty \exp\left(-\left(\frac{\zeta^2}{2}-\frac{k}{\zeta}\right)\bar{R}\dot{\bar{R}}\right)\frac{d\zeta}{\zeta^2}}\right\}. \tag{9}$$

Using the found temperature profile, we obtain the implicit equation for $\bar{R}\dot{\bar{R}}$:

$$\bar{R}\dot{\bar{R}}\int_1^\infty \exp\left(-\left(\frac{\zeta^2}{2}-\frac{k}{\zeta}-\frac{1}{2}+k\right)\bar{R}\dot{\bar{R}}\right)\frac{d\zeta}{\zeta^2} = \theta_R\{(k+1)\mathbf{Ku}\}^{-1}. \tag{10}$$

It follows from (10) that $\bar{R}\dot{\bar{R}}$ is a function of the temperature θ_R at the crystallization front and does not depend explicitly on time: $\bar{R}\dot{\bar{R}} = \bar{a}^*(\theta_R(\tau))/2$. Here $\bar{a}^*$ is the sought function, which can be found from the differential equation

$$\dot{\theta}_R = 2(1-\theta_R)^2\{(d\bar{a}^*/d\theta_R) + \bar{a}^*/(1-\theta_R)\}^{-1}. \tag{11}$$

Numerical calculations indicate that the temperature θ_R at the crystallization front increases quite slowly within the time scale of establishing the quasisteady state. This makes valid the assumption of a quasi-steady-state process (but only in terms of the variables τ and χ). Fig. 1 shows the time dependences of the crystal radius and supercooling at the crystallization front ($\mathbf{Ku} = 2$). Clearly, the found analytical solution agrees pretty well with the numerical solution.

As follows from the analysis of (9)–(11), the melt temperature at the crystallization front at $\mathbf{Ku} > 1$ asymptotically approaches the equilibrium phase transition temperature and the function $\bar{a}^*$ tends to some constant. The temperature field around the crystal becomes a function of just one variable χ, and the solution of the boundary value problem becomes self-similar: $\bar{R}(\tau) = \sqrt{\bar{a}^*\tau}$. This stage of crystallization can be characterized as the equilibrium one.

At $\mathbf{Ku} < 1$, the crystal grows according to nearly the same scenario as at $\mathbf{Ku} > 1$ with the only difference that the melt temperature at the crystallization front tends to $\theta_R^* = \mathbf{Ku}$ rather than to the equilibrium temperature of the phase transition. Hence, the crystal grows under essentially nonequilibrium conditions during the whole process. The time dependence of the crystal radius is linear at this stage: $\bar{R}(\tau) = (1 - \mathbf{Ku})\,\tau$.

2.2. Overall crystallization

Let us now consider the description of overall crystallization kinetics. In general the form of the probabilistic solution for the problem of finding the time of isothermal crystallization was given by Kolmogorov:

$$X(t) = 1 - \exp\left(- \int_0^t J(\tau)\, V(t,t')\, d\tau \right) = 1 - \exp\left(-(\pi/3) J_0 v_0^3 t^4\right),$$

where $V = (4\pi/3) v_0^3 (t - t')^3$ — volume of the crystal; J — nucleation rate; $J_0 = J(\Delta T_0)$; $v_0 = K\Delta T_0$. It is noteworthy that though the full crystallization time is infinite, the number of formed crystallites is finite $n^* \approx 0.9\,(J_0/v_0)^{3/4}$. Obviously, the theory based on the assumption that the process is isothermal can only be used as the zero-order approximation [1, 5]. In order to devise a more accurate description of overall crystallization kinetics, it is necessary to account for the change of the melt supercooling in the process of crystal formation and growth. It can be made in the following way. The heat released in the phase change process is uniformly distributed through the volume. During the solid phase growth the supercooling will decrease and the nucleation process will quickly stop. Though this approach is apparent, it can lead to an appreciable error. It is due to the fact that the temperature field is highly non-uniform in real melts. This is why another approach is suggested in this work.

We will use nucleation rate from classical kinetic theory (also we will regard nucleation as being purely homogeneous, this is acceptable for cases of significant supercooling and also for small volumes where the probability of finding a nucleation site is small):

$$J = B \exp\{-W_{cr}/(k_B T)\},$$

where B — kinetic coefficient; $W_{cr} = 16\pi\sigma^3 T_{mel}^2/(3\rho_s^2\, L^2 \Delta T^2)$ is the work of formation of a critical nucleus; σ — solid-liquid surface tension.

As we can see before, around the crystal, the heated area is forming. Therefore, the probability of a new crystal forming near the previously formed one will be much less, the farther away it is. This is due to a strong nucleation rate dependence on supercooling. To take this fact into account in equations describing the process of overall crystallization, real function $J(r)$ is substituted by the step function so that, near the crystal (in the "forbidden" zone), it equals zero and outside this zone it equals the nucleation rate at the initial supercooling. The size of the thermal boundary layer thickness r_T is defined by the equality of latency periods in the cell for the "step" approximation and using real dependency $J(r)$ respectively. It is worth noting that this substitution of real dependency $J(r)$ with the step function is accurate: $r_T = \left\{R^3 + 3\int_R^S \{1 - \bar{J}(r)\}\, r^2\, dr\right\}^{1/3}$, where $\bar{J} = J/J_0$. With the distance from the center of the crystal, $\bar{J}$ rapidly converges to 1. That is why the upper integral limit can be safely changed to infinity. Thus, we can diverge from the cellular model of crystal growth during the nucleation stage.

In the light of the above-mentioned the generation rate of new nuclei is given by the product of nucleation rate and the volume where nucleation can happen:

$$dN/dt = J(t)\{1 - X_T(t)\},$$

where N — number of nuclei per unit volume; X_T — total volume of "forbidden" regions per unit volume in an ensemble of the crystals can be found from the integral equation:

$$X_T(t) = \int_0^t V_T(t,t')\,dN = \int_0^t J(t')\{1 - X_T(t')\}\,V_T(t,t')\,dt',$$

where $V_T = (4\pi/3)r_T^3$ — volume of a single "forbidden" zone around one crystal. Thus, the time dependence of the crystal phase volume per unit of the initial volume is

$$X(t) = \int_0^t J(t')\{1 - X_T(t')\}\,V(t,t')\,dt',$$

where $V = (4\pi/3)R^3$ is the volume of one crystal.

At the moment t^*, when the "forbidden" zones will cover the whole volume, the nucleation of new crystals will cease. The subsequent phase transition reduces merely to growth of the already nucleated crystal growth and it stops when the supercooling is annulled by the latent transition heat. The crystal phase volume reaches the value of $X^* = \mathbf{Ku}^{-1}$ if $\mathbf{Ku} > 1$. Full solidification can be achieved if $\mathbf{Ku} < 1$.

When, most of the time, the crystal growth occurs near the equilibrium (if $\mathbf{Ku} > 1$) and the setting time for the stationary nucleation rate is negligible, the problem can be simplified significantly. In this case, the temperature profile does not depend on time and, so: $r_T = \varkappa R$, i. e. the boundary layer thickness is in the direct ratio with the crystal size and proportionality coefficient $\varkappa$ does not depend on time. Using dimensionless variable $\tau = t/t_0$, where $t_0 = \left((4\pi/3)\varkappa^3 J_0\, a^{*3/2}\right)^{-2/5}$; $a^* = a_l\,\bar{a}^*$, we get

$$X_T(\tau) = \int_0^\tau \{1 - X_T(\tau')\}\,(\tau - \tau')^{3/2}\,d\tau'; \quad X(\tau) = \varkappa^{-3} X_T(\tau); \tag{12}$$

$$N(\tau) = J_0 t_0 \int_0^\tau \{1 - X_T(\tau')\}\,d\tau'. \tag{13}$$

The solution of (12), (13) can be found using the method of successive approximations. Our analysis shows that even the second iteration is sufficient to provide a good accuracy in determining function $X_T(\tau)$:

$$X_T(\tau) \approx (2/5)\,\tau^{5/2} - (3\pi/640)\,\tau^5; \quad N^* = N(\tau^*) \approx 0.56\,(J_0/(\varkappa^2 a^*))^{3/5}.$$

From this equation it is easy to find the full nucleation time: $\tau^* \approx 1.5$; $t^* = \tau^* t_0$. The crystal phase volume at the moment of nucleation completion is $X(\tau^*) = \varkappa^{-3}$. It is clear that the volume is small even for a relatively small $\varkappa$. It shows that the duration of the nucleation stage is significantly shorter than that of the post-nucleation stage of the phase transition. This fact can be used to significantly simplify the description of phase transition. If additional cooling is applied to the melt until fully crystallized, the mean size of crystallization grains will be $R^* = (4\pi(k+1)N^*/3)^{-1/3} \approx 0.75(k+1)^{-1/3}(J_0/(\varkappa^2 a^*))^{-1/5}$.

Let us consider another extreme case important for high supercooling. Under this conditions, most of the crystals grow under non-equilibrium conditions with supercooling at the interface equal to the initial one. In this case $\dot{R} = v_0 \approx K\Delta T_0$; $R(t) = v_0 t$ and

$$X(\tau) = \int_0^\tau \{1 - X(\tau')\}\,(\tau - \tau')^3\,d\tau', \tag{14}$$

where $\tau = t/\tilde{t}_0$; $\tilde{t}_0 = \left((4\pi/3)J_0\,(\varkappa v_0)^3\right)^{-1/4}$. The solution of (14) can be found analytically:

$$X(\tau) = 1 - \cos((3/2)^{1/4}\tau)\,\mathrm{ch}((3/2)^{1/4}\tau) \approx \tau^4/4; \quad N^* \approx 0.79\,(J_0/(\varkappa v_0))^{3/4}.$$

From it one can see that the nucleation period lasts for $\tau^* \approx \sqrt{2}; t^* = \tau^* \tilde{t}_0$. The average size of crystal grains in a solidified material is $R^* \approx 0.67\,(k+1)^{-1/3}(J_0/(\varkappa v_0))^{-1/4}$.

The dependence of final crystal grains number on the initial supercooling is illustrated in Fig. 2. It is shown that the Kolmogorov theory is justified only under high supercooling when the process is close to isothermal.

The model can be modified to be used in the case of gradual formation of metastability. Let us consider the volume of melt V_0 cooling with rate $q = -dT/dt$. The condition for the formation of the first nucleus is: $V_0 \int_{T_{\rm mel}}^{T_0} (J/q)\, dT \simeq 1$, where T_0 is the temperature when the intensive phase transition starts. Due to the transience of the nucleation process, we can safely say that most nuclei form under temperature T_0 as if the melt is cooled instantly. It considerably simplifies the model of nucleation process, but the factors responsible for the change of metastability during the following phase transition should be correctly taken into account.

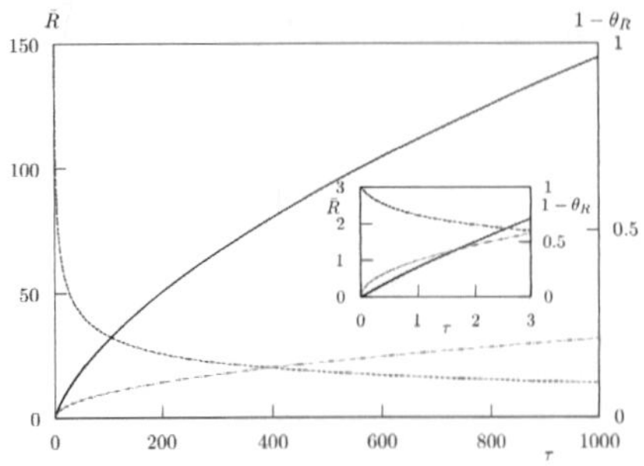

Figure 1. Time dependencies of the crystal radius and melt supercooling at the crystallization front. The dash-dotted line is the quasi-steady-state solution.

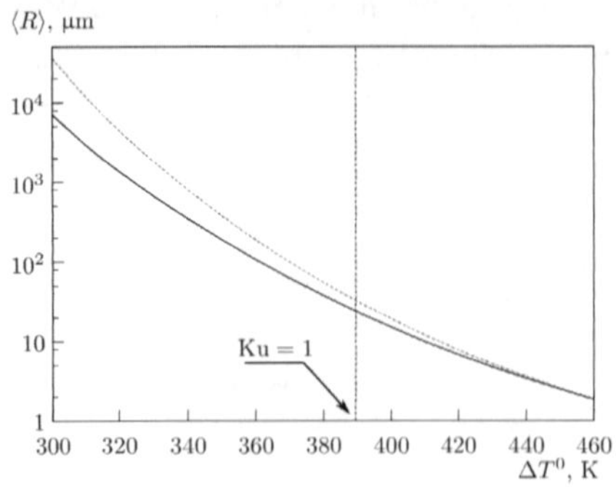

Figure 2. Dependence of size of crystal grains (nickel) on initial supercooling of the melt: solid line — developed model; dashed line — Kolmogorov's theory.

3. Conclusion

To summarize, the found solution fully describes the kinetics of the growth of a spherical crystal under strongly nonequilibrium conditions in a wide range of supercoolings taking into account shrinkage of the substance under solidification. The suggested model of overall crystallization allows us to find the nucleation time which, by itself, is a non-trivial problem and it has never been addressed in classical contributions. Also the model allows us to find the final number of crystal grains and, therefore, their size in a solidified material.

This work was supported by the Russian Science Foundation, project # 15-19-10025.

References

[1] Chernov A A 2009 *Int. J. Heat Mass Transfer* **52** 3290–3296
[2] Chernov A A and Pil'nik A A 2015 *J. Exp. Theor. Phys. Letters* **102** 526–529
[3] Chernov A A 2009 *J. Eng. Thermophys.* **18** 238–241
[4] Chernov A A, Kedrinsky V K and Pil'nik A A 2014 *Phys. Fluids* **26** 116602
[5] Chernov A A, Pilnik A A and Islamov D R 2016 *J. Cryst. Growth.* **450** 45–49

Gas hydrate formation at cyclic boiling-condensation of hydrate-forming gas in a closed volume of liquid

D S Elistratov [1,2], I V Mezentsev[1,2], A V Meleshkin[1,2], S L Elistratov[2], A Yu Manakov[3]

[1]Novosibirsk State University, Novosibirsk, 630090, Russia

[2]Kutateladze Institute of Thermophysics SB RAS, Novosibirsk, 630090, Russia

[3]Nikolaev Institute of Inorganic Chemistry SB RAS, Novosibirsk, 630090, Russia

E-mail: mezn@itp.nsc.ru

Abstract. A new method of the cyclic process of boiling-condensation of hydrate-forming gas in a closed volume was developed at Kutateladze Institute of Thermophysics based on the previous studies. This method compares favorably with previously developed techniques by its simplicity because it does not require a constant supply of gas and there is no need for a mixing device.

Introduction

Since the end of the 19th century, the properties of gas hydrates have been studied. In [1-5] and many other works, the properties of hydrates of different gases are described in detail: the structure, conditions and regularities of their formation and storage. Since the beginning of 2000s, the studies of gas hydrate production using various modeling gases have been performed at Kutateladze Institute of Thermophysics [5-9]. In 2006, the institute researchers have received a patent for the method of gas hydrates production [11]. The essence of invention is as follows: the compressed and cooled gas is fed into a reaction vessel filled with water under the static pressure and temperature below the equilibrium temperature of hydrate formation at this static pressure. The gas pressure is slightly higher than the pressure in the reaction vessel, and the gas temperature equals the temperature of water in the vessel. Gas is mixed with water. The pressure shock waves with the amplitude of up to hundreds atmospheres are generated in the gas-liquid medium. The shock waves can be generated by the electromagnetic pulse emitters, pneumostrikers or other devices. When the shock wave propagates in the gas-liquid medium, the pressure increases and the gas phase splits in the entire reaction vessel due to low wave dissipation, and, therefore, the degree of medium metastability increases as well as the number of hydrate nucleation centers; the size of gas inclusions decreases; the interface becomes larger; the relative velocity of gas inclusions in liquid increases; and liquid motion becomes turbulent. All these phenomena lead to acceleration of mass transfer process at the interface and, hence, intensify the process of hydrate formation. This method uses the impact effect on the gas-liquid mixture. The continuation of this research was the development of explosive boiling method, whose essence is as follows: gaseous Freon is fed under the pressure into a sealed vessel, half-filled with water, cooled to 2-4°C. Freon enters the vessel, increases the pressure, and being cooled to the ambient temperature, it is liquefied. Since liquefied Freon is much heavier than water, it descends and forms a layer on the vessel bottom. When supplied Freon is on the bottom, the pressure is bled down. As a result, liquid Freon on the vessel bottom boils up explosively and mixes with water. The boiling process is accompanied by mass formation of the shock waves, crushing the gas bubbles; as a result, a developed interface is formed. The bubbles move relative to water, and surrounding water is cold. Combination of these factors leads to mass formation of Freon hydrate throughout the water volume. It should be noted that the rate of hydrate formation is determined by heat transfer and it is not limited by diffusion of gas into liquid.

Together with gas phase crushing, heat transfer from the interface, where gas hydrate is formed, is of considerable importance. According to the studies of the Japanese scientists [10], enthalpy of hydrate formation of Freon R-134A equals 124 kJ/mole. Thus, when one kilo of gaseous Freon is converted into gas hydrate, 310 kJ of heat is released.

This paper describes a new method of Freon gas hydrate production based on cyclic boiling-condensation of hydrate-forming gas in a confined space. This method is not as fast as those described above, however, it differs favorably from them by its simplicity and low energy consumption.

1. Experimental setup and methods

The working section is a sealed vessel of 740-mm height and 150X150-mm cross-section (Fig. 1). Two walls of this vessel are made of an optically transparent material, and they allow visualization of the inside processes. The required temperature is maintained in the setup by the in-built cooling system (water jacket). The lower lid of setup is not thermally insulated, and it is heated by the ambient medium. The temperature is measured by the DTS204-RT100 sensor, and the pressure is measured by the PD-100 probe. This setup allows investigations in a wide range of temperatures (from -10 to 60°C) and pressures (from 0.1 to 10 MPa).

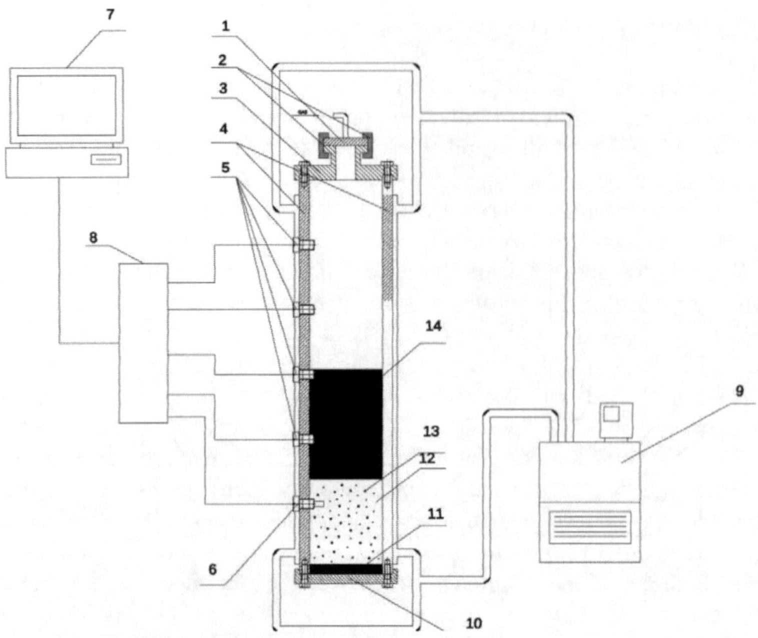

Figure. 1. The scheme of experimental setup of the "autoclave" type: 1 — Plug with a hole for gas introduction; 2 — Clamp; 3 — Upper lid; 4 — Lateral walls with in-built cooling system; 5 — Pressure sensors; 6 — Temperature sensor; 7 — PC; 8 — Data acquisition system; 9 — Cryostat; 10 — Lower lid; 11— Liquefied Freon R-134A; 12 — Distilled water; 13 — Freon bubbles; 14 — Freon hydrate.

The essence of experiment was as follows. The reaction vessel was half-filled with water. Water temperature was cooled to 2-4°C, then, 0.5 kg of gaseous Freon R-134A was fed into the chamber. Since the temperature of the vessel walls was negative, Freon was quickly liquefied and settled on the bottom. When gas supply into the working section was stopped, the pressure was leveled and boiling on the lower lid of the vessel started. Due to intense boiling, a large number of refrigerant vapor bubbles were formed. Rising to the surface, vapor in bubbles was cooled, and the conditions for Freon hydrate formation were formed. The hydrate film started growing on the bubble walls. Reached the free surface of liquid, the bubbles collapsed, leaving hydrate flakes. Freon vapors, which did not convert to hydrate, were condensed on the cooled walls and descended to the bottom of the vessel, where they were involved again into the boiling process. It is evident from the description that the process is cyclic and continues until all liquid Freon turns to hydrate at boiling (Fig. 2). Process intensity depended on heat removal from the lateral walls and heat supply to the bottom of setup.

Figure. 2. Gas hydrate obtained in experiment.

Conclusion

This work describes the new cyclic method for producing hydrate of modeling gas. The experimental studies were carried out and the processes occurring in the setup during gas hydrate formation were described.

Acknowledgments

This work was supported by the Russian Science Foundation (project no. 15-19-10025). We thank Novosibirsk State University for granting us the access to libraries.

References

[1] Gudmundsson J, Mork M, Graff O 2002 *Proc. Of the 4th Intern. Conf. on gas hydrates* pp 997-1002

[2] Makagon Y F 1997 *Hydrates of Hydrocarbons* (Tulsa, Oklahoma. Pennwell publishing company)

[3] Yakushev V S, Istomin V A 1993 Gas-hydrates self-preservation effect *Physics and Chemistry of Ice* pp 136-139

[4] Sloan E D, Koh C A 2008 Clathrate hydrates of natural gases 3rd ed *Chemical industries series* p 721

[5] Misura S Y 2013 Effect of Heat Transfer on the Kinetics of Methan hydrate Dissociation *Chem. Phys. Lett* vol. 583 pp 34-37

[6] Nakoryakov V E, Tsoi A N, Mezentsev I V and Meleshkin A V 2014 Explosive Boiling of Liquid Nitrogen Jet in Water *Journal of Engineering Thermophysics* vol.23 № 1 pp 1-8

[7] Nakoryakov V E, Dontsov V E, Chernov A A 2006 Gas hydrate formation in a gas-liquid mixture behind a shock wave Doklady Physic vol. 51 № 11 P 621-624

[8] Dontsov V E,. Chernov A A 2009 Dilution and hydrate forming process in shock waves Int. J. Heat Mass Transfer vol. 52 № 21-22 P 4919-4928

[9] Chernov A A, Dontsov V E 2011 The processes of dissolution and hydrate forming behind the shock wave in the gas–liquid medium with gas mixture bubbles Int. J. Heat Mass Transfer vol 54 № 19-20 P 4307-4316

[10] T. Akiya, T. Shimazaki, M. Oowa, M. Matsuo, and Y. Yoshida 1999 Formation Conditions of Clathrates Between HFC Alternative Refrigerants and Water International Journal of Thermophysics. Vol. 20, No. 6, 1999

[11] RF patent 2270053, 2006, B01F3/04.

Journal of Physics: Conference Series **754** (2016) 042006 doi:10.1088/1742-6596/754/4/042006

Water droplet evaporation and dynamics in a mini-channel under action of the gas flow

E A Isachenko[1,2], E V Orlik[1] and E F Bykovskaya[1]

[1]Kutateladze Institute of Thermophysics SB RAS, 1 Lavrentyev Ave., 630090, Novosibirsk, Russia

[2] Physics Department, Novosibirsk State University, 2 Pirogova Str., 630090, Novosibirsk, Russia

E-mail: isach-e@yandex.ru

Abstract. An experimental setup was developed to study the vaporization and dynamics of liquid droplets, blown by the gas flow in a mini-channel. The shadow method was the main method of measurement; a drop was also observed from the top. A series of experiments was carried out with single water drops with volumes varying from 60 to 150 μl in the channel of 6 mm height on the polished stainless steel substrate. The experiments have resulted in the dependences of evaporation rate in the temperature range of the substrate surface from 25 to 70°C and Reynolds numbers of the gas flow from 0 to 2500. The advancing and receding contact angles were measured depending on the Re number of the gas flow. The gas flow rate at which the droplet motion over the substrate starts was determined depending on the surface temperature at different drop volumes.

1. Introduction

The development of microelectronics is closely related to the problem of heat removal. Heat generation rate in microprocessors of modern computers reaches 100-200 W/cm^2, which is the limit of air cooling systems. Therefore, the most perspective appear to be liquid, two-phase and evaporative cooling systems. To date, the most affective are the systems based on impact jets and sprays, two-phase flows in mini-channels and fluid flows in micro-channels. At cooling microelectronic equipment, these systems enable removal of heat fluxes of up to 500-800 W/cm^2 [1, 2, 3, 4, 5]. However, already now the electronic industry is ready to produce components where the heat flux density can reach values of 1-2 kW/cm^2 and higher [6]. Even the most effective systems that use the two-phase flows are unable to cool such components, which is a technological barrier to the further development of microelectronic systems. One of the technical solutions to achieve significant intensification of heat transfer and, accordingly, an affective cooling of microelectronic equipment with local heat generation is a device, forming near-wall droplet flows in micro - and mini-channels. The transition from a continuous film flow to the droplet flow with an increase in the length of the contact lines leads to the intensification of heat transfer at evaporation [7]. Droplet evaporation has been widely studied numerically and experimentally, but mostly in still air [8,9, 10]. There are a few works studying the dynamics of single liquid droplets in rather high channels [11], but the existing data are not sufficient.

The objectives of this work are as follows:

•the evaporation of a single water droplet placed on a heated substrate in a flat channel under the influence of the laminar air flow;
•the dynamics of a single water droplet, moving on cold and hot surfaces.

2. Describing the experimental setup and measurement methods

The scheme of the experimental setup is shown in figure 1. It includes a channel with height varying from 4 mm to 20 mm, removable substrates, the substrate temperature control, and the air supply system. The side walls and the top cover are equipped with optical windows for visualization. The aim of this work is to study the influence of substrate temperature and Reynolds number of the air flow on the drop evaporation rate in a channel of 6 mm height. The used working fluid was ultrapure water, obtained with the use of Milli-Q system. The drop was placed into the channel with a syringe. The research was carried out at the initial drop volume of 60-150 μl. All experiments were realized on a substrate of stainless steel with polished surface. The Reynolds number of the gas flow was varied from 0 to 2500. The gas temperature was maintained equal to 25°C using a system of thermoregulation. The substrate was heated by Peltier elements, installed in contact with the substrate. The substrate temperature during the experiment was regulated with a precision of 0.2° using the PR-59 controller.

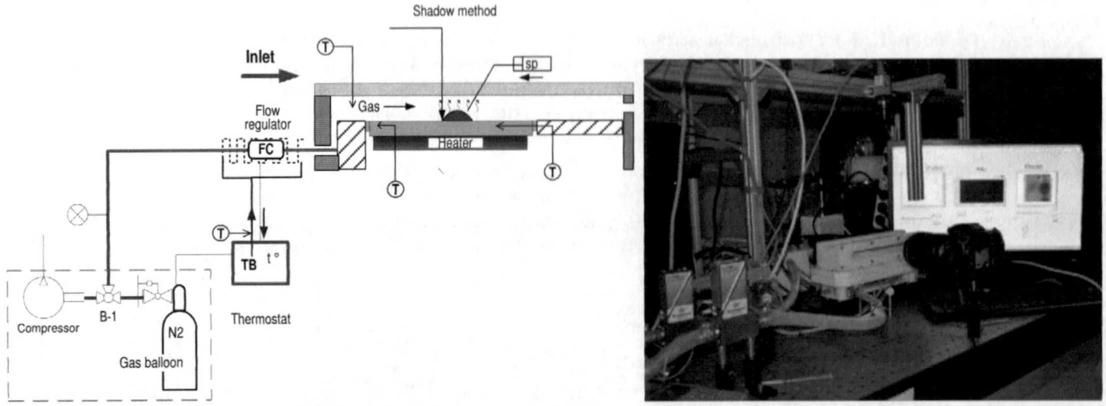

Figure 1. Scheme and photograph of the setup.

To register the contour of a sessile drop on the solid substrate the shadow method was used. Its principle is based on the fact that the physical object is illuminated by a parallel light beam, and its shadow is recorded by the camera, as shown in Fig. 2. Optical equipment allows obtaining images with a resolution of 6 μm/pixel (see Fig. 3 and 7). The obtained images were processed by different methods with the help of software (the Drop Shape Analysis by KRÜSS). The most suitable for measuring the contact angle of wetting in the experiments was the tangential method, since the drop had an asymmetric shape due to the impact of air flow. The principle of the tangential method is based on adapting the contour to the equation of the conic sectors. Then, the derivative of this equation equal to the wetting angle was taken at the point of intersection of the contour line and the baseline. Before the experiments, the hysteresis of the contact angle on the substrate was measured. The advancing angle of wetting was around 80 degrees, and the receding angle was around 50 degrees.

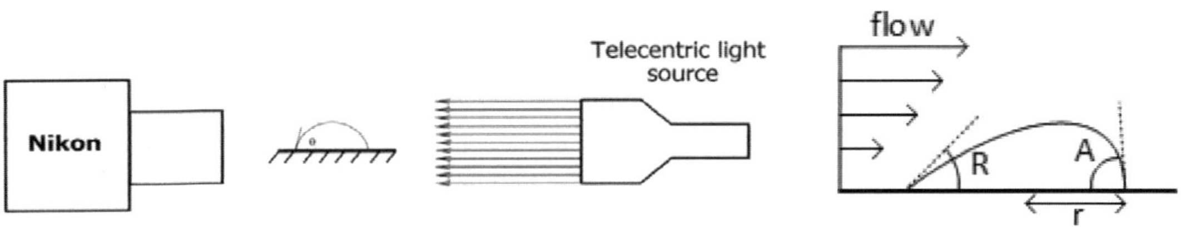

a) b)

Figure 2. Shadow method –a), schematic drawing of the droplet in the channel –b).

3. Results

3.1. Evaporation of a sessile drop

Figure 3 shows photographs of droplets at different substrate temperatures and times. Images were registered with an interval of 10 seconds. Analyzing the results of the experiment given in Figs. 4, 5 and 6, we can draw the following conclusions: the rate of evaporation is determined by the substrate temperature; the gas flow rate also affects the evaporation, and this affect is maximal at the initial moment of time. The change of drop volume is well approximated by a parabolic profile (see Fig. 4). Figure 5 shows data for the evaporation rate J related to the droplet surface area depending on the droplet size. The graph shows that the evaporation rate changes several times and has a local minimum. The explanation for this is the fact that in the first moment of time the drop remains similar, i.e. the ratio of height to diameter of the droplet remains unchanged (see Fig. 6). Since the gas velocity in a flat channel has a parabolic profile, the velocity of drop blowing decreases with reduction in the drop height; this explains the local minimum. As soon as the drop becomes flat, i.e. the ratio of the drop contact area to the substrate surface area from which evaporation begins starts growing. The same tendency of increasing of J was observed in [10] for sessile droplet in motionless air (see ref [10]).

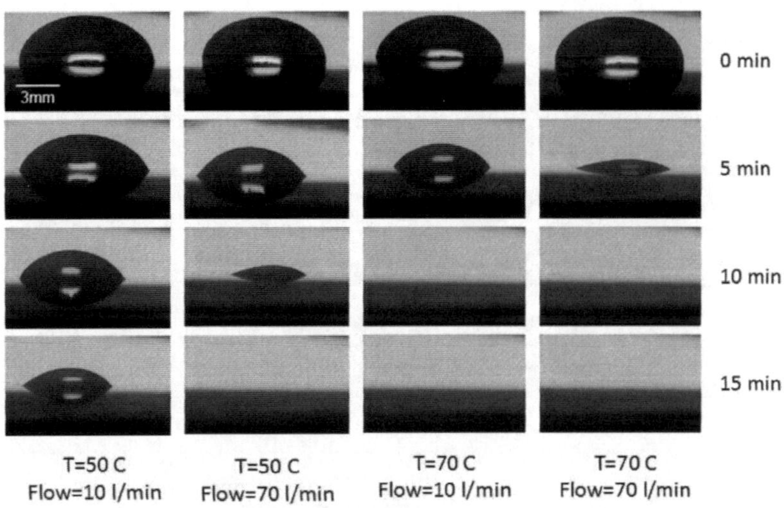

Figure 3. Droplet evaporation depending on the substrate temperature and gas flow rate.

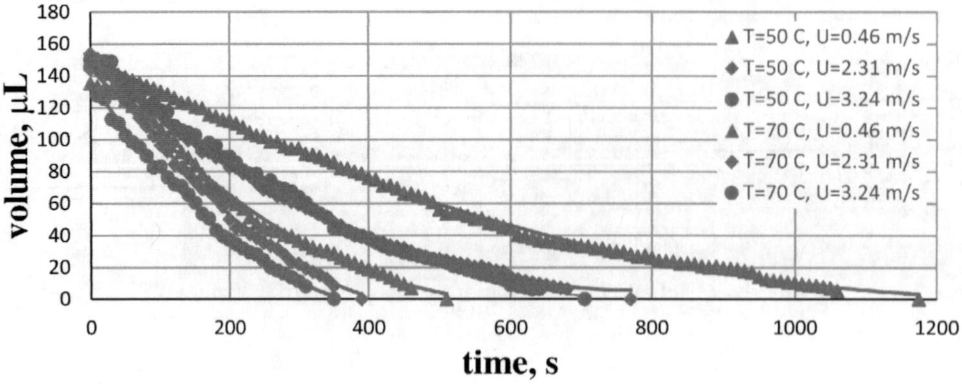

Figure 4. The change of drop volume during evaporation for different experimental conditions.

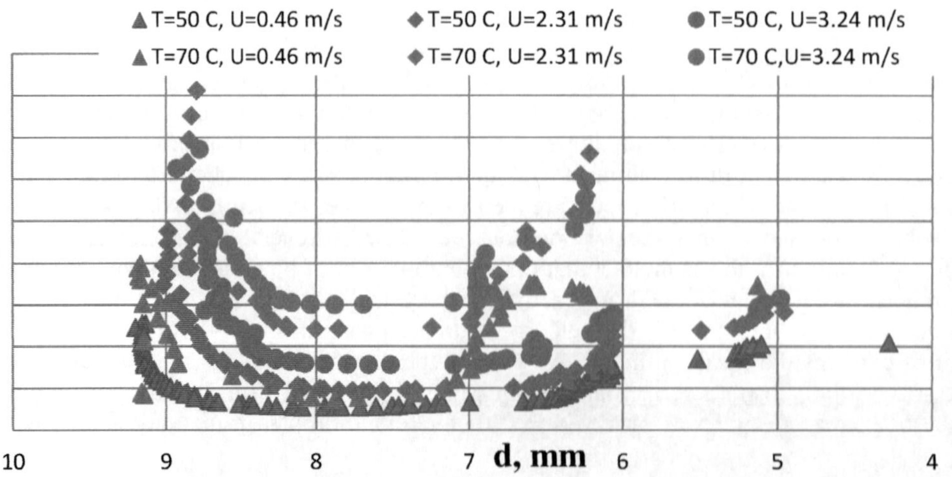

Figure 5. The rate of evaporation versus the droplet diameter.

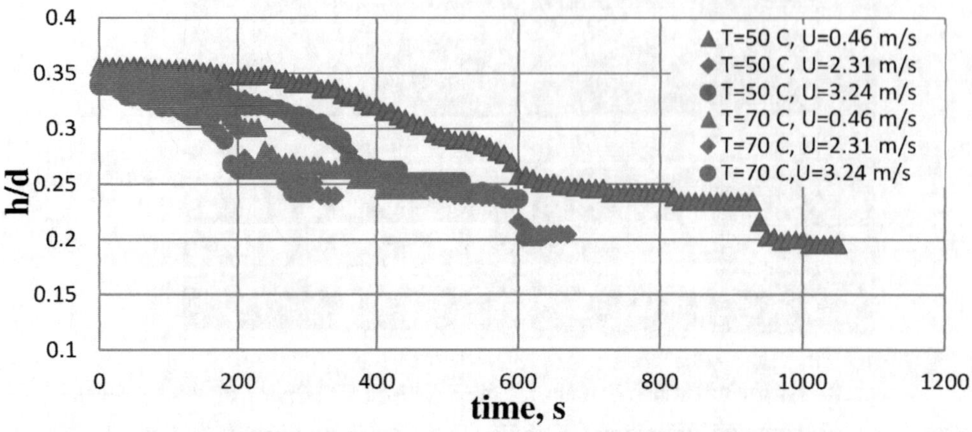

Figure 6. The ratio of height to the droplet diameter versus time.

3.2 Droplet dynamics

At an increased velocity of the gas flow, a drop lost symmetry (see Fig. 7), and at some velocity values it started moving over the substrate (see Fig. 8). It should be noted that increasing temperature of the substrate surface, leads a decrease in the receding contact angle; in other words, the stack to the surface. One of the explanations that must be taken into consideration is that the viscosity and the surface tension of water significantly depend on temperature.

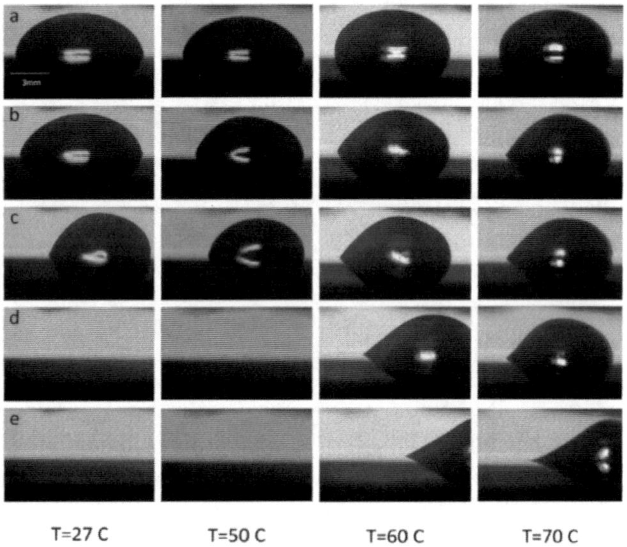

Figure 7. Droplet dynamics Flow rate: (a) 0 l/min, (b) 90 l/min, (c) 100 l/min, (d) 120 l/min, (e) 135 l/min.

Figure 8 shows the friction force affecting the contact line. It was determined by the formula $\sigma r(\cos(A)-\cos(R))$, where σ is the surface tension, and the angles A and R, r are shown in Fig. 2. The larger symbols correspond to the values at which the drop began its motion. The work [11] studied the dynamics of different droplet on the silane substrates. The values of velocity required to move droplet and correspondent friction forces in [11] were very similar to those obtained in our work. The effect of heating on the dynamics of droplets in [11] was not taken into account. In our work it was found that with increasing temperature the drop motion starts at somewhat larger gas flows, and the friction force affecting the drop on the heated surface, calculated by the above formula, is a few times larger than in the case without heating.

4. Conclusion

In this paper, we have carried out a detailed study of the evaporation of a single droplet in the 6 mm channel. With this view, the images of the drops were recorded with an interval of 10 sec and analyzed. It has been experimentally found out that the evaporation rate is determined by the temperature of the substrate surface, and the velocity of the gas flow blowing the drop also affects evaporation. It is found local minimum on evaporation rate curves. New data on the dynamics of water drops on a heated surface have been obtained. It has been found that with increasing temperature, the droplet starts to stick to the substrate surface.

The authors express their gratitude to the Russian Science Foundation for the support of this work (project No. 14-19-01755).

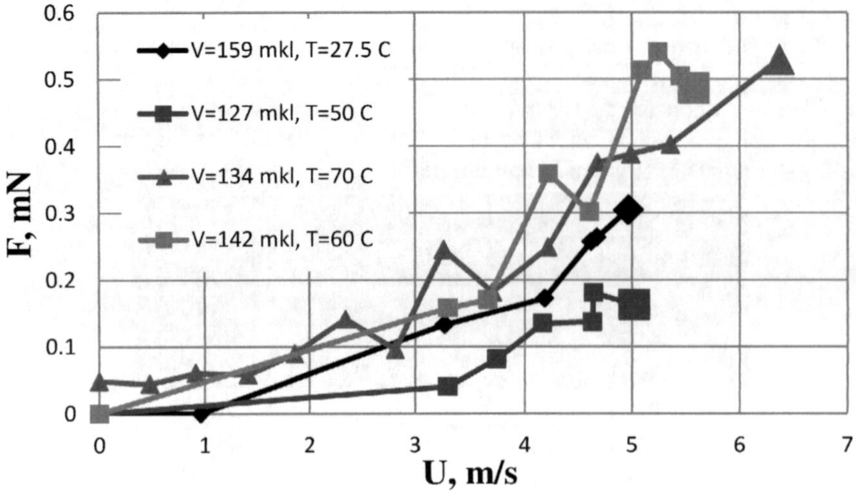

Figure 8. Friction force affecting the drop contact line, depending on the air flow rate and substrate temperature.

References

[1] Chinnov E.A and Kabov O.A 2004 The Affect of three-dimensional deformations on local heat transfer to a nonuniformly heated falling film of liquid *High Temperatures* **42**, No 2 pp 267–277 DOI: 10.1023/B:HITE.0000026159.25885.55

[2] Zaitsev D.V, Kabov O.A, Cheverda V.V and Bufetov N.S 2004 The affect of wave formation and wetting angle on the thermocapillary breakdown of a falling liquid film *High temperatures 42* No.3 pp 450–456 http://dx.doi.org/10.1023/B:HITE.0000033882.23387.19

[3] Tomoki Hirokawa, Haruhiko Ohta and Oleg A. Kabov 2015 Experimental Investigation on Behaviors and Heat Transfer in Shear-Driven Liquid Film Flow, *Interfacial Phenomena and Heat Transfer* **3** No 3 pp 303–317 DOI: 10.1615/InterfacPhenomHeatTransfer.2016015813

[4] Zaitsev D.V, Rodionov D.A and Kabov O.A 2009 Critical Heat Flux in a Locally Heated Liquid Film Driven by Gas Flow in a Minichannel, *Technical Physics Letters*, **35** No. 7 pp 680–682 DOI: 10.1134/S106378500907027X

[5] Jungho Kim 2006 Spray cooling heat transfer: The state of the art *International Journal of Heat and Fluid Flow* **28** Issue 4 pp 753-767

[6] Bar-Cohen A and Wang P 2012 Thermal Management of On-Chip Hot Spots *ASME Journal Heat Transfer* **134** (5) 051017-28 DOI 10.1115/1.4005708

[7] An Zou, Ashish Chanana, Amit Agrawal, Peter C. Wayner and Jr & Shalabh, C Maroo1 2016 Maroo1 Steady State Vapor Bubble in Pool Boiling *Scientific Reports* **6** Article number: 20240, doi: 10.1038/srep20240

[8] Kabov O Zaitsev D 2013 The affect of wetting hysteresis on drop spreading under gravity *Doklady Physics* **58** (7) pp 292–295 DOI: 10.1134/S1028335813070045

[9] Semenov A, Feoktistov D, Zaitsev D, Kuznetsov G and Kabov O 2015 Experimental investigation of liquid drop evaporation on a heated solid surface *Thermophysics and Aeromechanics* **22**(6) pp 771–774 DOI: 10.1134/S086986431506013X

[10] Gatapova E, Semenov A, Zaitsev D and Kabov O 2014 Evaporation of a sessile water drop on heated surface with controlled wettability *Colloids and Surfaces A: Physicochemical and Engineering Aspects* **441** pp 776–785 DOI: 10.1016/j.colsurfa.2013.05.046

[11] Fan J, Wilson M C T and Kapur N 2011 Displacement of liquid droplets on a surface by a shearing air flow *Journal of Colloid and Interface Science* **356** pp 286–292 doi:10.1016/j.jcis.2010.12.087

Journal of Physics: Conference Series **754** (2016) 042007 doi:10.1088/1742-6596/754/4/042007

Multiphase systems: Wave processes, state dynamics, and mathematical models

V K Kedrinskii

Lavrentyev Institute of Hydrodynamics SB RAS, 15 Lavrentyev Prosp., Novosibirsk, 630090, Russia

E-mail: kedr@hydro.nsc.ru

Abstract. Phenomena that accompany the processes of shock wave loading and bubble cavitation are briefly reviewed. The mechanisms of spalling due to shock wave interaction with the free surface, formation of "laser radiation" by bubble clusters, and bubble systems in volcanic conduits (problems of their transformation to a gas-particle system and destruction) are considered.

1. Introduction

High-velocity hydrodynamics of multiphase media as an individual research field appeared and started its intense development actually in the second half of the last century. It started from solving problems of practical significance, which were based on the fact of drastic enhancement of medium compressibility at the instant when a gaseous component was formed in the liquid carrier phase. Almost simultaneously with experimental investigations of the dynamics of interaction of gas bubbles and bubble screens with shock waves, two mathematical models were developed, which later became the classical tools. The first model was developed by Rakhmatulin [1] in the form of a system of conservation laws for each phase, which are related through force interaction and joint deformation. The second one was the Iordanskii-Kogarko-van Wijngaarden (IKW) gas-dynamic model [2-4], where the bubble medium is considered as a homogeneous phase with averaged characteristics. Multiphase media with unique properties and wave processes have recently turned out to be "on demand" in solving many problems in interdisciplinary fields: fluid mechanics and solid state physics, Earth sciences, acoustics, and medicine [5]. Here we can mention a large number of physical formulations aimed at solving problems of the so-called "acoustic laser," where active and passive bubble media are used as working media. Such media are capable of absorbing incident disturbances, amplify them, and re-emit the energy in a specified direction. We mean here free bubble systems shaped as a sphere, torus, or cord-type clusters [6-8]. In spherical clusters, significant retardation of the shock wave (SW) refracted into this cluster as compared to the external SW initiating the process in the cluster is often used. This effect allows one to control the focusing process, including the SW focal point in the cluster, its enhancement, and directed secondary radiation on the interface with the ambient liquid, by means of changing the gas phase concentration. Excitation of the bubble torus by the external SW leads not only to SW amplification in this cluster, but also to its focusing onto the axis of symmetry; as a result, a nonlinear process is observed at a certain distance from the torus plane: a series of Mach disks is formed. A cord-shaped cluster is something like a waveguide with focusing onto the axis of symmetry and amplification of the wave propagating along the axis of symmetry. Naturally, these effects can become much more pronounced if a chemically active mixture is used as a gas phase. As was shown in [9], here is often no need to consider Todes' bimolecular kinetics or calculate thermodynamic parameters of the chemical mixture. It is sufficient to use the model of an instantaneous adiabatic explosion in a constant volume at the instant of reaching the temperature of gas ignition in the bubble, according to which the new pressure in the bubble is determined on the basis of data on the heat generated in the explosion and the ratio of specific heats of the explosion products becomes different. Investigations in a different direction, namely, shock-induced cavitation, which was developed and formulated in [10], will be also considered in the presentation. It is based on the idea of the multiphase character of cavitating media and the adequacy of the IKW model to the processes in such media, regardless which phase of loading is considered (shock wave or expansion wave (EW)).

2. Mathematical model of a multiphase medium

2.1. Mathematical IKW model

$$\frac{d\rho}{dt}+\rho\cdot\operatorname{div}\mathbf{v}=0,\qquad \frac{dv}{dt}=-\frac{1}{\rho}\frac{\partial p}{\partial z}-g+\frac{1}{\rho}\frac{\partial}{\partial z}\left(\mu\frac{\partial v}{\partial z}\right),$$

$$p=p_0+\frac{\rho_0 c_0^2}{n}\left[\left(\frac{\rho}{\rho_0(1-K)}\right)^n-1\right],\qquad K=K_0\left(\frac{R}{R_0}\right)^3,\qquad \frac{4}{3}\pi R^3 p_s=\frac{m_g}{M}kT,\tag{1}$$

$$R\frac{d^2R}{dt^2}+\frac{3}{2}\left(\frac{dR}{dt}\right)^2=\frac{p_s-p}{\rho_l}-\frac{4\mu}{\rho_l}\frac{\dot R}{R}.$$

These equations are the mass and momentum conservation laws, equation of state in form of Taite taking into account the bubbles, equation of gas state, and Raleigh equation.

2.2. Kinetics

$$\mu=\mu_0\exp\{E_\mu^* K_\mu(C_0-C)/kT\},\qquad J=J_0\exp\{-G[(p_{ch}/\Delta p)^2-1]\},\tag{2}$$

$$\frac{dm_g}{dt}=4\pi R^2\rho D\left.\frac{\partial C}{\partial r}\right|_{r=R}.$$

Here, μ is the viscocity of magmatic melt, J is the nucleation frequency, D is the diffusion coefficient.

3. Low-depth underwater explosion, spalling effects

One of the classical examples of the development of intense bubble cavitation near the free surface is the reflection of an SW generated by an underwater explosion (UE) from this surface. Figure 1 shows an axisymmetric formulation of this problem; the charge weight is 1.2 g, and the charge is located at a depth of 5 cm. A typical frame of high-speed filming (t = 64 μs, a trace of the SW front is visible) shows that the cavitation region structure differs from the expected one. The calculation is performed for conditions as close to the experimental conditions as possible and adequately reflecting available data on the state of microroughness in settled water: R_0 = 0.5 μm and volume fraction of the gas phase $k = 10^{-11}$, (IKW model). Obviously, high-speed cameras capture only resolvable objects during the process (here $R \geq 100$ μm). Imposing and satisfying this condition in calculations of the dynamics of the cavitation region structure, we found that it is this "resolution" factor that is responsible for the resultant shape of this region.

Figure 2 shows a typical frame of high-speed filming of one of the later instants of cavitation region evolution, when spalling processes begin (figure 2, 'O'). A plane formulation was studied, where the charge (TNT, cylindrical charge, d = 6.4 mm) was located perpendicular to the slot walls. Destruction of the foam phase into spalling structures in the case of SW loading of a liquid layer has not been explained for a long time.

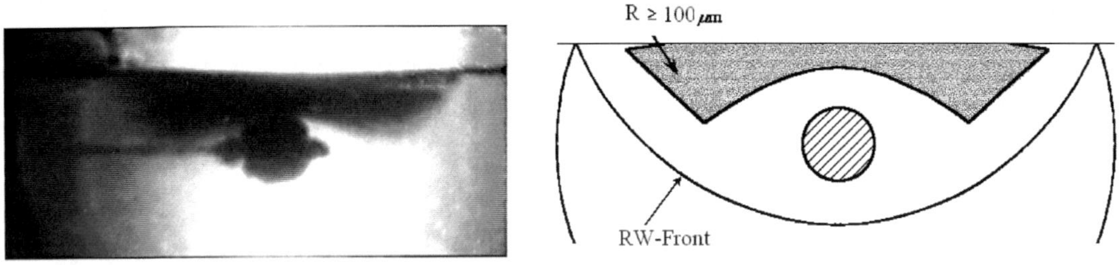

Figure 1. Underwater explosion: experiment Q_{sph} = 1.2 g, h = 5 cm; the numerical result is predicted by the IKW model.

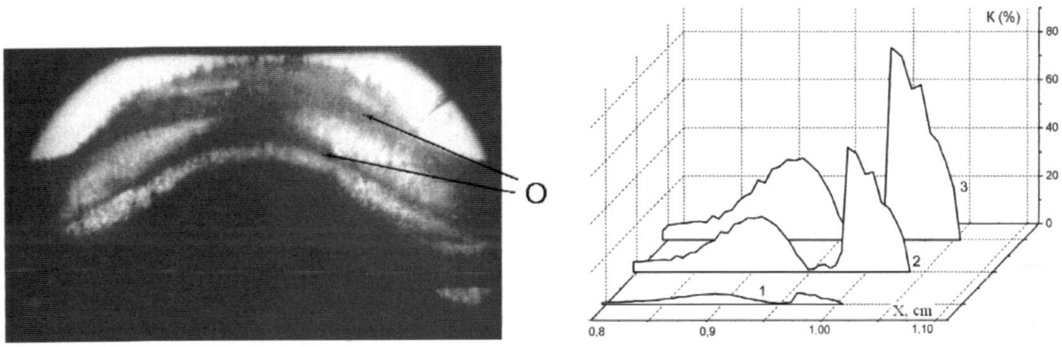

Figure 2. Plane formulation, charge Q_{cyl} = 1.2 g, h = 10 cm; the structure of the cavitation regions is described by the IKW model.

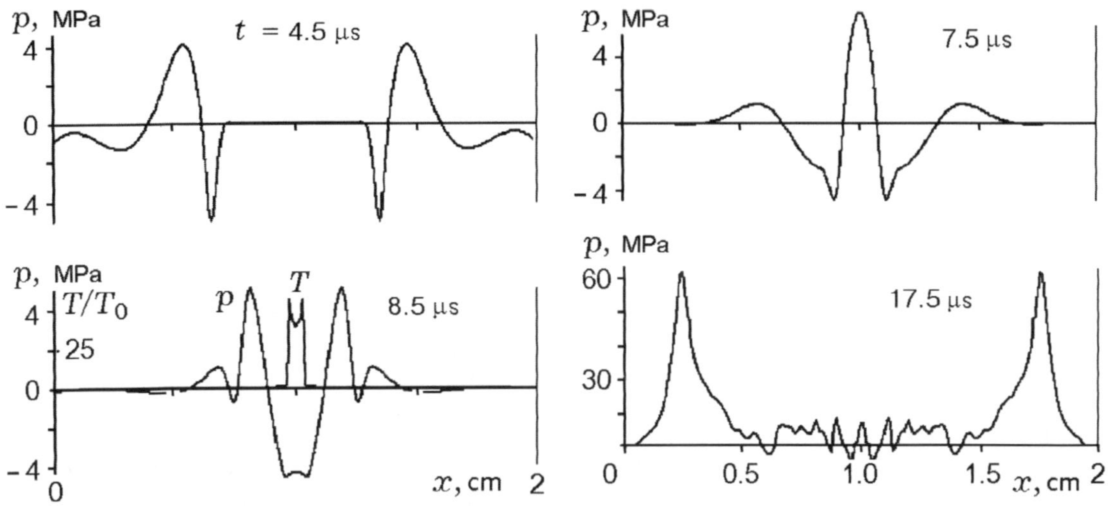

Figure 3. Model of detonation initiation in the case of a collision of weak expansion waves within a short distance of 2 cm .

An explanation was given when an unexpected fact was discovered: after its reflection from the free surface, the SW transforms into an expansion wave with an oscillating structure if SW reflection occurs in a medium with microroughness elements, which serve as cavitation nuclei behind the EW front. The calculated results in figure 2 clearly demonstrate the trend to separation of the region into individual spalling layers. This property of EWs can serve as a basis for one possible mechanism of initiation of explosive processes in readily ignited liquefied gases due to sudden depressurization of the gas holder shell, which is accompanied by generation of expansion waves. The latter can initiate cavitation processes with formation of bubbles with the chemically active mixture inside. It turned out that a collision of positive phases of even weak EWs propagating in such a medium can initiate the formation of bubble detonation waves in a bubble system, which can lead to serious consequences (figure 3). A special situation is observed for the equation of state – the model of instantaneous detonation and Todes' bimolecular model proposed in the 1930s. In the first case, when the initiation temperature is reached, the pressure in the corresponding bubbles instantaneously increases to a value determined from the transition condition $p_{expl} = \rho_{det}(\gamma_{det} - 1)\cdot Q_{expl}$.

4. SW interaction with bubble systems in the HST

We consider a horizontally located hydrodynamic shock tube (HST) with a coaxially aligned spherical bubble cluster (figure 4). An SW propagates from left to right; when the SW reaches the spherical cluster, it is refracted into the cluster. As a result, another SW is formed in the cluster; the parameters of this SW are determined by the volume fraction of bubbles in the cluster and by the parameters of the SW refracted into the cluster (figure 4): $R_{\rm cl}$= 4.5 cm, P_b=3 MPa, k_0=0.01, R_{ST}=15 cm, $R_{\rm bub}$=0.1 cm.

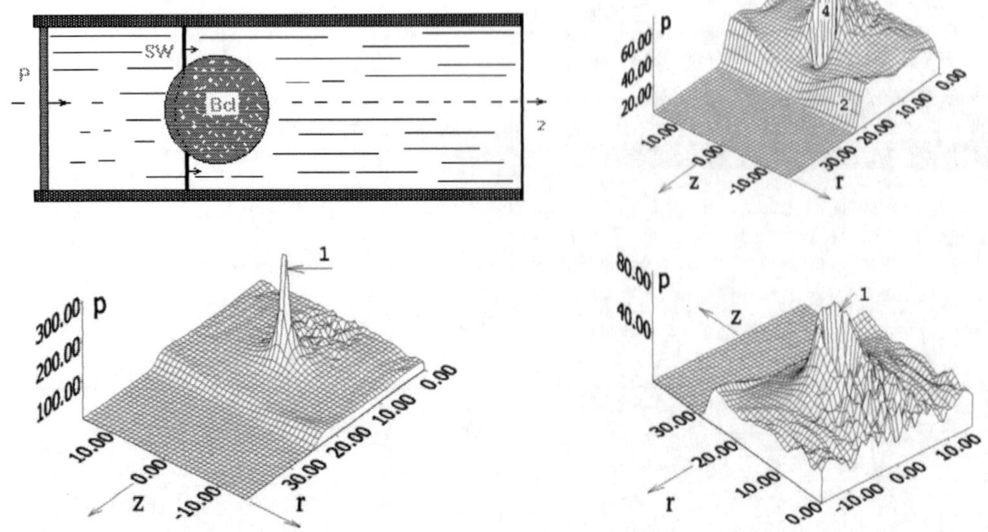

Figure 4. Interaction between the SW and the bubble cluster; dynamics of the SW field structure: SW focusing (4) in the cluster, focus (1) and SW (2) – result outside the cluster [6].

The figure demonstrates essentially nonlinear focusing of the SW (4) with a strong gradient of pressure along the front inside the cluster on its interface with the ambient medium. The HST with jumps of the cross-sectional area along the axis (figure 5) filled by a liquid with an active mixture inside gas bubbles offers a certain approach to solving new problems of wave amplification. The pattern shown in figure 5 contains two cross sections in the SW propagation direction, where the SW enters the annular channel from the left and forms a bubble detonation wave (BDW). The annular channel length is sufficient for BDW formation. In section 2, the BDW becomes expanded and reflected from wall 2-3. Section 3 is filled with a pure liquid. The right part of figure 5 illustrates the instant of focusing of the wave with the amplitude of about 15 MPa onto the z axis; the wave amplitude in the vicinity of the axis is from 50 to 60 Pa. Domain 2, where the main wave field is formed ahead of the entrance to domain 3, is located directly under the camera. The final result is shown in the previous frame: about 50 MPa in channel 3.

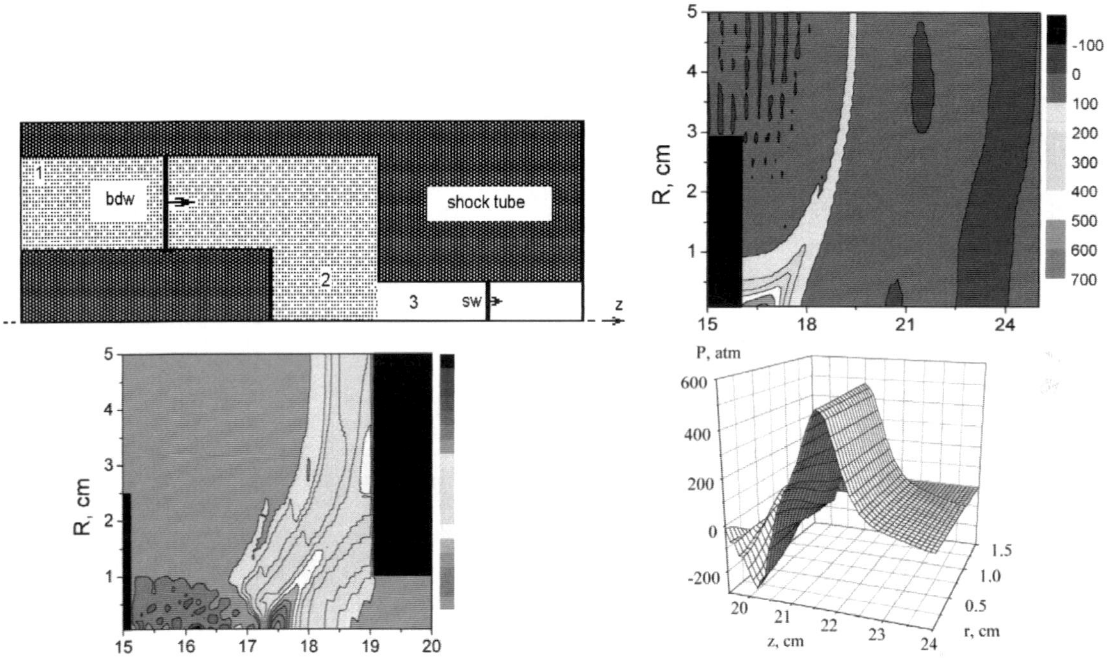

Figure 5. Hydrodynamic shock tube with internal jumps of the cross sections, the z axis is the axis of symmetry. The calculation is performed under the instantaneous detonation condition at $k_0 = 0.01$, $R_0 = 0.1$ cm, and $P_{bd} = 15$ MPa ($t = 100$ μs) [11].

5. Dynamics of the magma flow structure behind the front of explosive decompression waves

Many investigations of the last decades are aimed at finding natural mechanisms of formation of discontinuities in the flow of intensely cavitating magma in volcano conduits, which actually pose the problem of a cyclic behavior. Its solution has to be sought first of all in the dynamics of the magma flow structure and its state. We consider a 1-km vertical column of the magma connected by its bottom with the explosive chamber of the volcano (pressure of 170 MPa); on the top, the conduit is covered by a plug separating the magma melt and the atmosphere. The magma melt is saturated with the gas (5-6%) and microcrystalites. At $t = 0$ s, the plug disappears, and a decompression wave starts to form on the free surface and propagate downward.

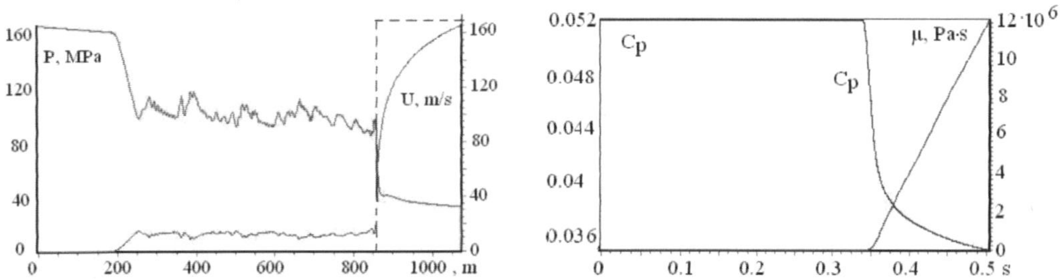

Figure 6. Anomalous zone P–U (200 m) with a sharp jump in velocity and drastic reduction of the gas phase C_p (t).

Cavitation processes develop vigorously behind the front of this decompression wave. Depending on the density of cavitation nuclei, an anomalous zone with a jump of the mass velocity and the basic parameters determining the state of this zone starts to form near the free surface (figure 6). In a short time, the inflow of the diffusion gas ($C_p(t)$) drastically decreases, and the main parameters reach their asymptotic values: the anomalous zone is ready to separate along the line of the mass velocity jump [12].

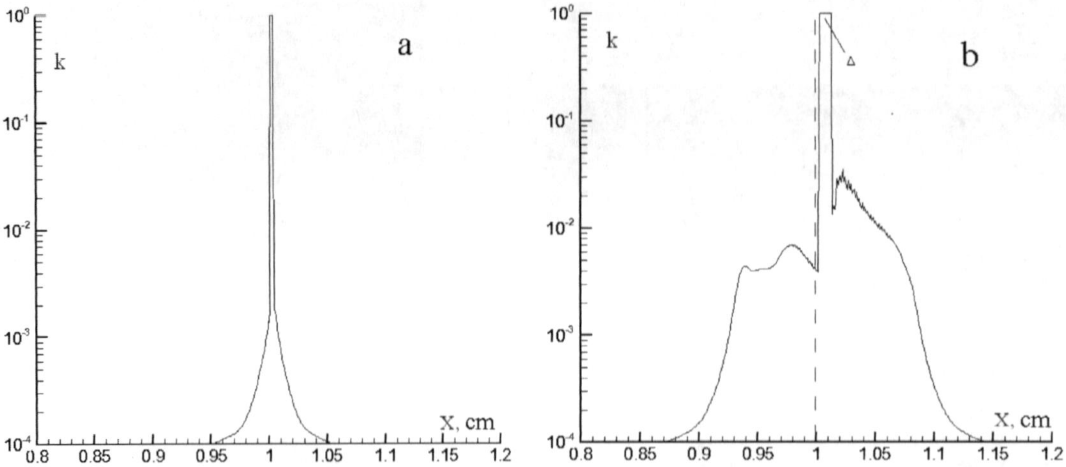

Figure 7. a) $t = 0.2$ μs, initial stage of the discontinuity; b) $t = 2$ μs, the discontinuity is already formed.

The process of formation of a discontinuity and its subsequent dynamics in a layer of a bubble liquid in the region of the mass velocity jump is calculated for the first time [13] within the framework of a general formulation of the problem.

Acknowledgments
The work was financially supported by the Russian Foundation of Basic Research (Grant No. 15-05-03336) and Lavrentyev Institute of Hydrodynamics of the Siberian Branch of the Russian Academy of Sciences (Project No. III.22.3.1).

References
[1] Rakhmatulin Kh A 1956 *Appl. Math. Mech.* **20**(2) 184–95
[2] Iordanskii S V 1960 *J. Appl. Mech. Tech. Phys*. No. 3 102–10
[3] Kogarko B S 1961 *Dokl. Akad. Nauk SSSR* **137**(6) 1331–3
[4] Wijngaarden Leen, van. 1968 *J. Fluid Mech.* **33** 465–74
[5] Kedrinskii V K 1997 *J. Shock Waves* **7**(2) 63–76
[6] Kedrinskii V K , Shokin Yu I, Vshivkov V A, Dudnikova G I and Lazareva G G 2001 *Dokl. Phys.* **46**(12) 856–9
[7] Kedrinskii V K, Vshivkov V A, Dudnikova G I, Shokin Yu I and Lazareva G G 2004 *J. Exp. Theor. Phys*. **98**(6) 1138–45
[8] Kedrinskii V K, Vshivkov V A and Lazareva G G 2005 *J. Appl. Mech. Tech. Phys*. **46**(5) 652–7
[9] Kedrinskii V K, Mader Ch. 1987 In: H. Gronig (ed.) *Proc. 16th Intern. Symp. on Shock Tube and Waves* pp 371–6
[10] Kedrinskii V K 2007 In: Shock Wave Science and Technology Reference Library vol. 1 Multiphase Flow I. (ed. Marinus van Dongen) – Chapter 3, Shock Induced Cavitation pp 67–97
[11] Kedrinskii V K, Maslov I V and Taratuta S P 2002 *J. Appl. Mech. Tech. Phys*. **43**(2) 256–63
[12] Kedrinskii V K 2014 *J. Appl. Mech. Tech. Phys*. **55**(2) 276–81
[13] Vshivkov V A, Kedrinskii V K, Dudnikova G I and Shokin Yu I 2015 *Dokl. Phys.* **60**(9) 392–5

Evaporation of liquid microdroplets levitated above a solid surface heated below the saturation temperature

D P Kirichenko[1,2], **D V Zaitsev**[1], **O A Kabov** [1]

[1]Kutateladze Institute of Thermophysics, Siberian Branch of Russian Academy of Sciences, Novosibirsk, Russia
[2]Novosibirsk State University, Novosibirsk, Russia

E-mail: dmitriy_kirichenko@mail.ru

Abstract. This paper presents a study of the interaction of liquid microdroplets falling on a solid surface whose temperature is varied from 75 °C to 155 °C. It has been shown for the first time that droplet levitation above a solid surface is possible at a temperature below the saturation temperature. It has been found that for levitated droplets, the specific evaporation rate is constant in time, but for sessile droplets, it increases sharply. The evaporation rate for sessile droplet was found an order of magnitude higher than that for levitated droplets.

1. Introduction

Effective solutions for removing high-density heat fluxes are systems based on gas-droplet flows or sprays [1, 2, 3] as well as systems using two-phase flow in a microchannel [4, 5, 6]. The effectiveness of these methods can be impaired by the Leidenfrost effect [7]. Liquid microdroplets may not reach the heating surface, which can affect the cooling capacity. The behavior of liquid microdroplets falling on a solid surface heated below the Leidenfrost temperature (about 200 °C for water) is not well understood. At relatively low temperatures, droplet levitation [8, 9] can have a significant effect on the performance of two-phase cooling systems.

The objectives of this work were to study the properties of liquid microdroplets levitated above a solid surface heated below the saturation temperature and to compare the evaporation rate for levitated and sessile microdroplets.

2. Experimental setup and methods

The experimental setup is shown schematically in Fig. 1. The working section is made of stainless steel with an embedded copper rod with dimensions of 10x10 cm^2. The surface was heated by a Nichrome coil supplied with current from a power source. Substrate surface temperature was measured by thermocouples embedded in the copper rod. The surface temperature of the copper rod varied from 75 to 155 °C. The experiments were performed in the working section open to the atmosphere at an air temperature of 25–27 °C. Liquid microdroplets with a size of the order 10 μm were applied on the working surface by a spray device placed a few centimeters above the heating area. Droplets were examined over the surface of the copper rod. Room-temperature nanofiltered distilled water was used as the working fluid. The radius of droplets was varied from 5 to 23 μm and was measured from images obtained with a FASTCAM SA1.1 high-speed camera (5600 fps at a resolution of 1024x1024 pixels) coupled with a microscope objective. The optical resolution of the system was 0.78 μm/pixel.

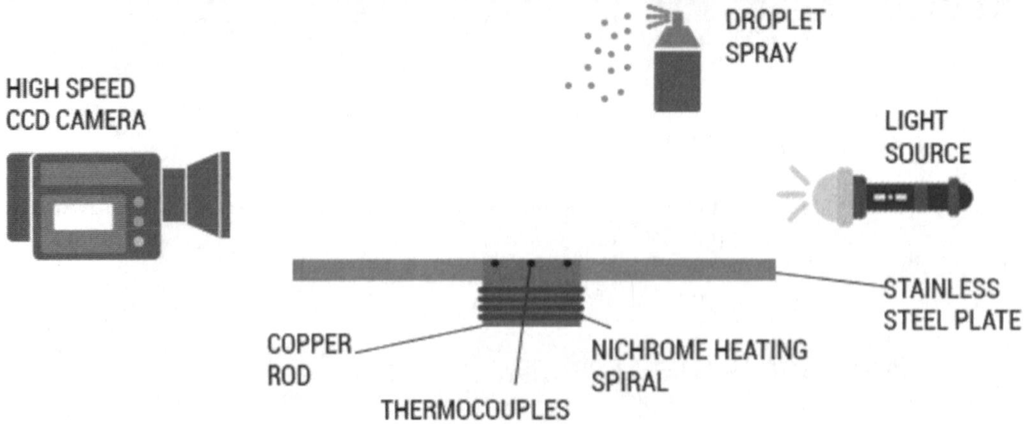

Figure 1. Schematic diagram of the experimental setup.

3. Results

It has been found that microdroplets with size of the order 10 μm can levitate above the surface heated below the saturation temperature. The minimum recorded substrate temperature at which droplet levitation occurred was 75 °C, which is far below not only the Leidenfrost temperature for water but also the saturation temperature. Figure 2 shows a photograph of levitated droplets at a substrate temperature of 75 °C. Due to evaporation, the droplet size decreases with time. Depending on the substrate temperature the droplet life changes from 0,01 to 1 second.

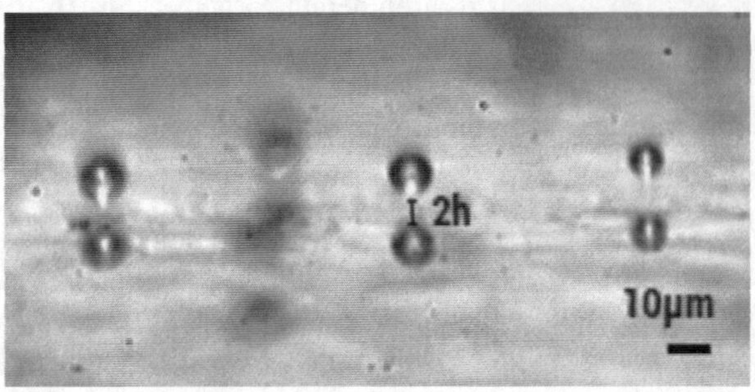

Figure 2. Levitated microdroplets and their reflection from the heated surface (for a droplet located at the center: r=5.4 μm, h=3.5 μm) at a substrate temperature of 75 °C.

Figure 3 shows plots of the radius of an evaporating droplet versus time for different surface temperatures and initial droplet sizes. The plots were well fitted by linear relations. It can be seen that increasing the substrate temperature increases the slope of the plot and decreases the droplet lifetime.

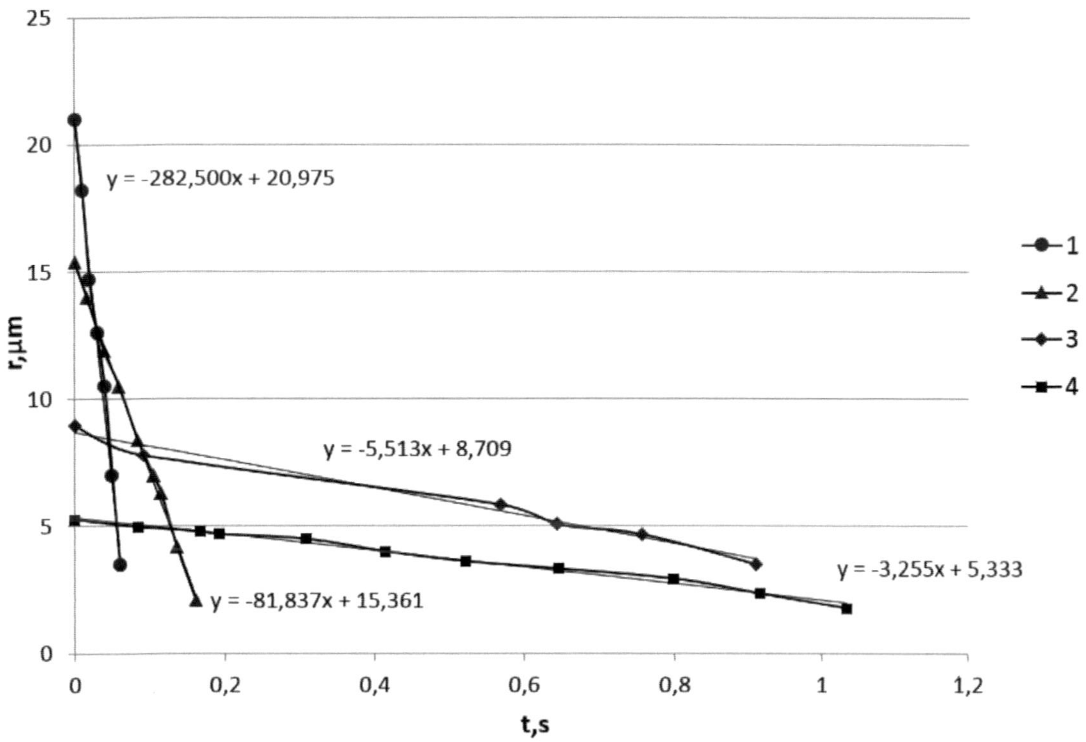

Figure 3. Radius of a levitated droplet versus time at a substrate temperature T= 155 °C (1), 135 °C (2), 93 °C (3), and 80 °C (4).

In Fig. 4, lines 2, 3, 4 and 5 show the specific evaporation rate of a droplet (droplet weight loss per unit time per unit surface area) versus time constructed from the approximating straight lines presented in Fig. 3. It can be seen that for levitated droplets, the specific evaporation rate does not change throughout the droplet evaporation process (for ease of comparison, the data in Fig. 4 are shown up to 0.01 second only). At the same time, the nominal value increases with increasing temperature. For comparison, the evaporation rate for a sessile droplet at the same surface temperature of 135 °C is plotted (line 1). It can be seen that the evaporation rate of the levitated droplet is an order of magnitude lower than that of the sessile droplet, which can significantly affect the performance of spray cooling systems. In the case of sessile microdroplets, the specific evaporation rate is not constant but increases rapidly with time. An increase in the specific evaporation rate was also observed in [10, 11] for relatively large sessile droplets (droplet size of several mm).

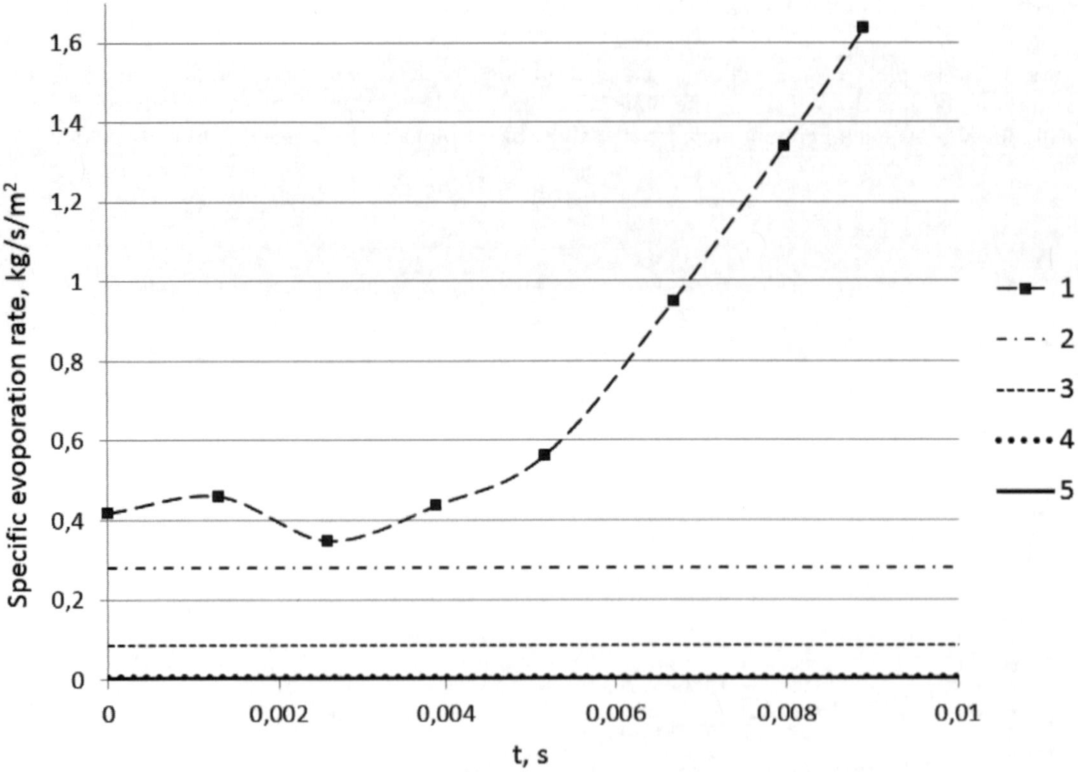

Figure 4. Specific evaporation rate of a droplet per unit surface area for a sessile droplet (1) at a substrate temperature of 135°C and for a levitated droplet at a substrate temperature of 155 °C (2), 135 °C (3), 93 °C (4), and 80 °C (5).

4. Conclusions

Liquid microdroplets falling on a solid surface have been studied experimentally and the following results have been obtained.

(1) Microdroplets with a size of order 10 μm can be levitate above a solid surface heated below the saturation temperature. The minimum temperature at which levitated droplets were observed was 75 °C, which is far below not only the Leidenfrost temperature, but also saturation temperature.

(2) For sessile microdroplets on the surface, the specific evaporation rate increases rapidly with time, while for levitated droplets, it is constant in time. The evaporation rate of levitated droplets is an order of magnitude lower than that of the sessile droplet.

The work was realized within the Russian Foundation for basic research (project № 16-38-50059). The shadow optical system was elaborated under the support of Russian Science Foundation (project № 14-19-01755).

References

[1] Kim J. *International Journal of Heat and Fluid Flow*, Volume 28, Issue 4, Pages 753-767, 2006. http://dx.doi.org/10.1016/j.ijheatfluidflow.2006.09.003

[2] Srikar R., Gambaryan-Roisman T., Steffes C., Stephan P., Tropea C., Yarin A.L. *International Journal of Heat and Mass Transfer*, Vol. 52, Issues 25–26, 2009, Pages 5814–5826 http://dx.doi.org/10.1016/j.ijheatmasstransfer.2009.07.021

[3] Kandlikar S. G., Bapat A. V., *Heat Transfer Engineering* Volume 28, Issue 11, 2007 pages 911-923 http://dx.doi.org/10.1080/01457630701421703

[4] Kabov O.A., Lyulin Yu.V., Marchuk I.V. and Zaitsev D.V. *International Journal of Heat and Fluid Flow*, Vol. 28, pp. 103-112, 2007. http://dx.doi.org/10.1016/j.ijheatfluidflow.2006.05.010

[5] Zaitsev D.V., Rodionov D.A. and Kabov O.A. *Technical Physics letters*, Vol. 35, No 7, pp. 680-682, 2009. http://dx.doi.org/10.1134/S106378500907027X

[6] Kabova Yu.O., Kuznetsov V.V., Kabov O.A. *Microgravity sci. technol.*, Vol. 20, No. 3-4, p. 187-192, http://dx.doi.org/200810.1007/s12217-008-9032-5

[7] Shirota M., van Limbeek M. A.J., Sun C., Prosperetti A., and Lohse D. *Phys. Rev. Lett.* 116, 064501, 2016. http://dx.doi.org/10.1103/PhysRevLett.116.064501

[8] Fedorets A.A., Marchuk I.V. and Kabov O.A. *Technical Physics Letters*, Vol. 37, No. 2, pp. 116-118, 2011. http://dx.doi.org/10.1134/S1063785011020064

[9] Fedorets A. A., Marchuk I.V., P.A. Stryzhak, Kabov O. A. *Thermophysics and Aeromechanics*, Vol.22, No.4, pp. 535-538, 2015. http://dx.doi.org/10.1134/S0869864315040137

[10] Gatapova E.Ya., Semenov A.A., Zaitsev D.V., Kabov O.A. *Colloids and Surfaces A: Physicochemical and Engineering Aspects*, vol. 441, pp.776-785, 2014. http://dx.doi.org/10.1016/j.colsurfa.2013.05.046

[11] Semenov, A.A., Feoktistov, D.V., Zaitsev, D.V. et al. *Thermophysics and Aeromechanics* (2015) 22: 771. http://dx.doi.org/10.1134/S086986431506013X

Kinetics of powder layer shrinkage during electron-beam treatment

A Knyazeva[1,2]

[1] Tomsk Polytechnic University, 30 Lenina Avenue, Tomsk, 634050, Russia
[2] Institute of Strength Physics and Materials Science, 2/4 Academicheskii Prosp., Tomsk, 634055, Russia

E-mail: anna-knyazeva@mail.ru

Abstract. When layerwise synthesis of new material occurs, only thin surface area, thin melt film or thin powder layer undergo essential changes. To describe the behaviour of thin layer where temperature and composition change, a model similar to shallow-water theory is suggested. The equations for thin layer behaviour are deduced for the Maxwell model of viscoelastic body in the approximation of incompressibility. The hydrodynamic part of the problem is coupled with thermal-kinetical one.

1. Introduction

Electron-beam (EB) technologies are interesting for development of manufacturing and for investigation in the directions involving the modification of material properties and its surfaces. Recent applications are connected with additive manufacturing [1]. Thermophysical phenomena determining the additive manufacturing results are the same [2] that take a place in the processes of laser and electron beam welding, surfacing and surface treatment [3]. However, there are the features in the laser and electron beam energy absorption by material. When EB acts on the powder layer, practically total energy has been used for the heating. Because EB-treatment happens in the vacuum chamber (without inert atmosphere or air), the result of the technological process is a more predictable. Generally, there are several types of technologies of EB- material synthesis: powder layer has been deposited preliminarily on the substrate and then it is treated by EB; powder enters in the melting pool traversing EB; the metallic wires have been used instead powder jet; EB acts locally to initiate the melting and intermixing between different materials or new phase formation in the heated area, and etc. In any case, the evolutions of the composition and the structure of treated material are very complex.

In additive manufacturing applications, it is necessary to know and predict the possible shrinkage of a powder layer during Selective Laser or Electron Beam Melting. This phenomenon in different technologies can be linear [4] or nonlinear in connection with nonlinear thermal conductivity or phase transitions [5]. There are many physical factors that affect the densification of the products during EB - sintering of metallic powders. It is not possible practically to take into account all phenomena in generalized model because different physical phenomena are characterized by different spatial and temporal scales. The set of limiting phenomena depends on the materials under study. Thermal-kinetical models of surface EB-treatment were suggested in [6-8]. Even in the simplest approximation (without hydrodynamics of melting pool and without rheological features of heterogeneous media), similar models give reliable change if powder layer thickness with the temperature at the variation of EB-parameters. The kinetic laws can be constructed based on Phase-field method [9] and lattice Boltzmann equation method [10].

In this paper, new approach is suggested to describe the evolution of thin surface layer under traveling EB-action.

2. Key equations

When layerwise synthesis of new material occurs, only thin surface area, thin melt film or thin powder layer undergo essential changes. In this thin area, the structure, composition and properties form. To describe the behavior of this thin area, one can use the approximation typical for the shallow-water theory [11]. However, the pure (classical) hydrodynamic model is not suitable here because solid and liquid phases and their mixture present in this traveling area simultaneously.

We assume that the behavior of the mixture consisting of powders, melt, and new solid and liquid phases and containing the pores can be described on the base of Maxwell model of viscoelastic body [12]. In this model, the components of deviators of stress and strain tensors are connected by relations

$$e_{ij} = \frac{1}{2\mu}\frac{ds_{ij}}{dt} + \frac{1}{2\eta_v}s_{ij}, \tag{1}$$

and the usual equality

$$\sigma_{kk} = 3K(\varepsilon_{kk} - \omega) \tag{2}$$

takes a place. Here, $e_{ij} = \varepsilon_{ij} - \frac{1}{3}\varepsilon_{kk}$, $s_{ij} = \sigma_{ij} - \frac{1}{3}\varepsilon_{kk}$, μ is the shear module, K is the bulk module, η_V is the viscosity, the function ω depends on the temperature, composition and porosity.

Substituting (2) into (1), we obtain

$$\frac{d\sigma_{ij}}{dt} + \frac{\mu}{\eta_v}\sigma_{ij} = 2\mu\frac{d\varepsilon_{ij}}{dt} + \delta_{ij}\left[\lambda\frac{d\varepsilon_{kk}}{dt} - K\frac{d\omega}{dt} + \frac{K\mu}{\eta_v}(\varepsilon_{kk} - \omega)\right]. \tag{3}$$

Here, $\lambda = K - \frac{2}{3}\mu$. Viscosity depends on the temperature, composition and porosity. If $\eta_V \to \infty$, we come to the model of solid body. If $\eta_V \to 0$, we obtain ideal liquid.

Motion and continuity equations are usual for hydrodynamics

$$\rho\frac{d\mathbf{v}}{dt} = \mathbf{F} + \nabla\cdot\boldsymbol{\sigma}, \tag{4}$$

$$\frac{d\rho}{dt} + \rho\nabla\cdot\mathbf{v} = 0. \tag{5}$$

Energy equation in the form of thermal conductivity equation contains terms connecting with viscous dissipation, coupling and sources due to chemical reactions

$$\sigma\cdot\cdot\frac{d\varepsilon}{dt} + c_\varepsilon\rho\frac{dT}{dt} = \nabla\cdot\lambda_T\nabla T + W - 3KT\alpha_T\frac{d\varepsilon_{kk}}{dt}. \tag{6}$$

Here, c_ε is the heat capacity for constant strains, λ_T is effective thermal conductivity, α_T is the linear thermal expansion coefficient, ρ is the density, $\mathbf{v}$ is the velocity vector.

Further we take into account that metals have high thermal conductivity, only thin layer can melt, the *OX* and *OY* axes of Cartesian coordinate system are directed along the treated surface, the OZ axis is perpendicular to this surface and is directed upward.

3. Equations for thin layer

Firstly, we construct thermal conductivity equation for metallic substrate with the preliminary deposited coating. The properties of the substrate with the thickness h_A have the index "A". We use the index "B" for the coating with small thickness $h_B = h_B(t, x, y)$. The contact between the substances is assumed as ideal. In the first approximation, the substrate stays solid and immovable. The external heat source connecting with travelling EB inter into the boundary condition

$$\lambda_B \frac{\partial T}{\partial z} + c_B\rho_B w(T - T_0) = q_0 f(x, y; t) = W_{ext}.$$

Due to high thermal conductivity of metals and small thickness of the coating, we integrate the equation (6) along OZ axis. As a result, we obtain the equation

$$U\frac{\partial T}{\partial t} + c_B\rho_B h_B\left[u'\frac{\partial T}{\partial x} + v'\frac{\partial T}{\partial y}\right] = \frac{\partial}{\partial x}\left(W\frac{\partial T}{\partial x}\right) + \frac{\partial}{\partial y}\left(W\frac{\partial T}{\partial y}\right) + W_{ext} + h_B W(T) - H(x, y, t), \quad (7)$$

where

$$U = c_A\rho_A h_A + c_B\rho_B h_B;\; W = \lambda_A h_A + \lambda_B h_B,$$

and convective velocities contain additive terms, connecting with coating thickness evolution:

$$u' = u + \frac{\lambda_B}{c_B\rho_B}\frac{1}{h_B}\frac{\partial h_B}{\partial x};\; v' = v + \frac{\lambda_B}{c_B\rho_B}\frac{1}{h_B}\frac{\partial h_B}{\partial y}.$$

The term $H(x, y, t)$ depends on the solution of mechanical part of the problem.

Hence, secondly, we should find the equations describing the evolution of the coating thickness. Thermal-kinetical model [8] uses the kinetic equations for thickness h_B, porosity θ and the relation

$$\rho = \rho_S(1-\theta).$$

Taking into account the relations

$$\frac{d\varepsilon_{ij}}{dt} = \frac{1}{2}\left(\frac{\partial v_i}{\partial x_j} + \frac{\partial v_j}{\partial x_i}\right),$$

where $v_1 = u, v_2 = v, v_3 = w$; $x_1 = x, x_2 = y, x_3 = z$, we can write

$$\frac{d\varepsilon_{kk}}{dt} = \nabla\cdot v = -\frac{1}{\rho}\frac{d\rho}{dt}.$$

For porous substance, we can assume

$$\varepsilon_{kk} = ln\frac{\rho_0}{\rho} = ln\frac{1-\theta_0}{1-\theta},$$

where index "0" relates to initial state.

For EB-sintering powder mixtures, the medium models used for hot pressing [13] are suitable.

For viscoelastic body of Maxwell type, if medium is incompressible, we come to equations

$$\frac{d\sigma_{ij}}{dt} + \frac{\mu}{\eta_v}\sigma_{ij} = 2\mu\frac{d\varepsilon_{ij}}{dt} + \left[\frac{K\mu}{\eta_v}(\varepsilon_{kk} - \omega) - K\frac{d\omega}{dt}\right]\delta_{ij}. \quad (8)$$

Neglecting the velocity component w in comparison with u, v and using usual dynamical condition, we can find

$$\sigma_{zz} = \rho g(\zeta - z), \quad (9)$$

where $\zeta = \zeta(t; x, y)$ is the coordinate of the free boundary.

Then substituting (9) into (8) for $i = j = z$, we obtain

$$\rho g\frac{d\zeta}{dt} + \frac{\mu}{\eta_v}\rho g(\zeta - z) = 2\mu\frac{\partial w}{\partial z} + \left[\frac{K\mu}{\eta_v}(\varepsilon_{kk} - \omega) - K\frac{d\omega}{dt}\right].$$

Now we can integrate this equation with z using the kinematical conditions:

$$w = 0 \text{ for } z = 0 \text{ and } \frac{\partial\zeta}{\partial t} + u\frac{\partial\zeta}{\partial x} + v\frac{\partial\zeta}{\partial y} = w \text{ for } z = \zeta. \quad (10)$$

As a result, we have

$$\rho g\zeta\frac{\partial\zeta}{\partial t}+\frac{\mu}{\eta_v}\rho g\frac{\zeta^2}{2}=2\mu\left[\frac{\partial\zeta}{\partial t}+u\frac{\partial\zeta}{\partial x}+v\frac{\partial\zeta}{\partial y}\right]+\left[\frac{\mu}{\eta_v}\zeta p-K\zeta\frac{d\omega}{dt}\right], \tag{11}$$

where $p=\frac{\sigma_{kk}}{3}$.

Continuity equation $\nabla\cdot\mathbf{v}=0$ together with (10) gives known equation

$$\frac{\partial\zeta}{\partial t}+\frac{\partial}{\partial x}(\zeta u)+\frac{\partial}{\partial y}(\zeta v)=0. \tag{12}$$

The final equation system ncludes (12) for ζ, (11) for p, the equations

$$\frac{d\sigma_{xx}}{dt}+\frac{\mu}{\eta_v}\sigma_{xx}=2\mu\frac{\partial u}{\partial x}+\frac{\mu}{\eta_v}p-K\frac{d\omega}{dt},$$

$$\frac{d\sigma_{yy}}{dt}+\frac{\mu}{\eta_v}\sigma_{yy}=2\mu\frac{\partial v}{\partial y}+\frac{\mu}{\eta_v}p-K\frac{d\omega}{dt},$$

$$\frac{d\sigma_{xy}}{dt}+\frac{\mu}{\eta_v}\sigma_{xy}=\mu\left(\frac{\partial u}{\partial y}+\frac{\partial v}{\partial x}\right),\ \sigma_{xy}=\sigma_{yx}$$

for stress tensor components and equations

$$\rho\left(\frac{\partial u}{\partial t}+u\frac{\partial u}{\partial x}+v\frac{\partial u}{\partial y}\right)=\frac{\partial\sigma_{xx}}{\partial x}+\frac{\partial\sigma_{xy}}{\partial y},$$

$$\rho\left(\frac{\partial v}{\partial t}+u\frac{\partial v}{\partial x}+v\frac{\partial v}{\partial y}\right)=\frac{\partial\sigma_{yx}}{\partial x}+\frac{\partial\sigma_{yy}}{\partial y}$$

for velocities.

For particular cases, we can simplify the equations.

The more rigorous dynamical conditions should include the force of surface tension.

It is obvious that $h_B=\zeta$. In this approximation the density ρ is the function of the temperature, composition and porosity similarly to the function ω. To study the composition evolution, it is necessary to formulate special subproblem usual for macro kinetics and chemical hydrodynamics.

4. Conclusion

So, the model is formulated for the description of thin layer behavior during the electron beam treatment. It is shown that mechanism of heat transfer appears connecting with treated layer evolution. It is necessary further to determine the laws for viscosity evolution with the temperature especially near the melting temperature, where rheological behavior of the mixture changes. For various technological conditions, the model could be modified similarly to [7,14].

Acknowledgments

The study was supported by the Russian Foundation for Basic Research (grant No. 16-58-00116).

References

[1] Gong X, Anderson T and Chou K 2014 Review on powder-based electron beam additive manufacturing technology *Manufacturing Rev.* **1**(2) 1–12

[2] Thomas M, Baxter G J and Todd I 2016 Normalized model-based processing diagrams for additive layer manufacture of engineering alloys *Acta Materialia* **108** 26–35

[3] Rykalin N N, Zuev A A and Uglov A A 1978 *The Foundation of Electron-Beam Treatment of Materials* (Moscow: Mashinostroenie) (in Russian) p 239

[4] Sercombe T B and Hopkinson N 2006 Process Shrinkage and Accuracy during Indirect Laser Sintering of Aluminium *Advanced Engineering Materials* **8**(4) 260–4

[5] Ranogajec J, Djuric M, Radeka M and Jovanic P 2000 Influence of particle size and furnace

atmosphere on the sintering of powder for tiles production *Ceramics – Silikáty* **44**(2) 71–7
[6] Knyazeva A G, Pobol I L Gordienko A I., Demidov V N, Kryukova O N and Oleschuk I G 2007 Simulation of thermophysical and physico-chemical processes occurring at coating formation in electron-beam technologies of surface modification of metallic materials *Physical Mesomechanics* **10**(3-4) 207–8
[7] Knyazeva A G, Pobol I L and Demidov V N 2016 Mathematical modelling of thermal and kinetic phenomena in electron-beam technologies – International Seminar on Interdisciplinary Problems in Additive Technologies *IOP Conf. Series: Materials Science and Engineering* **140** 012021
[8] Knyazeva A G and Bookrina N V 2003 Three-dimensional model of thermal treatment of surface by external energy source taking into account the shrinkage of powder layer *Materials of 12th Int. Conf. on Radiation Physics and Chemistry of Inorganic Materials* (Tomsk: TPU) 167–4
[9] Sahoo S and Chou K 2016 Phase-field simulation of microstructure evolution of Ti-6Al-4V in electron-beam additive manufacturing process *Additive Manufacturing* **9** 14–24
[10] Attar E and Korner C 2009 Lattice Boltzman method for dynamic wetting problems *J. Colloid and Interface Science* **335** 84–93
[11] Lyapidewskii V Yu and Teshukov V M 2000 *Mathematical Models of Long Waves Propagation in Nhomogeneous Liquids* (Novosibirsk: SB RAS) (in Russian)
[12] Boley B A and Weiner J H 1960 *Theory of Thermal Stresses* (Wiley, New York)
[13] Kovalchenko M S 1980 *Theoretical Basis of Hot Treatment of Porous Materials* (Kiev: Naukova Dumka,) (in Russian) p 240
[14] Gordienko A I, Knyazeva A G and Pobol I L 2016 Coating synthesis controlled by electron-beam heating - International Seminar on Interdisciplinary Problems in Additive Technologies *IOP Conf. Series: Materials Science and Engineering* **140** 012018

Dynamics of the wall thermal boundary layer at the initial stage of the nucleate boiling

A A Levin, A S Safarov and P V Khan

Melentiev Energy Systems Institute Siberian Branch of the Russian Academy of Sciences
664033, Irkutsk, Lermontov St., 130, phone +7 (3952) 424-700, 500-646

E-mail: lirt@mail.ru

Abstract. This paper presents the results of the experimental study of the water nucleate boiling on the surface of the cylinder heater. We investigated the initial stage of formation of the thermal boundary layer with rise and departure of bubbles with the departure diameter of 50 μm. To build the field temperature at the moment prior to the start of boiling, we used a numerical model implemented in the COMSOL multiphysics software. The experimental data on bubbles dynamics was compared to the numerical calculation of the temperature, and the relationship between the temperature growth rate, the bubble departure diameter and the Jacob number was established. The outcome results demonstrated that the non-transient heat influences the departure diameter and we need to employ the modified Jacob number to build predictive models.

1. Introduction

At present, it is impossible to quantitatively estimate the nucleation centers density [1,2], which is the decisive value for the heat transfer coefficient and intensive vaporization start time calculation. Therefore, the extension of the experimental data about the onset of critical heat flux conditions is as relevant as ever. So far, only a few number of the research papers dealing with the problem of experimental investigation of transient boiling on thin wire-wound heaters and in micro-channels [3-5] have been published. However, the data for the large-size heater is lacking.

Although the individual bubble size and lifetime are stochastic values, their scales can be described and predicted in terms of dimensionless criteria. Kim and Kim [6] published their experimental results together with a review of the partial nucleate pool boiling data, which they have been accumulating for more than 40 years. According to their study, many researchers included the bubble radius at departure, R_d, in the Bond number that sets the balance between the buoyancy and the inter-facial tension forces

$$Bo = g(\rho_l - \rho_v)(2R_d)^2 / \sigma, \quad (1)$$

and expressed the Bond number in terms of the Jacob number that sets the ratio of the superheat and the latent heat:

$$Ja = \rho_l C_{pl} \Delta T / (\rho_v h_{fg}) = \rho_l C_{pl} (T_{wall} - T_s) / (\rho_v h_{fg}) \quad (2)$$

$$Bo = \gamma Ja^{\alpha} \tag{3}$$

Here T_w stands for the wall temperature, T_s denotes the saturation temperature, ρ_l is the liquid density, ρ_v is the vapor density, and h_{fg} denotes the latent heat.

Kim et el [7] have shown that the effect of various pool temperature conditions can be taken into account if the characteristic superheat includes the pool temperature T_0

$$Ja = \rho_l C_{pl} \Delta T^* / (\rho_v h_{fg}) \tag{4}$$

$$\Delta T^* = (T_w - T_s)\beta + (T_0 - T_s)(1-\beta) \tag{5}$$

As shown in [7], the data for the constant temperature and the constant heat flux conditions can be fitted with β=0.7 . In case of the fast temperature growth, the wall thermal boundary layer can be much thinner than that one for the stationary flow, affecting the departure diameters and the departure time of bubbles. In the present study, the experimental data on bubbles dynamics is compared with numerical calculation of the temperature field in order to clarify the relationship between the temperature growth rate, the bubbles diameter, the coefficient β, the Jacob number and physical meaning of the characteristic temperature.

2. Experimental study of bubbles dynamics

Experiments with rapidly increasing heating power were conducted on flowing water in a channel with a cylindrical type 321 stainless steel heater in the center. The heater has the following characteristics: 12 mm outer diameter, 10 mm inner diameter, 4 μm roughness. The channel has optically transparent windows and is equipped with the temperature, voltage and pressure measuring probes.

The channel is mounted into a closed contour designed for studies of transitional processes during boiling. The contour comprises the following parts: channel with a high power pulse heater, pump, cooler, additional heater with automatic control of inflow temperature, bypass line for adjustment of pressure and flux.

Heating power was generated by conducting three-phase rectified electric current through the heater with the pulse duration of 60–300 ms and the temperature growth rate of 1000–7000 K/s. The water flowed along the channel upwards with the average velocity of 0.2 m/s, the inflow pressure of 0.11 MPa, and the inflow temperature of 30 °C, 60 °C and 90 °C, corresponding to the 13–73K subcooling.

The heater wall temperature, the nucleation, the bubble growth and the collapse were studied considering such parameters as the inflow temperature and the heating rate.

The wall temperature was measured by a thermocouple, the dynamics of nucleation was recorded using a speed videography.

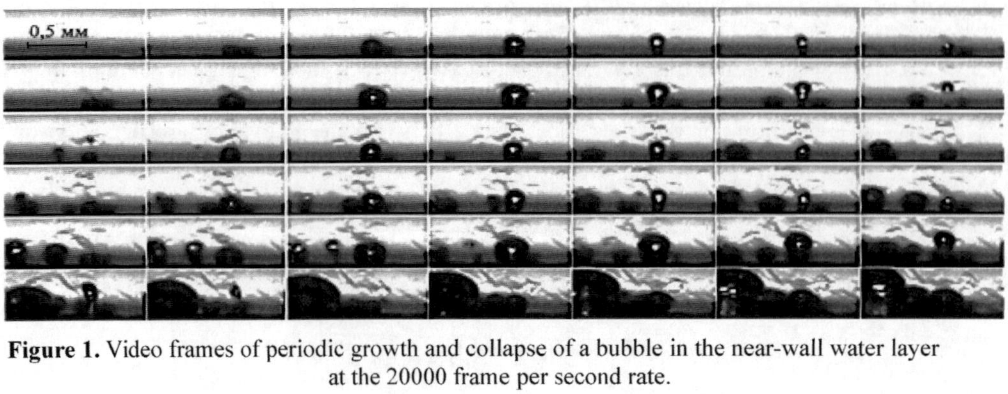

Figure 1. Video frames of periodic growth and collapse of a bubble in the near-wall water layer at the 20000 frame per second rate.

The lighting scheme proposed in [8] was applied to obtain a high speed and high quality video. The green light-emitting CVT-120 diode with nominal power of 77 Watt was subjected to the short pulse current with a 4–5 times excess of the nominal amplitude. Such a short overload does not destroy the light-emitting diode and provides the high brightness light pulse. The short exposition of 5–10 μs improves sharpness of moving objects in the video frame. The videography rate was equal to 20000 frames per second. The diode emitting surface was small enough (12 mm^2) to form a parallel light beam by means of a convex lens.

As one can see at the video frames (Fig.1), processes of nucleation, bubbles growth and collapse repeat each 200–400 μs at the same location. It was established that during the first stage of boiling, from the first micro-bubbles to the beginning of intensive boiling, the wall thermal boundary layer increased by 25% only. At the same time, the water is heated at the bigger distance from the wall by means of the vapor condensation at the bubbles top. Hence, the conditions for transient explosive boiling are formed in the same way as it happens during large over-cooling.

3. Numerical estimation of the temperature field

Using the COMSOL Multiphysics software, we performed the numerical analysis and obtained the water temperature near the heater surface. We only modeled heating of the homogeneous liquid phase, and therefore the temperature field remained adequate until the first bubbles appeared.

The axisymmetric 2D model of the steel heater and the water tank was build and verified against a full 3D model and thermocouple data. The laminar flow, the convective and the conductive heat transfer in water, the conductive heat transfer and the joule heat in solid are analyzed using temperature-dependent material properties. The metal temperature at the bottom and at the top was the same as the inflow temperature, taking three values of 30 °C, 60 °C and 90 °C, corresponding to the 73 K, 43 K and 13 K subcooling.

Both in the physical experiment and in the numerical model, the heater temperature was increased due to electric current. The actual voltage history was measured during the physical experiment and subjected as an input data during the numerical simulation.

4. Results and discussions

In each experiment, we considered from one to five of the first collapsed bubbles and calculated the wall temperature T_w and the water temperature $T(x)$ as a function of distance from the heater. We selected only the bubbles with the biggest diameter ± 2.5 μm, which collapsed in 0.5 ms or less after the collapse of the first bubble.

The examples of heating history for two experiments are shown at Fig. 2.

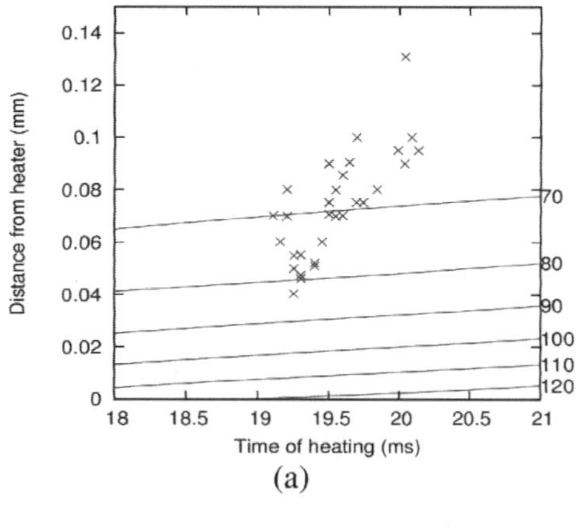

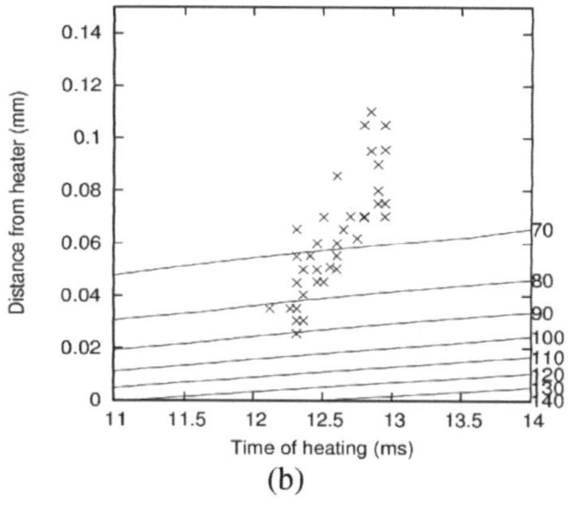

Figure 2. Temperature growth and first bubbles collapse times and diameters for experiments with T_0=60°C. (a) - low heating rate, (b) – high heating rate.

The overheat at different levels with respect to the bubble diameter was plotted and fitted against the characteristic overheat (5) and the diameter of each selected bubble. According to Fig. 3, the temperatures at x=0.1÷0.2D give the most relevant representation of the characteristic overheat, because at smaller distances the short time temperature fluctuations distort the relationship, and at bigger distances the temperature might go below the saturation level, resulting in the negative overheat. Then the overheat $[T(x)–T_s]$ correlates to the formula (5) with β=0.79±0.01. At this level, the overheat is 20–40 K.

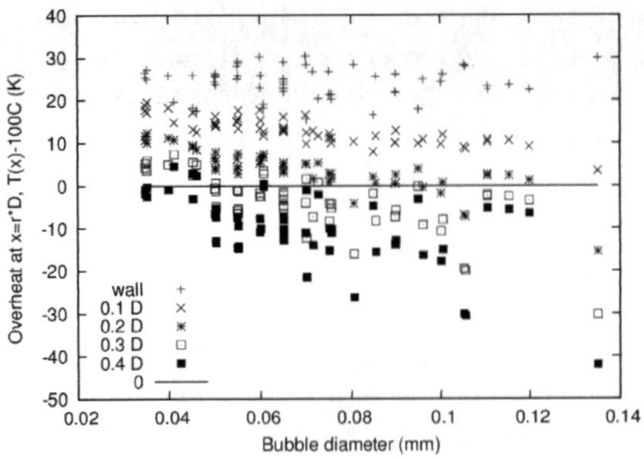

Figure 3. Calculated overheat at the distances, proportional to the bubble diameter.

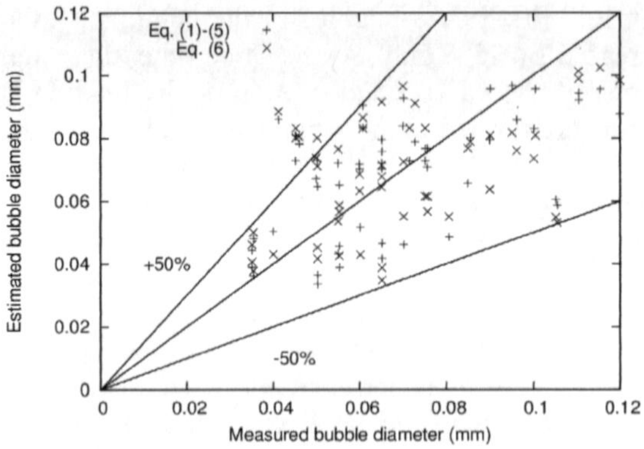

Figure 4. Comparison of the measured and estimated bubble diameter.

Then, according to (3), (4) and (5), the biggest bubble diameter was estimated by the formula

$$d(T_w, T_0, \mathrm{d}T/\mathrm{d}t) = (\gamma_1 + \gamma_2 \mathrm{d}T/\mathrm{d}t)[(T_w - T_s)\beta + (T_0 - T_s)(1-\beta)]^{\alpha} \qquad (6)$$

It fitted the data for the pool temperature T_0 from 30°C to 90°C and the average heating rate dT/dt from 2.5 to 5.5 K/ms, giving α=0.69±0.01, γ_1=0.013±0.002, γ_2=–0.0013±0.0003.

Therefore, we conclude that the influence of subcooling is described well by the equation (5) proposed by Kim et el. [7], and the relation (6) can include the value of the transient heater temperature as a coefficient. It has been revealed that for the faster heating rates the bubbles become smaller, although the characteristic temperatures correspond to the temperatures in the liquid at the distance from the wall measured as 0.15 of the bubble diameter. Thus, for the bigger heat flows the bubbles have smaller departure diameters due to the thinner thermal boundary layer.

5. References

[1] Kim J 2009 *Int. J. Multiphase Flow* **35** 1067–76.
[2] Yagov V V 2008 Teploenergetika **3** 58–64.
[3] Glod S, Poulikakos D, Zhao Z and Yadigarogly G 2002 *Int. J. Heat Mass Transfer* **45** 367-79.
[4] Liu D, Lee P S and Garimella S V 2005 *Int. J. Heat Mass Transfer* **48** 5134-49.
[5] Deev V I, Kutsenko K V, Lavruhin A A and Belyh K O 2010 *Proc. 4Th Russian Nat. Conf. Heat Transfer* (Moscow: Nat. R. Univ. "MPEI") **8** 81–4.
[6] Kim J and Kim M H 2006 *Int. J. of Multiphase Flow* **32** 1269-86
[7] Kim J, Oh B D and Kim M H 2004 *Proc. Int. Conf. I Multiphase Flow* CMF-2004-293.
[8] Willert C, Stasicki B, Klinner J and Moessner S 2010 *Meas. Sci. Technol.* **21** 075402.

Intensification of hydrate formation by means of explosive boiling incipience of rarefied gas in a bulk of water

I V Mezentsev[1,2], A V Meleshkin[1,2], D S Elistratov[1,2], S L Elistratov[2], M N Mutali[1]

[1]Novosibirsk State University, Novosibirsk, 630090, Russia

[2]Kutateladze Institute of Thermophysics SB RAS, Novosibirsk, Russia

E-mail: mezn@itp.nsc.ru

Abstract. The experiments on obtaining gas hydrate of refrigerant 134a were carried out by the method, based on explosive boiling-up of a layer of liquefied gas in a bulk of water at decompression. It is shown that this method combines several factors, leading to intensification of hydrate formation process, resulting in the fast gas hydrate growth. The effect of the decompression rate on the volume of produced hydrate was studied experimentally.

Introduction

As far back as in the beginning of the 20th century, the technological offers on the storage and transport of natural gas in the hydrate state appeared. This method attracts the experts' attention for quite long time due to the property of gas hydrates to concentrate large volumes of gas at relatively low pressures. Without a pipeline a promising way of gas transportation is its conversion into the gas hydrate (solid) state and its transportation at the atmospheric pressure and low temperature (-10 ... -20°C) [1]. In [2 - 5] and some other monographs, the properties of gas hydrates, basic conditions and features of their formation are described; the types of their crystallization, mechanisms of gas hydrate formation and decomposition are presented.

Previously, we have carried out the experiments on introduction of liquid nitrogen jet [6]. These studies were focused on a new method of gas hydrates production, based on the shock-wave method. The maximal pressure amplitude in experiments was 53 bar, and the rate of pressure rise was 567 bar/s. The advantage of the shock-wave method as compared to the conventional analogues is significant (by an order and more) intensification of the process of gas hydrate formation [7-9].

1. Experimental setup and methods

The studies were carried out using the setup in the form of a high-pressure reactor of up to 25 MPa. The working section was made as a stainless steel vessel (diameter of 100 mm, height of 300 mm) with a water jacket for thermal stating (Fig.1).

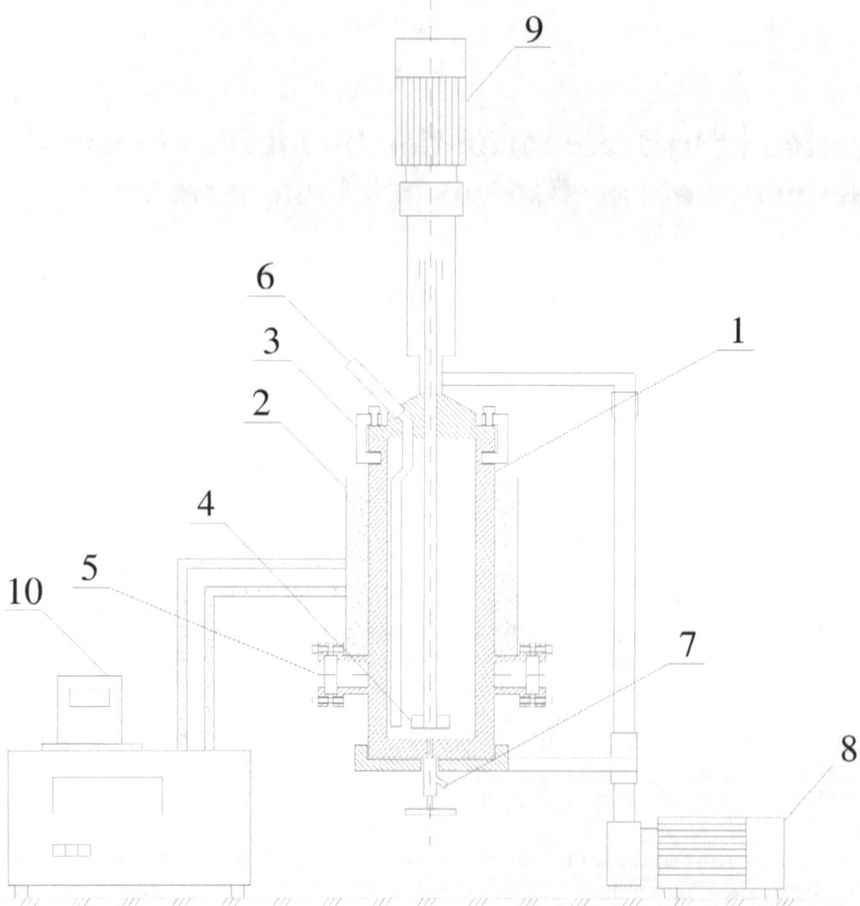

Figure 1. Scheme of experimental setup: 1 – vessel; 2 – water jacket; 3 – lid; 4 – mixer; 5 – viewing windows; 6 – bottom tube; 7 – bottom valve; 8 – drive for working section lowering/lifting; 9 – mixer drive; 10 – cryostat.

The experiment has been performed in the following manner. Water in the amount of 200 ml was into the reaction vessel and cooled by the cryostat through the outer walls to the temperature of hydrate formation. Further, 200 g of Freon 134a from the tank was fed into the water-filled vessel (with the temperature of 2-8°C). The pressure in the tank is much higher than the pressure in the autoclave, and the temperature of gas in the tank is higher than the water temperature. Entering the autoclave, gas started pressurization. Gas was cooled to the temperature of water, causing it liquation and accumulation in the form of a layer on the autoclave bottom. Then the pressure was decreased, liquefied gas under the water layer started boiling over the entire volume, so that there was an increase in interfacial surface. On the bubble walls, the growth of the gas hydrate layer started. Due to active boiling, the bubbles with the hydrate film, formed on the interface, collided constantly and deformed, causing the hydrate film peeling, and the rate of hydrate formation did not decreased. The heat released during hydrate formation was compensated by heat absorption during boiling. Due to superposition of these processes, there was a fast increase in the mass of gas hydrate.

Analysis of results

To perform the balance calculation, the following experiment was performed. Immediately after completion of refrigerant boiling, causing leveling the pressure in the working section with the atmospheric one, the system was brought to the state to be in the zone of hydrate stability. Further, the system was heated; to do this, the temperature in the thermostat was raised by 1°C every 10 minutes. The pressure in the system remained almost unchanged until the hydrate equilibrium line was not passed, and gas hydrate decay into water and gaseous Freon started. Freon increases the pressure in the system (Fig. 2).

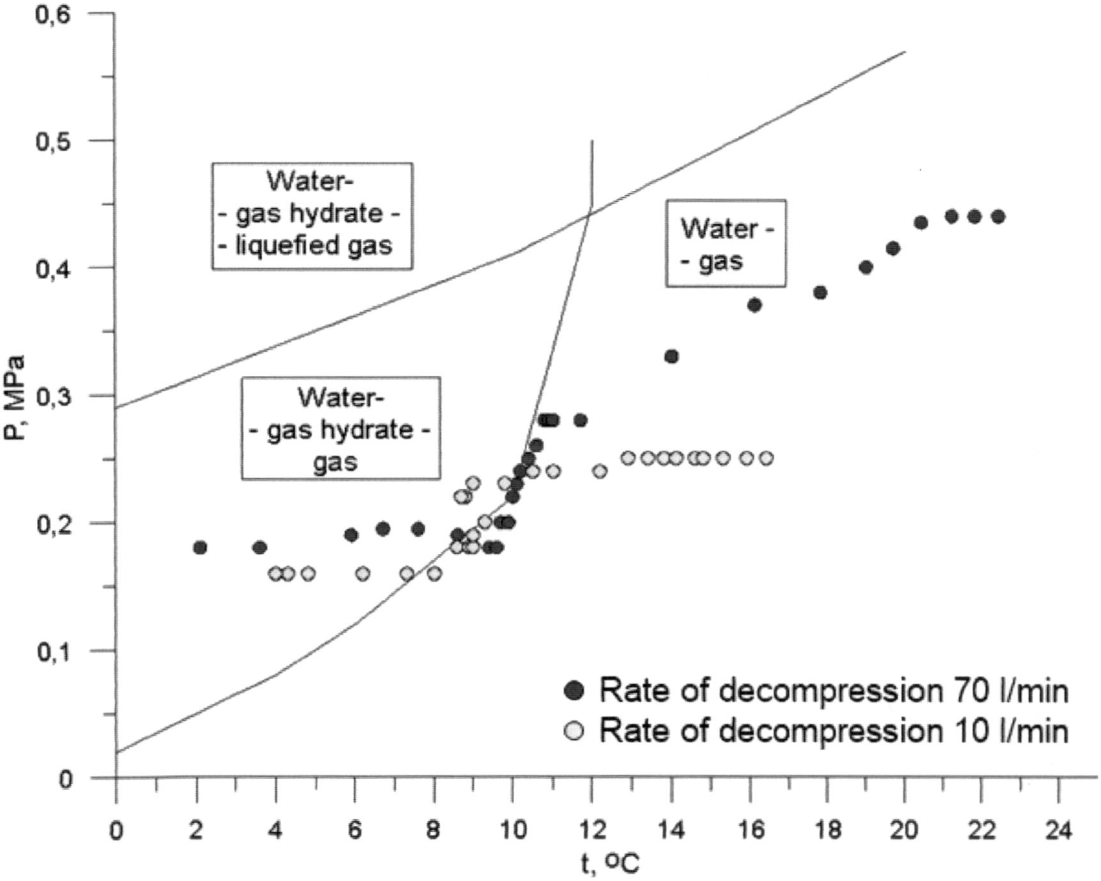

Figure 2. Dynamics of pressure change in the system at gas hydrate decay

The mass of gas released at gas hydrate decay is determined using data obtained:

$$v_g = \frac{\Delta PV - v_0 R \Delta T}{RT}$$

$$m_g = v_g M$$

where v_g is the amount of released gas matter, ΔP is the pressure jump in the system after gas hydrate decomposition, V is the volume of the gas layer in the autoclave, v_0 is the amount of substance before gas hydrate decomposition, ΔT is heating temperature, m_g is mass of gas, M is molecular mass of gas.

Taking this into account, we find the mass of produced gas hydrate:

$$m_h = \frac{V_g}{\lambda_n}$$

where λ_n is the coefficient of gas content in gas hydrate.

Since in experiments, all water converted into gas hydrate and ice, we find the mass of ice using the equation of mass balance m_s.

$$m_s + m_h = m_l + m_g$$

According to calculations, since the rate of pressure drop affects the mass of produced gas hydrate, at the rate of pressure drop of 10 l/min, the mass share of hydrate relative to ice is 15%, and for 70 l/min, it is 40,7%.

Conclusion

In this paper, the new method of gas hydrate production, based on liquefied gas boiling under the water surface due to pressure relief, was checked experimentally. The process of gas hydrates production was described, the advantages of this method over the other were determined. The diagrams in this paper show the presence of gas hydrate in the system due to an increase in pressure. The high rate of discharge increases the amount of produced hydrate.

Acknowledgments

This work was supported by the Russian Science Foundation (project no. 15-19-10025). We thank Novosibirsk State University for granting us the access to libraries.

References

[1] Gudmundsson J, Mork M, Graff O 2002 *Proc. Of the 4th Intern. Conf. on gas hydrates* pp 997-1002
[2] Makagon Y F 1997 *Hydrates of Hydrocarbons* (Tulsa, Oklahoma. Pennwell publishing company)
[3] Yakushev V S, Istomin V A 1993 Gas-hydrates self-preservation effect *Physics and Chemistry of Ice* pp 136-139
[4] Sloan E D, Koh C A 2008 Clathrate hydrates of natural gases 3rd ed *Chemical industries series* p 721
[5] Misura S Y 2013 Effect of Heat Transfer on the Kinetics of Methan hydrate Dissociation *Chem. Phys. Lett* vol. 583 pp 34-37
[6] Nakoryakov V E, Tsoi A N, Mezentsev I V and Meleshkin A V 2014 Explosive Boiling of Liquid Nitrogen Jet in Water *Journal of Engineering Thermophysics* vol.23 № 1 pp 1-8
[7] Nakoryakov V E, Dontsov V E, Chernov A A 2006 Gas hydrate formation in a gas-liquid mixture behind a shock wave *Doklady Physic* vol. 51 № 11 P 621-624
[8] Dontsov V E,. Chernov A A 2009 Dilution and hydrate forming process in shock waves *Int. J. Heat Mass Transfer* vol. 52 № 21-22 P 4919-4928
[9] Chernov A A, Dontsov V E 2011 The processes of dissolution and hydrate forming behind the shock wave in the gas–liquid medium with gas mixture bubbles *Int. J. Heat Mass Transfer* vol 54 № 19-20 P 4307-4316

Features of liquid mixtures separation in large-scale distillation columns with structured packing. New ideas and approaches

A N Pavlenko[1], V E Zhukov[1], N I Pecherkin[1], X Li[1,2] and H Sui[1,2]

[1] Kutateladze Institute of Thermophysics SB RAS, Novosibirsk, Russia

[2]School of Chemical Engineering and Technology, Tianjin University, Tianjin, China

pavl@itp.nsc.ru; zhukov@itp.nsc.ru

Abstract. Negative vapor stratification along the height of distillation column caused by different density of vapor mixture components and higher temperature at the column bottom, leads to formation of large-scale maldistribution of temperature and mixture composition over the column cross-section even at uniform irrigation of the structured packing. Experimental results concerning the dynamic effect of packing irrigation on separation efficiency of the two-component mixture of R-21 and R-114 are presented in this paper. The structured packing Zulser 350Y was installed in the distillation column with the diameter of 0.9 m. Experiments were carried out on the 10- and 19-layer packing with an overall height of 2.1 and 4 m, respectively. The liquid distributor with independently controlled 126 valves for each irrigation point, developed by the authors, was used for packing irrigation. The experiments showed that the periodic impact of the irrigation system on the large-scale non-uniformity of mixture composition, formed in the packing, could significantly affect the distribution of flow parameters over the cross-section and height of the mass transfer unit. Essentially non-uniform periodic irrigation of the packing can improve the separation efficiency of the column within 20%, if the switching periods are comparable with the times of formation of large-scale non-uniformity.

1. Introduction

Distillation columns are widely used by petroleum, chemical and food industries to separate the mixtures of various liquids. The basic requirement for the effective operation of distillation column is uniform distribution of the liquid film over the packing and the counter-current vapor flow over the column cross-section. To achieve this, the packing of metal sheets with deposited structures of different scales and highly permeable porous packing were designed [1]. In recent years, in order to increase the production capacity and improve the separation efficiency at distillation, the scholars have performed a lot of study on the structured packing [2–10]. These researches were mainly focused on the improvement of packing structure, special treatment of packing surface and organization of the optimal structure of irrigation points in the liquid distributor. The use of the modern structured packing allows more uniform distribution of a liquid film over the packing surface, but it does not completely solve the problem of uniform distribution of the flows along the entire height of the column.

As it is mentioned in [4], negative vapor stratification along the packing height caused by different densities of vapor mixture components and higher temperature at the column bottom can lead to formation of large-scale maldistribution of temperature and mixture compositions over the column cross-section even at uniform packing irrigation by liquid. In our experiments, we have implemented the idea of non-stationary use of the liquid distributor for destruction of such stationary large-scale non-uniformities of local liquid and vapor flow rates, and mixture composition over the column cross-section.

2. Set-up and method description

The experiments were carried out on the large-scale research setup "Large Freon Column", described in detail in [2-4]. The structured packing Zulser 350Y was installed in the distillation column with the diameter of 0.9 m. The experiments were performed on the 10- and 19-layer packing with the overall height of 2.1 and 4.0 m, respectively. The mixture of Freon R-21 and R-114 was used as the working liquid, chosen by its characteristics to model the separation of cryogenic mixtures. The liquid distributor specially designed by the authors was used for packing irrigation. This distributor included 126 electromagnetic valves overlapping the holes of 5-mm diameter. Each valve was controlled independently by a PC using specially developed software allowing manual control of each valve, control of an arbitrary group of valves and programmed valve switching by any predetermined algorithm with time discretization of 1 s. A response of the column to the influence of liquid distributor was observed in real time by indications of three groups of thermometers mounted in 3 different cross-sections of the column along the packing height. Each group consisted of 16 thermometers uniformly distributed over the packing cross-section. Their readings were displayed on the computer monitor as the topograms. The groups of thermometers were installed in the bottom (two layers from the packing bottom), middle and upper (two layers from the packing top) packing cross-sections. The on-screen information allowed evaluation of both the structure of large-scale temperature non-uniformity and the value of this non-uniformity.

The experiments were carried out for different irrigation point patterns of the liquid distributor, keeping the total number of irrigation points. The maximal number of experimental data was obtained for two patterns of irrigation points, namely: *Pulse* pattern and *Half_Column (HC)* pattern (*Fig.*1). *Pulse* pattern provides packing irrigation through the holes distributed uniformly over the cross-section during time t_1, then all holes are closed for period t_2 (*Fig.1a*). Then, the cycle is repeated. *HC* pattern provides irrigation of a half of the packing cross-section during time t_1 (*Fig.1b*), the second half of cross-section stays closed. Then, the valves are switched, and the irrigated and dry halves of cross-section change for period t_2 (*Fig.1c*). Then, the cycle repeated periodically. The main data array for the *HC* pattern is obtained for equal values of t_1 and t_2 within 5 – 160 s. The pictures of packing irrigation by *HC* pattern are shown in *Fig. 2*.

a) 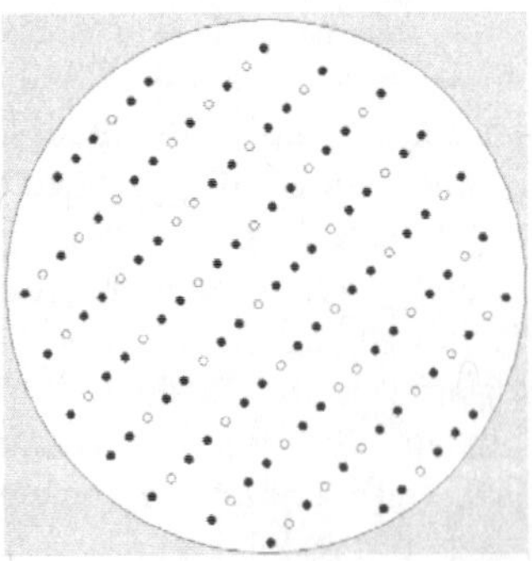 b) 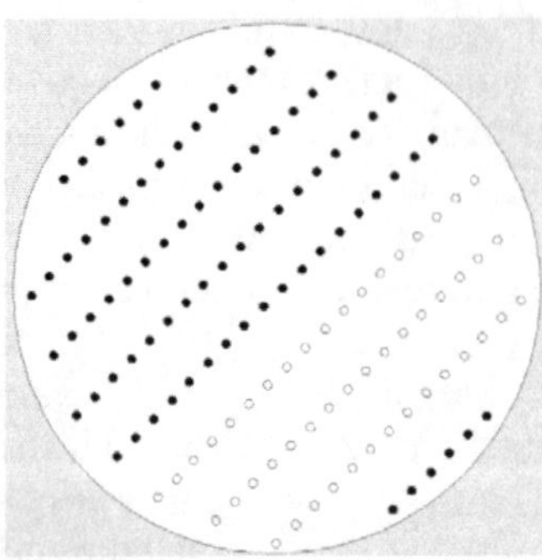 c) 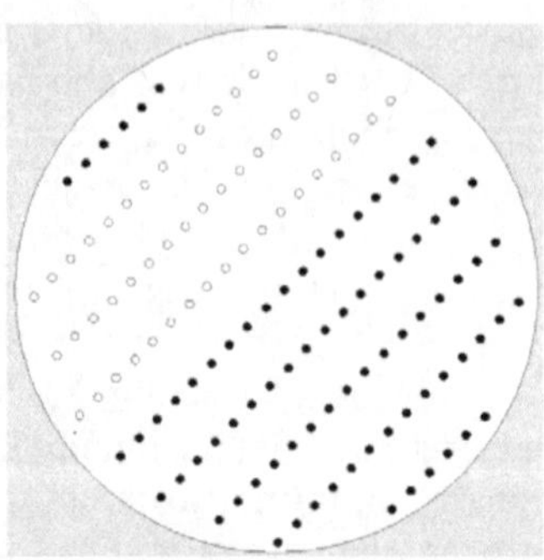

Figure 1. Irrigation point patterns in different experiments. *a*) – uniform irrigation by *Pulse* pattern; *b*) – first semi-period of irrigation by *HC* pattern; *c*) – second semi-period of irrigation by *HC* pattern.

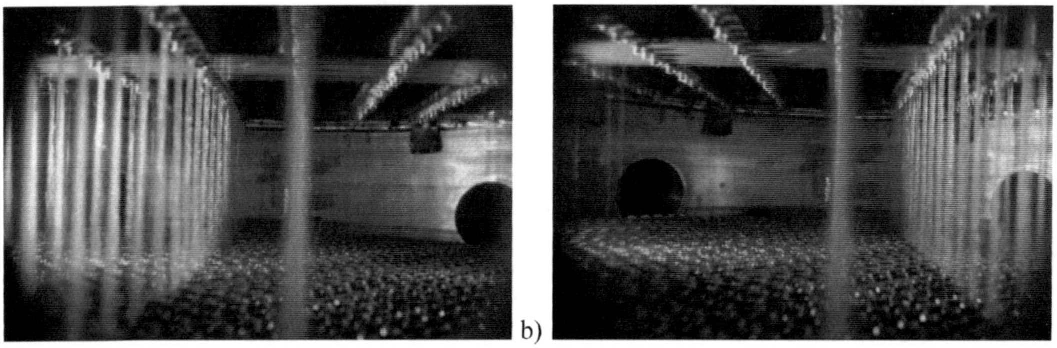

Figure 2. The pictures of packing irrigation by *HC* pattern. *a*) – first semi-period of irrigation; *b*) – second semi-period of irrigation.

3. Results and disscussions

Dependences of the height of transfer unit (*HTU*) on the period of valve switching for the *HC* pattern, determined as a sum of times t_1 and t_2, are shown in *Fig. 3*. The experiments were carried out on the packing with the heights of 2.1 and 4.0 m at the same superficial velocity of vapor K_v = 0.024 m/s. The zero period corresponds to the conditions of uniform packing irrigation.

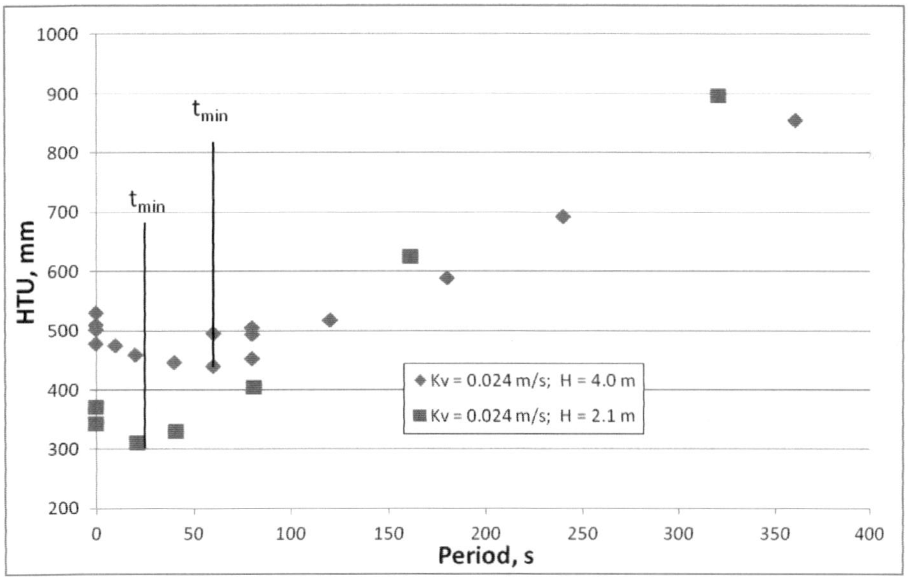

Figure 3. Height of transfer unit (HTU) vs. period of valve switching on the liquid distributor for the *HC* pattern.

As it can be seen from the diagram, the efficiency of mixture separation in a short packing (10 layers, H = 2.1 m) is significantly higher than in the packing of 4 m (19 layers). This fact is well known and it is explained by formation of large-scale maldistribution of local flow parameters along the packing height [4]. In engineering practice, the liquid redistributors are usually installed in 10 packing layers to align these maldistributions. In our experiments, at periodic packing irrigation by the *HC* pattern, we observed the improvement of mixture separation efficiency in the range of switching periods of up to 60 s for the packing height of 4 m and up to 20 s for the height of 2.1 m.

The following increase in the switching period led to deterioration in the efficiency of mixture separation. When switching with a short period, separation efficiency was close to separation efficiency at uniform irrigation because during irrigation of a half of the packing cross-section liquid did not have time to penetrate through a significant part of the packing height. Switching irrigation to another half of the column cross-section of the column, in fact, created the conditions of uniform irrigation. An increase in the period of irrigation led to the fact that during one semi-period of irrigation liquid spread over the packing height, and the large-scale non-uniformity, corresponding to non-uniform irrigation of the first semi-period, started forming in the packing. Switching at this moment to the second semi-period of irrigation, whose structure is diametrically different to the first semi-period led to rearrangement of the forming large-scale non-uniformity in the opposite direction. As a result, at switching period t_{min}, the maximal separation efficiency was achieved; it was 20% higher than separation efficiency at uniform irrigation. The value of optimal period t_{min} is determined by the time of liquid flow along the packing height, and it depends on both the packing height and superficial velocity of liquid and vapor flows.

The experiments in the regime of periodic irrigation (*Pulse* pattern) were carried out on the packing with the height of 2.1 m at superficial vapor velocity K_v = 0.024 and 0.035 m/s. Dependence of the relative height of transfer unit is shown in *Fig. 4* at different ratios of irrigation time (t_1) and time of irrigation stop (t_2).

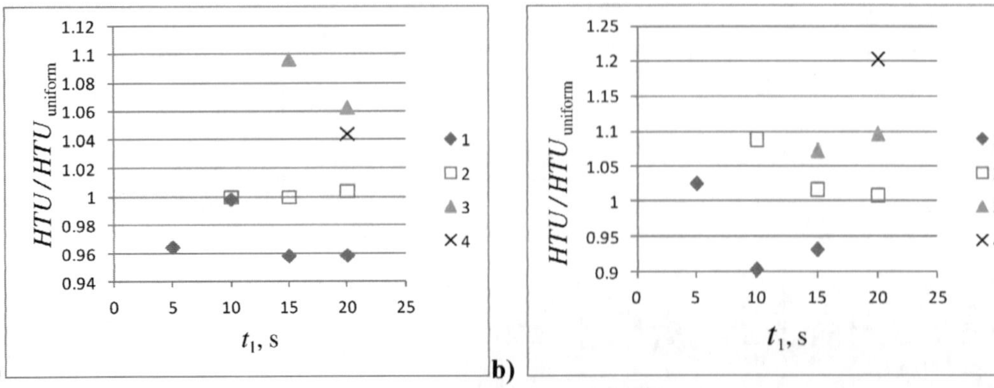

Figure 4. Relative height of transfer unit, normalized by *HTU* at uniform irrigation. *a)* – K_v = 0.024; *b)* – K_v = 0.035 m/s. t_2: *1* – 5 s; *2* – 10; *3* – 15; *4* – 20.

According to the diagrams, short-term (within the time of liquid flow on the packing) complete periodical stop of the liquid flow irrigating the packing does not lead to deterioration of mixture separation; in some cases, we can even observe the improvement of separation efficiency within 5 - 10%.

4. Conclusions

According to the experiments, the influence of the periodic irrigation on the large-scale non-uniformity of mixture composition, formed in the packing, can affect significantly the structure of flow parameter distribution over the cross-section and height of the mass transfer surface. Essentially non-uniform periodic irrigation of the packing can improve separation efficiency of the column within 20%, if the switching periods are comparable with the times of formation of large-scale non-uniformity. Short-term (within the time of liquid flow on the packing) complete periodical stop of the liquid flow irrigating the packing does not lead to deterioration of mixture separation; in some cases, we can even observe the improvement of separation efficiency within 5 – 10%.

Acknowledgment

This research has been performed at IT SB RAS with the support of the Russian Science Foundation (project No. 14-49-00010).

References

[1] Pavlenko A N, Zeng J, Pecherkin N I, Zhukov V E and Volodin O A 2016 *Journal of Engineering Thermophysics* **25** 1–14
[2] Pavlenko A N, Pecherkin N A, Chekhovich V Yu, Zhukov V E, Sunder S and Houghton P 2009 *Theoretical Foundations of Chemical Engineering* **43** 1–11
[3] Pavlenko A N, Zhukov V E, Pecherkin N I, Chekhovich V Yu, Sunder S and Houghton P 2010 *Theoretical Foundations of Chemical Engineering* **44** 869–76
[4] Pavlenko A N, Zhukov V E, Pecherkin N I, Volodin O A, Shilkin A and Grossmann C 2014 *AIChE J.* **60** 690–705
[5] Soulaine C, Horgue P, Franc J and Quintard M 2014 *AIChE J.* **60** 3665–74
[6] Ehler C and Fieg G 2014 *AIChE J.* **60** 3833–47
[7] Pavlenko A N, Li X, Zhukov V E, Pecherkin N E, Volodin O A, Surtaev A S, Gao X, Zhang L, Sui H and Li H 2015 *Journal of Engineering Thermophysics* **24** 210–21
[8] Oluji´ca Z, Rietfortb T, Jansenb H and Zich E 2015 *Chemical Engineering Research and Design* **99** 14–19
[9] Yao Y, Pavlenko A N and Volodin O A 2015 *Journal of Engineering Thermophysics* **24** 222–36
[10] Stepykin A, Sidyagin A and Ulyanov V 2015 *Transactions TSTU* **21** 445–52

Journal of Physics: Conference Series **754** (2016) 042013 doi:10.1088/1742-6596/754/4/042013

Visual study of propagation of self-sustained evaporation front within the thickness of a thermal liquid layer

V E Zhukov, M I Moiseev and D V Kuznetsov

Kutateladze Institute of Thermophysics SB RAS, Novosibirsk, Russia

zhukov@itp.nsc.ru

Abstract. In the range of low reduced pressures, the development of self-sustained evaporation front along the heat-releasing surface at non-stationary heat release is an important factor that determines possible transition to film boiling at heat fluxes, significantly lower than the critical heat fluxes at stationary heat release. This paper presents the experimental results on the scale of a leading part of the interface of self-sustained evaporation front at stepped heat release. The scale of the leading part of the interface of the evaporation front is compared with the thermal layer thickness, registered using the shadow method of visualization at high-speed video shooting with up to 25,000 frames per second. Experiments were carried out in Freon R21 under the conditions of free convection at relative pressures of 0.032 - 0.068. It is shown that self-sustained evaporation front spreads along the heated wall within the thickness of a liquid layer, superheated relative to the saturation temperature. Dependence of the front velocity on wall superheating relative to the saturation temperature does not change with significant subcooling to the temperature of liquid saturation in the volume.

1. Introduction

Non-stationary heat release within a local region of the heat exchanger can create a metastable layer of a contacting liquid coolant and lead to formation and propagation of a vapor film. The vapor front propagates along the heat-releasing surface due to transfer of heat accumulated in a thin thermal layer during heating the single-phase liquid. In the case of non-stationary heat release, when the times of evaporation front formation are considerably less than the times of convection development, the thermal layer thickness increases by the root law, accumulating heat from the heat-releasing surface by the mechanism of heat conduction. Several studies demonstrate that the average velocity of evaporation front propagation is constant for the same temperature of the heat-releasing wall in the case of stationary heat release [1], and the front propagates with acceleration at non-stationary heat release and increasing temperature of the heat-releasing wall [2]. It is shown in [3] that the velocity of evaporation front interface pulsates in a vicinity of the average front velocity. The mode of pulsations is low-frequency; it is determined by the heater scale comparable with the scale of the capillary constant. The high-frequency mode is determined by the thickness of the vapor layer of evaporation front. There are some models that describe propagation of undisturbed self-sustained evaporation front [1,4,5], as well as the single attempts to model the dynamics of front propagation dynamics with consideration of small-scale interface perturbations under the conditions of intensive evaporation [6]. Despite a long history of studying such a phenomenon as the self-sustained evaporation front, to date there is no model that describes the dynamics of front propagation with consideration of all processes occurring at the interface of the evaporation front.

The aim of this work is to obtain new experimental data on the process of evaporation front propagation within the thickness of a thermal layer at non-stationary heat release. It is necessary to develop the model description of this phenomenon.

2. Setup and method description

Experiments were performed in a large volume of liquid Freon R-21 in the range of reduced pressures of 0.032 - 0.068. The detailed description of setup is given in [2]. The tube with the diameter of 3 mm made of steel 12X18H10T was used as a working section. Heat release was organized by the pass of a rectangular current pulse with the amplitude of 420 A and predetermined duration of 10 - 90 ms through the working section. Due to the high thermal inertia of the working section, the wall temperature increased almost linearly with time (the initial exponential region). The wall temperature was determined by dependence of electrical resistance of the working section on temperature, as well as by numerical calculation using the heat conductivity equation. In addition to the wall temperature, the temperature profile in liquid near the heat-releasing wall was determined by numerical calculation at specified duration of heat release and the value of electric current through the heater. The use of the heat conductivity equation in this case is admissible, as the process of front formation and propagation was observed for the short times, significantly less than the time of convection development. Laminar convection was observed in experiments during the times of the order of 300 ms. To visualize the evaporation front within the thickness of the thermal layer illumination in the transmitted light was used. A laser level with the wavelength of 650 μm or LED SMD assemblies was used as a light source. In the case of laser use in the working section of 3-mm diameter, the thermal layer was observed as a shaded area; when using LED lighting, the upper border of the thermal layer was observed as a dark line, and the space of the thermal layer near the heating wall was observed as the clear liquid.

In experiments, there was high-speed recording that allows observation of evaporation front propagation along a length of about 30 mm. According to this video (frequency of 25,000 frames per second), the propagation velocity of evaporation front was determined depending on wall superheating relative to the saturation temperature. There was also macrorecording, which allowed us to consider the structure of the front interface in detail with simultaneous observation of the thermal layer using the shadow techniques. The maximal video resolution obtained in macrorecording is about 10 μm per pixel.

3. Results and disscussions

Time dependences of thermal layer thickness from the beginning of heat release, measured by video records and calculated are presented in *Fig. 1*. The thickness of the thermal layer calculated numerically for the cylindrical heater from the condition of 98-% temperature difference between the wall and undisturbed liquid (line *6*) and calculated for a flat layer (line *3*) by formula (*1*) coincide satisfactorily with the thicknesses measured in experiments by the shadow method.

$$\delta = 2.4\sqrt{a\tau} \tag{1}$$

Lines *4* and *5* determined though numerical calculation and corresponding to the thickness of a thermal layer, limited by the temperature of saturation in experiments with subcooling, are also shown in *Fig. 1*.

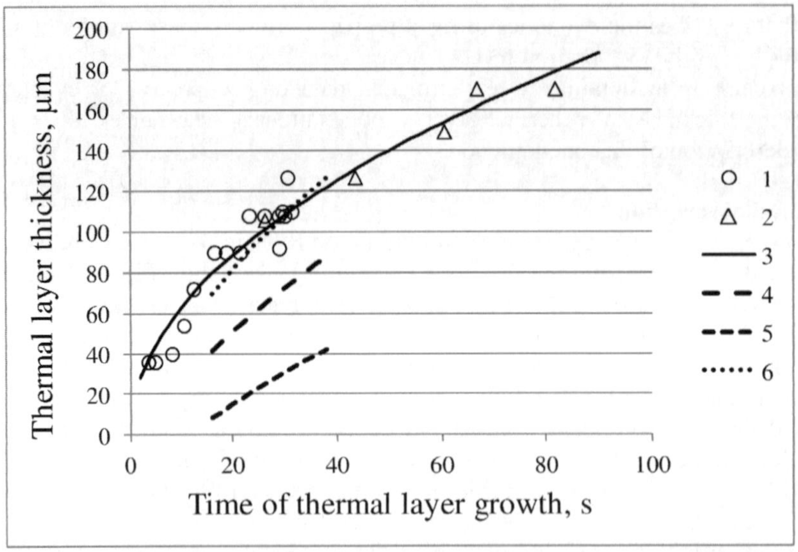

Figure 1. Thermal layer thickness measured in the video frames. *1* –saturated liquid; *2* –subcooled liquid, subcooling $\eta = 24$ K; *3*–calculation by formula (1); *4,5,6* – numerical calculation of superheated layer, $\eta = 4.2$ K; 24; 0, respectively.

The fragments of evaporation front propagation in the saturated and subcooling liquids at the same temperature 22.3 °C are shown in *Figs. 2 – 4*.

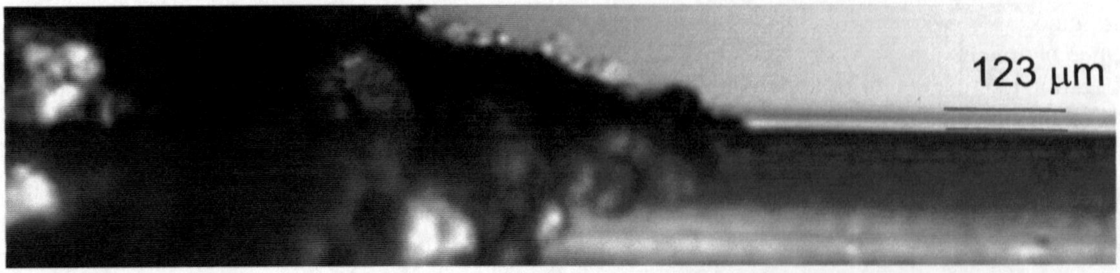

Figure 2. Fragment of evaporation front propagation in saturated liquid. $P = 0.167$ MPa. Thermal layer thickness is 123 μm.

In *Fig. 2*, the thermal layer boundary is observed as a washed-out shaded line at some distance from the cylindrical wall. The vapor front propagated along the heater within the thermal layer thickness due to heat accumulated in the superheated layer. At that, only a part of heat from the thermal layer is spent for evaporation in the leading part of the vapor layer; another part of the superheated layer is pushed off by the propagating front and, then, it is spent for vapor layer thickening. In liquid significantly subcooled to the saturation temperature the front propagates in a thin part of the thermal layer adjacent to the heated wall and superheated relative to the saturation temperature under the given pressure (*Fig. 3*).

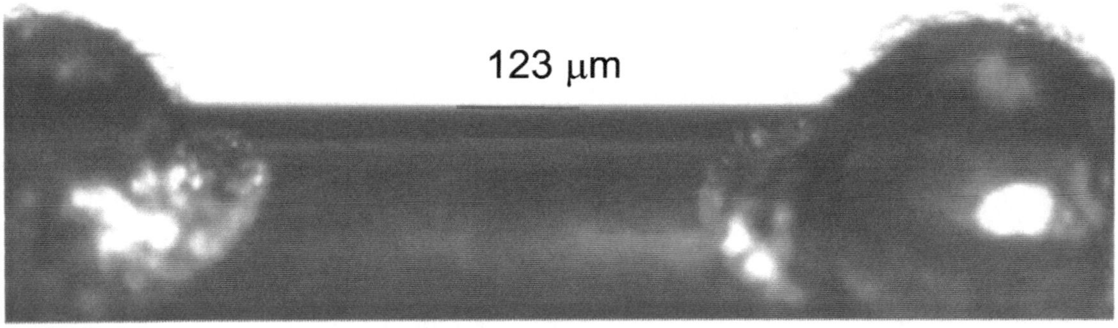

Figure 3. Fragment of evaporation front propagation in subcooled liquid. The thermal layer thickness to the saturation temperature is 38 μm, total thermal layer thickness is 123 μm. $\Delta T_s = 46$ K; $T_s = 46$ K; subcooling $\eta = 24$ K; $P = 0.353$ MPa.

It can be seen in *Fig. 3* that the thermal layer thickness in the frontal part on the upper generatrix of the working section is about a third part of the thermal layer thickness observed as a dark shaded area with the thickness of about 120 μm. The calculated value of the layer adjacent to the wall and having a temperature above the saturation temperature under the conditions of our experiment is 38 μm. As we can see, the thickness of the frontal part of the vapor layer thickness correlates with the thickness of the liquid layer superheated relative to the saturation temperature. The fragment of evaporation front propagation under the similar conditions, but at a lower subcooling value ($\eta = 4.2$ K) is shown in *Fig. 4*. The thickness of the thermal layer with the temperature higher than the saturation temperature in this experiment is 84 μm. As it can be seen in the video frame, the thickness of the frontal part of the vapor layer is about 0.5 - 0.7 of the total thickness of the thermal layer, observed as a shaded area with the calculated thickness of 123 μm. Also on the scene observed pushing off the remaining part of the thermal layer from the heated is also observed in this frame.

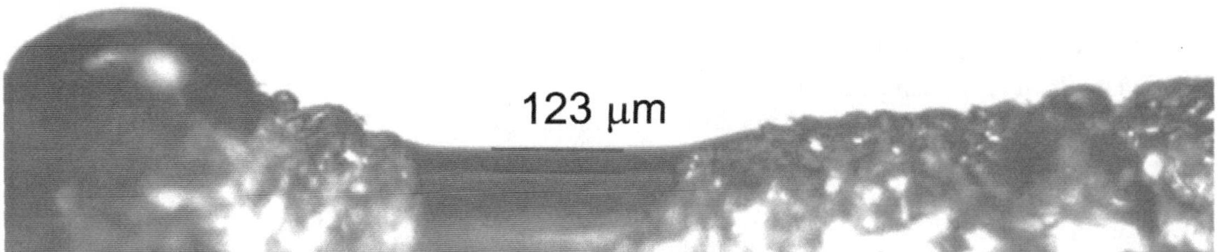

Figure 4. Fragment of evaporation front propagation in subcooled liquid. The thermal layer thickness to the saturation temperature is 84 μm, total thermal layer thickness is 123 μm. $\Delta T_s = 65.7$ K; $T_s = 26.5$ K; недогрев $\eta = 4.2$ K; $P = 0.19$ MPa.

Dependence of propagation velocity of a self-sustained evaporation on wall superheating relative to the saturation temperature is shown in *Fig. 5*. Data are shown for liquid subcooled to the saturation temperature in a varying degree at three different pressures. As it can be seen from the diagram, the front velocity is virtually independent of the value of liquid subcooling, and it is determined by the superheating degree of near-wall liquid layer with respect to the saturation temperature.

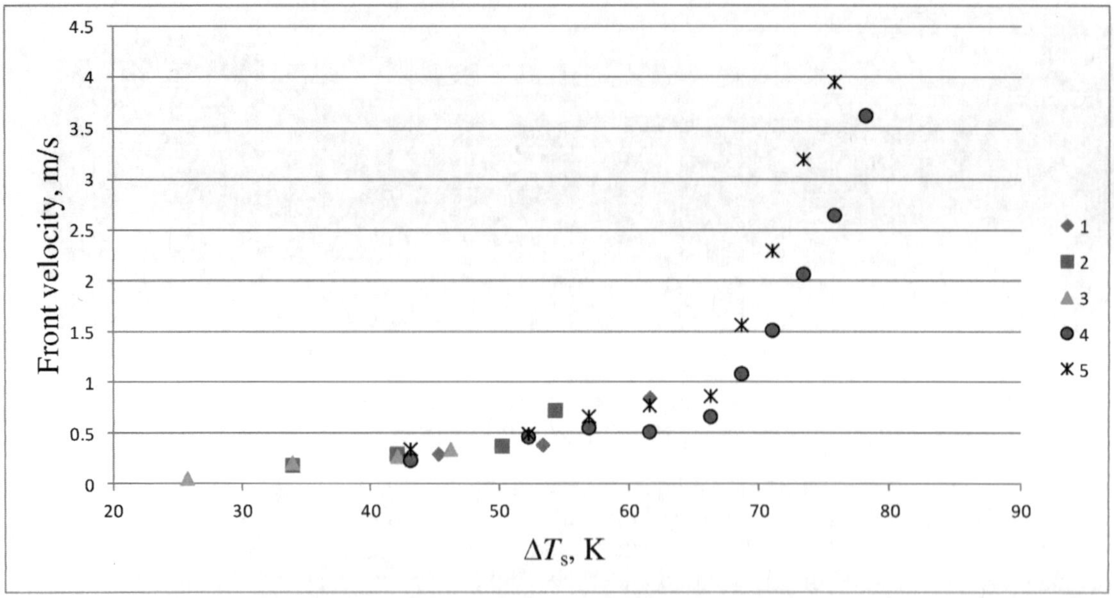

Figure 5. Propagation velocity of self-sustained evaporation front vs. wall superheating relative to the saturation temperature. *1* – η = 4.2 K; *P* = 0.19 MPa; *2* – η = 15.6 K; *P* = 0.272 MPa; *3* – η = 24 K; *P* = 0.353 MPa; *4* – η = 0 K; *P* = 0.167 MPa; *5* – η = 0 K; *P* = 0.353 MPa.

4. Conclusions

The experiments on propagation of self-sustained evaporation front along the cylindrical heat-releasing surface with simultaneous visualization of the thermal layer have been carried out. The results of measuring the thermal layer thickness according to the video agree satisfactorily with the calculated value of the layer. The experiments showed that the self-sustained evaporation front propagates along the heated wall within the thickness of the liquid layer superheated relative to the saturation temperature. Dependence of the front velocity on the saturation temperature does not change at significant subcooling to the temperature of liquid saturation in the volume.

Acknowledgment

This research has been performed at IT SB RAS with the support of the RFBR (project No. 15-08-01359).

References

[1] Aktershev S P and Ovchinnikov V V 2008 *J. Appl. Mech. Tech. Phys.* **49** 47-55
[2] Pavlenko A N, Tairov A A, Zhukov V E, Levin A A and Tsoi A N 2011 *Journal of Engineering Thermophysics* **20(4)** 1–27.
[3] Pavlenko A N, Tairov A A, Zhukov V E, Levin A A and Moiseev M I 2014 *Journal of Engineering Thermophysics* **23**(3) 173–193.
[4] Avksentyuk B P 1995 *Russian Journal of Engineering Thermophysics* **5** 1-8
[5] Pavlov P A and Vinogradov V E 2010 *High Temperature*, **48** 683-90
[6] Pavlenko A N and Lel' V V 1999 *Thermophysics and Aeromechanics* **6** 105–17

On peculiarities of near-threshold initiation of powder density explosive by air shock wave and by solid impactor

A O Kashkarov[1,2], A P Ershov[1,2] and E R Pruuel[1,2]

[1] Lavrentyev Institute of Hydrodynamics SB RAS, 630090, 15 Ac. Lavrentieva ave., Novosibirsk, Russia

[2] Novosibirsk State University, 630090, 2 Pirogova Str., Novosibirsk, Russia

E-mail: kashkarov@hydro.nsc.ru

Abstract. The features of near-threshold mode of initiating by gas-tight piston and high-enthalpy gas flow was evaluated for a powder density explosive PETN. Both methods lead to the development of detonation in about 10 μs time. The synchrotron radiation diagnostics have shown that the initial stages of the process were significantly different, that diversity being caused by the influence of the gas flow in the pores of the charge. In this work, the effect of the gas flow on the mode of initiation was studied experimentally.

1. Introduction

There are two basic methods of detonation initiation in high explosives (HEs). The impact of the shock wave of sufficient amplitude leads to the development of the detonation in a microsecond time interval, or even less. However, this method requires a strong action of some impactor, usually accelerated by the pressure of detonation products. The most common case of shock initiator is conventional detonator. To initiate detonation in the base charge, first you need to arrange the same process in the detonator, typically using sensitive, and therefore dangerous primary explosives.

The second way of initiation is deflagration to detonation transition (DDT). This comparatively slow process may take milliseconds. In contrast to the shock initiation, the initial pulse may be rather weak. Typical HE ignited at the initial normal conditions burns slowly, but the gradual increase of pressure leads to acceleration of reaction rate. Now, fast (hundreds m/s) convective combustion eventually develops in the permeable for gaseous products charge – the hot gases penetrate into the pores of fresh explosive and ignite new portions of material. Rapid gas emission causes the compression of the HE, and finally the detonation wave develops with further pressure rise. Obviously, such process is possible when HE is confined in a shell, for example, in a strong metal tube. DDT is relatively less determined process, what is natural at longer time needed.

In our works [1, 2, 3, 4], the different way to initiate detonation was studied in detail, which can be regarded as an intermediate between the two classical methods described above, and which, to certain extent, combines their advantages. This is the impact of high enthalpy gas flow with the powdered HE. There are several ways to create a flow with the required parameters–the

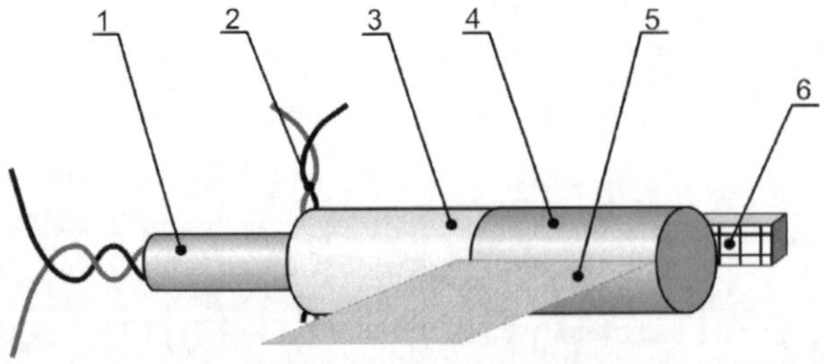

Figure 1. The experimental scheme. 1 – generator of the initiating flow, 2 – contact sensor, 3 – air gap, 4 – sample, 5 – SR beam, 6 – detector of passed radiation.

electric explosion of wires, breakthrough of the combustion products from a separate chamber, gas detonation, etc. [5, 6, 7]. In the blasting practice, high enthalpy flow occurs when the detonation is transmitted through the air gap. At given sufficient flow intensity, the detonation develops within the 10 μs time range. This is fast enough to discard the requirement of the strong casing since inertial confinement is sufficient enough. As a result, the initiation can be realized in size of standard detonators.

Physical prerequisite for the high enthalpy initiation is quite intense activation of explosive powder by the hot gas flow. As a result, the rapid convective burning develops in the charge immediately, thus the slow layer-to-layer combustion stage which is the weak link of standard DDT is avoided.

In this study, three variants of impulse action on powder PETN are compared experimentally:

- near-critical high-enthalpy initiation which leads to successful development of detonation;
- subcritical, weaker impact of the gas flow, which results in a failure;
- the impact by the metal plate.

We found that the detonation occurs under conditions of moderate compression of powder PETN, whereas in the case of failure the compression is much higher. The reason of this paradox is the rapid reaction with appreciable generation of the gas phase in the case of successful initiation, which reduces the compressibility of the material. As a result, the powdered explosive remains permeable for the gas, and the detonation develops through two-phase flow stage. On the contrary, in the case of failure, the gas generation due to chemical reaction inside the charge is insignificant. Thus, the explosive is easily compressed similarly to an inert powder of comparable mechanical properties. Pores in the charge are closed and convective combustion has no chance to penetrate into the layers of undisturbed material.

2. Experiment

The processes within the explosive sample were studied using the synchrotron radiation (SR) diagnostics [8, 9] which provides a shadow slit X-ray film with spatial resolution down to 0.1 mm and the time interval between frames is 0.496 μs. Thus, it was possible to trace the initiation from the stage of undisturbed state of the material to the development of the detonation. Calibration of measuring equipment allowed us to convert the beam attenuation data into the quantity of mass along the beamline (measured in g/cm^2). This is simply the product of density and the sample diameter for uniform one-dimensional axisymmetric case. The real flow is not uniform and essentially two dimensional which complicates recovering of the density distribution. However, in our case, the direct qualitative data on the evolution of the quantity of mass along the beamline during the first microseconds of the process were found to be informative enough.

Journal of Physics: Conference Series **754** (2016) 052001 doi:10.1088/1742-6596/754/5/052001

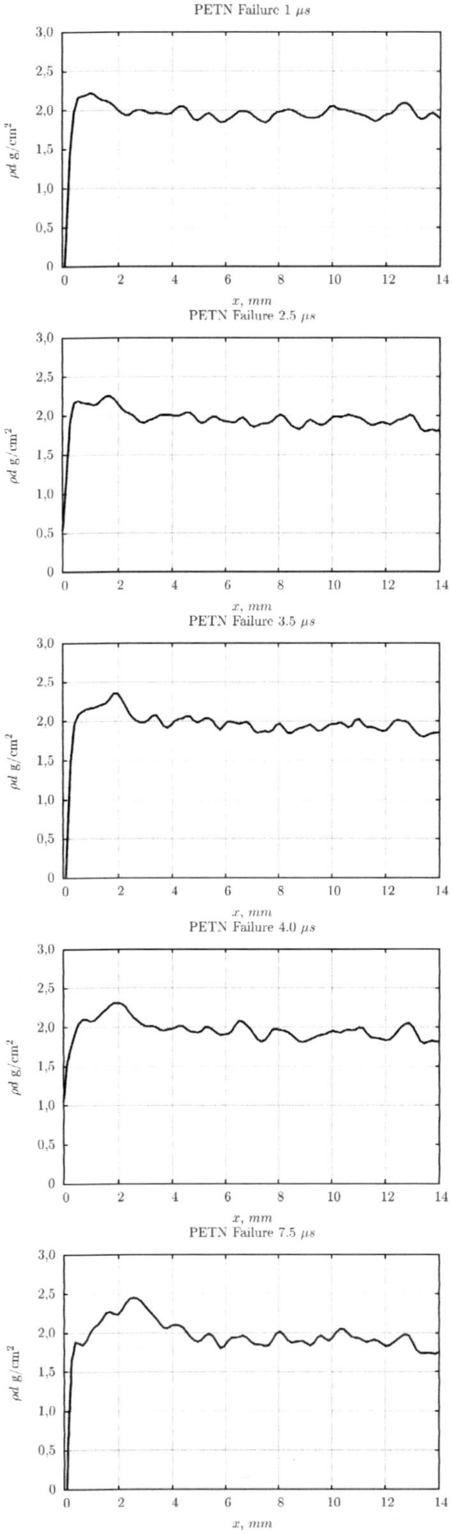

Figure 2. Failure, subcritical regime. The speed of the air shock wave is 2.4 km/s.

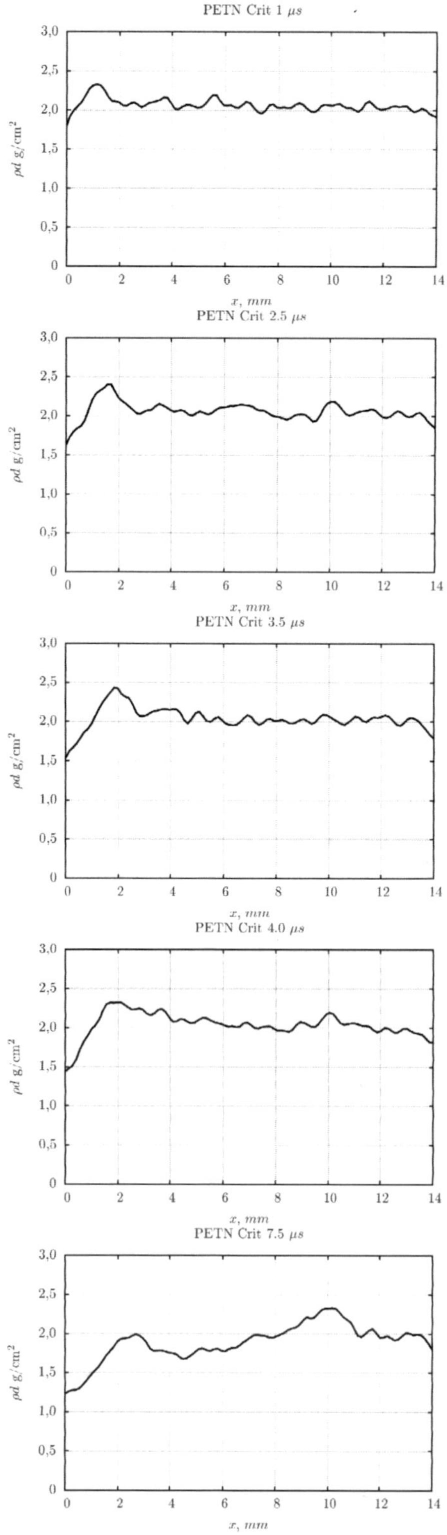

Figure 3. Near-critical regime. The speed of the air shock wave is 2.6 km/s.

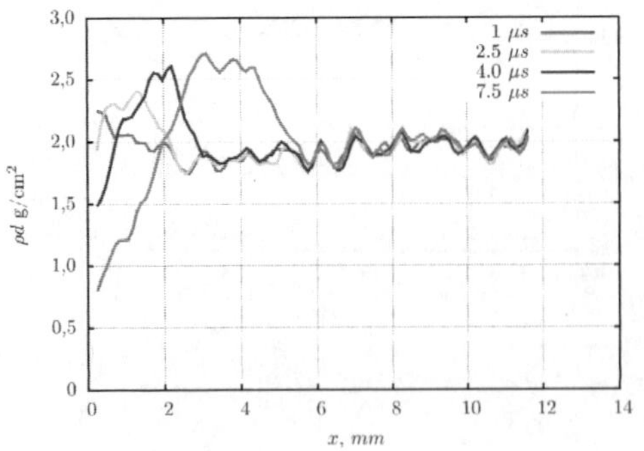

Figure 4. Impact of high enthalpy flow on the inert powder.

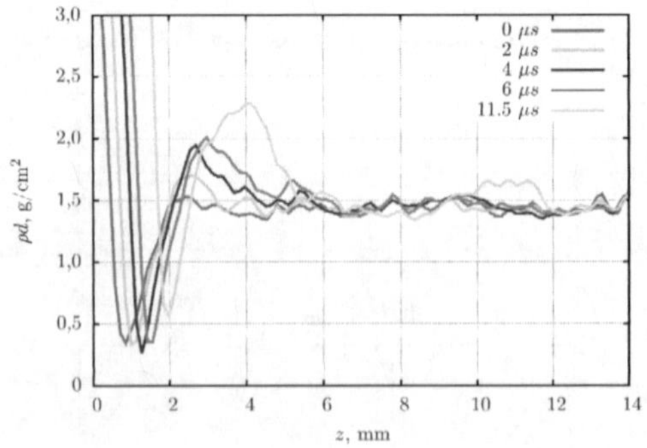

Figure 5. Initiation of powder density PETN by Al impactor.

The scheme of experiments is shown in figure 1. The explosive generator of the initiating flow (1) produced the hot gas behind the air shock. The same primary charge closed the contact sensor (2) for the synchronization of the measuring equipment. The air gap (3) was adjusted to regulate the flow parameters. The flat SR beam (5) is directed onto the test sample (4). Passed radiation is registered by the linear detector (6).

Near-critical regime of initiation PETN charge of powder density by high-enthalpy gas flow occurred if the velocity of the incident air shock wave was 2.6 km/s. If the air shock velocity was reduced to 2.4 km/s, the detonation was not registered during the observation time that was more than 10 μs. We also compared the experimental data on HE with the results obtained for inert powder (sugar) of similar density and dispersity.

The dynamics of the mass on the beamline for the case of failure in which the hot gas impact does not produce detonation is shown in figure 2. Figure 3 shows the data for the near-critical initiation regime. One may note that during the start of both processes the behaviour of samples was almost identical. The discrepancies begin at approximately fourth microsecond from the moment of impact. In the case of the successful initiation the compression of the charge slows down and acceleration of the wave front was observed which further develops into the detonation wave. For the case of failure, the radiographed mass continues to increase to greater level than that achieved with more intensive impact, without any disturbances propagating ahead of the compressed region. The velocity of compression wave was significantly slower than in the case of near-critical initiation.

Clearly, the discrepancies in the development of the process are the manifestation of the reaction induced at the surface of explosive grains in the case of a successful detonation. The influence of the gas produced in the reaction on the movement of the condensed phase becomes visible after about 4 microseconds.

The results for the more intensive impact on the inert powder are shown in figure 4. One can note more rapid formation of a dense region without any disturbances ahead of the main wave front. The dynamics of the process at an early stage is similar to powdered charge initiation by solid impactor (figure 5), 2 mm thick aluminium plate, details see in [4]. Here the charge was initiated by compression wave instead of penetration of hot gases. The absence of reaction on the particle's surfaces leads to formation of the thick compressed region. In this case, the reaction starts at the front of the compression wave and propagates into the undisturbed explosive, with eventual transition to detonation, while the region of compressed explosive remains inert.

3. Conclusions

Impact of high-enthalpy gas flow with the porous charges lead to quite different results depending on its intensity. If the hot gas flow is too weak to initiate the fast combustion of the condensed phase, then a powder is compacted into a dense plug near the charge interface. This plug blocks the gases from penetrating into the main charge. In general, such a process is similar to initiation by solid impactor. On the contrary, more intense gas flow filtering into the pores leads to ignition of explosive particles. The gas produced in the reaction prevents the further compression of the material, so the porous bed retains its permeability. Moreover, the reaction gases filtrate into the explosive in the initial state producing the convective combustion wave. The visible difference in the wave dynamics was observed at 4 μs from the start of the process. This gives an upper bound for the PETN ignition time in a near-threshold regime. Actually, the ignition occurs much earlier since the gas produced in the reaction needs time to affect the movement of dense solid phase. To conclude, the effect of gas released in the combustion process is crucial.

It was found recently that in coarse grained loose packed explosives even the developed detonation wave may depend upon the convective processes [10]. We would like to underline that these convective mechanism appeared to be most important in particular for PETN, studied in the present paper. Thus, there exists a possibility of the common nature of accelerating wave in PETN, from the start of the high enthalpy initiation process to the final "normal" detonation regime.

Acknowledgments

This work was supported by Russian Foundation for Basic Research (Grant No. 15-03-01039).

References

[1] Luk'yanchikov L A, Pruuel E R, Kashkarov A O and Ten K A 2010 *J. Appl. Mech. Tech. Phys.* **51**(4) 453–62 doi: 10.1007/s10808-010-0061-7

[2] Ershov A P, Kashkarov A O, Lukyanchikov L A and Pruuel E R 2013 *Combust., Expl., Shock Waves* **49**(1) 79–81 doi: 10.1134/S0010508213010097

[3] Ershov A P, Kashkarov A O and Pruuel E R 2015 *Combust., Expl., Shock Waves* **51**(6) 700–9 doi: 10.1134/S0010508215060118

[4] Ershov A P, Kashkarov A O and Pruuel E R 2016 *Combust., Expl., Shock Waves* **52**(1) 96–101 doi 10.1134/S0010508216010135

[5] Andreev V V and Luk'yanchikov L A 1974 *Combust., Expl., Shock Waves* **10**(6) 818–23 doi: 10.1007/BF01463829

[6] Luk'yanchikov L A 2000 *J. Appl. Mech. Tech. Phys.* **41**(5) 806–17 doi: 10.1007/BF02468725

[7] Grigor'ev V V, Luk'yanchikov L A, Pruuel E R and Vasil'ev A A 2001 *Combust., Expl., Shock Waves* **37**(5) 572–9 doi: 10.1023/A:1012349305278

[8] Titov V M, Pruuel E R, Ten K A, Luk'yanchikov L A, Merzhievskii L A, Tolochko B P, Zhulanov V V and Shekhtman L I 2011 *Combust., Expl., Shock Waves* **47**(6) 3–15 doi: 10.1134/S0010508211060013

[9] Pruuel E R, Ten K A, Tolochko B P, Merzhievskii L A, Luk'yanchikov L A, Aulchenko V M, Zhulanov V V, Shekhtman L I and Titov V M 2013 *Doklady Physics* **58**(1) 24–8 doi: 10.1134/S1028335813010035

[10] Ershov A P, Kashkarov A O, Pruuel E R, Satonkina N P, Sil'vestrov V V, Yunoshev A S and Plastinin A V 2016 *J. Appl. Phys.* **119** 075903 doi: 10.1063/1.4942359

On the mechanisms and criteria of deflagration-to-detonation transition in gases

A D Kiverin, I S Yakovenko, M F Ivanov
Joint Institute for High Temperatures of the Russian Academy of Sciences (JIHT RAS)
125412, Izhorskaya st. 13 Bd.2, Moscow, Russia

E-mail: alexeykiverin@gmail.com

Abstract. The origins of different supersonic combustion regimes in channels are discussed. It is shown that the evolution of supersonic flames including the formation of quasi-steady supersonic flames, transition to detonation directly at the flame front and auto-ignition inside the kernels at some distance ahead the flame front are fully determined by the same gasdynamical effects and peculiarities of combustion kinetics. Three basic criteria are formulated aiming to establish correlation between the initial state of the gaseous combustible mixture and the possibility of deflagration-to-detonation transition. Formulated criteria are in an adequate agreement with the data on combustion of hydrogen-based mixtures.

1. Introduction

One of the most important unresolved questions of contemporary theory of combustion and detonation is the determination of criteria for deflagration-to-detonation transition (DDT), which is of primal interest for perspective technical systems design as well as for the safety control of their operational regimes. It is known that in most active gaseous mixtures such as hydrogen-oxygen, acetylene-oxygen, ethylene-oxygen etc. the flame propagating through the channel with acceleration is able to cause detonation onset on relatively small spatial-time scales even in short and narrow channels (see e.g. [1, 2, 3]). In less reactive mixtures, including hydrogen-air and hydrocarbon-air mixtures, the sufficient flame acceleration becomes possible only inside wide obstructed channels and only on rather large distance [4, 5]. However, it should be noted that even in such conditions the DDT is not intrinsic to all the mixtures. The resulting regime alternative to detonation onset is a quasi-steady flame propagation with trans- or supersonic speed. Such regimes are experimentally observed in lean and rich mixtures [4, 5] as well as in mixtures under low pressure [6]. In addition, in low-reactive mixtures [7] and high-reactive mixtures at low pressure [8] one can also observe the formation of auto-ignition kernels at the distance ahead the propagating flame. In number of cases such localized auto-ignition becomes origin of detonation wave formation that is one of the mechanisms of DDT together with the mechanism of detonation onset directly on the flame front observed in high-reactive mixtures. Earlier in [9] the mechanism of DDT directly on the flame front was formulated and described on the example of stoichiometric hydrogen-oxygen mixture on the basis of available experiments interpretation and high-precision numerical simulations. Further investigations partly reviewed in [10] allowed us to understand physical and chemical mechanisms determining the evolution of combustion waves propagating through the channels, including flame acceleration up to sonic and supersonic speeds and two mechanisms of DDT. The present paper represents the basics of quantitative DDT criteria elaboration on the basis of recent work. All the considerations concerning DDT criteria discussed below are proved by means of high-precision numerical simulations. The calculations are carried out using one- and two-dimensional reactive CFD codes elaborated and approved in our previous papers (see e.g. [10, 11, 12] and citations within).

2. On the scenarios and criteria of DDT

Consider the primary features of flame propagation through the channel filled with combustible mixture under given temperature and pressure. Independently on the channel geometry, the main mechanism of flame acceleration is the combustion products expansion that cause the formation of compression waves transferring the energy and momentum to the flow. The accelerating flow drags the flame and cause its corresponding acceleration and stretching, that in turn determine the increase of the mass of combustible mixture burning per time unit. Independent on the channel geometry and the concrete mechanism of flame stretching, the positive feedback between flow acceleration and flame acceleration is establishing, that determine averagely exponential character of acceleration. However, according to the Vulis law of "inverse actions" [13] the acceleration of the flow inside the channel, with the continuous energy release inside the reaction zone, is limited by the velocity value equal to the sonic speed. Thus, on the basis of fundamental gasdynamical concepts it can be concluded that the maximal flow velocity of the reactive flow cannot exceed the sonic speed in the combustion products ($\max(u_f) = a_b$). The corresponding upper limit for the flame speed (relative to the channel walls) is $\max(U_{f,L}) = U_f + a_b$, where U_f is the burning velocity (or the flame speed relative to the fresh mixture). It should be noted that at corresponding time instant the state of fresh mixture ahead of the flame is determined by the set of compression waves irradiated from the propagating flame during the whole process of flame acceleration before reaching the discussed limit ($\max(u_f) = a_b$). Herewith depending on the mixture the burning velocity $U_f(T, p)$ can vary sufficiently compare with one at initial conditions (T_0, p_0).

As soon as the flame speed relative to the channel walls becomes supersonic ($U_{f,L} > a_f$, where a_f - sonic speed in the fresh mixture), the acoustic perturbations occur to be chocked by the supersonic flow formed ahead. Due to this, such a regimes of flame propagation with supersonic speed are usually called "chocked flames". After a chocked regime is established, the flow begins to decelerate according to the Vulis law, and this in turn provides additional compression of the fresh mixture inside the flame front. In such conditions the burning velocity changes ($U_f = U_f(T, p)$), that can cause either additional flame acceleration or flame deceleration depending on the mixture chemical features. Generally the burning velocity U_f increases with temperature and is a non-monotonic function of pressure $U_f \sim p^{(n-2)/2}$, where n is a global reaction order changing with mixture composition, pressure and temperature. Therefore, an additional compression of the fresh mixture can cause progressive or intermediate flame acceleration or even deceleration of the flame. In first case, the additional acceleration causes corresponding compression on the scales of the reaction zone. A shock wave is forming, which together with a reaction front forms a detonation wave as it was reported in [9]. In case of insufficient burning velocity growth or its decrease, the gasdynamical mechanisms play the leading role and the flow velocity stagnates in the vicinity of gasdynamical limit ($\max(u_f) = a_b$). Thereby experimentally observed oscillations of the flame speed are determined by the competition between gasdynamics and chemistry. This non-steady behavior of the flame propagation causes discrete generation of the compression waves driven by the pockets of compressed gas, outrunning from the flame front on the deceleration stages of chocked flame evolution [12]. On the x-t diagrams illustrating this regime, one can observe sequence of shocks and contact surfaces propagating ahead the flame front [7]. In number of cases one can also observe the auto-ignition kernels arising at these contact surfaces.

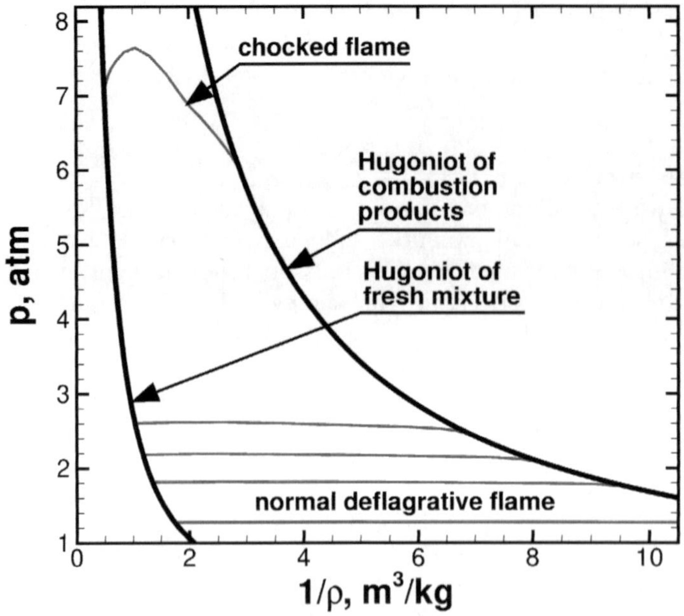

Figure 1. Hydrogen-oxygen flame front in p-V coordinates at different stages of flame acceleration (red lines, presented at equally distant time instants). Thick black lines represent the Hugoniot curves for fresh mixture and combustion products.

The overall pattern described above includes all the regimes observed experimentally such as quasi-steady supersonic flames, DDT and auto-ignition kernels formation ahead the flame front. According to this, it can be concluded that all these phenomena have the same nature and can be described by the same basic mechanisms. Consider now the evolution of fresh mixture state during the whole process of flame acceleration. Figure 1 shows the calculated evolution of the hydrogen-oxygen flame acceleration through the two-dimensional channel in p-V coordinates. The presented profiles illustrate the flame structures on the stages of normal deflagration and chocked flame propagation through the compressed medium. One can see, that the compression of the fresh mixture during the whole process takes place along the Hugoniot curve that can be also observed experimentally. Thus in [14] a rather wide review of experimental measurements of fresh mixture state for different mixtures is presented, that confirms the known fact, that the state of medium compressed by the sequence of compression waves corresponds to the state at Hugoniot curve. As the state of fresh mixture during the whole process of flame acceleration is described adequately by the shock-wave compression, we have assumed that it is rather convenient to elaborate the DDT criteria along the Hugoniot curve that will allow us to get correlation between the initial state (T_0, p_0) of the considered mixture and the probability of DDT.

As it was already discussed above, the additional compression of the fresh mixture ahead of the chocked flame causes the change of the burning velocity and as a result the additional acceleration of the flow. Herewith, additional compression of the fresh mixture results in increase of local sonic speed that in turn determines the critical flow velocity. Taking this issue into account it can be shown that the criterion exists beyond which the flame acceleration mechanism due to chemical reaction begins to play the leading role in contrast with the gasdynamical one. If

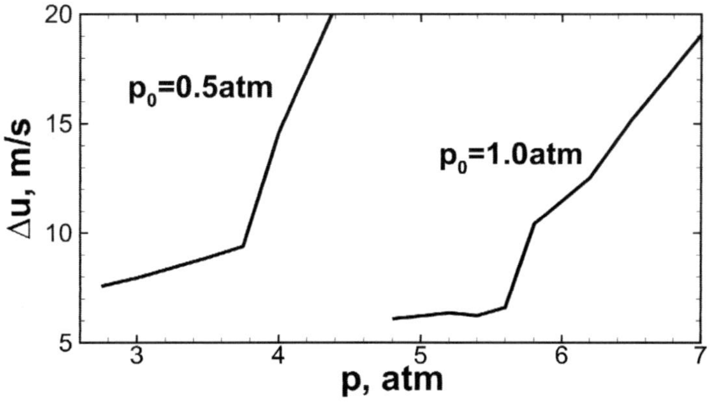

Figure 2. Values of flow velocity variation due to combustion kinetics depending on the pressure characterizing the state on the Hugoniot curve. Two curves represent data for two different initial conditions (p_0=0.5 and 1.0atm).

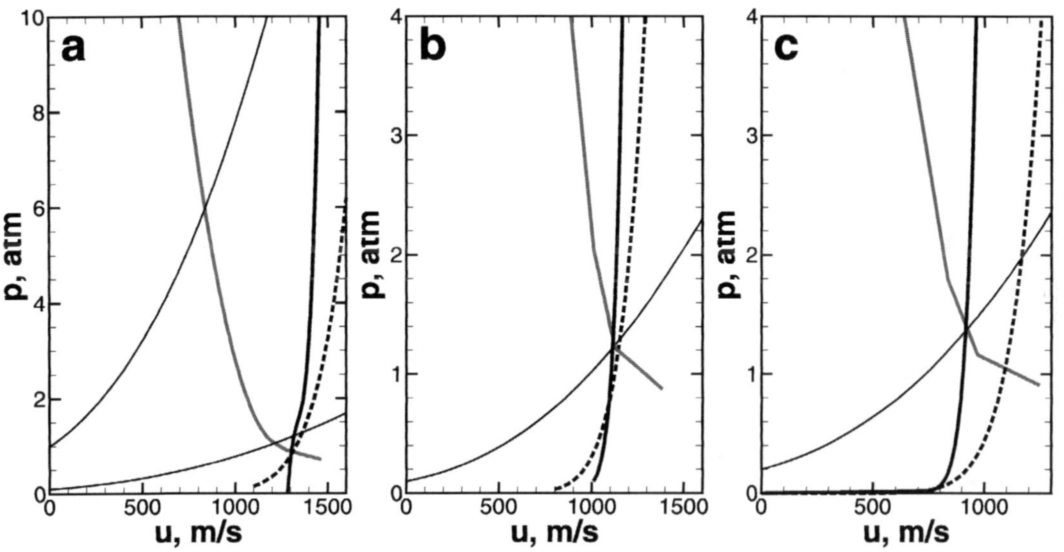

Figure 3. Diagram for determination the possibility of different supersonic combustion regimes of stoichiometric hydrogen-oxygen (a), equimolar hydrogen-oxygen (b) and stoichiometric hydrogen-air (c). Thick black lines - gasdynamical velocity limits. Red lines - criteria of additional supersonic flame acceleration due to peculiarities of chemical kinetics. Dashed lines - "crossover". Thin lines - Hugoniot curves corresponding to different initial pressures: (a) - 1.0 and 0.1atm, (b) 0.11atm, (c) 0.2atm.

the variation of burning velocity (ΔU_f) under additional compression becomes greater than the corresponding variation of sonic speed (Δa_f), the leading role in the following flame acceleration will switch to the mechanism of reaction intensification with compression. If this criterion is not satisfied, the flow will accelerate via gasdynamical mechanism up to the critical gasdynamical velocity scale equal to a_b. Note that due to the additional compression the sonic speed is rising with temperature as $a_f \sim \sqrt{T}$, while the burning velocity is defined by non-monotonic function of temperate and pressure. Thus, for example, in hydrogen-oxygen mixture compressed via shock Hugoniot from the initial normal conditions ($p_0 = 1$atm, $T_0 = 300$K) as well as from the state with lower pressure the function $U_f(T,p)$ changes its rate of increase at some critical state realized along the Hugoniot compression curve (see figure 2). Therefore, along every Hugoniot curve built for different initial conditions one can obtain the criterion beyond which the features of chemical reaction determine additional flame acceleration even if the flow velocity has reached the gasdynamical threshold. To obtain this criterion quantitatively, it was proposed to solve the following problem. We studied numerically the stability of converging initially supersonic flow with combustion front propagating in it. The numerical setup was as following. A spherically symmetric domain was initially divided in three regions: (1) $r < R_0$ - fresh mixture at initial temperature and pressure (T_0 and p_0), (2) $R_0 < r < R_1$ - ignition kernel containing the fresh mixture at elevated temperature, (3) $r > R_1$ - high pressurized driver gas at pressure p_1 and temperature T_0. Discontinuity decay at $r = R_1$ provided a converging flow of compressed gas. A reaction front was formed in the vicinity of the contact surface and propagated from the ignition kernel towards the fresh mixture. The solution of considered problem qualitatively reproduces all the features intrinsic to the supersonic reactive flow in channel. In both cases, the flow is continuously decelerating, that provides compression of the fresh mixture before burning inside the flame front. Criteria obtained numerically in such a way for hydrogen containing mixtures are presented in figure 3 by thick red lines. Thin lines represent some characteristic Hugoniots and thick black lines represent the gasdynamical criterion $u_f = a_b$. According to the obtained diagrams it can be concluded, that there are two basic opportunities: 1) if the flow accelerates up to the gasdynamical criterion prior the moment of reaching the state for which kinetics induces further acceleration of the chocked flame than a quasi steady supersonic combustion regime will establish, 2) otherwise DDT can occur.

An important criterion for gaseous combustibles is the transition from so-called non-degenerate explosion to the degenerate one characterized by sufficiently fast induction period compared with the period of exothermal stage. In literature this criterion is also known as a "crossover" determining the balance between the chain branching and chain termination reactions. As the mixture state transits into the region of degenerate explosion, the burning rate of the fresh mixture can become of the order or even higher than that inside the flame front. In such a case, the mechanism of flame propagation switches from the normal deflagration to the combustion of the mixture reacting at the initial conditions [15]. Set of such "crossover" states is presented in fig. 3 by the dashed lines. According to the presented diagrams, there are two basic opportunities of supersonic combustion in the vicinity of "crossover". In the first case, the supersonic flow continues to accelerate due to the features of chemical kinetics and achieves the state beyond the "crossover". The independent ignition of the mixture ahead the flame front becomes possible, that can be observed experimentally as a local "spontaneous" acceleration of the flame front sheet. The auto-ignition kernel in such conditions is located in the gas directly ahead the contact surface pushing the flow, where the mixture occurs to be the most heated. In the considered case the role of this contact surface belongs to the flame front, and therefore the ignition visually almost cannot be separated from the primal flame front. The second scenario is realized when the flow first achieves the "crossover" state and then the gasdynamical velocity limit, however the chemical activity of the mixture is not enough to provide additional acceleration of the supersonic flow. In such conditions a quasi-steady regime of supersonic flame propagation takes place which is characterized by the downwind propagating shock waves followed by contact surfaces. As the state of mixture corresponds to the so-called degenerate explosion the contact surfaces driven by the additionally compressed gas can become the sources of auto-ignition kernels at some distance from the primal flame. Both regimes were obtained experimentally in [8] in the equimolar hydrogen-oxygen mixture at lowered pressure (~0.11atm).

It is interesting to note, that the regime with DDT on the flame front was also observed in this mixture that defines some instability which was not observed in stoichiometric hydrogen-oxygen mixture at normal initial conditions, also considered in [8], and for which only regime with DDT on the flame front was observed. As it can be concluded from fig. 3b, in case of equimolar hydrogen-oxygen mixture at pressure of 0.11atm, the flow achieves the gasdynamical velocity limit in the vicinity of intersection of all the three limits discussed above. Therefore because of experimental errors the possibility of three different regimes becomes possible. In stoichiometric mixture at higher pressures this instability was not observed experimentally [1, 2, 8, 16], that correlates with the results presented in fig. 3a. It should be also noted, that in less reactive mixtures and even in hydrogen-air mixture, there is no such a region in which the intersection of all the three criteria is observed. Thus in fig. 3c one can clearly see the limit between two possible regimes of DDT and quasi-steady supersonic flame propagation. According to fig. 3c quantitatively criterion equals to 0.2atm, that from our point of view agrees adequately with available experimental data. For example, in [17] an asymptotic increase of the pre-detonation distance was observed with the pressure decrease down to 0.3atm.

3. Conclusion

Three basic criteria were formulated aiming to establish correlation between the initial state of the gaseous combustible mixture and the possibility of DDT and the mechanism of its arising in channel. The first one is the gasdynamical velocity limit - the maximal flow velocity which can be achieved due to the flame acceleration through the channel. The second one is the critical state in which the features of combustion kinetics allow additional acceleration of the supersonic flow even if it has already exceeded the gasdynamical velocity limit. The third one is the so-called "crossover" determining the critical state in which the pre-flame kinetics starts to play a sufficient role, and alternative mechanisms of DDT become to be possible. Formulated criteria have demonstrated an adequate compliance with the data on combustion of hydrogen-based mixtures.

References

[1] Salamandra G D, Bazhenova T V and Naboko I M 1959 *Proc. Comb. Inst.* **7** 851-55
[2] Kuznetsov M, Alekseev V, Matsukov I and Dorofeev S 2005 *Shock Waves* **14** 205-15
[3] Wu M, Burke M, Son S and Yetter R 2007 *Proc. Comb. Inst.* **31** 2429-36
[4] Peraldi O, Knystautas R and Lee J H S 1986 *Proc Comb. Inst.* **21** 1629-37
[5] Ciccarelli G and Dorofeev S 2008 *Progress in Energy and Combustion Science* **34** 499-550
[6] Manzhalei V I 1992 *Combustion, Explosion and Shock Waves* **28** **3** 296-302
[7] Smirnov N N and Tyurnikov M V 1995 *Comb. and flame* **100** **4** 661-8
[8] Utriew P A and Oppenheim A K 1966 *Proc. Roy. Soc. A.* **295** **1440** 13-53
[9] Ivanov M F et. al. 2010 *Dokl. Akad. Nauk.* **434** **6** 756-9
[10] Ivanov M F et. al. 2014 *Uspekhi Physics* **184** **3** 247-264
[11] Ivanov M F, Kiverin A D and Liberman M A 2015 *Combustion and Flame* **162** **10** 3612-21
[12] Kiverin A D and Yakovenko I S 2015 *Journal of Physics: Conference Series* **653** 012061-7
[13] Vulis L A 1946 *Reports of the Academy of Sciences of the USSR* **54** **8** 669-72
[14] Silvestrini M et. al. 2008 *J. Loss Prevent Proc.***21** 555-62
[15] Zeldovich Ya B 1980 *Comb. and flame* **39, 3** 219-24
[16] Meyer J W, Urtiew P A and Oppenheim A K 1970 *Comb. and flame* **14** 13-20
[17] Vasiliev A A 2012 *Combustion, Explosion and Shock Waves* **48** **3** 25-34

Modeling of the flame propagation in coal-dust- methane air mixture in an enclosed sphere volume.

A Yu Krainov and K M Moiseeva

Tomsk State University, Tomsk, 634050 Russia

E-mail: Moiseeva_KM@t-sk.ru

Abstract. The results of the numerical simulation of the flame front propagation in coal-dust-methane-air mixture in an enclosed volume with the ignition source in the center of the volume are presented. The mathematical model is based on a dual-velocity two-phase model of the reacting gas-dispersion medium. The system of equations includes the mass-conversation equation, the impulse-conversation equation, the total energy-conversation equation of the gas and particles taking into account the thermal conductivity and chemical reactions in the gas and on the particle surface, mass-conversation equation of the mixture gas components considering the diffusion and the burn-out and the particle burn-out equation. The influence of the coal particle mass on the pressure in the volume after the mixture burn out and on the burn-out time has been investigated. It has been shown that the burning rate of the coal-dust methane air mixtures depends on the coal particle size.

1. Introduction

The experimental studies on the influence of the coal-dust on the ignition characteristics of gas mixtures have been conducted for several decades. In the studies the mixture proportions (the methane and coal particle mass content) and the reaction places (a spherical bomb, a tube, dust-laden flame) have been varied, the influence of the additives (deterrents and retarders) on the burning characteristics of the hybrid mixtures has been analyzed.

It has been shown in [1] that the reactive gas and coal dust mixture is able to explode under the low concentration of the coal dust in the air. The data in [1] show that the coal dust in the air and the methane air mixture in separate trials under the chosen concentrations in the air are not able to explode, but its mixture can be highly explosive. The flame propagation velocity in the methane air mixture increases with the presence of the small coal dust particles [2]. The experimental study on the burning of the coal dust and 6.5% methane-air mixture in a spherical volume has been carried out in [3]. The authors in [3] indicate the significant influence of the coal dust particles on the burning characteristics of the reactive gas mixtures. The rate of the pressure growth in the volume with the presence of the coal dust particles in the mixture is higher than that without the particles. In [4] for the coal dust-air mixture it has been shown theoretically that under the normal conditions the mixture burning become possible by adding a small amount of methane or by preheating the burner walls. The influence of the coal dust particles on the hybrid mixture burning rate has been investigated in [5]. The results have shown that for the laminar flames the small size particles decrease the normal velocity of the flame front.

In [6] the authors proposed the physical-mathematical model of the ignition and the detonation burning of the coal dust suspension in the air. The research on the comparison of the obtained results from one-velocity model and dual-velocity models has been carried out. It has been observed in [7] that at the low concentrations of the fuel in the mixture the presence of the reacting particles increase the velocity of flame front propagation.

The experimental and theoretical researches have shown that the burning characteristics of the hybrid mixtures significantly depend on the size of the reacting particles. The burning process is affected by the surrounding conditions. One of the methods determining the burning characteristics of the coal dust and reactive gas mixture is the burning in a spherical volume. In the present paper the numerical simulation of the flame front propagation in the coal-dust methane–air mixture in the enclosed spherical volume has been carried out. The aim of the study was to determine the mass and the particle size influence on the rate of the pressure growth.

2. Mathematical model

It is supposed that the monodisperse suspension of the coal dust particles which mass is m_{pyl}, and the particle radius is r_k and 1% methane-air mixture are located in the enclosed spherical volume V with the radius r_S. The ignition of the mixture is executed by the hot spot in the center of the volume with the radius r_0. The heat transfer to the surrounding environment is neglected. The diffusion and the heat transfer coefficients depend on the temperature [8]. The gas motion caused by the heat expansion due to the rising temperature is taken into account. There are two parallel reactions in the mixture, the first is the exothermic chemical reaction in the gas, and the second one is the heterogeneous reaction on the particle surface. The reaction rate in the gas is determined by the second-order kinetics (the first-order kinetics for methane and the first-order kinetics for oxygen). The specific reaction rate of the gas depends on the temperature under the Arrhenius law. The heterogeneous reaction rate is limited by the mass-transfer coefficient β. The friction between the particles and the gas is taken into account.

The mathematical model is based on the dual-velocity two-phase model of the reacting gas-dispersion medium. The system of equations includes the mass-conversation equation, the impulse-conversation equation, the total energy-conversation equation of the gas and the particles taking into account the thermal conductivity and the chemical reactions in the gas and on the particle surface, the mass-conversation equation of the mixture gas components considering the diffusion and the combustion and the particle radius and concentration equations. The mathematical formulation of the problem under the made assumptions has the following form:

The gas continuity equation:

$$\frac{\partial \rho_g}{\partial t}+\frac{\partial \rho_g u_g}{\partial r}=-\frac{2\rho_g u_g}{r}+G. \tag{1}$$

The gas impulse–conversation equation:

$$\frac{\partial(\rho_g u_g)}{\partial t}+\frac{\partial(\rho_g u_g^2+p)}{\partial r}=-\frac{2\rho_g u_g^2}{r}-\tau_{tr}+Gu_k. \tag{2}$$

The gas energy-conversation equation:

$$\frac{\partial \rho_g\left(\varepsilon_g+\frac{u_g^2}{2}\right)}{\partial t}+\frac{\partial\left[\rho_g u_g\left(\varepsilon_g+\frac{u_g^2}{2}\right)+pu_g\right]}{\partial r}=-\frac{2\left[\rho_g u_g\left(\varepsilon_g+\frac{u_g^2}{2}\right)+pu_g\right]}{r}+Gc_{p,k}T_k-u_k\tau_{tr}+$$
$$G\frac{u_k^2}{2}+\alpha_k n_k S_k\left(T_k-T_g\right)+\frac{1}{r^2}\frac{\partial}{\partial r}\left(r^2\lambda(T_g)\frac{\partial T_g}{\partial r}\right)+Q_1 k_{01}\rho_{CH4}\rho_{O2}\exp\left(-\frac{E_1}{R_u T_g}\right). \tag{3}$$

The methane mass balance equation:

$$\frac{\partial \rho_{CH4}}{\partial t}+\frac{\partial \rho_{CH4}u_g}{\partial r}=-\frac{2\rho_{CH4}u_g}{r}+\frac{1}{r^2}\frac{\partial}{\partial r}\left(r^2 D(T_g)\frac{\partial \rho_{CH4}}{\partial r}\right)-k_{01}\rho_{CH4}\rho_{O2}\exp\left(-\frac{E_1}{R_u T_g}\right). \tag{4}$$

The oxygen mass balance equation:

$$\frac{\partial \rho_{O2}}{\partial t}+\frac{\partial \rho_{O2} u_g}{\partial r}=-\frac{2\rho_{O2} u_g}{r}+\frac{1}{r^2}\frac{\partial}{\partial r}\left(r^2 D\left(T_g\right)\frac{\partial \rho_{O2}}{\partial r}\right)-\alpha_1 G-\alpha_2 k_{01}\rho_{CH4}\rho_{O2}\exp\left(-\frac{E_1}{R_u T_g}\right). \quad (5)$$

The particle mass balance equation:

$$\frac{\partial \rho_k}{\partial t}+\frac{\partial \rho_k u_k}{\partial r}=-\frac{2\rho_k u_k}{r}-G. \quad (6)$$

The particle impulse–conversation equation:

$$\frac{\partial\left(\rho_k u_k\right)}{\partial t}+\frac{\partial \rho_k u_k^2}{\partial r}=-\frac{2\rho_k u_k^2}{r}+\tau_{tr}-G u_k. \quad (7)$$

The particle energy-conversation equation:

$$\frac{\partial \rho_k\left(\varepsilon_k+\frac{u_k^2}{2}\right)}{\partial t}+\frac{\partial \rho_k u_k\left(\varepsilon_k+\frac{u_k^2}{2}\right)}{\partial r}=-\frac{2\rho_k u_k\left(\varepsilon_k+\frac{u_k^2}{2}\right)}{r}-\alpha_k S_k n_k\left(T_k-T_g\right)+Q_2 G-G c_{p,k} T- G\frac{u_k^2}{2}+\tau_{tr} u_k. \quad (8)$$

The particle number concentration equation:

$$\frac{\partial n_k}{\partial t}+\frac{\partial n_k u_k}{\partial r}=-\frac{2 n_k u_k}{r}. \quad (9)$$

The particle radius equation:

$$r_k=\left(\frac{3\rho_k}{4\pi \rho_k^0 n_k}\right)^{1/3}. \quad (10)$$

The perfect-gas law:

$$p=\rho_g R_g T_g. \quad (11)$$

The particle temperature equation:

$$T_k=\frac{\varepsilon_k}{c_k}. \quad (12)$$

The initial condition:

$$T_g\left(r,0\right)=\begin{cases}T_z,\ 0\le r\le r_0\\ T_b,\ r_0<r\le r_s\end{cases},\ T_k\left(r,0\right)=T_b,\ \rho_{CH4}\left(r,0\right)=\rho_{CH4,b},\ \rho_{O2}\left(r,0\right)=\rho_{O2,b},\ \rho_k\left(r,0\right)=\rho_{k,b},\ u_g\left(r,0\right)=u_k\left(r,0\right)=0,\ p\left(r,0\right)=p_b,\ n_k\left(r,0\right)=n_{k,b}. \quad (13)$$

The boundary conditions:

$$\frac{\partial \rho_{CH4}\left(0,t\right)}{\partial r}=\frac{\partial \rho_{O2}\left(0,t\right)}{\partial r}=\frac{\partial T_g\left(0,t\right)}{\partial r}=0,\ u_k\left(0,t\right)=u_g\left(0,t\right)=0. \quad (14)$$

$$\frac{\partial \rho_{CH4}\left(r_S,t\right)}{\partial r}=\frac{\partial \rho_{O2}\left(r_S,t\right)}{\partial r}=\frac{\partial T_g\left(r_S,t\right)}{\partial r}=0,\ u_k\left(r_S,t\right)=u_g\left(r_S,t\right)=0. \quad (15)$$

Where : $\varepsilon_g=\frac{p}{\rho_g(\gamma-1)}$ – the internal gas energy, $\varepsilon_k=c_k T_k$ – the internal particle energy, $\lambda=\lambda_{st}\left(\frac{T}{T_v}\right)^s$ – the thermal conductivity [8], $D=D_{st}\left(\frac{T}{T_v}\right)^s$ – the diffusion coefficient, $\alpha_k=\frac{Nu_k\lambda_g}{2r_k}$ – the gas-particles heat transfer coefficient, $\gamma=\frac{c_p}{c_v}$ – the adiabatic exponent, $\beta_{m,i}=\frac{\lambda_g(T)Nu_D}{c_g\rho_g r_{k,i}}$ – – the particles mass-transfer coefficient [10], $G=\alpha_5 n_k S_k j_1\rho_{O2}$ – the rate of mass changing,

$j_1 = \dfrac{\beta_m k_{02} \exp(-E_2/R_u T_k)}{\beta_m + k_{02} \exp(-E_2/R_u T_k)}$ – the heterogeneous reaction rate, $\tau_{tr} = n_k F_{tr}$, – the gas-particles interacting force, $F_{tr} = \dfrac{12(1+0.15\mathrm{Re}^{0.682}) S_m \rho_g (u_g - u_k)|u_g - u_k|}{\mathrm{Re}}$, $\mathrm{Re} = \dfrac{2\rho_g r_k |u_g - u_{k,i}|}{\eta}$ – Reynolds number, $Nu_k = 2 + (Nu_l^2 + Nu_t^2)^{1/2}$, $Nu_l = 0.664\mathrm{Re}^{0.5}$, $Nu_t = 0.037\mathrm{Re}^{0.8}$, $\alpha_1 = \dfrac{\mu_{O2}\nu_{O2}}{\mu_C \nu_C}$ – stoichiometric coefficient in coal-oxygen reaction, $\alpha_2 = \dfrac{\mu_{O2}\nu_{O2}}{\mu_{CH4}\nu_{CH4}}$ – stoichiometric coefficient in methane-oxygen reaction, , ρ – the density, ρ_{CH4} – the methane partial density, ρ_{O2} – the oxygen partial density, u – the velocity, t – the time, r – the radius coordinate, r_k – the particle radius, p – the pressure, λ – the thermal conductivity, D – the diffusion coefficient, Q – the reaction heat, k_0 – pre-exponential factor in the Arrhenius law, T – the temperature, E – energy of activation, R_u – the molar gas constant, R_g – gas constant, r_0 – the radius of the high-temperature region (the hot spot), r_S – the radius of the spherical volume. Indexes: b – the initial conditions of the methane-air mixture state parameters, g – gas parameters; k – particle parameters. For Q, k_0, E index 1 is the parameters of the reaction in the gas; index 2 is the parameters of the heterogeneous reaction.

The problem (1) – (15) was solved numerically by S.K. Goduvov's method [11]. The summands on the right side of equations which determine the diffusion and the heat transfer processes were explicit approximated by the three-point stencil. To solve the equations (1) – (5) the breakdown breakup of an arbitrary discontinuity scheme of gas parameters was used [11]. To solve the equations (6) – (8) the method from [12] was used. The spatial step was set to $\Delta h = 10^{-5}$ m to provide the sufficient number of different scheme points in the region of the reaction and warm-up zone in front of the burning front. The value of the step provides at least 30 points of difference scheme in the warm-up zone. The time step was calculated by Courant's stability criterion $\Delta t = \dfrac{0.8\Delta h}{\max[c] + \max[|u|]}$, where c is the sound velocity.

To verify the computer program the test problem about the thermal explosion in an enclosed volume has been solved. The determination error of the adiabatic temperature after the explosion was equal to 1 %. The value of the hot spot energy was larger than the critical ignition energy of the mixture.

3. Results and discussion

The calculations were held under the following parameters: $Q_1 = 55.7$ MJ/kg, $Q_2 = 29$ MJ/kg, $k_{01} = 1.125\cdot10^{12}$ $m^3/(kg\cdot s)$, $k_{02} = 7.9\cdot10^4$ m/s, $E_1 = 239$ kJ/mol, $E_2 = 135$ kJ/mol, $c_{p,g} = 1065$ $J/(kg\cdot K)$, $c_{v,g} = 768.2$ $J/(kg\cdot K)$, $c_k = 1464.4$ $J/(kg\cdot K)$, $\rho_k^0 = 1400\ kg/m^3$, $\rho_{CH4,b} = 6.4\cdot10^{-3}\ kg/m^3$, $\rho_{O2,b} = 0.264\ kg/m^3$, $\lambda_{st} = 0.025$ $W/(m\cdot K)$, $\eta = 2\cdot10^{-5}$ $Pa\cdot s$, $D_{st} = 1.992\cdot10^{-5}$ m^2/s, $p_b = 0.1$ MPa, $s = 2/3$, $R_u = 8.31$ $J/(mol\cdot K)$, $R_g = 288.95$ $J/(kg\cdot K)$, $\alpha_1 = 2.67$, $\alpha_2 = 4$, $\gamma = 1.39$, $r_0 = 2\cdot10^{-3}$ m, $T_b = 300$ K, $V = 10^{-2}$ m^3, $r_s = 0.134$ m, $r_{k,b} = 10^{-6}$ m, $T_z = 1500$ K. The value of particle density ρ_k was varied in the range of 0.065 ÷ 0.085 kg/m^3. The results of the calculations are shown in Pic. 1 – 3.

The distribution of the temperature and the gas velocity at different moments of time under coal particle mass $\rho_{k,b} = 0.068$ kg/m^3 are shown in Pic 1. The curves 11 correspond to the state of the temperature and the gas velocity after the burn out of the whole mixture in the volume. It has been determined from the calculations that the particle velocity was almost the same as gas velocity. We have obtained the pressure-coal particle mass curve (Pic. 2) and the time-coal particle mass curve (Pic. 3) at the end of the burning. The increase of the coal particle mass leads to the increase of the pressure in the volume (Pic. 2) and to the decrease of the mixture total burn out time (Pic. 3).

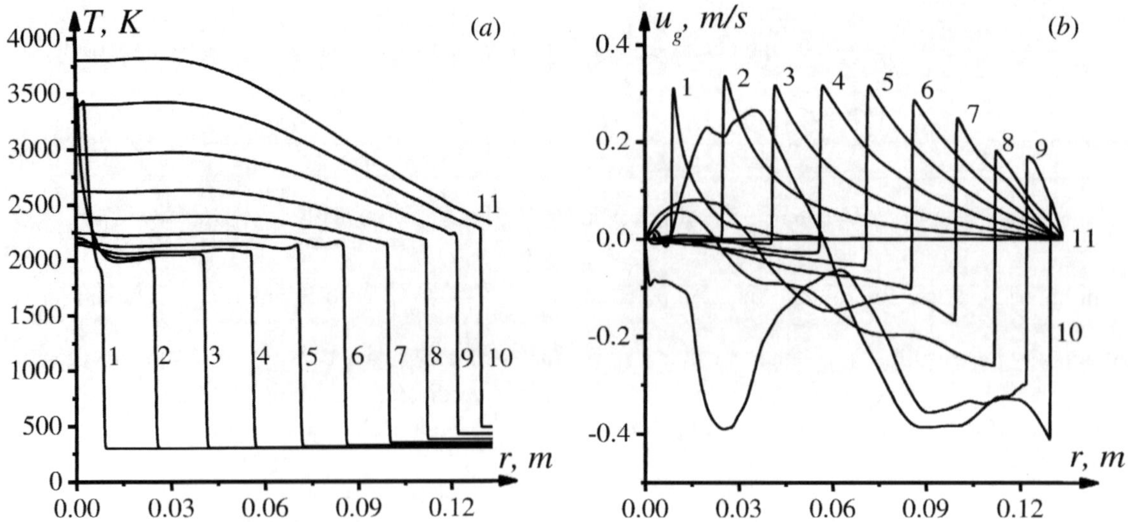

Pic. 1. The temperature distributions (*a*) and the gas velocity distributions (*b*) at the fixed moments of the time. $\rho_{k,b} = 0.068$ *kg/m*^{3}; $r_{k,b} = 10^{-6}$ *m*; 1 – *t* = 0.02 *s*, 2 – *t* = 0.06 *s*, 3 – *t* = 0.1 *s*, 4 – *t* = 0.14 *s*, 5 – *t* = 0.18 *s*, 6 – *t* = 0.22 *s*, 7 – *t* = 0.26 *s*, 8 – *t* = 0.3 *s*, 9 – *t* = 0.34 *s*, 10 – *t* = 0.38 *s*, 11 – *t* = 1 *s*

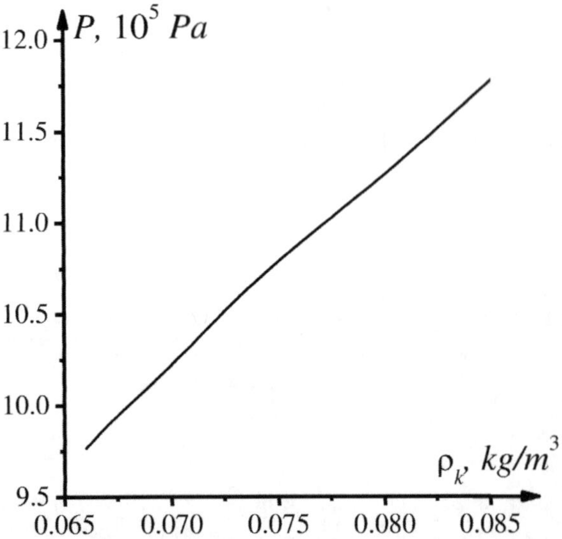

Pic. 2. The pressure-coal particle mass curve at the end of the burning

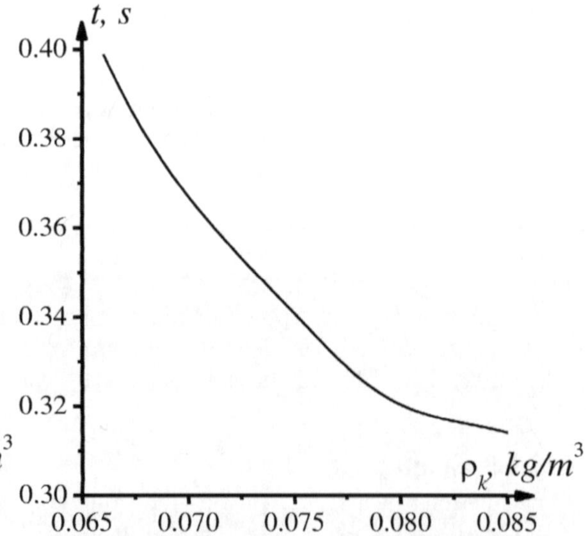

Pic. 3. The time-coal particle mass curve at the end of the burning

The hot spot ignites the mixture in the center of the volume. The amount of the methane in the mixture is not enough to support the burning therefore the flame front propagation depends on the amount of the coal dust in the volume. According to the obtained results the flame front doesn't propagate along the volume when the coal dust mass is less than 0.065 *kg/m*^{3} and the particle size is more than $r_{k,b} = 10^{-6}$ *m*. The particles don't have enough time to react and the burning is extinguished. But if the particle size is less than $r_{k,b} = 10^{-7}$ *m* and the coal dust mass is less than 0.065 *kg/m*^{3} the burning of the coal dust methane-air mixture becomes possible. It is largely due to the quick burning of the small particles which are able to support the burning front.

Pic. 4*a* shows the pressure growth-time curves under $\rho_k = 0.068$ *kg/m*^{3}. Pic. 4*b* shows the curves of the flame front propagation in time. The flame front coordinate is *r* in which the content of the oxidizer is half of the initial value. The rate of the pressure growth increases with the decrease of the coal particles size (Pic. 4*a*).

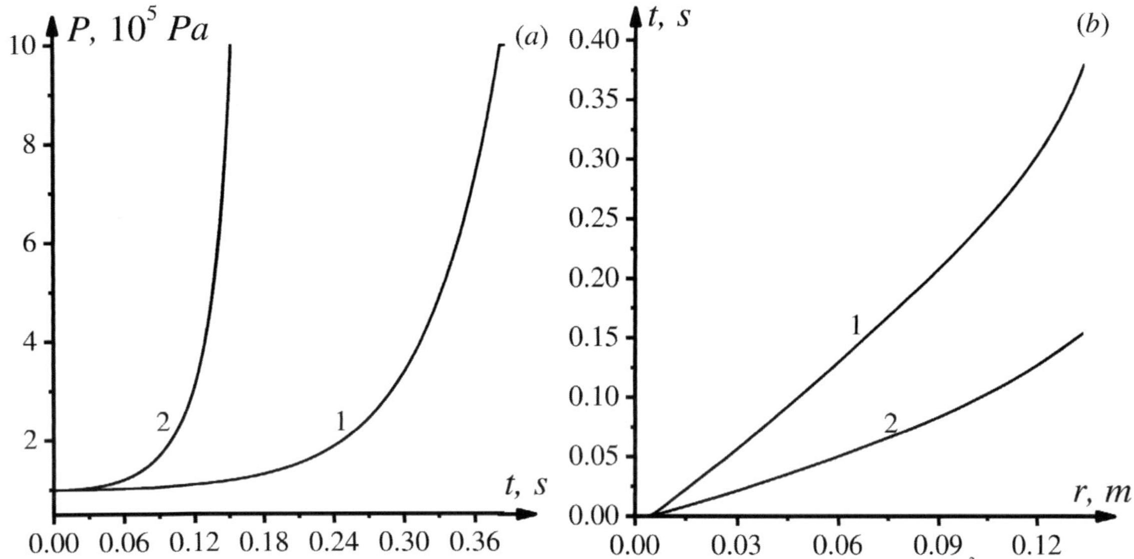

Pic. 4. The pressure change in time (*a*), flame front coordinate (*b*). $\rho_k = 0.068$ *kg/m*^{3};
$r_{k,b} = 10^{-6}$ *m* (curve 1), 10^{-7} *m* (curve 2)

4. Conclusions

We have carried out the numerical simulation of the coal dust methane-air mixture flame propagation in the enclosed spherical volume with the hot spot in the center. The influence of the coal particle mass on the pressure in the volume after the mixture burn-out and on the burn-out time has been investigated. We have obtained the pressure growth rate-coal particle size dependences.

Acknowledgments

This work was supported by the Russian Science Foundation grant No. 16-33-60091 mol_a_dk.

References:

[1] Kalyakin S, Kostenko V, Zav'yalova E and Shtrokh L 2012 *Aktualne problemy zwalczania zagrożeń górniczych: II konferencja naukowo-techniczna (Brenna)* No 7-9 (Brenna) p 176
[2] Krainov A Yu 2000 *Combust., Expl.,Shock Waves* **36 (2)** 157-63
[3] Baklanov A M, Valiulin S V, Dubtsov S N, Zamashchikov V V, Korzhavin A A, Onischuk A A, Purtov P A, Klishin V I, Paleev D Y and Kontorovich A E 2015 *Doklady Phys. Chem.* **461 (1)** 57-60
[4] Bradley D, Lawes M, Scott M J and Usta N 2001 *Combust. Flame* **124 (1-2)** 82–105
[5] Rockwell S R and Rangwala A S 2013 *Combust. Flame* **160** 635-40
[6] Fedorov A and Khmel T 2005 *Combust., Expl.,Shock Waves* **41 (1)** 78-87
[7] Dement'ev A A, Krainov A Yu and Moiseeva K M 2015 *Tomsk State University Journal of Mathematics and Mechanics* **5(37)** 55-64
[8] Shkadinskii K G and Barzykin V V 1968 *Combust., Expl.,Shock Waves* **4 (2)** 100-4
[9] Nigmatulin R I 1990 *Dynamics of Multiphase Media* (CRC Press; Rev Sub edition)
[10] Frank-Kamenetskiy D A 1987 *Diffusion and Heat Transfer in Chemical Kinetics* (Moscow: Nauka)
[11] Godunov S K, Zabrodin A V, Ivanov M Ya, Kraiko A N and Prokopov G P 1976 *Numerical solution of multi-dimensional problems of gas dynamics* (Moscow: Nauka)
[12] Kraiko A N 1979 *Journal of applied mathematics and mechanics* **43 (3)** 539-49

Journal of Physics: Conference Series **754** (2016) 052004 doi:10.1088/1742-6596/754/5/052004

The growth of carbon nanoparticles during the detonation of trinitrotoluene

I A Rubtsov[1,4], K A Ten[1,4], E R Pruuel[1,4], A O Kashkarov[1,4], B P Tolochko[2,4], V V Zhulanov[3,4], L I Shekhtman[3,4] and P A Piminov[3,4]

[1] Lavrentyev Institute of Hydrodynamics of the Siberian Branch of the Russian Academy of Sciences, 15 Lavrentyev Prosp., Novosibirsk, 630090, Russia
[2] Institute of Solid State Chemistry and Mechanochemistry of the Siberian Branch of the Russian Academy of Sciences, 18 Kutateladze Str., Novosibirsk, 630090, Russia
[3] Budker Institute of Nuclear Physics of the Siberian Branch of the Russian Academy of Sciences, 11 Lavrentyev Prosp., Novosibirsk, 630090, Russia
[4] Novosibirsk State University, 2 Pirogova Str., Novosibirsk, 630090, Russia

E-mail: rubtsov@hydro.nsc.ru

Abstract. In this work we present experimental data on measuring distributions of small-angle X-ray scattering (SAXS) during cast trinitrotoluene (TNT) detonation of 30 and 40 mm in diameter. Dynamics of average size of condensed carbon nanoparticle inkreases has been restored from experimental SAXS data. The work was carried out at the SYRAFEEMA (**Sy**nchrotron **Ra**diation **F**acility for **E**xploring **E**nergetic **Ma**terials) station at accelerator complex VEPP-4M (Budker Institute of Nuclear Physics). We observe minimal size of particles of order of 2 nm directly behind the detonation front. Later, the average size of carbon nanoparticles increases within 4–12 μs and reach values of 6 nm.

1. Introduction

In the early 2000s, studying of triaminotrinitrobenzene (TATB) in the nuclear centers of Russia and USA renewed interest of carbon condensation at detonation of high explosives (HEs) with negative oxygen balance. The results [1, 2] have shown that the assumption of the additional allocation of energy behind a chemical reaction zone better describes the experimental data on acceleration of the metal plates. The results of experiments can be explained by the additional energy released due to the exothermic coagulation of carbon clusters.

At the same time there were the first experiments on the dynamic registration of small-angle X-ray scattering (SAXS) distributions, which showed a long signal growth in the detonation of a mixture of trinitrotoluene (TNT) and hexogen (RDX) 50/50 [3].

The SAXS signal carry the information about the variations of density inside the investigated sample. This method is widely used for the static analysis of structure of dispersion sistems. In [3–5], it was shown that the integral SAXS signal increases with increasing negative oxygen balance. At raying of explosion products by synchrotron radiation (SR), the condensed carbon nanoparticles plays role the density variations.

The ability to use short high-periodical SR flashes of high-energy accelerators allowed us to reveal the time evolution of the signal at detonation of HEs. One can restore the dynamics of the average particle size of the condensed carbon at detonation of HEs from the SAXS data [6].

Presently, registration of SAXS signal is the only way to register experimently the dynamics of the size of the condensed carbon nanoparticles during detonation of HEs [7–9].

A new experimental station on accelerator complex the VEPP-4M put into operation in 2015. These new SYRAFEEMA station give much higher SR intensity than the similar station "Extreme States of Matter" on the VEPP-3 accelerator complex. It allowed us to X-ray examination of explosive charges of larger diameter (till 40 mm) and increase in the mass of the charges studies by ten times till 200 grams [6].

Similar experiments was carried out at Advanced Photon Source (Argonne National Laboratory, USA). In these studies, the hexanitrostilbene detonation produces carbon particles with radius of gyration of 2.7 nm, which corresponds to the 7 nm diameter of spherical particles. That particles was registered in 400 ns and remained constant in several next microseconds. It should be noted, that small charges of ≈ 6 mm in diameter were used in their study [9]. Dynamics of carbon condensation in big charge are still remains an actual problem.

Here we present the experimental data on particle size dynamics during TNT detonation of charge of different diameter. The study of TNT is important because this explosive is the main component of mixtures used in the commercial production of ultrafine diamonds.

2. Experimental setup

The SAXS measurements were carried out at SYRAFEEMA station on the VEPP-4M. For dynamic experiments on registration SAXS, the scheme of measurement is shown in figure 3. The Kratky collimator was installed ahead the explosion chamber. It consist of lower K_1 and upper K_2 knives which formed initial SR beam. Straight beam was cut off by knife K_3 ahead the detector. The scattered rays detected by the detector DIMEX [10] with an angular resolution of $3 \cdot 10^{-5}$ rad. Changing the position of all three knives was tuned by the micropositioner with accuracy 3 μm. After that, the SR beam size was 0.5 mm in thickness (figure 4 (1)) and 25 mm in width. The distance from knives K_2 and K_3 to the detector was 4660 and 231 mm, respectively. The time between frames was 600 ns.

The position of knife K_3 was chosen such that the angular measurement range to be 2–200 detector channels (0.06–5.8 mrad). This SAXS measurement range allowed to register particles of carbon of a diameter of D: $d_{min} \leq D \leq d_{max}$. $d_{min} = \pi/q_{max} = \lambda/4\sin(\theta_{max}) \approx 2$ nm, $d_{max} = \pi/q_{min} = \lambda/4\sin(\theta_{min}) \approx 100$ nm.

Figure 1 presents the spectral characteristics of the radiation of SYRAFEEMA station (real spectrum). The real spectrum of SR at the station depends on the wiggler emission spectrum, the absorption spectrum of the charge, and the detection efficiency of the DIMEX-3 detector. Calculations have shown the possibility of using such a polychromatic spectrum to restore the size of the scattering centers by replacing the real radiation spectrum to effective photon energy. The effective photon energy was determined by the least squares method so that the deviations between the SAXS signal with the real spectrum and the SAXS with the effective energy was minimal. Thus, we obtained an effective X-ray energy of 38 keV (see figure 2). At this energy, the SR wavelength is equal to $\lambda = h\frac{c}{E} = 0.0326$ nm. The monochromatic radiation energy that most closely describes the SAXS from the real spectrum at the station is considered as the effective energy.

3. SAXS measurement experiments

The sizes of the scattering centers were determine using the Guinier approximation – $I(q, R) = I_0 \exp(-q^2 R_g^2/3)$, where R_g is the radius of gyration of the particle, $q = 4\pi\sin(\theta)/\lambda$ is the scattering vector, λ is the scattered radiation wavelength and 2θ is the scattering angle [11,12]. For a spherical homogeneous particle of radius R, $R_g = \sqrt{3/5}R$.

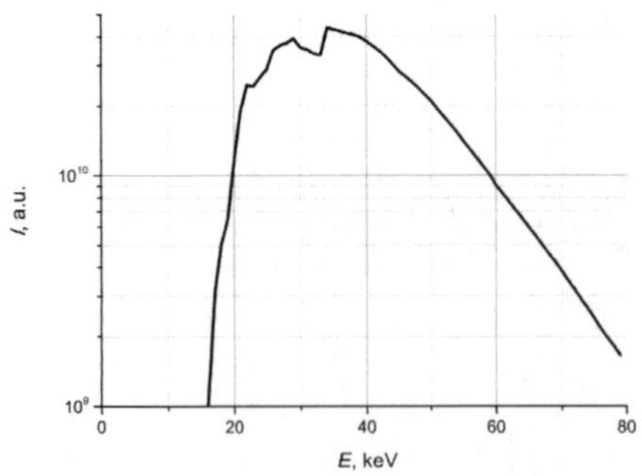

Figure 1. Real radiation spectrum at the SYRAFEEMA station.

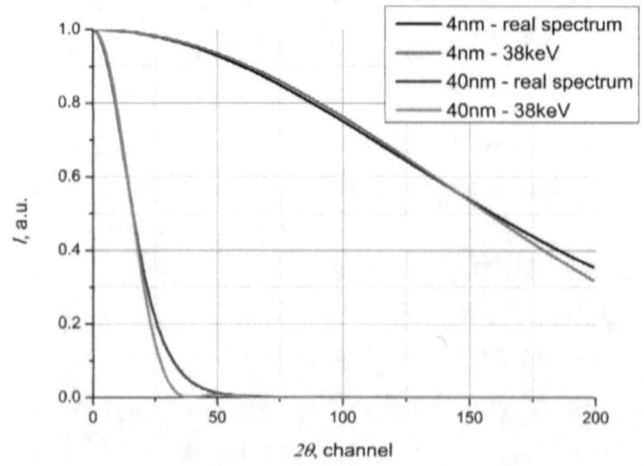

Figure 2. SAXS with consideration of the spectrum and effective energy.

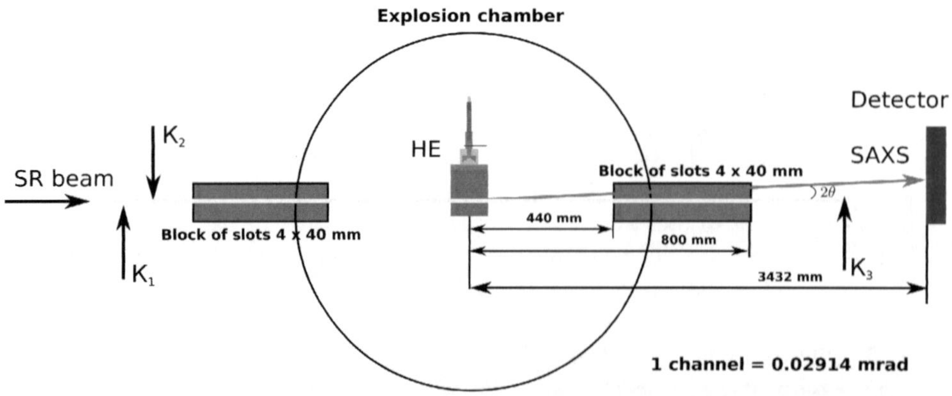

Figure 3. Scheme of experiments on SAXS measuring.

Taking the logarithm of the intensity $\ln(I(q,R)) = \ln(I_0) - q^2R^2/5$, we obtain a function which decreases linearly versus q^2. We can determine the size of the spherical particle $D = 2R = 2\sqrt{-5k}$ using the slope k of this line [11, 12].

Testing the possibility of using SR in dynamic explosive experiments was tested under static conditions on the scattering of the ultra-fine diamonds (UFD) (figure 4(2)). UFD particles were placed in the center of the explosion chamber instead of explosion charge. Transmission electron microscopy shows the average size of these particles is 5 nm. Analysis of SAXS distributions from UFD was determined using Guinier approximation. The slope was k $\approx$ -1.4 (figure 5), this slope corresponds to the particle size of D=5.4 nm. It can be seen that the graph of $\ln(I)$ versus q^2 is different from a straight line. This means that the scattering centers have different sizes. The approximation by a straight line is applicable for the average particle size.

Test experiments have shown the ability to determine the particle size distribution obtained by treating the SAXS signal on the VEPP-4M. During TNT detonation scatter occurs on the particles of condensed carbon in the explosion products.

In our experiments we used cast samples of TNT of the same size (40 mm in diameter and 60 mm in length) and of density 1.6 g/cm^3. The SAXS intensity distributions during the detonation of TNT are shown in figure 6. Each colored line represents the distribution of SAXS signal at a certain time. The sequence of the angular distributions gives the time dependence of dynamics

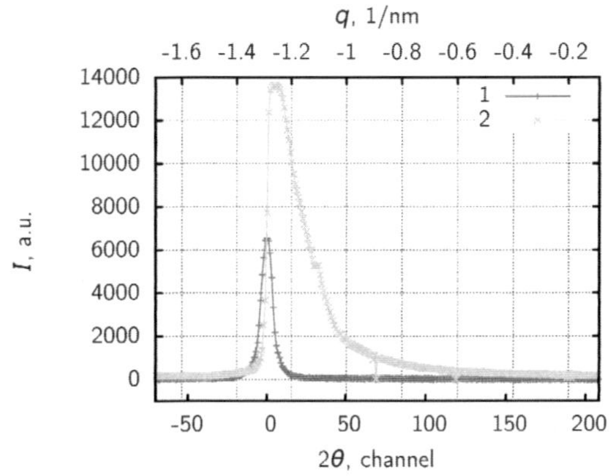

Figure 4. Setting: 1 – attenuated straight beam, 2 – scattering on particles of ultra-fine diamonds (UFD).

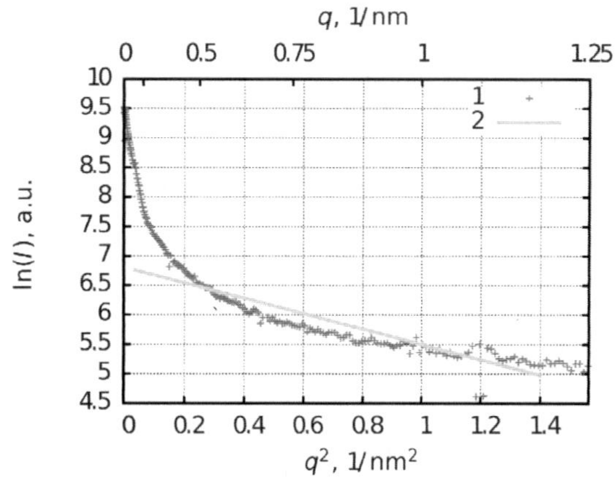

Figure 5. The dependence of $\ln(I)$ on q^2: 1 – $\ln(I)$, 2 – approximation by straight line for UFD (k ≈ -1.4, D ≈ 5.4 nm).

of the process. Time is measured from the moment of passage of the detonation front.

Figure 7 represented the Guinier approximations at the 10 frame in one experiment on detonation of TNT charge of 40 mm in diameter. The slope of straight line approximation is k ≈ -0.99 corresponds to the scattering particles of the diameter of D ≈ 4.4 nm. After processing of all frames was obtain dynamics of average particle sizes for one experiment. Figure 8 presents the dynamics of the average size of nanoparticles for all experiments during the detonation of TNT charge of 40 mm in diameter. Figure 9 shows the same information for the charge of 30 mm in diameter.

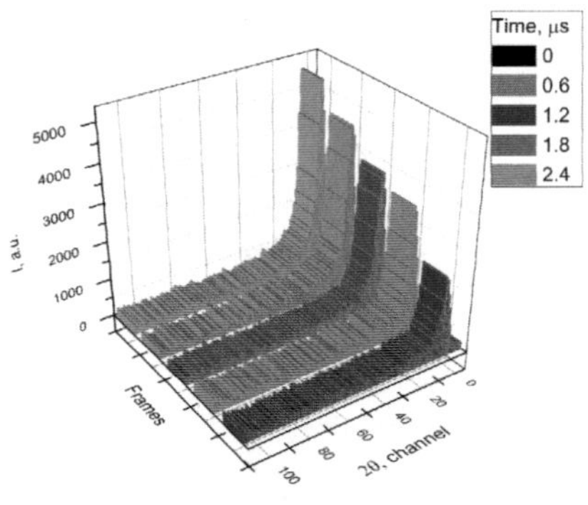

Figure 6. SAXS data during TNT detonation in the first 2.5 μs for one experiment.

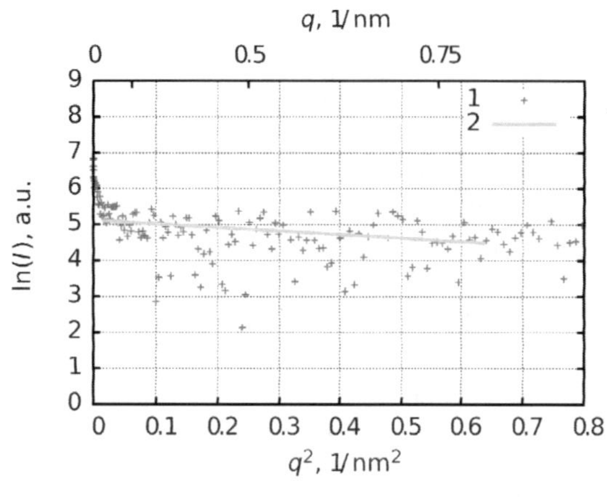

Figure 7. 1 – SAXS for TNT detonation at 6 μs behind the front, 2 – Guinier approximation.

The dynamics of nanoparticle sizes obtained under the following assumptions. The scattering centers are a sphere with uniform density (excluding the phase state). Carbon in the diamond form gives more intensive SAXS signal. Replacement of real radiation spectrum to

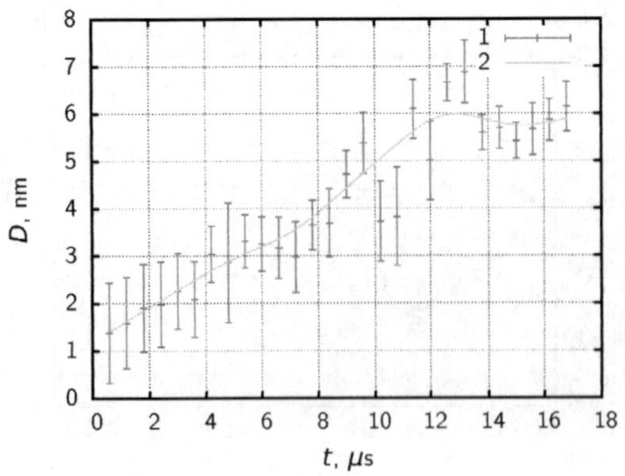

Figure 8. Average size of carbon particle versus time in detonation of a cylindrical TNT charge of 40 mm diameter: 1 – experimental data, 2 – smooth spline.

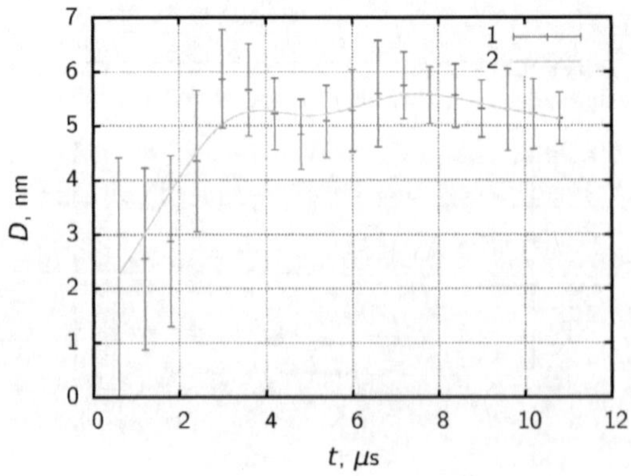

Figure 9. Average size of carbon particle versus time in detonation of a cylindrical TNT charge of 30 mm diameter: 1 – experimental data, 2 – smooth spline.

a monochromatic, as shown previously does not lead to large error. Replacement the size distribution of nanoparticles by average size introduces the greatest error in mind a very strong dependence of the SAXS intensity from the particle size. Therefore, further efforts will be made to determine the size distribution of the scattering centers.

4. Conclusions

The SAXS dynamics in detonation of cast TNT charges (of different diameter) was measured. Dynamics of the average size of the condensed carbon nanoparticles behind a chemical reaction zone recovered from SAXS distributions in the Guinier approximation. The minimall particle size registered in experiments is 2 nm. Particles of this size are registered immediately behind the detonation front. Then the average size of particles increases. We observe, that the time of nanoparticles growth behind the chemical reaction zone depends on size of initial charge. It increase with increase in diameter of charge.

The experiments do not allow us to speak about the phase state of nanoparticles. However, a further increase in the intensity of SR will enable to measure the diffraction at large angles, and, consequently, to determine the phase composition of condensed carbon in the detonation products.

Acknowledgments

This work was supported by Russian Foundation for Basic Research (project No. 16-29-01050).

References

[1] Tarver C M, Kury J W and Breithaupt R D 1997 *Journal of Applied Physics* **82** 3771–82 URL `http://scitation.aip.org/content/aip/journal/jap/82/8/10.1063/1.365739`

[2] Grebenkin K, Taranik M and Zherebtsov A 2006 *Proc. 13th International Detonation Symposium* 496–505

[3] Aleshaev A N, Zubkov P I, Kulipanov G N, Lukyanchikov L A, Lyakhov N Z, Mishnev S I, Ten K A, Titov V M, Tolochko B P, Fedotov M G and Sheromov M A 2001 *Combustion, Explosion, and Shock Waves* **37** 585–93

[4] Evdokov O, Fedotov M, Kulipanov G, Luckjanchikov L, Lyakhov N, Mishnev S, Sharafutdinov M, Sheromov M, Ten K, Titov V, Tolochko B and Zubkov P 2001 *Nuclear Instruments and Methods in Physics Research Section A: Accelerators, Spectrometers, Detectors and Associated Equipment* **470**

236 – 9 ISSN 0168-9002 proceedings of the 13th National Synchrotron Radiation Conference URL `http://www.sciencedirect.com/science/article/pii/S0168900201010579`

[5] Titov V, Tolochko B, Ten K, Lukyanchikov L and Pruuel E 2007 *Diamond and Related Materials* **16** 2009 – 13 ISSN 0925-9635 proceedings of the Joint International Conference: Nanocarbon and Nanodiamond 2006 URL `http://www.sciencedirect.com/science/article/pii/S0925963507003810`

[6] Ten K A, Titov V M, Pruuel E R, Kashkarov A O, Tolochko B P, Aminov Y A, Loboyko B G, Muzyrya A K and Smirnov E B 2014 *Proc. 15th International Detonation Symposium* 369–74

[7] Titov V M, Pruuel E R, Ten K A, Lukyanchikov L A, Merzhievskii L A, Tolochko B P, Zhulanov V V and Shekhtman L I 2011 *Combustion, Explosion, and Shock Waves* **47** 3–15

[8] Pruuel E R, Ten K A, Tolochko B P, Merzhievskii L A, Lukyanchikov L A, Aulchenko V M, Zhulanov V V, Shekhtman L I and Titov V M 2013 *Doklady Physics* **58** 24–8

[9] Bagge-Hansen M, Lauderbach L, Hodgin R, Bastea S, Fried L, Jones A, van Buuren T, Hansen D, Benterou J, May C, Graber T, Jensen B J, Ilavsky J and Willey T M 2015 *Journal of Applied Physics* **117** 245902 URL `http://scitation.aip.org/content/aip/journal/jap/117/24/10.1063/1.4922866`

[10] Shekhtman L I, Aulchenko V M, Bondar A E, Dolgov A D, Kudryavtsev V N, Nikolenko D M, Papushev P A, Pruuel E R, Rachek I A, Ten K A, Titov V M, Tolochko B P, Zhilich V N and Zhulanov V V 2012 *Journal of Instrumentation* **7** C03021 URL `http://stacks.iop.org/1748-0221/7/i=03/a=C03021`

[11] Feigin L A and Svergun D I 1987 *Structure Analysis by Small-Angle X-Ray and Neutron Scattering* (NY, Plenum Press.)

[12] Schnablegger H and Singh Y 2013 *The SAXS Guide* (Anton Paar GmbH)

Journal of Physics: Conference Series **754** (2016) 052005 doi:10.1088/1742-6596/754/5/052005

Comparison of coal reactivityduring conversion into different oxidizing medium

A G Korotkikh[1], K V Slyusarskiy[1], K B Larionov[1] and V I Osipov[1]

[1]Department of Nuclear and thermal power plants, Tomsk polytechnic university, Tomsk, 634050, Russia

E-mail: konstantinsv@tpu.ru

Abstract. Acoal conversion process of different coal samples into three different types of oxidizing medium (argon, air and steam) were studied by means of thermogravimetry. Two coal types with different metamorphism degree (lignite and bituminous coal) were used. The experimental procedure was carried out in non-isothermal conditions in temperature range from 373 K to 1273 K with 20 K/min heating rate. Purge gas consisted of argon and oxidizer with volumetric ratio 1:24 and had 250 ml/min flow rate.The ignition and burnout indexes were calculated to evaluate sample reactivity at different oxidizing mediums. The highest reactivity coefficient values in same atmosphere were obtained for lignite. It was caused by higher particle special surface area and volatile matter content.

1. Introduction

The different coal conversion processes are very important for all fuel-using technologies as it defines many properties: ignition temperature [1], char porosity and surface area [2], pyrolysis gas composition [3] etc. For combustion the pyrolysis and volatile matter release define process characteristics [2-3]. That's why the information on coal conversion in specific conditions for combustion (low pressure, excess amount of air with low water and carbon dioxide content, high heating rate) is widely represented in scientific literature [4]. On the other hand, the development of energy-efficient technologies like gasification or oxy-fuel combustion creates need for data on pyrolysis process characteristics in other conditions.Gasification is process of turning solid fuel into combustible gas for further application [5]. The gasification is realized in oxidizing medium with low oxygen and high steam and carbon dioxide content [6] as well as at high pressures and relatively low temperatures [7]. The conclusion could be made that process properties in combustion and gasification conditions are significantly different. The lower process intensity determine higher interest for particle pore structure formation and surface area evolution during pyrolysis.

The coal conversion processes for two coal samples in medium of argon, air and steam were studied by means of thermogravimetry as well as sample chemical composition and physical properties. Thetwo reactivity indexes – ignition and burnout indexes – were calculated.

2. Experimental section

2.1 Sample characterization

Samples of two coal types were used: bituminous coal from Kuznetskiy deposit and lignite from Kansk-Achinsk deposit. Raw material was grinded in ball-mill and sieved through cribble with mesh size 80 μm. Fractional composition of produced powder was investigated via laser diffractometryanalyzer HELOS (Sympatec, Germany). As long as powder was produced by sieving through 80 μm the larger fraction was considered to have cylindrical form with diameter lower than 80 μm. The coal sample specific surface area values were defined by means of BET analysis usingnitrogen via 3Flex surface characterization analyzer (Micrometrics, USA) and also presented in table 1.The volumetric version of sorption method was used. Surface specific area was defined by nitrogen vapor low-temperature isotherm.

Table 1. Coal samples properties.

Solid fuel sample	**Sample surface specific area, m^2/g**	**Particle average diameter, μm**
Bituminous coal	2,6	69,31±0,29
Lignite	6,3	69,67±0,09

The coal proximate and ultimate analysis resultsare presented in table 2. The proximate and ultimate analysis data are given for sample dry mass.

Table 2. Coal sample composition

Solid fuel sample	Proximate analysis, mass. %				Ultimate analysis, mass. %				
	M	A	V	FC	C	H	N	S	O
Bituminous coal	3	21	20	66	68	4	3	<1	4
Lignite	7	16	32	45	50	4	2	<1	28

*The M refer to moisture, A – to ash, V – volatile matter, FC – fixed carbon, C – carbon, H – hydrogen, N – nitrogen, S – sulfur, O – oxygen.

2.2 Experimental procedure

The 20 mg coal powder samples thermal analysis was made using simultaneous TG-DSC analyzer Netzsch STA 449 F3 Jupiter (Netzsch, Germany). The samples were heated into furnace with rate 20 °C/min to 1000 °C. To eliminate particle drying stage the process was studied in temperature interval from 200 to 1000 °C.The heating rate was chosen empirically to supply reaction in wide temperature range while ensuring completeness of conversion at estimated temperature. Signal registration was started from 100 °C to eliminate water evaporation phase from measurement results. The purge argon flow rate for all experiments was 10 ml/min (it was needed for correct operation of analyzer). For experiments in pure argon medium the purge gas flow was supplied with additional 40 ml/min of argon (at standard conditions), for experiments in steam medium – 240 ml/min (or 140 mg/min) of steam, in air medium – 240 ml/minof atmospheric air (at standard conditions).

2.3 Reactivity evaluation

The coal reactivity was determined using ignition index as long as coal ignition appears on this stage according to [8]. To evaluate the further influence of formed by pyrolysis particle pore structure on oxidation kinetics the burnout index was calculated [8, 9]. The higher value the ignition and burnout indexes have the higher coal reactivity. The ignition index was calculated by equation as follows [8]:

$$D_i = \frac{v_{\max}}{t_p \cdot t_e}$$

Here $v_{\max}$ is maximum reaction rate, mass. %/s; t_p is the moment of time corresponding to maximum reaction rate, s; t_e is ignition time, s.

The burnout index was defined by following equation [9]:

$$D_f = \frac{v_{\max}}{\Delta t_{1/2} t_p t_f}$$

Here $\Delta t_{1/2}$ is time period of reaction rate values which are higher than $v_{\max}/2$, s; t_f is burnout time, s.

The ignition time values were defined according to presented in [9] methodic by TG-curves (fig.1). The ignition temperature and time corresponds to crossing point (B in fig.1) of tangent lines to TG-

curve at maximal rate (point A in fig.1) and initial level. Other time values were defined using DTG-curves (fig.2).
These indexes characterize coal reactivity, the lower the ignition temperature and thehigher the ignition index and the burnout index, the better the reactivity. Calculation were made for the most developed stage.For better illustration of experimental results, relative reaction rate was calculated.

3. Results and discussion

3.1 TG and DTG curves

The experimental TG and DTGcurves for lignite and bituminous coal in different mediums are presented in figures 1 and 2, respectively. The solid line refer to experiments in air medium, dash line – for experiments in steam medium, das-dotted line – for experiments in argon medium.

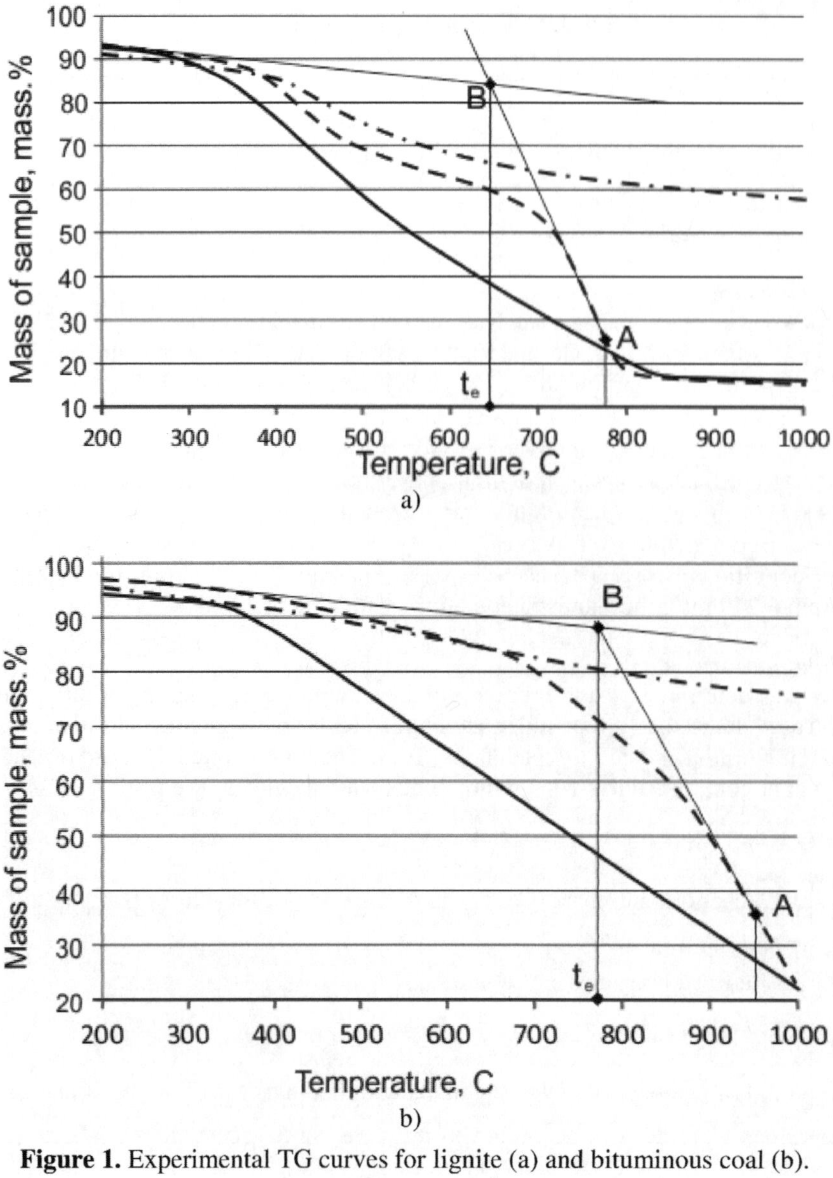

Figure 1. Experimental TG curves for lignite (a) and bituminous coal (b).

IOP Publishing
doi:10.1088/1742-6596/754/5/052005

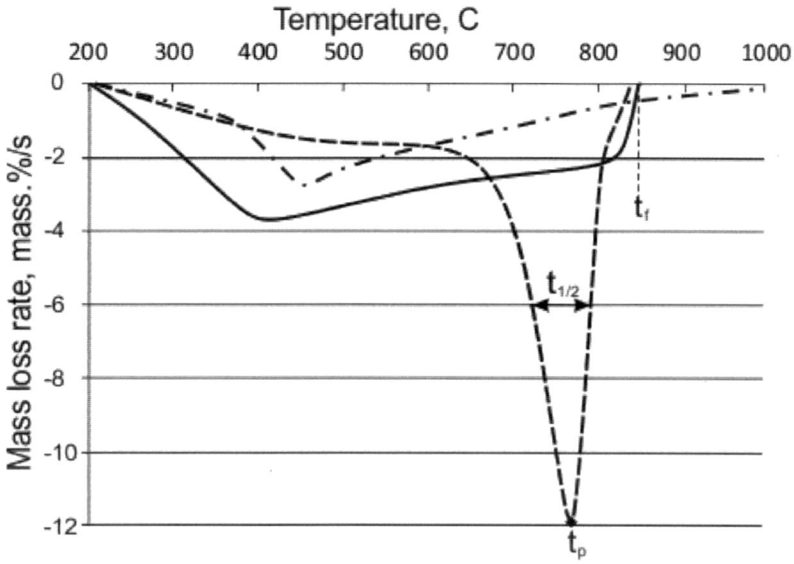

a)

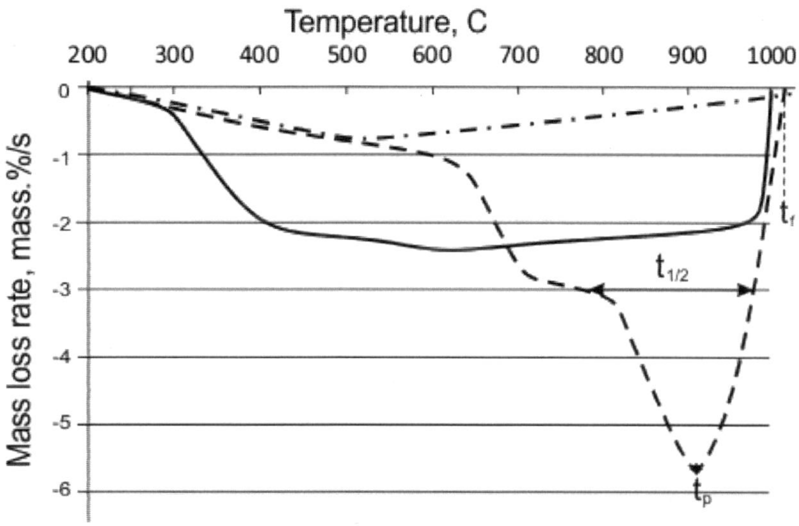

b)

Figure 2. Experimental DTG curves for lignite (a) and bituminous coal (b).

The TG-curve analysis for sample oxidation in different mediums have the same form. It has one stage in argon and air atmosphere and two stages in steam.

The argon curve situated above others and has the lesser slope that indicates a low reaction rateat all temperatures.

The average rate for steam and air experiments are close to each other while the form is completely different. The curve for oxidation in air medium has one clear stage resulted into one curve slope. For steam oxidation it has two stages with lesser slope at temperatures lesser than 700 °C and high slope within temperature range 700-1000 °C. It is well seen on DTG-curves as well.

Reaction rate for lignite samples in same conditions is higher than it for bituminous coal. It is connected to lesser carbon content in lignite samples. One more consequence is lower temperature of coal burnout.

3.2 Reactivity evaluation

Coal reactivity was evaluated according to presented in paragraph 2.3 methodology. Results are presented in table 3.

Table 3.Coal oxidation reactivity in different oxidizing medium.

Coal sample	Oxidizing medium	v_{max}, mass.%/s	$\frac{v_{max}}{m_{current}} 10^2$, 1/s	t_p, s	t_e, s	$\Delta t_{1/2}$, s	t_f, s	$D_i \cdot 10^3$, 1/s^2	$D_f \cdot 10^9$, 1/s^3
Lignite	Argon	2,69	4,4	1233	1065	333	2541	7,37	55,69
	Air	3,53	5,5	1020	795	1875	2397	15,67	16,63
	Steam	11,96	58,3	2175	1815	195	2370	10,91	257,01
Bituminous coal	Argon	0,74	1,1	1437	1125	1404	2865	1,65	2,77
	Air	2,32	3,6	1665	945	2055	3000	5,31	4,88
	Steam	5,66	87,1	2700	2175	870	3075	3,47	16,93

The reactivity for lignite conversion processes is lower than for bituminous coal. It is connected to more developed pore structure of lignite particles and higher volatile matter content. It is true for both ignition and burnout indexes: their values for lignite are higher than for bituminous coal. This is further proved by close maximal reaction rate values for lignite in argon and air mediums.

Different maximal reaction rate value points out its complex nature which consisted of carbon oxidation reaction and volatile matter release. The second process has larger contribution and caused higher burnout characteristics because of low carbon oxidation reaction impact and low $\Delta t_{1/2}$ values. The situation for bituminous coal is similar but low volatile matter content resulted into lower burnout index. Because of different nature the processes in argon medium has ignition index values two times higher than for processes in air. This tendency could be seen for both coals. Burnout characteristics are better for processes with high substance content. That's why it is high for processes in argon with lignite and for chemical reactions with bituminous coal.

In general, ignition temperature for both lignite and bituminous coal in argon and air atmospheres are close to each other. In steam they are significantly higher as well as maximal reaction rate. It's connected to higher activity of water molecules in interaction with carbon.

4 Conclusion

The coal pyrolysis process in three different mediums (inert argon, air and steam) was investigated by means of thermogravimetry in temperature range from 200 to 1000 °C with heating rate 20 K/min. The gas flow during experiments was 250 ml/min for steam and air, 60 ml/min for inert argon. The sample surface area, average particle size and coal composition were determined.

The ignition and burnout indexes for all samples and processes were determined in order to evaluate coal reactivity in different processes. The lignite reactivity was higher than bituminous coal for all processes because of calculated indexes higher values . It may be connected to higher volatile matter content and more complex particle pore structure. The burnout index of samples during steam oxidation was higher than its during other experiments, especially in high temperature area, while ignition index for lignite was higher than for bituminous coal. It may be applied to improve fuel conversion process performance.

5 Reference

[1] Gorlov EG, Andrienko V G, Nefedov K B, Lutsenko S V andNefedov BK2009 *Solid Fuel Chem.* **43** 94-98
[2] Syrodoi SV, Kuznetsov GV andSalomatov VV 2015*Solid Fuel Chem.* **49** 365-71
[3] Yang J, Chen H, Zhao W and Zhou J 2016*J. Therm. Anal. Calorim.***124** 519-28
[4] Hees J, Zabrodiec D, Massmeyer A, Habermehl M andKneer R 2016 *Flow Turbul. Combust.***96** 417-31
[5] Korotkikh A G and Slyusarskiy K V 2015 *MATEC web conf.***23** 010201-6

[6] Zhang B, Ren Z, Shi S, Yan S and Fang F 2016 *Chem. Eng. Sci.* **152** 227-38
[7] Kelley MA, Jakulewicz MS, Dreyer CB, Parker TE, Porter J M2015 *Rev. Sci. Instrum.* **86** 055106
[8] Kang Y, Yu-Ming Z, Qing-Zhao Y, Cheng F and Ze-Wu Z 2012 *Reac. Kinet. Mech. Cat.* **106** 369-77
[9] Li XG, Ma BG, Xu ZW and Wang XG2006 *Thermochim. Acta* **441**79–83

Acknowledgement
The study was realized in "National research Tomsk polytechnic university" in framework of federal target program "Research and development in prior directions of scientific-technological complex development in Russia in 2014-2020 year", unique identifier of R&D project RFMEFI58114X0001.

Numerical modelling of continuous spin detonation in rich methane-oxygen mixture

A V Trotsyuk

Lavrentyev Institute of Hydrodynamics, Siberian Branch of the Russian Academy of Science, Novosibirsk 630090, Russia

E-mail: trotsyuk@hydro.nsc.ru

Abstract. A numerical simulation of a two-dimensional structure of the detonation wave (DW) in a rich (equivalence ratio φ=1.5) methane-air mixture at normal initial condition has been conducted. The computations have been performed in a wide range of channel heights. From the analysis of the flow structure and the number of primary transverse waves in the channel, the dominant size of the detonation cell for studied mixture has been determined to be 45÷50 cm. Based on the fundamental studies of multi-front (cellular) structure of the classical propagating DW in methane mixtures, numerical simulation of continuous spin detonation (CSD) of rich (φ=1.2) methane-oxygen mixture has been carried out in the cylindrical detonation chamber (DC) of the rocket-type engine. We studied the global flow structure in DC, and the detailed structure of the front of the rotating DW. Integral characteristics of the detonation process – the distribution of average values of static and total pressure along the length of the DC, and the value of specific impulse have been obtained. The geometric limit of stable existence of CSD has been determined.

1. Introduction

Detonation waves (DWs) in reacting gaseous mixtures have the complex three-dimensional and time-dependent multifront (cellular) structure. A geometrical parameter of this structure is the transverse size of the elementary cell of the DW – a_0. Based on this value, it is possible to determine such critical parameters of detonation as the critical conditions of detonation combustion, the critical energy of initiation, geometrical limits of steady detonation propagation, etc., i.e. to estimate the detonation hazard of gaseous mixture and to evaluate the possibility of using detonations in various schemes of advanced detonation-based propulsion systems.

A very regular cellular DW structure in hydrogen–oxygen mixtures was studied in our previous 2D numerical simulation [1]. A good agreement was obtained between the numerical results and experimental data on the size a_0 over a wide range of initial pressures and degrees of mixture dilution by argon. In the subsequent studies we have simulated the irregular detonation structure in stoichiometric methane–air [2] and methane– oxygen [3] mixtures.

The objectives of this work are further development of the proposed approximate model of chemical kinetics of methane to describe the detonation combustion of rich methane based gaseous mixtures, to perform a numerical simulation of the 2D irregular cellular structure of propagating DW in a rich methane–air mixture, and to study the flow structure in cylindrical DC with continuous rotating DW in a rich methane-oxygen mixture.

2. Governing equations and model of chemical kinetics

The dynamics of the compressible chemically reactive medium is described by the two-dimensional Euler equations.

The chemical reaction in the DW is described according to the two-stage model of the detonation kinetics (induction period and main heat release stage) first proposed in [4]. In the framework of this idea, a two-stage model of the kinetics of detonation combustion of methane-based mixtures $CH_4+a_1H_2+a_2O_2+a_3N_2+a_4Ar+a_5H_2O$ has been developed and described in more detail in our paper [5]. The duration of the induction period is determined by known empirical formula [6] for methane mixtures. Constants of the approximate kinetic model have a clear physical meaning and are calculated from the tabulated thermochemical parameters of the mixture before the two-dimensional numerical calculation of the DW structure.

In this study, we adopt proposed model to describe a rich methane mixtures. The preliminary thermodynamic calculations of Chapman-Joguet (CJ) point and Von Neumann spike parameters by this kinetics in rich (equivalence ratio φ=1.5) methane-air mixture $1.5CH_4 + 2Air$ reveal that this model is accurate. The comparison has been done with CEARUN (NASA) numerical code calculations. The model is useful for multi-dimensional numerical simulations of detonation processes and is currently the only simple model kinetics with high accuracy to calculate the irregular structure of the detonation wave.

The system of governing equations was closed by the well-known thermal equation of state for an ideal gas.

3. Numerical method

The resultant systems of equations were solved numerically using the code based on the Godynov-type finite-volume scheme [7] with the fourth-order MUSCL TVD reconstruction [8] and the advanced HLLC algorithm [9] for an approximate solution of the Riemann problem. In implementation of HLLC algorithm for the case of a chemically reacting mixture, the "energy relaxation method" [10] was used. This method eliminates the problem of numerical solution of the Riemann problem for a medium with a complicated nonlinear equation of state (including that with a variable ratio of specific heats). Integration in time was performed with second-order accuracy by using additive semi-implicit Runge-Kutta methods [11]. The time step was determined at each time layer of the solution from the stability condition [7]. In the present simulations, the values of the Courant number were CFL=0.3–0.4.

In numerical simulations of propagating 2D detonation wave in the straight channel we use an adaptive moving grids technique in x direction with local refinement in the vicinity of a leading shock front. In more details it was described in [1-3].

The codes are parallelized with MPI library using the domain decomposition technique.

4. Physical background: 2D structure of DW in a rich methane–air mixture in channel

Numerical simulation was performed for a rich methane–air 1.5 CH_4 + 2 Air at normal initial conditions with the equivalence ratio φ=1.5. This value of φ corresponds to the mole fraction of methane in mixture c= 0.136, which is on the upper (rich) limit for detonability and near the upper limit for combustion of the methane–air mixtures [12]. The DW propagation in a 2D straightline channel was studied. The boundary conditions and the method of DW initiation are described in [1-3].

In this study we have modelled the DW structure in channels with a height H = 40 cm, 45 cm, 50 cm, 60 cm, 65 cm, 70 cm. The total number of numerical cells in the x and y directions were N_x=2000, N_y=2000 respectively. Calculations with varying H showed that the cellular structure of the DW with two main (primary) transverse waves exists in sufficiently wide range of H. This range is much wider than for hydrogen-oxygen mixtures [1].

To study grid convergence of the numerical solutions the calculations have been performed with twice as many number of numerical cells N_x=4000, N_y=4000 for channel height H = 45 cm, 50 cm, and 65 cm as in regular simulations. The simulations revealed a flowfield with two approximately equal primary transverse waves for H = 45 cm and 50 cm. Based on these results of the calculations, we conclude that grid convergence of the numerical solutions was obtained in our study.

From the analysis of the flow structure with varying H, the transverse size of the dominant detonation cell for rich (φ = 1.5) methane-air mixture determines to be a_0 = 45÷50 cm. The dominant cell is the cell formed by a pair of primary transverse waves that are nearly identical intensity and symmetric. For other values of the channel height H the two primary transverse waves have different intensity and reveal a chaotic uncoordinated movement. Empirical data on the size of the cell a_0 for a given mixture composition are not known to the author, is necessary to conduct large-scale experiments for this mixture.

5. Simulation of continuous rotating DW in a rich methane-oxygen mixture

Numerical investigation of continuous spin detonation (CSD) in a rich methane-oxygen 1.2 CH_4 + 2 O_2 (φ = 1.2) in the annular cylindrical detonation chamber (DC) has been conducted according physical and mathematical 2D model, see [13,14]. The described above kinetics model of detonation combustion of methane mixtures has been used. In order to compare numerical simulation results with experiments in annular cylindrical DC (average diameter d_c=3,5 cm, the annular gap Δ = 0.5 cm, DC length L_c=9 cm) [18], two-dimensional time-dependent calculations have been done under the same geometric and input governing physical parameters, as in the experiments [14], see Chapter 2, Table 2.1. Total mass flow rate of a fresh 1.2 CH_4 + 2 O_2 mixture from manifold (receiver) in the experiment was G_0 = 85 g/s, which ensure the specific mass flow rate in the annular channel of the DC $g_{\Sigma 0} = G_0/S_c = \rho \cdot u_y$ = 154.6 kg/s·m^2 on the inlet boundary y = 0. In the experiments the stagnation pressure and temperature in injection manifold (receiver) were p_m=6.81 atm, T_m=293.15 K, and S_*/S_c=0.1026. Here S_c – DC annular channel cross-sectional area, S_* – the total throat area of Laval micro-nozzles at the inlet wall of DC [13], ρ – mixture density, u_y – velocity on y axis. In experiments [14] with this mixture a single rotating DW has been obtained in the given DC, respectively a two-dimensional computational domain width has been selected $W = \pi \cdot d_c$=10.9956 cm. A periodic boundary conditions have been set at the left and right boundaries x=0 and x=W [13]. On the outlet boundary y= L_c a value of counterpressure p_{out}=1 atm has been set.

Numerical simulations have shown that for given manifold parameters (p_m=6.81 atm, T_m=293.15 K) a stable CSD regime with a single rotating transverse DW (TDW) cannot exist in DC with predetermined geometrical dimensions. After the numerous numerical calculations it has been found that self-sustained continuous rotating DW can exist with the following parameters in the manifold p_m=40 atm, T_m=1000 K, see Figures 1 and 2.

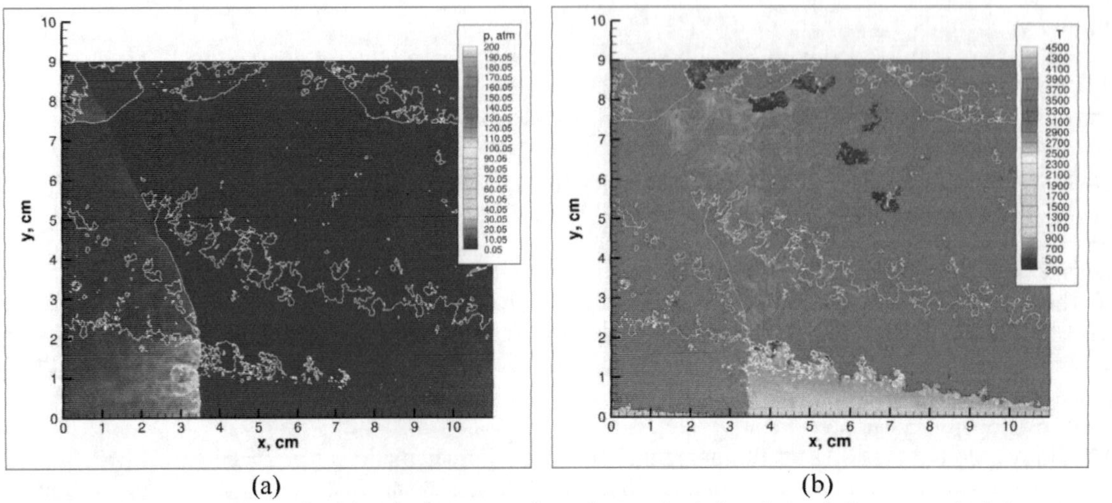

Figure 1. Structure of rotating detonation wave in a detonation chamber of diameter d_c=3,5 cm at p_m=40 atm, T_m=1000 K: (a) pressure (atm) flow field; (b) temperature (K) flow field.

Figure 1 shows the flow structure with a single TDW in this DC after the establishing of steady spinning regime. The pressure and temperature flow fields are shown for time moment t=844.6 μs, see Fig.2a. A white curve on these flow fields shows an instant position of Mach number $M=u_y/c=1.0$ isoline, where c is a speed of sound. Figure 1 clearly shows the cellular (multifront) structure of the front of this TDW, that is similar to a multifront structure of TWs on the front of propagating DW. On Figure 1b one can see a number of the unburned pockets of gas. These pockets lead to variation in time of the value of the specific flow rate $g_{\Sigma,Out}$ on the outlet boundary.

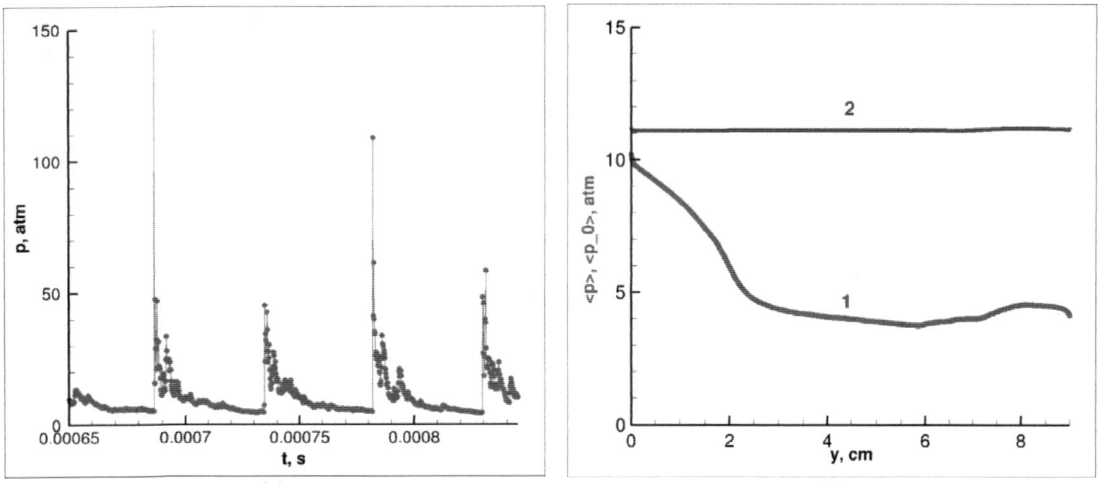

(a) (b)

Figure 2. Spin detonation parameters: (a) static pressure p (atm) history at the point x=0, y=0, see Figure 1; (b) distribution of average static pressure $\langle p \rangle$ (curve 1) and total pressure $\langle p_0 \rangle$ (curve 2) along the axial length of a detonation chamber.

Figure 2a shows the dependence of the static pressure (atm) on time (s) on the inlet boundary of the DC. This is a virtual pressure gauge data at a fixed point (x = 0, y = 0). Analysis of the obtained pressure history shows, that within 194.6 μs of the detonation process the TDW is on a steady rotation regime with an average time period of rotation $\langle \Delta t \rangle$=47.6 ± 0.2 μs.

Figure 2b presents the distribution of calculated values of the averaged on t from 650 μs to 844.6 μs and on $0 \le x \le W$ the static $\langle p(y) \rangle$ and total $\langle p_0(y) \rangle$ pressures along the length of DC channel. The data for the drawing of these curves has been taken from the simulation of detonation process shown on Figure 2a. It is seen that the static pressure (curve 1) decreases along the length of the chamber. Moreover, in the main heat release zone ($0 < y < 2.0$ cm) the rapid decline of pressure from about 10 atm to 6 atm is observed, and then it decreases slowly with minor fluctuations to a value of about 4 atm at the outlet of the channel ($y=L_c$=9 cm). The length of the main heat release zone, obtained from these graphics correlates very well with the average height of TDW $\langle h \rangle$, see Figure 1. The curve 2 shows, that the averaged total pressure determined as $p_0=p+\rho \cdot u_y^2$ is constant (with very minor fluctuations) along the length of the DC. This total pressure behavior is completely consistent with the laws of gas dynamics.

Numerical simulations have shown that a reduction of the stagnation temperature to T_m=900 K at p_m=40 atm leads to a complete breakdown of detonation combustion process and disappearance of TDW in DC. The reduction of stagnation pressure to p_m=35 atm at T_m=1000 K still gives stationary in general detonation regime with rotating TDW. The TDW shape is significantly changing over time and has greatly distorted form with 1 or 2 transverse waves on its front. This mode can be considered as marginal, insofar further decrease of stagnation pressure to p_m=30 atm at T_m=1000 K leads to the complete destruction of the CSD regime.

6. Discussion and Conclusions

The two-stage model of the kinetics of detonation combustion of methane-based mixtures with an oxygen and air has been developed. The proposed model of the kinetics of detonation combustion of methane is highly accurate and consistent with the second law of thermodynamics. The use of this kinetic model for the simulation of the two-dimensional cellular structure of detonation wave provides realistic results. The dominant size of the detonation cell for stoichiometric methane-air mixture determines to be 34±1 cm, and 0.3÷0.35 cm for methane-oxygen mixture [2,3]. These values are in good agreement with all available experimental data. A numerical simulation of 2D structure of the DW in rich (equivalence ratio φ=1.5) methane-air mixture has been performed. From the analysis of the flow structure, the size of the detonation cell for this mixture has been determined to be a_0=45÷50 cm. This value is in a good agreement with the result of the analytical model of the detonation cell by Vasil'ev and Nikolaev [15] a_0=38 cm. The experimental values of the cell size for this rich mixture composition are not available in literature up to now.

The 2D simulations for these three gas mixtures were reproduced the DW irregular cellular structure with all its main features observed in the experiments [16,17] with stoichiometric methane-air and methane-oxygen mixtures: a chaotic uncoordinated movement of the main transverse waves; the presence of a fine (secondary) cell structure at the transverse waves themselves; the existence of numerous secondary transverse waves at the leading shock front, forming a hierarchy of the decreasing the size DW front perturbations; and a significant number of pockets of the unburned mixture at a considerable distance behind the DW front, etc.

The use of the proposed model of chemical kinetics and the same two-dimensional numerical code to calculate CSD in a rich methane-oxygen mixture in the annular cylindrical DC produced very interesting results. It has been found that it is impossible to obtain a stable CSD regime with the parameters in the manifold, at which in experiment was obtained stably rotating single TDW. In [14] from the analysis of experimental data it was obtained that the primary control parameter, which determines the boundaries of the existence of the CSD regime and the main characteristics of detonation flow in DC, is the specific mass flow rate $<g_{\Sigma In}>$ of fresh mixture at inlet boundary of DC. In experiments [18] this value was $<g_{\Sigma In}>$=154.6 kg/s·m^2 for the manifold parameters p_m=6.81 atm, T_m=293.15 K. In the numerical simulations the stable CSD regime with single TDW has been obtained at the substantially greater (in several times) value of $<g_{\Sigma In}>$=470 kg/s·m^2 (p_m=40 atm, T_m=1000 K) in the same DC, and the marginal regime has been realized at $<g_{\Sigma In}>$=411.4 kg/s·m^2 (p_m=35 atm, T_m=1000 K).

Numerical simulation also showed that in addition to the specific flow rate $<g_{\Sigma In}>$, the second governing parameter is the stagnation temperature T_m. Indeed, reducing the stagnation temperature from T_m=1000 K to T_m=900 K at p_m=40 atm leads to termination of the CSD. Although for this case, the specific flow rate $<g_{\Sigma In}>$=494.45 kg/s·m^2 is even more than for the steady CSD mode $<g_{\Sigma In}>$=470 kg/s·m^2.

A significant discrepancy has been obtained for the TDW velocity. In the numerical simulation the value of the average velocity is $<D>/D_{CJ}$=0.931 (see Figure 2a), and in experiment [14] this value is $<D>/D_{CJ}$=0.75. In our opinion, such low experimental value of the TDW velocity raises the question whether this shock is in a reality the detonation wave in the classical consideration according to the ZND theory.

However, the height of the TDW is in the very good agreement with the experimental data. In 2D simulation the height is $<h>$=18.7 mm, $<h>$=15÷20 mm has been obtained in experiment [14].

Thus, the developed two-stage kinetic model and 2D numerical code give excellent results in the modelling of irregular cellular structures in methane-based mixtures in the case of DW propagation in the straight 2D channel. However, in the case of simulations of the rotating DW in the annular channel of cylindrical DC the significant discrepancies have been obtained between the calculated and experimental values of governing parameters under which the stable CSD regime is realized.

Acknowledgements
The authors gratefully acknowledge the financial contribution from the Russian Foundation for Basic Research (Grant No. 14-01-00406 and Grant No. 16-01-00102). The computations were performed on an MVS-100K supercomputer at the Joint SuperComputer Center (JSCC) of the Russian Academy of Sciences (Moscow).

References
[1] Trotsyuk A V 1999 Numerical simulation of the structure of two-dimensional gaseous detonation of an H_2–O_2–Ar mixture *Combust. Explo. Shock Waves* **35** 549–58
[2] Fomin P A, Trotsyuk A V and Vasil'ev A A 2015 Numerical study of cellular detonation structures of methane mixtures *J. Loss. Prevent. Proc.* **36** 394–403 DOI: 10.1016/j.jlp.2015.03.012
[3] Trotsyuk A V and Vasil'ev A A 2014 Numerical study of cellular structure of detonation of a methane–oxygen mixture *Modern Science. Researches, Results, Technologies (ISSN 2076-6866)* № 1 (14) 139–45
[4] Korobeinikov V P, Levin V A, Markov V V and Chernyi G G 1972 Propagation of blast waves in a combustible gas *Acta Astronaut.* **17** 529–37
[5] Fomin P A, Trotsyuk A V and Vasil'ev A A 2014 Approximate model of chemical reaction kinetics for detonation processes in mixture of CH_4 with air *Combust. Sci. Technol.* **186** 1716–35 DOI: 10.1080/00102202.2014.935643
[6] Soloukhin R I 1969 Measurement methods and basic results of shock-tube experiments *Proc. 7th Int. Symp. on Shock Tubes* (Novosibirsk: SB Academy of Sciense USSR Press)
[7] Godynov S K 1976 *Numerical Solution of the Multi-Dimensional Problems of Gaseous Dynamics* (Moscow: Nauka Press) 400 p in Russian
[8] Yamamoto S and Daiguji H 1993 Higher-order-accurate upwind schemes for solving the compressible Euler and Navier-Stokes equations *Comput. Fluids* **22** 259–70
[9] Batten P, Leschziner M A and Goldberg U C 1997 Average-state Jacobians and implicit methods for compressible viscous and turbulent flows. *J. Comput. Phys.* **137** 38–78
[10] Coquel F and Perthame B. 1998 Relaxation of energy and approximate Riemann solvers for general pressure laws in fluid dynamics *SIAM J. Numer. Anal.* **35** 2223–49
[11] Shen J W and Zhong X 1996 Semi-implicit Runge-Kutta schemes for non-autonomous differential equations in reactive flow computations *AIAA 27th Fluid Dynamics Conf., New Orleans, LA, June 17-20, 1996* AIAA Paper No.96–1969
[12] Vasil'ev A A Valishev A I Vasil'ev V A Panfilova L V and Topchian M E 2000 Detonation hazards of methane mixtures *Archivum Combustionis* **20** 31–48
[13] Zhdan S A Bykovskii F A and Vedernikov E F 2007 Mathematical modeling of a rotating detonation wave in a hydrogen-oxygen mixture *Combust. Explo. Shock Waves* **43** 449–59
[14] Zhdan S A and Bykovskii F A 2013 *Continious Spin Detonation* (Novosibirsk: Lavrentyev Institute of Hydrodynamics Press) 422 p in Russian
[15] Vasil'ev A A and Nikolaev Yu A 1976 Model of the nucleus of a multifront gas detonation *Combust. Explo. Shock Waves* **12** 667–74
[16] Manzhalei V I and Mitrofanov V V 1973 The stability of detonation shock waves with a spinning configuration. *Combust. Explo. Shock Waves* **9** 614–20
[17] Manzhalei V I 1977 Fine structure of the leading front of a gas detonation *Combust. Explo. Shock Waves* **13** 402–4

Initiation of explosive mixtures having multi-sized structures

A A Vasil'ev[1], V A Vasiliev[1], A V Trotsyuk[1]

[1] Lavrentyev Institute of Hydrodynamics SB RAS, 630090 Novosibirsk, Russia

E-mail: gasdet@hydro.nsc.ru

Abstract. Theory of strong blast was used as the basis for the experimental method of determining of the energy of source which provides the initiation of combustible mixture. For mono-fuel mixtures the following parameters were experimentally determined at testing: the critical initiation energy of a cylindrical detonation wave in mixtures 2H2+O2 and C2H2+2.5O2 (exploding wire); the critical initiation energy of a spherical detonation in a mixture of C2H2+2.5O2 (electrical discharge). Similarly, for the double-fuel mixtures of acetylene – nitrous oxide – oxygen (having bifurcation cellular structures) the critical initiation energy of spherical wave was determined also. It was found that for the stoichiometric mixture on both fuel components the critical energy of mixture with the bifurcation structure was undervalued by several times in comparison with the value of the critical energy for the mono-fuel mixture, in which the cell size at a given pressure is determined by the large scale of bifurcation cells. This result shows the decrease of the critical energy with an increase of the number of "hot spots", which are the numerous areas of collision of the transverse waves of large and small scales in a mixture with bifurcation properties.

1. Introduction

The effect of excitation of combustion and detonation processes in the fuel-oxidizer mixture usually demonstrates a "threshold" character ("yes" - "no") for each initiator (Figure 1). In an idealized model of strong blast in an inert medium (for example, [1-2]) the energy of the explosion is defined as the parameter that identify the function of the velocity of the blast wave on distance (or time). By analogy,

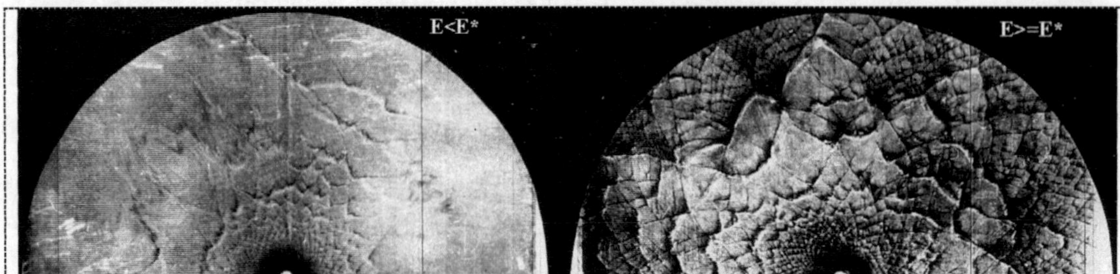

Figure 1. Threshold character of initiation: left – no initiation, right – successful initiation of detonation in monofuel mixture C2H2+2.5O2.

the minimum energy of an initiator, providing 100-percent excitation of burning or detonation regimes, is commonly known as the critical energy for combustible mixture. The critical ignition energy E_{ign} at the spark ignition has traditionally been treated as the primary parameter of the flammability of mixture. The critical energy of detonation initiation E^* by the ideal source (in terms of spatial and temporal characteristics of the initiator) is a measure of the detonability of mixtures: the smaller E^*, the more dangerous is the mixture.

The formulation of the initiation condition on the basis of a single parameter - the critical energy - is very attractive. However, the multi-front structure of a real detonation wave (instead of its idealized one-dimensional model with a smooth front) does not allow to consider the problem of initiation of waves of combustion and detonation as completely solved: up to now the following questions are poorly studied: the influence of the spatial and temporal characteristics of the initiator on the excitement of the combustible mixture; the formation of self-supporting wave (detonation or combustion), including the case of an initiation of charges of intermediate symmetry [3].

Indeed, each mixture at fixed conditions (pressure, temperature, composition) has some characteristic spatial r^* and temporal t^* scales (for example, the sizes of the induction zone or the reaction zone and the corresponding induction or reaction periods). At the same time, under these conditions, the explosive mixture absorbs the energy E_ν from the initiator for the finite period of time t_0 in some limited area of space $V_0 = f(r)$ (E_ν is part of the energy E_0, initially stored in the initiator):

$$E_\nu = \int_0^{t_0} \cdot \int_0^{V_0} \varepsilon(t,V) \cdot dt \cdot dV = \eta E_0 \qquad (1)$$

$\varepsilon(r,t)$ is a function, which describes the spatial-temporal law of the input energy, ν is the index of symmetry ($\nu = 1,2,3$ for planar, cylindrical and spherical case, respectively). It is evident that for different initiators (electric or laser spark, exploding wire, explosive charge, thermal igniter, flow of hot and active particles, etc.) the function $\varepsilon(r,t)$ has its own specific type, and, in general, E_ν is a complex function of the characteristic scales of the initiator and the mixture (more precisely – of their ratio). Influence of temporal and spatial factors of the energy transfer from the initiator to combustible mixture and the features of excitation and development of a chemical reaction in the mixture are crucial in initiating tasks, and above all – in the correct experimental determination of the critical energy and its optimization.

Note that recently the new type of gas detonation structure was experimentally observed: the fixed structures of various sizes (large and small) [4-5] were detected simultaneously on the smoked imprint. In connection with the discovery of two-sized (bifurcation) cellular structures the following question becomes fundamental: which the scale of the cells is responsible for the initiation of a mixture. This work is the world's first study devoted to the critical initiation of mixtures with multiscale structures.

It should be noted that the problem of determining the critical initiation energy of the multi-front detonation is fundamental in gas dynamics of reacting systems. A lack of reliable data on the initiation significantly hinders the creation of a perspective detonation engine for the aviation and aerospace industries, which are engaged in intensive development of the world's leading countries.

2. Energetic equivalence of different initiators

A wide range of combustible mixtures – from close to stoichiometric of fuel-oxygen mix (FOM), and fuel-air mix (FAM) – up to enriched or depleted mixtures, makes it impossible to apply any single type of initiator (a range of critical energy is about 15 orders of magnitude). The real initiators have individual spatial-temporal characteristics of its energy input in the explosive mixture. As a consequence the problem of comparison of different sources and identification of "useful" energy using a universal method equally suitable for various initiators and various mixtures becomes important.

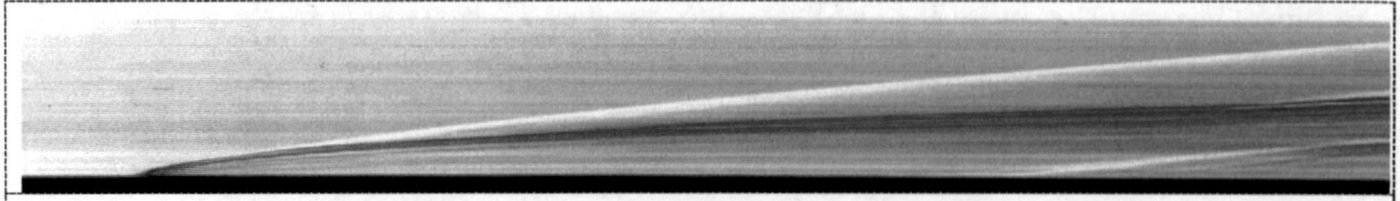

Figure 2. Trajectory of the blast wave in the inert mixture C2H2+2.5N2.

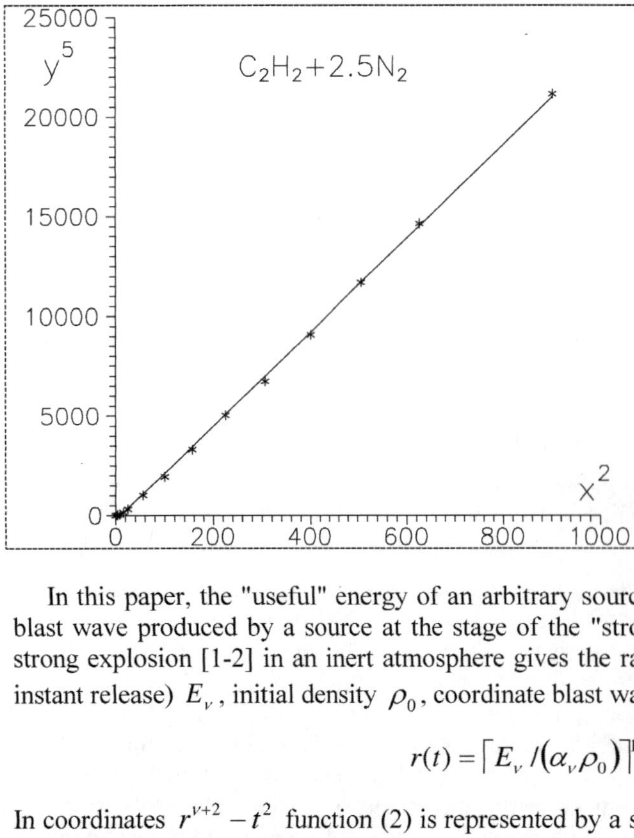

Figure 3. Trajectory of spherical blast wave in the inert mixture from the point of view of the strong blast theory ($y^5 - x^2$) with a large linear part. The similar dependence with the linear part is characteristic also for the cylindrical blast wave (in coordinates $y^4 - x^2$).

In this paper, the "useful" energy of an arbitrary source was determined using the intensity of the blast wave produced by a source at the stage of the "strong explosion", see Figure 2. The theory of strong explosion [1-2] in an inert atmosphere gives the ratio between the energy of the explosion (at instant release) E_ν, initial density ρ_0, coordinate blast wave r and time t in the following form:

$$r(t) = \left[E_\nu / (\alpha_\nu \rho_0)\right]^{1/(\nu+2)} \cdot t^{2/(\nu+2)}. \quad (2)$$

In coordinates $r^{\nu+2} - t^2$ function (2) is represented by a straight line, see Figure 3. The energy E_ν is determined by the slope of the linear part, which in this case can be attributed to the meaning of "useful" energy of source.

3. Experimental details and results

This technique has been successfully tested for the case of the cylindrical initiating of detonation wave (DW) using exploding wires [6]. The energy E_2^* in mixture C2H2+2.5O2 and 2H2+O2 was experimentally determined as a test.

This article presents the results of the hemispherical initiation of wave by pointed electrical discharge also. The mixture of C2H2+2.5O2 was used as the base for mono-fuel system and the mixture C2H2+(2.5–k/2)O2+k·N2O was choose as 2–fuel system with bifurcation properties. Initiation of the mixture was carried out using a high-voltage generator (at the capacitor $C = 1$ μF charged to a voltage of $U = 4.5$ kV (mode of R1 with the stored energy of 10 joules) or $U = 9$ kV

(R2 mode with 40 J of stored energy). The typical schlieren streak image of critical regime of initiation of detonation is demonstrated on Figure 4.

The energy E_3^* at using of electrical discharge in mixture C2H2+2.5O2 was experimentally determined as a test. Figure 2 shows the photographic scan of the blast wave trajectory in an inert mixture C2H2+2.5N2 as an analog of the combustible mixture (replacement of O2 to N2, see [6]). Figure 3 shows the dependence $r^5 = f(t^2)$, obtained by the recalculation of the trajectory of the blast wave from the point of view of the theory of the strong explosion (hereinafter the blast wave coordinate r and time t is replaced by x and y coordinates of the trajectory image of the blast wave on a film).

It is clearly seen that there is a large linear region, and its linearity over a large area of the trajectory allows one to determine the critical energy with a reasonable accuracy. The critical mode of initiation in R1 regime (10 J) was observed at the initial pressure $P_{R1}^* = 0.1$ atm, while in R2 regime (40 J) it was observed at $P_{R2}^* = 0.04$ atm. By the linear parts of graphs of $r^5 = f(t^2)$ the critical energy was defined: $E^* = 6.8$ J (68% of the stored energy) at $P_0 = 0.1$ atm and $E^* = 23.2$ J ($\eta = 58\%$) at $P_0 = 0.04$ atm.

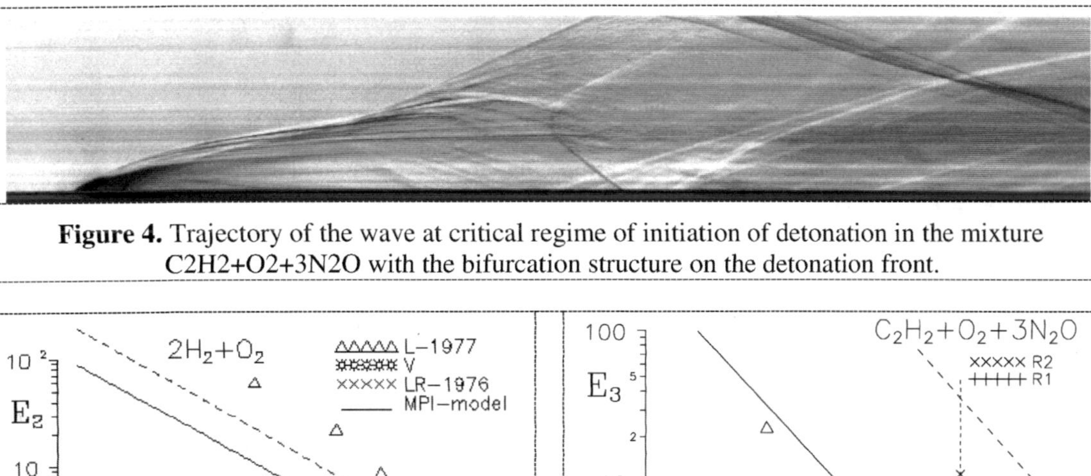

Figure 4. Trajectory of the wave at critical regime of initiation of detonation in the mixture C2H2+O2+3N2O with the bifurcation structure on the detonation front.

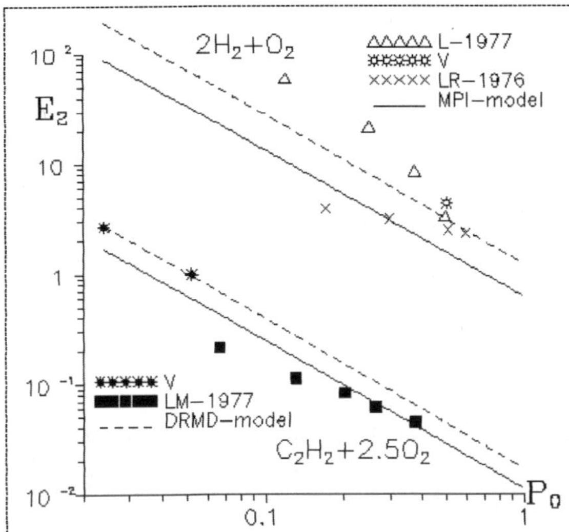

Figure 5. The experimental results about initiation of the cylindrical detonation in mono-fuel mixtures 2H2+O2 and C2H2+2.5O2.

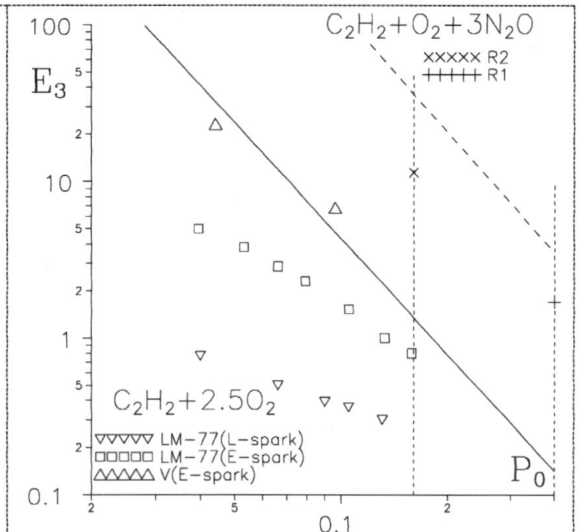

Figure 6. The experimental results about initiation of spherical detonation in the mono-fuel mixture C2H2+2.5O2 and in 2-fuel mixture C2H2+O2+3N2O with bifurcation structures of the detonation front.

Figure 5 presents data on the critical energy initiation of cylindrical DW together with data from the [7-9]: symbols - experimental data, lines – calculated results (the details of initiation models can be found in [10]). It should be noted a significant difference between the experimental data obtained at various initiators, both among themselves and in their slopes (dependence $E_2^* = f(P_0)$).

Figure 6 is a summary graph of an experimentally determined critical initiation energy of the spherical detonation for classical mixture C2H2+2.5O2 supplemented by the data of initiation energy E^* for the same mixture as in [9] when using electric (E-spark) and laser (L-spark) discharge as the initiator. One can see an appreciable spread of data in [9], not only in the values (several times smaller at the laser initiation), but also in the nature of tilting approximating lines $E_3^* = f(P_0)$ for each type of the initiator. Authors of [9] do not discuss the reason for such differences, the true nature of this is not clear and requires detailed studies (the same remark can be made for Figure 5). Experimental points presented in this paper for mixture C2H2+2.5O2 are located closer to the calculated line data in comparison with Canadian researchers.

Approximate line, the slope of which determines the critical energy of initiation, is defined by the equation $Y = AX - B$, i.e. it does not pass through the origin of coordinates, due to a non–instantaneousity of allocated initiated energy (see Eq. (1)). Let $X_i = B/A$ be the point of intersection of the continuation of the approximate line with the x–axis. Then the corresponding time t_i characterizes the delay of equivalent instantaneous initiation with respect to the real non-instantaneous energy release. In other words, in the case of non–instantaneous source (i.e., $E^* = f(t)$) the time t_i characterizes the critical time allocation of the "useful" energy. The energy released after the time t_i until the time of formation of a self-sustaining DW in the mix t_{form} does not affects the initiation process.

Similarly, for the dual-fuel mixtures of acetylene – nitrous oxide – oxygen, having the bifurcation properties of cellular structures, E_3^* was determined also. For example, in mixture of C2H2 + O2 + 3N2O the critical mode of initiation of the mixture with the bifurcation properties using R1 (10 J) mode was observed at the initial pressure $P_{R1}^* = 0.4$ atm, and in R2 (40 J) regime at $P_{R2}^* = 0.16$ atm. The processing path of the blast wave trajectory in an inert mixture from the standpoint of the model of a strong explosion on a straight-line portion of the graph of the dependence r^5 on t^2 allows one to determine the corresponding critical energy for the initiation of the spherical DW: $P_0 = 0.4$ bar $E_3^* = 1.7$ J ($\eta = 17\%$), while at $P_0 = 0.16$ atm $E_3^* = 11.4$ J ($\eta = 28\%$).

In Figure 6 the symbols on the vertical dotted lines mark the data on initiation of a mixture of C2H2 + O2 + 3N2O with bifurcation properties. Inclined dotted line is the calculated data for the mixture of C2H2 + 3N2 + 2.5O2 as its non-bifurcation analogue. It is evident that the experimental values for a mixture of C2H2 + O2 + 3N2O are located significantly below the estimated line for the mixture of C2H2 + 2.5O2 + 3N2 (about 3 times), although the sizes of the cells were unexpectedly close.

4. Conclusion

It was found that for stoichiometric mixture (on both fuel) the critical energy for mixture with bifurcation structure is understated by several times in comparison with the value for mono-fuel mixture, in which the cell size at a given pressure is determined by the large scale bifurcation cells. This result shows a decline of E_3^* with an increase in the number of "hot spots", which are the numerous areas of collisions of large and small-scale transverse waves in a mixture with bifurcation properties.

5. Acknowledgments

This investigation was partially financially supported by the Russian Foundation of Basic Researches (grant № 14-01-00406).

References

[1] Sedov L I 1993 *Similarity and dimensional methods in mechanics*. (CRC press)
[2] Korobeinikov V P 1991 *Problems of point blast theory*. (Springer Science & Business Media)
[3] Vasil'ev A A 2015 The cellular structure of the multi-detonation and initiation (review) *Comb. Expl. & Shock Waves* **51** 1 9–30
[4] Presles H N, Desbordes D, Guirard M and Guerraud C 1996 Gaseous nitromethane and nitromethane-oxygen mixture, a new detonation structure *Shock Waves* **6** 111–14
[5] Vasil'ev A A 2014 Monopropellant as a source of bifurcation properties of multi-systems *Comb. Expl. & Shock Waves* **50** 2 14–23
[6] Vasil'ev A A 1983 Study of critical initiation of gas detonation *Comb. Expl. & Shock Waves* **19** 1 121-31
[7] Lee J H 1977 Initiation of gaseous detonation *Ann. Rev. Phys. Chem.* **28** 75–104
[8] Lee J H and Ramamurthi K 1976 On the concept of the critical size of a detonation kernel *Comb. Flame* **27** 331–40
[9] Lee J H and Matsui H 1977 A comparison of the critical energies for direct initiation of spherical detonations in acetylene-oxygen mixtures *Combust. Flame* **28** 61–66
[10] Vasil'ev A A 2012 Dynamic parameters of detonation *Shock Waves Science and Technology Reference Library, Volume 6: Detonation Dynamics* ed F Zhang (Berlin-Heidelberg: Springer-Verlag) pp 213-79

Journal of Physics: Conference Series **754** (2016) 062001 doi:10.1088/1742-6596/754/6/062001

Simulation of liquid cube fracture with SPH

M N Davydov

Lavrentyev Institute of Hydrodynamics of the Siberian Branch of the Russian Academy of Sciences, Novosibirsk, Russia

E-mail: davydov@hydro.nsc.ru

Abstract. The dynamics of fracture of liquid cube with high-pressure cavity is studied numerically in a three-dimensional formulation using the Smoothed Particle Hydrodynamics method (SPH). It is shown that the propogation of shock wave from growing cavity and reflection from the free side is due to relaxation of tensile stresses in the liquid volume and formation of quasistationary mass-velocity field which provides fracture of the fluid.

1. Introduction

The problem of liquid breakdown in intense rarefaction waves is directly related to the notion of strength, which does not have a precise and clear definition in the mechanics of liquids under dynamic loading, unlike in solid mechanics [1]. A typical examples of this process is the ejection fraction of magmatic melt from the channel of the volcano during the exposive eruption [2] or the formation of sultans out on the free surface at shallow underwater explosions.

Numerical methods based on finite differences are unable to solve such problems, therefore it is required to use methods that can simulate of flows in the complex variable geometry region. In this paper, a model problem of an underwater explosion in the cube of pure (one-phase) liquid using the SPH method [3] is considered.

2. SPH method

The smoothed particle hydrodynamics (SPH) method [4, 3] is an effective meshless Lagrangian numerical method used to calculate flow structures with an unknown free boundary, in particular, high-velocity processes in media with the modeled object topology appreciably changed under intense dynamic loading.

In this method N spherical particles are placed into the modeled physical space. Each particle has a mass m_i, internal energy e_i, and velocity $\vec{v}_i$. The particles move in accordance with the laws of mechanics. If these physical quantities are known at a certain time at all points $j = 1, \ldots, N$ to which N particles are placed, i.e., if a certain function $f(r_j)$ is specified, then its value at an arbitrary point of the modeled space can be obtained by means of discretization of the interpolation formula

$$\langle f(r) \rangle = \int f(r')W(r - r', h)dr,$$

where h is the smoothing radius and $W(r - r', h)$ is the smoothing function (kernel) for which the following conditions should be satisfied:

$$\int W(r,h)dr = 1, \qquad \lim_{h\to 0} W(r,h) = \delta(r)$$

In this work we use the known definition of the kernel based on third-order piecewise-spline functions [4].

As the smoothing function W is not equal to zero only in a certain (small) neighborhood of the point with the coordinate r, the result of interpolation is affected only by those nodes that are located in the neighborhood of smoothing of this point. Therefore, it is possible to avoid summation over all known values of the function and to perform it only over the neighboring nodes (particles) located at distances smaller than $2h$.

It should be noted that the procedure of sampling of all particles in space has a quadratic order of complexity, and it is next to impossible to implement it in practice. As the particles arbitrarily move in space and can freely mix, the task of effective (time-efficient) searching for the neighboring particles is extremely important for the SPH method. One possible approach to solving this problem is to divide the space into cells and to perform summation only over particles in the neighboring cells. It should be noted that the mesh constructed for this purpose is an auxiliary tool used for accelerating the process of searching for the neighboring particles rather than for approximation; therefore, it does not affect the resultant solution, and the SPH method is still a meshless one.

In the SPH method the gas dynamic equations have the form

$$\frac{\partial \rho_i}{\partial t} = -\rho_i \sum_{j=1}^{N} \frac{m_j}{\rho_j}(v_j - v_i)\nabla W(r_i - r_j, h)$$

$$\frac{\partial v_i}{\partial t} = -\sum_{j=1}^{N} m_j \left(\frac{p_i}{\rho_i^2} + \frac{p_j}{\rho_j^2}\right)\nabla W(r_i - r_j, h).$$

We use the known definition of the liquid state equation:

$$p = p_0 + \frac{\rho_0 c_0^2}{\gamma}\left[\left(\frac{\rho}{\rho_0}\right)^{\gamma} - 1\right],$$

where $\gamma = 7.15$, ρ_0 and p_0 is a reference density and pressure, c_0 is a speed of sound (approximate 1500 m/s).

3. Results of numerical simulations

Initially, the liquid is a cube with sides of 1 cm, located on the XY plane, OZ axis passes through the center of the cube. Inside the liquid, at a distance of 2.5 mm from the top side of the cube, there is a spherical cavity radius of 1 mm, the center of which is located on the axis OZ. The pressure in the cavity is 10000 atm, the rest of the liquid is at atmospheric pressure p_0. Figure 1–3 shows the distribution of pressure and particle velocities in the central section of the cube (the plane $y = 0$).

The shock wave propagates from the high-pressure cavity. The amplitude of shock wave decreases as it propagates. Figure 1 shows the begin of reflection the shock wave from the top of the cube at time 1 μs. Near to the cavity layer of liquid starts to move. Only particles in which $\vec{v} \neq 0$ is painted on the figure. It is clearly seen that the velocity distribution is symmetric with respect to the cavity.

Figure 2 shows that the wave reaches the faces of the cube and is reflected from its borders, at first it occurs on the side face, then on the bottom. By this time, the upper bound is significantly

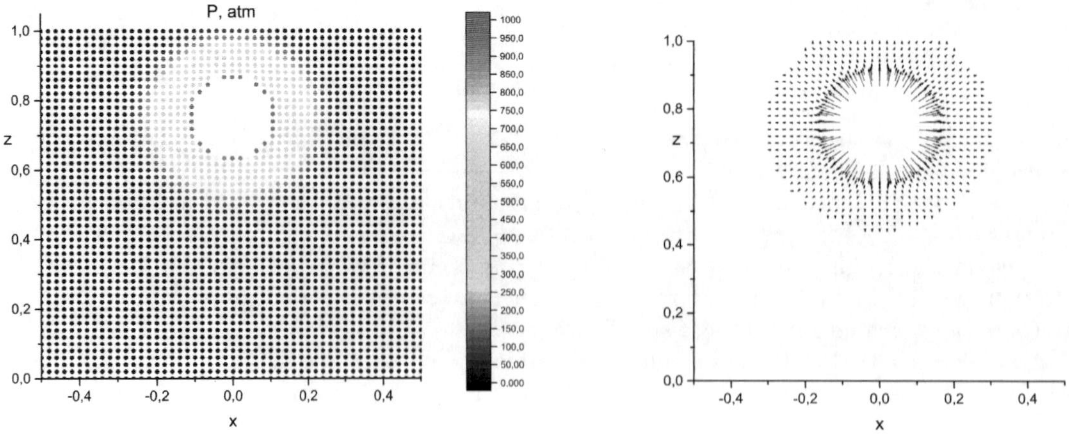

Figure 1. Distribution of pressure and particle velocities in the central section of the cube (1 μs)

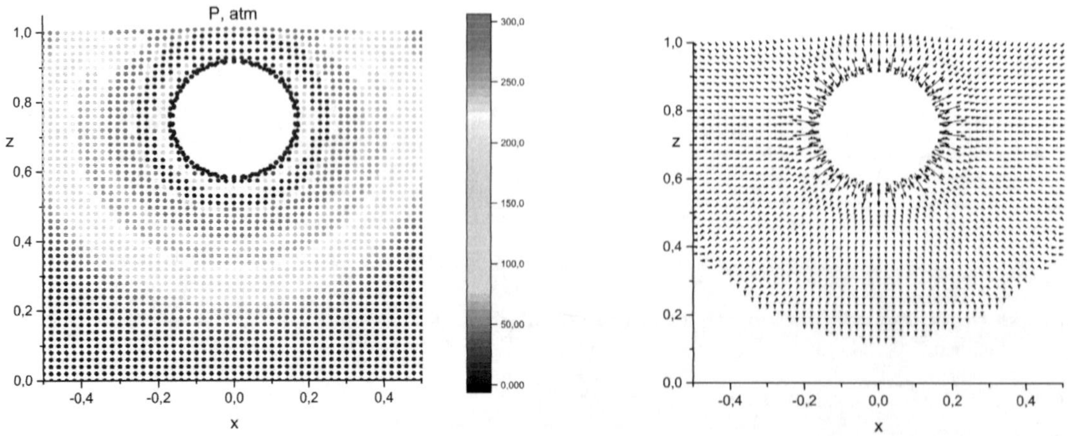

Figure 2. Distribution of pressure and particle velocities in the central section of the cube (5 μs)

deformed, curling up. Almost all particles are in motion, it is clear that the particles at the boundary of the cavity "push" other particles.

After a few tens of microseconds pressure field substantially relaxes, only individual particles have the pressure greater then p_0, and scored the particle velocity continues to scatter out of inertia, thereby simulating the fraction of the liquid (Fig. 3). One clearly sees a deformation of the side and bottom faces of the cube. The particle size distribution shows that inside a cube formed low-density zones, which lead to the growth medium fracture into fragments.

Figure 4 shows the development of process in 3D view, it is an isosurface of the density field for the half of the cube, the value of the density is $\rho^* = 0.45$ g/cm^3. It means that the isosurface bounds the area where the density of the liquid is greater than or equal to the ρ^*. Clearly that by the time 500 μs (Fig. 4a) the cavity has increased in size by several times, "stretched" up and deformed the top face of the cube, the hole is formed. Bottom and side faces of the cube is

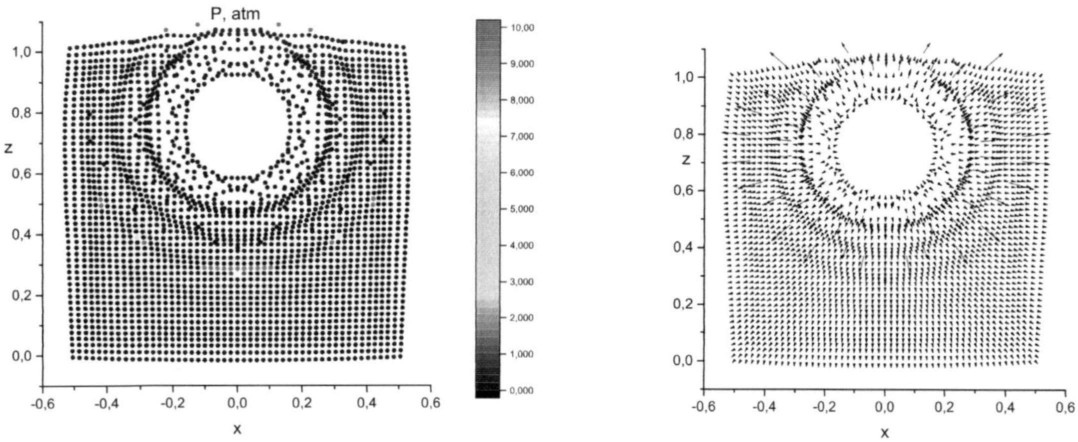

Figure 3. Distribution of pressure and particle velocities in the central section of the cube (25 μs)

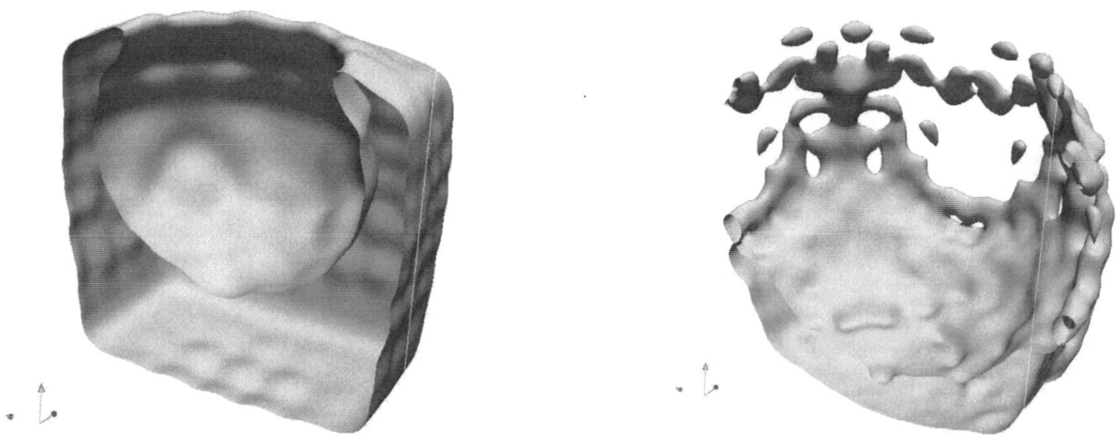

Figure 4. Isosurface $\rho = 0.45$ g/cm^3 for the half cube in times 500 and 10000 μs

slightly deformed.

It should be noted that the same liquid particles spreads to distances greater than the smoothing radius, which means the loss of bonding in the SPH formulation and is interpreted as the medium fracture [5].

In the future, the vertical jet of liquid is formed on the top face of the cube, which gradually destroyed. Flying off individual spalls are formed on the side and bottom faces, where the destruction is not so strongly expressed (Fig. 4b). In the future and in these areas the destruction of the liquid will occur and the emergence of a breakaway from the main volume of "layers". It is evident that the environment, flying away, loses consistency and disintegrates into fragments. Further development of the process leads to the dissipation of the particles and the final destruction of the liquid.

4. Conclusion

The results of the numerical analysis of the processes of formation and dynamics of the structure of spalling zones after reflection of the shock wave generated by an underwater explosion from the free surface are presented. It shows that the SPH method can be used to study the dynamics of the liquid cube fracture. A numerical simulation shows that the shock wave reflection from the free surface leads to relax of a high pressure and subsequent decomposition into individual fragments and clusters of almost free particles.

Acknowledgments

This work was supported by the Russian Foundation for Basic Research (Grant No. 15-05-03336).

References

[1] Davydov M N and Kedrinskii V K 2008 *J. Appl. Mech. and Tech. Phys.* **49** 209–216
[2] Davydov M N 2012 *J. Appl. Mech. and Tech. Phys.* **53** 324–332
[3] Monaghan J J 2005 *Reports on Progress in Physics* **68** 1703
[4] Monaghan J J 1992 *Annual Review of Astronomy and Astrophysics* **30** 543–574 (*Preprint* `http://www.annualreviews.org/doi/pdf/10.1146/annurev.aa.30.090192.002551`)
[5] Davydov M N and Kedrinskii V K 2013 *J. Appl. Mech. and Tech. Phys.* **54**

Comparative study of different combustion models for turbulent gas flames

A A Dekterev[1,2], Ar A Dekterev[1], A V Minakov [1,2]

[1] Kutateladze Institute of Thermophysics, Siberian Branch, Russian Academy of Sciences, Novosibirsk, Russia

[2] Siberian Federal University, Krasnoyarsk, Russia E-mail: dektererv@mail.ru

Abstract. Several popular turbulent combustion models have been tested in a computational study of three experimentally well-documented non-swirling and swirling jet flames. Different combinations of turbulence, combustion and reaction mechanisms models were considered. It is shown that the eddy- dissipation concept (EDC) and the probability-density function (PDF) of flamelet combustion models with detailed kinetics mechanisms provide the best results for all flames examined. For some cases, a combination of RANS turbulence models and less costly combustion approaches (Hybrid or EDC with 4 reactions) also gives acceptable results.

Combustion processes are very common in nature and technology. Most of the technological processes that ensure the life necessities of people are based on combustion processes such as energy, transport, metallurgy, petrochemicals, and others. The optimization and development of such systems is impossible without detailed study and modelling of the combustion process. Computational fluid dynamics (CFD) methods are becoming currently increasingly common in industry to solve multi-physical problems. They allow accelerating design and optimization processes of products, thereby reducing the financial costs of the project. At that, CFD allows partial elimination of the need to conduct expensive experiments, as provides an opportunity to optimize the product design based on its virtual model. The gas combustion is a complex process, which is determined by the joint effects of chemical, heat and mass transfer processes in turbulent flows of multicomponent gas mixtures. To calculate gas flames, it is necessary to implement a combination of these processes in computational model. The present paper is dealing with the development and testing of numerical simulation techniques for gas combustion process, as well as its practical use when solving applied problems.

In the calculations we used in-house SigmaFlow software package and Fluent commercial package. The flow parameters and turbulence characteristics were determined on the basis of solutions to both the Reynolds averaged Navier-Stokes equations based on k-ε and MSST models, and using eddy-resolving LES and hybrid RANS/LES turbulence models. To describe the radiation heat transfer we used the discrete ordinates model with the application of the weighted gray gases model to calculate the absorption coefficient.

The calculation of gas combustion was carried out using different approaches and models. In particular, a simple hybrid model, in which the reaction rate is selected as the lower of the rate calculated on the basis of 2-to-4-stage reaction mechanisms ($Ri_{,KIN}$) (Tables 1 and 2) and the turbulent mixing rate of components ($R_{i,EBU}$).

$$R_{i,KIN} = -AM_i X_i^{\nu_i} X_{OX}^{\nu_{OX}} T^{\beta} e^{-E_a/RT} \quad (1)$$

where X_i, R_i - molar concentration and reaction rate of i-th reagent
A_i, E_i - pre-exponential factor and activation energy
ν_i, ν_{OX}, β - empirical constants
X_{OX} – molar concentration of oxidant

$$R_{i,EBU} = -\frac{\rho\varepsilon}{k} A \cdot MIN(Y_i, \frac{Y_{OX}}{S_{OX}}, B\frac{Y_{pr}}{S_{pr}}) \quad (2)$$

where A и B – coefficients, equals 4.0 and 0.5;
Y_i, Y_{OX}, Y_{PR} - mass concentrations of reagent, oxidant and product of combustion
S_{OX}, S_{PR} - stoichiometric coefficients

$$R_i = -MIN(|R_{i,KIN}|, |R_{i,EBU}|) \quad (3)$$

Table 1. Two-stage methane reaction mechanism [1].

No.	Reaction	E,J/kmol	A, kmol/(м^3·s)	B	Components degree γ
1	$2CH_4+3O_2 \rightarrow 2CO+4H_2O$	$2.03\cdot10^8$	$2.8\cdot10^{12}$	0	γ_{CH4}=-0.3, γ_{O2}=1.3
2	$2CO+O_2 \rightarrow 2CO_2$	$1.67\cdot10^8$	$2.91\cdot10^{12}$	0	γ_{CO}=1, γ_{O2}=0.25

Table 2. Four-stage methane reaction mechanism [2]

No.	Reaction	E, J/kmol	A, kmol/(M^3·s)	β	Components degree γ
1	$CH_4+0{,}5O_2 \rightarrow CO+2H_2$	$1.26\cdot10^8$	$4.4\cdot10^{11}$	0	γ_{CH4}=0.5, γ_{O2}=1.25
2	$CH_4+H_2O \rightarrow CO+3H_2$	$1.26\cdot10^8$	$3.1\cdot10^8$	0	γ_{CH4}=1, γ_{H2O}=1
3	$CO+0{,}5O_2+0H_2O \rightarrow CO_2$	$6.69\cdot10^7$	$2.5\cdot10^8$	0	γ_{CO}=1, γ_{O2}=0.3
4	$H_2+0{,}5O_2 \rightarrow H_2O$	$1.46\cdot10^8$	$7.9\cdot10^{10}$	0	γ_{H2}=1, γ_{O2}=0.5

For more complex generalized eddy-dissipation model (EDC) we considered different reduced reaction mechanisms, as well as the PDF technique with the application of laminar flamelets for the detailed GRI 3.0 reaction mechanism.

The verification of the methodology was carried out using experimental data on the direct-flow combustion of methane in a bounded domain [3] and in the open space (Flame D) [4]. When modeling these flames, we used both the k-ε turbulence model and an anisotropic Reynolds stress model (RSM).

The results of the direct-flow combustion simulation in a bounded domain showed that for this problem the use of a 2-stage mechanism for both hybrid model and EDC model gives an overestimation of the temperature. Though, the 4-stage mechanism gives a noticeable improvement. Its application for both models gives a fairly close agreement with the experimental data. The use of more complex reaction schemes with EDC model gives the results for temperature, CH4 and O2 very close to those obtained using the 4-stage scheme. Small differences are observed only in the CO and CO_2 concentration profiles.

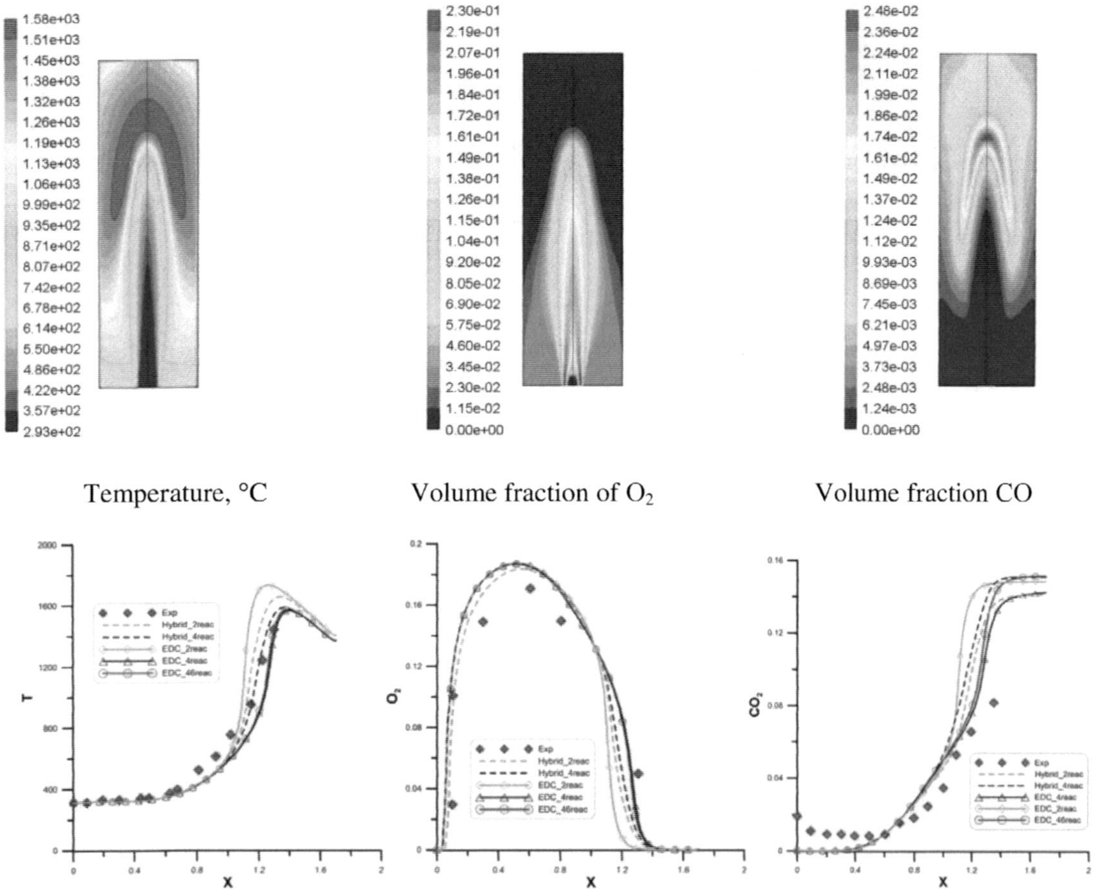

Figure 1. Direct-flow flame in a bounded domain. Temperature, O_2 and CO concentration fields. Temperature, O_2 and CO_2 concentration profiles along the axis of the burner.

The Flame D simulation results showed that the EDC model with 4-stage mechanism gives the result, which is far from the experimental data in terms of all the evaluation criteria. However, the transition to the use of more detailed chemistry (46 reactions) drastically changes the outcome. We have obtained close agreement between calculated and experimental data. The PDF Flamelet model with detailed GRI 3.0 reaction mechanism showed good results as well. It is worth noting that for the considered direct-flow flames the influence of the used turbulence model on the simulation result is less pronounced than the effect of the chosen reaction mechanism.

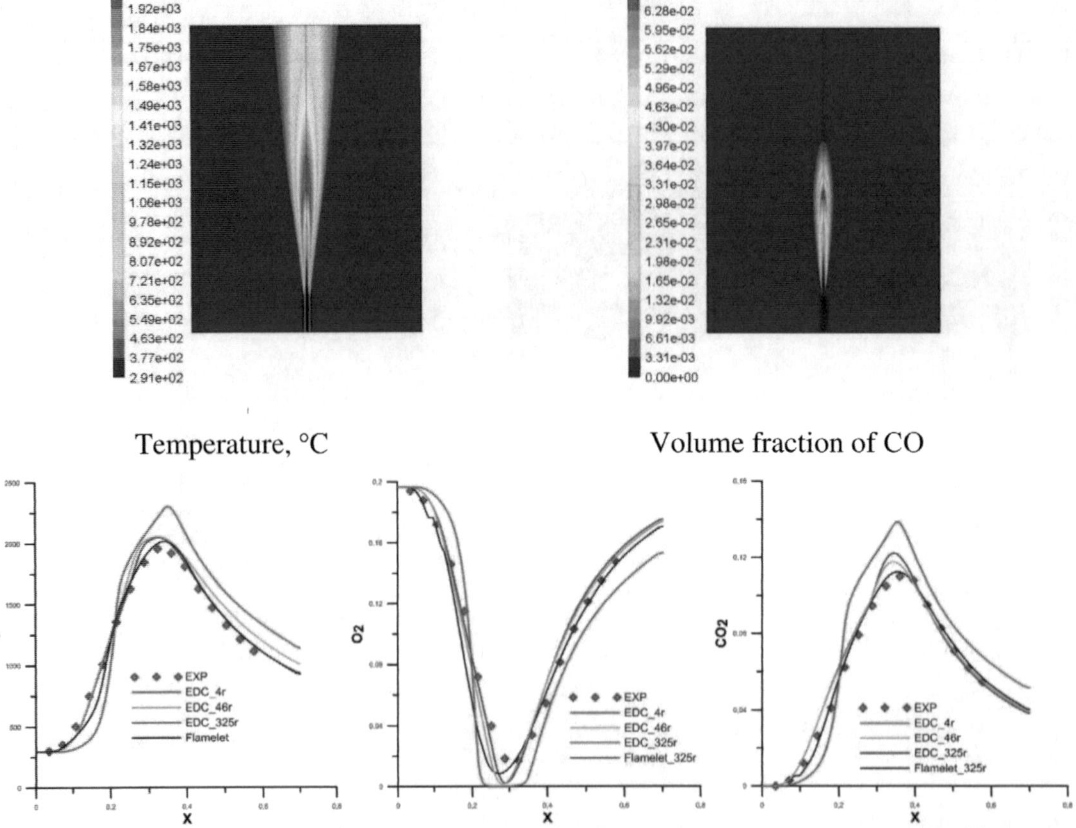

Figure2. Direct-flow flame in an open space (Flame D). Temperature and CO concentration fields. Temperature, O_2 and CO_2 concentration fields along the axis of the burner.

To test the swirling flame models we used the data obtained from the laboratory-scale burner at University of Sydney [5]. Various modes were studied in terms of the mixture composition and twist parameter. For simulation of unsteady flow, we used the URANS and LES approaches, while the EDC model and flamelet technique with different reaction mechanisms were used as combustion models. The close agreement with experimental data was obtained in terms of the temperatures and the gas mixture components (Figure 3,4).

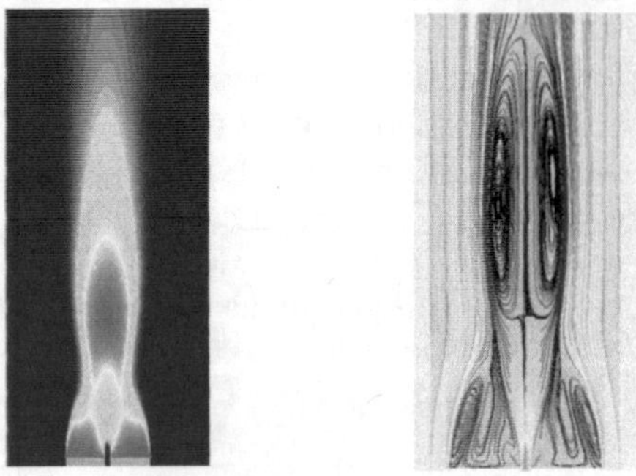

Figure 3. A swirling flame in the burner of the University of Sydney (SM1 regime). The temperature field and the streamlines.

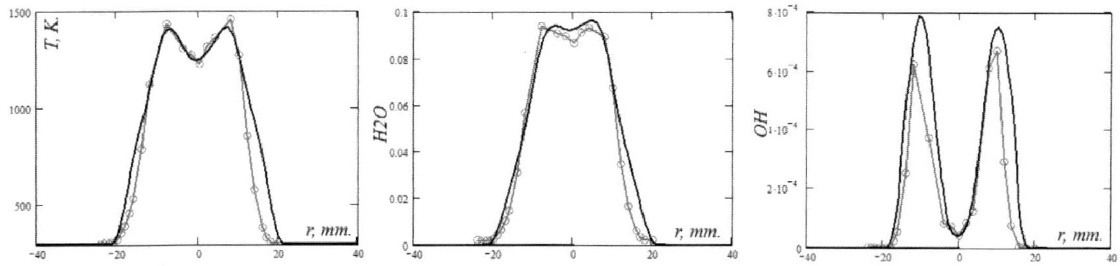

Figure 4. A swirling flame in the burner of the University of Sydney (SM1 regime). Comparison of temperature, H_2O and OH concentration profiles with the experimental data in section x=40 mm. Red line – experiment, black one – calculation

The fine-tuned calculation technique was applied successfully in calculations of burners for the combustion of wet oil-associated gas in oil and gas production, in modelling of operation modes of the gas turbine combustion chamber, optimization of the afterburning process of anode gases from electrolytic cells, and for the evaluation of heat fluxes to the ground in the event of spontaneously igniting oil and gas torches under natural conditions (Figure.5).

The oil-associated gas combustion system. Temperature isosurface of 1200°C.

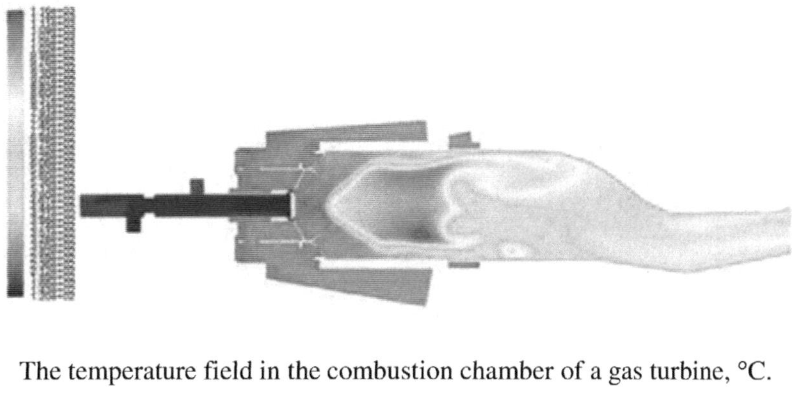

The temperature field in the combustion chamber of a gas turbine, °C.

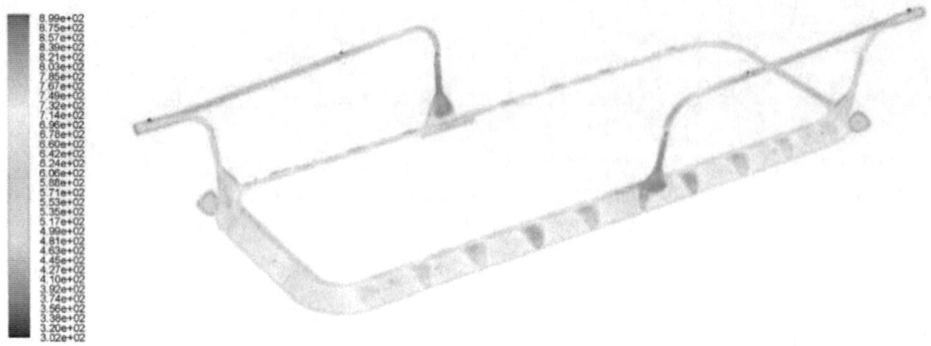

The temperature field in the gas-collecting dome of the electrolysis unit
Figure 5. Examples of calculations used for various combustion applications.

Overall, the research results have shown that the use of the EDC and PDF combustion models with flamelets, when using detailed kinetic mechanisms in combination with the eddy-resolving turbulence models provides the best results for all flames. However, such simulations require large computer resources. In some cases, more simple combustion models can be applied in combination with RANS/URANS turbulence models to obtain acceptable results with a much lower computational cost.

To obtain the correct simulation results, when performing applied calculations, it is necessary to carry out the preliminary analysis of possible correct use of the RANS turbulence models and simple reaction mechanisms.

[1] Da Silva, C.V., et al. Numerical Simulation of the Combustion of Methane and Air in a Cylindrical Chamber. Engenharia Térmica (Thermal Engineering), Vol. 5 - Nº 01 - July 2006, pp.13-21.

[2] Jones, W.P. and Lindstedt, R.P., 1988, Global reaction schemes for hydrocarbon combustion. Combust. Flame, 73, pp. 222–233.

[3] Garreton, D., Simonin O. Aerodynamics of steady state combustion chambers and furnaces // ASCF Ercoftac CFD Workshop. – Chatou, 1995

[4] Barlow, R.S., Frank, J.H., Karpetis, A.N., and Chen, J.-Y., 2005, Piloted methane/air jet flames: scalar structure and transport effects.Combust. Flame, 143, pp. 433-449.

[5] Kalt, P.A.M., Al-Abdeli, Y.M., Masri, A.R., and Barlow, R.S., 2002, Swirling turbulent non-premixed flames of methane: flow field and compositional structure. Proc. Combust. Inst. 29, pp. 1913-1919.

Application of software complex turbo problem solver to rayleigh-taylor instability modeling

S V Fortova, P S Utkin, V V Shepelev

Institute for Computer Aided Design of the Russian Academy of Sciences, 19/18 2nd Brestskaya str., Moscow 123056, Russian Federation

E-mail: sfortova@mail.ru

Abstract. The dynamic processes which take place during high-speed impact of two metal plates with different densities are investigated using three-dimensional numerical simulations. It is shown that as a result of the impact the Rayleigh-Taylor instability forms which leads to the formation of three-dimensional ring-shaped structures on the surface of the metal with smaller density. The comparative analysis of the metals interface deformation process with the use of different equations of state is performed. The numerical study is carried out by means of special software complex Turbo Problem Solver developed by the authors. The software complex Turbo Problem Solver implements generalized approach to the construction of hydrodynamic code for various computational fluid dynamics problems. Turbo Problem Solver provides several numerical schemes and software blocks to set initial, boundary conditions and mass forces. The solution of test problem about Rayleigh-Taylor instability growth and development for the case of very rapid density growth is also presented.

1. Introduction

The problem of explosive hardening and explosive welding is the actual issue for production of high-strength bimetallic compounds and constructions with the desired characteristics [1]. Explosive welding is a process of metallic compound production as a result of high-speed impact due to the energy of condensed explosives detonation products. The process is followed by the complex dynamic effects which are the subject of a number of experimental and numerical studies, see [2] for example. In [2] on the example of high-speed impact of lead plate with the plates of different metals the features of the processes in the near interface regions of the colliding plates are investigated experimentally.

The scheme of the experiment [2] is the following. The lead plate is thrown by the products of condensed explosive detonation to the steel plate under some angle with the speed several hundred meters per second. The plates become deformed with the formation of solid compound in some cases. The partial melting of the colliding surfaces takes place and the transition of the metals to the elastoplastic state occurs due the high energy release. So during some time (about 10 μs) after an impact the metals are in the elastoplastic state and behave as pseudo-fluids before the backward transition starts. Circumstantial proof of that fact is the existence of crateriform splashes on the surface of steel plate in the direction of lead plate. The explanation of the splashes was given in [2] on the assumption of the Rayleigh-Taylor instability (RTI) development. Such type of instability can be realized in the rarefaction wave (RW) which moves through the plates interface from the lead plate free side if the initial disintegration of discontinuity occurred with the formation of two shock waves (SW). Acceleration in the RW is directed to the side of the metal with smaller density that is the necessary condition of the RTI development (the velocity and density gradients have different signs).

It should be noted that there are a lot of difficulties in experimental observations of the process in consideration due to the extremely high temperatures and pressures and very short times. In such conditions numerical modeling provides the possibility to obtain very precise effects that could not be obtained in natural experiments. The paper is devoted to the numerical simulation of high-speed impact of two metal plates in the statement qualitatively similar to [2] using three-dimensional (3D) Euler equations and different equations of state.

2. Statement of the problem

Consider the interaction of the lead plate with the density $\rho = 11300$ kg/m^3 and thickness $h = 0.002$ m with the steel plate ($\rho = 7900$ kg/m^3) or the copper plate ($\rho = 8900$ kg/m^3) of the same thickness. The lead plate is thrown in the vertical direction with the speed $w = 500$ m/s. As the initial disturbance we set the point disturbance 500 m/s of the velocity on the throwing plate surface. The initial pressure is equal to $p = 10^{12}$ Pa, acceleration is equal to $g = 10^7$ m/s^2 and directed to the metal with smaller density. The computational area is the cube with the edge length 0.004 m. We set slip conditions on the upper (upper surface of the throwing plate) and lower (lower surface of the target plate) boundaries and periodic conditions of the side faces.

Note that due to the large computational costs we do not consider the whole development of the process during plates impact which is described above. Instead our statement qualitatively corresponds to the moment when the RW from the free surface of the throwing plate reaches the interface. Calculations are carried out up to the moment when the disturbances from the initial impact reach the upper and lower boundaries of the computational area.

3. Mathematical model and numerical algorithm

Mathematical model is based on 3D non-stationary two-component Euler equations [3] supplemented by different equations of state (EOS) (Cartesian frame, the notations are standard):

$$\frac{\partial \mathbf{U}}{\partial t} + \frac{\partial \mathbf{F}}{\partial x} + \frac{\partial \mathbf{G}}{\partial y} + \frac{\partial \mathbf{H}}{\partial z} = \mathbf{S},$$

$$\mathbf{U} = \begin{bmatrix} \rho_1 \\ \rho u \\ \rho v \\ \rho w \\ e \\ \rho_2 \end{bmatrix}, \mathbf{F} = \begin{bmatrix} \rho_1 u \\ \rho u^2 + p \\ \rho u v \\ \rho u w \\ (e+p)\, u \\ \rho_2 u \end{bmatrix}, \mathbf{G} = \begin{bmatrix} \rho_1 v \\ \rho v u \\ \rho v^2 + p \\ \rho v w \\ (e+p)\, v \\ \rho_2 v \end{bmatrix}, \mathbf{H} = \begin{bmatrix} \rho_1 w \\ \rho w u \\ \rho w v \\ \rho w^2 + p \\ (e+p)\, w \\ \rho_2 w \end{bmatrix}, \mathbf{S} = \begin{bmatrix} 0 \\ 0 \\ 0 \\ -\rho g \\ -\rho g w \\ 0 \end{bmatrix},$$

$$e = \frac{\rho\left(u^2 + v^2 + w^2\right)}{2} + \rho\epsilon\left(p, \rho\right), \ \rho = \rho_1 + \rho_2.$$

We use three different EOS, namely: (i) ideal gas, (ii) barotropic gas and (iii) wide-range EOS for real metals.

As the first approach to explain the experimental results the model of ideal gas with the specific heat ratio $\gamma = 5/3$ was considered. The second considered EOS was the barotropic EOS:

$$p = p_0 + E\frac{\rho - \rho_0}{\rho_0},$$

where p_0 corresponds to the curve of "cold" compression of the substance which coincides with the "cold" isentropic line, E — module of elasticity in the solid phase. As the third model we chose the wide-range semi-empirical EOS [4]. The last EOS not only well describes the condensed phase of the substance under the low temperatures but also takes into account the phase transition from the solid to the liquid one. Wide-range semi-empirical EOS are based on the extensive experimental data concerning shock compression of the solid and porous substances under normal conditions. At the extremely high pressures and temperatures they have the asymptotics of Tomas-Fermi and Debye-Huckel relations.

Numerical algorithm is based on the physical processes and spatial directions splitting techniques. The set of equations is written in the characteristics form. We solve three one-dimensional systems along each coordinate directions independently using Roe numerical flux function. Time integration is performed using explicit Euler scheme. Spatial approximation order of the scheme is increased up to second with the use of hybrid schemes approach. In this approach the switching between the central and upwind differences is realized for each characteristics based on the sign of characteristics. So the the numerical algorithm is characterized by theoretical first approximation order in time and the second approximation order in space on smooth solutions. The detailed description of the numerical algorithm could be found elsewhere [5]. The solutions of test cases including well-known Riemann problems could be found in [3]. As an example Figure 1 demonstrates the solution of the so called Sod problem [6]. The initial disintegration of discontinuity leads to the formation of "right" SW, contact surface and "left" RW. The SW smearing on three-four computational cells which is seen in Figure 1 is typical for the second approximation order schemes.

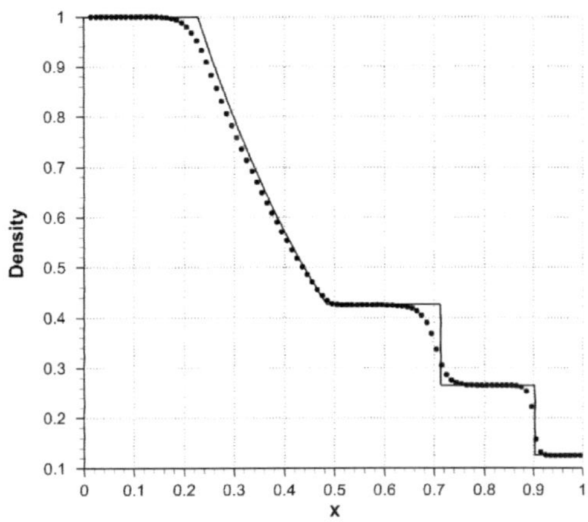

Figure 1. Predicted density distribution (dots) in comparison with the exact solution (solid line) in the Sod problem [6]

The typical size of the computational grid is $100 \times 100 \times 100$. For the numerical simulations of the problems which are described by the hyperbolic system of equations the authors developed computer code Turbo Problem Solver (TPS) [7]. TPS has modular structure and consists from the independent blocks responsible for different parts of the numerical method. TPS provides to the user the possibility to change numerical scheme, initial and boundary conditions and mass forces. The code is written in C++ and is parallelized using Message Passing Interface (MPI) package using domain decomposition approach.

4. RTI development and growth on the interface between fluids

As a test case we chose the problem very similar to the well-known experiments of K.I. Read [9]. Consider very thin squared-shape tank, approximately 15 × 13 mm size. The upper part of a tank is filled with dense fluid (ethanol) and the lower part is filled with air. The densities ratio is $\rho_1/\rho_2 = 600$. 2D approximation is used. Initial disturbance is applied to the between-fluid interface itself: $\xi = a_0 \sin(2\pi X/\lambda)$ is the disturbed boundary. For our simulation $a_0 = 0.5$ mm and $\lambda = 65$ mm are taken.

Large particles method [8] is used for the simulation up to 100 ms on the mesh of 150 × 130 cells. Figure 2 shows the development of the instability, trajectories of the mixing zone boundaries and their velocities evolution.

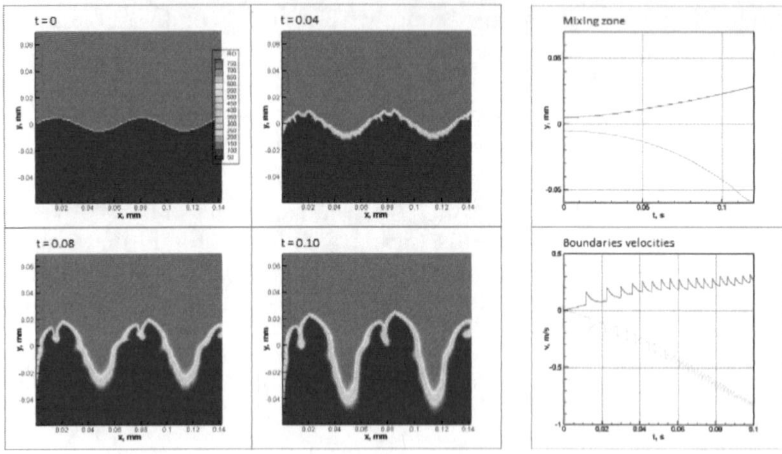

Figure 2. Growth and development of RTI on the interface between relatively dense (e.g. ethanol) and relatively sparse (e.g. air) fluids. $\rho_1/\rho_2 = 600$.

5. Results of numerical experiments

The presented results concern the ideal gas EOS and barotropic EOS for the case of steel target plate. For the ideal gas EOS with the point disturbance in the center of one of the plates we obtained the splash in the direction to the lead plate. The shape of the splash is similar to that observed in [2]. Figure 3 shows density isosurface at the time momet 5 μs after the impact. The domain $z > 0$ corresponds to the lead plate and $z < 0$ to the steel plate.

For the barotropic EOS we also get the splash in the direction to the lead plate. The splash has the crateriform shape. Figure 4 shows density isosurface at the time momet 5 μs after the impact. The results obtained using two different EOS are in qualitative agreement.

6. Conclusions

To conclude let us figure out the main features of the process obtained in the natural experiment and confirmed by our numerical study:

(i) The instability development on the interface of the metal plates is characterized by splashes from the metal plate with the smaller density to the plate with the greater density.

(ii) The increase of the throwing plate thickness with the fixed velocity leads to the disturbance wavelength growth on the interface.

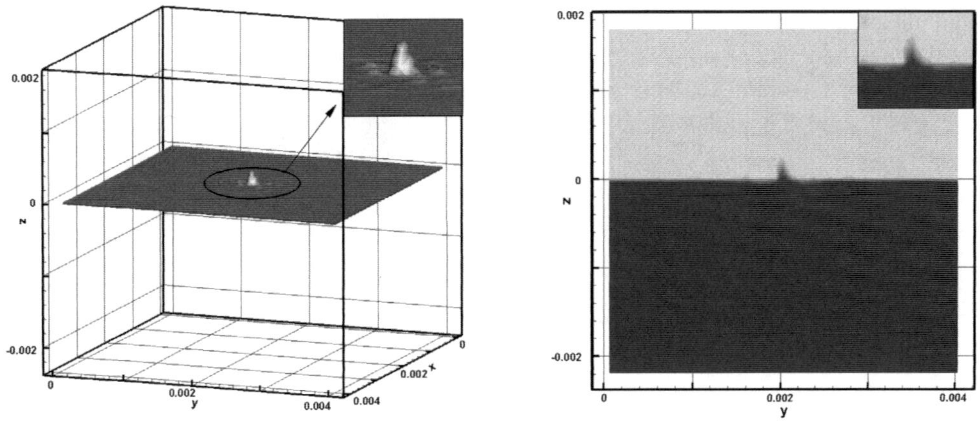

Figure 3. Density isosurface and cross section. The splash to the direction of lead plate. Calculation with the use of ideal gas EOS.

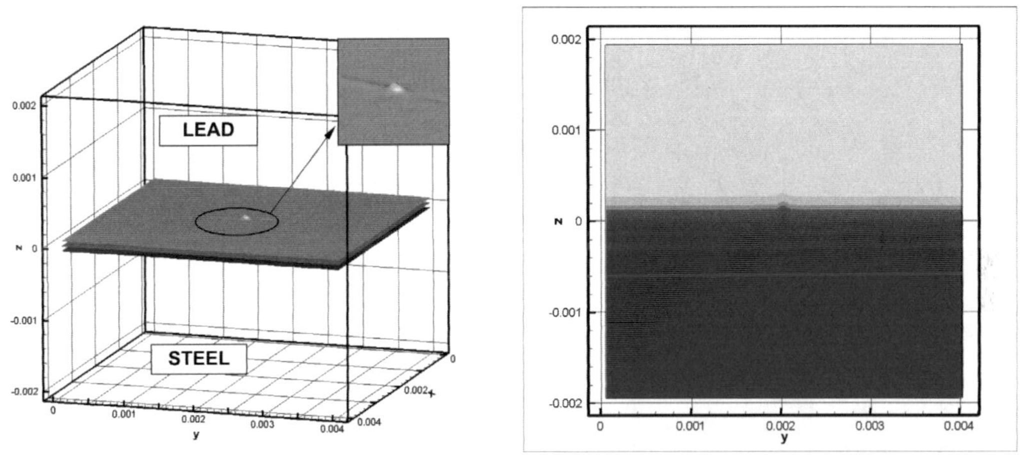

Figure 4. Density isosurface and cross section. The splash to the direction of lead plate. Calculation with the use of barotropic EOS.

The numerical simulations are carried out with the use of software complex Turbo Problem Solver developed by the authors and suitable for the three-dimensional numerical solution of the hyperbolic problems.

References

[1] Deribas A A 1972 *Physics of hardening and explosive welding* (Novosibirsk: Nauka)
[2] Yakovlev I V 1973 *Fiz. Gorenia i Vzryva* **9**(3) 447
[3] Belotserkovskii O M and Oparin A M 2001 *Numerical experiment in turbulence. From order to chaos.* (Moscow: Nauka)
[4] Bushman A V, Lomonosov I V and Fortov V E 1989 The models of wide-range equations of state under the high energy densities (*Preprint JIHT*) **6-287**
[5] Fortova S V 2014 *Comp. Math. and Math. Phys.* **54**(3) 553
[6] Toro E F 2009 *Riemann Solvers and Numerical Methods for Fluid Dynamics* (3d Eds., Springer)

[7] Fortova S V, Kraginkskii L M, Chikitkin A V and Oparina E I 2013 *Matem. Mod.* **25**(5) 123
[8] Belotserkovskii O M 1994 *Numerical modeling in the continuum media mechanics* (Moscow: Fizmatlit)
[9] Read K I 1984 *Physica D* **12** 45.

About the flowage of vortex chamber with outlet diaphragm and side slot swirler

M S Makarov and V S Naumkin

S S Kutateladze Institute of Thermophysics, Siberian Branch of the Russian Academy of Sciences, Ac. Lavrentiev ave., 1, Novosibirsk, Russia

naumkinv@mail.ru

Abstract. The vortex chambers are the widespread equipment in energy, chemical and mining industries. The upgrade of existing equipment and development of new models of vortex chambers require the detail analysis of swirling flow structure. This paper presents results of numerical modeling of the airflow dynamic in a vortex chamber with the outlet diaphragm and side slot swirler. The main attention is paid to the end boundary layers and to their influence on the flowage of vortex chambers. The study has been done for a wide range of swirl angles with Reynolds numbers of up to 1720.

1. Introduction

The vortex chambers have been used to organize various high-intensity physicochemical processes in gases, liquids or multi-phase flows. Wide-known cyclone separators for gas-liquid and dust-gas flows are applied in chemical, mining and metallurgical industries [1-3]. The vortex energy separators based on the Ranque-Hilsch effect are applied in local cooling systems of gas flows and conditioning systems [4-6]. In energy industry, the vortex burners are used to increase completeness of liquid and solid fuel combustion; the vortex cyclones are applied to clear combustion products from slag and harmful impurities [1, 7 and 8]. The vortex technologies allow increasing the specific productivity of devices and decreasing their mass and dimensions. At the same time, the absence of effective calculation methods for flow dynamic, heat and mass transfer analysis in vortex chambers and incompleteness criterion set for development of engineering calculation methods and methods of scaling this equipment slow down the progress of vortex technologies [9].

The studies of vortex chambers with side slot swirler are presented in [9, 10]. It is shown that the side slot swirler provides better uniformity of gas flow, lesser turbulent pulsations of axial jet, lesser hydraulic resistance of vortex chamber and swirler. In paper [9], the results of complete experimental study of the influence of the inlet slot area, swirl angle, diameter of outlet diaphragm, height of chamber and outlet branch pipe on the flow dynamic in the bulk of vortex chamber are presented. The influence of end boundary layers on the flow dynamic was not considered. All data was obtained by means of multichannel air pressure probes. The semiempirical method of calculation of the velocity and pressure fields in the bulk of vortex chamber was suggested.

The studies of flow dynamic in the end boundary layers are presented in papers [10, 11]. The experimental measurements of flow velocity were carried out both by probe method and non-perturbative optical Laser Doppler Anemometry (LDA) [12, 13]. The swirl angle of the inlet flow, where the flow regime is switch to new configuration, was obtained. For the new regime, the inlet mass of gas flows along the sidewall into the end boundary layers, than it turns out along the chamber axis and exits through the outlet diaphragm. The radial mass flow in the bulk of vortex chamber closes to zero and therefore the new regime was named "unflowage".

This paper presents the results of numerical modeling of air flow dynamic in the vortex chamber with the side slot swirler at various mass flow rates and swirl angles. The analysis of flow configuration is done based on obtained results. The mass flow rates through the bulk of chamber and the end boundary layers are calculated for a wide range of swirl angles. The influence of reverse flow into the chamber through the outlet diaphragm on it flowage is shown.

2. Statement of the problem and numerical approach

The air flow through the vortex chamber with the side slot swirler and outlet diaphragm is considered. The geometry of chamber is presented in figure 1. The diameter of vortex chamber is D_v = 160 mm, height is H = 196 mm, H/D_v = 1.225. The outlet diaphragm with diameter d_f = 32 mm is installed at the chamber axis, the clamping factor is d_f/D_v = 0.2. The heights of slots in side swirler are equal to the chamber height. The swirl angle α is varied from 0.5 to 90°. The width of outlet cross section for each slot l is constant to all swirl angles and equals s/sin 45° for slot width s = 1 mm. It is equivalent to varying the slot width as s=sin α /sin 45°. The number of slots n is 12. The slots are uniformly spaced over the circular of sidewall.

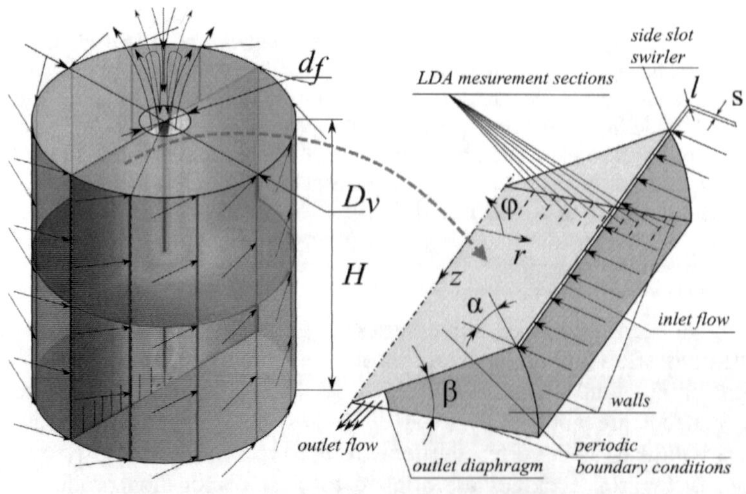

Figure 1. The geometry of vortex chamber (on the left) and calculation domain with boundary conditions (on the right).

The calculation domain for 3D modeling is the sector with opening angle β=360°/n. On the cylindrical face, the rectangular space with width l is separated as the outlet cross section of swirler slot or equivalent the inlet flow boundary (see figure 1 on the right). On the axis of top face, the sector with radius d_f /2 is separated as the outlet flow boundary. The lateral faces are joined by periodic boundary conditions.

The mesh is generated by in-home code GGN. The mesh consists of the hexahedral cells in the bulk and wedge sells near the axis. The exponent compression is used to the walls, periodic boundaries and radial section separating the diaphragm and bulk of the chamber. The number of nodes is 720000.

The mass rate of inlet flow G is 1, 2.33 and 4 gr/s. The pressure p_0 and temperature T_0 on the outlet boundary are 1 atm and 300 K, respectively. The air is supposed incompressible with constant viscosity μ = 1.85·10^{-5} Pa·s and density ρ = 1.17 kg/m^3. The Reynolds numbers built on the diameter of vortex chamber $\mathrm{Re}_v = 4G/\pi\mu D_v$ for studied mass flow rates equal 430, 1000 and 1720.

We solved the problem by the SIMPLE numerical method, based on the 3D RANS equations for the steady and unsteady flows with the second order scheme for pressure and third order scheme for momentum equations. ANSYS Fluent solver was licensed for SSCC SB RAS. For studying the

Reynolds number range, the flow regime in the bulk of chamber is laminar, but in the vortex decay area on the bottom wall the flow configuration is very complex. To consider the influence of possible turbulent effect for a small swirl angle, we used the RSM turbulence model with calculation of turbulence dissipation by means of k-ε model and standard wall functions. The convergence for all variables and mass flow rates are provided with maximum error 10^{-4}.

3. Result and discussion

3.1. Validation of numerical approach

Figure 2 shows the circumferential and radial velocity distributions across the bottom end boundary layer. For this case, the air is supplied to the vortex chamber through the 12th slots side swirler with geometrical swirl angle of 45°. The slots width *s* is 1 mm. The mass flow rate is 2.33 gr/s (Re_v =430).

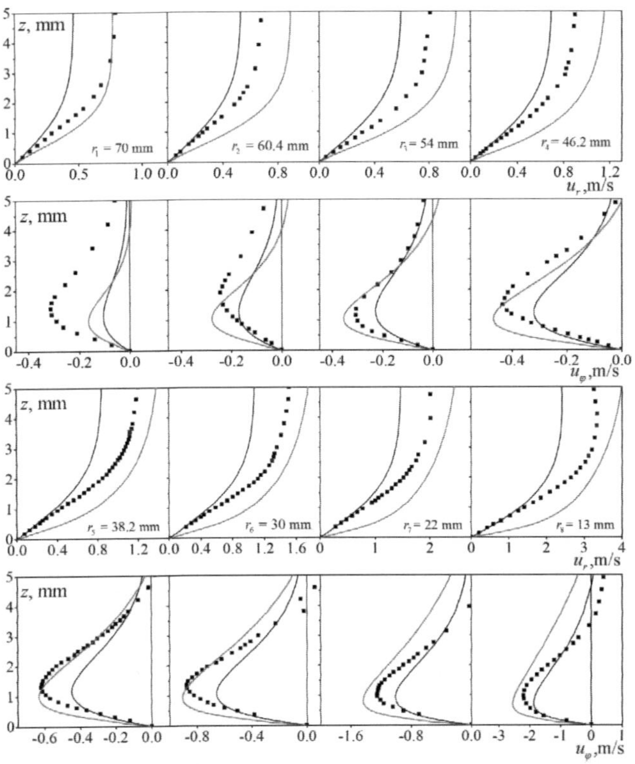

Figure 2. Circumferential velocity (u_r) and radial velocity (u_φ) distributions across the bottom end boundary layer in the sections at 70, 60.4, 54, 46.2, 38.2, 30, 22 and 13 mm from the axis of vortex chamber: points – experimental data; black lines – numerical modeling without turbulence model; red lines – numerical modeling with RSM turbulence model.

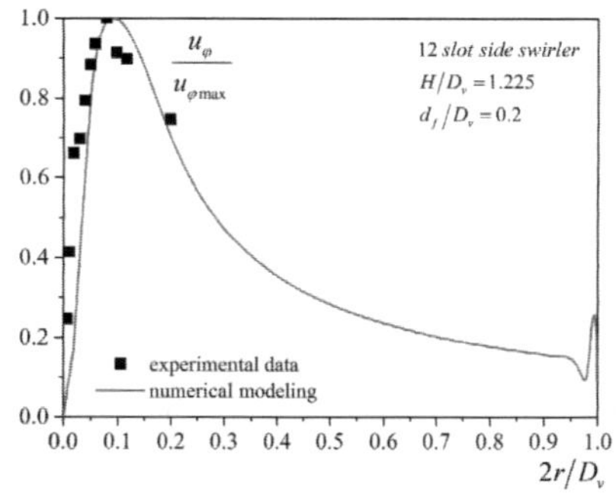

Figure 3. Dimensionless circumferential velocity depending on the dimensionless radius of vortex chamber: points – experimental data [9]; red line – numerical modeling.

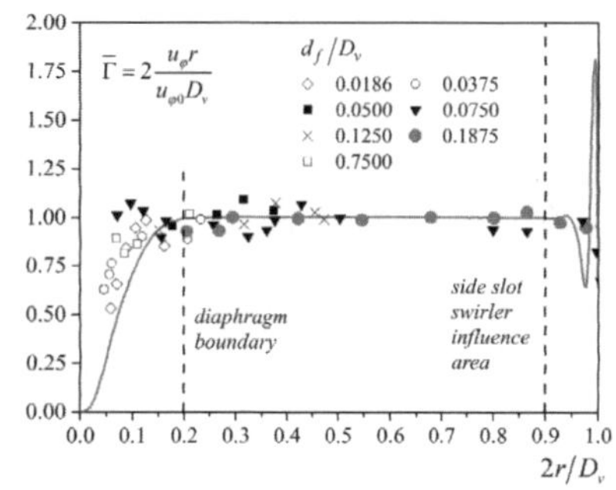

Figure 4. Dimensionless circulation depending on the dimensionless radius of vortex chamber: points – experimental data [9] for various vortex chambers with outlet diaphragm; red line – numerical modeling.

The points present the experimental data [15] obtained by LDA methods in the radial sections at 70, 60.4, 54, 46.2, 38.2, 30, 22 and 13 mm from the axis of vortex chamber. The black line shows the modeling data obtained without using the turbulence model, the red line show the data obtained with the use of above-described RSM turbulence model. As we can see, the results of modeling with RSM model correlate better with experimental data. A certain difference in the circumferential velocity distribution we can see near the side swirler and near the chamber axis. Near the side swirler, the radial velocity obtained in modelling has a lesser value than that obtained in experiment. It can be explained by the smooth edge of the slot in experiment and the sharp edge in the model. Different forms lead to the different widths and angles of flat jets on the slots outlet. Near the axis, the flow is unstable. It was noted in experiment description that the axial swirl jet in the vortex chamber has weak precession rotation around the geometrical axis with frequencies of few Hertz. The LDA measurements were not synchronized with the position of own axis of swirl jet, and therefore the experimental data were averaged over the random sample of flow configurations near the axis. The used numerical modeling method does not allow solving the problems with precessing axial jet. Figure 2 presents the results of steady solution.

Figure 3 shows the dimensionless circumferential velocity depending on the dimensionless radius of vortex chamber for the axial cross section at the distance from the bottom wall of 100 mm. The numerical data are compared with the experimental results [9]. The parameters of the flow are the same as in Figure 2. We can clearly see that the flat jet is pressed to the sidewall near the swirler. The circumferential velocity has the maximum in this area. The maximum of circumferential velocity near the axis lies at the half radius of diaphragm and it is agreed with the experiment.

The comparison of the modeling data with the generalized experimental data of dimensionless circulation obtained to the various vortex chambers with outlet diaphragm [9] is presented in Figure 4. The red points correspond to the same geometry of vortex chamber as we modeled. As we can see, circulation has constant distribution along the radius in the bulk area, tends to zero at the axis and has the maximum near the swirler. The slot swirler perturbs the flow near the sidewall, but its influence is significant only, if the distance from the axis of the chamber exceeds $0.45D_v$.

3.2. Flowage of vortex chamber with the outlet diaphragm and side slot swirler

A flowage criterion is the ratio of the summarized mass flow rate through the top and bottom end boundary layers G_b to the full mass flow rate through the vortex chamber G_r. The complexity of flowage analysis is associated with choosing a radial section, where the velocity both in the end boundary layers and in bulk of the chamber has the most characteristic distribution. Obviously, the full mass flow rate through the section on the sidewall is equal to the defined inlet mass flow rate, and the mass flow rate through the end boundary layers is equal to zero. As Figure 2 shows, the flow near the bottom wall is radially accelerated, therefore the gas flows into the bottom end boundary layer through the upper bound along the radius. The mass flow rate of gas in the end boundary layers increases from the sidewall to axis. If we choose the radial cross section near the axis or bound of outlet diaphragm, the mass flow rate through the end boundary layers is close to zero too. In this area, the flow is radially decelerated and axially accelerated; the axial swirl jet is formed. For the small swirl angle the recirculation area is formed near the axis, therefore the full mass flow rate through the vortex chamber is increased, often several times as compared with the defined inlet mass flow rate through the slots swirler. Obviously, the increasing full mass flow should be taken into account in flowage analysis.

In present study, the control section for calculation of flowage criterion is chosen as the radial section at the distance 60.4 mm ($0.38D_v$) from the axis. This section is located far from the axial swirl jet influence area and is characterized the constant circulation in the bulk of chamber and well-developed flow in the end boundary layers. As it's showed in previous section for the studied Reynolds number range the thickness of bottom end boundary layer is not exceed the 5 mm. The modeling data shows the top end boundary layer thickness is equal to 5 mm too. For the flowage analysis the mass flow rate through the end boundary layers is determined as the part of full mass flow rate through the parts of chosen axial section with the height of 5 mm to each part near the top and bottom walls. Figure 5 clearly shows the areas of end boundary layers in comparison with the bulk of the chamber.

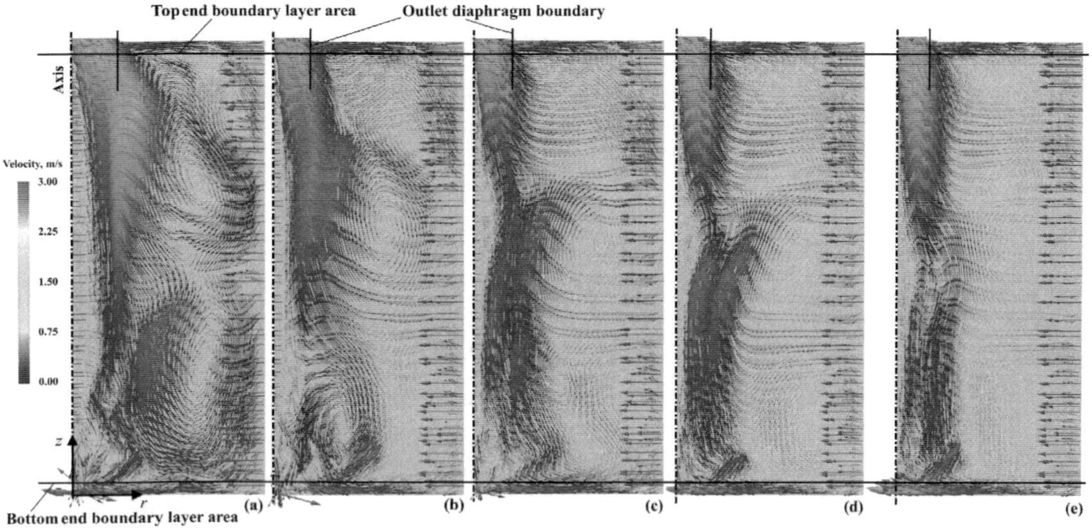

Figure 5. Vector fields of velocity project on (r, z) coordinate plane for various phases of flow in a vortex chamber: from flowage regime (a) to unflowage regime (e).

The results of calculation of flowage criterion for the studied vortex cambers G_b/G_r depending on the tangents of geometric swirl angle of side slot swirler are presented in Figure 6. The numerical modeling data for various inlet mass flow rates are shown by points. As we can see, the decreasing swirl angle from 90° to 30° leads to increasing mass flow rate through the end boundary layers, but the flowage criterion is still lesser than 1. The full mass flow rate through the control section is higher than the mass flow through the end boundary layers. The further decreasing the swirl angle leads to the switch in flow configuration. The air flow is pressed to the sidewall. The axial swirl jet is swelled and its diameter is increased. Two or more toroidal vortexes are formed in the bulk of the chamber. The flow through the end boundary layers is still non-intensive. The flow pattern is presented in Figure 5 (a). The further decreasing the swirl angle leads to formation of the recirculating flow area on the diaphragm axis. The axial swirl jet turns to the unstable flow regime. As we can see in Figure 6, in this regime the flowage criterion can be higher than 1, close to 0 or even lesser than 0. The data dispersion is presented in Figure 6 by grey field. The changing of flowage criterion is associated with the increasing full mass flow rate through the control section due to gas inflows through the diaphragm and unstable axial flow. The characteristic phases during flow reconfiguration from the flowage to unflowage regime are presented in Figure 5. As the conclusion, the flowage of vortex chamber for the small swirl angle and unsteady regime of the flow is the conditional criterion. At various time moments, the vortex chamber could be the flowage or unflowage. The determination of the characteristic frequencies of transition from flowage to unflowage regime and identification of the main factors influencing the transition is the interesting problem for the further studies.

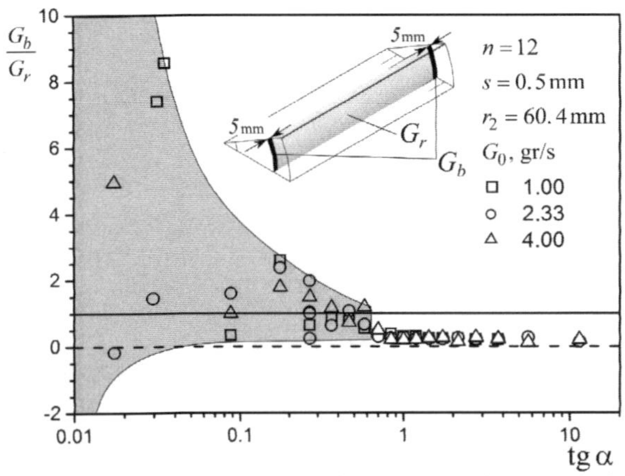

Figure 6. Flowage criterion depending on the geometric swirl angle of side swirler.

Acknowledgments

This work was supported by the Russian Science Foundation, grant 14-19-00402.

References

[1] Alekseenko S V and Okulov V L 1996 *Thermophysics and aeromechanics* **3** 2 97–128

[2] Balabekov O S and Balabaev L Sh 1991 *Gas cleaning in the chemical industry (processes and apparatus)* (Moscow: Chemistry)

[3] Bagryantsev V I and Terekhov V I 1985 *Izv. Sib. Otd. Akad. Nauk SSSR: Techn. nauki* **1** 87–93

[4] Eiamsa-ard S, Promvonge P 2008 *Renewable and sustainable energy reviews* **12** 1822–42

[5] Xue Y, Arjomandi M and Kelso R 2010 *Exp. thermal and fluid science* **34** 1367–74

[6] Piralishvilly Sh A, Polyaev V M and Sergeev M N *Vortex effect. Experiment, theory, technical solutions* 2000 (Moscow: UNPC Energomash)

[7] Reznikov M I and Lipov Yu M *Steam boilers of thermal power plants* 1981 (Moscow: Energoizdat)

[8] Pitsukha E A, Teplitskii Yu S and Borodulya V A 2014 *Journal of engineering physics and thermophysics* **87** 5 1017–21

[9] Smulsky J J *Aerodynamics and processes in vortex chambers* 1992 (Novosibirsk: Nauka)

[10] Kutateladze S S, Volchkov E P and Terekhov V I 1987 *Aerodynamics and heat and mass transfer in confined vortex flows* (Novosibirsk: IT SB RAS)

[11] Vochkov E P, Semenov S V, Terekhov V I 1991 *Exp. thermal and fluid science* **4** 546–57

[12] Dubnishev Yu V, Belousov V G, Pavlov V A and Meledin V G 1990 *Optica applicata* **20** 3 112–20

[13] Titkov V I and Lukashov V V 2006 *Optoelectronics, instrumentation and data processing* **4** 85–91

[14] Volchkov E P, Dvornikov N A, Lukashov V V and Abdrakhmanov R Kh 2013 *Thermophysics and aeromechanics* **20** 6 663–8

[15] Abdrakhmanov R Kh, Lukashov V V, Makarov M S and Naumkin V S 2015 *Proc. of Conf. Heat and mass transfer and hydrodynamics in swirling flows: 5th international conference (Russia, Kazan)* 302-15

Verification of a low Mach variable-density Navier-Stokes solver for turbulent combustion

R Mullyadzhanov[1,2], E Palkin[1,2], B Ničeno[3], L Vervisch[4], K Hanjalić[2,5]

[1] Institute of Thermophysics SB RAS, Novosibirsk, Russia
[2] Novosibirsk State University, Novosibirsk, Russia
[3] Paul Scherrer Institute, Villigen, Switzerland
[4] CORIA-CNRS and INSA Rouen, Normandie Université, Rouen, France
[5] Delft University of Technology, Delft, The Netherlands

E-mail: rustammul@gmail.com

Abstract. We describe the low Mach variable-density Navier-Stokes numerical iterative solution procedure implemented in the finite-volume unstructured T-FlowS code. As the test cases we use a number of analytic manufactured solutions and Rayleigh-Taylor instability problem from the literature for algorithm verification purposes. The tests show that the code is second-order accurate in agreement with the spatial discretization scheme. We outline the recent combustion ADEF model implemented in the program.

1. Introduction

During the last few decades rapid development of the computational technologies stimulated the research using computational codes for fundamental and applied studies. Increasing complexity of codes makes the verification process essential for research purposes [1]. In the present work we assess the accuracy of the implemented algorithm in the variable-density Navier-Stokes solver related to the combustion problems where quite a broad range of problems can serve as a benchmark [2–9]. As the test cases we use the analytic manufactured solutions derived by Shunn et al [10], the Rayleigh-Taylor instability problem employed by Desjardins et al [11] and describe the combustion modeling framework [12] used in Large-eddy simulations of the Cambridge stratified burner [13].

2. Governing equations and iterative procedure

At low Mach numbers variable-density reacting flows can be described be the following conservation equations for mass, momentum and scalars together with a suitable equation of state:

$$\frac{\partial \rho}{\partial t} + \frac{\partial \rho u_j}{\partial x_j} = \dot{Q}_\rho, \tag{1}$$

$$\frac{\partial \rho u_i}{\partial t} + \frac{\partial \rho u_i u_j}{\partial x_j} = -\frac{\partial p}{\partial x_i} + \frac{\partial}{\partial x_j}\Big(2\mu S_{ij}\Big) + \dot{Q}_{u_i}, \tag{2}$$

$$\frac{\partial \rho\phi_k}{\partial t} + \frac{\partial \rho\phi_k u_j}{\partial x_j} = \frac{\partial}{\partial x_j}\Big(\rho\alpha_k \frac{\partial \phi_k}{\partial x_j}\Big) + \dot{Q}_{\phi_k}, \tag{3}$$

$$\rho = f\Big(\phi_1, \phi_2, ..., \phi_k, ...\Big), \tag{4}$$

where

$$S_{ij} = \frac{1}{2}\Big(\frac{\partial u_i}{\partial x_j} + \frac{\partial u_j}{\partial x_i}\Big) - \frac{1}{3}\delta_{ij}\frac{\partial u_l}{\partial x_l} \tag{5}$$

is the rate-of-strain tensor. The variables ρ, u_i, p and ϕ_k denote the density, velocity vector components, pressure and additional scalar fields. The dynamic viscosity and kinematic diffusivity coefficients are denoted by μ and α_k. The source terms $\dot{Q}_\rho$, $\dot{Q}_{u_i}$ and $\dot{Q}_{\phi_k}$ in Eqs. (1)-(3) are used to construct the analytic manufactured solutions or denote the subgrid-scale terms appearing in the framework of Large-eddy simulations (LES). In the present work we consider only one additional scalar ϕ, thus, further we omit the subscript index 'k'.

The following iteration procedure similar to the one described by [14] is implemented into the finite-volume T-FlowS code [15] employing unstructured cell-centered collocated grids and featuring second-order accuracy in time and space. Below the superscript n and $n-1$ refers to solutions that are known from two previous time levels, the subscript 0 indicates the initial guess or the current provisional value of some quantity.

Step 0: The solution from the previous time step n for momentum, pressure and scalar are used as the initial best guess:

$$(\rho u_i)_0^{n+1} = (\rho u_i)^n, \quad p_0^{n+1} = p^n, \quad \phi_0^{n+1} = \phi^n. \tag{6}$$

The predictor ρ_0^{n+1} for the density is calculated from the continuity equation, Eq. (1), integrated over a control volume (CV):

$$\int_{CV} \frac{\partial \rho}{\partial t} dV = \frac{\Delta V}{\Delta t}\Big(\frac{3}{2}\rho_0^{n+1} - 2\rho^n - \frac{1}{2}\rho^{n-1}\Big) = -\int_{CV} \frac{\partial \rho u_j}{\partial x_j} dV = -\int_{faces} \rho u_j n_j dS, \tag{7}$$

where the Gauss theorem is used to go from the volume to the surface integration. The time derivative is approximated with a three-point backward-difference scheme (second order).

Step 1: The scalar equation is advanced so that a better estimate for the density can be obtained early in the iteration process. Time advancing Eq. (3) yields $(\rho\phi)^{n+1}$, from which a provisional estimate for ϕ_0^{n+1} is obtained using the current density predictor:

$$\phi_0^{n+1} = (\rho\phi)^{n+1} / \rho_0^{n+1}. \tag{8}$$

Step 2: Update the density from the equation of state using the provisional scalar values:

$$\rho^{n+1} = f(\phi_0^{n+1}). \tag{9}$$

Step 3: Update the scalar based on the new density:

$$\phi^{n+1} = (\rho\phi)^{n+1} / \rho^{n+1}. \tag{10}$$

Step 4: Advance in time the momentum equations, Eq. (2), to obtain provisional values for the momentum components, $g_i = \rho u_i$. Compute provisional velocity field:

$$u_{i,0}^{n+1} = g_{i,0}^{n+1} / \rho^{n+1}. \tag{11}$$

Step 5: A Poisson equation is solved to satisfy the continuity equation adjusting the values of momentum and pressure. We consider additive corrections, δg_i and δp, to the momentum and pressure such that

$$g_i^{n+1} = g_{i,0}^{n+1} + \delta g_i^{n+1}, \quad p^{n+1} = p_0^{n+1} + \delta p^{n+1}, \tag{12}$$

We then substitute these expressions into the momentum equation. Following Pierce [14] we group all other terms into a generic right-hand-side term, R_i,

$$\frac{\partial g_i^{n+1}}{\partial t} = \frac{\partial (g_{i,0}^{n+1} + \delta g_i^{n+1})}{\partial t} = -\frac{\partial (p_0^{n+1} + \delta p^{n+1})}{\partial x_i} + R_i. \tag{13}$$

Provided that the following equation has been satisfied on the previous step

$$\frac{\partial g_{i,0}^{n+1}}{\partial t} = -\frac{\partial p_0^{n+1}}{\partial x_i} + R_i, \tag{14}$$

we have to satisfy the rest of the equation

$$\frac{\partial \delta g_i^{n+1}}{\partial t} = -\frac{\partial \delta p^{n+1}}{\partial x_i}. \tag{15}$$

The time derivative is approximated using the three-point backward-difference to be consistent with Eq. (7). Taking the divergence of this expression and integrating over a control volume, we arrive to

$$\frac{1}{\Delta t}\int_{CV} \frac{\partial}{\partial x_i}\Big(\frac{3}{2}\delta g_i^{n+1} - 2\delta g_i^{n} + \frac{1}{2}\delta g_i^{n-1}\Big) dV = -\int_{CV} \frac{\partial^2 \delta p^{n+1}}{\partial x_i^2} dV. \tag{16}$$

Note that $\delta g_i^n = \delta g_i^{n-1} = 0$ provided the algorithm has satisfied both the momentum and continuity equations on previous time steps. The derivative of the momentum correction can be expressed as the residual of the continuity equation:

$$\frac{\partial \delta g_i^{n+1}}{\partial x_i} = \frac{\partial (g_i^{n+1} - g_{i,0}^{n+1})}{\partial x_i} = -\frac{\partial \rho}{\partial t}^{n+1} - \frac{\partial g_{i,0}^{n+1}}{\partial x_i} = -\delta(cont.)^{n+1}. \tag{17}$$

Going from volume to surface integrals, we obtain

$$\frac{3}{2\Delta t}\int_{CV} \delta(cont.)^{n+1} dV = \int_{faces} \frac{\partial \delta p^{n+1}}{\partial x_i} n_i dS \tag{18}$$

which is solved for the pressure correction derivatives in the cell centers.

Step 6: The momentum components, velocity and pressure at the cell centers are updated:

$$g_i^{n+1} = g_{i,0}^{n+1} - \Delta t \frac{\partial \delta p^{n+1}}{\partial x_i}, \quad u_i^{n+1} = g_i^{n+1}/\rho^{n+1}, \tag{19}$$

$$p^{n+1} = p_0^{n+1} + \delta p^{n+1}. \tag{20}$$

In the next section we describe a set of test cases to verify the described algorithm.

Problem 1	Value	Problem 2	Value	Problem 3	Value
ρ_0	20	ρ_0	20	ρ_0	5
ρ_1	1	ρ_1	1	ρ_1	1
k_1	4	u_F	1	$k = \omega$	2
k_2	2	v_F	0.5	$u_F = v_F$	0.5
w_0	5	a	0.2	$\rho\alpha_\phi = \mu$	0.001
$\rho\alpha_\phi = \mu$	0.03	b	20		
		k	4π		
		ω	1.5		
		$\rho\alpha_\phi = \mu$	0.001		

Table 1. Parameters of problems introduced by Shunn et al [10]

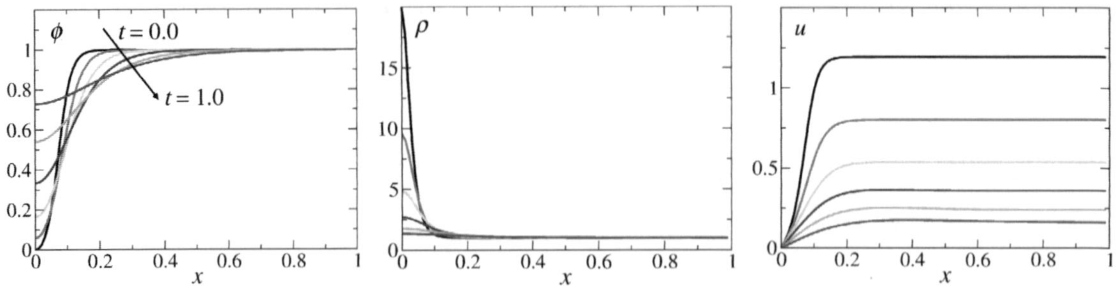

Figure 1. Profiles of ϕ, ρ and u for Problem 1 and different time instants (increment $\Delta t = 0.2$).

3. Laminar test cases

3.1. Problem 1: one-dimensional diffusion

We consider a one-dimensional problem described by the following relations:

$$\phi(x,t) = \frac{\exp(-k_1 t) - \cosh(w_0 x \exp(-k_2 t))}{\exp(-k_1 t)\left(1 - \frac{\rho_0}{\rho_1}\right) - \cosh(w_0 x \exp(-k_2 t))}, \tag{21}$$

$$\rho(x,t) = \left(\frac{\phi(x,t)}{\rho_1} + \frac{1-\phi(x,t)}{\rho_2}\right)^{-1}, \tag{22}$$

$$u(x,t) = 2k_2 \exp(-k_1 t)\frac{\Delta\rho}{\rho(x,t)}\left(\frac{\hat{u}x}{\hat{u}^2+1} + \frac{\left(\frac{k_1}{k_2} - 1\right)(\arctan\hat{u} - \frac{\pi}{4})}{w_0 \exp(-k_2 t)}\right), \tag{23}$$

where $\hat{u} = \exp(w_0 x \exp(-k_2 t))$ and $\Delta\rho = \rho_0 - \rho_1$, w_0, k_1 and k_2 are constants (see Table 1). The corresponding source terms in Eqs. (1)-(3) are explicitly derived by Shunn et al [10] (see their Appendices). The computational domain for this problem is $0 \leq x \leq 2$ and $0 \leq t \leq 1$ and the time step is $\Delta t = 0.00125$. The number of uniformly distributed cells is $n_x = 64$, 128 and 256. Figure 1 shows the profiles of ϕ, ρ and u for different time instants. The rate of convergence of the algorithm is assessed by calculating the L_2-error for different grid refinement levels, where

$$L_2 = \sqrt{\frac{\sum_{i=1}^{N}(\Phi_{ex} - \Phi_i)^2}{\sum_{i=1}^{N}\Phi_{ex}^2}} \tag{24}$$

and Φ is some test function. The subscript 'ex' denotes the exact solution while 'i' corresponds to the value in the i-th control volume with the total number N. Figure 2 shows the L_2-error

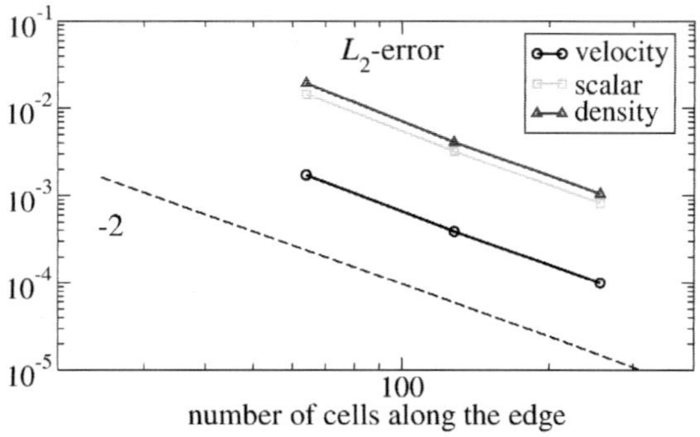

Figure 2. Log-log plot of L_2-error for ϕ, ρ and u for Problem 1 and $t = 1$. The dashed line shows the slope corresponding to the second-order accurate discretization.

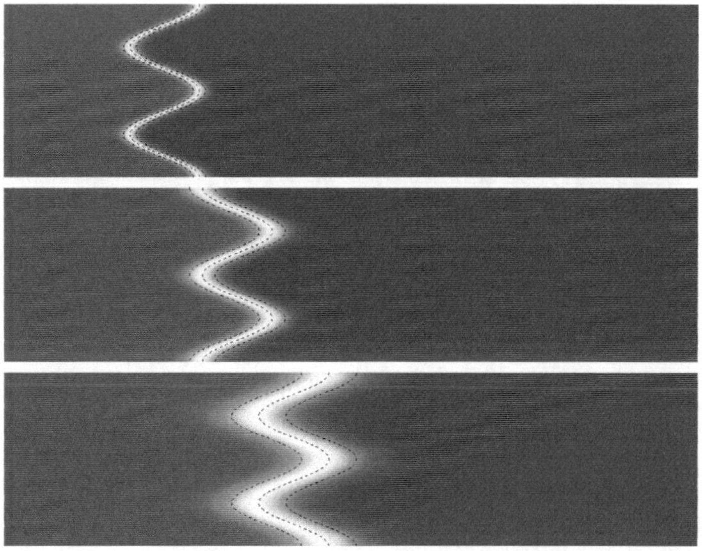

Figure 3. Contours of $\rho(t, x, y)$ for $t = 0$ (top), $t = 0.5$ (middle), $t = 1$ (bottom). Red color corresponds to $\rho = 1$ and blue color to $\rho = 20$.

for ϕ, ρ and u at the moment $t = 1$. Note that the decrease of the error when n_x is increased confirms the second-order accuracy of the algorithm.

3.2. Problem 2: two-dimensional advection and diffusion

The second problem describes the advection of a diffusing two-dimensional 'flame' front (see Fig. 3):

$$\phi(x, y, t) = \frac{1 + \tanh(b\hat{x}\exp(-\omega t))}{\left(1 + \frac{\rho_0}{\rho_1}\right) + \left(1 - \frac{\rho_0}{\rho_1}\right)\tanh(b\hat{x}\exp(-\omega t))}, \quad (25)$$

$$\rho(x, y, t) = \left(\frac{\phi(x, y, t)}{\rho_1} + \frac{1 - \phi(x, y, t)}{\rho_2}\right)^{-1}, \quad (26)$$

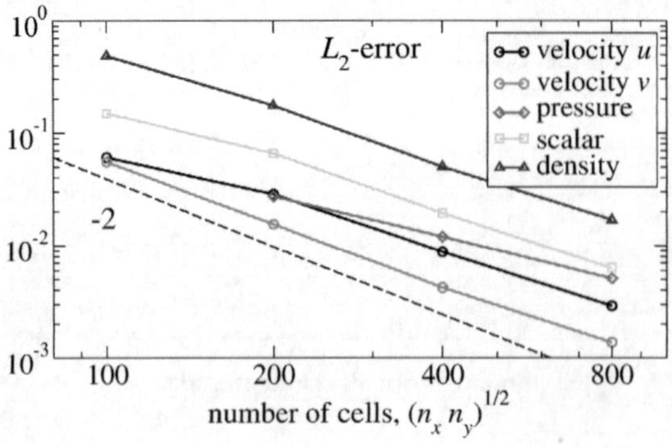

Figure 4. L_2-error for u, v, p, ϕ and ρ for Problem 2 and $t = 1$.

$$u(x, y, t) = \frac{\Delta\rho}{\rho}\Big(\omega\hat{x} - \frac{\omega\hat{x} - u_F}{\exp(2b\hat{x}\exp(-\omega t)) + 1} - \frac{\omega\log(\exp(2b\hat{x}\exp(-\omega t)) + 1)}{2b\exp(-\omega t)}\Big), \quad (27)$$

$$v(x, y, t) = v_F, \quad p(x, y, t) = 0, \quad (28)$$

where $\hat{x}(x, y, t) = u_F t - x + a\cos(k(v_F t - y))$, $\Delta\rho = \rho_0 - \rho_1$ and a, b, k, ω, u_F and v_F are parameters (see Table 1). The computational domain for this problem is $-1 \le x \le 3$, $-1/2 \le y \le 1/2$ and $0 \le t \le 1$. Four meshes are considered with the number of uniform cells in each direction $n_x \times n_y = 200 \times 50$, 400×100, 800×200 and 1600×400. The time step is $\Delta t = 0.00125$. Figure 4 shows the spatial convergence of the L_2-error with the grid refinement approximately indicating the second-order accuracy of the code.

3.3. Problem 3: two-dimensional oscillating density field

Further we consider a time-periodic solution of the following form (see Fig. 5):

$$\phi(x, y, t) = \frac{1 + \sin(\pi k\hat{x})\sin(\pi k\hat{y})\cos(\pi\omega t)}{\left(1 + \frac{\rho_0}{\rho_1}\right) + \left(1 - \frac{\rho_0}{\rho_1}\right)\sin(\pi k\hat{x})\sin(\pi k\hat{y})\cos(\pi\omega t)}, \quad (29)$$

$$\rho(x, y, t) = \left(\frac{\phi(x, y, t)}{\rho_1} + \frac{1 - \phi(x, y, t)}{\rho_2}\right)^{-1}, \quad (30)$$

$$u(x, y, t) = \frac{\rho_1 - \rho_0}{\rho(x, y, t)}\left(\frac{-\omega}{4k}\right)\cos(\pi k\hat{x})\sin(\pi k\hat{y})\sin(\pi\omega t), \quad (31)$$

$$v(x, y, t) = \frac{\rho_1 - \rho_0}{\rho(x, y, t)}\left(\frac{-\omega}{4k}\right)\sin(\pi k\hat{x})\cos(\pi k\hat{y})\sin(\pi\omega t), \quad (32)$$

$$p(x, y, t) = \frac{1}{2}\rho(x, y, t)u(x, y, t)v(x, y, t), \quad (33)$$

where $\hat{x} = x - u_F t$ and $\hat{y} = y - v_F t$. The computational domain for this problem is $-1 \le x \le 1$, $-1 \le y \le 1$ and $0 \le t \le 1$. Structured hexahedral and unstructured triangular grids have been used for calculations with four levels of refinement, i.e. $n_x \times n_y = 32^2$, 64^2, 128^2 and 256^2. The time step descreases from $\Delta t = 0.025$ for the coarsest mesh to $\Delta t = 0.003125$ for the finest. Figure 6 shows the L_2-error convergence for hexahedral and triangular grids in comparison with the results of Shunn et al [10]. While unstructured meshes demonstrate similar absolute level of the L_2-error, the hexahedral grids with the present algorithm bring lower error compared to the data from the literature.

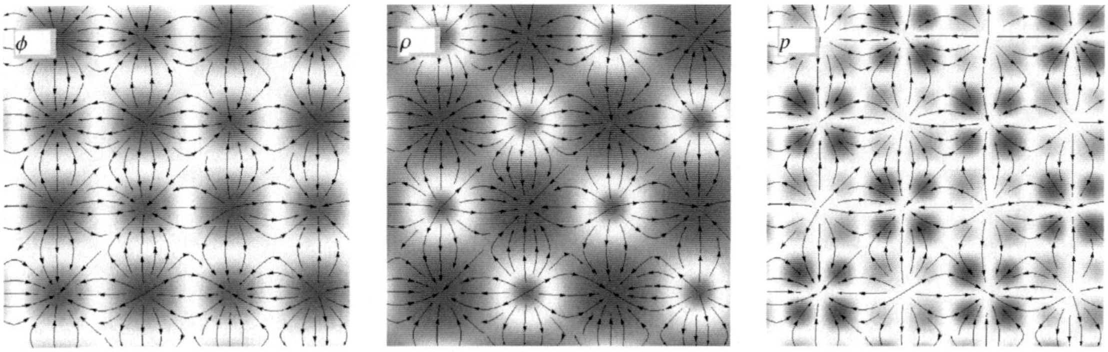

Figure 5. Colorplots of ϕ, ρ and p and streamlines at $t = 0$ for Problem 3.

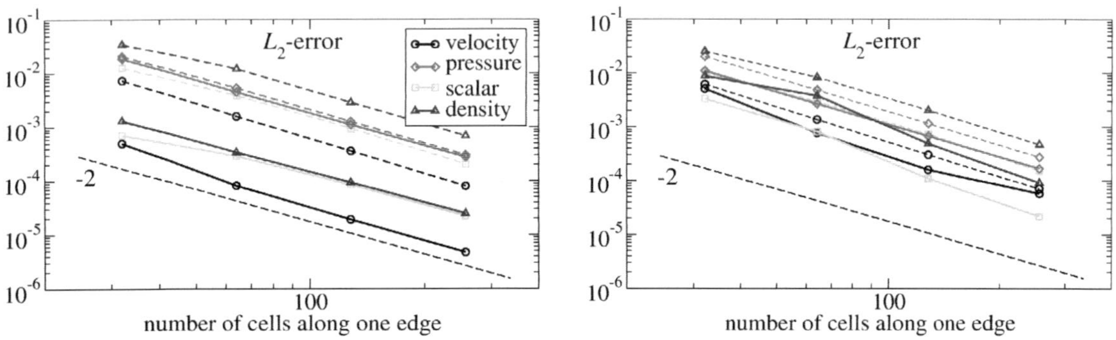

Figure 6. Left: uniform hexahedral mesh. Right: unstructured triangular mesh. Solid lines correspond to the present simulations while dashed lines denote results of Shunn et al [10].

3.4. Problem 4: Rayleigh-Taylor instability

Following Desjardins et al [11] we consider the two-dimensional Rayleigh-Taylor instability problem. A rectangular domain of the size $[-0.5, 0.5] \times [-0.5, 0.5]$ filled with two miscible fluids separated by a horizontal perturbed interface. The heavy fluid with $\rho_0 = 1$ is above the light fluid with $\rho_1 = 0.1$. The exact location of the interface is given by

$$y_{int}(x) = -\gamma \sum_{k=1}^{8} \cos(\omega_k \pi x), \tag{34}$$

where $\gamma = 0.001$ is the perturbations amplitude and $\omega_k = 4, 14, 23, 28, 33, 42, 51, 59$. The initial scalar field is set by the relation:

$$\phi(x, y, t = 0) = \frac{1}{2}\Big[1 + \tanh\Big(\frac{y_{int}(x) - y}{2\delta}\Big)\Big], \tag{35}$$

with $\delta = 0.002$. The density obeys the same equation of state used in previous problems:

$$\rho(x, y, t) = \Big(\frac{\phi(x, y, t)}{\rho_0} + \frac{1 - \phi(x, y, t)}{\rho_1}\Big)^{-1}. \tag{36}$$

The two fluids have identical kinematic viscosity $\nu = \mu/\rho = 0.001$ and kinematic diffusivity $\alpha_\phi = 0.0005$. The gravity body force is added to the momentum equation with the gravity

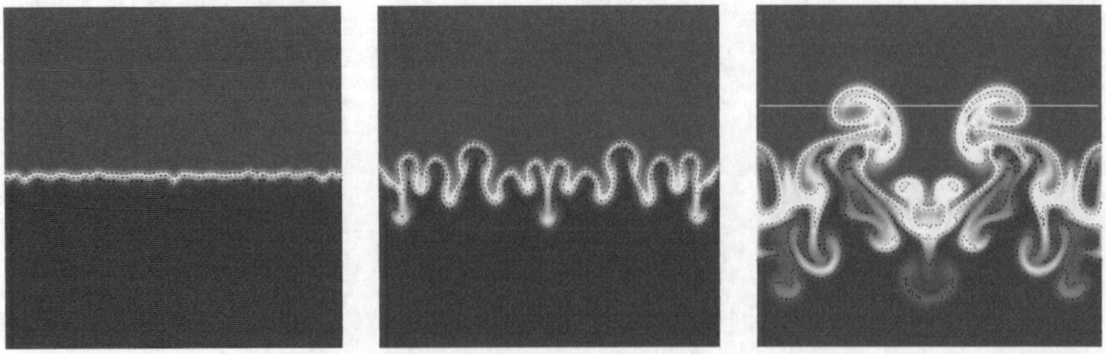

Figure 7. Contour plot of ρ at $t = 0.1$ (left), $t = 0.4$ (middle) and $t = 0.75$ (right) computed on a 128×128 grid. The dashed lines correspond to $\rho = 0.2$, 0.5 and 0.8 levels. Red color corresponds to the heavy fluid ($\rho = 1$) and blue color to the light one ($\rho = 0.1$).

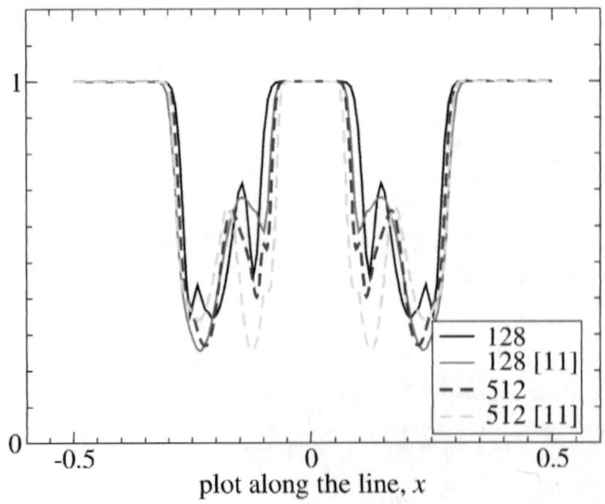

Figure 8. The plot of ρ at $t = 0.75$ and $y = 0.2$ (the line is shown in Fig. 7, right) for different mesh resolutions. The data from the literature [11] is also shown.

acceleration constant set to $g = 9$ so that the Reynolds number is $Re = \sqrt{gL_y}L_x/\nu = 3000$. Simulations have been performed on a uniform hexahedral mesh with $n_x \times n_y = 128^2$ and 512^2. The time step size is $\Delta t = 0.001$ for the coarse mesh and $\Delta t = 0.00025$ for the fine mesh. The comparison of our results with the data of Desjardins et al [11] shows good agreement although not excellent. In our case coarse and fine simulations robustly capture the same peaks in the profile of $\rho(x, y = 0.2, t = 0.75)$ while in case of Desjardins et al the results are more different with the mesh refinement. We imposed no-slip conditions on top and bottom wall and periodic conditions in x-direction for the side walls while Desjardins et al did not mention the boundary conditions. This issue could be a possible source of deviations.

4. Turbulent test case

In this section we introduce a recent approach to the Large-eddy simulation of premixed turbulent combustion called the ADEF model (Approximate Deconvolution and Explicit flame Filtering) for subgrid-scale modeling of scalar fields. First, let us define the spatial filtering procedure denoted by an overbar symbol of some test function ϕ:

$$\overline{\phi}(x,t) = \int_{-\infty}^{+\infty} \phi(x',t)G(x-x')dx', \tag{37}$$

where G is the filter kernel. The density weighted filtering (similar to Favre time-averaging) is introduced as follows

$$\overline{\rho}\widetilde{\phi}(x,t) = \overline{\rho\phi}(x,t) = \int_{-\infty}^{+\infty} \rho(x',t)\phi(x',t)G(x-x')dx', \tag{38}$$

Applying the filtering operation to the system of Eqs. (1)-(4) we obtain

$$\frac{\partial\overline{\rho}}{\partial t} + \frac{\partial\overline{\rho}\widetilde{u}_j}{\partial x_j} = 0, \tag{39}$$

$$\frac{\partial\overline{\rho}\widetilde{u}_i}{\partial t} + \frac{\partial\overline{\rho}\widetilde{u}_i\widetilde{u}_j}{\partial x_j} = -\frac{\partial\overline{p}}{\partial x_i} + \frac{\partial}{\partial x_j}\Big(2\mu(\widetilde{\phi})\widetilde{S}_{ij}\Big) - \frac{\partial}{\partial x_j}\Big(\tau_{ij}^{c,u} - \tau_{ij}^{d,u}\Big), \tag{40}$$

$$\frac{\partial\overline{\rho}\widetilde{\phi}}{\partial t} + \frac{\partial\overline{\rho}\widetilde{\phi}\widetilde{u}_j}{\partial x_j} = \overline{\dot{\omega}(\phi)} + \frac{\partial}{\partial x_j}\Big(\overline{\rho}\alpha(\widetilde{\phi})\frac{\partial\widetilde{\phi}}{\partial x_j}\Big) - \frac{\partial}{\partial x_j}\Big(\tau_j^{c,\phi} - \tau_j^{d,\phi}\Big), \tag{41}$$

$$\overline{\rho} = f(\widetilde{\phi}), \tag{42}$$

where

$$\tau_{ij}^{c,u} = \overline{\rho u_i u_j} - \overline{\rho}\widetilde{u}_i\widetilde{u}_j, \qquad \tau_{ij}^{d,u} = \overline{2\mu(\phi)S_{ij}} - 2\mu(\widetilde{\phi})\widetilde{S}_{ij}, \tag{43}$$

$$\tau_j^{c,\phi} = \overline{\rho\phi u_j} - \overline{\rho}\widetilde{\phi}\widetilde{u}_j, \qquad \tau_j^{d,\phi} = \overline{\rho\alpha(\phi)\frac{\partial\phi}{\partial x_j}} - \overline{\rho}\alpha(\widetilde{\phi})\frac{\partial\widetilde{\phi}}{\partial x_j}, \tag{44}$$

The subgrid-scale terms, Eq. (43), in the momentum equation are expressed with the dynamic Smagorinsky model:

$$\tau_{ij}^{c,u} - \tau_{ij}^{d,u} = -2\mu_t\widetilde{S}_{ij}, \tag{45}$$

where μ_t is the time and space-dependent parameter which is calculated using a standard dynamic routine [16]. The unclosed $\tau_j^{c,\phi}$ term in Eq. (41) can also be modeled using the Boussinesq hypothesis:

$$\tau_j^{c,\phi} = -\overline{\rho}\alpha_t\frac{\partial\widetilde{\phi}}{\partial x_j}, \tag{46}$$

At the same time according to the thickened flame approach [17]:

$$\overline{\rho\alpha(\phi)\frac{\partial\phi}{\partial x_j}} = \mathcal{F}\overline{\rho}\alpha(\widetilde{\phi})\frac{\partial\widetilde{\phi}}{\partial x_j}, \tag{47}$$

where

$$\mathcal{F} = \frac{\dot{\omega}(\widetilde{\phi})}{\overline{\dot{\omega}(\phi)}} \tag{48}$$

In the end, the equation for the filtered scalar field is the following:

$$\frac{\partial \overline{\rho}\widetilde{\phi}}{\partial t} + \frac{\partial \overline{\rho}\widetilde{\phi}\widetilde{u}_j}{\partial x_j} = \overline{\dot{\omega}(\phi)} + \frac{\partial}{\partial x_j}\Big[\overline{\rho}\Big(\mathcal{F}\alpha + \alpha_t\Big)\frac{\partial \widetilde{\phi}}{\partial x_j}\Big]. \tag{49}$$

Momentum and scalar equations contain unclosed terms which can be expressed if the inverse filtering operation is defined [12], for example, the chemical source term is

$$\overline{\dot{\omega}(\phi)} = \overline{\dot{\omega}(\widetilde{\mathcal{L}}^{-1}[\widetilde{\phi}])}, \tag{50}$$

where $\mathcal{L}^{-1}$ denotes the inverse of the filtering operation. For a Gaussian filter the inverse operator can be approximately described as (see [18])

$$\phi(x,t) = \mathcal{L}^{-1}[\overline{\phi}(x,t)] = \overline{\phi}(x,t) - \frac{\Delta^2}{24}\frac{\partial^2 \overline{\phi}}{\partial x_k^2}, \tag{51}$$

where Δ is the local filter width. The above described model further will be applied to the laboratory stratified burner [13] for the validation purposes.

5. Conclusions

In this work we described the numerical algorithm and assessed the accuracy of the low Mach variable-density Navier-Stokes solver on a number of analytic manufactured solutions and Rayleigh-Taylor instability problem. It was shown that the code reproduces second order of accuracy expected from the program. After the verification step we will validate the recent Approximate Deconvolution and Explicit flame Filtering combustion model on experimental data from a number of laboratory burners.

Acknowledgments

This work is funded by RFBR grants No. 15-08-06166 and No. 14-08-01233a. The computational resources are provided by Siberian Supercomputer Center SB RAS (Novosibirsk) and Joint Supercomputer Center RAS (Moscow).

References

[1] Oberkampf WL, Roy CJ 2010 *Verification and Validation in Scientific Computing* (Cambridge: Cambridge University Press)

[2] Shunn L, Ham F 2007 Method of manufactured solutions applied to variable-density flow solvers *Annu. Res. Briefs, CTR* 155-168

[3] Knikker R 2011 A comparative study of high-order variable-property segregated algorithms for unsteady low Mach number flows *Int. J. Numer. Meth. Fluids* **66** 403–427

[4] Vedovoto JM, da Silveira Neto A, Mura A, da Silva LF 2011 Application of the method of manufactured solutions to the verification of a pressure-based finite-volume numerical scheme *Comp. Fluids* **51** (1) 85–99

[5] Ferrer PJ, Buttay R, Lehnasch G, Mura A 2014 A detailed verification procedure for compressible reactive multicomponent Navier-Stokes solvers *Comp. Fluids* **89** 88–110

[6] Waltz J, Canfield TR, Morgan NR, Risinger LD, Wohlbier JG 2014 Manufactured solutions for the three-dimensional Euler equations with relevance to Inertial Confinement Fusion *J. Comp Phys.* **267** 196–209

[7] Tyliszczak A 2014 Projection method for high-order compact schemes for low Mach number flows in enclosures *Int. J. Numer. Meth. Fluids* **24** (5) 1141–1174

[8] Tyliszczak A 2016 High-order compact difference algorithm on half-staggered meshes for low Mach number flows *Comp. Fluids* **127** 131–145

[9] Choudhary A, Roy CJ, Dietiker JF, Shahnam M, Garg R, Musser J 2016 Code verification for multiphase flows using the method of manufactured solutions *Comp. Fluids* **80** 150–163

[10] Shunn L, Ham F, Moin P 2012 Verification of variable-density flow solvers using manufactured solutions *J. Comp. Phys.* **231** (9) 3801–3827

[11] Desjardins O, Blanquart G, Balarac G, Pitsch H 2008 High order conservative finite difference scheme for variable density low Mach number turbulent flows *J. Comp. Phys.* **227** (15) 7125–7159
[12] Domingo P, Vervisch L 2015 Large Eddy Simulation of premixed turbulent combustion using approximate deconvolution and explicit flame filtering *Proc. Comb. Inst.* **35** (2) 1349–1357
[13] Zhou R, Balusamy S, Sweeney MS, Barlow RS, Hochgreb S 2013 Flow field measurements of a series of turbulent premixed and stratified methane/air flames *Comb. Flame* **160** (10) 2017–2028
[14] Pierce CD 2001 *Progress-variable approach for large-eddy simulation of turbulent combustion* (PhD, Stanford University)
[15] Ničeno B, Hanjalić K. 2005 Unstructured large-eddy-and conjugate heat transfer simulations of wall-bounded flows in *Modeling and Simulation of Turbulent Heat Transfer (Developments in Heat Transfer)* (UK: WIT)
[16] Moin P, Squires K, Cabot W, Lee S 1991 A dynamic subgridscale model for compressible turbulence and scalar transport *Phys. Fluids A* **3** (11) 2746-57
[17] Colin O, Ducros F, Veynante D, Poinsot T 2000 A thickened flame model for large eddy simulations of turbulent premixed combustion *Phys. Fluids* **12** (7) 1843-63
[18] Katopodes FV, Street RL, Ferziger JH 2000 A theory for the subfilter-scale model in large-eddy simulation *Tech. Report, Stanford* K1

High-frequency rotary oscillations control of flow around cylinder at Re = 1.4×10^5

E Palkin[1,2], M Hadžiabdić[3], K Hanjalić[2,4]

[1]Institute of Thermophysics SB RAS, Novosibirsk, Russia
[2]Novosibirsk State University, Novosibirsk, Russia
[3]International University of Sarajevo, Sarajevo, Bosnia and Hercegovina
[4]Delft University of Technology, Delft, The Netherlands

E-mail: palkinev89@gmail.com

Abstract.
We perform a series of URANS simulations of the flow over a rotary oscillating cylinder at $Re = 1.4 \times 10^5$ to study the possible reduction of the drag and lift forces acting on the body when the rigid wall is rotary oscillating around the axis of symmetry. Two parameters of the external forcing are varied, i.e. the amplitude of rotation and the frequency. We find that the high-frequency and relatively high-amplitude forcing leads to the drag reduction of 78% compared to the non-rotating case. The oscillations (rms) of drag and lift coefficients are also significantly reduced.

1. Introduction

Most of the flow regimes over bluff bodies feature the unsteady quasi-periodic vortex shedding known as the Karman vortex street. These oscillations affect the drag and lift forces acting on the body. The control of the amplitude and the possible reduction of these forces, in general, is of utmost importance in a wide range of practical applications. In this work we investigate the effect of rotary oscillations of a cylinder of diameter D in a uniform fluid flow with velocity U_∞. The relatively high Reynolds number of interest $Re = U_\infty D/\nu = 1.4 \times 10^5$, where ν is the kinematic viscosity, corresponds to a sub-critical regime with the laminar boundary layer separation but is close to the drag crisis phenomenon ($Re \approx 2.0 \times 10^5$) where the drag coefficient reduces naturally.

Tokumaru and Dimotakis [1] studied the same flow at moderate $Re = 1.5 \times 10^4$ and showed that the drag coefficient can be reduced by 80% if high-frequency rotary oscillating motion of the form

$$U_\theta \Big|_{wall} = \frac{\Omega D}{2} \sin(2\pi f t)$$

is applied to the cylinder wall, where U_θ is the tangential velocity, Ω the oscillation amplitude, f the frequency and t the time. The physical reason behind the significant drag reduction is the transformation of the wake velocity profile which reorganizes and becomes more narrow (see Fig. 1). Shiels and Leonard [2] qualitatively confirmed the results of Tokumaru and Dimotakis with the two-dimensional simulations in the range of $150 \leq Re \leq 1.5 \times 10^4$ stressing that the effect of the drag reduction enhances with the Reynolds number increase. Relatively coarse three-dimensional large-eddy simulations (LES) performed by Du and Dalton [3] at $Re = 1.5 \quad 10^4$

also confirmed the results of Tokumaru and Dimotakis although with the drag reduction of 57% instead of optimistic 80% obtained in experiments due to inaccurate method for drag coefficient calculation.

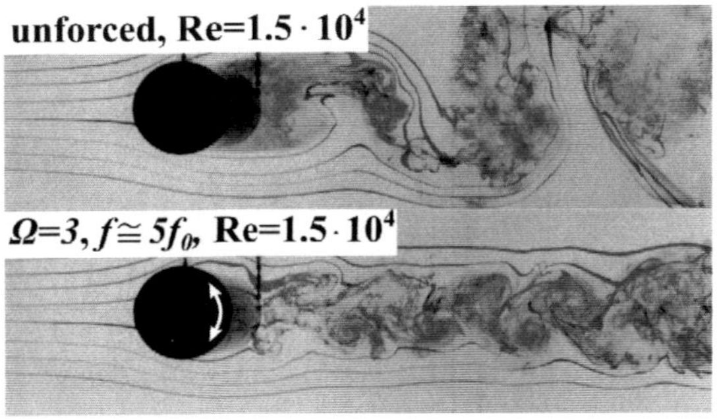

Figure 1. Visualization of the flow with and without the rotary oscillations from experiments [1]. The frequency f_0 corresponds to the natural (unforced) vortex shedding process.

In the present work we use a verified Unsteady Reynolds-averaged Navier-Stokes method with second moment closure Reynolds Stress Model of Jakirlić and Hanjalić [4] (URANS RSM JH) to investigate the effect of rotary oscillations on the drag and lift forces in a broad range of parameters. This model has been previously verified on LES of the same non-rotating case [5] and experimental data [6]. The relatively high Reynolds number $Re = 1.4 \times 10^5$ is of practical importance in contrast with most previous low Re studies. Our precursor work [7] in the low-frequency range ($f = f_0$, where f_0 corresponds to the natural/unforced vortex shedding process) did not reveal drag reduction in agreement with the work of previous investigators. Here we focus on higher-frequency oscillations with $f = 2.5f_0$ which are found optimal by Du and Dalton for $Re = 1.5 \times 10^4$.

2. Computational details

The simulations were performed using the TU Delft in-house unstructured finite-volume computational code T-FlowS, with the cell-centred collocated grid structure. The convective term in the transport equations was discretized by a second-order upwind-biased TVD differencing scheme. The diffusion term was discretized by the second-order central differences. The time-marching was performed using a fully-implicit three-level time scheme. The pressure-velocity coupling was handled by the SIMPLE algorithm. The mesh comprised 2.3×10^6 hexahedral cells. The computational domain was a box with $25D$ length in streamwise and $20D$ in crosswise directions while the spanwise length was $2D$ (see Fig. 2). The center of the cylinder was placed at $10D$ from the inflow boundary with uniform velocity U_∞ and zero free-stream turbulence. The slip conditions were set at the top and bottom boundaries, the convective outflow condition at the outlet and the no-slip condition on cylinder wall. Periodic conditions were imposed in the cylinder axis direction.

3. Results

The central parameter of the problem is the drag and lift coefficients:

$$C_D = 2F_x/(\rho U_\infty^2 S), \qquad C_L = 2F_y/(\rho U_\infty^2 S),$$

where ρ is the density of the fluid and S is the area of the cylinder wall. The force components F_x and F_y acting on the body include the pressure and viscous contributions. We compare the time-averaged value $\overline{C}_D = 1.27$[1] previously obtained for the non-rotating case using URANS

[1] LES [6] provided $\overline{C}_D = 1.24$, the same as the experimental data of Cantwell and Coles [5]

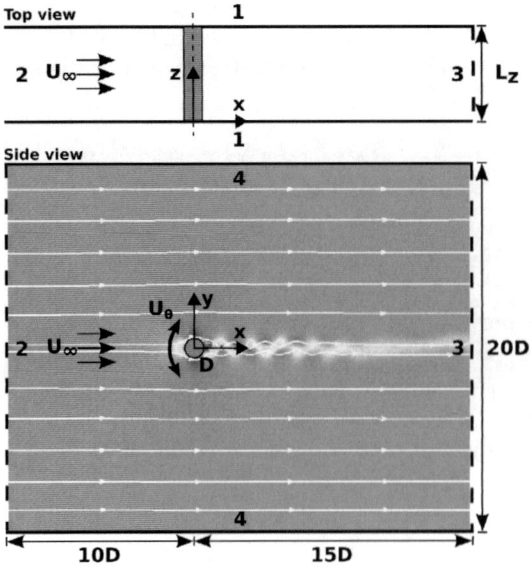

Figure 2. Computational domain. The numbers from 1 to 4 correspond to periodic, inflow, convective outflow and slip conditions.

RSM JH to low-frequency and high-frequency rotary forced regimes. The first one was forced at $f = f_0$ with $f_0 \approx 0.22$, investigated previously [7], did not show significant improvements of the drag characteristics. On the other hand, the second regime with $f = 2.5f_0$ showed substantial decrease of $\overline{C}_D$. All cases were forced with the non-dimensional amplitude $\Omega = 1$, 2 and 3 (value of $\Omega D/U_\infty$). Table 1 shows the value of $\overline{C}_D$ for various Ω and f values. The optimal among considered control parameters are $\Omega = 2$ and $f = 2.5f_0$ where the drag coefficient value ($\overline{C}_D = 0.28$) is reduced by 78% compared to the non-rotating case ($\overline{C}_D = 1.27$).

Table 1. Time-averaged drag coefficient $\overline{C}_D$ at $Re = 1.4 \times 10^5$ for various Ω and f.

	$\Omega = 1$	$\Omega = 2$	$\Omega = 3$
$f = 1.0f_0$	1.30	1.21	1.33
$f = 2.5f_0$	0.56	0.28	0.42

A sample of the time-history of the drag and lift coefficients for various Ω and $f = 2.5f_0$ is presented in Fig. 3. Note that at relatively low $\Omega = 1$ the signal of C_L resembles the one for unforced case with only modulations imposed by rotation. However, for higher Ω the signal is clearly $2.5f_0$-periodic. In practice not only the time-averaged value of C_D is important but also the amplitude of oscillations of the forces. Note that the optimal regime also exhibits the lowest rms values of C_D and C_L. Figure 4 shows the streamlines and axial velocity field (left) and vorticity field (right) for various forcing parameters. As previously mentioned the high-frequency forcing reorganizes the vortical structure of the wake reducing the velocity defect leading to the smaller drag coefficient. Future work will be devoted to extending the parameter $\Omega - f$ map.

4. Conclusion

We performed a series of URANS simulations of the flow over a rotary oscillating cylinder at $Re = 1.4 \times 10^5$. Low- and high-frequency oscillations were studied with various amplitudes of

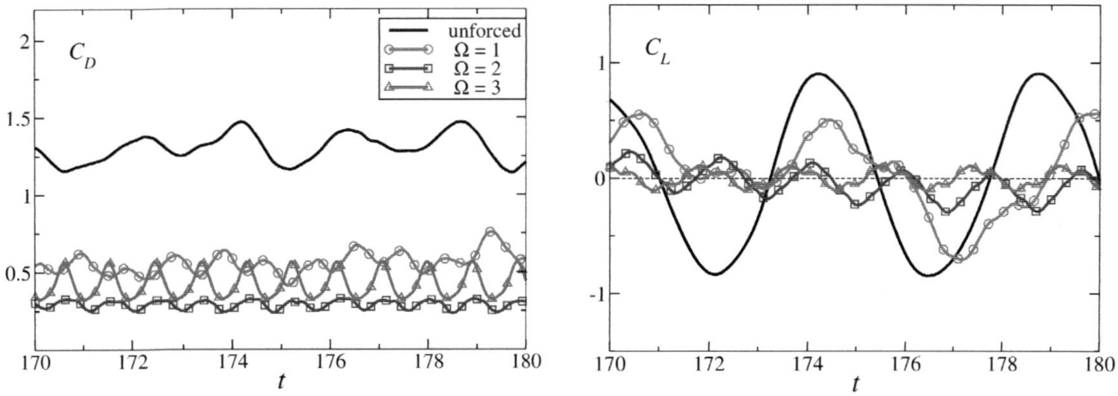

Figure 3. Drag (left) and lift (right) coefficients for simulations with various Ω and f.

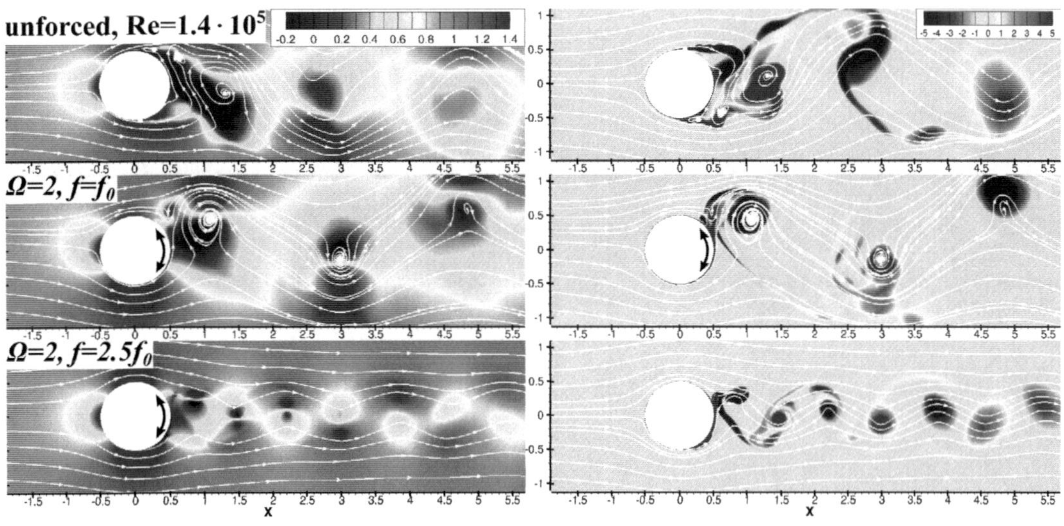

Figure 4. Axial velocity (left) and z-vorticity (right) for simulations with various Ω and f.

rotation in order to determine the optimal parameters when the drag and lift coefficients are significantly reduced. It was found that the high-frequency oscillations can lead to at least 78% $\overline{C}_D$ reduction in comparison with the non-rotating case. Future work will focus on a wide range $\Omega - f$ parameteric study for a complete optimization map.

Acknowledgments
This work is funded by Russian Science Foundation under the grant No. 14-29-00203. The computational resources are provided by Siberian Supercomputer Center SB RAS (Novosibirsk) and Joint Supercomputer Center RAS (Moscow).

References
[1] Tokumaru PT, Dimotakis PE 1991 Rotary oscillation control of a cylinder wake *J. Fluid Mech.* **224** 77–90
[2] Shiels D, Leonard A 2001 Investigation of a drag reduction on a circular cylinder in rotary oscillation *J. Fluid Mech.* **431** 297–322

[3] Du L, Dalton C 2013 LES calculation for uniform flow past a rotationally oscillating cylinder *J. Fluids Struct.* **42** 40–54
[4] Jakirlić S, Hanjalić K 2002 A new approach to modelling near-wall turbulence energy and stress dissipation *J. Fluid Mech.* **459** 139–166
[5] Cantwell B, Coles D 1983 An experimental study of entrainment and transport in the turbulent near wake of a circular cylinder *J. Fluid Mech.* **136** 321–374
[6] Palkin E, Mullyadzhanov R, Hadžiabdić M, Hanjalić K 2016 Scrutinizing URANS models in shedding flows: the case of the cylinder in cross flow *Flow Turb. Comb.*
[7] Palkin E, Mullyadzhanov R, Hadžiabdić M, Hanjalić K 2016 Rotary oscillations control of flow around cylinder at $Re = 1.4 \times 10^5$ *ICMAR Conf.*

Numerical and experimental study of low-frequency pressure pulsations in hydraulic units with Francis turbine

D Platonov, A Minakov, D Dekterev, A Sentyabov, A Dekterev

Siberian Federal University, Russia 660041 Krasnoyarsk, Svobodny ave., 79
Institute of Thermophysics SB RAS, Russia, 630090 Novosibirsk, Acad. Lavrentyev ave., 1

platonov-08@yandex.ru, tov-andrey@yandex.ru

Abstract. The paper presents the numerical simulation method of three-dimensional turbulent flows in the hydraulic turbine. This technique was verified by means of experimental data obtained on a water model of the Francis turbines. An aerodynamic stand, which is a miniature copy of the real hydraulic turbine, was designed. A series of experiments have been carried out on this stand and the corresponding calculations were performed. The dependence of the velocity and pressure pulsations profiles for different operation regimes are presented.

1. Introduction

The development of hydraulic turbo machinery have long reached a mature stage. At design conditions, the turbines usually perform with high efficiency and reliability. However, at part loads and in transient regimes the stability of the system can seriously be impaired leading to a decrease in efficiency, mechanical damage and system failure. The intrinsic unsteadiness at suboptimal conditions leads often to vortex breakdown and precessing helical vortices in form of a single or twin rope behind the runner, which cause intense flow pulsations and vibrations of the turbine structure that pose a serious risk of the system reliability and safety. These problems have long been in focus of the experimental researches and simplified analytics, but the full three-dimensional time-dependent dynamics has remained beyond the most advanced diagnostics techniques. Computer simulations have been seen as a potential tool for capturing the details of the instabilities and the vortical and turbulence structures. Such information can be obtained by direct and large-eddy simulations (DNS, LES) over a range of important scales. However, the accuracy of DNS and LES depends much on the numerical resolution. For complex configurations the Reynolds-averaged Navier-Stokes (RANS) approach, being computationally less demanding, has been considered as a more rational option for industrial purposes. However, because if its empirical contents, the RANS have not been regarded as a trustful research tool, and moreover, their credibility depends much on the choice of the turbulence model to close the time- or ensemble-averaged conservation equations.

Despite the extensive literature (e.g. Doerfler [1], Alligne et al. [2]), comprehensive experimental data for hydro turbines are still insufficient, apart from mean velocity and second- moment turbulence statistics measured at selected locations in laboratory models of some popular turbine types [3, 4]. On the other hand, there are still many open questions regarding the practically relevant numerical simulation of the flow in turbines of realistic configuration. Because of typically very high Re numbers, DNS and even LES, requiring enormous numerical mesh and computer resources, are usually beyond the capabilities of the designing. The URANS or hybrid RANS/LES methods remains the available approaches of the simulation.

Here the first challenge is related to selecting the appropriate RANS model that can reproduce satisfactorily the unsteady flow and the dynamics of at least the dominant large-scale vortex structures subjected to strong swirls, curvature and adverse pressure gradients in complex geometric shapes. The popular linear eddy-viscosity modes (k-ε, k-ω, and their many "improved" variants) lack of the sufficient physics and have shown to perform poorly in describing such flows.

2. Mathematical model and numerical method

Describing the flows in the flow path of water turbines one have to face several problems. The first challenge is related to modeling of turbulence in channels of complex geometric shapes and with strong swirl flow. At the same time for engineering calculations turbulence models, not only quite accurate describing of the average fields, but also large-scale flow pulsations is required. Widely spread in engineering calculations k-ε and k-ω turbulence models poorly describe such flows. To improve the modeling of turbulent swirling flows scientists are trying to either modify existing URANS turbulence models or use techniques, which resolve large-scale turbulent eddies (LES, DES).

The precession of the vortex core can accompany the swirling flow in hydraulic turbines. For modeling such phenomenon it is necessary to use non stationary simulation methods such as the method of large-eddy simulation. However, it requires a very detailed grid, especially near the walls. At the same time, RANS models are quite economical and describe the boundary layers well. To combine the advantages of these approaches the method of detached eddy simulation (DES) was proposed in [5].

In the simulation of hydraulic turbines it is necessary to consider the rotation of the runner. There are many approaches for modeling flows with rotating bodies, they include: dynamic grid, sliding mesh, moving mesh and method based on a moving reference frame. The most common and simple way to model the runner rotation is to use a rotating reference frame. The rotating coordinate system allows simulating flow in an approximation in which the runner is fixed and swirl flows pass through it. In this paper, modeling of the runner rotation was performed with rotated reference frame for runner zone. Numerous test calculations show the correctness of this approach, both in the description of integral characteristics of flow and pulsation [4, 6-7].

Below, the basic equations of the mathematical models in the rotating coordinate system are presented. The continuity equation (conservation of mass):

$$\nabla \mathbf{v} = 0 .$$

Momentum equation (conservation of momentum) in a rotating frame for relative velocities:

$$\frac{\partial \rho \mathbf{v}}{\partial t} + \nabla(\rho \mathbf{v}\mathbf{v}) = -\nabla p + \nabla(\boldsymbol{\tau}''' + \boldsymbol{\tau}') - \rho(2\boldsymbol{\Omega}\times\mathbf{v} + \boldsymbol{\Omega}\times(\boldsymbol{\Omega}\times\mathbf{r})),$$

where: $\mathbf{v}$ - velocity vector, $\boldsymbol{\tau}$ - stress tensor, Ω - angular velocity of runner rotation, p - static pressure, ρ - density. On the right side of the equation of conservation of momentum Coriolis' force and the centrifugal force are written.

Two turbulence modeling approach were used. The first one is second order closure, i.e. Reynolds stress model (RSM) and the second one is detached eddy simulation (DES).

One should say a few words about the numerical approach. Discretization of the transport equation was carried out by the control volume method on unstructured grids. Velocity and pressure fields for incompressible fluid were obtained by means of the SIMPLE procedure. For the approximation of the convective terms in the equation for momentum components, second order scheme was used. For the approximation of convective terms in the equations for turbulent characteristics, up-wind scheme was used.

3. Experiment

An experimental study was performed on an open aerodynamic stand. The configuration considered is a 4:1 scaled-down laboratory model of a Francis turbine (RO-230) of an LMZ (Leningrad Metal

Industry). Air supply in the draft tube model was carried out using a high performance air blower with the maximum flow rate Q_{max} =170 m^3/h and the excess pressure of Δp=0.4 bar. An ultrasonic flow meter «IRVIS-Ultra» and a frequency converter of the blower with feedback controlled air flow rate were used. For capturing the LDA signal the air flow was seeded with tracers produced by a standard-type generator of the paraffin oil aerosol with a mean particle diameter of 1-3 μm. Figure 1 provides a photo of the experimental turbine model. The draft tube model, runner, spiral case and guide vanes geometry were provided by the company LMZ and were manufactured using the technology of 3D printing.

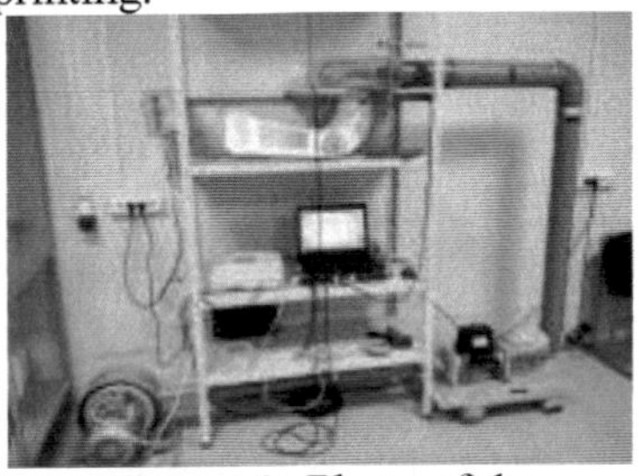

Figure 1: Photo of the experimental stand

Servo drive SPSh10-3410 ensured precise rotation frequency of the runner in the range from 0 to 3000 rpm. A computer ensured the control of the experiment. Using the original software, it was possible to maintain the flow regime with an accuracy of 1.5% and 0.5% of the flow rate and the runner rotation frequency respectively.

The experiments included measurements of velocity fields in the conical part of the draft tube using a system of two-component LDA "LAD-06i" [8], and measurements of the pressure pulsations with special acoustic sensors. The acoustic sensors were based on high precision microphones attached with tiny tips for capturing pressure pulsations at the local points [9]. Two of these sensors were used to conduct point-to-point measurements [9]. The perturbation mode that corresponds to the precessing vortex core was identified by analysing the signal from the sensors located in diametrically opposite angular positions. Note that the possibility of using well-proven acoustic technology is another advantage of air model stand.

Pressure fluctuations were recorded using a microphone Bruel & Kjaer Type 2250. The signals from the microphones were digitized by an analog-to-digital converter L-card E-440. The experiments were conducted at the runner rotation frequency from f= 0 to 2000 rpm with a step 500 rpm and a flow rate, varing in the range from Q =100 to 150 m^3/h with a step of 50 m^3/h. At each point, the signal detected by the acoustic sensors was digitized for 5 seconds with a sampling rate of 2 kHz.

The LDA measurements of the axial and tangential fluid velocity components were performed at a cross-section in the draft tube cone.

4. Francis-99 result

The section presents the results of simulation of the flow in the turbine Francis-99. Three operation point proposed for the Francis-99 workshop were calculated [4]. Geometry data provided by the workshop committee were used. Computational domain included the runner and the draft tube. The mesh consisted of 6 mln. control volumes. Constant radial and tangential velocity components were set at the runner inlet. Angle between velocity direction and the runner inlet corresponded to guide vane angle. Outflow boundary conditions were set at the draft tube outlet.

Three operation points were calculated with of Reynolds stress model. Part load and high load operation points were also calculated with DES method. Figure 2 shows the velocity magnitude in central cross-section for two operation points. At the best efficiency point, narrow concentrated vortex is formed under the runner tip. On the other hand, at part load operation point, axial flow is close to the draft tube wall. At this operation point, wide recirculation zone is formed at the axis in the draft tube cone.

Tables 1 present parameters of the turbine at the operation points (net head in calculations was defined between runner inlet and the draft tube outlet). Computational results were obtained with steady Reynolds stress model. The results of calculations agree with experimental results qualitatively. Part load operation point shows the largest discrepancy of the efficiency.

Fig. 3 – 4 show velocity components along the top line in the draft tube for two operation points. The results of the calculations are in close agreement with experimental data at two operation points both for Reynolds stress model and DES method. Part load operation point shows the smallest discrepancy.

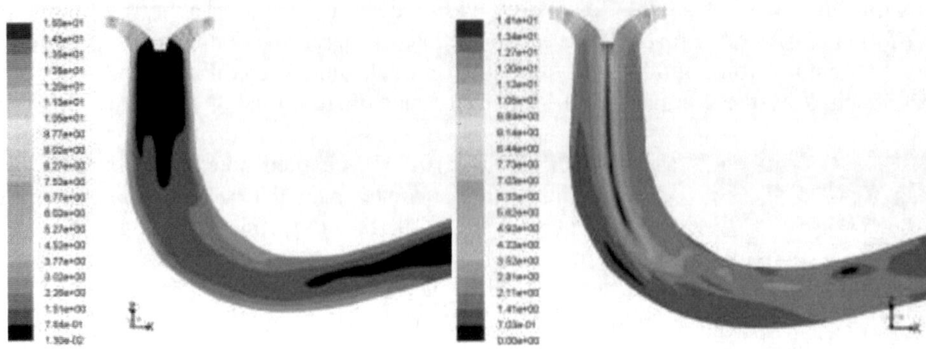

Figure 2: Velocity magnitude in central plane; part load and the best efficiency points

Table 1. Part load and the best efficiency point

Parameter	Part load		Best efficiency point	
	RSM	Experiment	RSM	Experiment
Net head [m]	11.30	12.29	11.61	11.91
Pressure torque [N m]	165.7	137.52	650.8	619.56
Friction torque [N m]	-3.5	6.54	-6.9	8.85
Total torque [N m]	162.2	–	643.9	–
Efficiency [%]	80.60	71.69	95.34	92.61

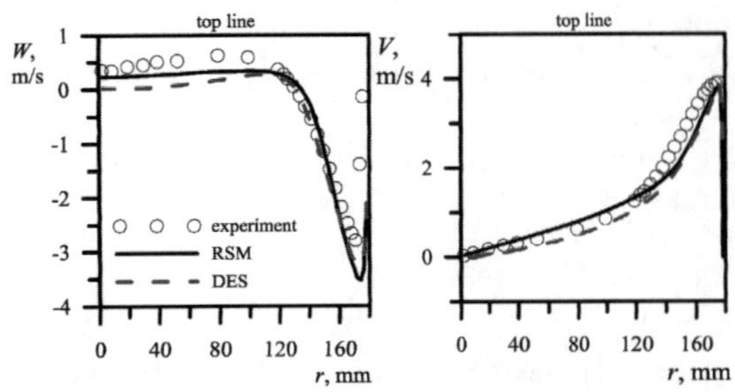

Figure 3: Velocity components in the draft tube, part load point

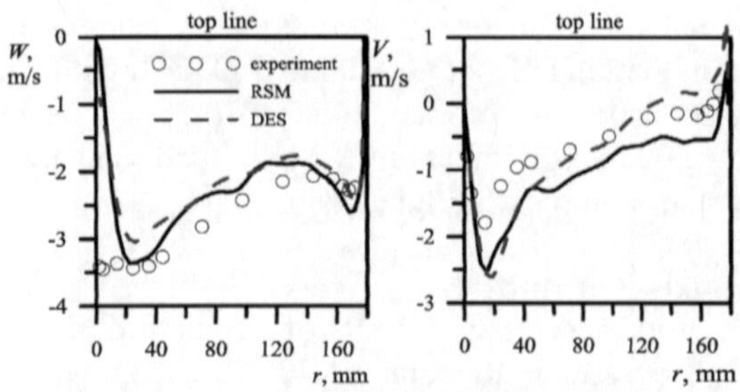

Figure 4: Velocity components in the draft tube, high load point

5. Aerodynamic stand

5.1 Experimental results

The structure of the flow downstream the runner was visualized and analyzed using the laser knife and flow tracers. The forming of the precessing vortex core in the draft tube diffuser was observed. Further instrumental and numerical study of the flow confirmed the presence of the PVC.

Measuring of the flow velocity downstream the runner was carried out using a LDA (Laser-Doppler anemometer) LAD-06I. Figure 6 shows velocity profiles for different flow rates (Q = 100, 150, 200 m^3/h). In addition, velocity profiles were measured for modes with different rotation speed of the runner including stationary rotor mode. As can be seen from the figure the stream in the draft tube is near the wall. A powerful recirculation zone is formed in the center and causes the PVC forming.

Using the sound-level meter (Bruel & Kjaer Type 2250) and piezo-electric sensors an amplitude and frequency of the pressure pulsations were measured. The analog-to-digital converter processed signals of sound-level meter and piezo-electric sensors. For example, spectra of the pulsations for mode with flow rate $Q = 200$ m^3/h and rotation speed $\omega = 2000$ rpm shown at figure 7.

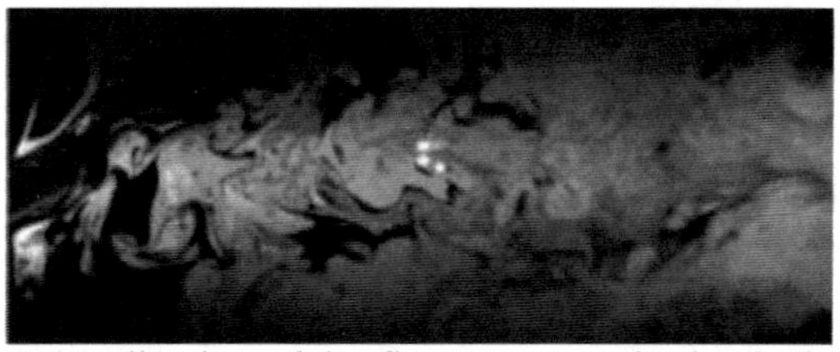

Figure 5: Visualization of the flow structure in the draft tube cone

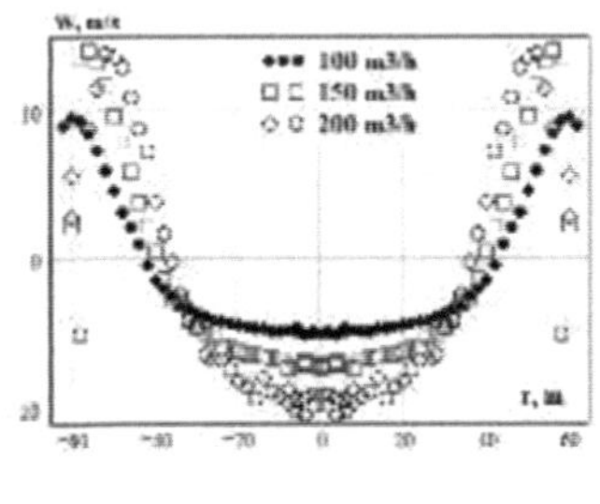

Figure 6: Profiles of the axial velocity in the draft tube diffuser, $\omega = 1000$ rpm

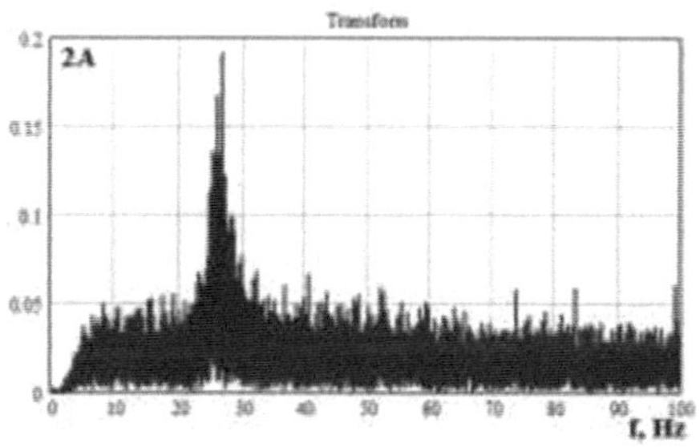

Figure 7: Spectra of the pressure pulsations

It is evident that increasing the flow rate leads to increasing the pulsations frequency. At the same time with increasing the runner speed the pulsations frequency reducing. It is quite logical because with increasing runner speed the flow swirling is reduced due to the fact that runner takes the whole flow swirling created by the stator and guide vanes.

5.2 Numerical research

The section presents the results of the simulation of the flow in the aerodynamic turbine stand. The calculations used unstructured grids of varying refinement. Figure 8 shows a comparison of the calculated and experimental axial velocity components. There is a satisfactory agreement. As has already been noted in the experiment there is a wide area of reverse flow, which indicates the presence of the precessing vortex core. This assumption is confirmed by the calculated data. Figure 9 shows the instantaneous pressure field in the cone of the draft tube. As seen in the diffuser of the draft tube is formed sufficiently strong PVC.

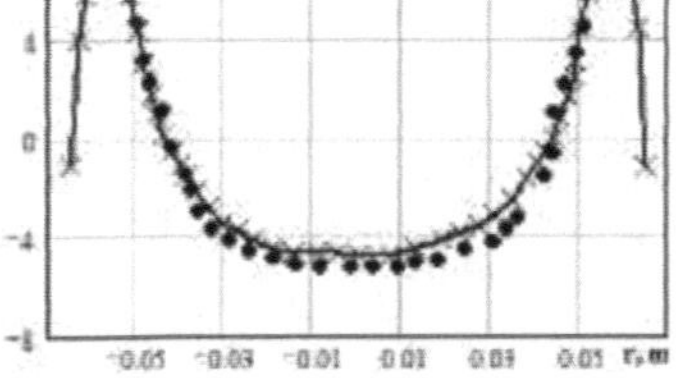

Figure 8: Profiles of the axial velocity in the draft tube diffuser

Figure 9: Vortex flow structure in the draft tube of the aerodynamic turbine (DES, RSM)

6. Conclusion

The paper proposed a numerical method for the simulation of flow in hydraulic turbines that was verified by means of experimental data. A unique aerodynamic stand, which is a miniature copy of the actual hydroelectric power, was designed. At the booth, a series of experiments simulating different modes of hydroelectric works. For this stand the velocity components and the pressure pulsation in the draft tube diffuser were measured. Using the numerical simulation technique the flow in the aerodynamic stand was calculated. The comparison of the calculated and experimental data showed close agreement that suggests the using possibility this numerical methods for the engineering applications.

We thank Russian Science Foundation for financial support (project No. 16-19-00138).

References

[1] Doerfler P. System dynamics of the Francis turbine half load surge. 1982, Proc. *IAHR Symp. on Hydraulic Machinery and Systems*, Amsterdam, Netherlands.Another reference

[2] Alligne S., Maruzewski P., Dinh T., Wang B., Fedorov A., Iosfin J. and Avellan F., 2010, Prediction of a Francis turbine prototype full load instability from investigations on the reduced scale model. Proc. *IAHR Symp. on Hydraulic Machinery and Systems*, Timisoara, Romania.

[3] Cervantes M.J., Engstrom T.F., Gustavsson L.H., 2005, Proc. *The third IAHR/ ERCOFTAC workshop on draft tube flows Turbine 99*. Sweden, Porjus,. 193 p.

[4] Minakov, A.V. Sentyabov, D.V. Platonov, A.A. Dekterev, A.A. Gavrilov, 2014, Numerical modeling of flow in the Francis-99 turbine with Reynolds stress model and detached eddy simulation method. *Journal of Physics: Conference Series* Volume 579, doi:10.1088/1742-6596/579/1/012004.

[5] Spalart P.R., Jou W.-H., Strelets M., Allmaras S.R. Comments on the feasibility of LES for wings and on a hybrid, RANS/LES approach / In Lue, C. and Lue, Z., (eds)// Advances in DNS/LES, Proceedings of 1st AFOSR International Conference on DNS/LES, Ruston, LA, August, 4-8, Greyden Press, Columbus, OH. – 1997. – pp. 137-147.

[6] Gavrilov A., Dekterev A., Sentyabov A., Minakov A., Platonov D. Application of hybrid methods to calculations of vortex precession in swirling flows // *Notes on Numerical Fluid Mechanics. 2012*. T. 117. P. 449-459.

[7] Kuznetsov I., Zakharov, A., Orekhov, G., Minakov, A., Dekterev, A., Platonov, D. Investigation of free discharge through the hydro units of high head Francis turbine. *IOP Conference Series: Earth and Environmental Science* 15 (PART 5). 2012. doi: 10.1088/1755-1315/15/5/052002

[8] Meledin V., Anikin Yu., Bakakin G., Glavniy V., Dvoinishnikov S., Kulikov D., Naumov I., Okulov V., Pavlov V., Rakhmanov V., Sadbakov O., Ilyin S., Mostovskiy N., Pylev I. Laser Doppler Diagnostic of Flow in Draft Tube behind Hydroturbine Runner // Turbomachines: Aeroelasticity, Aeroacoustics and Unsteady Aerodynamics, Edited by V.A. Skibin, V.E. Saren, N.M. Savin, S.M. Frolov. Moscow: Torus Press Ltd., 2006. –pp. 446-457.

[9] Litvinov I. V., Shtork S.I., Kuibin P.A., Alekseenko S. V., Hanjalic K., 2013, Experimental study and analytical reconstruction of precessing vortex in a tangential swirler, *Int. J. Heat Fluid Flow* 42: 251–264.

Heat transfer in a free liquid layer under action of additional tangential stresses: numerical modeling

E Rezanova

Altai State University, Lenin pr. 61, Barnaul, Russian Federation

Institute of Thermophysics SB RAS, Lavrentyev pr. 1, Novosibirsk, Russian Federation

E-mail: katerezanova@mail.ru

Abstract. The dynamics and heat transfer processes in a planar layer of viscous incompressible heat-conducting liquid with free boundaries under condition of zero gravity are investigated. The influence of thermocapillary forces and additional tangential stresses on the boundaries caused by the environment is taken into account. A special class of solutions of the Navier-Stokes equations is used to model the flows and layer deformations. The numerical results of investigation of the layer dynamics and temperature field in the three-dimensional case are presented.

1. Introduction

Problems related to the study of the infinite liquid layers with free boundaries have been studied for quite a long time. Modeling of such flows requires consideration of additional shear stresses on the free boundary and the study of the interaction of different mechanisms, causing fluid motion [1, 2]. Formulation of the problem in three dimensions for the study of the deformation of a viscous incompressible liquid layer under the influence of thermocapillary forces is carried out in [3, 4]. The first numerical modeling of deformation of the viscous liquid layer with free boundaries is given in [5]. The problem becomes more complicated, if one need to find the temperature of the liquid in the infinite deformable layer [6, 7].

In this paper the three-dimensional nonstationary liquid flow in an infinite plane-parallel layer is investigated under conditions of zero gravity. Thermocapillary forces and additional shear stresses are considered to have an influence on the dynamics of the liquid flow. Deformation of the layer and the heat transfer process are modeled on the basis of the exact solutions of the Navier-Stokes equations [3]. A numerical algorithm to find the position of layer boundaries, the velocity field and temperature distribution and some results of numerical studies are presented.

2. Statement of the problem

Consider the liquid occupying the infinite plane-parallel layer of incompressible viscous thermally conducting fluid $\Omega = \{(x, y, z)\colon -\infty < x < +\infty, -\infty < y < +\infty, -Z(t) < z < Z(t)\}$ in zero gravity. Free surfaces $\pm Z(t)$ remains rigid and parallel at all points in time. The coordinate system is chosen so that the Ox and Oy axis are directed along surfaces $\pm Z(t)$, and Oz axis is perpendicular to them (see figure 1). The unit vectors of external normals and tangential vectors to the free boundaries are defined as follows: $\mathbf{n} = (0, 0, \pm 1)$ and $\mathbf{s}_1 = (1, 0, 0)$, $\mathbf{s}_2 = (0, 1, 0)$.

Journal of Physics: Conference Series **754** (2016) 062008 doi:10.1088/1742-6596/754/6/062008

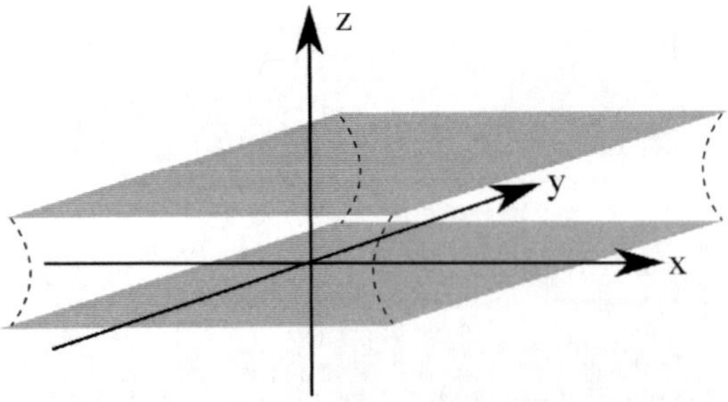

Figure 1. Geometry of flow domain.

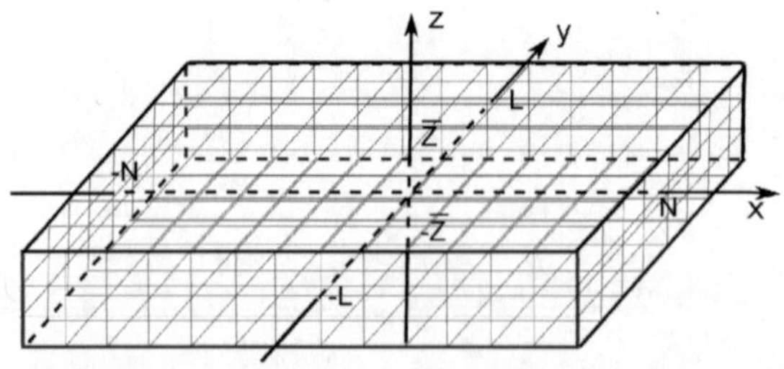

Figure 2. Scheme of the computational domain.

Unknown function (the velocity vector $\mathbf{v} = (u, v, w)$, the pressure p, the temperature T) satisfy the Navier-Stokes and heat transfer equations:

$$u_t + uu_x + vu_y + wu_z = -p_x + \frac{1}{Re}(u_{xx} + u_{yy} + u_{zz}),$$

$$v_t + uv_x + vv_y + wv_z = -p_y + \frac{1}{Re}(v_{xx} + v_{yy} + v_{zz}),$$

$$w_t + uw_x + vw_y + ww_z = -p_z + \frac{1}{Re}(w_{xx} + w_{yy} + w_{zz}),$$

$$u_x + v_y + w_z = 0,$$

$$T_t + uT_x + vT_y + wT_z = \frac{1}{RePr}(T_{xx} + T_{yy} + T_{zz}).$$

Here Re is the Reynolds number ($Re = \frac{v_* l}{\nu}$), Pr is the Prandtl number ($Pr = \frac{\nu}{\chi}$), ν and χ are the kinematic viscosity and thermal diffusivity coefficients, l is the characteristic length of the flow domain, v_* is the characteristic velocity, $t_* = \frac{l}{v_*}$ is the characteristic time.

The the kinematic and dynamic conditions on the free boundaries are determined as follows:

$$\mathbf{v} \cdot \mathbf{n} = \pm w|_{z=\pm Z(t)} = \frac{dZ}{dt}, \tag{1}$$

$$-p + \frac{2}{Re}\mathbf{n} \cdot D(\mathbf{v})\mathbf{n}|_{z=\pm Z(t)} = -P_g, \tag{2}$$

$$\mathbf{2s}_1 \cdot \mathbf{n} \cdot \mathbf{D(v)n}|_{z=\pm Z(t)} = \tau_1(x,t) - \frac{Ma}{RePr}T_x, \tag{3}$$

$$\mathbf{2s}_2 \cdot \mathbf{n} \cdot \mathbf{D(v)n}|_{z=\pm Z(t)} = \tau_2(y,t) - \frac{Ma}{RePr}T_y. \tag{4}$$

Temperature and shear stresses on free boundaries are the given functions of time and longitudinal coordinate:

$$T(x, y, \pm Z(t), t) = \frac{1}{2}A(t)x^2 + \frac{1}{2}B(t)y^2 + \Theta(t), \quad \tau_1(x,t) = x\tilde{\tau}(t), \quad \tau_2(y,t) = y\tilde{\tau}(t). \tag{5}$$

Here $\tau_1(x,t)$ and $\tau_2(y,t)$ are the shear stresses induced by the external environment, $A(t)$, $\Theta(t)$, $\tilde{\tau}(t)$ are the arbitrary functions that depend on time, $D(\mathbf{v})$ is the tensor of deformation rate, $P_g = \overline{P_g} - \dfrac{2}{Re}\overline{\rho}\overline{\nu}\mathbf{n}\cdot D(\mathbf{v}_g)\mathbf{n}|_{z=\pm Z(t)}$, $\overline{P_g}$ is the external pressure, $\overline{\rho}$, $\overline{\nu}$ is the relation of the densities and kinematic viscosity coefficients of gas and liquid, Ma is the Marangoni number ($Ma = \dfrac{\sigma_T T_* l}{\rho\nu\chi}$), T_* is the characteristic temperature, $p_* = \rho v_*^2$ is the characteristic pressure, ρ is the liquid density, v_g is the gas velocity, σ_T is the temperature coefficient of surface tension ($\sigma = \sigma_0 - \sigma_T(T - T_0)$, $\sigma_T > 0$). It is assumed that the action of normal stresses from the external environment may be neglected.

Components of the velocity vector will be determined in the form [3]

$$u(x,z,t) = (f+g)x, \quad v(y,z,t) = (f-g)y, \quad w(z,t) = -2\int_0^z f(\alpha,t)d\alpha,$$

where the functions $f(z,t)$ and $g(z,t)$ are the solutions of the system of equations

$$f_t + f^2 + g^2 - 2f_z\int_0^z f(\alpha,t)d\alpha - \frac{1}{Re}f_{zz} = 0, \tag{6}$$

$$g_t + 2fg - 2g_z\int_0^z f(\alpha,t)d\alpha - \frac{1}{Re}g_{zz} = 0. \tag{7}$$

The position of the free boundary $Z(t)$ satisfies the following integro-differential equation:

$$\frac{dZ}{dt} = -\int_0^{Z(t)} f(z,t)dz. \tag{8}$$

Dynamic conditions (3), (4) for the functions $f(z,t)$, $g(z,t)$ and $Z(t)$ are written in the form

$$f_z(Z(t),t) = \tilde{\tau}(t) - \frac{Ma}{RePr}(A(t)+B(t)), \quad g_z(Z(t),t) = \tilde{\tau}(t) - \frac{Ma}{RePr}(A(t)-B(t)).$$

If we are considering a symmetric flow then the line of symmetry must satisfy the condition $f_z(0,t) = 0$. The initial position of the layer is determined from the following equations:

$$f(z,0) = f_0(z) = 0, \quad g(z,0) = g_0(z) = 0 \ (0 \le z \le Z_0); \quad Z(0) = Z_0 \ (Z_0 > 0).$$

Also the problem has been supplemented by conditions at infinity.

The unknown functions of the problem are determined numerically. Functions $Z(t)$, $f(z,t)$, $g(z,t)$ can be found using the "predictor-corrector" algorithm. The position of the free boundary $Z(t)$ is determined using the following algorithm:

$$Z^{k+1} = Z^{k-1} - 2\Delta t\int_0^{Z^k} f^k(z)dz, \tag{9}$$

$$\bar{Z}^{k+1} = Z^k - \Delta t\left[\int_0^{Z^{k+1}} f^{k+1}(z)dz + \int_0^{Z^k} f^k(z)dz\right], \tag{10}$$

where (9) is the predictor and (10) is the corrector scheme [5, 7]. The new spatial grid is determined on the each temporal layer.

The predictor-corrector algorithm of second-order accuracy is used to calculate functions $f(t,z)$, $g(t,z)$:

$$q^{k+1/4} = q^k + 0.5\Delta t[-\Lambda_1 q^{k+1/4} + \Phi^k],$$

$$q^{k+1/2} = q^{k+1/4} + 0.5\Delta t[-\Lambda_2 q^{k+1/2}],$$

$$q^{k+1} = q^k + \Delta t[-\Lambda q^{k+1/2} + \Phi^k].$$

Here q^k are the the values of the unknown function f or g at the grid points, $\Lambda = \Lambda_1 + \Lambda_2$, Λ_i are the finite-difference operators corresponding to differential operators $(-1/Re)\partial^2/\partial z^2$ and $-F\partial/\partial z$, functions F and Φ^k are determined from (8). The integrals in the equations (6)-(8) are determined by quadrature formulas. The interpolation formulas are used for the transition to a new temporal layer.

3. Numerical algorithm for heat transfer simulation

The values of temperature distribution in the layer can be determined in the following parallelepiped (figure 1):

$$\Omega = \{-N < x < N, -L < y < L, -\overline{Z} < z < \overline{Z}\}. \tag{11}$$

The conditions (1)-(6) will be imposed at $z = \pm\overline{Z}$. "Soft" conditions resulting conditions at infinity will be imposed on the "lateral walls" of a parallelepiped. As such conditions the conditions of the following type can be used:

$$T_{xx} = 0, \qquad T_{yy} = 0.$$

These expressions are the result of conditions $T \to T_\infty$ ($T_\infty = const$) at infinity on the assumption $T_x \to 0$, $T_y \to 0$.

Numerical investigation of heat transfer process is based on the stabilizing corrections method. Finite-difference scheme of second order approximation for the solution of the heat transfer equation can be represented as follows [8]:

$$\frac{T^{k+1/3} - T^k}{\Delta t} = K_1 T^{k+1/3} + K_2 T^k + K_3 T^k + S^k,$$

$$\frac{T^{k+2/3} - T^{k+1/3}}{\Delta t} = K_2(T^{k+2/3} - T^k),$$

$$\frac{T^{k+1} - T^{k+2/3}}{\Delta t} = K_3(T^{k+1} - T^k). \tag{12}$$

Here $T^k(x, y, z) = T(x, y, z, t^k)$, K_i are the finite-difference analogues of the corresponding differential operators ($K_1 \sim \frac{1}{RePr}\partial^2/\partial z^2$, $K_2 \sim \frac{1}{RePr}\partial^2/\partial x^2$, $K_3 \sim \frac{1}{RePr}\partial^2/\partial y^2$), S^k is the finite-difference analog of convective term ($S \sim -u\frac{\partial T}{\partial x} - v\frac{\partial T}{\partial y} - w\frac{\partial T}{\partial z}$), Δt is the time step. The grid (x_n, y_l, z_m) is introduced to implement the presented finite-difference scheme: $x_n = (n-1)h_x (n = 1, ..., \bar{N}+1)$, $h_x = N/\bar{N}$, $y_l = (l-1)h_y (l = 1, ..., \bar{L}+1)$, $h_y = L/\bar{L}$. This grid is moving along the vertical coordinate: $z_m = (m-1)h_z (m = 1, ..., \bar{M}+1)$, $h_z = \bar{Z}/\bar{M}$, $\bar{Z} = Z^{k+1}$ is the new position of the free boundary on the time layer $k+1$. We assume that the flow is symmetric with respect to OZ (see [1]-[5]). Then the mathematical modeling can be devoted to construction of the solution in the half-layer $0 < z < Z(t)$. The conditions $T_x = 0$, $T_y = 0$ on the boundary $z = 0$ will be imposed in this case.

The second-order finite-difference approximations for second derivatives and the central difference scheme for approximation of convective terms are applied The numerical scheme (12) can be represented as a system of linear algebraic equations. At each time sublayer the systems of equations are implemented by the sweep method. The transition to the new spatial grid at the new time step is performed using the Newton interpolation formula [7].

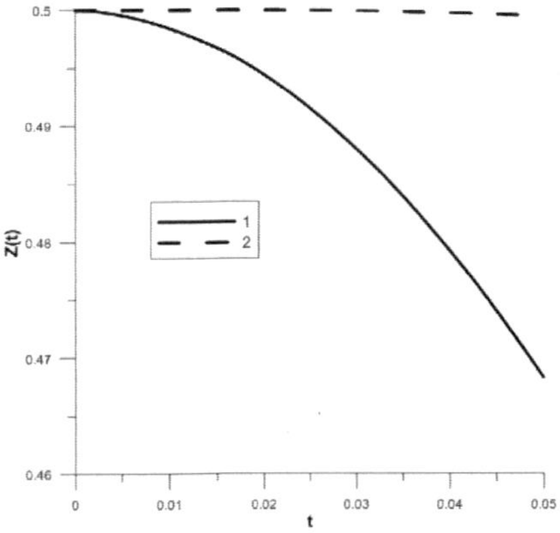

Figure 3. Changing of the free boundary position over time; 1 — the case when $A(t) = A_0/(t^2+1)$, $B(t) = B_0/(t^2+1)$, $\tilde{\tau}(t) = A_\tau \tau_0/(t^2+1)$, 2 — the case when $A(t) = A_0 texp(t)$, $B(t) = B_0 texp(t)$, $\tilde{\tau}(t) = A_\tau \tau_0 texp(t)$.

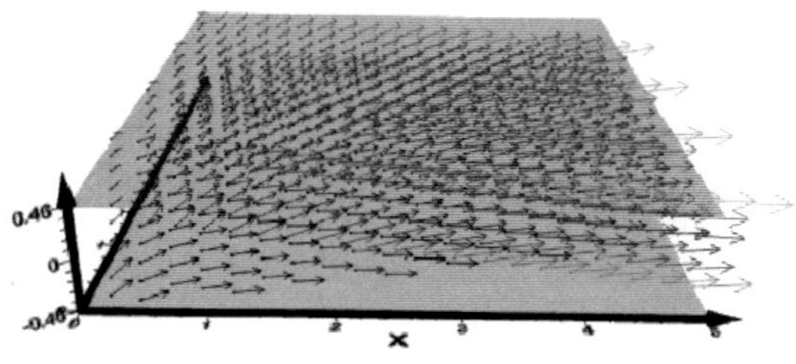

Figure 4. Spreading of the layer, $A(t) = A_0/(t^2+1)$, $B(t) = B_0/(t^2+1)$, $\tilde{\tau}(t) = A_\tau \tau_0/(t^2+1)$, $t = 0.05$.

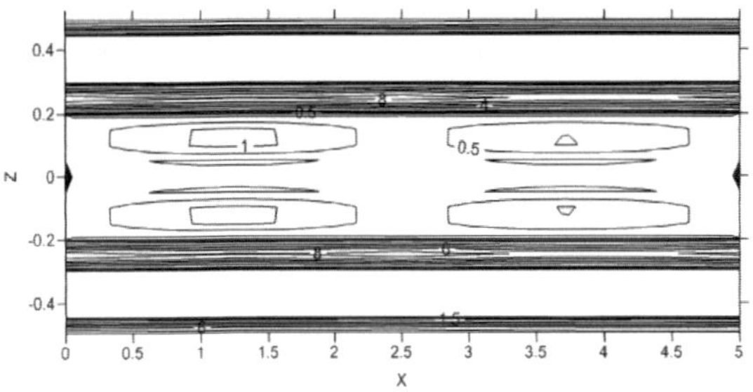

Figure 5. The temperature distribution in the XZ plane (y=2.5), t=0.02.

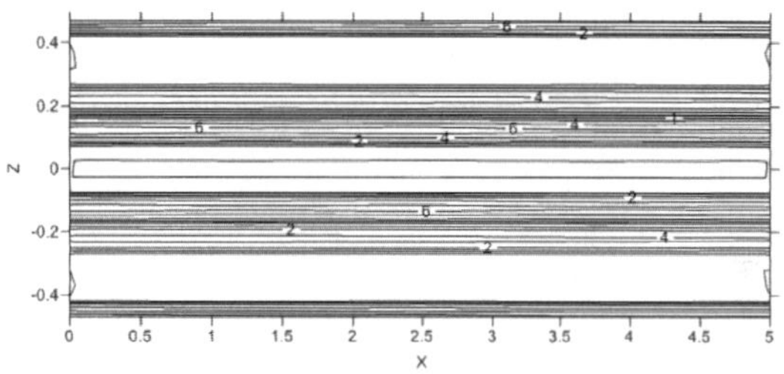

Figure 6. The temperature distribution in the XZ plane (y=2.5), t=0.05.

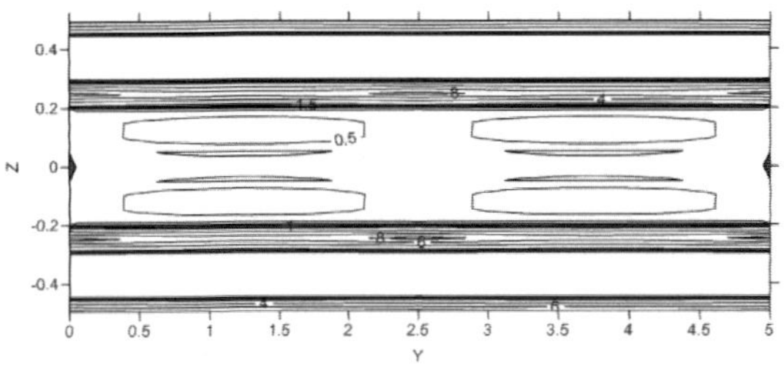

Figure 7. The temperature distribution in the YZ plane (x=5), t=0.02.

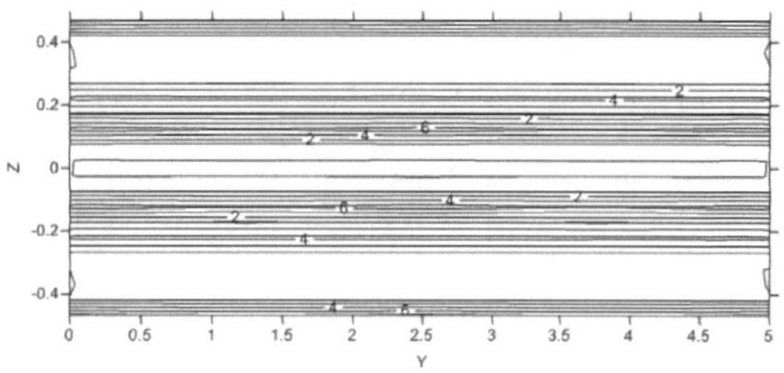

Figure 8. The temperature distribution in the YZ plane (x=5), t=0.05.

4. Numerical results

Numerical results characterize the spreading of an incompressible viscous layer and the liquid temperature distribution in the considered domain. Figure 3 shows the change of the free boundary position over time in the case of different time dependent functions $A(t)$, $B(t)$ and

$\tilde{\tau}(t)$. Position of the free liquid boundary at the initial time is defined as the $Z_0 = 0.5$, Θ is assumed to be 10. In the case when the dependencies $A(t)$, $B(t)$ and $\tilde{\tau}(t)$ have the form $A(t) = A_0 texp(t)$, $B(t) = B_0 texp(t)$, $\tilde{\tau}(t) = A_\tau \tau_0 texp(t)$ the layer is spreading much slower than in the case when $A(t) = A_0/(t^2+1)$, $B(t) = B_0/(t^2+1)$, $\tilde{\tau}(t) = A_\tau \tau_0/(t^2+1)$ (here $A_0 = -0.1$, $B_0 = -0.01$, $A_\tau = 400$, $\tau_0 = -0.1$). Figure 4 demonstrates an example of three-dimensional spreading of a liquid layer at time t=0.05.

The temperature distribution is defined in the parallelepiped where $N = L = 5$ (see (11)). Figures 5 and 6 demonstrates temperature distribution in the spreading liquid layer in the XZ plane when the dependencies $A(t)$, $B(t)$ and $\tilde{\tau}(t)$ have the form $A(t) = A_0/(t^2+1)$, $B(t) = B_0/(t^2+1)$, $\tilde{\tau}(t) = A_\tau \tau_0/(t^2+1)$ at time t=0.02 and t=0.05, respectively. In figures 7 and 8 shown the "lateral wall" $x = 5$ (YZ plane) in the case of the same dependency functions $A(t)$, $B(t)$ and $\tilde{\tau}(t)$ of time as the figures 5 and 6. Note that the temperature redistribution in the layer is observed over time. At the parallelepiped boundaries the temperature lower than in the central part of the investigated area. The qualitative distribution of the temperature is retained within the whole layer despite the decrease in temperature at the boundaries of the parallelepiped.

Numerical investigations of the temperature field in a liquid layer were also carried out for the case when boundary conditions at the interfaces are coordinated with the boundary conditions on the artificial boundaries. The conditions $T_{xx} = A(t)$ and $T_{yy} = B(t)$ were used as the conditions on the lateral walls. The temperature distribution in the layer remains qualitatively be the same; there are only the small quantitative differences (approximately 3%) in the values of the temperature functions.

5. Conclusion

The processes of deformation of a viscous incompressible liquid layer with free boundaries and heat transfer in the layer are modeled using exact solutions of the Navier-Stokes equations. A numerical algorithm for solving the problem of the heat distribution in the parallelepiped domain with moving boundaries in three dimensions. The change of the heat distribution in a spreads liquid layer is investigated.

Acknowledgments

The research was supported by the Ministry of Education and Science of Russia (Agreement No 14.613.21.0011, project identifier RFMEFI61314X0011). The author is grateful to Goncharova O.N. for helpful discussions.

References

[1] Goncharova O and Kabov O 2010 *Microgravity Sci. Technol.* **22** 407-414
[2] Goncharova O, Kabova Yu and Kabov O 2011 *Computational Thermal Sci.* **3(5)** 389-396
[3] Pukhnachov V 2000 Model of a viscous layer deformation by the thermocapillary forces *Max-Planck-Institut fuer die Mathematik in den Naturwissenschaften, Leipzig. Preprint* **50**
[4] Pukhnachov V 2002 *European J. of Applied Math.* **13(2)** 205-224.
[5] Pukhnachova T 2000 *Symmetry and diff. equations* (Krasnoyarsk) 183-186
[6] Goncharova O and Kondratenko O 2011 *Izv. AGU* **1-2(69)** 23-31
[7] Voevodin A, Goncharova O and Kondratenko O 2013 *Izv. AGU* **1-2(77)** 16-21
[8] Goncharova O 2007 *Izv. AGU* **1(53)** 39-44

Journal of Physics: Conference Series **754** (2016) 062009 doi:10.1088/1742-6596/754/6/062009

Simulation of heat and mass transfer in turbulent channel flow using the spectral-element method: effect of spatial resolution

V Ryzhenkov[1,2], V Ivashchenko[1,2], R Vinuesa[3], R Mullyadzhanov[1,2]

[1]Institute of Thermophysics SB RAS, Novosibirsk, Russia

[2]Novosibirsk State University, Novosibirsk, Russia

[3]Linné FLOW Centre, KTH Mechanics, Stockholm, Sweden and Swedish e-Science Research Centre (SeRC), Stockholm, Sweden

E-mail: rustammul@gmail.com

Abstract. We use the open-source code nek5000 to assess the accuracy of high-order spectral element large-eddy simulations (LES) of a turbulent channel flow depending on the spatial resolution compared to the direct numerical simulation (DNS). The Reynolds number $Re = 6800$ is considered based on the bulk velocity and half-width of the channel. The filtered governing equations are closed with the dynamic Smagorinsky model for subgrid stresses and heat flux. The results show very good agreement between LES and DNS for time-averaged velocity and temperature profiles and their fluctuations. Even the coarse LES grid which contains around 30 times less points than the DNS one provided predictions of the friction velocity within 2.0% accuracy interval.

1. Introduction

The turbulent regime of a fluid motion is often encountered in many natural phenomena and practical applications. The complexity of a turbulent flow is due to a wide spectrum of scales interacting in a non-linear fashion [1]. The problem becomes even more complicated when the flow is constrained by a rigid surface which substantially changes the properties of turbulence by introducing inhomogeneity in the wall-normal direction. The control of wall-bounded turbulent characteristics is a great scientific challenge where numerical simulations can effectively contribute.

The fluid motion is described by the Navier–Stokes equations which are nowadays possible to simulate directly resolving all the relevant scales. The obvious drawback of this approach is the high computational cost since the discretization and time step should be comparable with the size and turnover time of the smallest eddy. In a Large-eddy simulation (LES) one applies a low-pass filter to the Navier–Stokes equations. A well-known source of a possible error into the final solution is the common assumption that the spatial filtering operation commutes with spatial derivatives. The resulting equations describe the dynamics of the vortices which are larger than the filter size with a new unclosed term on the right-hand side representing the influence of the 'small' scales on the 'larger' ones. The subgrid scale model is a second source of error. The last issue in LES is the numerical scheme of discretization which brings the third kind of error. All three types of errors can interact with each other. Various approaches and

Journal of Physics: Conference Series **754** (2016) 062009 doi:10.1088/1742-6596/754/6/062009

criteria have been suggested in the literature to assess the quality of LES, such as Refs. [2–7] among others. In the present work we employ a high-order spectral element method [8] to discretize the equations, thus, significantly reducing the discretization error in the final solution, and investigate the performance of the widespread dynamic Smagorinsky model for the subgrid stress tensor and heat flux vector. We perform a set of Large-eddy simulations with different levels of grid refinement using the open-source spectral element code nek5000 [9] and compare the results with DNS.

2. The governing equations and computational details

The governing equations are the following:

$$\frac{\partial \boldsymbol{u}}{\partial t} + (\boldsymbol{u}\nabla)\boldsymbol{u} = -\nabla p + \frac{1}{Re}\nabla^2 \boldsymbol{u}, \qquad \nabla \cdot \boldsymbol{u} = 0, \tag{1}$$

$$\frac{\partial \theta}{\partial t} + (\boldsymbol{u}\nabla)\theta = \frac{1}{Pe}\nabla^2 \theta, \tag{2}$$

where $\boldsymbol{u}$, p and θ are the dimensionless velocity vector, pressure and temperature fields, respectively. The Reynolds number $Re = U_b H/\nu$ is defined using the bulk velocity U_b, half-width of the channel H and constant kinematic viscosity of the fluid ν. Another non-dimensional criterion is the Péclet number $Pe = U_b H/\alpha$, where α is the constant thermal diffusivity. All simulations are performed for the Prandlt number $Pr = Pe/Re = \nu/\alpha = 1$. Equations (1) and (2) are used to perform direct numerical simulations (DNS).

Applying the spatial low-pass filtering [10] denoted by tilde

$$\widetilde{\boldsymbol{u}}(\boldsymbol{x}, t) = \int G(\boldsymbol{x} - \boldsymbol{r})\boldsymbol{u}(\boldsymbol{r}, t)d\boldsymbol{r}, \tag{3}$$

where G is the filter function, to equations (1) and (2) we arrive to the filtered Navier–Stokes equations:

$$\frac{\partial \widetilde{\boldsymbol{u}}}{\partial t} + (\widetilde{\boldsymbol{u}}\nabla)\widetilde{\boldsymbol{u}} = -\nabla \widetilde{p} + \frac{1}{Re}\nabla^2 \widetilde{\boldsymbol{u}} - \nabla \cdot \tau, \qquad \nabla \cdot \widetilde{\boldsymbol{u}} = 0, \tag{4}$$

$$\frac{\partial \widetilde{\theta}}{\partial t} + (\widetilde{\boldsymbol{u}}\nabla)\widetilde{\theta} = \frac{1}{Pe}\nabla^2 \widetilde{\theta} - \nabla \cdot \boldsymbol{q}, \tag{5}$$

where τ denotes the subgrid stress tensor and $\boldsymbol{q}$ represents the subgrid heat flux vector:

$$\tau = \widetilde{\boldsymbol{u} \cdot \boldsymbol{u}} - \widetilde{\boldsymbol{u}} \cdot \widetilde{\boldsymbol{u}}, \qquad \boldsymbol{q} = \widetilde{\boldsymbol{u}\,\theta} - \widetilde{\boldsymbol{u}}\,\widetilde{\theta}, \tag{6}$$

The subgrid stress tensor and heat flux vector are expressed using the Boussinesq hypothesis:

$$\tau_{ij} - \tau_{kk}\delta_{ij}/3 = -2\nu_t \widetilde{S}_{ij}, \qquad \nu_t = (C_s\Delta)^2|\widetilde{S}|, \qquad q_i = -Pr_t^{-1}\nu_t\, \partial_i \widetilde{\theta}, \tag{7}$$

where δ_{ij} is the Kronecker delta, ν_t, C_s and Pr_t are the turbulent viscosity, Smagorinsky constant and turbulent Prandtl number which are computed dynamically as described below, Δ is the filter width, and $|S| = \sqrt{2S_{ij}S_{ij}}$ is the magnitude of the strain-rate tensor

$$S_{ij} = \frac{1}{2}(\partial_j u_i + \partial_i u_j), \tag{8}$$

where a compact notation for the derivative is adopted, $\partial_j u_i = \partial u_i/\partial x_j$. The Smagorinsky constant and turbulent Prandtl number are defined as follows [11, 12]:

$$C_s^2 = -\Big\langle \Big(\widehat{\widetilde{u}_i\widetilde{u}_j} - \widehat{\widetilde{u}}_i\widehat{\widetilde{u}}_j\Big)\widetilde{S}_{ij}\Big\rangle \Big/ \Big\langle 2\Big(\widehat{\Delta}^2|\widehat{\widetilde{S}}|\widehat{\widetilde{S}}_{ij} - \Delta^2\widehat{|\widetilde{S}|\widetilde{S}_{ij}}\Big)\widetilde{S}_{ij}\Big\rangle, \tag{9}$$

$$Pr_t = C_s^2\left[\Big\langle\Big(\widehat{\widetilde{u}_i\widetilde{\theta}} - \widehat{\widetilde{u}}_i\widehat{\widetilde{\theta}}\Big)\partial_i\widetilde{\theta}\Big\rangle \Big/ \Big\langle\Big(\widehat{\Delta}^2|\widehat{\widetilde{S}}|\partial_i\widehat{\widetilde{\theta}} - \Delta^2\widehat{|\widetilde{S}|\partial_i\widetilde{\theta}}\Big)\partial_i\widetilde{\theta}\Big\rangle\right]^{-1}, \tag{10}$$

Journal of Physics: Conference Series **754** (2016) 062009 doi:10.1088/1742-6596/754/6/062009

where the angular brackets ($\langle\ .\ \rangle$) denote the averaging along the homogeneous directions (x and z in our case, see Fig. 1 below). Thus, C_s and Pr_t are the functions of time and one spatial (wall-normal) direction only. The value of the Smagorinsky constant is clipped to zero in case the negative values occur. The test filter $\widehat{\Delta} = 2\Delta$ in the position (x_i, y_i, z_i) is defined as $\sqrt[3]{\Delta x_i \Delta y_j \Delta z_k}$, where $\Delta x_i = x_{i+1} - x_{i-1}$, $\Delta y_j = y_{j+1} - y_{j-1}$, $\Delta z_k = z_{k+1} - z_{k-1}$ with i, j, k being the structured indices of the point inside each spectral element. The above described governing equations are implemented in the nek5000 code [9] with a spatial discretisation based on the spectral-element method using Lagrange polynomials. The Navier–Stokes equations are cast in a weak form and discretised in space by means of Galerkin approximation using Nth-order Lagrange polynomial interpolants on the Gauss-Lobatto-Legendre points for the velocity field and $(N-2)$th-order for the pressure ($\mathbb{P}_N - \mathbb{P}_{N-2}$ formulation). In the present work the polynomial order N is set to 7. The semi-implicit time-stepping scheme is of the third-order accuracy.

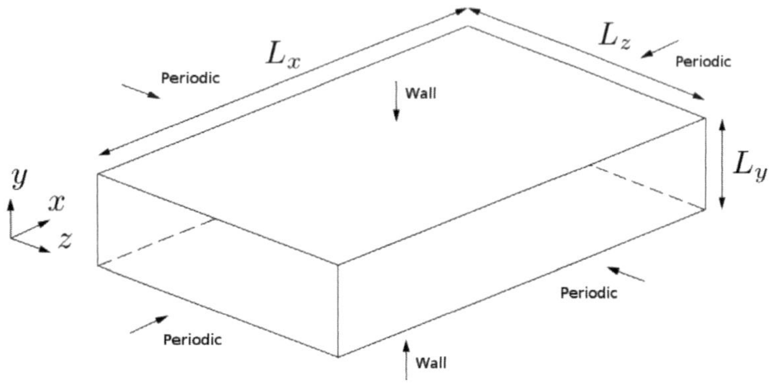

Figure 1. Geometry of the channel and boundary conditions

We study a turbulent flow between two infinite parallel walls at a Reynolds number of $Re = 6800$. A rectangular computational domain of the size $L_x \times L_y \times L_z = 2\pi H \times 2H \times \pi H$ is used where x is the streamwise coordinate (along the flow), y is the wall-normal axis and z is the spanwise coordinate, Fig. 1. The planes $y = \pm H$ correspond to the position of parallel walls. The infinite flow domain is modeled by periodic boundary conditions in x and z directions. No-slip condition ($\boldsymbol{u} = 0$) and constant temperature for hot upper ($\theta = \theta_h$) and cold lower ($\theta = \theta_c$) wall are imposed. A dynamically adjusted body force is added in equations (1) and (3) to keep a constant mass flux through the plane $y - z$. Note that all the quantities below are non-dimensionalized using U_b, $\Delta\theta = \theta_h - \theta_c$ and H unless otherwise specified.

The initial conditions of the simulation are set using the analytic profiles $u_x = 5(1 - y^4)/4$, $u_y = u_z = 0$ and $\theta = (1 + y)/2$. The random noise with a 5% amplitude of the bulk velocity is added to all velocity components to accelerate the transition to turbulence. The statistical analysis is performed for the flow only after a sufficient period of computational time to minimize the influence of initial transients on time-averaged characteristics. Figure 2 shows the instantaneous temperature field of the flow going from left to right from the DNS visualizing the near-wall vortical structures. Table 2 shows the characteristics of the mesh used for DNS and LES, where N_{se} is the number of spectral elements along each direction (x, y, z), N_{points} is the number of nodes along each direction and N_{tot} is the total number of nodes. We used structured hexahedral grids for the present calculations.

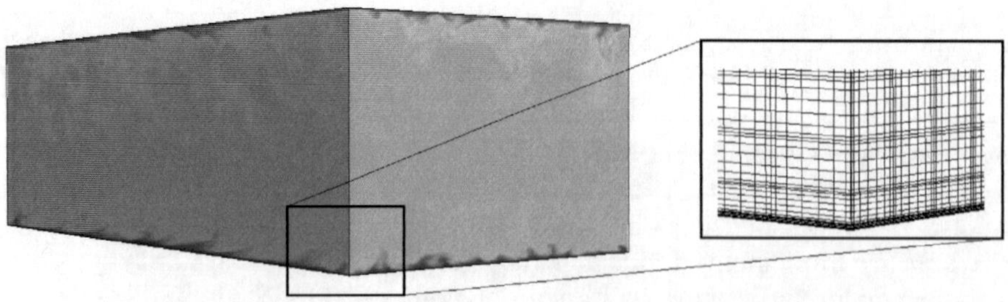

Figure 2. Instanteneous temperature field from DNS. The flow goes from left to right. The top wall is hot (red), the bottom one is cold (blue). The inset on the right shows a small fragment of the computational hexahedral mesh

Table 1. Mesh characteristics

	DNS	LES1	LES2	LES3
N_{se}	$40 \times 30 \times 30$	$26 \times 20 \times 20$	$20 \times 16 \times 16$	$14 \times 10 \times 10$
N_{points}	$320 \times 240 \times 240$	$208 \times 160 \times 160$	$160 \times 128 \times 128$	$112 \times 80 \times 80$
N_{total}	18.4×10^6	5.3×10^6	2.6×10^6	0.72×10^6

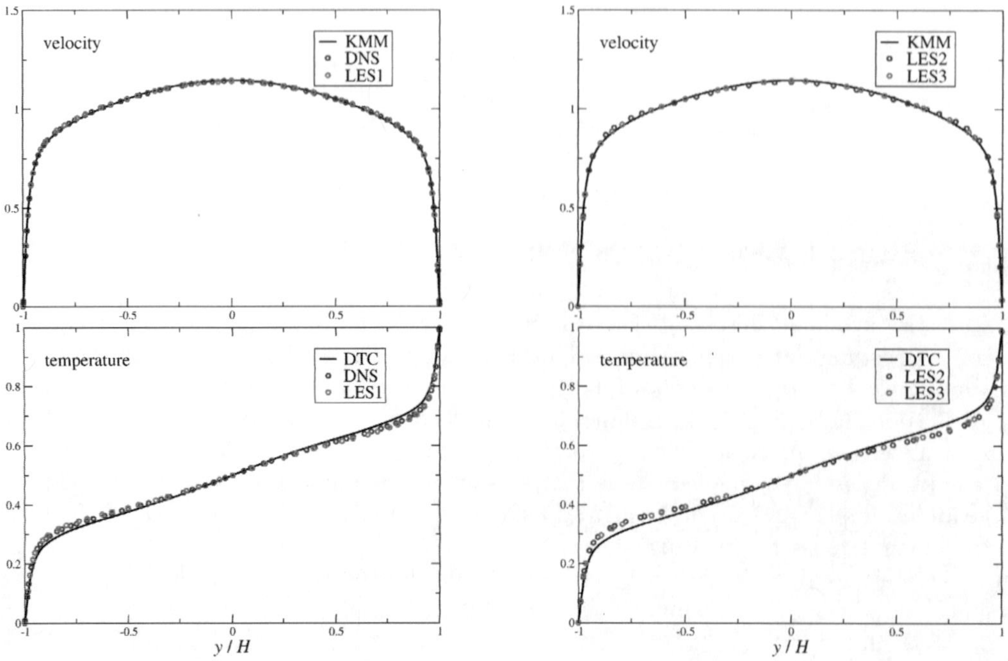

Figure 3. Comparison of the time-averaged velocity and temperature profiles with the data from KMM [13] and DTC [14]

3. Results

We compare time-averaged profiles of velocity, temperature and its fluctuations from the present DNS, LES and DNS data from the literature [13, 14]. The present data is also averaged along the homogeneous x and z directions. Figure 3 shows the comparison of $\overline{u}_x$ and $\overline{\theta}$ where the

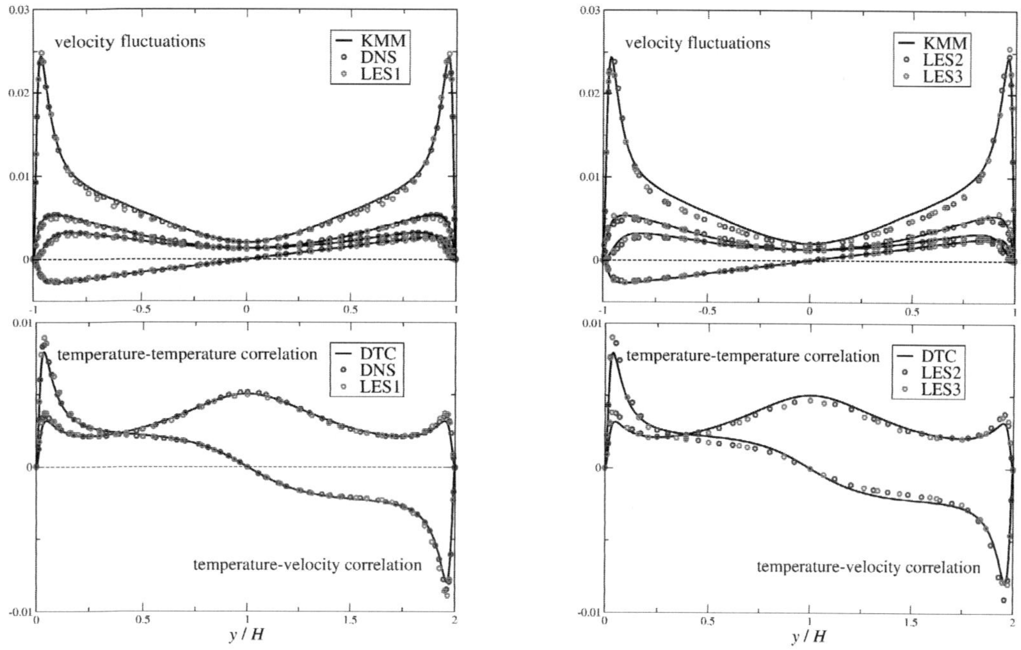

Figure 4. Comparison of the time-averaged velocity and temperature fluctuations with the data from KMM [13] and DTC [14]

overline denotes the time-averaging. The velocity profile is in excellent agreement with the data of [13] who used a fully spectral code based on Chebyshev polynomials in y and Fourier in x and z direction. The temperature profile also shows good agreement, however, slightly deviating from the data [14]. Note that in the present calculations $Pr = 1.0$ while $Pr = 0.71$ in [14] who used second order accurate discretization.

Figure 4 shows the comparison of different components of $\overline{u_i'u_j'}$ and $\overline{u_i'\theta'}$. DNS shows excellent agreement, while the LES data may slightly deviate from the results of [14]. The important fact is a good prediction of the fluctuations peak near the wall even for the very coarse mesh (LES3). This peak is the result of the dynamics of streaky structures in the near-wall region. Figure 5 shows the instantaneous field of u_x near the wall ($y/H = 0.95$) in the $x-z$ plane, where this peak of fluctuations appear. Visualizations for all meshes look very similar despite the differences in the spatial resolution. Table 3 shows the comparison of the non-dimensional friction velocity u_τ ($u_\tau^2 = Re^{-1}\partial_y\overline{u}_x|_{wall}$) and friction temperature $\theta_\tau = (Pe\ u_\tau)^{-1}\partial_y\overline{\theta}|_{wall}$. All simulations show excellent agreement with data from the literature within 2.0% accuaracy interval. Note that relatively high u_τ deviation calculated for LES1 mesh is probably caused by the effect of transients and insufficient averaging time. This issue is currently investigated. Note also the robust prediction of the correct θ_τ value for all grids even when u_τ deviates (LES1). The larger value of θ_τ in [14] is due to the lower Pe (0.71 against 1.0).

4. Conclusion

We performed a series of Large-eddy simulations of a turbulent channel flow at $Re = 6800$ to assess the accuracy of the numerical method coupled with dynamic Smagorinsky model for subgrid stress tensor and heat flux vector. An excellent agreement for the time-averaged velocity,

Table 2. Comparison of u_τ and θ_τ (percentage is given as a relative error in comparison with the present DNS data)

	[13, 14]	DNS	LES1	LES2	LES3
u_τ	0.05699	0.05701	0.05588 (-1.98%)	0.05663 (-0.67%)	0.05634 (-1.18%)
θ_τ	0.02306	0.01945	0.01933 (-0.62%)	0.01962 (+0.87%)	0.01981 (+1.85%)

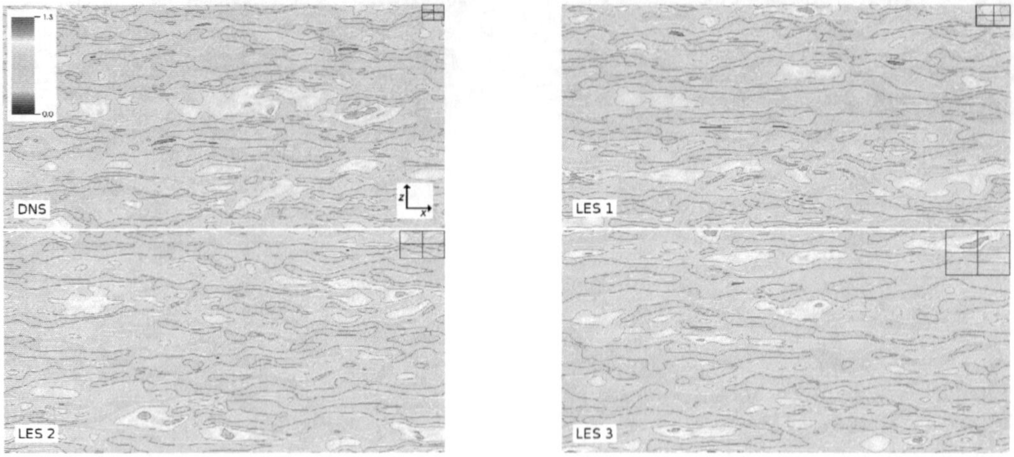

Figure 5. Instantaneous axial velocity field in $x - z$ plane near the wall at $y/H = 0.96$ ($y^+ = yu_\tau/\nu \approx 16$) for DNS and LES is shown. In the right top corner four spectral elements are shown schematically

temperature and their fluctuations is obtained. We also compared the value of friction velocity and temperature indicating the near-wall solution accuracy. Even the coarse LES grid which contains around 30 times less points than the DNS one provided predictions of the friction velocity within 2.0% accuracy interval which is encouraging for engineering applications. The next step is to assess the accuracy of LES in variable-density flows. Spectral-element method is well suited for procedures such as spatial filtering and deconvolution employed in reacting flow models [15].

Acknowledgments

This work is funded by the Russian Science Foundation grant No. 14-19-01685. The computational resources are provided by Siberian Supercomputer Center SB RAS (Novosibirsk).

References

[1] Landau LD, Lifshitz EM 1982 *Fluid Mechanics* (New York: Pergamon)
[2] Geurts BJ, Fröhlich J 2002 A framework for predicting accuracy limitations in large-eddy simulation *Phys. Fluids* **14**(6) 1070-6631
[3] Meyers J, Geurts BJ, Baelmans M 2003 Database analysis of errors in large-eddy simulation *Phys. Fluids* **15**(9) 1070-6631
[4] Klein M 2005 An attempt to assess the quality of Large Eddy Simulations in the context of implicit filtering *Flow Turb. Comb.* **75** 131–147
[5] Celik IB, Cehreli ZN, Yavuz I 2005 Index of resolution quality for Large Eddy Simulations *J. Fluids Engrg.* **127**(5) 949–958
[6] Meyers J, Geurts BJ, Sagaut P 2007 A computational error-assessment of central finite-volume discretizations in large-eddy simulation using a Smagorinsky model *J. Comp. Phys.* **227** 156–173

[7] van der Bos F, Geurts BJ 2010 Computational error-analysis of a discontinuous Galerkin discretization applied to large-eddy simulation of homogeneous turbulence *Comput. Methods Appl. Mech. Engrg.* **199** 903–915
[8] Deville MO, Fischer PF, Mund EH 2002 *High-order methods for incompressible fluid flow* (Cambridge University Press)
[9] Fischer P, Lottes J, Kerkemeier S Nek5000 web page (http://nek5000.mcs.anl.gov)
[10] Sagaut P 2006 *Large Eddy Simulation for Incompressible Flows: An Introduction* (Springer Science & Business Media)
[11] Germano M, Piomelli U, Moin P, Cabot WH 1991 A dynamic subgridscale eddy viscosity model *Phys. Fluids A* **3**(7) 1760-5
[12] Moin P, Squires K, Cabot W, Lee S 1991 A dynamic subgridscale model for compressible turbulence and scalar transport *Phys. Fluids A* **3**(11) 2746-57
[13] Moser RD, Kim J, Mansour NN 1999 Direct numerical simulation of turbulent channel flow up to $Re_\tau = 590$ *Phys. Fluids* **11**(4) 943–945
[14] Dharmarathne S, Tutkun M, Araya G, Castillo L 2016 Structures of scalar transport in a turbulent channel *Eur. J. Mech. B/Fluids* **55** 259–271
[15] Domingo P, Vervisch L 2015 Large Eddy Simulation of premixed turbulent combustion using approximate deconvolution and explicit flame filtering *Proc. Comb. Inst.* **35** (2) 1349–1357

The numerical study of energy separation in a two-cascade Leontiev tube

M S Makarov, S N Makarova and A A Shibaev

Kutateladze Institute of Thermophysics Siberian Branch of Russian Academy of Science, 1 Lavrentiev ave., Novosibirsk, Russia

alex-shib@yandex.ru

Abstract. This paper presents results of numerical modeling of energy separation for helium-xenon gas mixture flow in a two-cascade Leontiev tube with central and outer (annular) supersonic nozzles. The Mach number and stagnation temperature distributions in longitudinal section and the heat-transfer intensity from the subsonic to supersonic flow have been obtained. The dependences of the cooling effect, the temperature efficiency factor and the adiabatic efficiency on a stagnation pressure in the receiver have been investigated.

1. Introduction

It is known that various physical effects lead to energy separation in the gas flow without heat transfer between the gas and external heat-sources (to temperature stratification). The most famous energy-separation effect is the Ranque-Hilsch effect in the vortex tube [1–3]. This effect has been used for building the aircraft air-conditioning system, cooling systems with a swirl gas flow, and small refrigerators [4, 5].

In the Leontiev tube, the heat flux appears due to the temperature differences between the subsonic and supersonic parts of the flow when the expanding supersonic part of the flow is cooled. The dissipation of kinetic energy of the stagnating supersonic flow near the wall increases its temperature but for the low-Prandtl gas mixture the recovery temperature on the wall is lower than the temperature of the subsonic part of the flow. As a result, the heat transfers across the heat-conducting separating wall from the supersonic part of the flow to the subsonic part even if the stagnation parameters for both parts of the flow are equal.

The energy-separation method in a compressible flow was suggested and theoretically analyzed in [6, 7]. There it was shown theoretically that the heat transfer across the heat-conducting separating wall depended on the recovery temperature on the wall from the side, flown by the supersonic part of gas under the adiabatic conditions. To decrease the recovery temperature and increase the energy separation effect, it was suggested to use the low-Prandtl gas mixtures (helium-xenon, hydrogen-xenon etc.) [8–10], to blow some part of gas into the high-speed flow, to suck some part of gas from the low-speed flow [7, 11–14], or to use the wall with various surface reliefs [15].

The experimental investigations of heat transfer in the single Leontiev tube with a central cylindrical channel [16, 17] show the energy separation in the airflow with Prandtl number equal to 0.71. The authors obtain the static pressure distribution on the nozzle wall, and temperature and Mach number distributions in the outer cross section of the tube for the subsonic and supersonic parts of the flow. The results of numerical modeling of energy separation in the trans- and supersonic external flows are presented in monograph [18]. The study was performed for various gas mixtures with

Prandtl number from 0.2 to 0.8 and adiabatic factor from 1.3 to 1.67. It was shown that the subsonic part of the flow was cooled more effectively if the Mach number of the supersonic part of the flow varied from 1.9 to 2.4. For the transonic flows in the flat channel it was obtained that the adiabatic efficiency of the Leontiev tube, working on the mixture of helium-xenon with Prandtl number equal to 0.23, is higher than the adiabatic efficiency of the vortex tube, working on air with Prandtl number equal to 0.71 if other conditions being equal [19].

This paper presents the results of numerical modeling of energy separation for the helium-xenon gas mixture flow in the two-cascade Leontiev tube with central and outer (annular) supersonic nozzles. This configuration leads to an increase in the area of the cooled wall for the subsonic part of the flow and allows increasing the mass flow rate of the cooled gas.

2. Statement of the problem and numerical approach

A steady turbulent heat transfer is considered in the air (Pr=0.71) flowing through the nozzle with central cylindrical channel, used to test the numerical approach (see figure 1(a)), and in the helium-xenon mixture (Pr=0.23) flowing through the two-cascade Leontiev tube with central and outer (annular) supersonic nozzles, used in this study (see figure 1(b)).

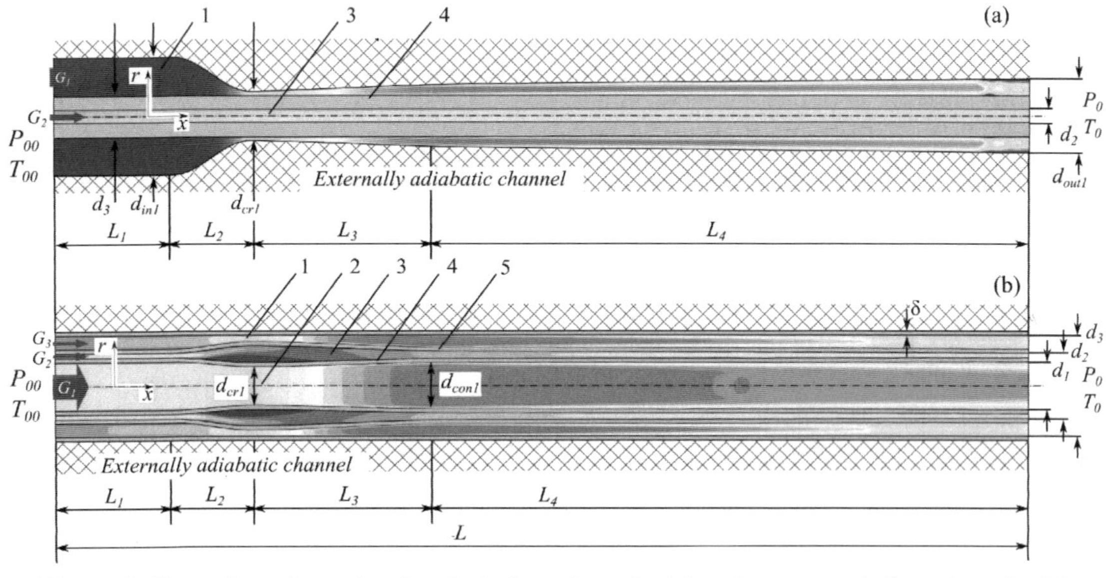

Figure 1. Channel configuration for single Leontiev tube (a) and two-cascade Leontiev tube (b): 1 – supersonic annular nozzle, 2 – supersonic central nozzle, 3 – external heat-conducting wall, 4 – internal heat-conducting wall.

The single Leontiev tube geometry experimentally studied in [16, 17] is used for the test configuration. The supersonic flow is formed in the outer annular Laval nozzle with outer diameter at the inlet d_{in1} = 30 mm, critical diameter d_{cr1} = 12.5 mm and outer diameter at the outlet d_{out1} = 19 mm. The inner diameter of the annular nozzle is equal to the outer diameter of the cylindrical channel, which is installed on the nozzle axis, d_3 = 10.4 mm. The expanding part of the Laval nozzle has the conical form (L_4 = 150 mm) and cubic conjugation in the critical section with converging part (L_3 = 50 mm). The expanding angle of the conical tube is 1.15°. The converging part of the Laval nozzle has the form of sinus half-period (L_2 = 20 mm). The length of the cylindrical inner part of the nozzle L_1 is equal to 30 mm. Thus, the whole length of the single Leontiev tube is 250 mm.

The two-cascade Leontiev tube has outer diameter d_3 = 27.2 mm and heat-conducting walls with thickness of 1 mm. The inner diameter of the central nozzle at the inlet and outer section d_1 is equal to

12.1 mm. The critical diameter d_{cr1} = 9.8 mm. The diameter of the critical section of the annular nozzle is 20 mm. The outer diameter of the subsonic channel d_2 = 16.5 mm. The expanding parts of the nozzles have conical form (L_4 = 150 mm) and cubic conjugation in the critical section with converging parts (L_3 = 50 mm). The converging parts and the cylindrical inner part of the nozzles have the same forms and lengths as in the single Leontiev tube. The whole length of the two-cascade tube is 250 mm.

We solved the problem by the implicit numerical method, based on the 2D axisymmetric RANS equations for the steady compressible flow with heat transfer. ANSYS Fluent solver licensed for SSCC SB RAS and self-made code were used for mesh generation and data analysis. The S-A turbulence model and the Kays-Crawford model [20] for the turbulent Prandtl number were used for modeling. As it was obtained in [18], this numerical model allows solving the convective heat transfer problems for the turbulent low-Prandtl gas flows. The turbulent intensity was 2 %, and the turbulent length scale was 1 m at the channel inlet. The adhesion conditions for motion equations, the temperature and heat flux coupling for energy equation were set as boundary conditions on the heat-conducting walls. The meshes contained from 90000 to 300000 nodes. The convergence for all variables and mass-flow rates in each channels were provided with maximum error 10^{-5}.

3. Result and discussion

3.1. Validation of numerical approach

The figure 2 shows the static pressure distribution on the outer wall, obtained by numerical modeling, in comparison with experimental data. In the test configuration, the air flows through the nozzle with central ebonit or cuprum cylindrical channel. The stagnation pressure at the inlet P_{00} is equal to 7.5 atm, and the stagnation pressure T_{00} = 21.8 °C.

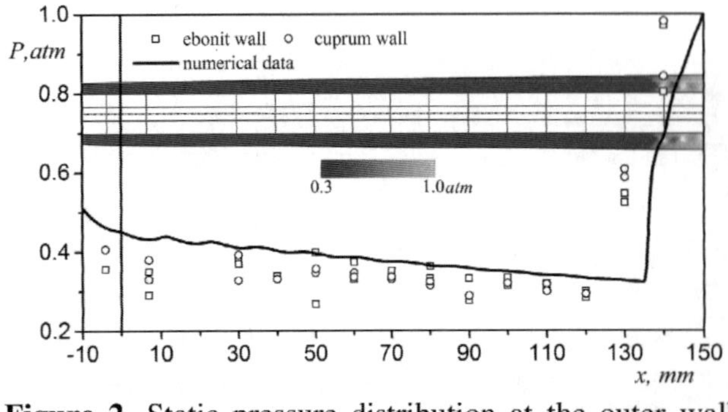

Figure 2. Static pressure distribution at the outer wall (line, points) and at the longitudinal section (color field) of the supersonic part of the flow: square is experimental data for low-heat-conducting wall (ebonit), circle is experimental data for high-heat-conducting wall (cuprum), and line is numerical data.

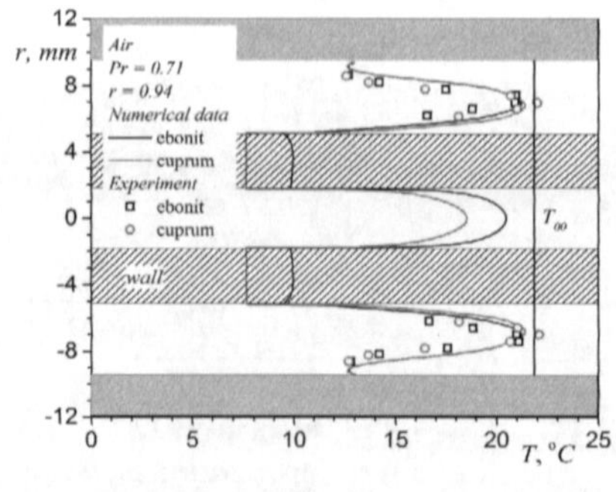

Figure 3. Recovery temperature distribution for the single Leontiev tube at the 7 mm distance from the outer cross section: vertical line is stagnation temperature.

As we can see, the numerical data are agreed with the experimental data. The static pressure is decreased down to 0.3 atm. Near the outlet section the system of pseudo-shocks is generated, and the supersonic flow stagnates with the pressure increase up to P_0 = 1 atm. The longitudinal coordinate of the first shock attachment that is obtained by numerical modeling lies 10 mm lower downstream compared with the experiment. We suppose that this disagreement is explained by the inaccuracy in an expanding angle of the conical part of the nozzle. It may be noted, that the energy separation has a weak influence on the flow dynamic in the Laval nozzle.

Figure 3 shows the experimental data for temperature, measured by thermo-probe at a distance of 7 mm from the nozzle outlet section. As it is noted in [16], the used thermo-probe allowed measuring the slightly less value of temperature than the stagnation temperature. The temperature obtained by the used numerical approach is indeed higher than the experimental data. For comparison with the experiment we calculate the recovery temperature distributions with recovery factor r = 0.94 based on numerical results. As we can see on figure 3, the numerical data for recovery temperature agree with the experimental data both on the form of distribution across the section and on the influence of heat-conductivity of the separated wall.

3.2. Energy separation in Leontiev tubes

Figure 4 presents the Mach number and stagnation temperature distributions for the helium-xenon gas mixture flow with the stagnation pressure in receiver 7.5 atm, stagnation temperature in receiver 21.8 °C for the single Leontiev tube and 327 °C for two-cascade Leontiev tube. The figure demonstrates the heating of supersonic flows by subsonic flow cooling.

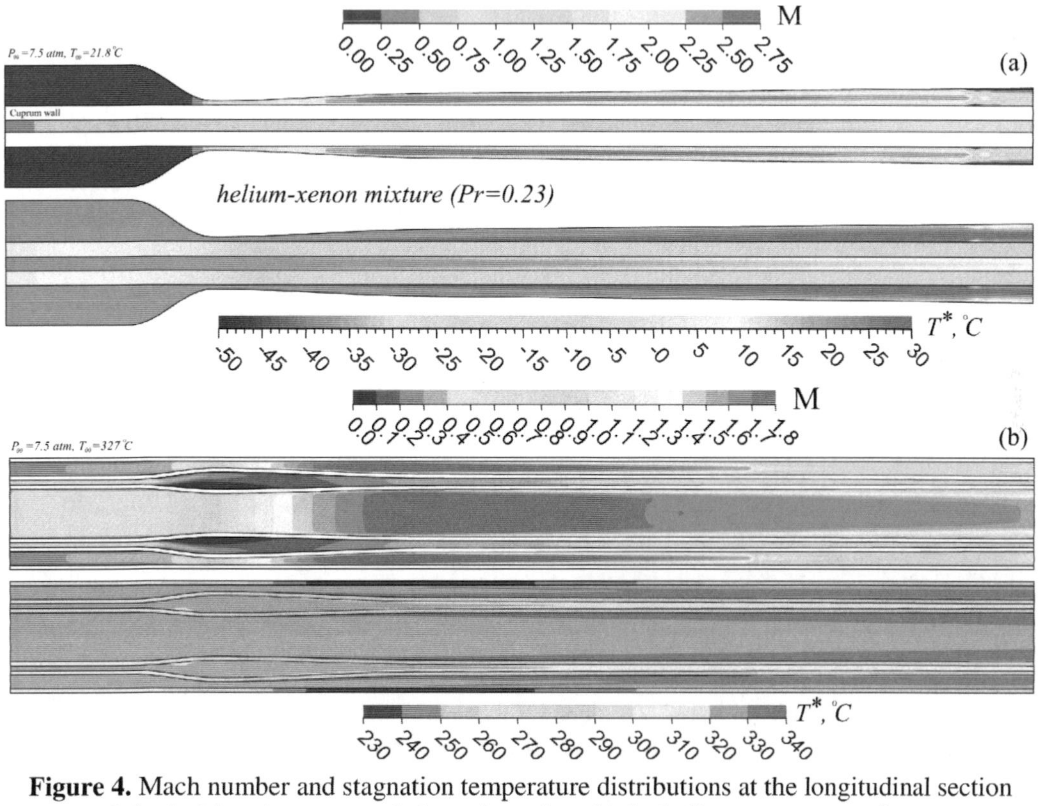

Figure 4. Mach number and stagnation temperature distributions at the longitudinal section of single (a) and two-cascade Leontiev tubes (b) for helium-xenon gas mixture.

Figure 5 presents the cooling effect $(\overline{T}_2^* - T_{00})$, the temperature efficiency factor η_T and adiabatic efficiency η depending on the stagnation pressure in the receiver for the single and two-cascade Leontiev tubes:

$$\eta_T = \frac{\overline{T}_2^* - T_{00}}{\Delta T_s^*}; \eta = \frac{G_2}{G}\frac{\overline{T}_2^* - T_{00}}{\Delta T_s^*}; \quad \Delta T_s^* = T_{00}\left(\left(P_{00}/\overline{P}_2^*\right)^{\frac{1-\gamma}{\gamma}} - 1\right).$$

G_2 is the mass flow rate of the cooled gas (in subsonic flow). G is the total mass flow rate through the tube. P_{00} and T_{00} are the stagnation pressure and temperature in the receiver. $\overline{P}_2^*$ and $\overline{T}_2^*$ are the mass average total pressure and total temperature of cooled gas at the outlet section. γ is the adiabatic exponent of gas.

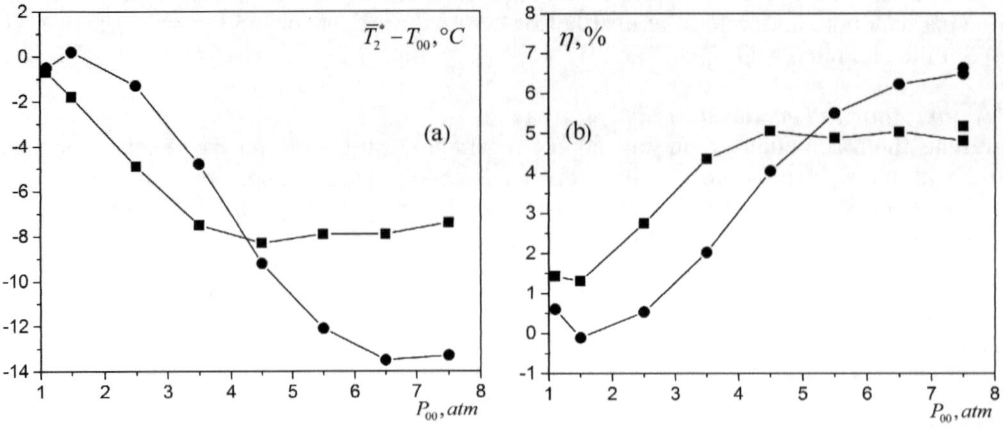

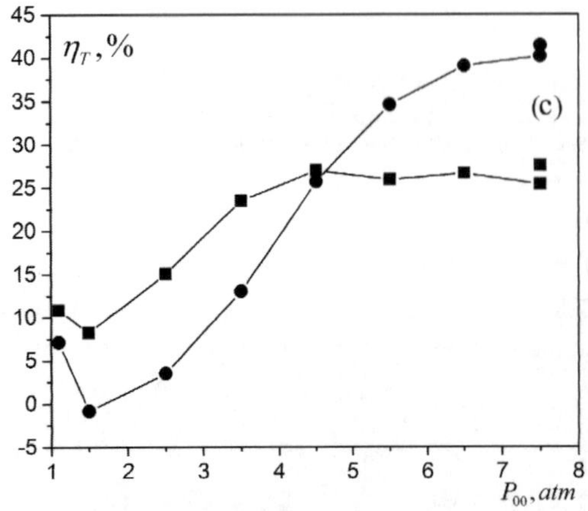

Figure 5. Cooling effect (a), temperature efficiency factor (b) and adiabatic efficiency (c) depending on the stagnation pressure in receiver for single Leontiev tube (circle points) and for two-cascade Leontiev tube (squire points).

For the two-cascade Leontiev tube it is shown that the maximum of temperature efficiency factor is achieved for the lesser value of stagnation pressure in the receiver (about 4.5 atm), and for the higher value it does not change. The adiabatic efficiency for the two-cascade Leontiev tube is higher compared to the single tube configuration if the stagnation pressure is less than 4.5 atm and achieves 25%. If the stagnation pressure is higher than 4.5 atm the single Leontiev tube is more effective and η_T achieves 40%. The temperature efficiency factor for the studied configurations of Leontiev tube reaches 5 – 6 %. In whole the efficiency of the two-cascade Leontiev tube is less than the efficiency of the single tube, but the cooling power of the two-cascade configuration is 2.5 time higher as a result of 6 time increase in the mass flow rate of the cooled gas. It is noted, that the efficiency of energy separation in the Leontiev tube does not depend on the stagnation temperature in the receiver. The obtained results for T_{00} = 21.8 °C and 327 °C are very close to each other.

4. Conclusions

Energy separation of the compressible gas flow in the single and two-cascade Leontiev tube has been analyzed. It is found that the parameters of energy-separation efficiency (temperature efficiency factor and adiabatic efficiency) for the two-cascade Leontiev tube are less than the same factors for the single Leontiev tube if the stagnation pressure in the receiver is near to maximum. If pressure in the receiver is less than 4.5 atm (for the studied configurations), the two-cascade Leontiev tube turns out to be a more effective energy separation device. For the same geometry parameters (length of tube and outer diameter) the two-cascade Leontiev tube is more powerful due to the high mass flow rate of the cooled part of the gas mixture.

Acknowledgments
The reported study was funded by RFBR according to the research project № 15-08-04203.

References

[1] Piralishvili Sh A, Polyaev V M and Sergeev M N 2000 *The Vortex Effect: Experiment, Theory, Engineering Solutions* (Moscow: UNPTs Energomash)
[2] Leontiev A I 2002 *Fluid Dynamics* **37** 4, 512–29
[3] Dubnishchev Yu N, Meledin V G, Pavlov V A and Yavorsky N I 2003 *Thermophysics and Aeromechanics* **10** 4 567–78
[4] Veretennikov S V 2010 *Vestnik RGATA* **17** 2 23–8
[5] Noskov A S, Lovtsov A V and Khait A 2012 *Computational continuum mechanics* **5** 3 313–21
[6] Leontiev A I 1997 *High Temperature* **35** 1 157–9
[7] Burtsev S A and Leontiev A I 2000 *Izv. Ross. Akad. Nauk: Energ.* **5** 101–13
[8] Leontiev A I, Lushchik V G and Yakubenko A E 2006 *High Temperature* **44** 2 234–42
[9] Vinogradov Yu A, Ermolaev I K, Zditovets A G and Leontiev A I 2005 *Izv. Ross. Akad. Nauk: Energ.* **4** 128–33
[10] Vigdorovich I I and Leontiev A I 2010 *Fluid Dynamics* **45** 3 434–40
[11] Leontiev A I, Lushchik V G and Yakubenko A E 2006 *Izv. Ross. Akad. Nauk: Energ.* **2** 12–8
[12] Leontiev A I, Lushchik V G and Yakubenko A E 2011 *Fluid Dynamics* **46** 6 935–41
[13] Leontiev A I, Lushchik V G and Makarova M S 2012 *High Temperature* **50** 6 739–43
[14] Leontiev A I, Lushchik, V G and Yakubenko A E 2007 *High Temperature* **45** 4 488–96
[15] Zditovets A G and Titov A A 2007 *Izv. Ross. Akad. Nauk: Energ.* **2** 111–7
[16] Vinogradov Yu A, Zditovets A G and Strongin M M 2013 *Fluid Dynamics* **48** 5 687–96
[17] Zditovets A G, Vinogradov Yu A and Titov A A 2014 *Proc. 15th Int. Heat Trans. Conf.* (Kyoto, Japan, IHTC15-8965)
[18] Nakorjakov V E, Makarov M S, Petukhov Yu I, Vitovsky O V and Elistratov S L 2015 *Heat transfer in a low-Prandtl gas mixtures flows* (Novosibirsk: Akademizdat)
[19] Makarov M S and Makarova S N 2013 *Thermophysics and Aeromechanics* **20** 6 777–87
[20] Weigand B, Ferguson J R and Crawford M E 1997 *Int. J. Heat Mass Transf.* **40** 17 4191–6

PIV and OH PLIF study of impinging propane-air jet-flames

Z D Kravtsov[1,2], Z D Tolstoguzov[1,2], L M Chikishev[1,2], V M Dulin[1,2]
[1]Kutateladze Institute of Thermophysics, Siberian Branch of RAS, 1 Ak. Lavrentyeva Avenue, Novosibirsk, 630090, Russia
[2]Novosibirsk State University, 2 Pirogova Street, Novosibirsk, 630090, Russia

E-mail: l.chikishev@gmail.com

Abstract. The paper reports on the study of flow structure of a reacting propane-air jet, issued from contraction nozzle and impinged on flat metallic surface, by using the particle image velocimetry technique. Flows with different nozzle-to-surface distance H/d were studied. The Reynolds number Re was in the range of 500÷5500, equivalence ratio Φ was varied from 0.8 to 1.4. Velocity field was measured for a conical premixed flame for Re = 1500 and Φ = 0.9. A region with flow recirculation was detected between the flame cone and the impingement surface for the case of $H/d = 4$. This flow feature may result in a reduced local heat transfer.

1. Introduction

The problem of heat and mass transfer during cooling and heating of surfaces by jet flows is relevant for a number of technical applications (e.g., paper drying, glass heating, cooling of microelectronics or turbine blades). One of the major shortcomings in the organization of impinging isothermal and reacting flows is the heterogeneity of the local heat transfer. Some applications, such as chemical vapor deposition, require homogeneity of heat and mass transfer. Vortex structures rolling-up in shear layer of jet flows influence on local heat transfer for small distances between the nozzle and surface [1-4]. Manipulation by these vortices (e.g., by external forcing) can be used to affect local intensity of heat transfer. Particle image velocimetry (PIV) and planar laser-induced fluorescence (PLIF) are useful tools to investigate influence of unsteady flow structures on heat transfer in impinging jets [5]. In impinging jets with premixed combustion, self-induced instabilities can arise without external forcing [6, 7]. Influence of this factor on heat transfer requires further studies.

Besides, near-wall combustion phenomena influence on the efficiency and formation of pollutions in many technical applications. This aspect is of a growing interest for modern downsized internal combustion engines with increased surface-to-volume ratios and higher power densities. Furthermore flame-wall interactions also affect the combustion in jet engines and gas turbine combustors. Since flame-wall interaction processes are not yet understood in details, the complex interactions between fluid motion, transport phenomena and chemical reactions during these processes need further investigations. Experimental studies of near-wall combustion in simplified geometry are required.

The present work focuses on PIV/PLIF measurements in a reacting jet, interacting with a flat wall in a simplified geometry. Namely, a conical premixed flame covered from the top with a flat water-cooled surface is studied. The confinement effect on the flow pattern has been investigated.

2. Apparatus and setup

The reacting jet flow of propane-air mixture issued from a contraction axisymmetric nozzle (with the outlet diameter d of 15 mm). The jet impinged normally on flat surface of the bottom of water-cooled cylindrical tank. The axis of the cylinder was aligned with the nozzle axis. The cylinder diameter was 80 mm. An electric traversing device was used to move the tank along its axis with the precision of 0.1 mm. The scheme of the experimental setup is shown in Figure1. A lighter was used to ignite the flame.

Flow rates of the air and propane were controlled by flow meters. The details about the burner can be found in the previous study [8]. Titanium dioxide particles of approximately 0.5 µm in diameter were used to seed the flow for the PIV measurements. The particles were contained in a vessel and were introduced to the passing through air flow by a mechanical mixer. The vessel was connected to the air supply line via a bypass scheme.

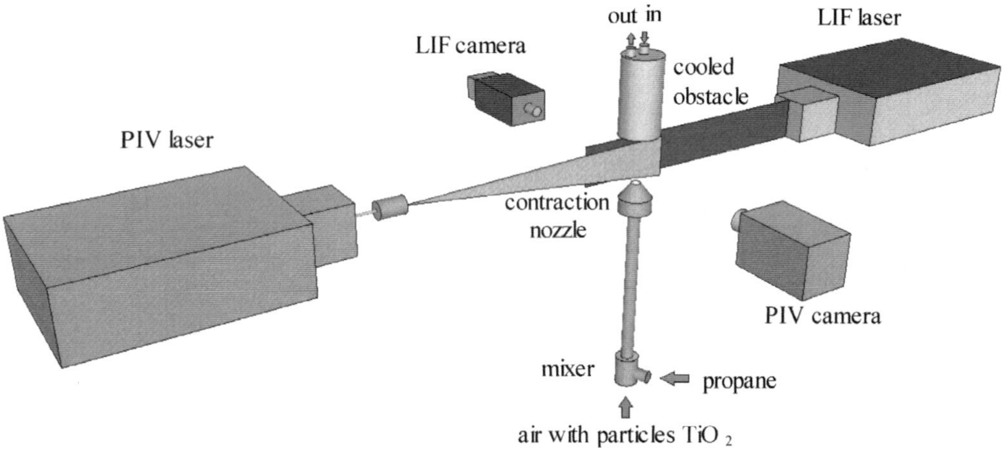

Figure 1. Sketch of PIV/PLIF experimental setup

Particle image velocimetry (PIV) was used for the measurement of velocity fields. Measuring area size was 80×80 mm. Double-head Nd:YAG laser (Quantel, Ever Green, up to 70mJ per each 7 ns pulse) was used to illuminate the particles in the measurement plane. The laser beam was converted into a laser sheet by using a system of cylindrical and spherical lenses. The opening angle of the laser sheet was 10 degrees. The energy of each laser pulse was measured by a power meter (Coherent LabMax). On average it was 53 mJ after the laser sheet optics.

Particles images were captured by a CCD camera (Bobcat ImperX), oriented normally to the laser sheet plane. A narrow band-pass optical filter (532 nm ± 10 nm) was used to reduce intensity of background and flame luminosity in the camera image. The laser and camera were synchronized via TTL signal. The time delay between two pulses for each image pair was 100 µs and corresponded to maximal displacement of the particles up to 15 pixels. The PIV images (2048×2048 pixels is size) were processed by using in-house software ActualFlow. The images were preprocessed to remove background (minimal intensity for each pixel) and impingement surface. During four iterations of an adaptive cross-correlation algorithm the interrogation area size was reduced from 64×64 to 32×32 pixels. The spatial overlap rate between the interrogation areas was 50%. The velocity fields were validated by signal-to-noise criterion.

A PLIF system (LaVision) was used to measure temperature distribution in the longitudinal cross-section of the jet-flame from OH* fluorescence. The second harmonic (532 nm) of a pulsed Nd:YAG laser (Quanta-Ray, 10Hz, 1 J per pulse) was used to pump a tunable dye laser (Sirah dye laser) with Rhodamine 6G. Frequency of the output radiation was doubled by using a BBO crystal. A small portion of the laser radiation was directed to a power meter using a semi-transparent mirror. The measurement of the energy of each laser pulse was necessary for correct comparison of the instantaneous fluorescence images with each other. The energy of the laser radiation in the wavelength range 283-284 nm was approximately 10 mJ. The laser beam was transformed into a collimated sheet with a width of 50 mm and thickness less than 0.8 mm (in the measurement region). The laser sheet passed through the axis of symmetry of the nozzle.

Image intensifier (LaVision IRO) based on a UV-sensitive photocathode S20 (multialkali) and multi-channel amplifier, was used to register fluorescence signal of the OH*. The quantum efficiency of the photocathode was about 25% for the considered wavelength range (280-320 nm). The image intensifier was equipped with a quartz lens (LaVision UV-lens, f # 2.8, 100 mm) and optical band-pass filter (300-320 nm). The use of the optical filter was necessary to remove unwanted intensity at the wavelength of the lasers, including reflections by the nozzle. The signal was recorded by sCMOS camera (16-bit images with resolution of 2560 × 2160 pixels). The images were processed by FlameMaster software from LaVision.

3. Results and discussion

3.1. PIV and PLIF data for unconfined flame

Temperature of the combustion products were detected by comparing intensity of OH* fluorescence, excited by two different transitions.

Figure 2. Examples of OH* PLIF intensity during excitation of two different transitions

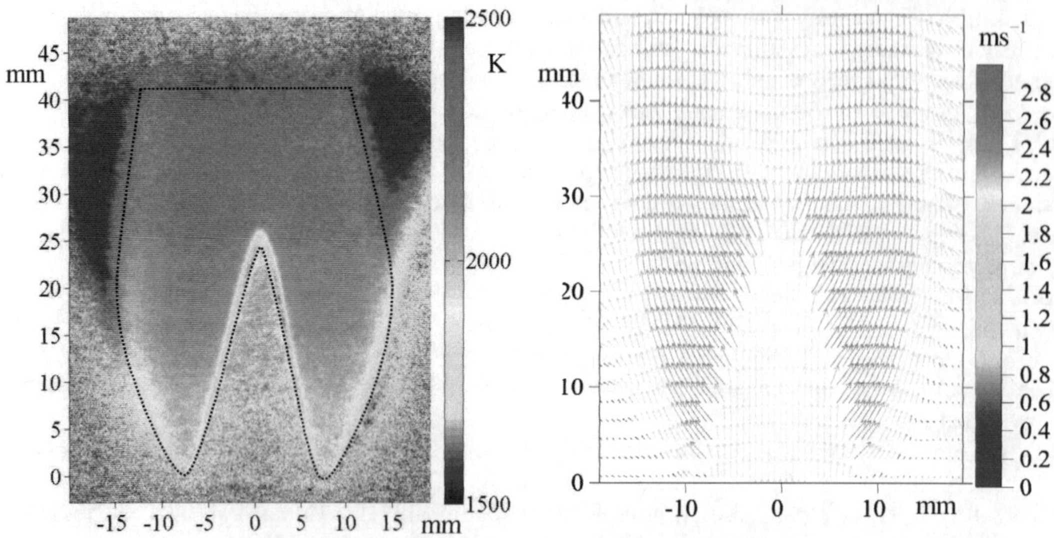

Figure 3. OH* PLIF and PIV results for unconfined conical flame ($\Phi = 0.9$, Re = 1500)

A pair of lines $R_1(14)$ и $P_1(2)$ ($v' = 1 \leftarrow v'' = 0$, $A^2\Sigma^+ \leftarrow X^2\Pi$) was selected to evaluate gas temperature. The signal is shown in Figure 2. Figure 3 shows the results of PLIF and PIV measurements for the unconfined propane-air flame. The temperature of the products is in the range of

2000÷2200 K, whereas the adiabatic temperature of the propane-air flame ($\Phi = 0.9$, 1 atm. and reactants temperature of 300 K), calculated using GRI-MECH 3.0 chemistry, is 2179 K. According to the analysis in [9], the uncertainty of the temperature measurement is expected to be roughly 10%.

3.2. *Visualisation of impinging flames*

Figure 4 shows the typical combustion regimes, which can be separate into two groups, namely, the seated flames, attached to the nozzle rim, and detached (lifted) flames, stabilized at a certain distance from the nozzle rim (similar flames were observed in [7]). The former type of the flames was observed for a wide range of H/d and Φ values. The lifted flames, stabilized near the impingement surface, were typical for fuel-rich mixtures and a wide range of H/d. Besides, lean lifted flames could exist near the impingement surface for $H/d = 1$ and 2, whereas unconfined flames were blown away for these conditions.

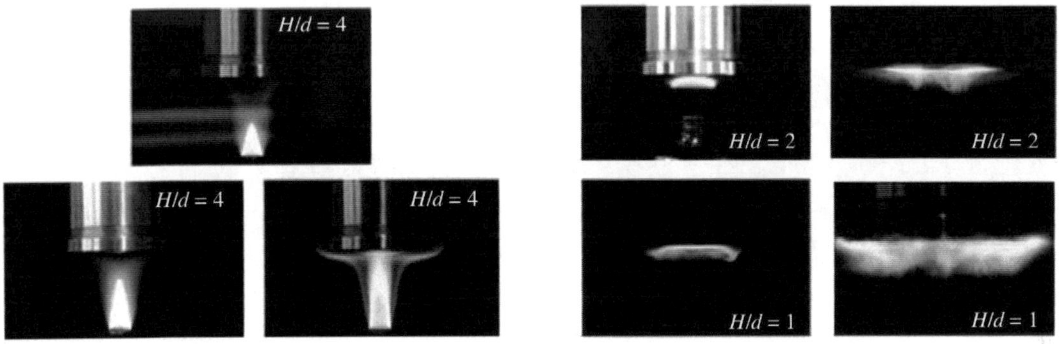

Figure 4. Photographs of seated (left) and detached (right) flame types

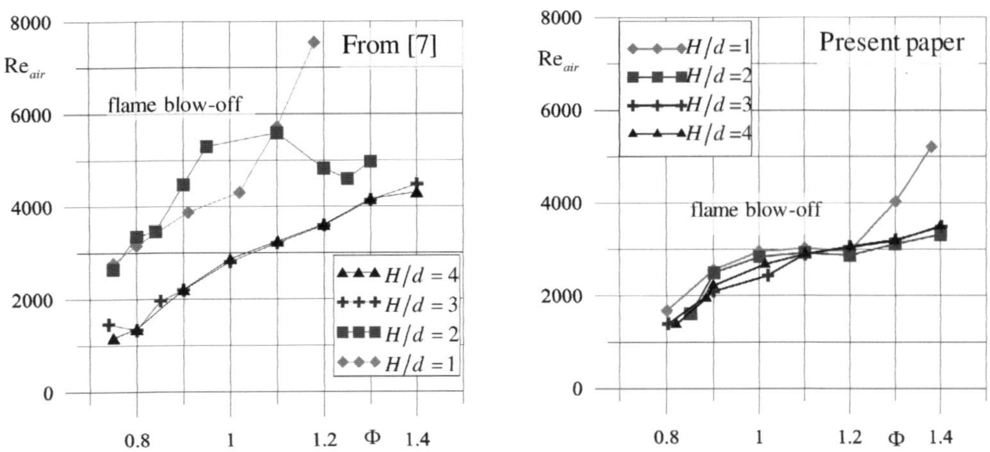

Figure 5. Blow-off curves for propane-air impinging flames for different nozzle-to-surface distances

Blow-off curves for different values of H/d are shown in Figure 5 on Re-Φ diagram. The Reynolds number is defined as $Re = 4Q_{air}/\pi\nu_{air}d$, where Q_{air} and ν_{air} are the flowrate and kinematic viscosity of the air, respectively. The blow-off curves separate the conditions, for which the flame stabilized on the nozzle or near the impinging surface, from region of Re and Φ values for which the flame was blown away. For the cases of $H/d = 4$ and 8 the flame was attached to the nozzle rim before the blow-off (for the studied range of Φ between 0.8 and 1.4). For $H/d = 1$ and 2, mixture with $\Phi > 1.2$ could burn as a detached (lifted) flame from the nozzle and stabilized near the impingement surface (see Figure 4).

Journal of Physics: Conference Series **754** (2016) 072001 doi:10.1088/1742-6596/754/7/072001

Blow-off limits obtained in [7] for another nozzle with diameter of 18 mm are shown in Figure 5. Presence of the impingement surface for $H/d = 1$ and 2 provides extended range of Re and Φ for combustion without blow off.

3.3. *PIV data for impinging flames*

The average velocity fields measured for the reacting impinging jets ($\Phi = 0.9$, $Re_{air} = 1500$) are shown in Figure 6. The bulk velocity of the mixture U_0 is approximately 1.6 ms^{-1}. The focus of this sub-section is on the impingement surface effect on the velocity field for the case of conical premixed flame. The data for the cases of $H/d = 1$, 2, 3, and 4 are presented. Except for $H/d = 1$, the flame had shape of a cone. The gas velocity increased downstream the flame front due to combustion and local gas density decrease. For the case $H/d = 2$ the cone top was located almost at the impingement surface. For the case $H/d = 1$ the flame shape was constricted by the obstacle. Important finding is a not intensive recirculation zone, present between the cone and the impingement surface for the case $H/d = 4$. Negative values of the axial velocity component are also detected for $H/d = 3$.

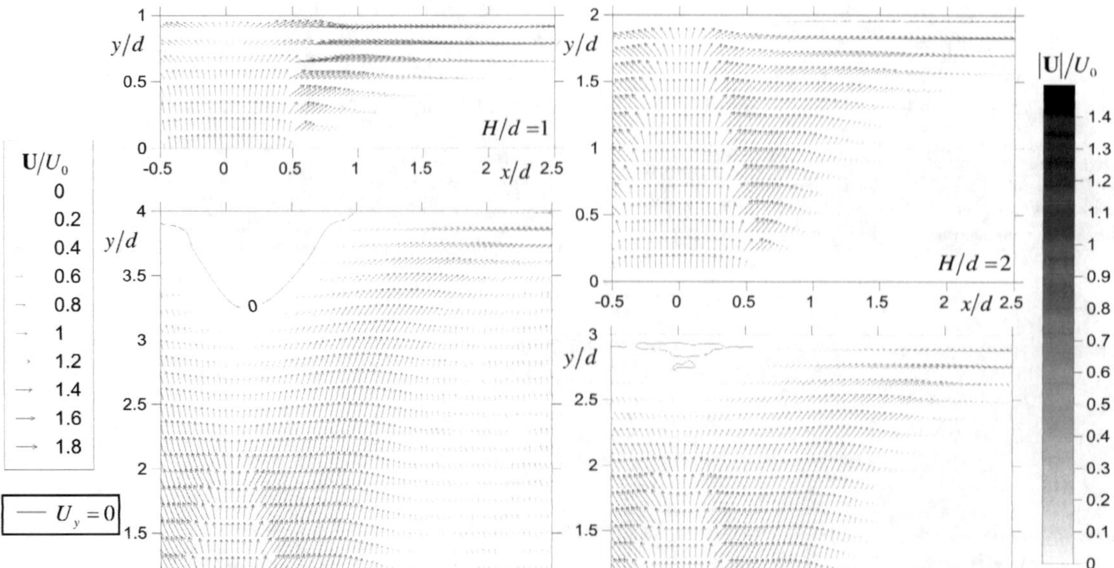

Figure 6. Mean velocity field for impinging propane-air flames ($\Phi = 0.9$, $Re_{air} = 1500$) for different H/d

4. Conclusions

Flow structure of premixed propane-air impinging jet-flames was studied by planar optical tehcniques. Depending on values of Re and Φ and also on the way of the flame ignition, the observed flames corresponded to two different types. Flames, attached to the nozzle rim, were observed for a wide range of Φ and H/d. Fuel-rich mixture could also burn as a lifted flame. For $H/d = 1$ and 2 the impingement surface provided a wider range of Re and Φ (fuel-rich lifted flames with $\Phi > 1.2$) for combustion without blow-off. Moreover, lean lifted flames could also burn near the surface for these values of H/d.

The velocity field in the conical lean flames was measured by using PIV. The flows for $H/d = 1$, 2, 3, and 4 were compared. It was found that a weak recirculation zone was formed between the flame cone and the impingement surface for the case of $H/d = 4$. This finding is important because local recirculation provides a stagnation region which may reduce local heat transfer between the solid surface and combustion products.

5. Acknowledgments

This work was funded by Russian Science Foundation (grant No. 16-19-10566).

6. References

[1] Kataoka K, Suguro M, Degawa H, Maruo K, Mihata I 1987 The effect of surface renewal due to largescale eddies on jet impingement heat transfer. *Int. J. Heat Mass Transfer* **30** 559–67

[2] Meola C, de Luca L, Carlomagno G 1996 Influence of shear layer dynamics on impingement heat transfer. *Exp. Therm. Fluid Sci.* **13** 29–37

[3] Hadziabdic M, Hanjalic K 2008 Vortical structures and heat transfer in a round impinging jet. *J. Fluid Mech.* **596** 221–60

[4] Roux S, Fenot M, Lalizel G, Brizzi L, Dorignac E 2011 Experimental investigation of the flow field and heat transfer of an impinging jet under acoustic excitation. *Int. J. Heat Mass Transfer* **54** 3277-90

[5] Sakakibara J, Hishida K, Maeda M 1997 Vortex structure and heat transfer in the stagnation region of an impinging plane jet (simultaneous measurements of velocity and temperature fields by digital particle image velocimetry and laser-induced fluorescence. *Int. J. Heat Mass Transfer* **40** 3163–76

[6] Renard P-H, Thevenin D, Rolon J, Candel S 2000 Dynamics of flame/vortex interactions. *Progr. Energy Combust. Sci.* **26** 225-82

[7] Fernandes E, Leandro R 2006 Modeling and experimental validation of unsteady impinging flames. *Combust. Flame.* **146** 674–86

[8] Abdurakipov S, Dulin V, Markovich D, Hanjalic K 2013 Expanding the stability range of a lifted propane flame by resonant acoustic excitation. *Combust. Sci. Technol.* **185** 1644–66

[9] Devillers R., Bruneaux G., Schulz C 2008 Development of a two-line OH-laser induced fluorescence thermometry diagnostics strategy for gas-phase temperature measurements in engines. *Appl. Optics*, **47** 5871-85

Artificial neural network for bubbles pattern recognition on the images

I E Poletaev[1,2], K S Pervunin[1,2], M P Tokarev[1]

[1] Kutateladze Institute of Thermophysics, the Russian Academy of Sciences, Siberian Branch
1, Lavrentyev Ave., Novosibirsk, 630090, Russia
[2]Novosibirsk National Research State University
2, Pirogova St., Novosibirsk, 630090, Russia

E-mail: poletaev.igor.phys@gmail.com

Abstract. Two-phase bubble flows have been used in many technological and energy processes as processing oil, chemical and nuclear reactors. This explains large interest to experimental and numerical studies of such flows last several decades. Exploiting of optical diagnostics for analysis of the bubble flows allows researchers obtaining of instantaneous velocity fields and gaseous phase distribution with the high spatial resolution non-intrusively. Behavior of light rays exhibits an intricate manner when they cross interphase boundaries of gaseous bubbles hence the identification of the bubbles images is a complicated problem. This work presents a method of bubbles images identification based on a modern technology of deep learning called convolutional neural networks (CNN). Neural networks are able to determine overlapping, blurred, and non-spherical bubble images. They can increase accuracy of the bubble image recognition, reduce the number of outliers, lower data processing time, and significantly decrease the number of settings for the identification in comparison with standard recognition methods developed before. In addition, usage of GPUs speeds up the learning process of CNN owning to the modern adaptive subgradient optimization techniques.

1. Introduction

Developing of such an important industrial sector as oil and gas, aerospace, energy inevitably linked with the usage of modern equipment. Unsteady two-phase bubble flows can be found in industrial apparatus and technical devices applied in these areas. In order to successfully tackle with technical challenges in optimizing of processes in existing setups and creating of the new equipment fundamental knowledge is needed on which such solutions can be based. It is necessary to develop precise and effective experimental methods for research of the complex processes.

During the analysis of experimental data in a two-phase flow, the separation of the obtained information on each phase is usually carried out. This helps to analyze a phase behavior independently from each other as they show different properties and study their mutual influence on the mixing efficiency and the turbulence level inside a flow [1].

Among the used non-intrusive methods for studying two-phase flows we can identify the two most promising shadow photography and planar laser techniques, for example as Planar Fluorescence for Bubbles Imaging (PFBI) [2]. The shadow technique has less spatial resolution along the depth, because a measurement area is recorded throughout the full depth and the localization depth is determined by optics' depth of field [3]. The second technique PFBI has a better localization depth and the spatial resolution is defined by a laser sheet thickness used for the illumination of the measurement plane. This gives an opportunity to increase the available range of the measured volume void fraction inside a flow up to 3% compared to the shadow technique where the upper limit can be assessed in 1-2%.

In order to obtain such parameters of a bubble flow as a velocity field and a distribution of a void fraction, we need numerical methods that are able to extract information on the spatial distribution of bubbles within the flow and their size distribution from experimental data.

The first approach is based on bubble search by edge identification on images and further recognition of the unilaterally connected area for the calculation of a bubble center. The second method is based on correlation search of the bubbles using a bubble template. For both methods, we can highlight common shortcomings: a large number of adjustable parameters (close to 20) and the specialization of algorithms to a particular kind of experimental data. Additionally, we can point out inability of the algorithm based on the edge identification to deal with the overlapped bubbles and the high computational cost inherited to the correlation analysis, though it is able to identify overlapped bubbles as separate objects.

Based on the above arguments, the aim of this study can be formulated as to create a fast and easy to use processing tool of data that was obtained during an experiment by optical diagnostics of two-phase bubble flows. The tool should have the ability to adapt itself to the variation of parameters of the registered bubble images and to deal with the overlapped bubble images.

For the first time to achieve the aims stated above we propose to use a method of artificial neural convolution networks in this study. This requires solving the following tasks:

- creation of algorithmic neural network approaches for recognition of the images of the disperse phase inside bubble flows, which will be easy to adjust to different experiment modifications and mainly they will be able to process overlapped bubble images;
- simulation of the realistic bubble images mimicking experimental data for creation of synthetic images that are needed for training of neural networks;
- implementation of the algorithms for processing of experimental images obtained by PFBI method of quantitative laser visualization.

Today many technical achievements such as speech recognition and face recognition in photographs, well-known machine translator, robots that can redraw paintings - all of this and more is made possible through a deep learning [4]. This is one of the directions of machine learning of an artificial intelligence, which has spread in the last decade thanks to the rapid development of computers and technology.

Briefly, the concept of deep learning can be described as a set of algorithms that attempt to simulate low and high-level abstractions in data by using architectures composed of multiple successive nonlinear transformations.

To extract features from images there have been developed a huge variety of methods, such as SIFT descriptors [5] or histogram of oriented gradients (HoG) [6]. However, all these methods without exception have a large number of settings that greatly complicated the work of a researcher. The concept of deep learning proposes to replace all these complicated and long procedures with learning algorithms. Thus, the basic principle of the method of deep learning can be formulated as the use of neural network architectures that are "trained" how to extract information from data.

The section 2 of this paper provides a brief theoretical presentation of the neural network approach and the methods used. In subsection 2.2, we describe architecture used in this study of the of the so-called convolutional neural networks that are the most advanced technologies for deep learning these days. Further in the next section 3 we present a description of an experimental setup and parameters for data acquisition. The last section 4 refers to the obtained results and their analysis.

2. Neural networks

2.1. *Simple neural network*

Figure 1.a shows a simplest implementation of a modern perceptron [7], where the signals $(y_1, \dots y_k)$ of each output neurons are generated by consistent transmission of the input signals $(x_1, \dots, x_d)$ from the neurons in the input layer the l neurons of the hidden layer towards the output layer. The artificial neuron itself has structure [8], shown in the figure 1.b. Each j-th neuron calculates the weighted sum of its inputs v_j with considering the bias b_j through activation function φ. Thus, the signal obtained by neuron goes to his single output. Artificial neurons are combined in a network so, that the outputs of some neurons are connected to the inputs of others, as it shown in figure 1.a.

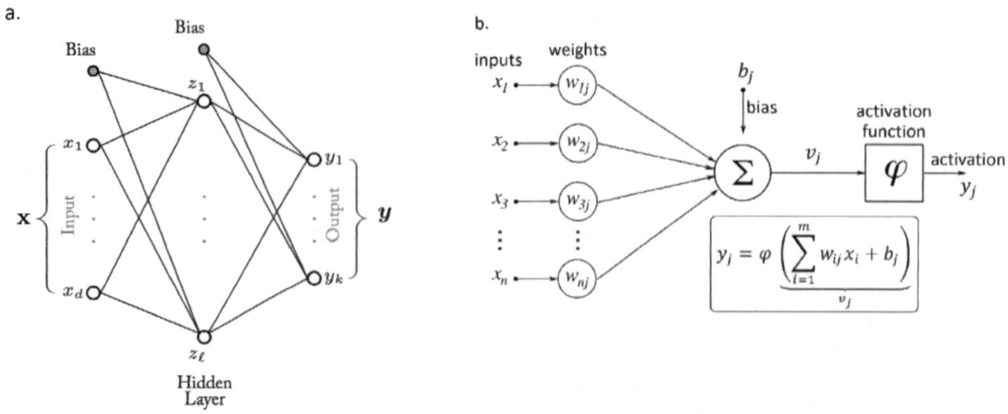

Figure 1. (a.) Single-layer perceptron, and (b.) the structure of an artificial neuron.

2.2. *Convolution neural networks*

One of the major deep-learning [4] technology, the *convolutional neural networks* (CNN), is aimed at processing graphic information: facial recognition on photos, handwritten digits, etc. Namely because of this basis, to address the problem of identifying bubbles in the images the CNN technology was used in this work. The idea of the convolutional networks was developed together with a group of neuroscientists and the programmer Yann LeCun. The work is based on the study of the principles of operation of the human's visual cortex [9-12]. Like the brain, the convolutional network passes information through several hierarchical layers of pre-processing where each layer performs a specific task. There are three main layers that constitute the CNN.

The convolution layer (figure 2.a) is constructed in such a way that graphically encodes some feature, such as the presence of a sloping line at a certain angle. Then the next layer, formed because of convolution operations with a weights matrix, shows the presence of the sloping line in the treating layer and its position, forming so-called feature maps. The convolutional neural network does not have one set of such weights but a whole range of them, encoding all kinds of lines and arcs from different angles. However, such convolution kernels are not given in advance rather than formed independently through training the network by the classic method of backpropagation [13,14,15,11].

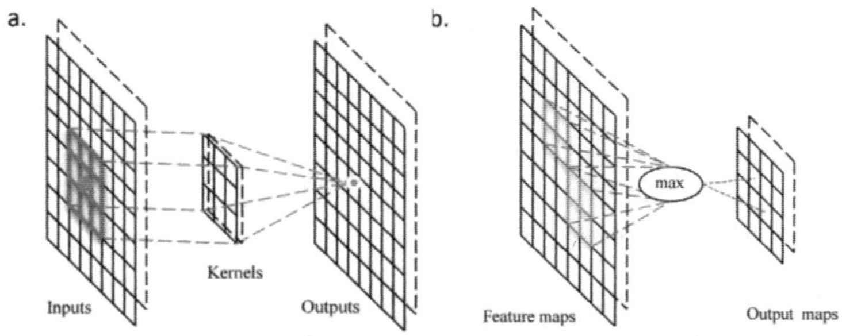

Figure 2. (a) Illustration of the operation principle of the convolution kernel convolutional layer and (b) the subsampling layer. On the left on each of the figures, the feature maps of $n-1$st layer are displayed and on the right the feature maps of n-th layer are depicted, obtained after the using of the corresponding operation.

Subsampling operation layer (figure 2.b) performs a reduction of a dimension of the formed feature maps. The subsampling refers to the process when from the several neighboring neurons of a feature map one is selected that has the maximum output value (max-pooling). Later, it forms one neuron of the next feature map with the reduced dimensions. Due to this operation, in addition to accelerating the further calculations, the network becomes more invariant to the scale of the input image.

Ordinary neuron layer – it is the fully-connected feedforward hidden layers that were previously described in subsection 2.1.

Thus, such architecture of CNNs is interleaving of convolutional, subsampling and ordinary layers of neurons, often namely in that order (see figure 3). Also, sometimes instead of a single layer, several consecutive layers of convolution or sub-sampling are used. Next, neural networks will be designated according to their main hyperparameters. For example, the convolutional network architecture in figure 3 will be marked as *CNN_8_16_50*, where the values 8 and 16 denote the numbers of the feature maps in the first and second convolutional layer respectively, and the number 50 corresponds to the size of the penultimate layer of fully-connected neurons.

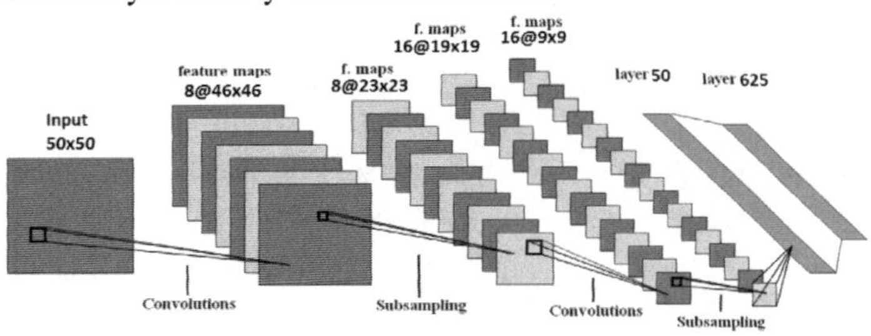

Figure 3. An example of the architecture of the convolutional neural network used in this work.

3. Experiment and used processing image techniques

3.1. Experiment setup

In order to train and test the developed artificial neural network on real experimental data, we visualized a bubbly free jet by PFBI technique (for details refer to [16, 17]). The bubbly jet flow was organized in a hydrodynamic rig. The experimental setup is of closed type and continuous operation. Its test section is made of organic glass with the following dimensions (HxWxL): 425x300x300 mm. The operating fluid was distilled water, which was driven by a centrifugal pump equipped with a control module to vary the rotor speed. The flow rate was measured by an ultrasonic flowmeter. The water temperature was maintained constant at 30±0.1 °C in the test section by means of a thermostatic regulator filled with tap water as a heat-transfer agent and containing a tubular electrical heater, cooling circuit and aquarium pump to permanently blend the heat carrier. Cooling and heating in the thermostat was implemented by a PID-control system activating an electromagnetic valve to start/stop the coolant supply to the cooling circuit and a switching relay to loop/break the electric line to the heater. The liquid temperature was measured by heat-variable resistors placed in the test section, heat exchanger and settling tank.

Figure 4. An example of the architecture of the convolutional neural network used in this work.

The submerged free jet was formed by a round contraction nozzle of 110 mm height with an inner diameter at its outlet edge D_N = 15 mm (see figure 4). The ratio of the inlet and outlet cross-section areas of the nozzle was 16. The Reynolds number based on the mean flow velocity $U_0 = 4\,Q_W/(\pi \cdot D_N^2) = 0.663$ m/s, where Q_W is the volume flow rate of the water (measurement error is 2%), and D_N equaled to 12,500. In order to saturate the flow with bubbles, an air-water mixer of a special design was used to provide a quasi-monodisperse size distribution of the bubbles (e.g., see figure 9.a). The air was supplied to the mixer by a compressor through a couple of filters of rough (5 μm) and fine (0.01 μm) cleaning. The air overpressure before the mixer was specified in the range of 9.8–19.6 kPa depending on the air volume flow rate Q_A by a high-precision reducer. The air temperature, pressure and flow rate were measured by a thermal mass flowmeter (TSI model 4140) with the measurement uncertainties 0.1 °C, 0.1 kPa and 2%, respectively. The length of the pipe with 20 mm inner diameter between the nozzle and mixer was 0.56 m. Different volume gas fractions $\beta = Q_A/(Q_W + Q_A)$ in the jet were achieved by changing Q_A using a needle valve. In experiments, β possessed four values: 0, 1, 2 and 3%. The mean diameter of the bubbles D_B was approximately 0.75 mm for all β.

For planar fluorescence imaging (PFBI approach), the water was merged with Rhodamine 6G as a fluorescent dye, the concentration of which was very low (about 20 μg/l). So, the water properties, especially viscosity and surface tension, can be considered unchanged. In order to illuminate and register the bubbles were suspended in the flow. PIV-system with a high temporal resolution consisting of a pulsed Nd:YAG Photonics Industries DM-532-50 laser (wavelength 532 nm, pulse, pulse energy 5 mJ and pulse duration 160 ns at repetition rate 10 kHz), Photron FASTCAM SA5 CMOS-camera (digit capacity 12 bits, resolution 1,024x1,024 px, acquisition rate 7 kHz) equipped with a Nikon AF Micro-Nikkor 60 mm f/2.8D lens and a low-pass optical filter (transmission edge at 570 nm) and Berkeley Nucleonics Corporation pulse/delay generator (model 575) for external synchronization was used. The thickness of a laser light sheet was about 0.8 mm in the measurement region. The distance between the camera and the measurement plane was roughly 360 mm. In the experiment, the recording rate was 2 kHz. The raw data (a series of 10,000 PFBI snapshots for each regime) was obtained continuously during 5 seconds.

3.2. Algorithm for training images generation

As it was described in subsection 2.2, in order to train a neural network, it is necessary to have training samples. In this work we propose to use the synthetic images, simulating the experimental data, for training the CNNs to search bubbles in images and determine their geometric centers. It is possible to generate the required number of test samples automatically with the known correct responses in advance. For training, the appearance of a bubble image in a square interrogation area of an image was simulated (see figure 5). A situation of hitting of the center of a bubble within an interrogation area was modeled in order that the algorithm could give an accurate assessment of the position of the bubble center. In case of the image of the bubble center went outside of the area, so that part of the bubble image was within the area or totally out of it, the algorithm should give a negative answer. The current situation with the overlapped bubble images was not modeled in this work. The image of a bubble was drawn as an ellipse with the specified center, values of the semi-axes and the rotation angle of semi-major axis with respect to the horizontal direction. The rim of a bubble was added to a black background and superimposed with Gaussian noise and images of tracer particles in the form of uniformly distributed impulse noise, smoothed by Gaussian filter with the kernel size equaled to 1. The intensity distribution profile of the bubble rim by its radius was generated based on the physical model described in [2].

Then a normalized brightness map of the generated image is used as the input data. As for the output data, we used a value map of the following function:

$$P(x,y) = e^{-(a\ x-x_0)^2-2b(x-x_0)(y-y_0)+c(y-y_0)^2)} \cdot \{\mathbf{1}\ if\ image\ has\ bubble, \mathbf{0}\ -\ otherwise\} \qquad \mathbf{1)}$$

- the function of the normal Gaussian distribution (normalized to unity), which reflects the probability of finding a pixel close to the geometric center of the bubble x_0, y_0). All of the coordinates in 1) are measured in pixels px) of a Cartesian coordinate system with the origin at the lower left corner. The coefficients in (1) are responsible for the rotation of the distribution by an angle θ (the angle corresponding to the rotation of the ellipse on the generated image), and are calculated as follows:

$$a = \frac{\cos(\theta)^2}{2\sigma_x^2} + \frac{\sin(\theta)^2}{2\sigma_y^2}; b = \frac{\sin(2\theta)}{4}\left(\frac{1}{\sigma_y^2} - \frac{1}{\sigma_x^2}\right); c = \frac{\cos(\theta)^2}{2\sigma_y^2} + \frac{\sin(\theta)^2}{2\sigma_x^2} \qquad 2)$$

The values of the dispersion parameters σ_x, σ_y in 2) were chosen so that the function 1) satisfyied the condition:

$$P(\tilde{x},\tilde{y}) \leq 0.01\ \forall\ \tilde{x},\tilde{y}) \in\ the\ rim\ of\ the\ drawn\ bubble\ (\mathbf{3})$$

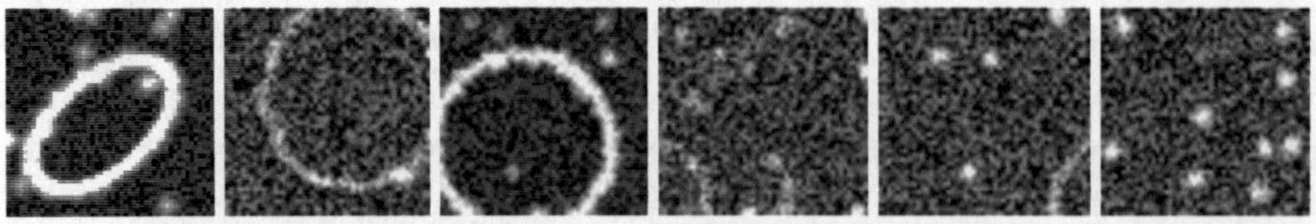

Figure 5. Examples of the generated synthetic images for training of the neural network.

3.3. *Bubble detection algorithm*

The processing algorithm of experimental images operated as described below in the algorithm 1. A square window moved through the experimental image, scanning by the trained neural network. In each window, if a bubble was found, then the most probable coordinates of its center were marked. Then after a pass through the entire image, obtained nearby centers were united, according to their probability weights, as belonging to the single physical bubble. Moreover, centers of a bubble $\mathbf{x_1} = x_{11}, y_{12})$ и $\mathbf{x_2} = (x_{21}, y_{22})$ were considered close to each other if $\| \mathbf{x_1} - \mathbf{x_2} \|_{L_2} \leq R$, where R is a parameter, depending on gas volume void fraction of the flow. On average, for the gas volume void fraction of $3 - 4\%$, the optimal value, which was obtained during several computational experiments, equals $R \approx 15\, px$.

The algorithm also uses probability thresholds $tresh_1$, $tresh_2$, characterizing the probability p_0 and p_{xy} to consider that object in the current window is a bubble and consider whether that (x, y)-th pixel is inside the bubble shape, respectively. Generally speaking, the optimum values of these threshold parameters (which was confirmed by the experiment) can be defined numerically on the basis of characteristics of the probability distribution (3) as $\overline{tresh_1} \approx 0.2; \overline{tresh_2} \approx 0.23$ (considering that the average bubble radius is about $15 - 20$ pixels).

Additionally, the option of using regression models (a nonlinear least squares method and its modifications) in order extract the unknown sizes and the rotation angle of a bubble by fitting the distribution of the output of the neural network was tested. However, this method was not used further because of the necessity to specify starting parameters of an approximating distribution, long calculation time (on the order of seconds for a window of 25x25 units), and, most importantly, due to the high probability the algorithm's divergence.

1. In an input image, a window with the size 50×50 px is moved;
   A parameter s (in px) of the window shift is fixed;
2. In the w_i-th window the p_0^i threshold is calculated by the CNN's outputs:

$$p_o^i = \frac{1}{N} \sum_{x,y)\in W_i}^{50\times 50} p_{xy}^i \cdot \begin{cases} 1, p_{xy}^i \geq \mathbf{tresh_2} \\ 0, p_{xy}^i < \mathbf{tresh_2} \end{cases}; N = \sum_{x,y)\in W_i}^{50\times 50} \begin{cases} 1, p_{xy}^i \geq tresh_2 \\ 0, p_{xy}^i < tresh_2 \end{cases}$$

3. ***If*** $p_0 > \mathbf{tresh_1}$, ***then*** the W_i contains a bubble with center $x_0, y_0) = argmax_{\,x,y)\in W_i}(p_{xy}^i)$;
4. Connection of k nearby centers $\{(x_{0j}, y_{0j}), j = 0 \ldots k\}$ into the single one:

$$x_{0i}, y_{0i}) = \frac{1}{k \cdot p_{max}^i} \sum_{j=1}^{k} (x_{0j}, y_{0j}) \cdot p_0^j; p_{max}^i = \max_{j=0\ldots k}(p_0^j);$$

Algorithm 1. The bubble detection algorithm.

3.4. Description of the finally chosen method

In this work for both of the above tasks, about 50 various architectures of the CNNs and the MLPs was tested in each group respectively. Selection of hyperparameters (layers' sizes, learn rate etc.) conducted automatically, using the Bayesian optimization method [18]. The experimental framework is implemented on C/C ++. In order to simplify the program's user interface, the Python wrapper has been written, and all "mathematics" - has been implemented using C ++ linear algebra libraries BLAS/LAPACK to improve performance. To prevent overfitting, it used a variety of methods [19] - it managed to obtain the absolute additive in accuracy in the test on the synthetic images about 7-9%.

The experimental data set consists of 12,000 images of 50 × 50 px, generated as it was described in subsection 3.2. Input vectors for both networks $\left(x_{11}^{d}, \ldots, x_{5050}^{d}\right)$ are a two-dimensional array of brightness values of each pixel: $x_{lm}^{d} \leftrightarrow I(x_l, y_m)$. The output vectors $\left(t_1^{d}, \ldots, t_{625}^{d}\right)$ are an unwrapped one-dimensional array of probabilities: $t_i^{d} \leftrightarrow P(x_i, y_i)$. We used the following algorithms: *Adam, AdaDelta, AdaGrad, SGD* and their various modifications [15] during the computational experiments of determination the optimal settings for the neural network training as the optimizers of the functional quality. The error function was the log likelihood function, which showed, in average, slightly better convergence speed (5-10% faster) of minimization of the quality function [15], than the quadratic error function.

Table 1. Examples of the studied neural networks.

Architecture	Parameters ($\times 10^3$)	Weight (KB)
MLP_100_100_100[a]	335	1298
MLP_256	800	3100
CNN_8_16_50	100	380
CNN_8_16_100	196	754
CNN_10_20_100	230	880

[a] Multi-layer perceptron (or *DNN*-deep neural network) with 100 neurons in hidden layers.

4. Results

4.1. Processing of synthetic images

Hereinafter, the results of a pair of the best tested architectures are presented for each of the groups, respectively. The next two figures of neural networks tests were obtained on the test data set of 1000 synthetic images through the training with the following experimental parameters:

- CNN: SGD-Nesterov with momentum 0.9
- DNN: SGD with momentum 0.9
- Examples/batch = 10; learn rate = 0.0003; 1000 test samples and 11000 train samples
- *GPU*: Nvidia GeForce GTX 980 Titan Z, 6 Gb RAM
- *CPU*: Intel Core i5 - 2.7 GHz; SDD

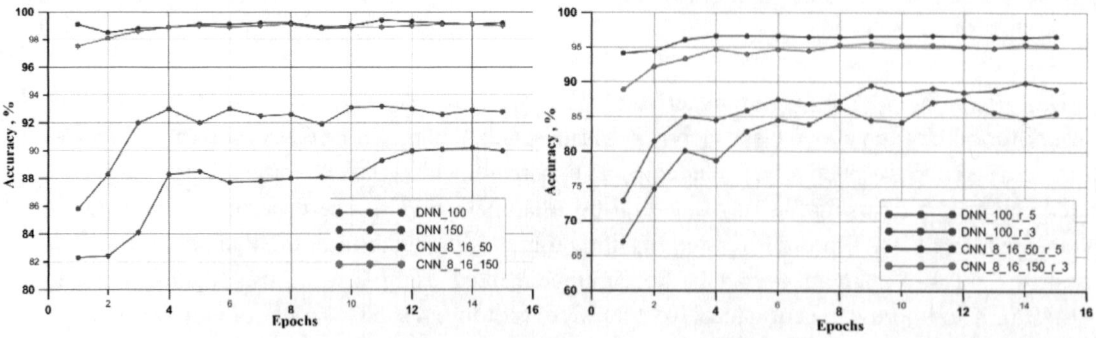

Figure 6. Correctly recognized bubbles on synthetic images (left) and correctly identified bubbles' geometrical centers with three pixel accuracy (right).

It is seen from figure 6 that the convolutional networks provide much better results than DNN networks. They show themselves better also in the problem of determining the exact center of the bubble. The accuracy of determining the size and orientation of the bubble (average for three parameters a, b and θ value in %) for different architectures almost completely (up $\approx 0.5\%$) coincides with the corresponding lines in figure 6 right. The following graph (figure 7) reflects the rate of convergence of different optimization methods for error of log-likelihood function for the first problem.

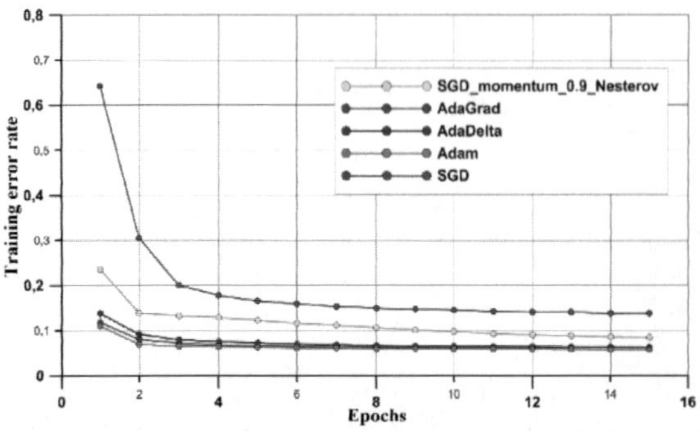

Figure 7. Minimization of the quality function by the various optimizers during the training of the CNN_8_16_50.

Algorithm	Epoch number[a]	Accuracy, %
Adam	**1**	**94.1**
AdaDelta	3	91.8
AdaGrad	2	90.3
SGD-Nesterov	7	93.2
SGD	11	91.6

Table 2. Efficiency of the optimization algorithms for CNN_8_16_50[a]. Epoch number, when the accuracy on the test data exceed 90%.

The Adam algorithm showed itself as the most effective (tables 2 and 3) during testing. The table 3 shows the training time in seconds of the network on CPU and GPU. It is clear from the graphs that on GPU the convolutional network learns very quickly, which allows the researcher to generate new types of training images and almost immediately verify the effectiveness of this new model.

Thus, after the numerous experiments the architecture CNN_8_6_50 (figure 3) was identified as the most effective. It is also worth to note, that on average the usage of GPU with parallelized calculations speeds up the training process up to 8-10 times. Subsequent analysis of the experimental data was carried out precisely by this neural network. Table 3 shows the best obtained the results for the trained CNN_8_16_50 model during synthetic image processing in the course of the experiments.

Table 3. Time of training the CNN_8_16_50 for various optimizers, seconds

Algorithm	CPU[a] (15 epochs)	CPU[b] (accuracy>90%)	GPU[a] (15 epochs)	GPU[b] (accuracy>90 %)
Adam	755	**48**	127	**9**
AdaDelta	740	100	124	30
AdaGrad	735	95	123	17
SGD-Nesterov	710	322	117	53
SGD	**690**	506	**114**	89

[a] Total time spent for the appropriate number of iterations;
[b] Total time necessary for reaching the respective accuracy.

Table 4. Final results on the synthetic images.

Identification accuracy of a bubbles, %	Center of bubble definition accuracy, %
99.8	**96.4**[a]

[a] With precision of 3 pixels;

4.2. *Comparison of the processing performance*

Table 5 shows average values of the total time required for processing of an image with the size of 1280x1024 pixels for the CNN_8_16_50 neural network and the correlation algorithm processing time. According to the tests, it was found that on average the proposed method realized in GPU is faster than correlation approach almost for 8-10 times.

Table 5. Processing time of the image with the size 1280 × 1024 px.

Processor	CNN[a] total time, sec	CNN[a] time for window, ms	The correlation algorithm total time, sec
CPU	**54.7**	**5**	≈80
GPU	**8.1**	**0.8**	not parallelized

[a] Window shift size = 10 pixels.

Table 6. Comparison of the correlation algorithm and the network CNN_8_16_50, trained with the accuracy 99.8% on 1000 test images.

Shift, px	$tresh_1$	$tresh_2$	Total number of bubbles	Number of errors[a]	Correctly recognized bubbles[b]
5	0.2	0.15		**5**	91-93
7	0.19	0.22		15	**94-96**
10	0.12	0.04	≈100	20	89-91
15	0.1	0.09		10	88-90
The correlation algorithm				30	80-85

[a] ≈number of unmarked bubbles and marked as bubbles blank areas: *false negative* + *false positive rates*

[b] ≈number of marked bubbles relative to the total number of bubbles in percentages: *true negative rate*

Journal of Physics: Conference Series **754** (2016) 072002 doi:10.1088/1742-6596/754/7/072002

Table 6 shows the performance of the correlation method and the convolutional network during the processing of the experimental image with the size 1280 × 1024 px (figure 8). Obviously, it is impossible to automatically assess this kind of results and table 6 gives approximate figures for several dozen different experimental images processed manually. According to the results, the neural network identifies more bubbles on average of 10-15% and makes less mistakes in 8-10% of cases. Moreover, accuracy of detection of a center of the bubbles for both methods is almost the same.

4.3. Processing of experimental images

Figure 8 shows the examples of how the same processed image section looks with the different shift values (subsection 3.3 algorithm 1). Only windows with bubbles chosen by the network are shown. The results obtained by the bubble detection algorithm are shown below.

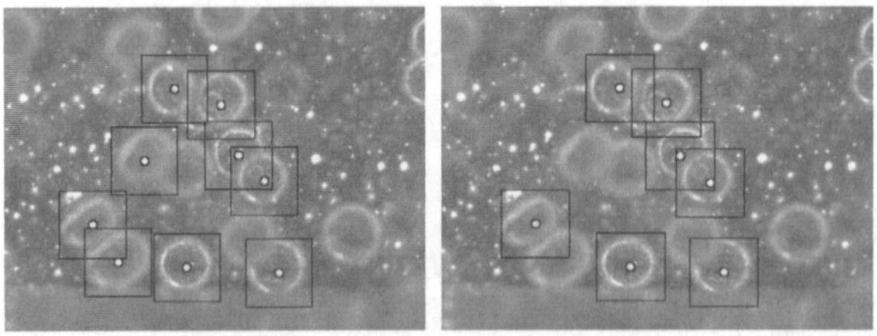

Figure 8. The results of the bubble detection algorithm. The marked centers refer to the interrogation area and not to the bubbles.

Left: $s = 5px$; $tresh_1 = 0.2$; $tresh_2 = 0.03$; *right:* $s = 10px$; $tresh_1 = 0.24$; $tresh_2 = 0.03$.

Figure 9 shows the examples of processing results of the experimental images by the convolution network and the correlation algorithm.

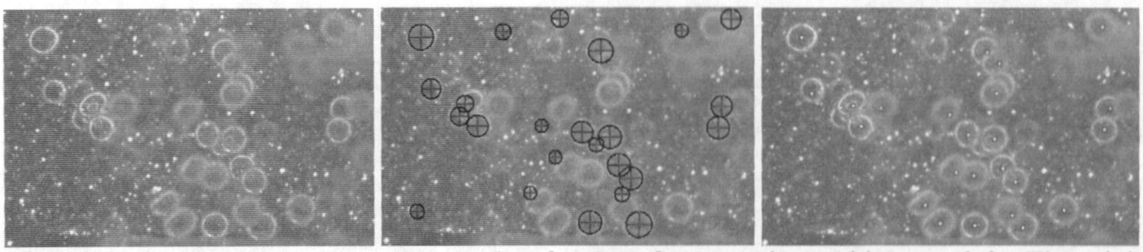

Figure 9. An example of the processing results of a part of an experimental image. *(left):* The original image and *(center):* The processing by the correlation algorithm. *(right):* The processing by the neural network

5. Conclusions

This paper demonstrates that modern technology of artificial neural networks can be successfully applied for experimental data processing of fluid dynamic experiments. As a result of this work, the software was developed and tested based on the deep learning algorithm for bubble center detection and synthetic bubble image generation for a neural network training. A number of different neural network architectures were studied. The CNN was chosen as the most efficient architecture. Preliminary tests of the bubble detection upon experimental images were conducted with the software.

A key aspect of our study is the end-to-end system including creating synthetic data set for training of the fully customizable convolution neural network model and implementation of methods for validation the trained models. Our machine learning based approach showed significantly faster processing time and better accuracy compared with existing approaches. Time for processing a standard image of 1280x1024 pixels takes about 8 seconds on GPU unlike 80 seconds to be taken by the CPU-based not parallelized correlation algorithm. Moreover, the neural network identifies 10-15% more bubbles on average and makes less mistakes in 8-10% of cases.

Future work will be focused on addition of the overlapped bubbles for synthetic training set and the improvement of physics of bubble simulation as well as developing reliable methods for detection of bubble sizes.

Acknowledgments
The work was financially supported by the Russian National Foundation (Grant No. 14-19-01685, guided by D.M. Markovich).

References

[1] Lance M, Bataille J (1991) Turbulence in the liquid phase of a uniform bubbly air–water flow. J of Fluid Mech. 222. pp 95–118

[2] V.M. Dulin, K.S. Pervunin, D.M. Markovich. A Technique for Round Bubbles Imaging via Planar Fluorescence // 17th International Symposium on Applications of Laser Techniques to Fluid Mechanics. Lisbon, Portugal, 07-10 July, 2012

[3] Lindken R, Merzkirch W (2002) A novel PIV technique for measurements in multiphase flows and its application to twophase bubbly flows. Exp Fluids 33. pp 814 – 25

[4] I. Goodfellow, Y. Bengio and A. Courville. Deep Learing. // An MIT Press book. 2016

[5] D. G. Lowe. Distinctive Image Features from Scale-Invariant Keypoints.// Computer Science Department. University of British Columbia, Vancouver, B.C., Canada. 2004

[6] N. Dalal and B. Triggs. Histograms of Oriented Gradients for Human Detection. // INRIA Rhone-Alps, Montbonnot 38334, France. 2004

[7] F. Rosenblatt Principles neurodynamics. Perceptron and the theory of brain mechanisms. // M.: Mir, 1965. p 480

[8] Osovsky C. Neural networks for processing information. Trans. from Polish I.D.Rudinskogo. // M .: Finance and Statistics, 2002. p 344

[9] Y. Le Cun, B. Boser, J.S. Denker, D. Henderson, R.E. Howard, W. Hubbard, L.D. Jackel, et al. Handwritten digit recognition with a back-propagation network. In Advances in neural information processing systems, 1990.

[10] Y. LeCun, F.J. Huang, and L. Bottou. Learning methods for generic object recognition with invariance to pose and lighting. In Computer Vision and Pattern Recognition, 2004. CVPR 2004. Proceedings of the 2004 IEEE Computer Society Conference on, volume 2, pages II–97. IEEE, 2004.

[11] Y. LeCun, K. Kavukcuoglu, and C. Farabet. Convolutional networks and applications in vision.

[12] Y. LeCun, L. Bottou, Y. Bengio, and P. Haffner. Gradient-based learning applied to document recognition.// Proceedings of the IEEE, 1998.

[13] Kingma, D., and Ba, J. "Adam": A method for stochastic optimization. // In International Conference for Learning Representations, p 201

[14] Nesterov Yu E. Methods of the convex optimization. // MTNMO, Moscow. 2010

[15] O. Dozat. Optimizing CNNs. Stanford // ArXiv preprint. 2016

[16] Akhmetbekov Ye.K., Alekseenko S.V., Dulin V.M., Markovich D.M., Pervunin K.S. Planar fluorescence for round bubble imaging and its application for the study of an axisymmetric two-phase jet // Experiments in Fluids. 2010. V. 48, No. 4. pp 615 – 29

[17] Dulin V.M., Markovich D.M., and Pervunin K.S. The optical principles of PFBI approach AIP Conf. Proc. 1428. 2012. pp 217–24

[18] Z. Brochu, V. M. Cora, N. de Freitas. A Tutorial on Bayesian Optimization of Expensive Cost Functions, with Application to Active User Modeling and Hierarchical Reinforcement Learning. // ArXiv preprint. 2010

[19] N.Srivastava, G. Hinton, A. Krizhevsky et al. Dropout: A Simple Way to Prevent Neural Networks from Overfitting. // Department of C.S., University of Toronto. Toronto, Ontario, M5S 3G4, Canada

Investigation of mass transfer in swirling turbulent flames

D Sharaborin[1,2], S Abdurakipov[1,2], V Dulin[1,2]

[1]Kutateladze Institute of Thermophysics, 1 Ak. Lavrentyeva Avenue, Novosibirsk, 630090, Russia
[2]Novosibirsk State University, 2 Pirogova Street, Novosibirsk, 630090, Russia

E-mail: sharaborin.d@gmail.com

Abstract. The present paper reports on analysis of flow structure and turbulent transport in swirling flames. The particle image velocimetry and spontaneous Raman scattering techniques were used for the measurements of 2D velocity and density distributions. The focus was placed on comparison between low- and high-swirl flows. A pronounced bubble-type vortex breakdown with strong flow precession took place in the latter case.

1. Introduction

Lean premixed combustion regimes are implemented in burners to achieve reduced emission of NO_x [1]. However, lean combustion regimes are known to be sensitive to different types of perturbations. In particular, a minor change in equivalence ratio can result in unwanted thermoacoustic instability [2], [3], [4]. Flow swirl is often used for flame stabilization, especially under severe conditions, such as combustion of ultra-lean mixtures with equivalence ratio close to the flammability limit. Vortex breakdown phenomenon is a feature of high-swirl flows. When the swirl rate (the axial flux of the angular momentum divided by the flux of the axial momentum and nozzle radius) of a non-reacting swirling jet exceeds a critical value of S_C (this value is found to be different in various papers: $S_C \approx 0.6$ in [5], $S_C \approx 1.1$ in [6], $S_C \approx 0.88$ in [7]), the vortex core breakdown takes place and a central recirculation zone often occurs [7].

Flow structure of swirling jets with combustion is more complex due to heat release, induced density gradients and other effects. It is known that shape of the vortex breakdown is sensitive to temperature differences between the jet and the surrounding fluid [8], [9]. Vortex breakdown can be delayed or, on the contrary, can be induced by the combustion. A number of studies were carried out by Cheng and others (e.g., [1], [10], [11], [12]) demonstrating that moderate swirl (without central recirculation zone) can be used for flame stabilization and provides less NO_x emissions in comparison to high-swirl flows with the recirculation zone. In this context it is clear that the flow organisation is important for flame stability, combustion efficiency, and emission levels for swirl burners. Comprehensive experimental data on flow structure and dynamics in swirl-stabilized combustors is necessary for deeper understanding of the flame behaviour and for development of more efficient burners. In the present study Particle image velocimetry (PIV) and spontaneous Raman scattering (SRS) are optical measurement tools were used to retrieve information about fluid structure, heat and mass transfer processes in swirling flames for moderate and high swirl rates.

2. Experimental setup

Sketch of the experimental setup for 2D measurements is shown in figure 1. The measurements were carried out in a combustion rig consisted of a swirling nozzle, flow seeding device, premixing pipe and section for the air and propane flow rate control. A vane swirler inside the contraction nozzle (exit diameter d = 15 mm) could be changed to vary the swirl rate. By using two swirlers with different inclination angle of blades, the swirl rate S (based on the definition in [5]) was 0.41 and 1.0. The Reynolds number Re_{air} (based on d, the bulk velocity U_0, and viscosity of the air) was 5 000. The equivalence ratio Φ of the propane-air mixture was 0.7.

In order to provide PIV measurements, the flows were seeded by TiO_2 particles with the average diameter of one micrometer. The seeding device instantly introduced solid particles to the air flow by using a stirring rod. The PIV system composed of a double-pulsed Nd:YLF Pegasus PIV laser and a pair of PCO 1200HS CMOS cameras. System was running at approximately 770 Hz frequency. A laser sheet, formed by the system of lenses, passed thru the central plane of the flows and had a thickness of 0.8 mm in the measurement region. The cameras were equipped with narrow-bandwidth optical filters admitting the light from the laser and suppressing the radiation of the flame. The system was operated by a computer with "ActualFlow" software. Stereo calibration was performed using the multi-level calibration target and a 3rd-order polynomial transform. For each flow case, 2400 instantaneous three-component velocity fields were measured.

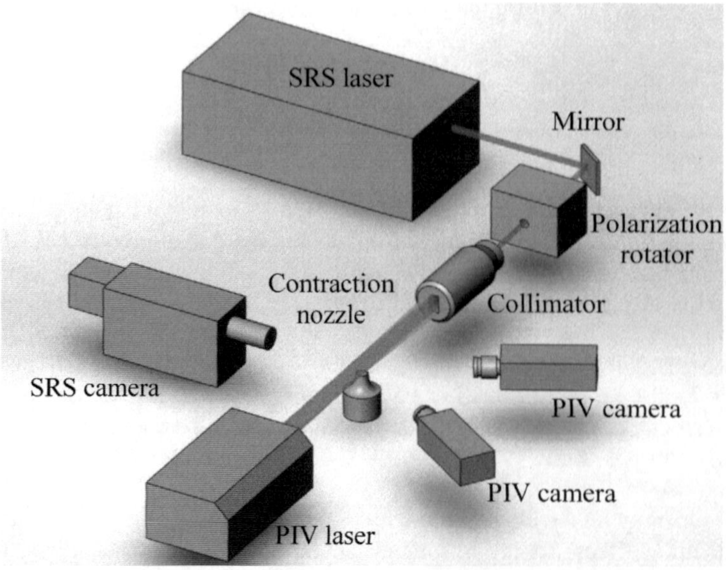

Figure 1. Sketch of the experimental setup for 2D SRS and stereo PIV system

The second (532 nm) harmonic of a pulsed Nd:YAG laser (Quanta-Ray) was used for the illumination of a sampling volume. The energy of 6 ns laser pulses was monitored by an energy meter (Coherent LabMax-TOP). On average, the energy was 690 mJ for 532 nm, with 5% RMS fluctuations. For registration of 2D images of the average SRS signal, a 16-bit ICCD camera (Princeton instruments PI-MAX-4) equipped with a tunable optical filter (VariSpec LC), based on liquid crystals, was used to collect the emission of the vibronic transitions (e.g., 607.3 nm and 473.3 for Stokes and anti-Stokes lines of ro-vibrational transition of nitrogen when excited by 532 nm). To provide planar illumination, the cylindrical laser beam was transformed into a laser sheet by using collimating optics. An additional multi-notch holographic filter was used to block the emission from Rayleigh scattering, which is about 10^4 times greater than the SRS and was not blocked completely by the tunable filter. The scattered light from 200 laser pulses was accumulated during acquisition of each image with 8×8 pixels binning.

The collection time of each pulse was 200 ns. Thirty frames were captured and averaged for each measured flame regime to reduce influence of CCD noise.

The data processing procedure of 2D SRS signal is similar to that of Rabenstein and Leipertz [13]. The laser sheet was passed through the room air and the flame. The room air was considered as the control volume (temperature of the air was monitored). Following Egermann et al. [14], the SRS signal were recorded for two perpendicular polarizations of the laser light. In order to minimize influence of the dark-current, reflections, and possible fluorescence the images for the P-polarization were subtracted from images for the S-polarization. Intensity of the Stokes lines depends on the temperature and volume fraction of the molecules. Intensity of the Stokes and anti-Stokes lines of SRS by nitrogen molecules was used to evaluate ratio between density in the flame and room air. According to [13], the dependence of the effective SRS cross-section on temperature was evaluated from calculations by RAMSES code [15].

3. Results

Average flow characteristics for the cases $S = 0.41$ and $S = 1.0$ are presented in figure 2. For the low-swirl jet ($S = 0.41$), the local minimum of the axial mean velocity is observed within the jet center. The positive value of the mean axial velocity indicates the absence of a permanent recirculation zone, and no clear vortex breakdown could be determined from the average velocity field. At the same time, weak local reserve flows were observed in realizations of the instantaneous velocity fields near the jet axis.

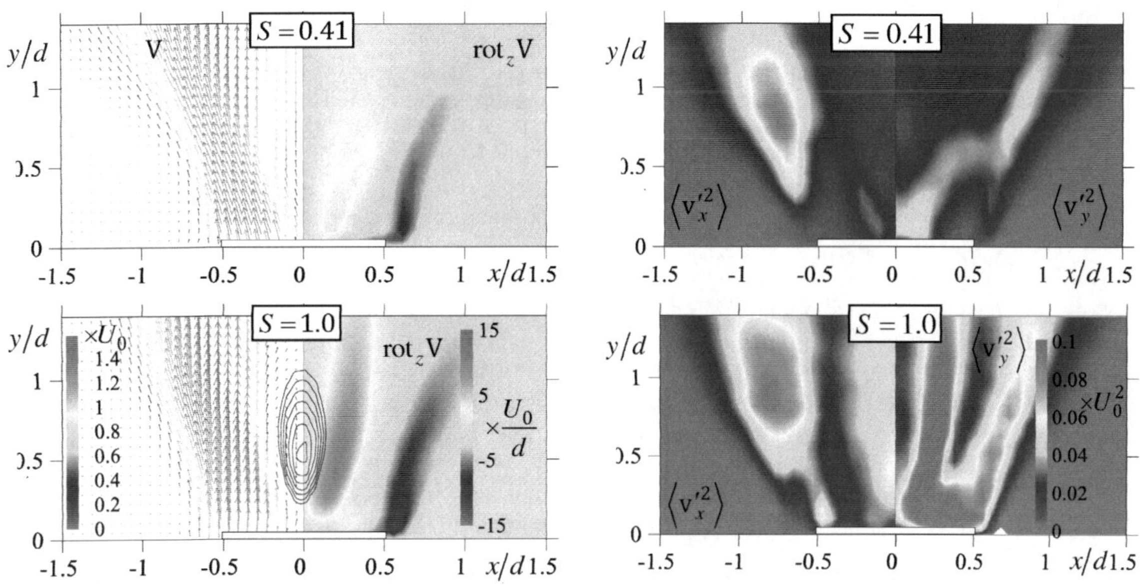

Figure 2. Mean velocity and vorticity distributions (left) and intensity of velocity fluctuations (right)

Shear of the axial velocity near the jet axis and outer mixing layer resulted in roll-up of eddies, which induced velocity fluctuations. For the high-swirl jet at ($S = 1.0$) a pronounced vortex breakdown took place with a bubble-type central recirculation zone. The mean velocity reached negative minimum value at the jet axis. From analysis of instantaneous velocity fields it was found that large-scale vortices emerged inside the recirculation zone and in the outer mixing layer were more intensive than in low-swirl case.

Figure 3 shows the photographs and distributions of the average concentration of molecular nitrogen in the premixed swirling flames measured by 2D SRS approach. Shape of the low-swirl flame corresponds to an inverted cone with the apex above the nozzle exit. The high-swirl flame is also an inverted cone with the apex inside the nozzle.

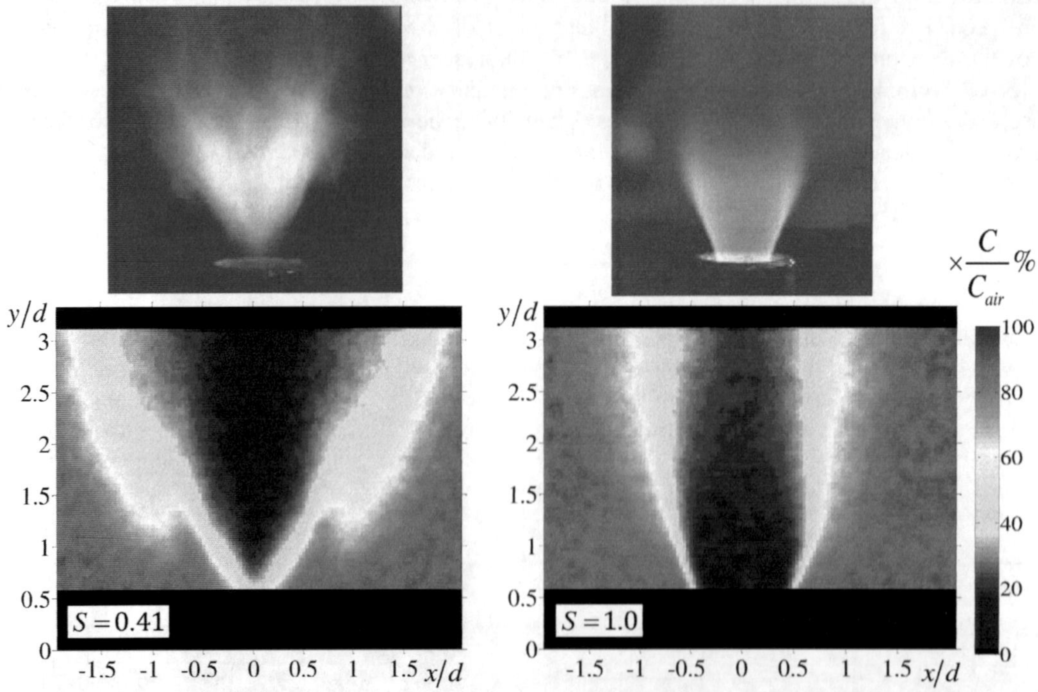

Figure 3. Photographs (top row) and distributions (bottom row) of the average density of nitrogen in low-swirl ($S = 0.41$) and high-swirl ($S = 1.0$) lean propane-air flames ($\Phi = 0.7$) measured by 2D SRS

Analysis of the spatial distributions of the density in the propane-air swirling flames revealed presence of a low-density region around the jet axis, inside the recirculation zone (inside the flame cone) and in the inner mixing layer around it. Low gas density was also detected in the outer mixing layer of the low-swirl flame due to the heat transfer from the center of the swirling flow. For analysis of the turbulent transport one can analyse the averaged equation of N_2 mass conversation [16]:

$$\left\langle \dot{\rho\omega}_{N_2}\mathbf{v}\right\rangle + \mathrm{div}\left(\left\langle \rho\omega_{N_2}\right\rangle\left\langle \mathbf{v}\right\rangle\right) + \mathrm{div}\left\langle \rho'\omega'_{N_2}\mathbf{v}'\right\rangle - \mathrm{div}\left(D_{N_2}\overline{\rho\mathrm{grad}\omega_{N_2}}\right) = \overline{r_{N_2}} \approx 0 \qquad (1)$$

Where ρ and $\mathbf{v}$ are the local gas density and velocity, correspondingly; ω_{N_2} is the local mass fraction of nitrogen molecules; r_{N_2} is production/consumption rate of N_2 during chemical reactions; D_{N_2} is the diffusion coefficient for nitrogen. Dot, brackets, and prime symbols denote temporal derivative, ensemble averaging, and fluctuating part, respectively. Local concentration of nitrogen molecules C_{N_2} is linearly proportional to $\rho\omega_{N_2}$. The first term in the left hand side of the equation vanishes for stationary turbulent flows. The second and third terms are convective and turbulent transports respectively. The fourth term is the diffusion, which is small for turbulent flows. The production/consumption rate of the nitrogen can be considered to be negligible. Thus, one can obtain:

$$-\mathrm{div}\left\langle \rho'\omega'_{N_2}\mathbf{v}'\right\rangle \approx \mathrm{div}\left(\left\langle \rho\omega_{N_2}\right\rangle\left\langle \mathbf{v}\right\rangle\right) \sim \frac{\partial C_{N_2}V_x}{\partial x} + \frac{\partial C_{N_2}V_y}{\partial y} \quad (2)$$

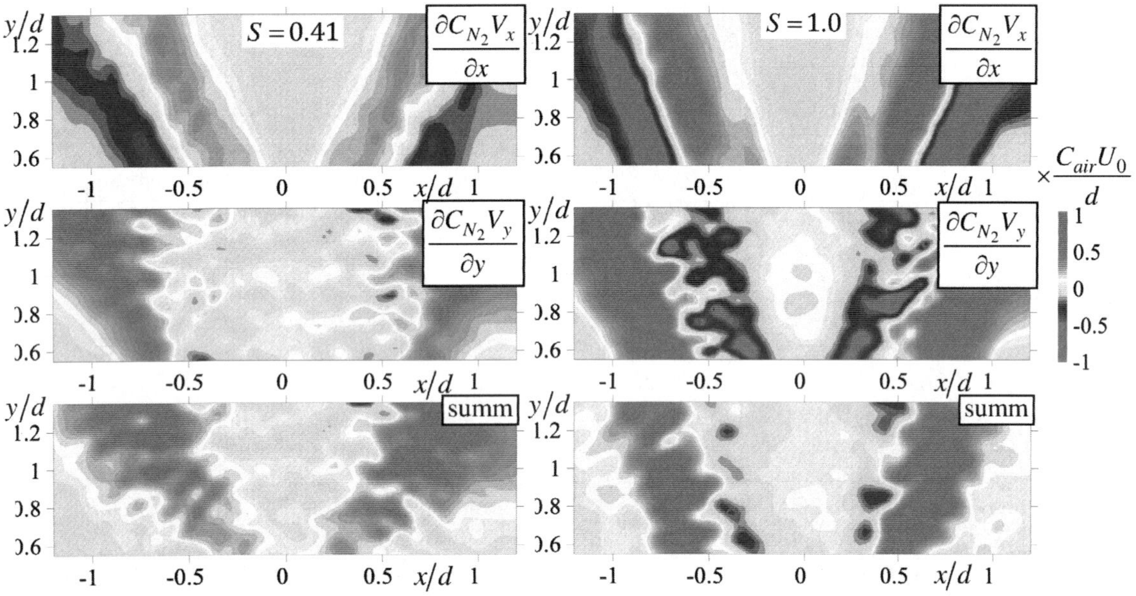

Figure 4. Convective transport terms of nitrogen molecules in low- ($S = 0.41$) and high-swirl ($S = 1.0$) flames

Figure 4 shows components of the convective transport term in the low- and high-swirl jets. Significant differences were observed between two cases near the jet axis. In the case of the high-swirl flame the presence of the permanent recirculation zone resulted in negative values of $\partial(C_{N2}V_y)/\partial y$ in the inner mixing layer and positive values around the jet axis.

4. Conclusion

2D spontaneous Raman scattering and particle image velocimetry techniques have been used to study flow structure and transport in swirling lean propane-air premixed flames. Average flow velocity and local density of nitrogen were analysed. The focus was placed on comparison between low- and high-swirl flows with swirl rate $S = 0.41$ and 1.0, respectively. The results showed that the low-swirl flow had a local minimum of the axial mean velocity at the jet axis. Positive value of the mean axial velocity indicated the absence of permanent recirculation zone. For the high-swirl flow, a pronounced vortex breakdown took place with bubble-type central recirculation zone. From the analysis of convective transport term it was concluded that the high-swirl flame was characterized by recirculation of the hot gas between the inner mixing layer and revers flow.

Acknowledgments

Financial support has been obtained from Russian Science Foundation (grant № 16-19-10566).

References

[1] Dunn-Rankin D 2008 *Lean Combustion: Technology and Control* (Amsterdam: Elsevier)

[2] Lieuwen T 1999 *Proc. Int. Gas Turbine and Aeroengine Congress and Exhibition (Indianapolis)* vol 2 *(ASME)* Paper 99-GT-003 pp 1-11

[3] Weigand P, Meier W, Duan X, Stricker W and Aigner M 2006 Investigations of swirl flames in a gas turbine model combustor: I. Flow field, structures, temperature, and species distributions *Combustion and flame* **144** 205-24.

[4] Meier W, Weigand P, Duan X and Giezendanner-Thoben R 2007 Detailed characterization of the dynamics of thermoacoustic pulsations in a lean premixed swirl flame *Combustion and Flame* **150** 2-26
[5] Gupta A, Lilley D and Syred N 1984 *Swirl Flows* (Tunbridge Wells, Kent: Abacus Press)
[6] Oberleithner K, Paschereit C, Seele R and Wygnanski I 2012 Formation of turbulent vortex breakdown: intermittency, criticality, and global instability *AIAA journal* **50** 1437-52
[7] Liang H and Maxworthy T 2005 An experimental investigation of swirling jets *J. Fluid Mech.* **525** 115-59
[8] Mourtazin D and Cohen J 2007 The effect of buoyancy on vortex breakdown in a swirling jet *J. Fluid Mech.* **571** 177-89
[9] Oberleithner K, Stöhr M, Im S, Arndt C and Steinberg A 2015 Formation and flame-induced suppression of the precessing vortex core in a swirl combustor: experiments and linear stability analysis *Combustion and Flame* **162** 3100-14
[10] Cheng R 2006 Low Swirl Combustion *The Gas Turbine Handbook* (Washington DC: DOE)
[11] Legrand M, Nogueira J, Lecuona A, Nauri S and Rodriguez P 2010 Atmospheric low swirl burner flow characterization with stereo PIV *Exp. Fluids* **48** 901-13
[12] Alekseenko S, Dulin V, Kozorezov Y, Markovich D, Shtork S and Tokarev M 2011 Flow structure of swirling turbulent propane flames *Flow, turbulence and combustion* **87** 569-95
[13] Rabenstein F and Leipertz A 1998 One-dimensional, time-resolved Raman measurements in a sooting flame made with 355-nm excitation *J. Appl. Opt.* **37** 4937-43
[14] Egermann J, Seeger T and Leipertz A 2004 Application of 266-nm and 355-nm Nd:YAG laser radiation for the investigation of fuel-rich sooting hydrocarbon flames by Raman scattering *J. Appl. Opt.* **43** pp 5564-74
[15] Hassel E 1993 Ultraviolet Raman scattering measurements in flames by the use of a narrow-band XeCl excimer laser *J. Appl. Opt.* **32** 4058-65
[16] Warnatz J, Maas U, Dibble R and Warnatz J 2001 *Combustion* vol. 3 (Berlin: Springer)

The 2D method for determining the temperatures field of the gas flow at the outlet of a multi-channel heat exchanger

E Yu Slesareva[1,3], S L Elistratov[2], and V V Ovchinnikov[1]

[1]SB RAS, Kutateladze Institute of Thermophysics, 630090 Novosibirsk, Russia

[2]Novosibirsk State Technical University, 630092 Novosibirsk, Russia

Corresponding author: styuardessa@yandex.ru

Abstract. The method for experimental determination of energy efficiency in the multi-channel heat exchanger was tested. The visualization of a temperatures field has been performed to determine the thermal structure of gas flows with the use of fast-response fine-meshed wire. Thermograms of the temperature fields of the multi-channels assembly at the outlet were registered by thermal imaging camera. Results show that the 2D method provides a sufficient time resolution for the temperature field for the steady-state gas flow regime, heat generation, and nonsteady regime. The 2D method allows us to determine the gas stream parameters at the channel outlet in real time, which are necessary for determining the efficiency of the heat exchanger. Qualitative and quantitative characters of temperature changes in the thermograms are consistent with modern physical understanding of the gas flow in channels.

1. Introduction

Mini and microchannel heat-exchanging apparatuses and gas reactors are the integral parts of innovative directions in the heat power engineering. Experimental data on the composition and temperature of gaseous reaction products, gained immediately at the outlet of such apparatuses, allow us to optimize their operation efficiency.

To visualize the temperature field in the medium, transparent for infrared rays, it is possible to use the heated slim tethers and meshes as indicators of temperature. The wire mesh is used in the thermographic method for experimental studying the thermal structures of the gas flow [1]. Using the thermographic method makes it possible to receive the thermogram of temperature changes inside the gas flow. In the present paper, the results of the 2D thermographic method [2] are presented for the multi-channel heat exchangers.

2. Measurement procedure

Experiments on measuring the nonstationary temperature fields at the outlet from six quasitriangular multi-channels were carried out on the model assembly, consisting of seven cylindrical heaters of equal diameter. The cross-section and internal passage section of the channel are presented in figure 1. Seven electric heaters (2) with the diameter of 5.9 mm and length of 0.5 m, which form six quasitriangular minichannels (3) (No. 1 ÷ No. 6) (triangle ABC consists of arcs AB, BC, CA), were positioned in a horizontal quartz tube (1) with the outer diameter of 23 mm and length of 0.5 m.

The heat losses from the packaging heaters to the atmosphere were decreased by using kaolinite wool (4) (7-mm layer) and non-flowing air spaces (5).

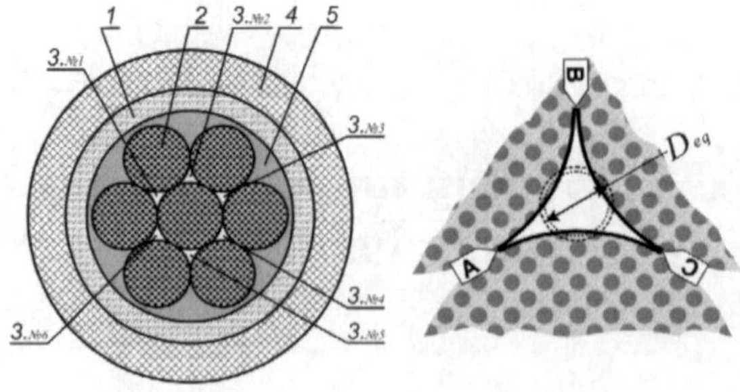

Figure 1. The cross-section (a) and shape (b) of the passage section of quasitriangular channel of multichannel assembly: 1 - quartz tube; 2 - heater; 3 - the passage channel (No. 1 ÷ No. 6) for gas; 4 - thermal insulation; 5 - air space; A, B and C - vertexes of the quasitriangular channel.

Table 1. Parameters of the passage channels in model assembly

No.	D_{eq}	S	Π	No.	D_{eq}	S	Π
	mm	mm^2	mm		mm	mm^2	mm
1	1.18	2.74	9.29	4	1.15	2.66	9.25
2	1.28	2.98	9.31	5	1.20	2.78	9.27
3	1.31	3.04	9.28	6	1.16	2.68	9.24

The total heat exchange area of the passage section of quasitriangular channels is 0.0275 m^2. Table 1 shows the parameters of a passage section in channels obtained by planimetry measurements in the pictures of the outlet of cross-sections (D_{eq} is hydraulic diameter, S is sectional area of the passage section of the channel, Π is length of wetted perimeter of the passage section of the channel). Figure 1b shows the visual relationship between the sizes of the passage section of the channel and its hydraulic diameter D_{eq}. The minimum (1.2 mm) and maximum (1.3 mm) values of D_{eq} in assembly are shown by the dashed lines. The difference in diameters is caused by inaccuracy of electric heaters in a bundle and ovality of cylindrical heaters. The thickness of the mesh-thermal detector wire is about 80 microns; cell size of the mesh is 0.3 × 0.3 mm. The wire is pressed close to the outlet surface of the assembly. Thermograms have been obtained by infrared imager NEC TH7100 with wavelength $\lambda = 8 \div 12$ microns. Time resolution is 300ms, spatial resolution is 100μm.

The dependence between a change in temperature T and time t was interpolated by dependence:

$$T(t) = T_0 + \Delta T \cdot \left[1 - \exp\left(\frac{t}{\tau}\right)\right] \tag{1}$$

where T_0 is initial temperature, ΔT is temperature difference, τ is typical scale time.

3. Discussion of results

The 2D thermographic method has been tested for determining the temperatures fields of the gas flow for stationary and non-stationary conditions of heating in cylindrical gas heaters at various air flow rates. Processing the thermograms allow us to receive the information about the temperature field in the gas flow at each passage section of minichannels.

Figure 2 shows four typical thermograms of samples at different times after the burst of thermal load of 4.8 W on the central heater with total air flow rate of 0.45 g/s and temperature-time graph of the average temperature in each of six channels. The temperature non-uniformity is caused by natural convection in the heating assembly and variation in individual gas flows in the channels, caused by differences in their areas of flow.

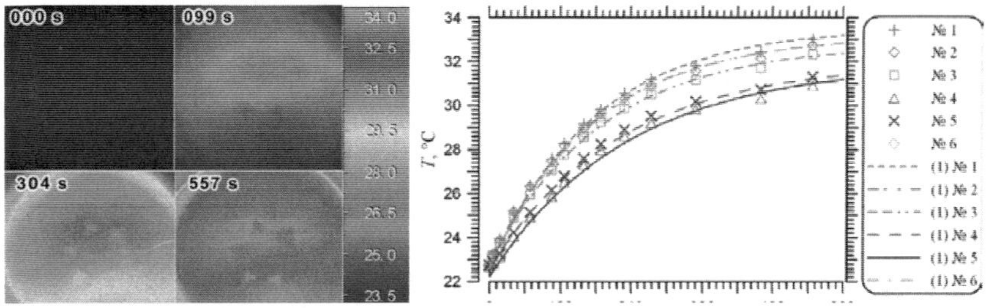

Figure 2. The thermograms of a wire mesh in the outlet cross-section after the heat load surge on central heater in the assembly and time graph of the average temperature in each of six channels.

Table 2. Parameter values for 'equation (1)' under the conditions in figure 2.

No. →	1	2	3	4	5	6
T_0, °C	22.6	22.6	22.6	22.6	22.6	22.6
ΔT, °C	11.0	10.4	10.0	9.4	9.7	10.6
τ, s	180	180	190	220	230	180

Figure 3 shows thermograms for the case of a decrease in the total air flow rate from 0.45 g/s to 0.26 g/s, under heat load of 33.6 W proportioned uniformly on all the heaters. Sampling starts from the moment of a decrease in the total flow rate of gas. It is easy to see that after the decrease in the total air flow rate the increase of temperature in the outlet cross-section takes place due to the gain of heating up of air in quasitriangular minichannels.

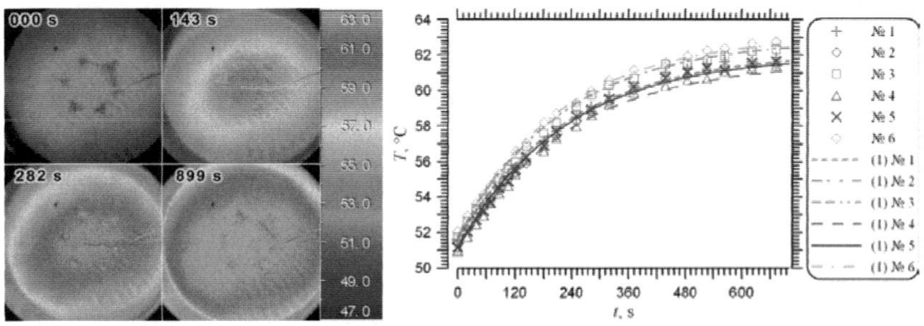

Figure 3. The thermograms of a wire mesh in the outlet cross-section after a decrease in the total flow rate of air and time graph of the average temperature in each of six channels.

Table 3. Parameter values for 'equation (1)' under the conditions in figure 3.

No. →	1	2	3	4	5	6
T_0, °C	51.6	51.2	51.7	50.9	51.1	51.8
ΔT, °C	10.6	10.8	11.2	10.7	10.9	11.3
τ, s	232	225	221	226	223	218

Figure 4 presents the variation in time of average temperature $T_{out}(t)$ in each of quasitriangular channels (marker) in comparison with calculated value of stationary temperature T_{out} (line). Calculation takes into account the redistribution of gas flow rate for each of the channel passage sections due to the difference of the area of passage section of the channels [3]. Thus, it takes into account the changing in the conditions of heat exchange along the length of the channel. In the present experiment the total air flow rate is 0.45 g/s, heat load on central heater is 4.8 W, surge of heat load on six peripheral heaters is 28.8 W. Sampling starts from the moment of the surge of heat load on peripheral heaters in the model assembly. Processing the thermograms has shown that after transition to the stationary conditions, the temperature of air at the outlet of assembly corresponds to the calculated value for the stationary flow.

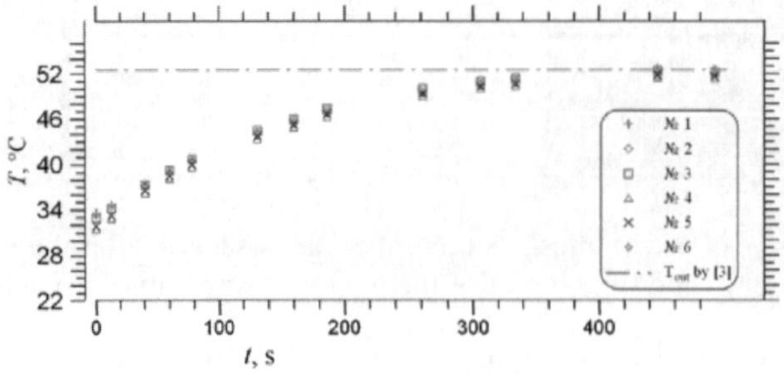

Figure 4. Time graph (upper axis) of the average temperature in each of six quasitriangular channels $T_{out}(t)$ after the heat load surge on peripheral heaters in the assembly. Calculated value of air temperature T_{out} by [3].

Table 4. Parameter values for 'equation (1)' under the conditions in figure 4.

No. →	1	2	3	4	5	6
T_0, °C	33.4	33.3	32.8	31.5	31.9	32.7
ΔT, °C	20.3	20.5	20.7	21.3	21.5	21.5
τ, s	173	175	157	160	175	180

The measured value of typical time scale τ is in good agreement with an estimate of time necessary for warming up the assembly to the stationary temperature field and it is determined by the heat release rate with the heat loss from the assembly taken into account. The thermograms allow us to see the 2D patterns of nonsteady preheating of gas in quasitriangular minichannels, and dynamics of preheating of model assembly in full detail. Let us note that it is necessary to use the infrared imagers with smaller wavelength λ for increasing accuracy of measurements in operations with channels of a smaller size.

4. Conclusion

Testing the 2D thermographic method for determining the thermal structure of gas flow using fast-response fine-meshed wire, at the outlet of the model of multi-channels assembly has shown the sufficient time resolution. Qualitative and quantitative characters of temperature changes in thermograms are consistent with modern physical understanding of the gas flow in channels.

The thermographic method is universal for studying the processes of stationary and non-stationary heat exchange. Accuracy of the method is determined by characteristics of the studied object and the equipment used for thermal imaging.

This work was supported by the Russian Foundation for Basic Research, project No. 16-38-00502 mol_a.

References

[1] Elistratov S L, Slesareva E Yu 2015 *EPJ Web of Conferences* **82** 01060

[2] Elistratov S L, Slesareva E Yu, Ovchinnikov V V 2015 *J. of Engineering Thermophysics* **24(4)** 346

[3] Sadik Kakac, Yaman Yener, Anchasa Pramuanjaroenkij 2013 *Convective Heat Transfer, Third Edition* (CRC Press) p 622.

Journal of Physics: Conference Series **754** (2016) 072005 doi:10.1088/1742-6596/754/7/072005

Non-zero helicity of a cyclonic vortex over localized heat source

A Sukhanovskii, A Evgrafova, E Popova

Institute of Continuous Media Mechanics - Academ. Korolyov, 1, Perm, 614013, Russia

E-mail: san@icmm.ru

Abstract. Experimental and numerical study of the steady-state cyclonic vortex from isolated heat source in a rotating fluid layer is described. The structure of laboratory cyclonic vortex is similar to the typical structure of tropical cyclones from observational data and numerical modelling including secondary flows in the boundary layer. Differential characteristics of the flow were studied by numerical simulation using CFD software FlowVision. It was found that helicity in a described system has non-zero value. Physical interpretation of helicity distribution is provided.

1. Introduction

Despite decades of research the problem of tropical cyclogenesis is unsolved and attracts close attention from many scientific groups. The complexity of the problem forces researchers to study tropical cyclogenesis step by step seeing as a main goal the theory that would describe all stages of the tropical cyclone formation. Laboratory model of hurricane-like vortex was proposed and studied in [1-3]. Where rotating layer of fluid with the localized heater in the bottom was considered. The important differences of this experimental approach were more viscous fluids (in comparison with water) and using of a shallow layer. Later, series of the experiments [4] were done for the same configuration using PIV system for velocity measurements. The main focus of [4] was on integral characteristics of the azimuthal flows such as angular momentum and kinetic energy. Detailed study of different constraints of the steady-state hurricane-like vortex were studied in [5]. The three main dimensional parameters that define the vortex structure for a fixed geometry – heating flux, rotation rate and viscosity were varied independently. It was shown that viscosity is one of the main parameters that define steady-state vortex structure. Increasing of kinematic viscosity may substantially suppress the cyclonic motion for fixed values of buoyancy flux and rotation rate. Strong competition between buoyancy and rotation provides the optimal ratio of the heating flux and rotation rate for achieving cyclonic vortex of maximal intensity. It was found that relatively small variation of the rotation rate for the fluids with low kinematic viscosity may remarkably change the cyclonic vortex structure and intensity.

Here we focused our attention on differential characteristics of the convective flow, mainly helicity H which is scalar production of velocity and vorticity vectors. It is known that helicity can be very important in a developed turbulent media. The flow in our experimental system is not fully turbulent but the structure of large-scale flow with high correlation of vertical vorticity and vertical velocity is very promising for helicity formation.

2. Experimental setup

Experimental model is a cylindrical vessel of diameter $D = 300$ mm, and height $H = 40$ mm (Fig. 1). The sides and bottom were made of Plexiglas with a thickness 3 mm and 20 mm respectively. There was no cover or additional heat insulation at the sidewalls. The heater is a brass cylindrical plate mounted flush with the bottom. The diameter of the plate d is 104 mm, and its thickness is 10 mm. The brass plate is heated by an electrical coil placed on the lower side of the disc. Cylindrical vessel was placed on a rotating horizontal table (Fig. 1). Silicon oils with different values of kinematic viscosity were used as working fluids. In all experiments, the depth of the fluid layer h was 30 mm and the surface of the fluid was always open. The room temperature was kept constant by air-conditioning system, and cooling of the fluid was provided mainly by the heat exchange with surrounding air on the free surface and some heat losses through sidewalls. For low values of kinematic viscosity it takes about 2 hours to obtain a steady-state regime. Temperature inside the fluid layer was measured at mid-height ($z = 15$ mm), near the periphery (about 3 cm from the sidewall) by copper-constantan thermocouple. It was used for the estimation of the mean temperature of the fluid. The velocity field measurements were made with a 2D particle image velocimetry (PIV) system Polis. The system included a dual pulsed Nd-YaG laser, a control unit, a digital CCD camera (11 megapixels), placed in a rotating frame, and a computer. The synchronization of the operation of the laser and the CCD camera, the measurement, and the processing of the results were performed using the software package Actual Flow. Cylindrical vessel works as a lens and narrow horizontal light sheet from the periphery to the center but all area of our interest in the central part of the vessel was illuminated. Also we need to note that due to strong optical distortions we did not make PIV measurements in a close proximity to the heater at height less than 2 mm. Iterative PIV algorithms [6] and decreasing of the size of the interrogation windows from 32×32 to 16×16 pixels provided a dynamic range of approximately 500 (the ratio of the maximum and minimum resolvable particle displacement). The PIV velocity measurements were accurate to within 5%, estimated from calibration experiments in solid body rotation and long time series.

Along with the dimensional parameters (heating flux, rotation rate and kinematic viscosity) we use the set of the non-dimensional parameters which are commonly used for similar problems and can help for comparison our results with results obtained by other researchers. These are the flux Grashof number Gr_f, non-dimensional rotation velocity Re, Ekman number E and Prandtl number Pr :

$$Gr_f = \frac{g\beta h^4 q}{c\rho\kappa\nu^2} \tag{1}$$

$$Re = \frac{\Omega h^2}{\nu} \tag{2}$$

$$E = \frac{\nu}{\Omega h^2} \tag{3}$$

$$Pr = \frac{\nu}{\kappa} \tag{4}$$

where g is the gravitational acceleration, h is the layer depth, β is the coefficient of thermal expansion, c is the thermal capacity, ρ is the density, ν is the coefficient of kinematic viscosity and κ is the thermal diffusivity, q is a heat flux ($q = P/S_h$, P is the power of the heater and S_h is the heater's surface area). The value of non-dimensional rotation velocity Re is equal to the inverse value of Ekman number E but because of different physical meaning of these parameters it is convenient to use them both.

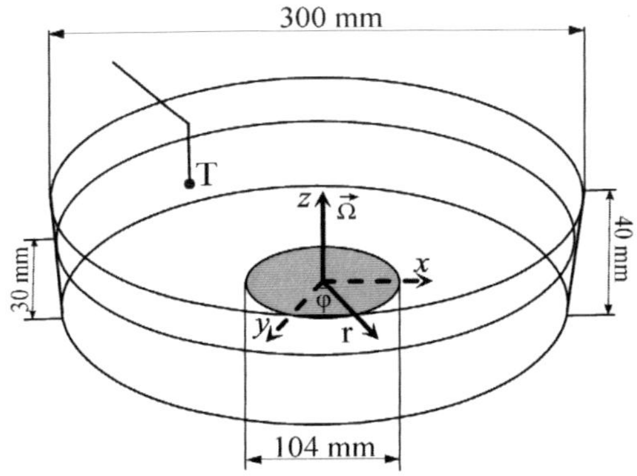

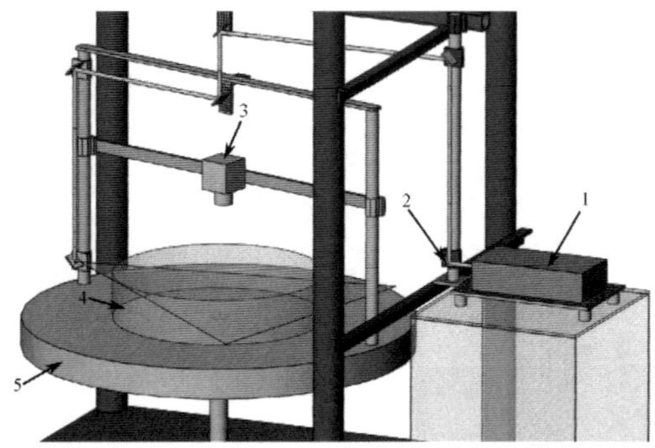

Figure 1. Experimental model, dimensions and location of the coordinate system. T - thermocouple for control of the mean temperature (left); Experimental setup: 1 - dual pulsed laser for PIV, 2 - laser sheet system, 3 - CCD camera, 4 - experimental model, 5 - rotating table (right).

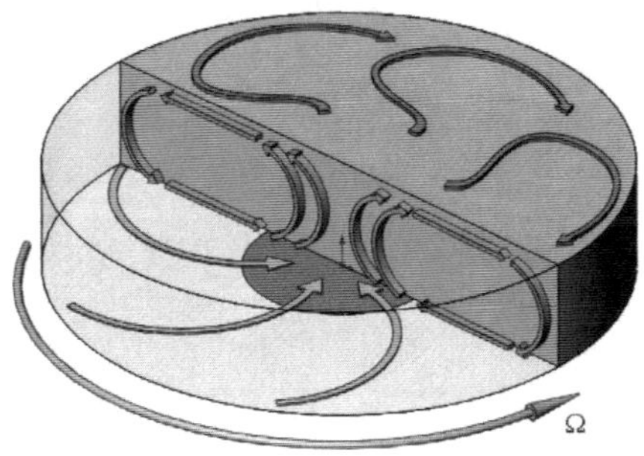

Figure 2. Scheme of large-scale circulation

3. General structure of the flow

The heat flux in the central part of the bottom initiates the intensive upward motion above the heater. Warm fluid cools at the free surface and moves toward the periphery where the cooled fluid moves downward along the side wall. After some time, large-scale advective flow occupies the whole vessel (Fig. 2, vertical cross-section). Experimental measurements of velocity fields in a non-rotating layer, in a vertical cross-section over the heating area showed that instantaneous fields are irregular and asymmetric. Along with the main updraft in the center there are less intensive but pronounced upgoing convective flows close to the periphery of the heater.

The large-scale advective flow in the lower part of the layer leads to the formation of boundary layers with potentially unstable temperature stratification above the heater and makes possible the generation of the secondary convective flows. Structure and specifics of secondary flows over the heater in the case of non-rotating cylindrical layer are described in details in [7].

The structure of the steady-state azimuthal flows (in a rotating frame) for different values of Grashof number is shown in Fig. 3. Positive (negative) values of velocity describe cyclonic (anticyclonic) motion. Distribution of azimuthal velocity is qualitatively similar to the one of a mature tropical cyclone [8]. The cyclonic vortex formation in the laboratory system can be described by following scenario. Large-scale radial circulation leads to the angular momentum transport and the angular momentum exchange on the solid boundaries. Convergent flow in the

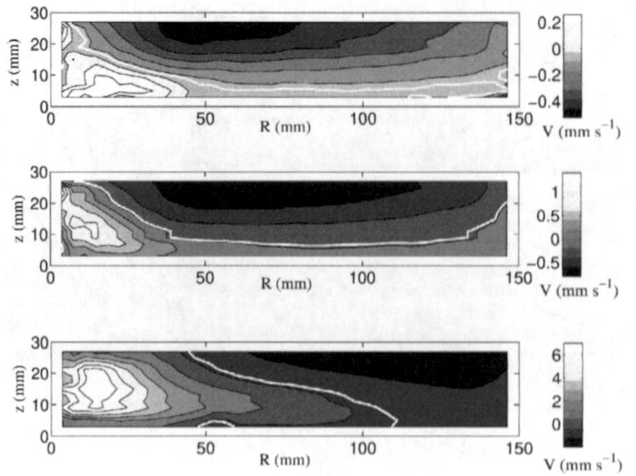

Figure 3. Mean azimuthal velocity fields for: (a) $Gr_f = 2.3 \cdot 10^5$, (b) $Gr_f = 7 \cdot 10^5$, (c) $Gr_f = 1.5 \cdot 10^7$, $\Omega = 0.069$ rad s^{-1}. Positive values describe cyclonic motion, negative - anticyclonic motion, thick solid white line shows the border between cyclonic and anticyclonic flows. Please note that for better representation of the flow structure the colour scale for each case is different.

lower layer brings the fluid parcels with large values of angular momentum from the periphery to the center and produces cyclonic motion (Fig. 2, lower horizontal cross-section). In the upper layer situation is the opposite - divergent flow takes the fluid with low values of angular momentum to the periphery resulting in anticyclonic motion (Fig. 2, upper horizontal cross-section). Friction in the viscous boundary layers leads to the sink of angular momentum in the part of the bottom occupied by cyclonic flow and produces source of angular momentum on the sidewalls when anticyclonic flow comes to the periphery. Zero net angular momentum flux on the solid boundaries is the necessary condition for the steady-state regime. Increasing of the heating flux produces more intensive radial circulation and cyclonic vortex (Fig. 3(b,c)) and pushed anticyclonic flow closer to the sidewalls.

4. Numerical simulation

Numerical simulations by computational fluid dynamics (CFD) software becomes powerful tool for studying of different hydrodynamic processes. For the numerical simulation presented here, we used the CFD package FlowVision. The fluid is assumed to be Newtonian and the flow is considered to be incompressible and laminar. The numerical finite volume code is used to solve the Boussinesq equations for thermal convection. Impermeable and no-slip velocity conditions are applied at the side wall and bottom. The upper boundary was stress-free. The bottom has a localized heat source in the central part defined through a heat flux. Geometry and physical properties of the working fluid were chosen similar to the experiment. In Table 1 values of non-dimensional parameters for presented numerical simulation are showed.

Table 1.

	$Gr_f \cdot 10^6$	Re	E	Pr
FlowVision	4.6	29	0.035	61
Experiment	4.5	27	0.036	60

There are two main mechanisms that can lead to the existence of non-zero helicity in a described system. Fig. 4 shows mean flow structure in a vertical cross-section. As a first mechanism we assume strong correlation between upward flow and vertical vorticity in a central area, over the heater. As a second mechanism we consider strong shear of radial and azimuthal velocities on the periphery which produces helicity. Fig. 5 shows mean helicity in the vertical cross-section (left figure) and mean helicity after azimuthal integration (right figure). As we assumed there is dominance of positive helicity in a central part and weak but integrally substantial negative helicity on the periphery in the upper layer. Integral helicity value is non-zero. As a next step we plan to study helicity distribution and its sources in detail.

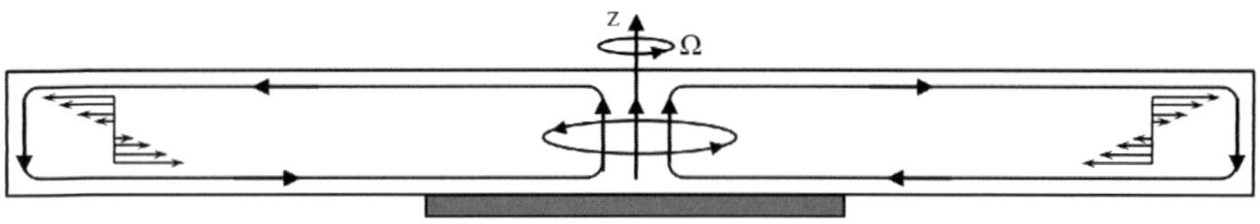

Figure 4. Mean flow structure in a vertical cross-section.

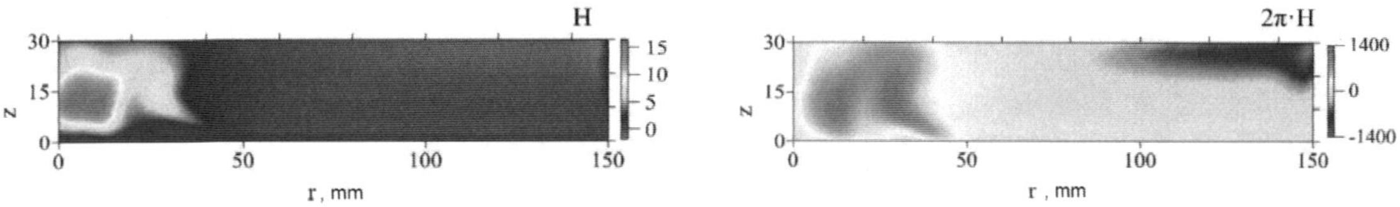

Figure 5. $Gr_f = 4.6 \cdot 10^6$, on the left - mean helicity, on the right - mean helicity after azimuthal integration.

5. Conclusions

Helicity is an invariant of the Euler equations of inviscid fluid flow. It is resulted from strong correlation between velocity and vorticity and its conservation imposes some restrictions on topology of the flow. It comes out that realization of the flow with substantial value of helicity in laboratory conditions is a complex problem. Experimental and numerical study of convection from isolated heat source in a rotating fluid layer showed that laboratory cyclonic vortex is similar to the typical tropical cyclone. Strong correlation between vertical velocity and cyclonic and anticyclonic vorticity makes this system very promising for helicity production. This paper presented new results concerning helicity distribution and its integral value in a laboratory system. Differential characteristics of the flow were studied by numerical simulation using CFD software FlowVision. There are two main mechanisms that produce helicity in a described system. As a first mechanism we assume strong correlation between upward flow and vertical vorticity in a central area, over the heater. As a second mechanism we consider strong shear of radial and azimuthal velocities on the periphery which also produces helicity. It was found out that there is dominance of positive helicity in a central part and weak but integrally substantial negative helicity on the periphery in the upper layer. The main result is that integral helicity of the described flow is non-zero.

This work was supported by the Russian Science Foundation (grant No. 16-41-02012).

References

[1] Bogatyrev GP 1990 *JETP Letters* **51** 557-559.

[2] Bogatyrev GP and Smorodin BL 1996 *JETP Letters* **63** 25-28.
[3] Bogatyrev GP et al. 2006 *Izvestiya Atmospheric and Oceanic Physics* **42** 423-429.
[4] Batalov et al. 2010 *Geophys. Astrophys. Fluid Dyn.* **104** 349368.
[5] Sukhanovskii et al. 2016 *Quart J.R.Met.Soc.* **142** 2214-2223.
[6] Scarano and Riethmuller 2000 *Exp. Fluids* **29** 51-60.
[7] Sukhanovskii et al. 2016 *Physica D* **316** 23-33.
[8] Smith et al. 2014 *Quart J.R.Met.Soc.* **140** 2638-2649.

Ultra dispersed mixture of PETN and RDX for explosive welding

K A Ten[1,5], E R Pruuel[1,5], A O Kashkarov[1,5], I A Rubtsov[1,5], I A Spirin[2], A L Mikhailov[2], L I Shekhtman[3,5], V V Zhulanov[3,5] and A A Deribas[4]

[1] Lavrentyev Institute of Hydrodynamics of the Siberian Branch of the Russian Academy of Sciences, 15 Lavrentyev Prosp., Novosibirsk, 630090, Russia
[2] RF-NC VNIIEF, 37 Mira Prosp., Sarov, 607190, Russia
[3] Budker Institute of Nuclear Physics of the Siberian Branch of the Russian Academy of Sciences, 11 Lavrentyev Prosp., Novosibirsk, 630090, Russia
[4] The Institute of Structural Macrokinetics RAS, 8 Academician Osipyan Str., Chernogolovka, 142432, Russia
[5] Novosibirsk State University, 2 Pirogova Str., Novosibirsk, 630090, Russia

E-mail: ten@hydro.nsc.ru

Abstract. Mixed formulations of ultrafine PETN+soda and RDX+soda have low detonation velocity and small critical diameters, which makes them attractive for application to new technological processes, such as welding explosion [1]. The above properties of these formulations are due to the use of nanopowders of PETN and RDX along with a phlegmatizing agent of sodium bicarbonate. The detonation parameters of these mixtures were studied using synchrotron radiation from the VEPP-3 accelerator (Budker Institute of Nuclear Physics). Techniques we developed were applied to the measurement of density distribution in the detonation front and the width of the reaction zone, as well as volume distribution of pressure, density and spread velocities in detonation of cylindrical charges.

1. Introduction

Synchrotron radiation (SR) as a source of X-rays has a number of unique features. The main one is the possibility of recording quick movies (with a frequency of up to 8 MHz) with a very small exposure ($\tau < 1$ ns). SR has small beam divergence, and use of a precision detector DIMEX enables recording of a multi-frame pattern of density distribution in shock waves and in a detonating high explosive, with good resolution (~ 100 μm).

Experiments to investigate bulk detonating mixtures of ultrafine PETN and RDX with soda (35/65) of a diameter from 10 to 20 mm were carried out at an experimental station on the VEPP-3 accelerator (BINP SB RAS). The difference between these mixed high explosives (HEs) and conventional ones is due to ultrafine PETN and RDX with a grain size of < 1 μm. To this end, the explosive component is dissolved in an organic solvent (a mixture of dimethylformamide, dimethyl sulfide and acetone) and precipitated in water. After stabilization, evaporation and drying, a nanopowder with a grain size of 1.0 to 0.01 μm is obtained [2]. When mixed with sodium bicarbonate (35/65), it is an HE with very small initial density ($\sim$ 0.5 to 1.0 g/cm^3) and low detonation velocity (2 km/s and less). With a small critical diameter ($\sim$ 2 mm), these compositions are very promising for use in explosion welding [1]. All samples of these mixed HEs have been patented and manufactured at VNIIEF (Sarov) [2].

The experiments on VEPP-3 were carried out in two set-ups: longitudinal and lateral measurement of X-ray absorption (a detector was placed along and across the direction of detonation). SR was recorded by a DIMEX detector of in-house development [3]. The measurement procedure is given in [4, 5]. In the first experiments, we obtained the detonation velocity of mixture versus the initial density, as well as density distributions in the detonation front of these compounds. The maximum density values in the Neumann peak were 1.15 g/cm^3 (PETN + soda) and 1.21 g/cm^3 (RDX + soda) with a detonation velocity of 2.3 km/s. The width of the chemical transformation zone of these compositions is 3.0 – 4.0 mm. In the lateral absorption measurement we obtained volume distributions of pressure, density and field of velocities of spread of detonation products.

2. Experimental set-ups and results

2.1. Measurement of passed SR along detonation propagation

The experimental set-up is described in [4, 5] and shown in figure 1.

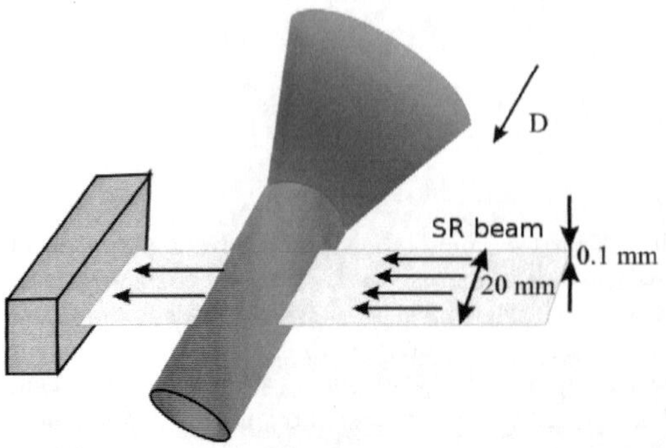

Figure 1. Experimental set-up for longitudinal measurement of absorption. D: direction of detonation in explosive charge (brown); yellow: SR.

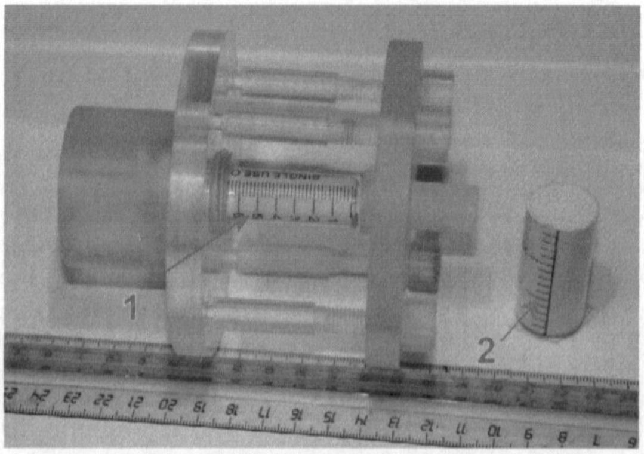

Figure 2. Assemblies with PETN+soda mixture prepared for experiment. 1: 15 ml syringe, 2: 20 ml syringe.

The charge preparation consisted in filling a syringe (20 ml) with the mixture and tapping the syringe until complete disappearance of cavities. The resulting density of the mixture varied from 0.50 to 0.81 g/cm^3 (figure 2).

In the longitudinal measurements, the mass distribution in the SR beam is recorded. In the detonation front there is observed a characteristic increase in the mass, caused by the compression of substance.

The radiation was recorded with a one-dimensional detector DIMEX-3 [3]. The detector has 512 channels with a width of 0.1 mm. The total length of the recording was determined by the width of the windows of the explosion chamber and was 18 mm. The resulting distributions of the passed radiation along the charge (frames) are shown in figure 3. Different colors show successive profiles with a step of 1.5 μs.

$z - t$ diagrams of the position of the front beginning were plotted for the mixtures under study. The resulting points were connected by an approximation line $z = Dt + c$, where z is the coordinate of the front, t is the time, D is the resulting front velocity, c is a constant. In the most cases, the error of determination of D was about 3 to 5%.

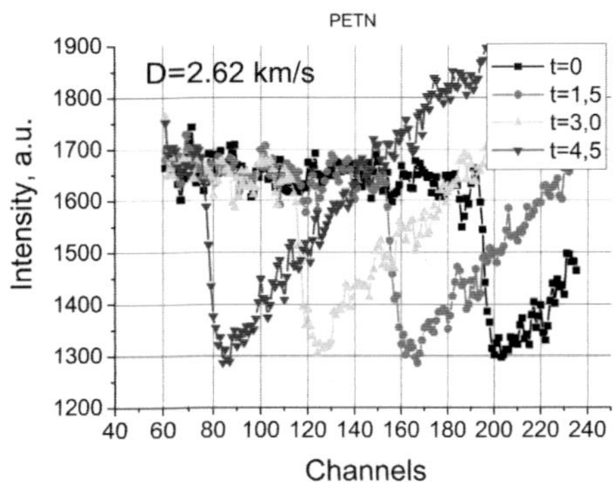

Figure 3. Relative intensity distribution (detector readings) vs. time. Profiles are given with interval of 1.5 μs. RDX + soda mixture; velocity of 2.35 km/s.

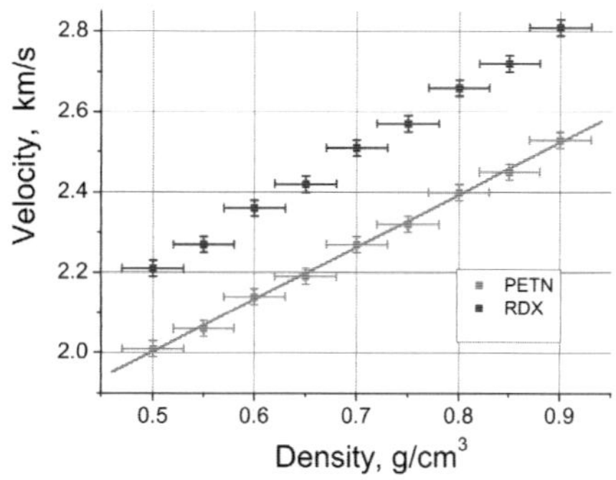

Figure 4. Detonation velocity vs. initial density of PETN+soda and RDX+soda mixtures.

When the initial density of the mixture is increased, the detonation velocity grows linearly (figure 4). The critical size, with which steady detonation in the mixtures was observed, was determined in experiments in plane geometry (figure 5). The cardboard box was 65 mm wide and 95 mm long. The height of the sides (the HE layer height) varied from 2 mm to 22 mm. The initiation was performed with an intermediate (booster) charge of plastic PETN.

Figure 5. General view of the box with PETN + soda mixture. HE layer height H = 3 mm.

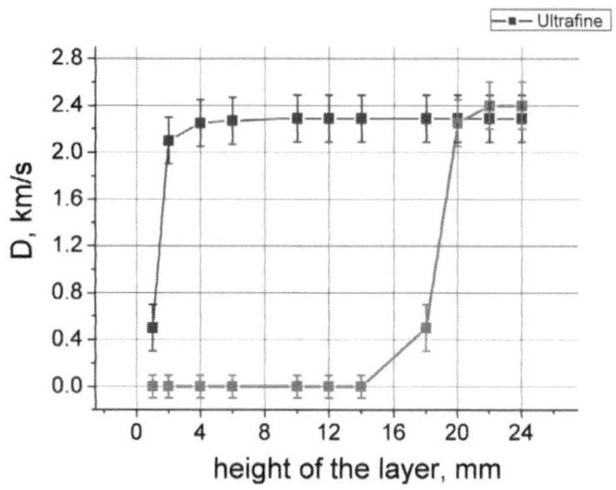

Figure 6. Detonation velocity vs. initial height of PETN + soda mixture layer.

Figure 6 presents the detonation velocity of ultrafine PETN versus the height of the explosive layer (blue line). The red line represents data for conventional PETN, with an average grain size of 30 μm.

The procedure of reconstruction of density distribution in the axial part of HE sample in experiments of this type is detailed in [5]. It consists of several stages. First, the distance dependence of the integral value $m(X) = 2\int_0^R \rho(X,r)dr$, the HE mass along the SR beam, which passes through the charge symmetry axis and perpendicular to it, is determined (R is the radius of the explosive charge). Since the m value varies considerably in the course of HE detonation and subsequent expansion of the explosion products, so the SR absorption spectrum changes. For this reason, each channel of the detector is subjected to pre-calibration, which enables determination of the dependence of the m value on the relative absorption of radiation by the HE under study, J/J_0, where J is the recorded SR flow and J_0 is the incident SR flow. That is done with HE samples of different thicknesses. The result of such calibration shown as the function $\ln(J/J_0)$ of m is close to a linear function and is interpolated with a parabola $\ln(J/J_0) = a_0a_1m + a_2m^2$.

The resulting smoothed dependences averaged over several experiments $\rho(X)$ are presented in figures 7 and 8, where $X = 0$ corresponds to the shock jump. Since no apparent inflections are observed on the resulting density distributions behind the Chapman-Jouget plane we mean the area where the parameters are 30% lower than those in the Neumann peak. Then, the width of the chemical reaction zone in the mixtures will be ~ 3 mm.

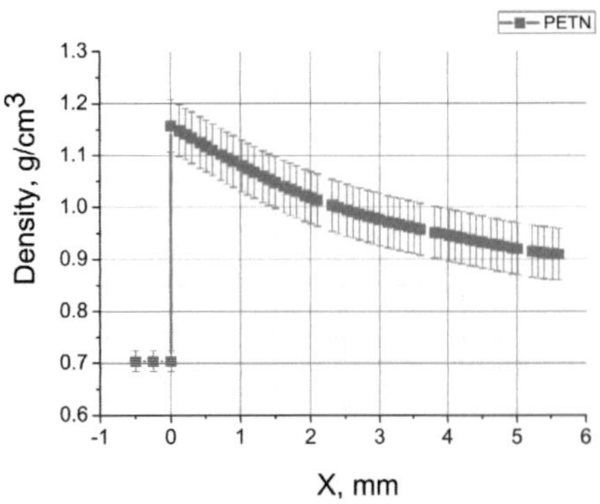

Figure 7. Density profile in PETN detonation front.

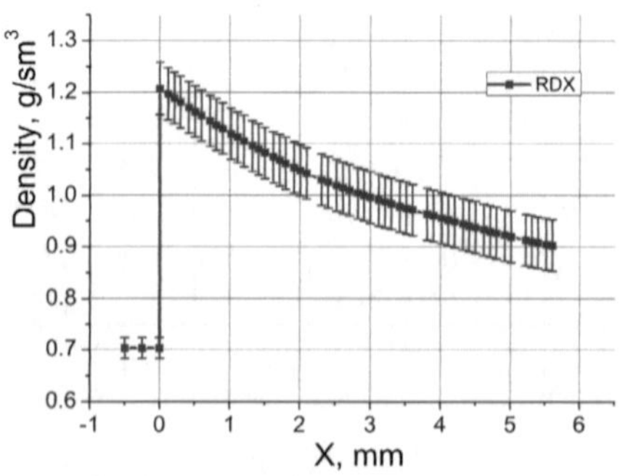

Figure 8. Density profile in RDX + soda detonation front.

An essential feature of the technique applied is the curvature of the detonation front in conjunction with its axial symmetry. When density is reconstructed from the results of recording of radiation passed through a sample, the curvature of the detonation front leads to an error. This error increases dramatically with the distance from the front, because of an assumption of spherical equidense layers in the front. For this reason, the area in which the density is reconstructed with acceptable accuracy is 3 mm at most.

2.2. Lateral measurement of absorption

We investigate the spread of products of detonation of a cylindrical PETN + soda charge in the set-up shown in figure 9. The experimental assembly (figure 2) was the same as in the longitudinal measurements. The lateral distribution of absorption is measured with the period between the SR pulses. The lateral distribution of mass is obtained after calibration of the absorption by the detector (figure 10). The set-up is described in detail in [4,5].

2.2.1. Reconstruction of density $\rho(r,t)$ with fixed z. Iterative method. The density reconstruction method consists in minimizing the mean square deviations of an experimental X-ray shadow (detector readings) from a density calculated from a distribution under study.

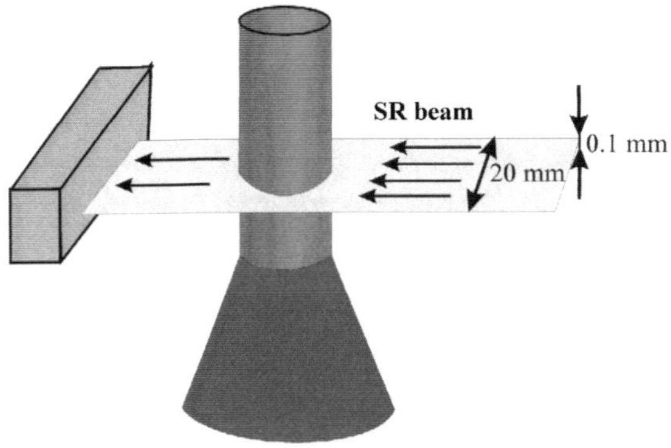

Figure 9. Set-up of experiments on measurement of absorption in spread of explosion products.

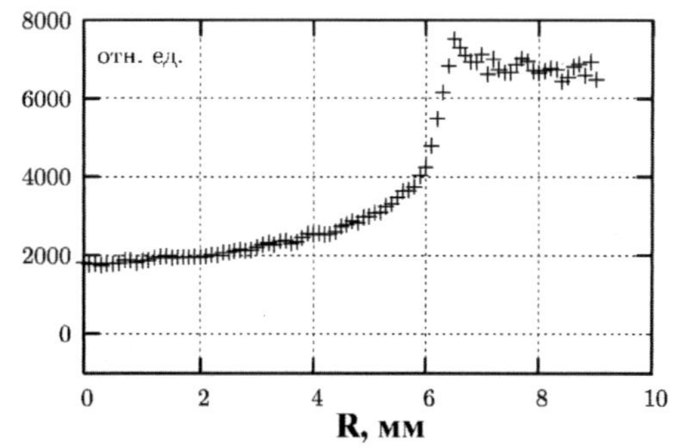

Figure 10. Intensity distribution on beam in detonation of PETN+soda mixture.

The density was sought in the form of a mesh function defined in nodes (figures 11 and 12). The density values $\rho(z,t)$ in the intervals were complemented with interpolating splines. For better accuracy, discontinuities in the desired density distribution were found and the splines were created along weak density gradients as far as feasible.

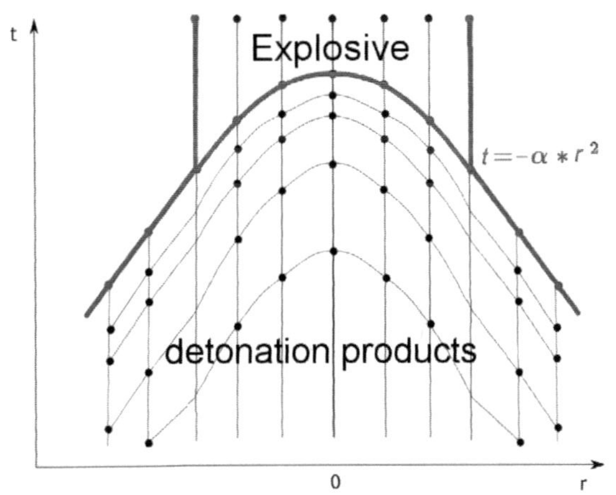

Figure 11. Scheme of approximation mesh generation for density reconstruction. Thick lines: density discontinuities; thin lines: curves along which interpolation splines are created; dots: areas where varying density value is given.

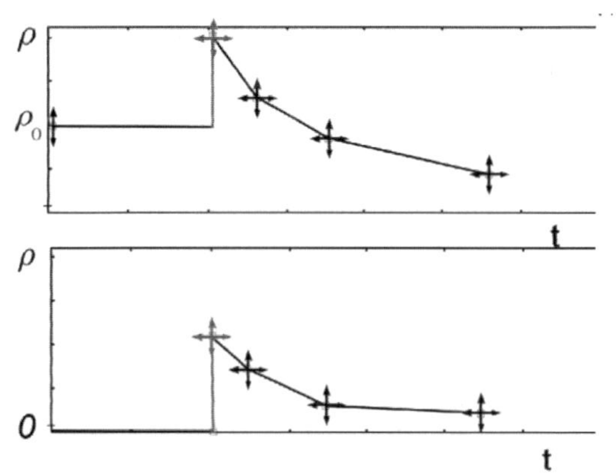

Figure 12. Approximation of density vs. time with fixed radius. Top: radius is inside charge; bottom: detonation products.

The density values in the nodes were determined from the conditions for minimum of the function $G(\vec{p})$

$$G(\vec{p}) = \sum_{i,j} (F(x_j,t_i) - F'_{\vec{p}}(x_j,t_i))^2,$$

$$F'_{\vec{p}}(x_j, t_i) = \int_{-\sqrt{R_0^2 - x_j^2}}^{\sqrt{R_0^2 - x_j^2}} \rho_{\vec{p}}(t_i, \sqrt{x_j^2 + y^2}) dy,$$

where $\vec{p}$ is the set of the desired density values in the nodes, $F(x_j, t_i)$ is the experimental shadow, $F'(x_j, t_i)$ is the shadow to test. The minimum was searched for numerically, using a simplex algorithm. The reconstruction results are presented in figures 13 and 14.

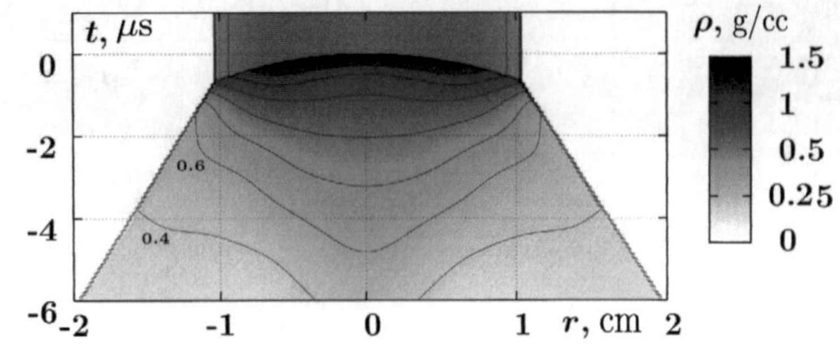

Figure 13. Reconstructed density distribution in PETN.

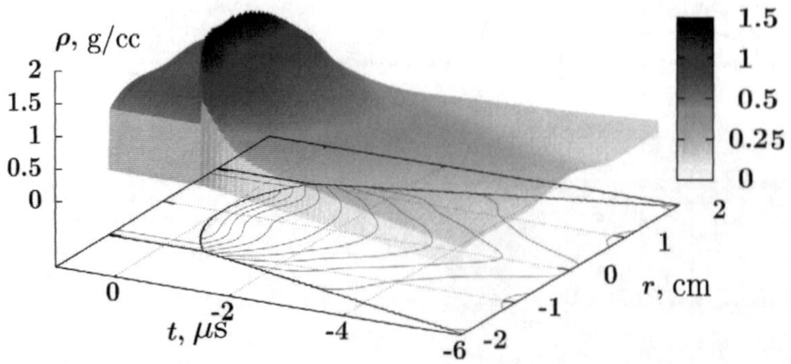

Figure 14. Reconstructed volume density distribution in PETN.

2.2.2. Reconstruction of equation of state and mechanical parameters of flow. The method of reconstruction of fields of gas-dynamic characteristics of detonation flow is based on numerical solution to the gas-dynamical problem in a set-up corresponding to the experiment. Let us consider the problem of a cylindrically symmetric gas flow. Then in the Euler coordinates, the equations of continuity and momentum look as follows

$$\frac{\partial r\rho u}{\partial r} + \frac{\partial r\rho v}{\partial z} = \frac{\partial r\rho}{\partial t},$$

$$\frac{\partial r\rho u^2}{\partial r} + \frac{\partial r\rho uv}{\partial z} + r\frac{\partial p}{\partial r} = \frac{\partial r\rho u}{\partial t},$$

$$\frac{\partial r\rho v^2}{\partial z} + \frac{\partial r\rho uv}{\partial r} + r\frac{\partial p}{\partial z} = \frac{\partial r\rho v}{\partial t},$$

where ρ is the density, p is the pressure, u and v are the axial and radial components of the velocity vector v, r and z are the radial and axial spatial coordinates, t is the time.

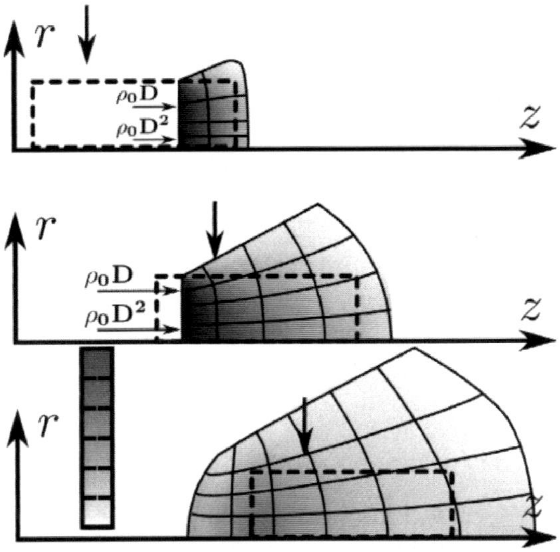

Figure 15. Scheme of set-up of gas-dynamical problem for calculation of detonation flow. Dotted line: initial borders of charge. Arrow: location of section to study.

Let us add the boundary conditions: a flow with known parameters $\rho_0 D$ is the mass flow; $\rho_0 D^2$ is the flow of the axial component of the momentum (figure 15) flows in through the flat surface of the detonation front. In the rest surface, the boundary conditions correspond to the decay of the discontinuity between the detonation products and the air.

Let us represent the unknown equation of state of the substance behind the detonation wave front in a parametric form,

$$p(\rho) = p_*(\rho/\rho_*)^{G(\rho)},$$

where $G(\rho)$ is a cubic spline based on nodes with fixed densities, ρ_i. The values of the adiabat G_i in these nodes are the sought for unknown parameters of the equation of state.

The above system was solved numerically by the Godunov method on the Lagrange mesh.

Having full information on the density, one can compare the X-ray shadow dynamics calculated numerically and measured experimentally, and thus choose a state equation that optimally describes the experimentally measured X-ray shadow.

With the results obtained one can detail the spatial distribution of the density of the expanding products of the PETN+soda mixture detonation. Figures 16 and 18 show the resulting pressure and density distributions, as well as their values along the charge axis (figures 17 and 19) for detonation of PETN+soda (the density $\rho = 0.75$ g/cm^3). Figure 20 shows the distributions of the mass velocity vectors.

With the obtained detailed information about the flow structure, one can reconstruct the unloading adiabat (ρ) from the experimental data (figure 21). The derivative $c = (\partial p/\partial \rho)^{1/2}$

Journal of Physics: Conference Series **754** (2016) 072006 doi:10.1088/1742-6596/754/7/072006

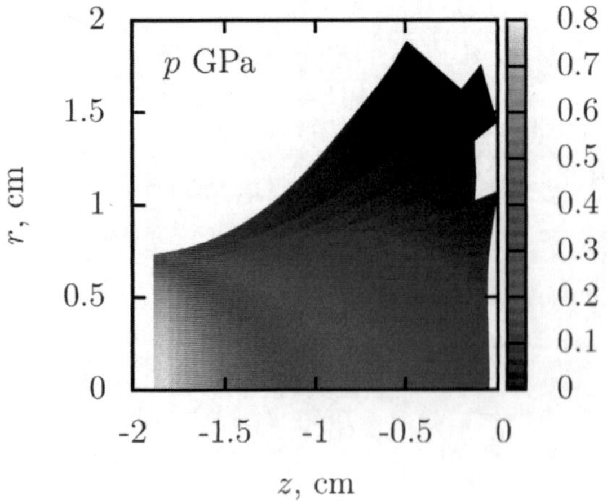

Figure 16. Volume distribution of pressure. ρ =0.75 g/cm^3.

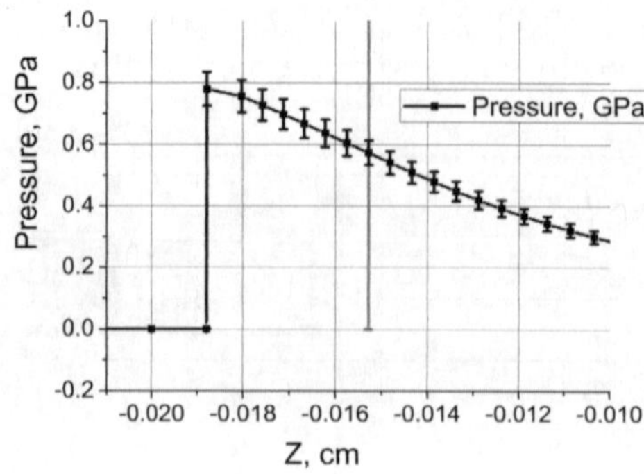

Figure 17. Pressure distribution along the charge axis.

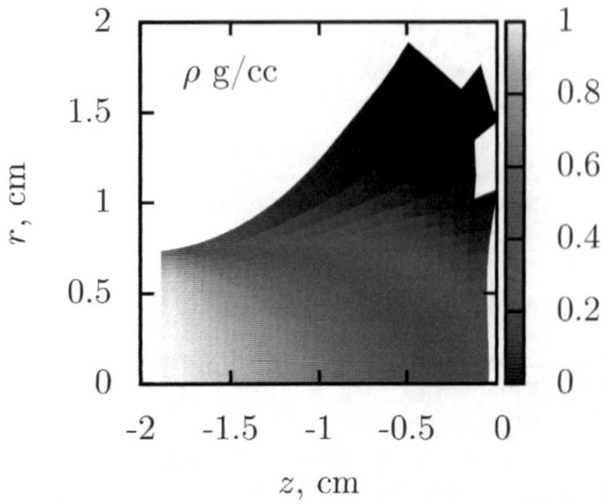

Figure 18. Volume distribution of density.

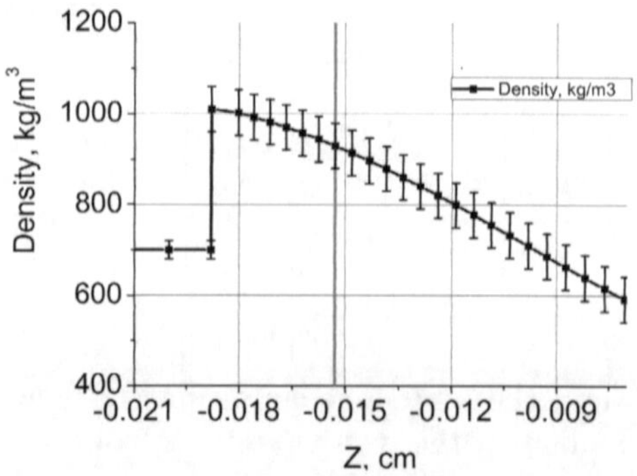

Figure 19. Density distribution along the charge axis.

will be interpreted as the speed of sound. After the reconstruction of the equation of state, it is calculated explicitly

$$c = (\frac{\partial p}{\partial \rho})^{1/2} = [p_0(\rho/\rho_{00})^{\gamma(\rho)}(\gamma(\rho)/\rho + \ln(\rho/\rho_{00})\gamma'(\rho))]^{1/2}.$$

Figure 22 shows charts of the speed of sound C, product expansion velocity U, and their sum along the charge axis. Equating the sum to the detonation velocity ($C+U=D$), we obtain the position of the Jouget plane (the blue dots in figure 22). The so-found Jouget plane is shown in figures 17 and 19 (the red line). In case of PETN+soda mixture and an initial density of 0.75 kg/m^3, the width of the chemical reaction zone is equal to 3.5 mm.

The spatial accuracy of the reconstruction of the flow characteristics is quite high; it is 1 to 2 detector channels, which is about 0.2 mm. The resulting time resolution, the statistics of the two experiments with an interval between frames of 0.125 μs taken into account, is about

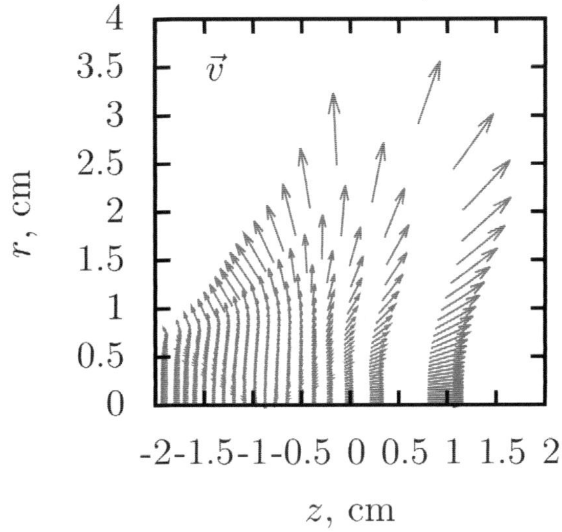

Figure 20. Distribution of velocity vectors of detonation products.

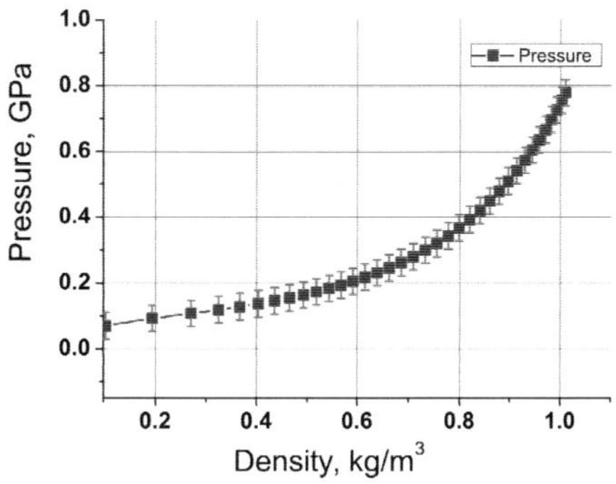

Figure 21. Reconstructed equation of unloading adiabat of detonation products. $p(\rho) = p_*(\rho/\rho_*)^{G(\rho)}$.

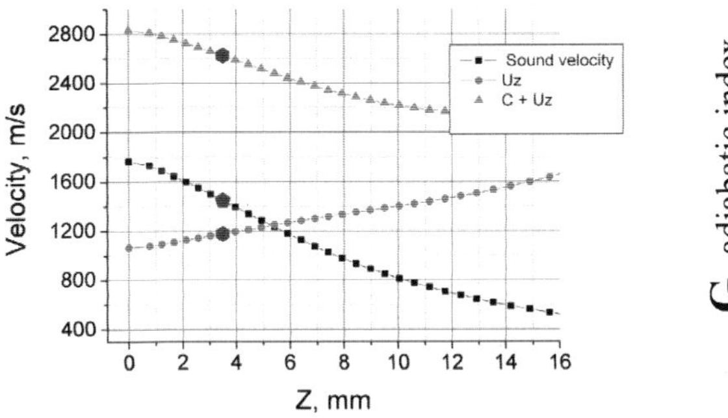

Figure 22. Charts of expansion velocity U_z, speed of sound C and their sum along charge axis. Blue dot: Jouget plane.

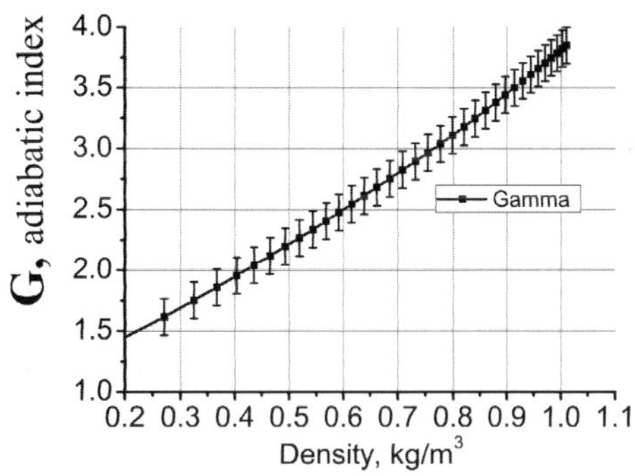

Figure 23. Chart of adiabatic index of expanding products.

0.08 μs. The accuracy of determination of the gas-dynamic parameters is adjusted using the conservation laws; it is estimated as 10% at least for a time scale of 0.5 μs.

3. Conclusions

The techniques using SR on VEPP-3 were applied to the measurement of the density distribution behind the detonation front, the reaction zone width, and dependence of detonation velocity on the initial density of ultrafine PETN + soda and RDX + soda (35/75) mixtures. The results of the measurements enabled reconstruction of volume distributions of density, pressure and flow velocities in expanding explosion products. The equation of state was derived for explosion products in the form of pressure versus density, $P(\rho)$.

Acknowledgments
This work was supported by Russian Foundation for Basic Research (project No. 16-29-01050).

References

[1] Andreevskikh L A, Dendenkov Y P, Drennov O B, Mikhailov A L, Titova N N and Deribas A A 2011 *Propellants, Explosives, Pyrotechnics* **36** 48–50 ISSN 1521-4087 URL `http://dx.doi.org/10.1002/prep.201000035`

[2] Drennov O B, Andreevskikh L A, Mikhajlov A L and Titova N N 2013 Method to manufacture composite explosive for explosion welding and composite explosive *Patent RU 2487108*

[3] Shekhtman L I, Aulchenko V M, Bondar A E, Dolgov A D, Kudryavtsev V N, Nikolenko D M, Papushev P A, Pruuel E R, Rachek I A, Ten K A, Titov V M, Tolochko B P, Zhilich V N and Zhulanov V V 2012 *Journal of Instrumentation* **7** C03021 URL `http://stacks.iop.org/1748-0221/7/i=03/a=C03021`

[4] Pruuel E R, Ten K A, Tolochko B P, Merzhievskii L A, Lukyanchikov L A, Aulchenko V M, Zhulanov V V, Shekhtman L I and Titov V M 2013 *Doklady Physics* **58** 24–8

[5] Titov V M, Pruuel E R, Ten K A, Lukyanchikov L A, Merzhievskii L A, Tolochko B P, Zhulanov V V and Shekhtman L I 2011 *Combustion, Explosion, and Shock Waves* **47** 3–15

Laws of the creep of metallic materials at high temperatures

I A Banshchikova[1], **B V Gorev**[1] **and M A Legan**[1,2]

[1] Lavrentyev Institute of Hydrodynamics SB RAS, 15 Prospekt Lavrentyeva, Novosibirsk, 630090, Russia
[2] Novosibirsk State Technical University, 20 Prospekt K. Marksa, Novosibirsk, 630073, Russia

E-mail: binna@ngs.ru

Abstract. The uniaxial tension with various strain rates for the alloy V95ochT1 at the room temperature and at the temperature of aging 165°C is investigated experimentally for finding of the rational modes of deformation. The range of deformation rates, close to strain rate at superplasticity, at the temperature of aging was found when the value of deformation at destruction is maximum. Increasing or decreasing of the deformation rate leads to decrease of the strain value at which the material is destroyed. Increasing the preheat period leads to a decrease in the strain value at fracture at temperature of aging. The elastoplastic characteristics are reduced in the alloy in the cold state after the full aging at 165°C without load.

1. Introduction

The processing of materials pressure in the quasi-static mode with deformation rates of the order of several percent per second is the most widespread method in industry. The treatment of new high-strength structural materials in this mode leads often to appearing of serious defects in the material and the product practically exhausts all resource already at the manufacturing stage before exploitation. The slow shaping modes due to irreversible creep strain and modes close to superplasticity are alternative to a fast shaping modes due to 'instant' plastic deformations [1-4].

The use of slow modes with the stress not exceeding the elastic limit with the strain rate about or less than $10^{-3}\,s^{-1}$ due to increasing temperature and increase the duration of force action leads to reduced [5-9]:

- level of stresses;
- damage of material at the manufacturing stage of a detail;
- level of residual stresses after removal of the load since the internal stresses relax to small values in the process of slow temperature-force action on the material.

2. Experiments and discussion

The treatment of the material in slow modes allows one to expand the range of a temperature and force loading regimes, including nonstationary modes of deformation, to combine the process of forming with the processes of heat treatment of the material. This significantly increases the ability of the material to deformation that leads to the preservation of residual service life (survivability) of details at the manufacturing stage [10-13]. For instance, bending plates of aluminum alloy ABT-102 with a thickness of 24 mm at the room temperature (in the foreground) and the hardening temperature of 470°C (in the background) are shown in figure 1.

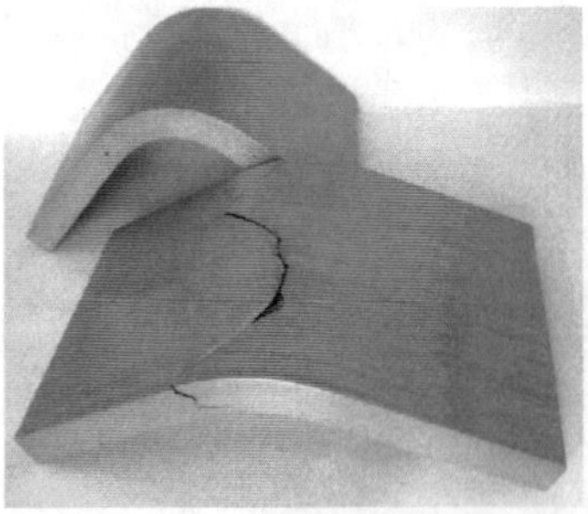

Figure 1. Bending plates of aluminum alloy ABT-102 with a thickness of 24 mm at the room temperature (in the foreground) and the hardening temperature of 470ºC (in the background).

Experimental studies in uniaxial tension for alloys based on aluminum V95ochT1 (very pure alloy V95 treated in mode T1) were carried out for the purpose of definition of rational modes of loading. The experiments were performed on testing machine Zwick/Roell Z100. The diameter of tested round specimens is 8 mm, working length L has value from 36 to 41 mm. Specimens were cut from the plate of thick 15 mm. The rate of the traverse displacement (one of the specimen end) has constant value during the experiment. The logarithmic strain is determined based on the traverse displacement. Since the strains obtained by measurements of the traverse displacement in the linear elastic stage have overestimated values, the strains are corrected based on measurements of the additional strain sensor in the elastic stage. The stresses are determined in view of the incompressibility. Since the cross-sectional area of the sample changes in the process of loading, the strain rate isn't constant value at a constant rate of the traverse displacement [14, 15]. For small deformations the strain rate can be assumed to be constant and equal to the rate at the initial stage of deformation with correction of the elastic component.

2.1. Elastic-plastic and creep properties of aluminum alloy V95ochT1 in tension at room temperature and the temperature of aging

The curves in figure 2 are diagrams 'stress σ - logarithmic strain ε' of uniaxial tension up to fracture at room temperature $T = 19$ºC (curves 1-4) and at aging temperature $T = 165$ºC (curves 5-6) with different strain rates. The strain rate at the initial stage of deformation, conditions of deformation and the fracture time t^* of specimens corresponding to the curves in figure 2 are given in table 1. The uniaxial tension 'σ-ε' diagrams at room temperature $T = 19$ºC after pre-deformation of 3% at $T = 165$ºC (curve 2) and after preheating for 18 h without any deformation at $T = 165$ºC (curve 4) are given for comparison. The preheating time at temperature $T = 165$ºC for curves 5 and 6 is 1.5 h.

Table 1. Modes of deformation corresponding to the curves in figure 2.

	$\dot{\varepsilon}$ (1/s)	T (°C)	Pretreatment at $T = 165$°C	t^* (h)
1	0.0004	19	—	0.11
2	0.0004	19	Pre-deformation of 3% during 0.4 h	0.09
3	0.000004	19	—	6.4
4	0.0004	19	Preheating 18 h without load	0.106
5	0.0004	165	Preheating 1.5 h without load	0.2
6	0.04	165	Preheating 1.5 h without load	0.003

Based on the results (curves 1-4), it is noted that after the full no-load aging (curve 4) the elastoplastic characteristics are reduced in the alloy in the cold state. The resulting elastic limit for the curve 6 is $\sigma_e = 446$ MPa. 'The resource plasticity' in the slow deforming at creep mode with strain rate $\dot{\varepsilon} = 0.0004$ 1/s at $T = 165$°C (curve 5 corresponding to elastic-creep deformations with stress not exceeding the elastic limit) is significantly higher than when cold forming (curve 4) and when quick elastic-plastic deformation (curve 6) at $T = 165$°C.

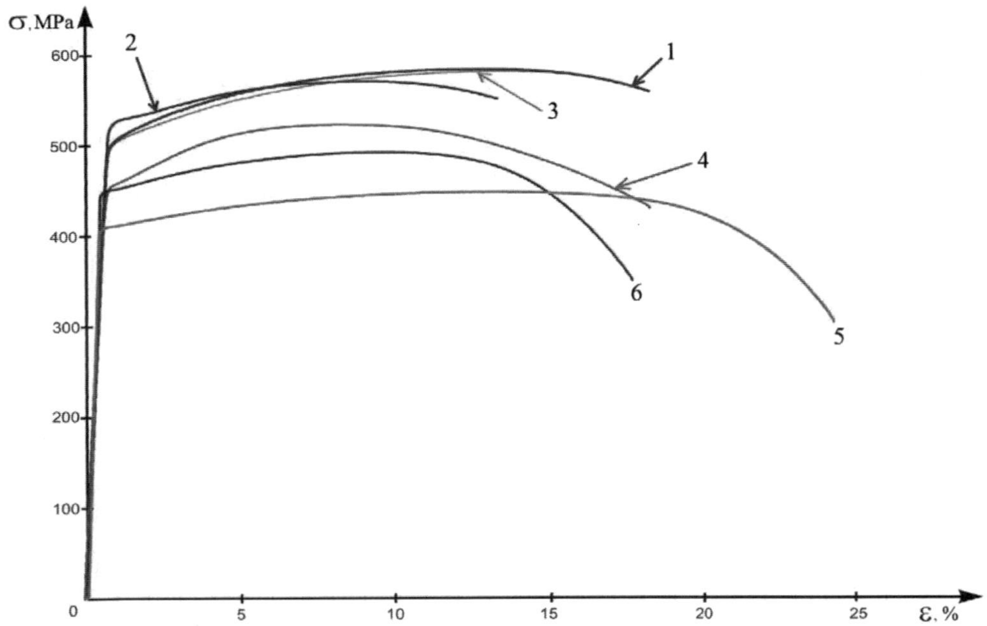

Figure 2. The diagrams 'σ-ε' of uniaxial tension at the room and at the aging temperatures.

2.2. *Rational modes at creep of aluminum alloy V95ochT1 at the aging temperature*

Curves 1-8 in figure 3 are experimental results in uniaxial tension with constant rate of the traverse displacement on machine Zwick/Roell Z100 for aluminum alloy V95ochT1 in creep modes with a short and long period of preheating at a temperature of aging $T = 165$°C. The rate, conditions of deformation and the fracture time t^* are indicated in the table 2. The heating time at the temperature $T = 165$°C before loading for curves 1, 3, 5, 6, 8 is 1.5 h. The long preheating time without loading for curve 2 is 10 h and for curve 4 is 20 h.

Table 2. Modes of deformation corresponding to the curves in figure 3.

	$\dot{\varepsilon}$ (1/s)	Preheating time (h)	t^* (h)
1	0.0004	1.5	0.2
2	0.0004	10.0	0.15
3	0.00004	1.5	1.49
4	0.0004	20.0	0.14
5	0.000004	1.5	9.1
6	0.04	1.5	0.003
7	0.002	2.5	0.04
8	0.002	1.5	0.04

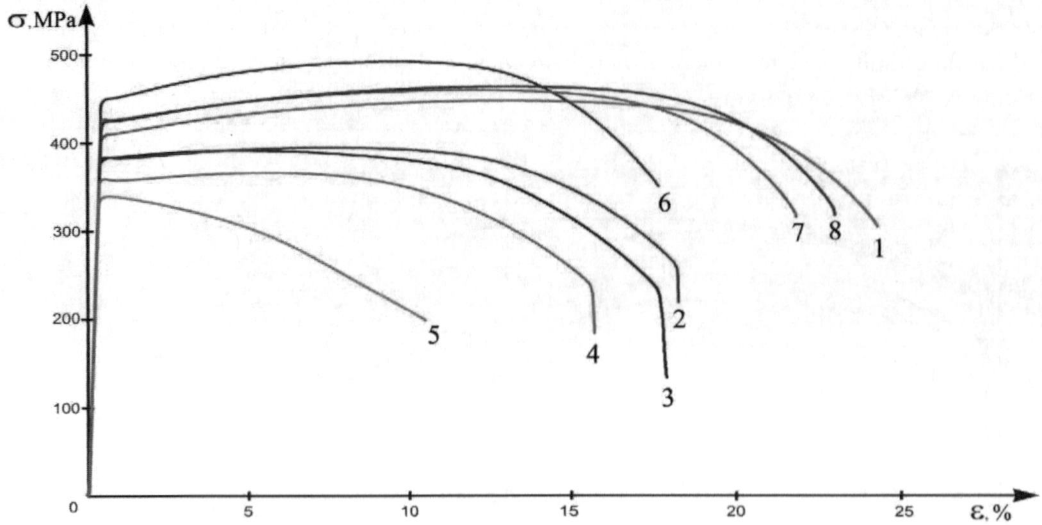

Figure 3. The diagrams 'σ-ε' of uniaxial tension in creep modes with a short and long period of preheating at the aging temperature T = 165°C.

From the presented results (curves 1, 3, 5, 6, 8), it follows that there is a range of strain rates (close to superplasticity, curve 1), when the value of deformation at fracture is the maximum. A comparative analysis of diagrams 1, 2 and 4 shows that increasing the preheating period and succeeding deformation at the strain rate $\dot{\varepsilon}$=0.0004 1/s also leads to lower value of deformation at fracture. Increased preheat period from 1.5 to 2.5 h without loading (the curves 8 and 7) does not lead to significant strain values reduction at which the specimen collapses.

3. Conclusion

From the analysis of experimental diagrams obtained at the room and at the aging temperature 165°C, it follows that a rational modes of deformation of the V95ochT1 alloy are the creep modes at the aging temperature 165°C with stresses that do not exceed the limit elasticity $\sigma_e = 446$ MPa at a rate of the order from 10^{-3} to 10^{-4} s^{-1}, which close to the superplasticity rate. Increasing or decreasing the rate of deformation leads to a decrease in the strain value at which the material is destroyed and to a reduction the residual service life.

The elastoplastic characteristics are reduced in the alloy in the cold state after the full aging at 165°C without load. The increase of the preheating period at the temperature of 165°C also leads to a decrease in the strain value at fracture. Thus the rational mode of deformation from the point of view of preservation of the residual service life is the mode of loading at creep with a rate of the order from 10^{-3} to 10^{-4} s^{-1} immediately after the workpiece preheating period from 1.5 to 2.5 h.

Acknowledgments
This work was partially supported by the Russian Foundation for Basic Research (project codes 15-01-07631-a and 16-08-00713-a).

References

[1] Padmanabhan K A, Vasin R A and Enikeev F U 2001 *Superplastic Flow: Phenomenology and Mechanics* (Berlin Heidelberg: Springer-Verlag) p 363
[2] Watanabe H and Fukusumi M 2014 *Journal of Materials Engineering and Performance* **23** 3551–7

[3] Gorev B V, Zakharova T E and Klopotov I D 2002 *Physical Mesomechanics* **5**(2) 17–22
[4] Enikeev F U 2008 *Russian Journal of Non-Ferrous Metals* **49**(1) 34–41
[5] Zolochevsky A A, Sklepus A N and Sklepus S N 2001 *Nonlinear Solid Mechanics* (Kharkov: Business Investor Group) p 720
[6] Banshchikova I A, Gorev B V and Tsvelodub I Yu 2007 *J. Appl. Mech. Tech. Phys.* **48** 751–54
[7] Banshchikova I A, Gorev B V and Sukhorukov I V 2002 *J. Appl. Mech. Tech. Phys.* **43** 448–56
[8] Banshchikova I A, Muraveva A E and Tsvelodub I Yu 2014 *Obrabotka Metallov — Metal Working and Material Science* (in Russian) **4**(65) 68–77
[9] Banshchikova I A and Tsvelodub I Yu 1996 *J. Appl. Mech. Tech. Phys.* **37** 876–83
[10] Gorev B V and Banshchikova I A 2008 *Vestn. Samar. Gos. Tekhn. Univ. Ser. Fiz.-Mat. Nauki* (in Russian) **2**(17) 110–7
[11] Klopotov I D, Ljubashevskaja I V, Raevskaja G A, Rublevskij L L and Sosnin O V 2002 Shaping apparatues *Patent RU 2251464*
[12] Pekarsh A I, Sarykov S E, Krupskij R F, Fizulakov R A, Stankevich A V, Olejnikov A I, Bojko V M and Krivenok A A 2001 Method of forming *Patent RU 2475322*
[13] Senatorova O G, Tkachenko E A, Sidelnikov V V, Antipov V V, Blinova N E, Shestov V V, Krasova E V and Girsh R I 2008 Procedure for production of item out of high-strength aluminium alloy *Patent RU 2396367*
[14] Panamaryov V A, Peretyat'ko V N, Gorev B V, Remorov V E and Bazaykin V I 2012 *Izvestiya Visshikh Uchebnykh Zavedenii. Chernaya Metallurgiya – Izvestiya. Ferrous Metallurgy* (in Russian) **55**(4) 24–6
[15] Panamarev V A, Karpov E V, Larichkin A Yu and Zhuravleva M I 2013 *Modern Science: Actual Problems of Theory and Practice. Series Natural and Technical Sciences* (in Russian) **5-6** 52–4

Thermo-mechanical relaxation of stresses in a glass-metal junction

O N Lyubimova and E P Solonenko

Department of mechanical and mathematical modeling, Engineering school, Far Eastern Federal University, Sykhanova 8, 690950, Vladivostok, Russia

E-mail: el-solonenko@yandex.ru

Abstract. This work deals with the evaluation of the stress-strain state in an infinite layered cylindrical junction of glass with steel. Structural changes in the glass behavior are described by the Tool-Narayanaswamy-Mazurin-Moynihan model. The relaxation processes in the glass have been presented by the temperature regime depending of the fictive temperature T_f, the relaxation time τ' or the viscosity η and the thermal expansion coefficient $\alpha^{(1)}$. The calculations of the stress components in the glass have been made by using the viscoelastic model. The presented model allows to find the optimal annealing conditions for the creation of the glass-metal composite material.

1. Introduction

The investigation of a layered cylindrical soldering of metal with glass (glass-to-metal seal) is associated with the creation and the optimization of the manufacturing mode of the glass-metal composite rod with a glass core [1]. The manufacturing method includes the heating process, the stable phase and the cooling stage. A technology is considered that the soldering metal onto glass is formed at the temperatures above the glass transition temperature, which the relaxation processes in the glass to metal seals are available, so, it is necessary to construct an algorithm for the calculation of the technological stresses in order to control the annealing mode and the residual stresses. The investigation of the relaxation processes in the glass is actual problem and many phenomenological models for describing of the glass transition kinetics have been presented in paper [2]. In this article, the evaluation of the stress-strain state of the seal is predicted by the Tool-Narayanaswamy model [3]. It is more appropriate way for the description of the experimental data on the vitrification of various amorphous materials, with using the Mazarin's algorithm for the calculation of the relaxation processes in plate seals [4]. The result of this work is the algorithm for the thermal stress calculations in the cylindrical seals with taking into account the changing structure of the glass and unloading the metal layer due to the relaxation processes in the glass.

2. Mathematical model

The deformation problem of the long two-layered cylinder under the influence of the varying temperature field is considered. The publication [5] has been shown that if the temperature changed uniform then the temperature distribution in the cross–over section of the cylinder was virtually unchanged for 20 mm cylinders in the diameter. It allows to consider that the temperature do not depend on the radius. So, under the assumption that there is a perfect contact between the glass to metal surfaces, the stresses at the junction are generated due to the significant differences in the mechanical characteristic of the joined materials, namely, the thermal expansion coefficients and the Youngs moduli. The dependencies linking the stresses and strains in the theory of thermo-elasticity including the plane strain hypothesis and axial symmetry have been well known [6]:

$$\begin{gathered}
\sigma_r^{(i)}(r,t) = \lambda^{(i)}(\varepsilon_\phi^{(i)}(r,t) + \varepsilon_r^{(i)}(r,t)) + 2\mu^{(i)}\varepsilon_r^{(i)}(r,t) - 2(\lambda^{(i)} + \mu^{(i)})\alpha^{(i)}(T(t) - T_0), \\
\sigma_\phi^{(i)}(r,t) = \lambda^{(i)}(\varepsilon_\phi^{(i)}(r,t) + \varepsilon_r^{(i)}(r,t)) + 2\mu^{(i)}\varepsilon_\phi^{(i)}(r,t) - 2(\lambda^{(i)} + \mu^{(i)})\alpha^{(i)}(T(t) - T_0), \\
\sigma_z^{(i)}(r,t) = \lambda^{(i)}(\varepsilon_\phi^{(i)}(r,t) + \varepsilon_r^{(i)}(r,t)) - 2\lambda^{(i)}\alpha^{(i)}(T(t) - T_0)), \\
\varepsilon_r^{(i)}(r,t) = \frac{\partial u_r^{(i)}(r,t)}{\partial r}, \quad \varepsilon_\phi^{(i)}(r,t) = \frac{u_r^{(i)}(r,t)}{r}, \\
T(t_0) = T_0, \quad \sigma^{(i)}(t_0) = 0, \quad \varepsilon^{(i)}(t_0) = 0, \quad u^{(1,2)}(t_0) = 0,
\end{gathered} \tag{1}$$

where i is the layer number, the 1st layer is the glass, the 2nd layer is the metal, $\lambda^{(i)}$ and $\mu^{(i)}$ are the Lame constants, $\alpha^{(i)}$ is the thermal expansion coefficients (figure 1). If the glass cylinder in the junction is solid then the stresses in the glass do not depend on the radius.

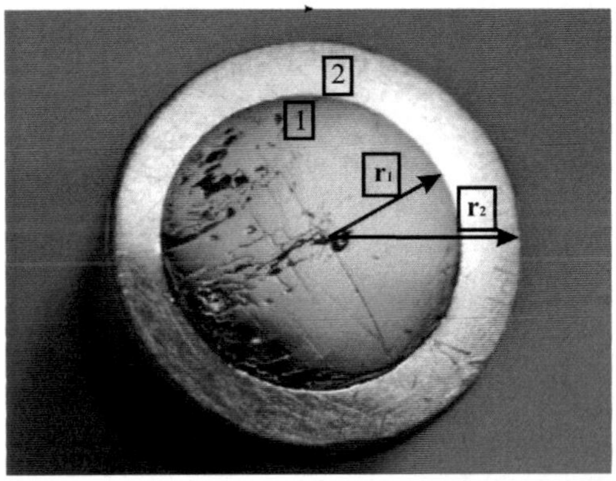

Figure 1. Glass-to-metal seal geometry.

The main characteristic of the structural changes (of the time and temperature) for the glass at the viscoelastic behavior in accordance with the Tool-Narayanaswamy-Mazurin-Moynihan (TNMM) model is the fictive temperature [3,4,7]

$$T_f(t) = T(t) - \int_0^t \left(\frac{dT}{dt}\right) M[\xi(t) - \xi(t')]dt', \tag{2}$$

$$\xi(t) = \int_0^t \left(\frac{dt'}{\tau'}\right), \tag{3}$$

$$M(\xi) = exp\left(-\xi^b\right), \tag{4}$$

where M is the relaxation kernel, ξ is the non-dimensional reduced time depending on the relaxation time τ' for the property at arbitrary thermal history, b is the kinetic model parameter.

The relaxation time can be determined as linearly depending on the viscosity in form [4]

$$\tau' = \frac{\eta(T,T_f)}{K},$$
$$\log\eta(T,T_f) = \log\eta(T_0) + (T_f^{-1} - T_0^{-1})B_e + (T^{-1} - T_f^{-1})B_m, \tag{5}$$

where K is the structural relaxation characteristic, $\eta(T_0)$ and T_0 are the initial values of the viscous and temperature. The initial temperature must satisfy the condition of metastable equilibrium, B_e and B_m are the parameters characterizing of the temperature dependencies of the viscosity at the equilibrium and frozen glass structure.

Equations (2) – (5), by using successive approximation algorithm, allow to calculate the fictive temperature T_f in every time step. The glass thermal expansion coefficient $\alpha^{(1)}$ depends on the temperature and can be determined by similarly specific heat [6] in form

$$\alpha^{(1)} = \alpha_g^{(1)} + (\alpha_m^{(1)} - \alpha_g^{(1)})\frac{dT_f}{dT}, \tag{6}$$

where $\alpha_m^{(1)}$ and $\alpha_g^{(1)}$ are the thermal expansion coefficients of metastable liquid phase and glass, respectively. The relaxation processes in the glass are determined by equations (2)–(6). If the stress relaxation is taken into account then it is proposed to use the discrete calculation algorithm presented in [3]. For each n time step, formulas for the residual stress calculations $\sigma_{res}^{(1)}(t)$ in the glass could be presented as

$$\sigma_{res\ r}^{(1)}(t_n) = \sigma_r^{(1)}(t_n) - \sigma_{rel\ r}^{(1)}(t_n), \quad \sigma_{res\ \phi}^{(1)}(t_n) = \sigma_\phi^{(1)}(t_n) - \sigma_{rel\ \phi}^{(1)}(t_n), \tag{7}$$

where $\sigma^{(1)}(t_n)$ are the elastic stresses found at the solution of the problem (1), $\sigma_{rel}^{(1)}(t_n)$ are the stresses relaxation by n time step and which are expressed as

$$\begin{aligned} &\sigma_{rel}^{(1)}(t_n) = \sum_{i=0}^{n-1} (1 - M(\xi(t_n) - \xi(t_i)))\, \Delta\sigma^{(1)}(t_i), \\ &\Delta\sigma^{(1)}(t_0) = 0, \quad \Delta\sigma^{(1)}(t_n) = \sigma^{(1)}(t_n) - \Delta\sigma^{(1)}(t_{n-1}). \end{aligned} \tag{8}$$

The solution of the problem (1) in elasticity theory is characterized by the integration constants which are also included in the calculations of the formulas (7) and (8) in the relaxation and residual stresses. The determination of these constants occurs at each time step from the boundary conditions on the free surface and the contact region

$$\begin{aligned} \sigma_{res\ r}^{(1)}(t_n)|_{r=r_1} &= \sigma_r^{(2)}(t_n)|_{r=r_1}, \\ \varepsilon_r^{(1)}(t_n)|_{r=r_1} &= \varepsilon_r^{(2)}(t_n)|_{r=r_1}, \\ \sigma_r^{(2)}(t_n)|_{r=r_2} &= f(t_n). \end{aligned} \tag{9}$$

The first boundary condition (9) allows to take into account the relaxation processes in the glass without breaking the ideal contact conditions in the junction.

3. Results and discussion

Calculations of the stress-strain state have been made for the cylindrical soldering of CH1 glass with steel 20. Material parameters using in the calculations have been taken as $r_1 = 4 \cdot 10^{-6}$ m, $r_2 = 5 \cdot 10^{-6}$ m, $T(t_0) = 600$ °C, $\eta(t_0) = 10^{10.25}$, $B_e = 28726.85$ °C, $B_m = 13726.85$ °C, $K = 10^{10.7}$, $\alpha_m^{(1)} = 210 \cdot 10^{-7}$ 1/°C, $\alpha_g^{(1)} = 52 \cdot 10^{-7}$ 1/°C, $b = 0.65$, $E_1 = 0.67 \cdot 10^5$ MPa,

$\alpha^{(2)} = 0.6 \cdot 10^{-6} \cdot T^{0.148}$ 1/°C [8] and $E_2 = 2 \cdot 10^5$ MPa. The temperature regime was slow cooling at a rate of 5 °C/min from 600 °C to 20 °C.

The analysis of the calculated results for the selected parameters showed that at the beginning of the cooling process the structural glass temperature coincided with the outside temperature and then stabilized (see figure 2), i.e. at the beginning there were changes in the melting glass structure, then the melt transformed into the glassy state and the glass structure became frozen. At this time, the relaxation time τ' was significantly increasing, respectively, the relaxation processes in the glass were almost stopping. For example, regarding the stress relaxation at the temperature 600 °C the relaxation time $\tau' = 1$ sec and at the temperature 550 °C the relaxation time $\tau' = 41$ min.

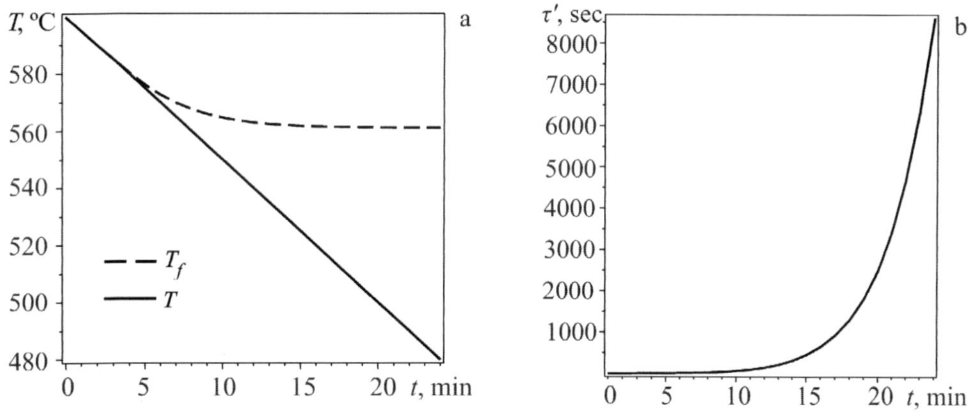

Figure 2. *a*) Time dependencies of the structural and outer temperatures of CH1 glass at a cooling rate of 5 °C/min. *b*) Time dependency of the relaxation time of glass at a cooling rate of 5 °C/min. Here, $T(t)$ is the outer temperature; $T_f(t)$ is the structural temperature; τ' is the relaxation time.

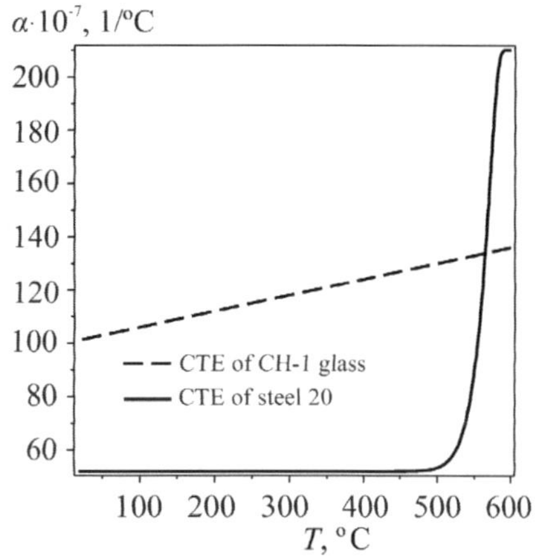

Figure 3. Temperature dependencies of thermal expansion coefficients (CTE).

The analysis of changes in the thermal expansion coefficients (figure 3) shows that, at cooling, the glass is exposed to the tensile stresses from 600 °C to 564.5 °C, there are no stresses in the composition at next temperature and after that, the problem could be actually solved without taking into account the relaxation processes in the glass, which almost do not affect the residual stresses (figures 4b and 5b).

At the beginning of the cooling process, the relaxation processes in the calculations significantly affects the technological stresses (figures 4a and 5a). Note that, the temperature

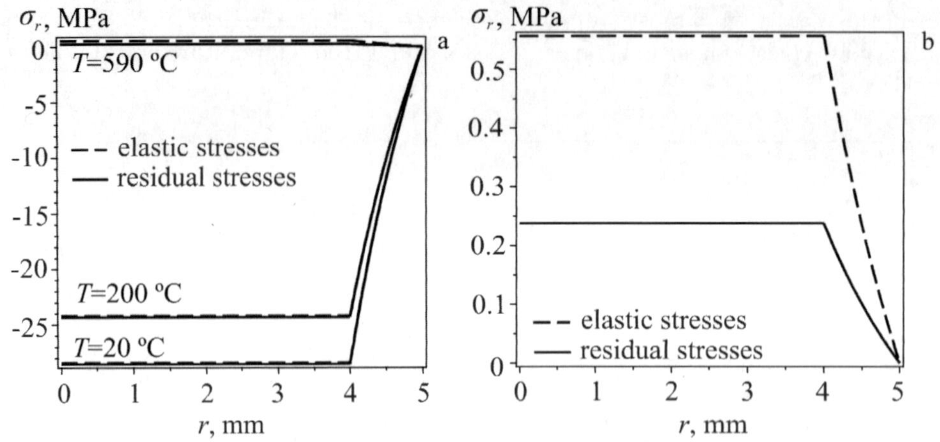

Figure 4. The distribution of the radial stresses in the glass-to-metal seal: *a*) the radial stresses at several temperature points; *b*) the radial stresses at temperature $T = 590$ °C.

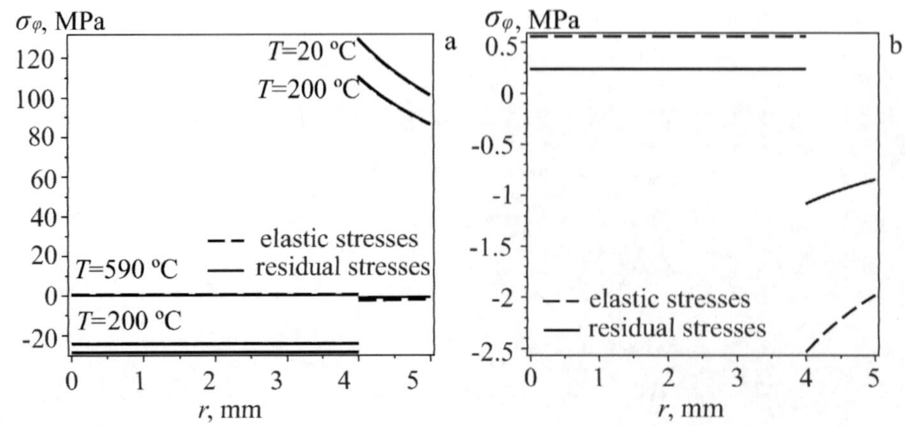

Figure 5. The distribution of the tangential stresses in the glass-to-metal seal: *a*) the tangential stresses at several temperature points; *b*) the tangential stresses at temperature $T = 590$ °C.

corresponding zero stress state (at slow cooling or heating) determines a possibility to solve the problem without consideration the stress relaxation into the junction. The junction could be reheated again without appearing of the additional stresses higher this temperature. Therefore, in order to make a layered cylindrical soldering of glass with metal, cooling at the temperatures above the temperature of the intersection curves (figure 3) should be slow, with taking into account time required for the stress relaxation and after this temperature, the cooling rate could be increased significantly.

4. Conclusion

The algorithm allows to describe the stress-strain state of the cylindrical junction, to simulate the optimal technological regime for the production of the junction with the specified residual stresses without the solution of the inverse problem. The simplicity of the algorithm allows to include the additional conditions considering the special mechanical properties of joined materials. For example, influence the temperature on the elastic modulus or the yield strength and the adhesion to the glass and metal surfaces.

Acknowledgements

This work began to make under support by Federal special purpose program event 1.2 Agreement 14.575.21.0009 ID number RFMEFI57514X0009.

References

[1] Lyubimova O N, Lyubimov E V, Gridasova E A and Nikiforov P A 2015 Sposob izgotovleniya steklometallokompozitnogo sterzhnya *RF Patent 2014120249/03(032372)* **5** (in Russian)

[2] Tropin T V, Schmelzer J W P and Aksenov V L 2016 Modern aspects of the kinetic theory of glass transition *Physics – Uspekhi* **59**(1) 42–66

[3] Narayanaswamy O S 1971 A model of structural relaxation in glass *J. Am. Ceram. Soc.* **54**(10) 491–8

[4] Mazurin O V 1980 *Otzhig Spaev Stekla s Metallom* (Leningrad: Energiya/Leningrad branch) p 140 (in Russian)

[5] Lyubimova O N, Pestov K N and Gridasova E A 2010 Mathematical modeling of the thermal process of diffusion welding of glass and metal *Sib. Zh. Ind. Mat.***13**(4) 52–63 (in Russian)

[6] Kudinov V A, Kartashov E M and Kalashnikov V V 2005 *Analytical Solutions of Problem of Heat and Mass Transfer and Thermoelasticity for Multilayered Structures* (Moscow:Vysshaya shkola) p 430 (in Russian)

[7] Moynihan C T, Macedo P B, Montrose C J et al. 1976 Structural Relaxation in Vitreous Materials *Ann. N. Y. Acad. Sci.* **279** 15–35

[8] Rous B 1969 *Glass in Electronics* (Moskow: Sovetskoe radio) p 356 (in Russian)

Journal of Physics: Conference Series **754** (2016) 092001 doi:10.1088/1742-6596/754/9/092001

Experimental study of liquid-liquid plug flow in a T-shaped microchannel

A V Kovalev[1,2], A A Yagodnitsyna[1,2] and A V Bilsky[1,2]

[1]Kutateladze Institute of Thermophysics, 1, Lavrentev Ave., Novosibirsk, 630090, Russia
[2]Novosibirsk State University, Novosibirsk, 2, Pirogova str., Novosibirsk, 630090, Russia

E-mail: yagodnitsinaaa@gmail.com

Abstract. Plug flow regime in immiscible liquid-liquid flow in a T-shaped microchannel is experimentally studied in the present work. For three different immiscible liquids plug length is measured using high-speed visualization of the flow. Micro-PIV technique is applied for instantaneous velocity field measurements in plugs. Velocity circulation in plugs is calculated for different bulk flow rates.

1. Introduction

Nowadays immiscible liquid-liquid flows in microchannels take place in a number of biological, chemical, medical and industrial processes [1,2]. Therefore the hydrodynamics investigation of immiscible liquid-liquid flows on a microscale is very important. A plug flow pattern is most suitable for conducting chemical reactions efficiently. In order to design and optimize liquid-liquid flow in microdevices the conditions required for plug flow formation, plug length, velocity and circulation inside plugs for liquids with different properties are needed to know.

Jovanovich et al. [3] measured a plug length depending on a plug velocity and a flow rate of dispersed and continuous phases for two different sets of immiscible liquids and two different channel diameters. The relationship between plug length and flow rate ratio was proposed by Garstecki et al. [4]. The corresponding model is valid for so called squeezing mode of plug formation when interfacial forces dominate shear stresses, i.e. Capillary number $Ca \ll 1$. This model was expanded by Xu et al. [5] who identified three different mechanisms of a plug formation process. There are the squeezing mode where a plug length is described by the correlation of Garstecki et al., the dripping mode where shear stresses dominate and a plug or droplet length is proportional to inversed Capillary number of a continuous phase, and the transient mode for which plug length dependence from flow rate ratio and Ca was also proposed in [5].

The circulation inside plugs is another important parameter required to determine a mixing efficiency. Kashid et al. [6] calculated circulation time for two models of plug flow: with film between a plug and a channel wall and without it. Velocity fields were obtained by means of CFD and validated by micro-PIV results in a narrow range of velocities. Dore et al. [7] investigated circulation inside water plugs in an ionic liquid for microchannels with circular cross section by micro-PIV technique. Three distinct circulation modes were found depending on plug velocity. Minimal circulation time and maximal velocity inside plugs were measured for different Ca and ionic liquid volume fractions.

In the present paper the peculiarities of the plug flow in a T-shaped microchannel were studied for different liquid-liquid sets. In order to access circulation inside plugs micro-PIV technique is used for velocity measurements for one of liquid-liquid sets.

2. Experimental setup

In the present work three different liquid-liquid sets were used for plug flow visualization in a T-shaped microchannel, i.e. kerosene - water, paraffin oil – water and paraffin oil – castor oil. The liquid properties measured are represented in Table 1. The T-shaped microchannel used is made of SU-8. The inlet channels have 200x200 um cross-section, outlet channel has 400x200 um cross-section. The flow was organized such that two opposing streams of liquids used were flowing horizontally in outlet channel. The length of inlet and outlet channels is 11.5 and 22.5 mm respectively which is sufficient to get steady plug flow. The photo of experimental setup for flow visualization and micro-PIV measurements inside plugs is presented in figure 1. There are pco.1200 hs high speed camera and inverted Zeiss Axio Observer.Z1 microscope with 5x magnification lens used for flow visualization. Frame rate varied from 5 to 1000 Hz. Halogen lamp is used for flow illumination. Flow is organized by KDS Gemini 88 double syringe pump. Flow visualization is done in two different channel areas: T-zone and at the end of the channel, which is 56 channel hydraulic diameters far from T-zone.

Table 1. Physical properties of liquids

Physical properties	Kerosene	Water	Paraffin oil	Castor oil
Density, kg/m^3	745	997	845	935
Viscosity, mPa•s	0.820	0.894	110	650
Interfacial tension, mN/m	45		48	14

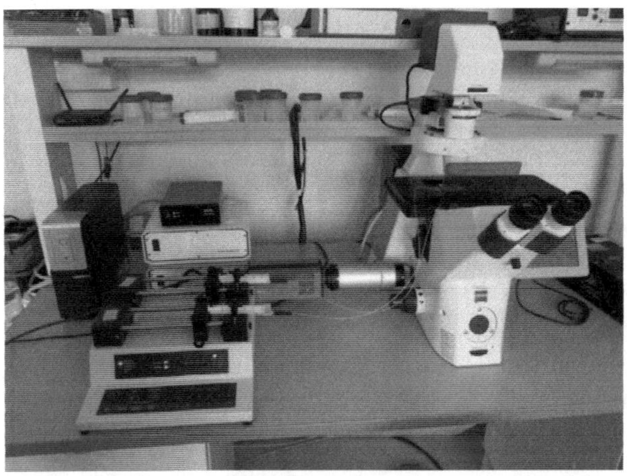

Figure 1. The photo of experimental setup for flow visualization and micro-PIV measurements.

For kerosene – water flow velocity field measurements are done inside water plugs by means of micro-PIV technique using POLIS PIV system. Water phase was seeded by 3 um fluorescent polystyrene particles. Flow was illuminated by double pulsed Nd:YAG laser. Particle images were captured using double exposure 4 Mpix CCD camera. Instantaneous velocity fields were calculated using iterative cross-correlation algorithm. The final spatial resolution of the measurements was 18 um for every velocity vector.

3. Results

3.1. Plug length measurements

ImageJ software was used for plug length measurements. Uncertainty of measurements was about 5 pixels, which corresponds to 0.01 mm. The averaging over a number of plugs was done for every value of volumetric flow rate. In figure 2 a plot of the dimensionless plug length versus flow rate ratio for all three sets of liquids is shown. To change the flow rate ratio the flow rate of a dispersed phase was changed while the flow rate of a continuous phase was constant. The standard deviation of the measured plug length is presented by error bars in figure 2. One could see that plug length increases with the flow rate ratio increase but dependence for castor oil – paraffin oil set differs from one for kerosene – water and paraffin oil – water sets. Moreover, the results for castor oil – paraffin oil do match well with the relation given in Garstecki et al. [4] while for two other sets of liquids there is no good agreement, in spite the fact that the squeezing mode of plug formation takes place in all cases. This difference between results and the prediction probably caused by adhesion forces which affect the plug formation process [8].

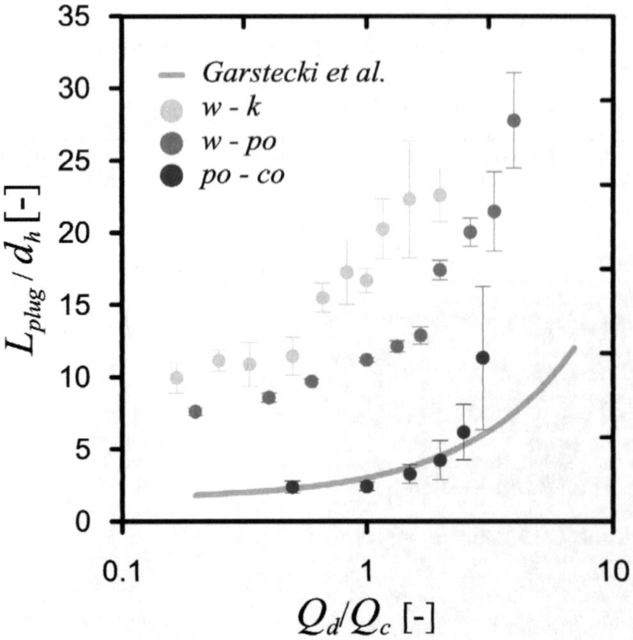

Figure 2. Non-dimensional plug length versus the ratio of rate of the flow for different liquid-liquid sets. Solid line is a plug length according to the scaling law by Garstecki et al. [4].

3.2. Velocity fields and circulation inside plugs

Velocity field measurements in water plugs are done in the area of plug formation and at the end of the channel. Figure 3 illustrates instantaneous velocity fields in central plane of the channel in different phases of water plug formation. One can see that after contact line formation between the plug and the channel wall velocity inside the plug increases significantly although the plug is still connected with the inlet channel by the dispersed fluid neck. This phenomenon could explain higher values of the plug length for kerosene – water and paraffin oil – water liquid sets.

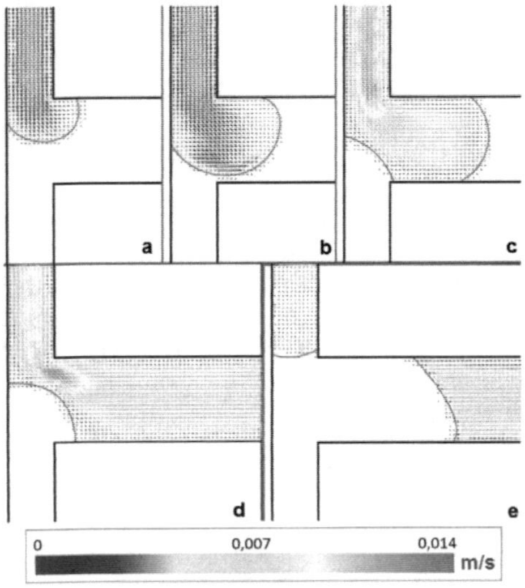

Figure 3. Instantaneous velocity fields in central plane of the channel in different phases of water plug formation.

In figure 4 velocity fields inside water plug at the end of the channel are presented. The maximum velocity is at the center of the plug and it decreases towards channel walls. In figure 4 c the difference between instantaneous velocity field and plug velocity is shown. One can see circulation inside plug with different sign left and right from plug axis.
Based on measured velocity fields velocity circulation inside plugs was calculated for different flow rates. Velocity circulation was calculated using Green's theorem by integrating vorticity over the surface bounded by closed contour shown in figure 4 c. In all cases the boundaries of integrating area were chosen nearby channel walls and channel center. Calculated circulation was normalized by integrating surface area. This allowed avoiding dependence of circulation on the length of integration surface. Circulation was calculated for two different contours shown in figure 4 c (black and red contours). The difference in velocity circulations for two contours did not exceed 10%.

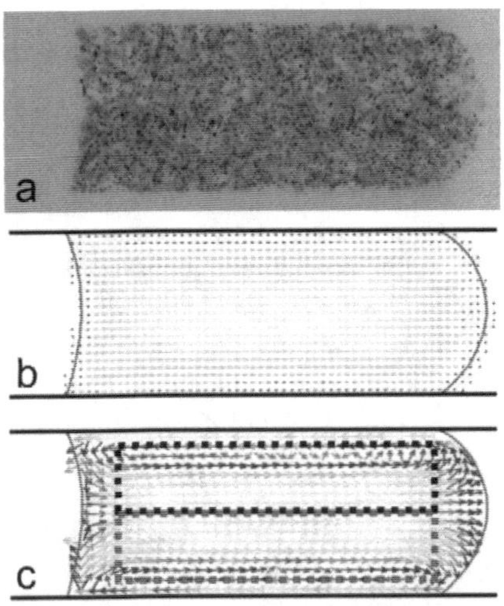

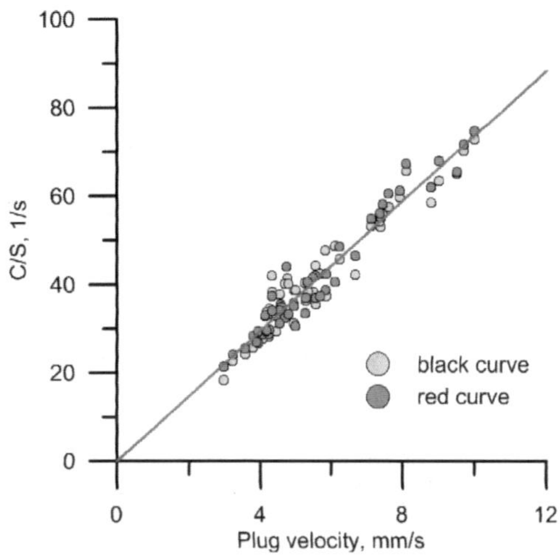

Figure 4. a) Inverted tracer particle image, b) instantaneous velocity field in water plug at the end of the channel, c) velocity circulations inside the plug.

Figure 5. Normalized circulation in water plugs versus plug velocity.

Normalized velocity circulation values in different flow rates are presented in figure 5. When plug velocity increase circulation growths too. It is obvious that circulation inside plugs defines intensity of convective mass transfer. In case of chemical reactions circulation is responsible for convective transfer of reagents. Thus circulation increase or, as follows from obtained result, plug velocity increase leads to decrease of reaction time. Nevertheless one should notice that reaction rate by microreactor unit length remains constant.

4. Conclusion

The plug flow in a T-shaped microchannel was studied in details. The experiments were carried out for three sets of immiscible liquids with different physical properties: kerosene – water, paraffin oil – water and castor oil – paraffin oil. The plug length was measured for all liquid-liquid sets. The plug length was compared with plug length scaling low proposed by Garstecki et al. [4]. It was found that for castor oil – paraffin oil set where adhesion forces for dispersed phase are negligible and plugs of paraffin oil do not wet channel walls the measured plug length matches well with the scaling law. But for kerosene – water and paraffin oil – water liquid sets there is mismatch between the measured plug length and the prediction given by scaling law. The discrepancy of the measured plug length with the model can be explained by existing of the contact line and an adhesion force which should be taken into account.

The plug formation process was studied by means of micro-PIV. The instantaneous velocity fields inside plugs were obtained for characteristic moments of plug formation in the T-junction. It is shown that after contact line formation between the plug and the channel wall velocity inside the plug increases significantly although the plug is still connected with the inlet channel by the dispersed fluid neck. This phenomenon could explain bigger values of the plug length for kerosene – water and paraffin oil – water liquid sets. The circulation inside the plugs was calculated from the velocity fields for different flow rates and shown to be linear with the plug velocity.

Acknowledgements

This work performed in the Kutateladze Institute of Thermophysics SB RAS by a grant from the Russian Science Foundation (project No 16-19-10519).

References

[1] Benavides J, Aguilar O, Lapizco-Encinas B H and Rito-Palomares M 2008 Extraction and Purification of Bioproducts and Nanoparticles using Aqueous Two-Phase Systems Strategies *Chem. Eng. Technol.* **31** pp 838–45

[2] Tran T M, Lan F, Thompson C S and Abate A R 2013 From tubes to drops: droplet-based microfluidics for ultrahigh-throughput biology *J. Phys. D. Appl. Phys.* **46** 114004

[3] Jovanovic J, Zhou W, Rebrov E V, Nijhuis T A, Hessel V and Schouten J C 2011 Liquid-liquid slug flow: Hydrodynamics and pressure drop *Chem. Eng. Sci.* **66** pp 42–54

[4] Garstecki P, Fuerstman M J, Stone A and Whitesides G M 2006 Formation of droplets and bubbles in a microfluidic T-junction — scaling and mechanism of break-up 2006 *Lab Chip* **6** pp 437–46

[5] Xu J H, Li S W, Tan J and Luo G S 2008 Correlations of droplet formation in T-junction microfluidic devices: from squeezing to dripping *Microfluid. Nanofluid.* **5** pp 711–17

[6] Kashid M, Renken A and Kiwi-Minsker L 2011 Influence of flow regime on mass transfer in different types of microchannels *Ind. Eng. Chem. Res.* **50** pp 6906–14

[7] Dore V, Tsaoulidis D and Angeli P 2012 Mixing patterns in water plugs during water/ionic liquid segmented flow in microchannels *Chem. Eng. Sci.* 80 pp 334-41

[8] Yagodnitsyna A A, Kovalev A V and Bilsky A V 2016 Flow patterns of immiscible liquid-liquid flow in a rectangular microchannel with T-junction *Chem. Eng. J.* **303** pp 547-54

Journal of Physics: Conference Series **754** (2016) 092002 doi:10.1088/1742-6596/754/9/092002

Investigation of forced convection ZrO_2 nanofluid in a channel with artificial roughness

A V Minakov[1,2], D V Guzei[2], K A Meshkov[2], V A Zhigarev[2], M I Pryazhnikov[1,2]

[1]Kutateladze Institute of Thermophysics SB RAS, 630090, Novosibirsk, Lavrentev av. 1, Russia.
[2] Siberian Federal University, 660041 Krasnoyarsk, Svobodnyi av. 79, Russia

E-mail: tov-andrey@yandex.ru

Abstract. The paper presents the results of experimental study of turbulent forced convection of water-based nanofluid with ZrO_2 nanoparticles. Volumetric concentration of the nanoparticles in the experiments was equal to 4%. The average particle size was 105nm. The study was carried out for a smooth round tube and round tube with knurling ring. The effect of additives of ZrO_2 nanoparticles on the average heat transfer coefficient and the pressure drop in the channel was investigated. Has been shown that the intensification of heat transfer by using nanofluid weakly dependent on the shape of the channel, and determined only by the properties nanofluid.

1. Introduction

The enhancement of convective heat transfer and the related experimental and theoretical research become at present an independent, important and rapidly developing field of heat transfer theory. High heat flux removal is a major consideration in the design of many machines, equipment and technologies, and can be accomplished using various kinds of heat transfer equipment. The urgency of this problem is determined by driving to enhance the performance of heat transfer devices, reduce energy costs and achieve maximum compactness with minimum material consumption. One of the solutions to the problem of heat transfer performance enhancement could be the use of so called nanofluids, which are fluids containing nanoparticles of various composition. The first experiments have shown that even very small additions of nanoparticles to the fluid (a fraction of volume concentration) may lead to increase in thermal conductivity and heat transfer of the nanofluids by tens of percent, whereas the critical heat flux may be increased several times [1-5]. Over the last two decades, a great number of works have been emerged in this area. Most of them were focused on the study of thermal conductivity and viscosity of nanofluids. Actually, the study of heat transfer was performed in a relatively few research works, and the results presented are extremely controversial [5]. Most works reveal an increase in heat transfer when using nanoparticles. The increase in heat transfer coefficient in nanofluids as compared to base fluids can range from a few percent to 350% for carbon nanotubes. However, there are publications showing the deterioration of heat transfer when adding nanoparticles. Most of these works are dealt with the study of convection of nanofluids in laminar flow regime [6-9]. Turbulent convection is explored to a substantially lesser degree.

The first work, in which the turbulent heat transfer of nanofluids was studied, is apparently the work of Pak and Cho [10]. They experimentally studied the turbulent heat transfer in water-based nanofluids, containing Al_2O_3 and TiO_2 nanoparticles, and flowing in a horizontal circular pipe with a constant wall heat flux. The results have shown that the Nusselt number in nanofluids increases with increasing both bulk concentration of the particles and the Reynolds number. However, in the same paper [10] it has been revealed also that at high concentrations of nanoparticles the heat transfer coefficient may be lower than that in pure water (by 12% in a nanofluid with particles concentration of 3%).

Convective heat transfer in turbulent flow regime of nanofluids containing copper nanoparticles was experimentally investigated in the subsequent work [11]. The experimental results have shown that adding nanoparticles to the base fluid significantly improved the heat transfer efficiency of base fluid (by 60% in a nanofluid with particles concentration of 2%), while the friction factor remained almost the same as for water.

In [12] the efficiency of heat transfer in water-based nanofluids containing titanium oxide nanoparticles was investigated in laminar and turbulent flow regimes in vertical tube under constant wall heat flux boundary condition. The results have shown that the heat transfer coefficient definitely increases with increase of nanoparticles concentration in both laminar and turbulent regimes at a fixed Reynolds number. The maximum intensification of heat transfer coefficient, recorded in the experiment, was 40% for 1.1% nanofluid. In this case, the pressure drop in nanofluids, when flowing in the channel, was very close to that in pure fluid. In addition, the authors investigated the effect of particles size, though no effect of size on the heat transfer coefficient was detected, possibly, because the particles were large enough (95-210 nm).

Duangthongsuk and Wongwises [13] experimentally studied turbulent heat transfer and pressure drop in water-based nanofluids with TiO_2 particles. And again the results were extremely controversial. They revealed a 32% enhancement in thermal performance at particles concentration of 1%, and 14% reduction in the heat transfer coefficient at particles concentration of 2% as compared with the pure fluid.

Fotukian and Nasr Esfahany [14] studied turbulent convective heat transfer in nanofluids with a very low concentration of Al_2O_3 nanoparticles in water flowing in a circular pipe. They recorded a clear increase in heat transfer coefficient and pressure drop with increasing concentration of particles. The maximum increase of heat transfer coefficient equal to 48% was reached at negligibly small volume concentration of the nanoparticles (0.054%). In a subsequent paper [15] the same authors studied turbulent heat transfer of water-based nanofluids containing CuO nanoparticles in a circular pipe. They achieved a 25% increase in the heat transfer coefficient and a 20% increase in pressure drop at a concentration of nanoparticles equal to 0.24%, as compared to pure water. Besides, they revealed that the heat transfer coefficient was almost independent of the concentration of nanoparticles within the observed range of volume percent change from 0.039 to 0.24.

Thus, the analysis of the literature shows that in the available works there is neither quantitative nor qualitative consensus in terms of turbulent heat transfer performance of nanofluids. In the meantime, almost all scientists note that the effect of nanoparticles on the heat transfer in the turbulent flow regime is much more complicated than that in laminar flow regime. This circumstance requires additional systematic experimental study of turbulent heat transfer of nanofluids.

The purpose of this paper - the study of the influence of artificial roughness on the turbulent forced convection and pressure drop for nanofluid.

2. Description of the experimental setup

The diagram of the installation to study the heat transfer coefficient is shown in figure 1. The installation is a closed loop with a circulating coolant [16]. The working fluid is pumped through the heated test section to the heat exchanger, where heat is removed by the thermostat. The flow rate of the working fluid in the loop is controlled by controlling the pump power by means of laboratory transformer. Power input to the pump is measured through Omix meter.

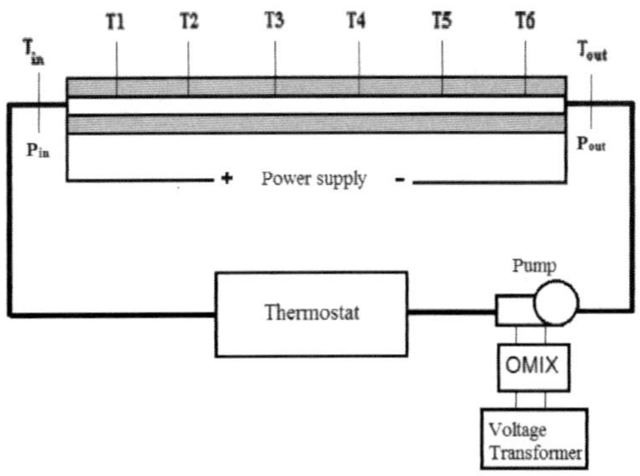

Figure 1. A schematic diagram of the measuring system.

The heated test section is a stainless steel tube 6 mm in diameter, 1m long with wall thickness of 0.5 mm. The tube was heated by supplying electric current directly to its wall. This heating technique allows obtaining a constant wall heat flux boundary condition. In addition, this heating technique is universal and easily applicable to tubes of any cross section. The tube is insulated by multi-layer insulation. The heating power is controlled by a transformer. Six chromel-copel thermocouples fixed on the tube wall at the equal distance from each other were used to measure the local temperature of the tube. Temperature measurements were carried out by TPM-200 meters. In addition, the temperatures at the inlet and outlet of the heated test section were measured using thermocouples. At that, the thermocouple designed to measure the fluid temperature at the outlet of the loop was located at a considerable distance from the end of the heated test section to ensure uniformity of the fluid temperature in the metering point. The loop section between the heater and the fluid temperature metering point was also insulated. The pressure drop measurements were carried out using a differential manometer OWEN PD200. The designed experimental setup was tested based on known empirical data for heat transfer of pure water. The water flow rate ranged from 0.65 to 2 l/min that corresponds to the range of Reynolds number from 2300 to 8000. Figure 2 shows the comparison of the experimental data on the average Nusselt number depending on the Reynolds number with the known empirical correlation [17]:

$$Nu = 0.021\mathrm{Re}^{0.8}\,\mathrm{Pr}^{0.43}.$$

Experimental Nusselt number – $Nu = \frac{\alpha d}{\lambda}$, where $\alpha = GC_P(T_{out} - T_{in})S^{-1}(T_w - \overline{T})^{-1}$ – is the average heat-transfer coefficient; G – mass flow rate measured by flowmeter, C_p – is the specific heat of the fluid; S – is the area of the lateral surface of the channel; d – is the diameter; T_{out}, T_{in} – are fluid temperatures at the channel inlet and outlet; $\overline{T} = \frac{(T_{in} + T_{out})}{2}$ – is the average fluid temperature; T_w – is the arithmetic mean of the channel wall temperature, obtained by averaging of the values of six thermocouples; Pr – is the Prandtl number; $\mathrm{Re} = \frac{4G}{\pi d \mu}$ – is the experimental Reynolds number, λ and μ – are fluid thermal conductivity and viscosity coefficients, respectively. The viscosity and thermal conductivity coefficients of this nanofluid also been measured and are equal to $\mu = 0.001047$ Pa×s and $\lambda = 0.657$ W/(m×K).

The viscosity of the nanofluid was measured using a rotational viscometer DV2T. Viscosity measurements were carried out within the range of shear velocity from 10 to 200 1/s at a temperature of 25°C. Before the measurements, the viscometer was calibrated with distilled water and ethylene glycol. The results obtained were in a good agreement with the reference data.

The coefficient of thermal conductivity of nanofluid was measured using a non-stationary hot wire method. The detailed description of the installation and its testing is given in [18].

As is obvious from the plots presented in figure 2, experimental data obtained in the turbulent flow regime, are in good agreement with the empirical correlation [17]. The discrepancies do not exceed 5% that is comparable to the accuracy of the correlation.

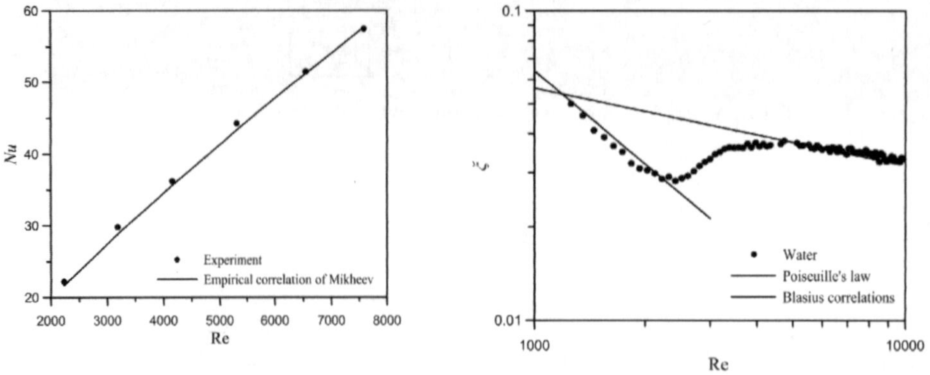

(a) (b)

Figure 2. Heat transfer coefficient (a) and friction factor (b) for pure water versus Reynolds number.

In addition to the heat transfer of pure water, pressure drop was measured as well. Figure 2b shows the measured correlation between friction factor and Reynolds number for pure water. The friction factor was calculated by known formula:

$$\xi = \frac{2d\Delta P}{\rho U^2 L}$$

where U – is the superficial velocity; d – is the diameter; L – is the test section length; and ΔP – is the measured pressure drop.

For comparison, the chart shows also calculations by Poiseuille's theoretical relationship $\xi = \frac{64}{\text{Re}}$ for laminar flow and Blasius correlation $\xi = 0.316\text{Re}^{-0.25}$ for turbulent flow. It is obvious that within the range of Reynolds numbers from 2300 to 3000 we observe flow transition from laminar to turbulent regime. The measured pressure drop values are consistent with the theoretical values with an accuracy of 5%.

3. The results of the experiments

The turbulent forced convection of water-based nanofluid with zirconia nanoparticles with the volume concentration of 4% was studied. This study was carried out for a round smooth tube and round tube with knurling ring. The experimental results are shown in figures 3-4. On the smooth round tube the ZrO_2 nanofluid allows to intensify the heat transfer coefficient by 27% compared to pure water at a fixed Reynolds number. The experiment was shown that the using of nanofluid can improve the heat transfer coefficient in the tube with knurling by 23% compared with pure water (figure 3). The difference in the intensification of heat transfer for smooth tube and tube with knurling due to nanofluid does not exceed the experimental error 5%.

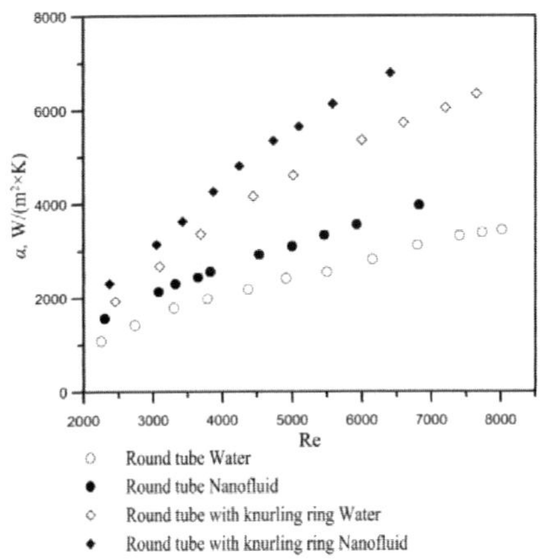

Figure 3. Average heat transfer coefficient versus Reynolds number.

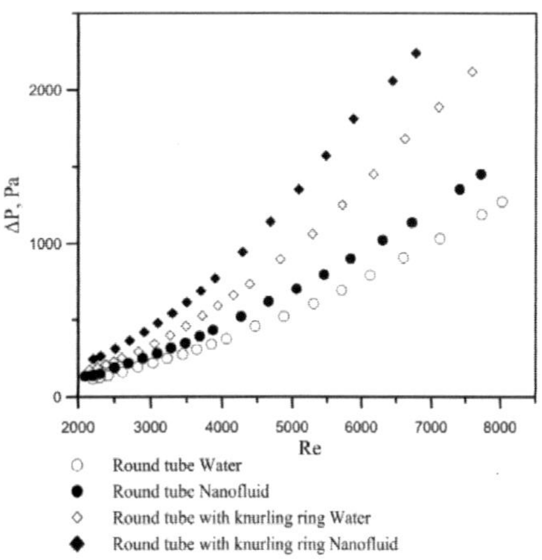

Figure 4. Pressure drop versus Reynolds number.

Figure 4 shows the pressure drop for both channels depending on Reynolds number for water and ZrO_2 nanofluid. Data analysis has shown that the nanofluid the increases pressure drop in the smooth and in the rough channels on the same value about 25%.

It is necessary to estimate of the thermal-hydraulic efficiency for the different channels. The heat transfer coefficient in the tube with a knurling ring to pure water is higher by 85%, than the heat transfer coefficient of smooth tube. However the pressure drop in the rough channel is also increased by 80% compared with a smooth tube. Nanofluid allows intensifying heat transfer in a smooth tube by 27% and increases the pressure drop is only 25% compared to pure water. To evaluate the thermal-hydraulic efficiency we used the ratio of heat transfer coefficients to the ratio pressure drops $\eta = \left(\dfrac{\alpha}{\alpha_{w,s}}\right)\left(\dfrac{dp}{dp_{w,s}}\right)^{-1}$. Here $\alpha_{w,s}$, $dp_{w,s}$ - average heat transfer coefficient and pressure drop in a smooth tube for pure water.

This dependence for different Reynolds numbers is shown in figure 5. The graph shows that the using of nanofluid in a round smooth tube has approximately the same efficiency, taking into account the measurement error. The applying of nanofluid in rough pipe is not effectively. The thermal-hydraulic for this case less than 1.

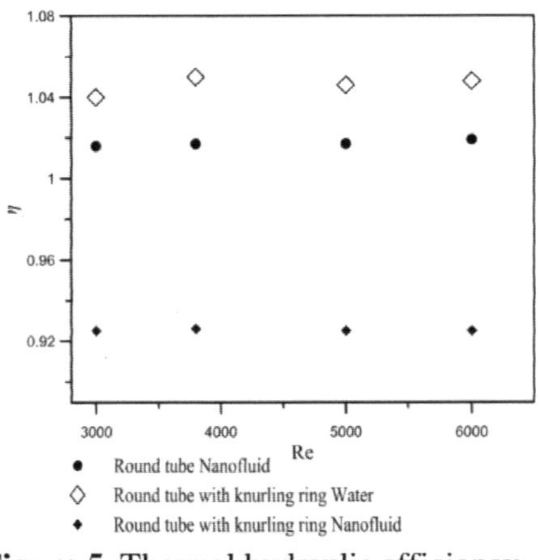

Figure 5. Thermal hydraulic efficiency versus Reynolds number

4. Conclusion

The turbulent forced convection of water-based nanofluid with zirconia nanoparticles was studied. The heat transfer coefficient and the pressure drop for pure water and nanofluid in the smooth and rough pipes been measured. The experiment showed that the heat transfer coefficient with using ZrO_2 nanofluid increases by about 25% in comparison with the pure water for both the smooth and rough channels. It was obtained that the nanofluid increases pressure drop in the smooth and in the rough channels on the same value about 25%. Thus, it has been shown that the intensification of heat transfer by using nanofluid weakly dependent on the shape of the channel, and determined only by the properties nanofluid.

A comparison of the thermal-hydraulic efficiency η of using nanofluid and using tubes with knurling ring has been done. It is shown that the using nanofluid in the smooth pipe has comparable performance with using tubes with knurling ring for pure water. The using of nanofluid in rough pipe is not effectively.

Acknowledgments

This work was financially supported by the Russian Foundation for Basic Research (Project 16-38-00569).

References

[1] Ahuja A S 1975 *J. Appl. Phys.* **46** 224
[2] Das S K, Choi S and Patel H 2006 *Heat Transfer Engineering* **20** (10) 3
[3] Das S K, Choi S, and Yu W and Pradeep T 2007 *Nanofluids Science and Technology* (New Jersey: Wiley-Interscience) p 397
[4] Wang X-Q and Mujumbar A S 2007 *Int. J. Thermal Science* **46** 1
[5] Yu W, France D M, Choi S and Routbort J L 2007 *Review and Assessment of Nanofluid Technology for Transportation and other Applications* (Illinois: Argonne National Laboratory) p 78
[6] Terekhov V I, Kalinina S V and Lemanov V V 2010 *Thermophysics and Aeromechanics* **2** 173
[7] Minakov A V, Lobasov A S, Guzei D V, Pryazhnikov M I and Rudyak V Ya 2015 *Applied Thermal Engineering* **88** 140-148.
[8] Minakov A V, Rudyak V Ya, Guzei D V, Lobasov A S 2015 *High Temperature* **53** (2) 246-253
[9] Guzei D V, Minakov A V, Rudyak V Ya, Dekterev A A 2014 *Technical Physics Letters* **40** (3) 203-206
[10] Pak B C and Cho Y I 1998 *Exp. Heat Transfer* **11** 151.
[11] QLi Y 2002 *Sci. China* **45** 408
[12] He Y, Jin Y, Chen H, Ding Y, Cang D and Lu H 2007 *Int. J. Heat Mass Transfer* **50** 2272
[13] Duangthongsuk W and Wongwises S 2010 *Int. J. Heat Mass Transfer* **53** 334–344
[14] Fotukian S M, Esfahany N 2010 *Int. J. Heat and Fluid Flow* **31** 606–612
[15] Fotukian S M, Esfahany N 2010 *Int. Communications in Heat and Mass Transfer* **37** 214–219
[16] Meriläinen A, Seppälä A, Saari K, Seitsonen J, Ruokolainen J, Puisto Rostedt N and Ala-Nissila T 2013 *Int. J. Heat and Mass Transfer* **61** 439–448
[17] Tsvetkov F F and Grigoriev B A 2005 *Heat and Mass Transfer* (Moscow: Moscow Power Engineering Institute) p 271
[18] Minakov A V, Rudyak V Ya, Guzei D V, Pryazhnikov M I, Lobasov A S 2015 *J. Eng. Phys. Thermophys* **88**(1) 149–162

Journal of Physics: Conference Series **754** (2016) 092003 doi:10.1088/1742-6596/754/9/092003

Study of the dependence the thermal conductivity of nanofluids on different parameters

M I Pryazhnikov[1,2,3], A V Minakov[1,2,3], D V Guzei[1,2], V Ya Rudyak[2]

[1] Siberian Federal University, 79 Svobodny pr., Krasnoyarsk, 660041, Russia
[2] Novosibirsk State University of Civil Engineering, 113 Leningradskaya Street, Novosibirsk, 630008, Russia
[3] Institute of Thermophysics SB RAS, 1 Acad. Lavrentiev pr., Novosibirsk, 630090, Russia

E-mail: arrivent@yandex.ru

Abstract. The paper presents the results of systematic measurements of the thermal conductivity coefficient of nanofluids at room temperature. It is shown that the thermal conductivity of all considered nanofluids depend on concentration, nanopartical size, as well as the base fluid properties. It was revealed that thermal conductivity coefficient of nanofluids increases with increasing concentration and size of nanoparticles.

1. Introduction

The first use of a fluid with microparticles for the purpose of heat-transfer intensification has been known since the mid-1970s (e.g., Ahuja, 1975 [1]). However, it failed to obtain substantial results. This is due to the fact that large disperse particles sediment quite rapidly and cause the channel walls to erode. Fluids in which the dispersed component is represented by nanoparticles are free from this drawback. The idea of using nanoparticles to enhance the heat conduction of a carrier fluid emerged about two decades ago. A two-phase medium whose carrier component is a regular liquid and whose dispersed component is represented by nanoparticles is currently referred to as a nanofluid (this term appeared for the first time in the work of Choi [2]). At present, one important fundamental issue is understanding mechanisms of transfer of heat in nanofluids. To explain the anomalous heat conduction of nanofluids one analyzes a few mechanisms: Brownian motion of nanoparticles (diffusion), the formation of a high-thermal-conductivity liquid layer on the "liquid–particle" interface, classification of nanoparticles, thermal diffusion (transfer of nanoparticles by the action of a temperature gradient), ballistic transfer of thermal energy in an individual particle and between nanoparticles occurring in the case of their contact, and others. The issue as to the contribution of these mechanisms to the effective thermal conductivity of various nanofluids remains open. Due to this, no general theory capable of predicting reliably the heat conduction of nanofluids has been created to date. Therefore, a reliable experimental procedure of measurement of the thermal-conductivity coefficient of nanofluids is required.

Despite the great number of works in which the heat conduction of nanofluids and their heat transfer are studied, the results obtained are often erratic. This is true of, e.g., heat transfer under natural-convection conditions [3]. Although an enhancement of the heat transfer in the case where nanoparticles are used is noted in most works [4], there are publications where a decrease in the heat transfer on adding them is demonstrated [5]. In addition to experimental investigations of the heat conduction of nanofluids, there are numerous theoretical works; however, a review of results of works by different researchers shows that it is impossible to closely predict the heat conduction of nanofluids at present. Therefore, comprehensive investigations of the heat transfer of nanofluids require a reliable procedure of measurement of the thermal-conductivity coefficient.

The present work we used hot-wire method. This method is widely employed to measure the thermal-conductivity coefficient of liquids.

2. Measurement technique

Thermal conductivity measurements were performed by non-stationary hot-wire method. Detailed description of the test bench and testing technique is given in [6]. Wheatstone bridge was used as the basis of the test bench instrumentation to measure an unknown electrical resistance of hot wire. A copper wire with a length of 150 mm and a diameter of 75 microns was used in the experiments. The wire serves one of the measuring bridge resistors R_w. The bridge includes also two other resistors: R_1=2 kOhm and R_3=1 Ohm as well as the resistance box R_2, needed to balance bridge. Initially the bridge is balanced and the output voltage does not exceed 10 μV. Balancing of the measuring circuit is carried out by a voltage of 0.1 V from a lab power source GWInstek GPC-3060D applied within a short period of time. Further, the circuit is energized by measuring voltage to record the out-of-balance voltage change of the bridge circuit over time. Voltage measurement is carried out using a precision voltmeter GWInstek GDM-78261 with increments of 10 ms. In addition, the test fluid temperature was measured by means of chromel-copel thermocouples connected to the TRM-138 meter.

Data processing was made in a following way. The initial resistance of the wire is found from the condition of the bridge circuit balance: $R_{w0} = R_1 R_3 / R_2$. The change in wire resistance is determined by the ratio

$$R_{wt} = R_3 \left[R_1 + \frac{V_o(R_1 + R_2)}{V_i} \right] \left[R_2 - \frac{V_o(R_1 + R_2)}{V_i} \right]^{-1},$$

where V_i – is the input voltage of the bridge, V_o – is out-of-balance bridge voltage. Given the dependence of electrical resistance of copper on temperature, we can determine the temperature change of wire:

$$\Delta T = (R_{wt2} - R_{wt1})(\alpha R_{w0})^{-1} = \Delta R_{wt} (\alpha R_{w0})^{-1},$$

where $\alpha = 0.0003831/\mathrm{K}$ – is the temperature coefficient, predetermined from measurement series of used copper wire resistance at different temperatures. The voltage drop on the wire is given by formula: $V_{Rw} = V_i R_{wt} / (R_{wt} + R_3)$. Then, the linear density of heat flux on the wire is defined as $q = V_{Rw}{}^2 / (L_w R_{wt})$: where L_w – is the length of the wire. In the experiments, a typical value of q was around 10 W/m.

Eventually, the heat conductivity coefficient of a fluid is defined as follows: $\lambda = \alpha q R_{w0} / (4\pi G)$, where G – is the angular coefficient of the linear section taken from the dependence of the wire resistance on the logarithm of time. The coefficient G was calculated by the least squares method. The value of thermal conductivity coefficient was obtained by averaging over ten measurements. During the measurements, the nanofluid temperature was maintained constant. In the presented measurements it was equal to 25°C. The resultant relative measurement error of fluid thermal conductivity coefficient does not exceed 3 %.

Preparation of nanofluids was carried out based on standard two step process. After adding to the base fluid the required amount of nanopowder, the nanofluid was first thoroughly mixed mechanically, and then was placed for the half-hour into an ultrasonic disperser "Sapphire" to destruct conglomerates of particles. The nanoparticles were purchased from "Plasmoterm" company (Moscow).

3. Dependence of thermal conductivity on particle concentration

The first question that arises in the study of thermal conductivity of any dispersed fluid is formulated quite simply: how thermal conductivity depends on the particle concentration φ. The answer to this question for coarse liquids with spherical particles was given by Maxwell [7], who obtained the following relation between the thermal conductivity coefficient of the suspension and base fluid:

$$\lambda = \lambda_f \frac{\lambda_p + 2\lambda_f + 2\varphi(\lambda_p - \lambda_f)}{\lambda_p + 2\lambda_f - \varphi(\lambda_p - \lambda_f)} \quad (1)$$

where λ_p, λ_f – is the thermal conductivity coefficient of particle material. Formula (1) is obtained for spherical particles non interacting with each other. Later Bruggemann [8] proposed a model, which took into account the interaction of the randomly distributed particles. This model, unlike (equation (1)), has no restrictions on particle concentration, though within the low concentration range it leads to the same results as the Maxwell model. Hamilton and Crosser [9] generalized the model (equation (1)) for the case of non-spherical particles. There are still quite a lot of models, a short list of which can be found, for example in [10], however, the model (equation (1)) satisfactorily describes numerous experimental data obtained in coarse fluids at not too high concentrations.

In all cases, increasing the particle concentration leads to significant enhancement of nanofluid thermal conductivity. The enhancement of nanofluid thermal conductivity over that of pure water at 6% concentration of nanoparticles reaches 5-28% that is greater than the values defined by equation (1). For illustration, figure 1 shows nanofluids thermal conductivity coefficients along with the values calculated by equation (1).

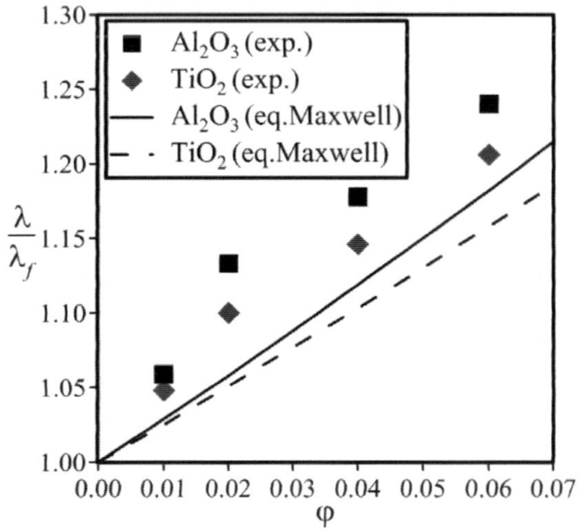

Figure 1. Relative thermal conductivity of water based nanofluids with 150 nm particles of TiO_2, Al_2O_3 depending on particle concentration.

4. Dependence of thermal conductivity on particle size

In order to clearly answer the question about the nature of this dependence, it was necessary to perform thermal conductivity measurements in nanofluids with particles of different size, though at the same volumetric concentration. We have performed these measurements using nanofluids with silicon, aluminum, zirconium, and titanium oxides particles. The particle sizes ranged from 10 to 150 nm. Pure water was used as the base fluid in all cases, while volumetric concentration of the nanoparticles was equal to 2%. The measurements were carried out at a temperature of 25°C. The data obtained are presented in figure 2 and allow us to unambiguously assert that the nanofluid thermal conductivity coefficient depends on the particle size; at that, the more particle size the higher thermal conductivity.

However, the degree of enhancement of thermal conductivity coefficient is very different. It is characteristic that for nanofluids containing SiO_2 and Al_2O_3 particles, the obtained relative thermal conductivity coefficients are lower than those defined by equation (1). This fact was noted in the literature before, though was not associated with the relationship between the thermal conductivity and particle size.

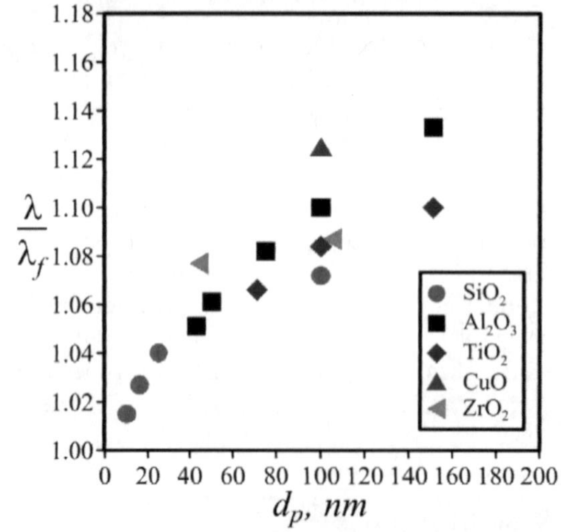

Figure 2. Relative thermal conductivity coefficients of nanofluids depending on nanoparticle size at volumetric concentration of 2%.

5. Dependence of thermal conductivity on the properties of the base fluid

To date, just a few works are known, where the effect of base fluid properties on nanofluid thermal conductivity was studied. This was first done in [11], which deals with nanofluids based on water, ethylene glycol, vacuum pump oil, and engine oil. It was noted that the highest thermal conductivity of the nanofluid was revealed in ethylene glycol based nanofluids. Later, the influence of the base fluid was studied by Xie and colleagues [12]. They developed a method to create a stable nanofluids, based on deionized water, ethylene glycol, and decene with multiwalled carbon nanotubes. It was found that the thermal conductivity coefficient enhancement decreases with increasing thermal conductivity of the base fluid. This conclusion is not consistent with the data presented in work [11], though, certainly, the studied nanofluids were quite specific. Therefore, it was extremely important to test the applicability of this conclusion with regard to the conventional nanofluids.

The complexity of studying the effects of the thermal conductivity of the base fluid on the thermal properties of nanofluids consists in the necessity to compare the data, at least of two nanofluids with the same concentration of nanoparticles and their size, though different base fluids. In the present work, the measurement series was carried out with nanofluids based on water and ethylene glycol, containing TiO_2 particles 150 nm in size. The data obtained differ systematically, though not much (see figure 3). In all cases, thermal conductivity enhancement of ethylene glycol based nanofluids is higher than that of water based nanofluids. At that, thermal conductivity coefficient of water is almost two and a half times higher than that of ethylene glycol.

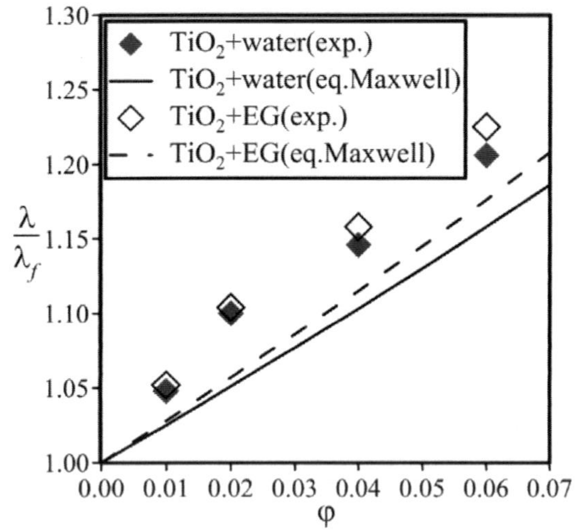

Figure 3. Relative thermal conductivity coefficient of water and ethylene glycol based nanofluids depending on TiO_2 particle concentration.

6. Conclusion

First of all, it should be noted that the nanofluid thermal conductivity is not described by the classical theories, including the Maxwell theory (equation (1)) for coarse fluids. The main difference is that the nanofluid thermal conductivity coefficient is a complicated function not only of the particle concentration, but also the particle size, material, and type of base fluid.

Measured thermal conductivity coefficients almost always exceed the values calculated by equation (1), though nanofluids with sufficiently small particles may have thermal conductivity coefficients even lower than those predicted by the Maxwell theory. However, in all cases, the nanofluid thermal conductivity coefficient enhances with increasing particle size. Starting with a certain particle size, the measured nanofluid thermal conductivity coefficients exceed the values predicted by classical theories (in particular, by equation (1)).

By definition, nanoparticles are particles sized from 1 to 100 nm. In the present work we studied the thermal conductivity of both nanofluids and dispersed fluids with larger particles. At that, in all cases we observed a monotonic enhancement of thermal conductivity with increasing particle size. However, as already mentioned, noted enhancement of thermal conductivity coefficient with increase in particle size cannot be unlimited. Determining the nature of dependence of the thermal conductivity coefficient in dispersed fluids on size of the particles larger than 100 nm requires further systematic study.

The base liquid also significantly influences the effective thermal conductivity of the nanofluid. In this work we have confirmed that the lower the thermal conductivity of the base fluid, the higher the relative thermal conductivity coefficient of the nanofluid. This is quite naturally explained by the fact that in the base fluid with the highest thermal conductivity the enhancement of thermal conductivity is weaker, at other conditions being equal. This, in particular, means that the supplement of the nanoparticles into the fluid will be most effective for the base fluid with low thermal conductivity. Nevertheless, the effect of base fluid properties on the effective thermal conductivity of nanofluid is still poorly studied.

Acknowledgments

The work is carried out at partial financial support of the Russian Scientific Foundation (Agreement No. 14-19-00312).

References

[1] Choi S 1995 *ASME.* **231** 99-105.
[2] Kumar D S, Choi S U S and Patel H E 2006 *Heat Transfer Eng*. **27**(10) 3–19.
[3] Godson L, Raja B, Mohan L D and Wongwises S 2010 *Renew. Sustainable Energy Rev.* **14** 629–641.
[4] Minakov A V, Lobasov A S, Pryazhnikov M I and Guzei D V 2014 *Defect Diffus Forum* **348** 123-138.
[5] Guzei D V, Minakov A V, Rudyak V Ya and Dekterev A A 2014 *Tech Phys Lett* **40** 203–206.
[6] Minakov A V, Guzei D V, Pryazhnikov M I, Lobasov A S and Rudyak V Ya 2015 *J. eng. phys. thermophys*. **88** 149-162.
[7] Maxwell J C 1881 *A treatise on electricity and magnetism*. Oxford, Clarendon Press.
[8] Bruggeman D A G . 1935 *Annalen der Physik*. **24** 636–679.
[9] Hamilton R L and Crosser O K 1962 *I&EC Fundam.* **1** 182–191.
[10] Rashmi W, Khalid M, Ong S S and Saidur R 2014 *Materials Research Express*. **1** 032001.
[11] Wang X, Xu X and Choi S U S 1999 *J. Thermophys. Heat Trans.* **13**(4) 474–480.
[12] Xie H, Lee H, Youn W and Choi M 2003 *J. Appl. Phys*. **94** (8) 4967–4971.

The morphology of carbon-metal composite synthesized in arc discharge

D V Smovzh, S Z Sakhapov and A V Zaikovskii

Kutateladze Institute of Thermophysics, Siberian Branch of Russian Scademy of Sciences, Novosibirsk

E-mail: smovzh@itp.nsc.ru

Abstract. The phase state of nanoparticles and function of distribution by size of particles, formed at joint electric arc spraying of metal (Ni/Cu/Ti/Pt/Zr) - carbon electrode, is studied. It is shown that the metal and carbide nanoparticles with the size of 2-9 nm are formed in the carbon matrix at spraying. When annealing the metal-carbon composites, the metal-containing nanoparticles oxidize and coagulate, forming the agglomerates of 100 nm or more.

1. Introduction

The arc discharge is one of the oldest, well-studied and promising methods of obtaining the nanomaterials [1], [2]. In particular, fullerenes were discovered in the arc discharge by Kretschmer in 1990 [3], and the CNT were firstly discovered at carbon deposition in the arc discharge by Iijima in 1991 [4]. Usually, in plasma of an arc discharge gap, the temperature varies from 5000 K in the channel center to 1000 K at the periphery [5]. Under the conditions of arc discharge, metal and carbon atomic spraying, followed by condensation at expansion to the buffer gas, takes place. Currently, when synthesizing nanoparticles in the arc discharges, the experimental efforts are focused on the study of mechanisms related to anode sputtering, deposition of material on the cathode, and current-voltage characteristics of the anode arc discharge [6], [7], [8], [9], [10], [11], and [12].

Due to simplicity of practical implementation of the plasma-arc method, the expansion of its capabilities is important for both scientific research and practical application. Synthesis of nanomaterials in the electric arc occurs under the non-equilibrium conditions. Moreover, the synthesis conditions (electric parameters of the arc, buffer gas and its pressure, geometry and composition of electrodes, etc.) influence kinetics of the synthesis. Therefore, the theoretical approaches to the study of processes occurring at synthesis are hampered, and the experimental studies go top.

In this work, we study the materials, formed in the gas phase at combined electric arc spraying of metal-graphite rods in the arc discharge. The powders of Ni, Cu, Ti, Pt, and Zr metals were used as the additions to the graphite electrode.

2. Experimental setup and methods

The metal-carbon black was synthesized in helium under the pressure of 25 torr and discharge current of 100 A, using the setup described in detail in [11]. The mole fraction of metal in the sprayed electrode was the same for all metals and equaled 6.4%. The synthesized material was annealed in air at 950°C during 2 hours. The materials were analyzed by the methods of transmission electron microscopy (JEOL – 2010) and X-ray diffractometry (Siemens D500).

3. Results

The TEM images of synthesized materials are shown in Fig. 1. The nanoparticles of metals integrated into the carbon matrix are obvious for each synthesized material. The images of materials annealed in the oxygen medium are also shown in Fig.1.

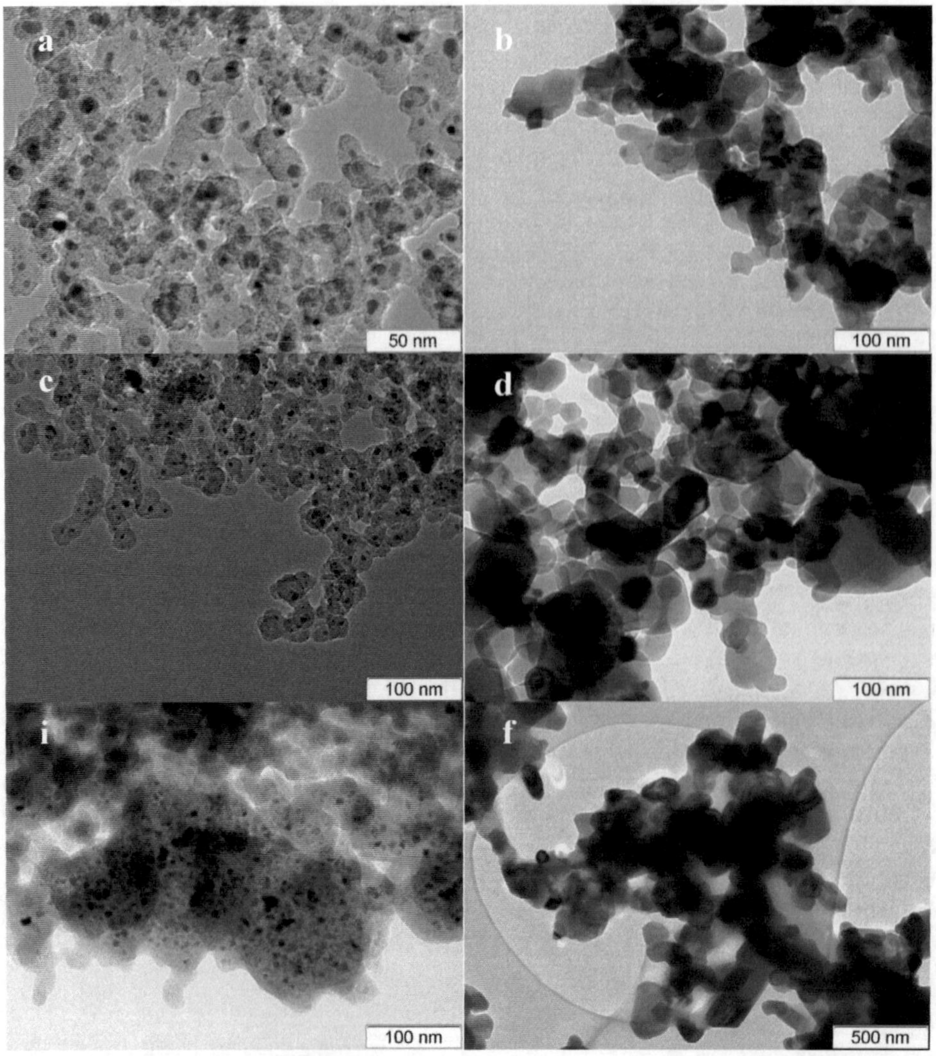

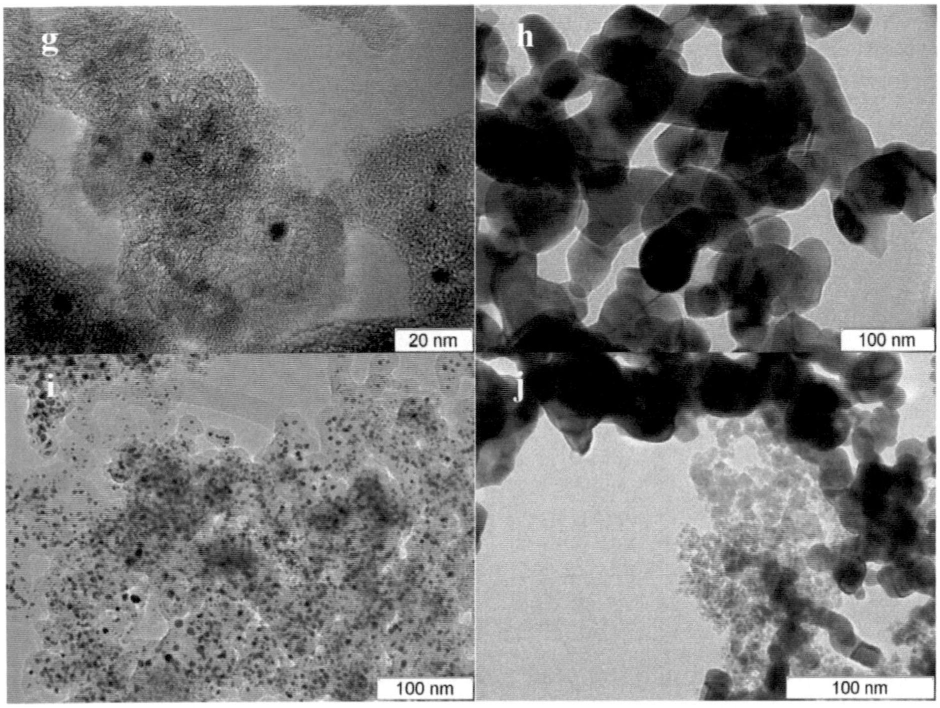

Figure 1. TEM images of Me-C composites. a) Cu-C, b) Annealed (a) at 950°C, c) Ni-C, d) Annealed (c) at 950°C, e) Ti-C, f) Annealed (e) at 950° C, g) Zr-C, h) Annealed (g) at 950°C, i) Pt-C, j) Annealed (i) at 950°C.

The function of size distribution of metal particles was measured by the method of statistical processing of a large array of TEM images. The data were described by lognormal distribution that corresponds to the function of cluster size distribution at agglomeration in the gas phase. After annealing, the nanoparticles fused, forming the agglomerates of the larger particles with the size of about 100 nm and higher; therefore, annealing these composites data does not allow obtaining the nanoparticles of metal oxides.

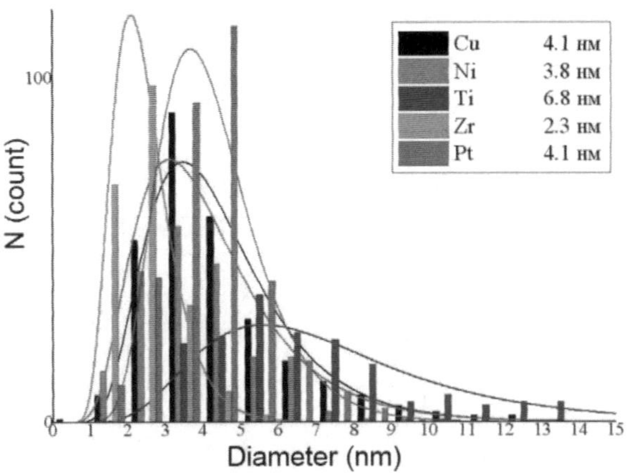

Figure 2. Size distribution of nanoparticles obtaining at sputtering the composite electrodes in the arc. The corresponding approximations by the lognormal distribution are presented with indication of the median.

The phase composition of synthesized samples is presented in Table 1; it can be seen that the crystalline phase is not observed in every sample. Formation of oxides of various stoichiometry and phase composition is characteristic of nanoparticle oxidation.

Table 1. Phase structure of samples.

Element composition of samples	Crystalline structure	Crystalline structure after annealing
Cu-C	none	Cu2CO5 (P 121/a1) Cu4O3 (I 41/amd)
Ni-C	none	None
Ti-C	TiC (Fm-3m)	Ti4O7 (A-1(-1)) TiO2 (Pbca) Ti6O11 (-1(1))
Zr-C	ZrC (Fm-3m)	ZrO2 (P 121/c1) ZrO2 (P bcm)
Pt-C	Pt (Fm-3m)	Pt (Fm-3m)

5. Conclusions

The method of electric arc spraying allows obtaining the nanoparticles of various metals with a narrow size distribution function. The absence of impurities (except carbon) makes it promising in comparison with the methods of solution chemistry. Additionally, in some applications, such as super capacitors, lithium-ion batteries, and fuel cells, it is necessary to obtain metal nanoparticles on a base matrix with high porosity. Various types of activated carbon are commonly used as these matrices. Thus, the proposed method can be considered as the one-satge synthesis of metal-carbon catalyst. Another important advantage of this technology is the possibility to get the joint composites of several metals.

In these works, the experimental data on the synthesis of metal-carbon composites were obtained. We investigated the size distribution functions for various metals in the carbon matrix. The average size of the formed nanoparticles is in the range of 2-9 nm. It is shown that for all studied metals, when the carbon matrix is removed, nanoparticles sinter and form the larger oxide particles of 100 nm and higher. We studied the phase state of particles; it is shown that when spraying the Pt electrodes, the particles of metal platinum are formed in the carbon matrix. When spraying Zr and Ti, the carbide particles are formed in the carbon matrix. When spraying Ni and Cu, the metal or carbide particles have no crystal structure.

6. Acknowledgements

The reported study was funded by RFBR, according to the research project No. 15-38-20454 mol_a_ved.

References

[1] Keidar, M., Factors affecting synthesis of single wall carbon nanotubes in arc discharge. Journal of Physics D: Applied Physics, 2007. 40(8): p. 2388.
[2] Ando, Y., et al., Mass production of single-wall carbon nanotubes by the arc plasma jet method. Chemical Physics Letters, 2000. 323(5–6): p. 580-585.
[3] Kratschmer, W., et al., Solid C60: a new form of carbon. Nature, 1990. 347(6291): p. 354-358.
[4] Iijima, S., Helical microtubules of graphitic carbon. Nature, 1991. 354(6348): p. 56-58.
[5] Waldorff, E.I., et al., Characterization of carbon nanotubes produced by arc discharge: Effect of the background pressure. Journal of Applied Physics, 2004. 95(5): p. 2749-2754.
[6] Fetterman, A.J., Y. Raitses, and M. Keidar, Enhanced ablation of small anodes in a carbon

nanotube arc plasma. Carbon, 2008. 46(10): p. 1322-1326.
[7] Shashurin, A., M. Keidar, and I.I. Beilis, Voltage-current characteristics of an anodic arc producing carbon nanotubes. Journal of Applied Physics, 2008. 104(6): p. 063311.
[8] Shashurin, A. and M. Keidar, Factors affecting the size and deposition rate of the cathode deposit in an anodic arc used to produce carbon nanotubes. Carbon, 2008. 46(13): p. 1826-1828.
[9] Keidar, M. and I.I. Beilis, Modeling of atmospheric-pressure anodic carbon arc producing carbon nanotubes. Journal of Applied Physics, 2009. 106(10): p. 103304.
[10] Smovzh, D.V., et al., Morphology of aluminium oxide nanostructures after calcination of arc discharge Al–C soot. Ceramics International, 2015. 41(7): p. 8814-8819.
[11] Kundrapu, M. and M. Keidar, Numerical simulation of carbon arc discharge for nanoparticle synthesis. Physics of Plasmas, 2012. 19(7): p. 073510.
[12] Scott J.H.J., Majetich S.A. Morphology, structure, and growth of nanoparticles produced in a carbon arc. Phys Rev B 1995, 52(17), 12564–12571Kraus A., Bar-Cohen A.A. 2015 Mechanical Engineers Handbook. IV:1:12:1-47

Non-linear memristor switching model

A. A. Chernov[1,2,3], D. R. Islamov[2,3], and A. A. Pik'nik[1,2,3]

[1]Kutateladze Institute of Thermophysics SB RAS,
Novosibirsk 630090, Russian Federation
[2]Rzhanov Institute of Semiconductor Physics SB RAS,
Novosibirsk 630090, Russian Federation
[3]Novosibirsk State University, Novosibirsk 630090, Russian Federation

E-mail: chernov@itp.nsc.ru

Abstract. We introduce a thermodynamical model of filament growing when a current pulse via memristor flows. The model is the boundary value problem, which includes nonstationary heat conduction equation with non-linear Joule heat source, Poisson equation, and Shockley-Read-Hall equations taking into account strong electron-phonon interactions in trap ionization and charge transport processes. The charge current, which defines the heating in the model, depends on the rate of the oxygen vacancy generation. The latter depends on the local temperature. The solution of the introduced problem allows one to describe the kinetics of the switch process and the final filament morphology.

1. Introduction

In recent years significant efforts are being made for implementation of fourth electronic element – resistor with memory called memristor. This element is a passive device whose resistance can be changed by passing through a certain electric current pulse. The memristor has a simple metal-high-k-metal structure.

Functionally the memristors can be divided into two-level (one-bit) and multilevel (multibit) ones. If the one-bit device may be used as a cell of non-volatile memory with a large storage time and high resistance to radiation, the multibit memristors on the one hand can increase the density of integration of the non-volatile memory, and on the other hand can be used to form the basis of adaptive neuromorphic (cognitive) computational systems. One of the most interesting and hitherto unsolved problems is the nature of the intermediate (discrete) multi-level states in a memristor.

The hypothesis that the effect of memristive switch is directly connected with the formation of a filament, which occurs when certain current pulse passes over a dielectric, is distributed among specialists. However, the kinetics of filament formation during the transition from the high resistance state of the memristor to a low resistive state is still not clear. Another unsolved problem in the way of developing memristor arrays is a description of the forming process of memristor elements – the first switch from the initial high resistance state to a low resistance at high voltages, accompanied by intense Joule heat release. Nature of the forming of memristor matrix is still unclear, and the absence of any plausible physical model constrains the practical implementation of these elements.

It is considered that charge transport in high-κ is associated with the presence of defects. The most probable and the most widespread defects are oxygen vacancies. Properties of these vacancies are actively studied both experimentally and theoretically. However, many contradictory data can be found in literature, and the role and characteristics of oxygen vacancies in the charge transport through dielectric process are not established yet. It is obvious that rate of change in structure of dielectric (namely the generation rate of oxygen vacancies associated with conductivity) in strong electric field is mainly determined by non-stationary processes of heat and mass transfer. Thus, the kinetics of such processes determines final morphology and properties of memristor.

2. The theory

Physics model of resistive switch in thin films of oxides of transition metals consists of 4 main subtasks: strongly nonlinear and non-stationary thermophysical problem, problem of finding charge distribution and charge transfer, problem of finding of the heat source function and problem of generation, recombination and migration of defects that act as traps for electrons and holes.

2.1. Thermophysical problem

Generation (recombination) of defects in active medium of ReRAM element is assumed to be induced by local heat release caused by electric current flowing through it. For modeling purposes the structure of memristor is assumed to consist of 3 layers: 2 half-infinite layers of metal electrodes with dielectric layer between them as shown in Fig 1. Layer thickness is d. Temperature field in the volume of memristor is described by heat conduction equation which we regard as having cylindrical symmetry:

$$\rho c\frac{\partial T}{\partial t} = \frac{1}{r}\frac{\partial}{\partial r}\left(\lambda r\frac{\partial T}{\partial r}\right) + \frac{\partial}{\partial z}\left(\lambda\frac{\partial T}{\partial z}\right) + Q, \quad (1)$$

where ρ is the medium density, c is the medium heat capacity, T is the medium local temperature, t is time, r and z are radius and height, respectively; λ is heat conductivity, Q is heat per unit volume (will be defined later on).

The boundary conditions are as follows: no heat flux on the axis of symmetry; the temperature far away from filament equals the initial temperature T_0; heat fluxes and temperatures at the dielectric-electrode border are equal.

2.2. Charge distribution and charge transfer problem

In solid bodies the charge carriers are electrons and electron holes — quasi-particles that represent collective reaction of electrons on valence atomic orbitals of elements that form the medium [1, 2]. The distinctive feature of such quasi-particles is that they have a charge equal to the charge of free electron q in absolute value and that they have an effective mass m^* defined by crystal structure of the substance. It is generally accepted that the charge of electron hole is positive and the charge of electron is negative. Later on we will assume that elementary charge q is positive.

Charge transport involves both «free» electrons and holes (1 in conduction band and 1' in valence band), and trapped quasi-particles (Fig. 2). The trapped quasi-particles can be ionized in respective zone i.e. become «free» (2, 2' in Fig. 2), «free» charge carriers can become trapped (3, 3' in Fig. 2), providing the closeness of traps trapped electrons and holes can tunnel between adjacent traps (4, 4' in Fig. 2). Also free electrons can recombine with trapped holes (5 in Fig. 2), free holes can recombine with trapped electrons (5' in Fig.2). Recombination of free electrons and free holes can be neglected because of low concentration of quasi-particles in carrier band and valence band of the dielectric. Abovementioned processes are described by Shockley-Read-Hall equations complemented by effect of strong electron-phonon interaction in the processes of ionization of traps and charge transport:

$$\frac{\partial n}{\partial t} = +\nabla\cdot(nv_n\boldsymbol{e}) - \sigma_n v_n n(N - n_t) + n_t P_{ion} - \sigma_r v_n n p_t, \quad (2)$$

$$\frac{\partial n_t}{\partial t} = \sigma_n v_n n(N - n_t) - n_t P_{ion} - \sigma_r v_p p n_t + \left(\frac{\partial n_t}{\partial t}\right)^{tun}, \quad (3)$$

$$\frac{\partial p}{\partial t} = -\nabla\cdot(pv_p\boldsymbol{e}) - \sigma_p v_p p(N - p_t) + p_t P_{ion} - \sigma_r v_n p n_t, \quad (4)$$

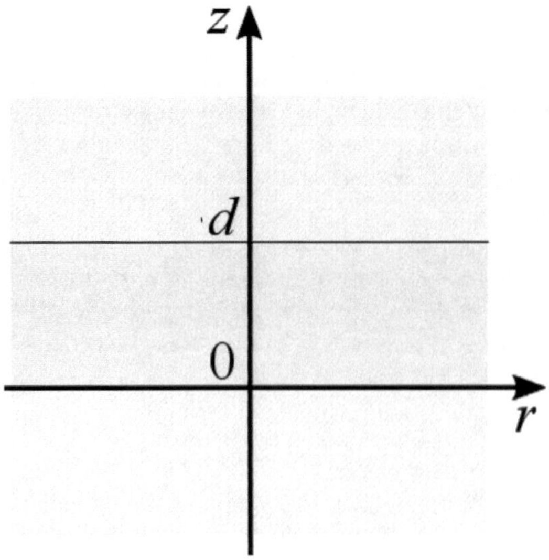

Figure 1. Model of the structure of Re-RAM element.

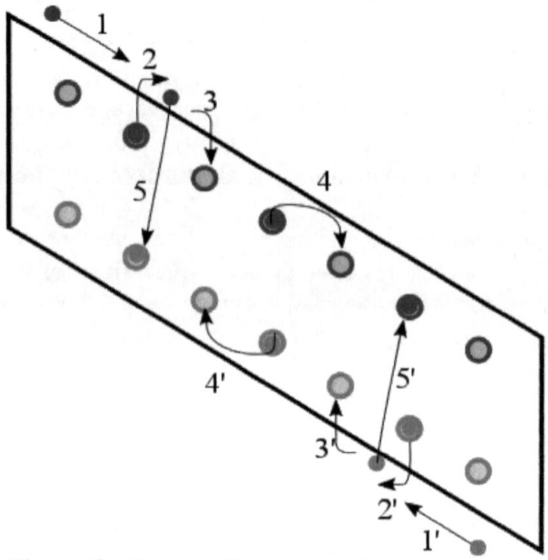

Figure 2. Energy diagram of the processes of localization, delocalization and charge transfer in dielectric medium.

$$\frac{\partial p_t}{\partial t} = \sigma_p v_p p(N - p_t) - p_t P_{ion} - \sigma_r v_n n p_t + \left(\frac{\partial p_t}{\partial t}\right)^{tun}, \tag{5}$$

where n is concentration of free electrons, v_n is drift velocity of free electrons, $\boldsymbol{e} = \boldsymbol{F}/F$ is a unit vector, $\boldsymbol{F}$ is electric field, σ_n is capture cross-section of electron on the trap, N is concentration of charge traps (assuming that the trap for electron and a trap for electron hole are the same defects, namely, oxygen vacancies), n_t is concentration of trapped electrons, P_{ion} is ionization rate of trapped charge carriers, σ_r is recombination cross-section for electron-hole pair, p_t is concentration of trapped holes, p is concentration of free holes, v_p is drift velocity of free holes, σ_p is capture cross-section of hole on trap. Terms $(\partial n_t/\partial t)^{tun}$ and $(\partial p_t/\partial t)^{tun}$ represent charge transfer caused by tunneling between traps:

$$\left(\frac{\partial n_t}{\partial t}\right)^{tun} = s\nabla \cdot \left(n_t\left(1 - \frac{n_t}{N}\right) P_{tun}\boldsymbol{e}\right), \left(\frac{\partial p_t}{\partial t}\right)^{tun} = -s\nabla \cdot \left(p_t\left(1 - \frac{p_t}{N}\right) P_{tun}\boldsymbol{e}\right), \tag{6}$$

where $s = N^{-1/3}$ – average distance between traps, P_{tun} is rate of tunneling between traps.

Process of ionization of trapped into conduction/valence band electrons/holes is described by the theory of multiphonon ionization of neutral isolated trap (2, 2' in Fig. 2):

$$P_{ion} = \sum_{n=-\infty}^{+\infty} \exp\left(\frac{nW_{ph}}{2kT} - \frac{W_{opt} - W_t}{W_{ph}} \coth\frac{W_{ph}}{2kT}\right) I_n\left(\frac{W_{opt} - W_t}{W_{ph}\sinh(W_{ph}/2kT)}\right) P_i(W_t + nW_{ph}),$$
$$P_i(W) = \frac{eF}{2\sqrt{2m^* W}} exp\left(\frac{-4}{3}\frac{\sqrt{2m^*}}{\hbar qF} W^{3/2}\right), \tag{7}$$

where W_{ph} is phonon energy (breathing mode), W_t is thermal trap energy, W_{opt} is optical trap energy, k is the Boltzmann constant, I_n are modified Bessel functions, $\hbar = h/2\pi$ is the Plank constant, P_i is probability of tunneling through the barrier W [3].

Jumping between adjacent traps is described by phonon-assisted tunneling between traps:

$$P_{tun} = \frac{2\sqrt{\pi}\hbar W_t}{m^* s^2 Q_0 \sqrt{kT}} \exp\left(-\frac{W_{opt} - W_t}{2kT}\right) \exp\left(-\frac{2s\sqrt{2m^* W_t}}{\hbar}\right) \sinh\left(\frac{eFs}{2kT}\right), \tag{8}$$

where $Q_0 = \sqrt{2(W_{opt} - W_t)}$ [4]. Filling of the first trap derives from contact and described by the following equations:

$$\frac{\partial n_t}{\partial t} = v_{inj} \cdot (N - n_t) - v_{ion} \cdot n_t + s\,\frac{\partial}{\partial z}\left(n_t\left(1 - \frac{n_t}{N}\right) P_{tun}^{-}\right), \tag{9}$$

$$\frac{\partial p_t}{\partial t} = v_{inj} \cdot (N - p_t) - v_{ion} \cdot p_t + s\,\frac{\partial}{\partial z}\left(p_t\left(1 - \frac{p_t}{N}\right) P_{tun}^{+}\right), \tag{10}$$

where coefficients of filling of the trap with electrons (holes) from the contact v_{inj} and back ionization of the trap *into* contact v_{ion} were found using results of Refs. [4, 5]. Ionization rate for phonon-bounded trap is described by following equation:

$$v_{ion} = \int \frac{V_{out}}{2a\sqrt{2\pi kT}} \exp\left(-\frac{(Q-Q_0)^2}{2kT} - \frac{3}{4}\frac{\sqrt{2m^*}((-\epsilon)^{3/2}-(-qFa-\epsilon)^{3/2})}{qF\hbar}\right) \times$$
$$\times \left(1+\exp\left(\frac{+\epsilon+qFa}{kT}\right)\right)^{-1} dQ, \quad (11)]$$
$$\epsilon = -Q_0(Q-Q_0) - W_{opt}.$$

Here Φ is energy difference between the energy of the bottom of conductive band (for electrons, the top of the valence *bands* for holes) of the dielectric and Fermi level of the contact, a is the distance from the trap to the contact, V_{out} is velocity of released from the trap electron (hole) in the contact. Injection velocity v_{inj} from contact to the trap can be found using the fact that in stationary conditions trap's filling complies statistics with Fermi level in the contact:

$$n_t(z=0,d) = \frac{N}{1+\exp\left(\frac{\Phi_n - W_t}{kT}\right)}; p_t(z=0,d) = \frac{N}{1+\exp\left(\frac{\Phi_p - W_t}{kT}\right)}. \quad (12)$$

Relationship between injection velocity and ionization velocity is found from stationary equations (9), (10) without charge transfer to adjacent traps:

$$v_{inj} = \exp\left(\frac{-\Phi_{n,p} - W_t}{kT}\right) v_{ion}. \quad (13)$$

Filling of the last trap can be calculated similarly with filling of the first with substitution of F with $-F$. Also, it is necessary to qualify that for electrons the first trap is the trap near the electrode with higher potential, for the holes — with lower one.

Total current $\boldsymbol{j}$ includes electron $\boldsymbol{j}_n$ and hole $\boldsymbol{j}_p$ components:

$$\boldsymbol{j} = \boldsymbol{j}_n + \boldsymbol{j}_p. \quad (14)$$

Electron component consists of free electron component and the tunnelling between traps component:

$$\boldsymbol{j}_n = qnv_n\boldsymbol{e} + \frac{q}{s^2}\left(\frac{n_t}{N}\left(1-\frac{n_t}{N}\right)P_{tun}\boldsymbol{e}\right). \quad (15)$$

Hole component has a similar form:

$$\boldsymbol{j}_p = qpv_p\boldsymbol{e} + \frac{q}{s^2}\left(\frac{p_t}{N}\left(1-\frac{p_t}{N}\right)P_{tun}\boldsymbol{e}\right). \quad (16)$$

Drift velocity of free carriers in weak electric fields is linearly dependent on field. If it reaches the thermal velocity v_T, the saturation occurs (dependency on field disappears):

$$v = \min(\mu F, v_T), v_T = \sqrt{\frac{8kT}{\pi m^*}}, \quad (17)$$

where $\mu \sim (0{,}1\text{-}10)$ cm^2/V·s is charge carrier mobility in dielectric layer.

It is necessary to note that free and trapped charge carriers in the volume of dielectric layer distort the external electric field, and their influence needs accounting for by solving Poisson equation:

$$\Delta\varphi = q\frac{(n-p)+(n_t-p_t)}{\varepsilon\varepsilon_0}; \boldsymbol{F} = -\boldsymbol{\nabla}\varphi, \quad (18)$$

where φ is electric field potential, ε is low-frequency permittivity of the medium.

2.3. *Heat source function*

Heat source function is caused by the release of heat due to inelastic scattering of free charge carriers (Joule heat) and recombination of electrons and holes. Generally speaking, the process of recombination can go with creation of photons or phonons. In the process of recombination of free electron with trapped hole (or free hole with trapped electron) the full momentum is nonzero and remains so after the recombination. It is known that photons have negligible momentum and phonons can have momentum comparable to ones of electrons and holes. So, in this model creation of photons can be considered highly improbable process and ignored. Thus, in recombination of opposite charges all energy releases in the form of phonons and should appear in the heat source function:

$$Q = (\boldsymbol{F}\cdot\boldsymbol{j}) + \sigma_r v_n n p_t(E_g - W_t^n) + \sigma_r v_p p n_t(E_g - W_t^p). \quad (19)$$

Here E_g is the band-gap energy of dielectric, W_t^n and W_t^p are thermal energies of electron and hole traps, respectively. The first member in Eq (19) represents Joule heat, the second one describes recombination of free electrons on trapped holes, the third one is recombination of free holes on trapped electrons. It is noteworthy that for values characteristic for this problem: electric field ~10^6 V/cm and current 10^{-6} A/cm^2, Joule heat have the value ~1 W/cm^3. For values $\sigma_t \sim 10^{-13}$ cm^2, $v_{n,p} \sim 10^7$ cm/s, $n, p \sim 10^7$ cm^{-3}, $n_t, p_t \sim 10^{18}$ cm^{-3}, $E_g \sim 6$ eV, $W_t \sim 1$ eV released energy from recombination can be ~(0.1÷10) W/cm^3. So, the heat release in recombination in this case cannot be neglected! The change in energy of charge carriers in tunneling between traps is taken into account in Eq (19) using equations (14)–(16).

2.4. Generation, recombination and migration of oxygen vacancies

As was mentioned above forming and resistive switch to low-resistance state is related to trap generation for charge carriers. Oxygen vacancies are acting as traps. In accordance with experimental data certain fraction of vacancies exists in dielectric immediately after synthesis. However, defects generated during forming are responsible for performance attributes. Switching to high-resistance state is caused by recombination and spatial redistribution of defects described by following system of equations:

$$\frac{\partial N}{\partial t} = (N - N_0)\frac{W_{ph}}{h}\exp\left(-\frac{W_{ox}-\sqrt{\frac{q^3F}{\pi\varepsilon\varepsilon_0}}}{kT}\right) - NN_{ox}v_{ox}\sigma_{ox} - s_0\nabla\cdot\left(N\left(1-\frac{N}{N_0}\right)\frac{W_{ph}}{h}\exp\left(-\frac{W-\frac{q^2}{\pi\varepsilon\varepsilon_0 s_0}}{kT}\right)\sinh\left(\frac{qFs_0}{2kT}\right)\boldsymbol{e}\right), \tag{20}$$

$$\frac{\partial N_{ox}}{\partial t} = (N - N_0)\frac{W_{ph}}{h}\exp\left(-\frac{W_{ox}-\sqrt{\frac{q^3F}{\pi\varepsilon\varepsilon_0}}}{kT}\right) - NN_{ox}v_{ox}\sigma_{ox} + \nabla\cdot(N_{ox}v_{ox}\boldsymbol{e}), \tag{21}$$

where N_0 is atomic concentration, $s_0 = N_0^{-1/3}$ is interatomic distance, W_{ox} is energy of interstitial-vacancy pair formation, N_{ox} is concentration of interstitial atoms of oxygen created in formation of interstitial-vacancy pair, v_{ox} is drift velocity of interstitial atoms of oxygen, σ_{ox} – recombination cross-section for interstitial-vacancy pair. The first term of the right side of (20) describes the generation rate for new interstitial-vacancy pair with lowering of energy of formation of the pair in strong electric field [6]. As a pre-exponential factor the concentration of oxygen atoms in their atom positions is used, frequency factor W_{ph}/h is calculated using characteristic oscillation frequencies of atoms near their equilibrium position i.e. phonon oscillation modes. The second term in (20) describes the recombination of oxygen vacancies with interstitial oxygen. The third one is associated with the drift of oxygen vacancies occurring as jumping of adjacent oxygen atoms to the vacancy position by Hill mechanism [7], at that initial vacancy recombines and in a place of atom new vacancy forms. Equation (21) describes generation/recombination of interstitial atoms of oxygen. The first and the second terms in (21) are similar by design with respective members of Eq. (20). The third one describes the drift of interstitial oxygen in electric and thermal fields. Dependence of drift velocity v_{ox} on temperature of the medium and electric field requires additional investigation.

To find the recombination cross-section of interstitial-vacancy pair following considerations were used. If interstitial oxygen atom gets to the area near the vacancy with attractive potential the interstitial-vacancy pair recombines. Within Frenkel model cross-section of such area is equal to the area of a circle with radius matching the distance between the center of the vacancy and the maximum of the potential:

$$\sigma_{ox} = \frac{q}{4\varepsilon\varepsilon_0 F}. \tag{22}$$

Equation (22) shows that area of recombination cross-section decrease with increase of electric field (voltage on the structure) for interstitial-vacancy pair. Thus, for linear dependence of drift velocity of interstitial-vacancy pair from electric field the probability of recombination does not depend on electric field. However, probability of generation grows exponentially with increase of electric field and temperature. So, in current problem the switching from high-resistance state to low-resistance state is anticipated in the area of local heating. At that the electric field will promote the transport of formed interstitial sites from the area of filament and migration of vacancies of oxygen to the active area. Under the electric field of opposite direction, the interstitial sites will migrate to the active area of filament and recombine with vacancies of oxygen. In the process oxygen vacancies will leave the area of filament. In conjunction the processes will lead to the switch to high-resistance state.

3. Summary

In conclusion, this study introduces thermodynamical model of filament growing when a current pulse of variable length and value flows. The model is the boundary value problem, which includes nonstationary heat conduction equation with non-linear Joule heat source, Poisson equation, and Shockley-Read-Hall equations taking into account strong electron-phonon interactions in trap ionization and charge transport processes. The charge current, which defines the heating in the model, depends on the rate of the oxygen vacancy generation. The latter depends on the local temperature. Kinetic parameters of our model are obtained from the transport and optical experiments and *ab initio* simulations. The solution of the introduced problem allows us to describe the kinetics of the switch process and the final filament morphology. The resulting filamentary structure in turn may explain the memristor switch.

End-to-end solution of the task allows us to develop recommendations for optimizing the memristor technology and predict the behavior of memristor arrays, including neuromorphic electronics and non-volatile large memory arrays

Acknowledgments

This work was supported by the Russian Science Foundation, grant #16-19-00002.

References

[1] Islamov D R, Gritsenko V A, Cheng C H and Chin A 2011 *Appl. Phys. Lett.* **99** 072109
[2] Gritsenko D V, Shaĭmeev S S , Lamin M A *et al.* 2005 *JETP Letters* **81** 587-589
[3] Makram-Ebeid S S and Lannoo M 1982 *Phys. Rev.* B **25** 6406-6424
[4] Nasyrov K A and Gritsenko V A 2011 *J. Appl. Phys.* **109** 093705
[5] Lundström I and Svensson C 1972 *J. Appl. Phys.* **43** 5045
[6] Frenkel J 1938 *Technical Physics of the USSR* **5** 685-695
[7] Hill R M 1971 *Phil. Mag.* **23**, 59-86

Dust particles in a spherical glow discharge

A V Fedoseev[1], S A Novopashin[1], S Z Sakhapov[1] and G I Sukhinin[1]
[1]Institute of Thermophysics SB RAS, Novosibirsk, 630090 Russia

E-mail: sakhapov@gmail.com

Abstract. The results of an experimental study of dusty plasma of a spherical gas discharge have been presented. It has been found out that the current density in the locations of dust particles remains unchanged in an unstratified glow discharge. The upper limit of the charge and size of particles have been evaluated.

1. Introduction

Synthesis of nanoparticles in low-temperature plasma plays an important role in fundamental research and in plasma technology [1,2]. Dust particles may be found in astrophysical experiments, in the ionosphere, and in fusion plasma. The possibility of formation of nanometer or micron size particles in laboratory plasma was described in [3-5]. The formation and growth of dust particles were observed in various types of gas discharges: RF discharges [3], DC discharges [4], and arc discharges [5].

In 2015, a new phenomenon was discovered [6]: in spherical strata, the clouds of dust particles are formed; they occasionally explode, split into two clouds and scatter in opposite directions. It turned out that the compact cloud entrained to the reactor wall decelerates many times slower than it could be expected from the evaluation on Epstein force of nanoparticles on a neutral background gas.

Spherical striations appear around a point anode (placed at the center of a grounded vacuum chamber, whose walls serve as cathode) at low pressures (10-50 Pa) of high-molecular gases (acetone, ethanol, methane, acetylene, etc.) and present a set of concentric luminous regions [7]. In discharge plasma in the high-molecular gas the complex plasma-chemical processes of dissociation, ionization and recombination result in the appearance of radicals, ions and small clusters. Their coagulation and recharge lead to the formation of negatively charged clusters and dust particles of nano or micro size. A special feature of the spherical gas discharge is absence of transverse and the presence of only radial flows of particles and energy, which allows concentrating high-density plasma in the central part of the discharge and enables considering discharge to be a one-dimensional problem.

2. Experimental setup

Experiments were carried out in a grounded steel cylindrical vacuum chamber (1) 60 cm in height and 50 cm in diameter (figure 1). An isolated electrode with a non-isolated tungsten end (2) 5mm in diameter was placed at the center of the vacuum chamber. The discharge occurred between the central electrode and grounded steel chamber wall. The gas discharge was supported by a high-voltage current source (3). Experiments were carried out under continuous pumping by a mechanical pump. Pressure was controlled by the pumping speed and by the inflow of ethanol vapor into the vacuum chamber. At positive potential of the central electrode, several spherical striations (4) around the central electrode were formed. The photo and video visualization was carried out through the glass windows. For observation, a quartz window (5) was placed perpendicular to the basic windows at half-height of the chamber, and a solid-state laser (6) with 35mW power and 532 nm wavelength was installed. Between the laser and the window a cylindrical lens (7) was placed. The vertical laser sheet formed by the laser beam and lens illuminated and highlighted the strata and space in the chamber between the anode and the window.

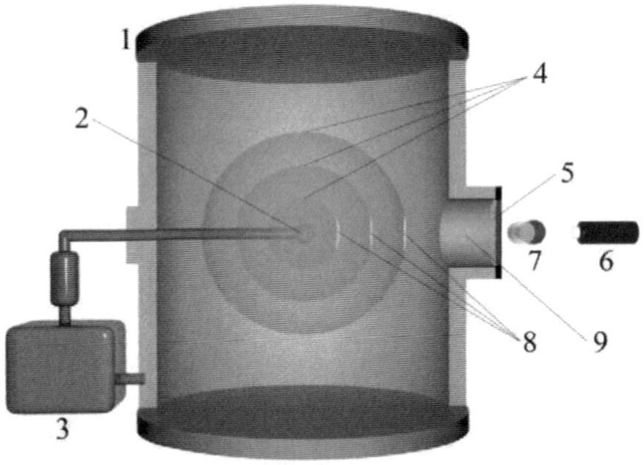

Figure 4. Spherical glow discharge experimental setup.
1 - vacuum chamber (cathode);
2 - central electrode(anode);
3 - high-voltage current source;
4 - striations;
5 - quartz window;
6 - laser;
7 - cylindrical lens;
8 - clouds of dust particles;
9 -hollow cathode.

3. Results and discussions

Below is a description of the condition of forming and the dynamics of dust particles depending on the conditions of a spherical gas discharge glow.

3.1. Glow discharge in ethyl alcohol.

Complex plasma-chemical processes in a spherical glow discharge in high-molecular gases result in formation a variety of molecules, active radicals and ions. It was discovered that the differential conductivity in the formed gas can be non-monotonic, which probably leads to the formation of spherical stratification [8]. At coagulation of these active particles the molecular clusters are formed; with discharge increase they grow to nano or micro size, acquiring a large negative charge $Q_d = -e_0 Z_d$ in plasma. Charge number Z_d of the dust particles in the plasma, as a rule, is directly proportional to the radius R_d of particles, and is determined by the equality of flows of plasma electrons and ions on the surface of dust particles.

In about a minute of glowing of a spherical stratified gas discharge particles are formed with sizes and in numbers that are sufficient to make scattering of laser radiation visible to the naked eye (figure 1(8)). In figure 2 it can be seen that the dust particles are concentrated at the boundaries of the strata towards the hollow cathode. The potential of the plasma discharge is shown in figure 3, where solid lines mark the location of dust particles. It is worth noting that since the Langmuir probe, having both a negative and a zero potential relative to the plasma, repels negatively charged dust particles [9], the measured potential is not affected by the cloud particles. Particles in the case of a stratified discharge are located in the region of the potential well, as it was previously proposed by the authors in [6]. However, in contrast to previous work, the cloud of particles did not leak in the direction of the cathode that is apparently related to the high electric field in the strata.

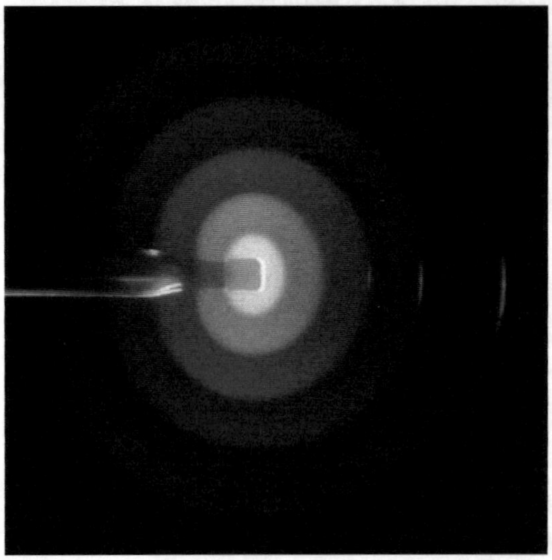

Figure 2. Photograph of the spherical striations and the clouds of dust particles. The pressure in the gas-discharge chamber is 0.1 Torr, the voltage and the discharge current are 538V and 15.3 mA, respectively.

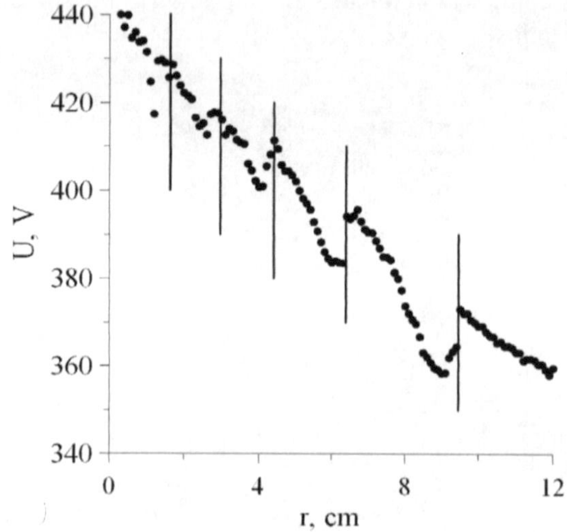

Figure 3. The radial distribution of the plasma potential in the discharge shown in Figure 2. The solid lines show location of dust particles.

3.2. Glow discharge in nitrogen

It was discovered that some of the particles formed in spherical stratified discharge have magnetic properties. This is due to the fact that the appearance of dust particles in gas discharge is not only caused by coagulation of chemically active molecules, but by sputtering of the electrodes as well. As a result, in a glow discharge in atomic and simple molecular gases there are formed particles consist of a substance of the electrodes, mainly the cathode. This is confirmed by the fact that dust particles are formed even in a glow discharge in simple molecular and atomic gases, for example nitrogen.

Photograph showing the discharge in nitrogen and scattering of laser radiation on dust particles is presented in figure 4. The appearance of particles occurs sooner with increasing discharge current (figure 5). Probe measurements show that the plasma potential outside of the anode and cathode regions is monotonous, and the electric field is almost constant and is approximately 2.5 V/cm. It has been found out that under rapid increase in the total current of the discharge the cloud of dust particles, whose boundaries are shown in figure 6 in solid lines, is moved from the anode so that the current density at the location of particle remains constant. The dependence of the current density j in the locations of the particles under various conditions is shown in figure 7.

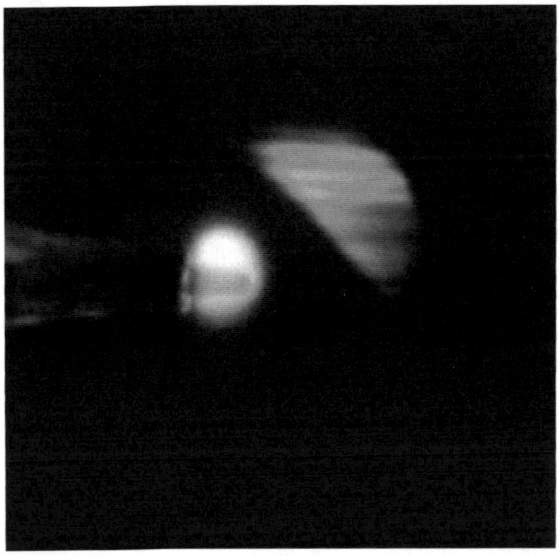

Figure 4. Photograph of the dust particles in the spherical gas discharge in nitrogen. The pressure in the gas-discharge chamber is 0.268 Torr, the voltage and the discharge current are 421V and 2 mA, respectively.

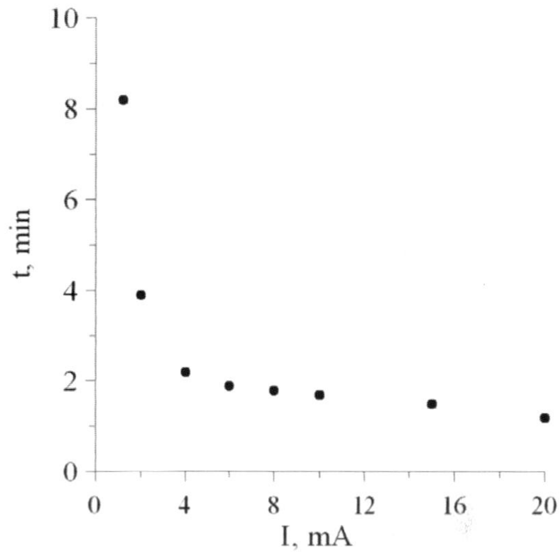

Figure 5. The dependence of time of appearance of dust particles on the discharge time.

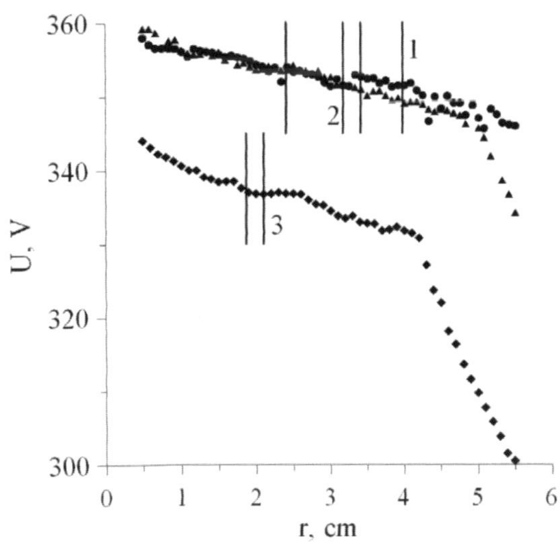

Figure 6. The radial distribution of the plasma potential in nitrogen. The solid lines show the location of dust particles. 1, 2, 3 are discharge currents of 20, 10, 6 mA, respectively.

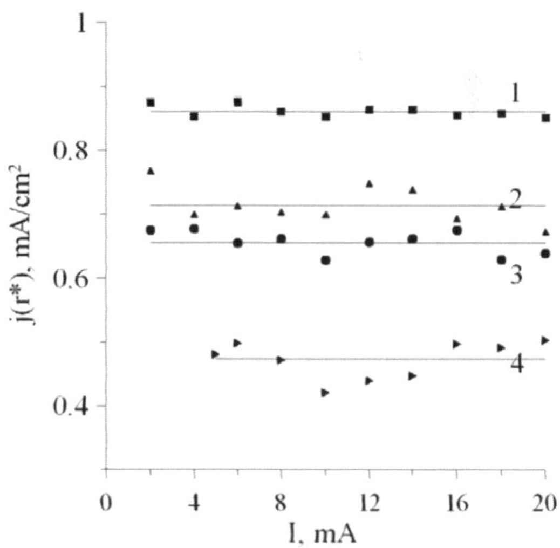

Figure 7. The dependence of the current density at the location r^* of the dust particles on the total discharge current. 1, 2, 3, 4 are nitrogen chamber pressures of 0.237, 0.268, 0.272, 0.421 Torr, respectively.

The location of the cloud of particles is determined by the forces acting on dust particles: gravity F_g, thermophoretic force F_{th}, the electrostatic force F_{el}, the drag force on the neutral gas F_{nd}, and the ion drag force F_{id} [6]. The force of neutral drag F_{nd} can be neglected because the gas at the center of the chamber is stationary. Evaluation of gravity $F_g = M_d \cdot g$ for particles with a $R_d = 1$ micron and $g = 9.8$ m/s^2 is approximately $F_g \approx 5 \cdot 10^{-15}$ N, i.e. by an order of magnitude smaller than typical values of other forces. Thermophoretic force $F_{th} \sim R_d^2 \cdot \nabla T_n$ is proportional to the square of particle radius and temperature gradient ∇T_n. The temperature gradient decreases rapidly with increasing distance from the anode due to spherical geometry, and for particles with radius $R_d = 1$ μm the thermophoretic force is small [6].

Thus, the particles are affected by two differently directed forces: the electric force F_{el} and the ion drag force F_{id}. The electric force in the field E of the discharge plasma acts on particle with a negative charge $-eZ_d$ in the direction of the anode:

$$F_{el} = -eZ_d E. \tag{1}$$

The ions in the field of the discharge move toward the cathode, dragging the particles with a force [6]:

$$F_{id} = \frac{4}{3}\left(\frac{2}{\pi}\right)^{1/2} \pi R_d^2 n_i \upsilon_{T_i} \left(\frac{e_0^2 Z_d}{R_d T_i}\right)^2 m_i u_i \Lambda(\beta_T), \tag{2}$$

where $\Lambda(\beta_t)$ is the Coulomb logarithm, and n_i, υ_{Ti}, u_i, T_i, m_i are concentration, thermal and drift velocity, temperature and mass of ions, respectively.

Let's evaluate the charge of the particle in the dust cloud in the experiment whose data are shown in figure 7 (1). Given the stationary nature of the gas discharge, forces (1) and (2) must be equal therefore:

$$Z_d = \frac{T_i^{3/2}}{\frac{4}{3}\left(\frac{2}{\pi} m_i\right)^{1/2} \pi n_i \mu_i e_0^3 \Lambda(\beta_T)} \tag{3}$$

where μ_i is the ion mobility, which at pressure $p = 0.237$ Torr is equal to $\mu_i = 6 \cdot 10^3$cm^2/(V·s). In the experiment, the current density of electrons $j_e = 0.861$ mA/cm^2, the drift velocity of the electrons $u_e(E/N) \approx 10^6$ cm/s at $E/N \approx 3$ Td. Given quasineutrality of plasma, we find the concentration of ions in the localization of dust particles $n_i \approx n_e = j_e/u_e \approx 5 \cdot 10^9$ cm^{-3}. Taking $\Lambda(\beta_t)=1$ and $T_i = 0.03$ eV (3), we find the charge number of an individual dust particle $Z_d \approx 4 \cdot 10^3$.

In the experiment ($n_i \approx n_e$ и $u_i/\upsilon_{Ti} < 1$) the absolute value of the dimensionless charge of the particle z, equal to the ratio of the potential energy of the particle to the energy of the electrons is approximately equal to 1[2]:

$$z = \frac{|Z_d| e^2}{R_d T_e} \approx 1. \tag{4}$$

Given (3) and (4), and taking the electron temperature T_e=2 eV [10], the radius of dust particles can be calculated to be about 3 microns.

It should be noted that the estimates given above did not take into account the thermophoretic force F_{th} which is directed towards the cathode and due to high heating of the anode may have a large value. Accounting for F_{th} can lead to a significant decrease in the value Z_d and R_d, so these are upper limits for the charge and size of the dust particles.

4. Conclusion

This paper presents the results of experimental studies of dusty plasma of spherical gas discharge, using optical and Langmuir probe methods. It is shown that dust particles are formed in a glow discharge in ethanol and nitrogen. In the case of the stratified discharge, the particles are concentrated at the border of strata. In unstratified discharge in nitrogen the dust particles gather in the cloud, which moves away from the anode along with the increase in the discharge current so that the current density in the cloud remains unchanged. The evaluation has showed that the particle size in the dust cloud in the glow discharge of nitrogen is about 5 microns under assumption that the thermophoretic force acting on the particles is small.

Acknowledgments
This research was funded by the Russian Science Foundation, Project No. 14-19-01379.

References

[1] Ostrikov K, Cvelbar U, Murphy A 2011 *J. Phys. D: Appl. Phys.* **44** 174001
[2] Fortov V E, Khrapak A G, Khrapak S A, Molotkov V I, Petrov O F 2004 *Physics-Uspekhi* **47** 447
[3] Selwyn G S, Singh J, Bennett R S 1989 *J. Vac. Sci. Technol.* **A7** 2758
[4] Scherbina A I, Burdyukh S V 2011 *IEEE Trans. Plasma Sci.* **39** 2746
[5] Ugarte D 1992 *Chem. Phys. Lett.* **198** 596
[6] Sakhapov S Z, Fedoseev AV, Sukhinin G I, Novopashin S A 2015 *EPL* **110** 15003
[7] Nerushev O A, Novopashin S A, Sukhinin G I, Radchenko V V 1998 *Phys. Rev. E.* **58** 4897
[8] Belikov A E, Sakhapov S Z 2011 *Phys. D: Appl. Phys.* **44** 45202
[9] Ussenov Y A, Ramazanov T S, Dzhumagulova K N, Dosbolayev M K 2014 *EPL* **105** 15002
[10] Novopashin S A, Radchenko V V, Sakhapov S Z 2008 *Journal of Engineering Thermophysics* **17** 71

Journal of Physics: Conference Series **754** (2016) 102003 doi:10.1088/1742-6596/754/10/102003

Application of dissipation-free numerical method CABARET for solving gasdynamics of combustion and detonation

M F Ivanov, A D Kiverin, S G Pinevich, I S Yakovenko
Joint Institute for High Temperatures of the Russian Academy of Sciences (JIHT RAS)
125412, Izhorskaya st. 13 Bd.2,Moscow, Russia

E-mail: yakovenko.ivan@bk.ru

Abstract. This paper discusses capabilities of the novel dissipation-free CABARET numerical algorithm to solve the range of complex non-stationary combustion problems. On the basis of detailed analysis of the obtained results and comparison with the data derived with the classic low-order coarse particles method it was shown that reactive flow evolution process may be strongly influenced by the artificial effects introduced by the numerical algorithm, numerical dissipation in particular. Revealed peculiarities of the flame propagation dynamics regimes taking place in considered tests allowed us to propose a number of requirements which should be taken into account when choosing numerical procedure suitable for modelling combustion processes in real technical environment.

1. Introduction

Increase in modern computational facilities performance continuously expands the range of problems that could be solved by means of numerical analysis and mathematical modeling. Modern high performance facilities allow solving the most resource intensive problems of computational science and engineering. Certainly one of the first places among them belongs to detailed numerical modeling of the non-stationary reactive flows and particularly the combustion phenomena in gaseous media. Most complex problems of combustion physics are related with evolution of the non-stationary and transient combustion regimes, which could be observed inside combustors of real technical and experimental facilities filled with highly reactive gaseous mixtures. Detailed analysis of these regimes features and peculiarities, identification of fundamental physical mechanisms and processes responsible for their dynamics and evolution is crucial for safety issues and developing novel propulsion devices and power stations, utilizing traditional or alternative energy sources more efficiently and environmentally safe. Due to the vast variety of factors and scales related to non-stationary gaseous combustion its theoretical and experimental analysis is limited to certain extent. And computational methods become the most insightful and powerful technique for combustion problems. Since the gaseous combustion process is highly sensitive to the adequacy of gas dynamic flow features reproduction, correct resolution of energy release zones and chemical transformations, results of modeling procedure are fully determined by the features of utilized computational method. Despite large variety of computational methods for modeling gas dynamic processes the problem of choosing the method appropriate for detailed combustion modeling is still controversial. Only some of developed computational algorithms are capable for simulating gas dynamic flows in compressible multicomponent mixtures of non-calorically perfect gases with varying thermodynamic properties and in presence of energy release induced by chemical transformations. Available computational resources make it possible to perform direct numerical simulations of combustion processes inside multidimensional chambers utilizing advanced physical models supplemented with detailed chemical kinetic schemes and taking into account real features of flammable media and variety of processes accompanying combustion. However despite continuous progress in increasing computational performance detailed numerical analysis of non-stationary combustion processes inside vessels of technical or laboratory scales remains very resource intensive task. Thereby numerical efficiency of the chosen numerical method is of great importance on a par with accuracy and robustness. Today two main approaches are commonly used to justify choice of numerical algorithm. First approach focuses on accuracy order of the numerical method. Most common examples are weighted essentially non-oscillating (WENO) finite difference and finite volume schemes and discontinuous Galerkin (DG) methods with high accuracy orders [1]. Being wide spread and repeatedly used for various problems of combustion phenomena modeling [2, 3, 4] these schemes have certain disadvantages. WENO schemes have wide spatial stencils increasing with accuracy order that determines large computational costs of WENO procedures and amount of interprocessor communication data required for performing parallel computations. DG methods are computationally demanding and tend to drop accuracy order so specific non-physical limiting procedures requiring user-defined parameters are needed to maintain high levels of accuracy. Described drawbacks could have major ineligible impact both on computational costs and obtained results accuracy crucial for non-stationary combustion detailed modeling on large spatial and temporal scales. Second approach that may be more perspective in a field of detailed combustion modeling consists in developing low-order dissipative-free computational methods. A vivid example of computational approach developed in a given ideology is CABARET method.

CABARET originates from work by Iserles on generalizaion of classical leapfrog numerical method for solving hyperbolic equations. In [5] Iserles proposed upwind leapfrog scheme with two-point spatial stencil and time discretization by the midpoint rule. It was also shown that upwind leapfrog scheme is conditionally stable for Courant number within [0,1] interval. Further detailed analysis of this scheme was performed in [6, 7], where it was called CABARET scheme with considering its antisymmetrical spatial stencil. Main advantages of CABARET scheme are compact spatial stencil, reversibility in time and capacity to provide exact solution with two different Courant numbers 1 and 0.5. These properties determine outstanding dissipative and dispersion features of the CABARET scheme [7]. Balance-characteristic CABARET version for numerical solution of the one-dimensional gasdynamics equations was elaborated in [8]. Proposed method with using flux variables correction necessary for providing monotonic solution has shown high accuracy on classical blast wave benchmark problem [9]. Further it was shown that scheme developed in [8] in case of special approximation of the initial conditions is monotonic [10] and strongly monotonic [11] for Courant numbers within (0,0.5] and non-monotonic for Courant numbers within (0.5,1). To avoid non-monotonic behavior two-step flux variables correction procedure was developed in [10] that ensures monotonic solution by means of CABARET scheme within entire stability region of Courant numbers in (0,1] interval. Lately two layer in time form of the CABARET scheme [12, 13] became widely used for solving multidimensional problems of gasdynamic [14] and hydraulic [15] flows. Monotonicity of this approach for one-dimensional and two-dimensional cases was studied in [16, 17].

CABARET method has a number of outstanding features among which absolute absence of numerical viscosity, most compact computational stencil, no need of any task specific parameters or physically precarious limiting procedures [13]. However CABARET method was not yet thoroughly tested in a field of multicomponent reactive flows modeling. This paper is devoted to description of the two layer in time form of the CABARET technique adaptation for solving combustion problems and to evaluation of its capabilities on a set of specific problems of non-stationary combustion evolution.

2. Mathematical model and CABARET numerical procedure

2.1. Mathematical model

To perform numerical simulation of multidimensional flame dynamics in chemically active gaseous mixture full system consisting of mass, momentum and energy conservation laws for viscous, compressible multicomponent gasdynamics was solved, taking into account effects of thermal conductivity, molecular diffusion, chemical transformations and heat release and equations of state of real gas for each mixture component. Fuel-oxidizer chemical reactions were defined by detailed chemical kinetics mechanism. Full set of governing equations in divergence form are presented below:

$$\frac{\partial \rho}{\partial t} + \frac{\partial (\rho u_i)}{\partial x_i} = \sum_k M_k \left[\frac{\partial}{\partial x_i} \left(\rho D_k (T) \frac{\partial Y_k}{\partial x_i} \right) \right]$$

$$\frac{\partial (\rho Y_k)}{\partial t} + \frac{\partial (\rho u_i Y_k)}{\partial x_i} = \left[\frac{\partial}{\partial x_i} \left(\rho D_k (T) \frac{\partial Y_k}{\partial x_i} \right) \right] + \left(\frac{\partial Y_k}{\partial t} \right)_{chem}$$

$$\frac{\partial (\rho u_j)}{\partial t} + \frac{\partial (\rho u_i u_j)}{\partial x_i} = \left[\frac{\partial}{\partial x_i} (\sigma_{ji} - \delta_{ij} p) \right]$$

$$\frac{\partial (\rho E)}{\partial t} + \frac{\partial (\rho u_i E)}{\partial x_i} = \left[\frac{\partial}{\partial x_i} (u_j \sigma_{ji} - u_i p) + \frac{\partial}{\partial x_i} \left(\kappa (T) \frac{\partial T}{\partial x_i} \right) \right] + \sum_k h_k \left(\frac{\partial Y_k}{\partial t} \right)_{chem}$$

Where viscous stress tensor

$$\sigma_{ij} = \mu (T) \left(\frac{\partial u_i}{\partial x_j} + \frac{\partial u_j}{\partial x_i} \right) - \frac{2}{3} \delta_{ij} \mu (T) \frac{\partial u_m}{\partial x_m}$$

System is enclosed with following real gas equations of state:

$$p = RTn = RT \sum_k Y_k$$

$$\varepsilon = c_V (T) T / \rho$$

Here x_i - spatial coordinates, p,ρ,u_i - pressure, density and components of mass velocity vector, Y_k - molar concentration of k-th species, M_k - molar mass of k-th species, $n = \sum_k Y_k$ $E = \varepsilon + u^2/2$ - specific total energy, ε - specific inner energy, R - universal gas constant, σ_{ij} - the viscous stress tensor, $c_V = \sum_k c_{V_k} Y_k$ - mixture constant volume volumetric heat capacity, c_{V_k} - the constant volume molar heat capacity of k-th species, h_k - the enthalpy of formation of k-th specie, $k(T)$, $\mu(T)$ - thermal conductivity and viscosity coefficients, $D_k(T)$ - diffusion coefficients of k-th specie, $(\partial Y_k/\partial t)_{chem}$ - variation of k-th specie concentration due to chemical reactions. Temperature dependence of the specific heats, heat capacities and enthalpies of each species were evaluated by interpolation of thermophysical JANAF data tables with fifth order polynomials [18]. The transport coefficients were calculated from the first principles of the gas kinetic theory [19]. Variations of mixture components concentrations in chemical reactions were defined by the solution of the ordinary differential equations system

$$\frac{dY_k}{dt} = F_k\left(Y_1, Y_2, Y_3, ..., Y_N, T\right), \quad k = 1, \ldots, N$$

The right hand parts of equation system contain the rates of chemical reactions, which depend on temperature according with Arrhenius law in a standard form [20].

2.2. CABARET numerical procedure

Detailed description of CABARET numerical algorithm can be found in monograph by Goloviznin et al. [13]. Here we provide general expressions of CABARET procedure for the system of governing equations described above. The initial governing system, excluding equations for component transfer, can be rewritten in simpler form:

$$\frac{\partial \vec{\varphi}}{\partial t} + A_i \frac{\partial \vec{\varphi}}{\partial x_i} = 0$$

Where $\vec{\varphi}^T = (\rho, u_i, p)$

Being of balance-characteristic type, CABARET numerical method requires derivation of the governing equations system in characteristic form. In general characteristic form can be written as follows:

$$\left(\Omega_i \cdot \frac{\partial \vec{\varphi}}{\partial t}\right) + \Lambda_i \cdot \left(\Omega_i \cdot \frac{\partial \vec{\varphi}}{\partial x_i}\right) = \Omega_i \cdot \vec{f} - \Omega_i \sum_{j \neq i} A_j \frac{\partial \vec{\varphi}}{\partial x_j}$$

Where in our case

$$\vec{f}^T = \left(F_\rho, \frac{F_{u_i} - u_i F_\rho}{\rho}, \frac{1}{\rho}\frac{\partial p}{\partial \varepsilon}\left(F_E + \left(u_i^2 - E\right) F_\rho - u_i F_{u_i}\right) + \frac{\partial p}{\partial \rho} F_\rho\right)$$

Ω_i matrix composed from left eigenvectors of matrix A_i, $\Lambda_i = E \cdot \lambda_i$, where E unity matrix and λ_i eigenvalues vector of matrix A_i.

$$F_\rho = \sum_k M_k \left[\frac{\partial}{\partial x_i}\left(\rho D_k\left(T\right) \frac{\partial Y_k}{\partial x_i}\right)\right]$$

$$F_{u_i} = \left[\frac{\partial}{\partial x_i}\left(\sigma_{ji}\right)\right]$$

$$F_E = \left[\frac{\partial}{\partial x_i}\left(u_j \sigma_{ji}\right) + \frac{\partial}{\partial x_i}\left(\kappa\left(T\right) \frac{\partial T}{\partial x_i}\right)\right] + \sum_k h_k \left(\frac{\partial Y_k}{\partial t}\right)_{chem}$$

Utilization of characteristic form of the governing equations system determines accurate localization of the solution features as shock waves and contact surfaces. At once approximation of the divergence form of the governing equation system performed during CABARET procedure ensures conservativity of the obtained solution.

Considering real equation of state pressure derivatives are obtained:

$$\frac{\partial p}{\partial \rho} = c^2 - \frac{(\gamma - 1)\, p}{\rho}, \quad \frac{\partial p}{\partial \varepsilon} = (\gamma - 1)\, \rho$$

Required adiabatic index and speed of sound for multicomponent mixture were found as follows:

$$\gamma = 1 + \frac{nR}{c_V}, \quad c = \sqrt{\gamma P / \rho}$$

To avoid negative values of species concentrations and preserve solution monotonicity convective species transport is obtained by using mass flow through cells faces calculated as a result of CABARET procedure.

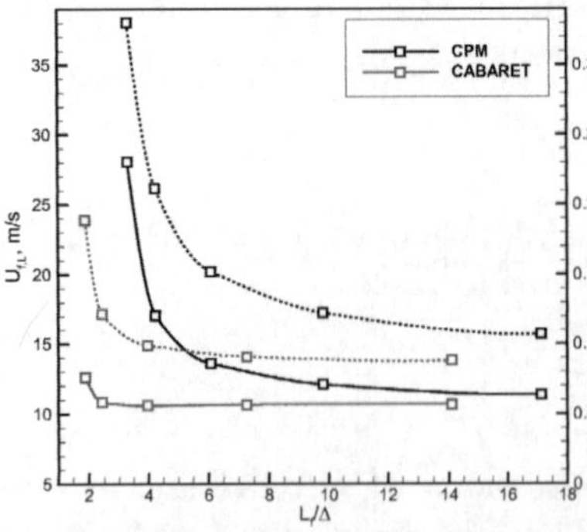

Figure 1. Resolution dependence of normal flame velocity (solid lines) and flame thickness (dashed lines).

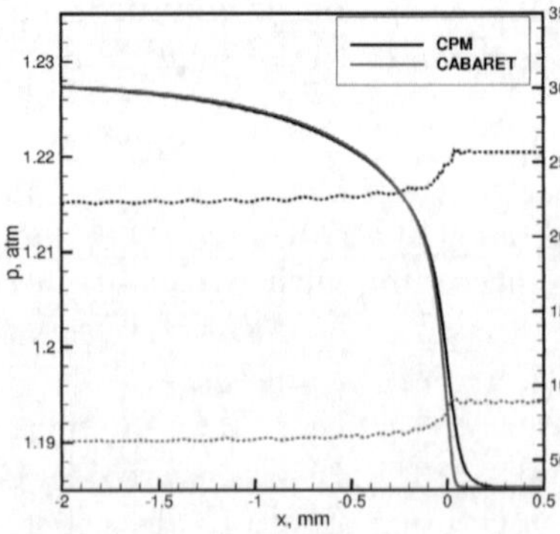

Figure 2. Structure of the steady-state laminar flame. Solid lines denote temperature profile, dashed lines denote pressure profile.

3. Non-stationary combustion processes

3.1. Laminar flames test problem

Developed approach to numerical modeling via CABARET method was tested on one-dimensional stationary laminar flame propagation problem. In absence of multidimensional factors flame ignited near closed end of channel filled with combustible mixture after passing through short transient stage proceeds in stable stationary regime characterized by laminar flame velocity which is defined merely by the features of chemical kinetics and background flow induced by transient flame ignition processes. Convergence properties of the problem solution with decreasing computational cell size provide an insight about accuracy of the flow reproduction provided by numerical method. Results of convergence test on laminar flame propagation problem in stoichiometric hydrogen-oxygen mixture are given on figure 1. Dependence of laminar flame velocity and flame front thickness on computational cell size obtained using CABARET method is shown on figure 1 together with results obtained with classic euler-lagrange gasdynamic computational method known as coarse particles method (CPM) [21]. Earlier CPM was intensively used by authors for solving various problems of combustion physics and demonstrated high stability and effectiveness [22, 23]. The major drawback of CPM is high artificial diffusion that can introduce some undesirable features of obtained results. Negative effect of numerical diffusion arising while solving laminar flame propagation problem appears in a lower rate of solution convergence. As seen from figure 1 CPM solution converges about four times slower than solution provided by CABARET. Figure 2 shows clearly wider reaction zone for CPM solution on the same cell size, although flame structure obtained by both CABARET and CPM is qualitatively similar that indicates correctness of developed CABARET technique implementation.

3.2. Confined flames evolution

In present research we were mainly focused on features of the flame evolution inside confined chambers. Problem of confined flames evolution which is very important from the point of view of numerous practical applications is still far from general theoretical description. Non-stationary confined flame evolution is defined by effects of development of multidimensional flame surface instabilities, intensive interactions between flame surface and flame induced acoustics and swirling flows, peculiarities of chemical kinetics and thermophysical properties of combustible mixture at abrupt conditions changes posed by presence of shock waves and

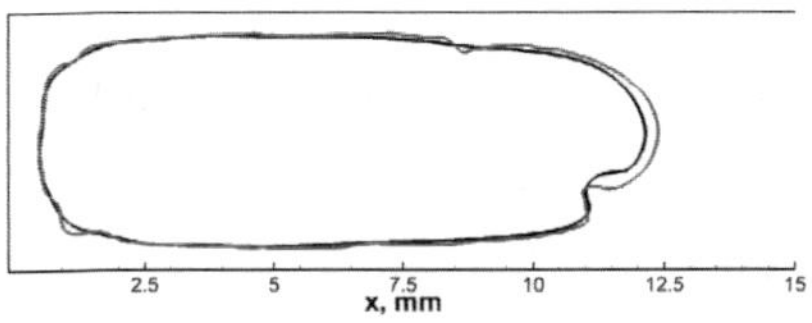

Figure 3. Flame front structure in narrow channel of 5 mm width. Solid lines represent isotherms 3000K. Red line denotes CABARET method and black line denotes CPM.

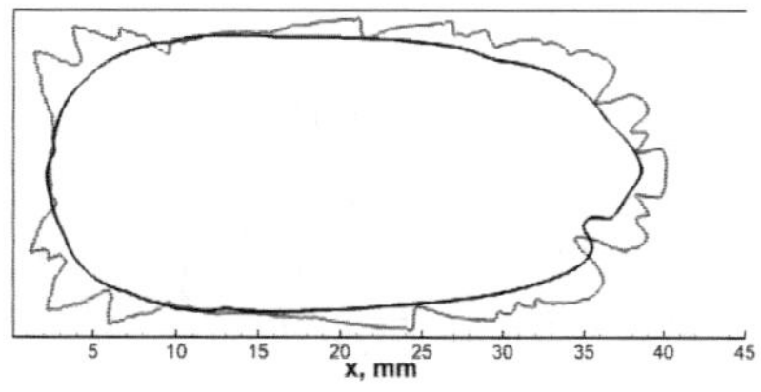

Figure 4. Flame front structure in wide channel of 2 cm width. Solid lines represent isotherms 3000K. Red line denotes CABARET method and black line denotes CPM.

tangent discontinuities. Gasdynamical nature of the most of the outlined effects indicates crucial importance of precise reproduction of gasdynamic flows by numerical algorithm. Gasdynamical benchmark tests performed by authors of CABARET method have shown outstanding efficiency and accuracy for long-term calculations of acoustic and vortex flows. These positive properties of CABARET method are evident while solving problem of flame propagation out from the ignition source inside closed channel.

First consider rectangular channel filled with stoichiometric hydrogen-oxygen mixture. Mixture is being ignited near the left channel end-wall by cylindrical heated region that resembles energy release by spark plug induced gas breakdown. Primal ignition area surface is slightly disturbed to create ignition conditions more similar to real experimental ones. Experimental investigations of confined flames ignition in wide tubes shows that flame propagation out from the ignition source is almost isotropic and the longitudinal propagation velocity is almost the same as transverse up to the very moment of flame achieving channel side walls. However this behavior is rarely reproduced numerically even with utilization of high order numerical schemes [24].

Flame expansion out from the ignition source can be divided into several stages. On the initial stage immediately after ignition flame propagates as freely expanding cylindrical flame, without any effect of confinement. Duration of flame propagation in this regime is restricted by the distance between flame surface and channel walls as the ignition and flame non-stationary flame expansion propagation induce acoustic waves that propagate with sonic speed in the fresh mixture, reflect from the channel walls and interact with the flame surface providing repelling effect. While flame is approaching channel walls interaction between flame and acoustics becomes more violent and other counteractive process is required to maintain spherical expansion. Expanding flame is subjected to various hydrodynamic instabilities with finite characteristic time scales required for their development. Development of hydrodynamic instabilities on the flame front results in local multidimensional strain and wrinkling of the flame front surface. Due to the flame curvature eddies are generated in the vicinity of the flame surface. On the initial stage of flame expansion eddies are growing while diffusing into combustion products area. As the flame advances near the channel walls counterflow starts to slow down flame front propagation and eddies could manage to penetrate into fresh mixture ahead of the flame due to diffusion and cause intense mixing and enhance flame propagation velocity.

However characteristic time is needed for instabilities to start having a noticeable impact on flame dynamics. Let us consider relatively narrow channel of 5 mm width. Results obtained with CABARET and CPM methods on the same time instance corresponding to flame surface achievement channel side walls via CABARET method are presented on figure 3. Here

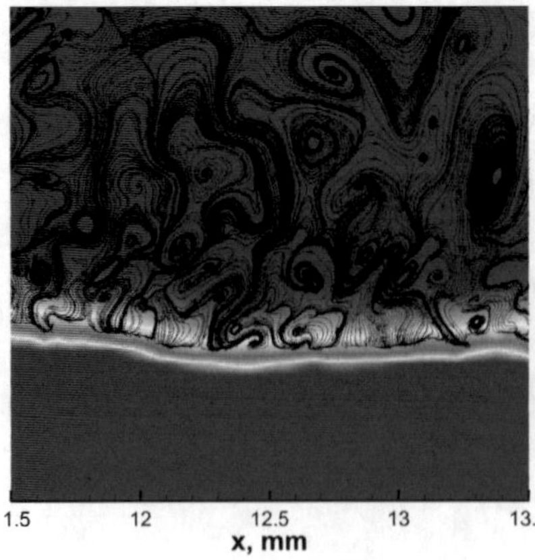

Figure 5. CABARET resolves intensive vortex generation on the spatial scales of reaction zone.

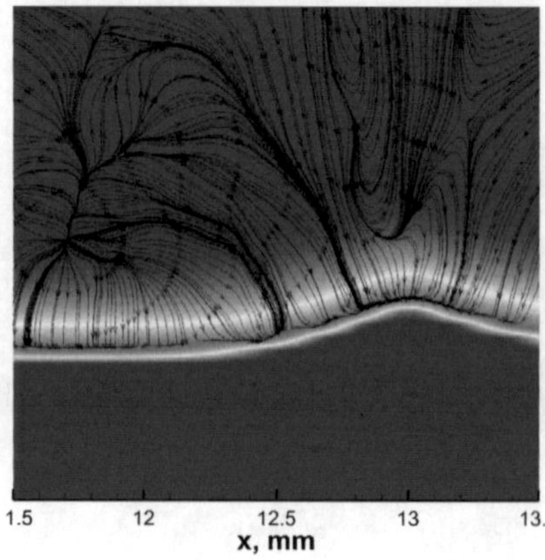

Figure 6. CPM prevents formation of small scale vortices due to artificial dissipation.

CABARET and CPM provide almost similar results that indicate that both methods reproduce interactions between acoustic waves and flame front contact discontinuity almost identically. Narrow channel width significantly limits duration of initial freely cylindrical flame propagation and instabilities development at this stage. Not being able to grow significantly flame surface instabilities are damped by intense acoustics reflected from the side walls. Thereby in narrow channel notable anisotropic expansion is observed with much stronger longitudinal compared with transverse flame surface strain. Situation drastically changes in wider channel of 2 m width results for which are presented in figure 4. Here we observe pronounced instability growth on the flame surface that results in faster transverse flame propagation and isotropic flame expansion in case of CABARET in comparison with CPM. Rugged flame front surface represent so called cellular flame structure that is more pronounced due to developed flame surface instability in CABARET solution. Examining flow structure near flame surface (figure 5) one could notice intensive vorticity of the flow in CABARET solution while vortex formation in CPM solution is suppressed by higher numerical diffusion (figure 6).

Described phenomenon may be crucial for detailed analysis of combustion wave acceleration process proceeding after flame ignition in semi-closed tubes. More pronounced transversal flame propagation restricts possible extension of the flame surface while adopting to the structure of the background flow [25, 26]. Gasdynamical flame acceleration after a certain time instance becomes caused by interaction between flame skirt and vortices generated in boundary layer behind leading shock wave (figure 7). However as it was shown in [27] flame acceleration due to gasdynamical factors cannot be a key mechanism on final stages of deflagration-to-detonation transition process since the restriction imposed on gasdynamical acceleration by Vulis law of inverse action [28]. Thus the interaction between flame front and vortices generated in boundary layer can be important only until flame front achieves velocities in laboratory reference system equal to sonic speed in combustion products. On the other hand obtained flame front flattening due to more accurate reproduction of flame interactions with acoustic waves and vortex generation may be essential for theoretical foundation of the tulip flame structure formation in opened tubes that was observed in recent experiments [29].

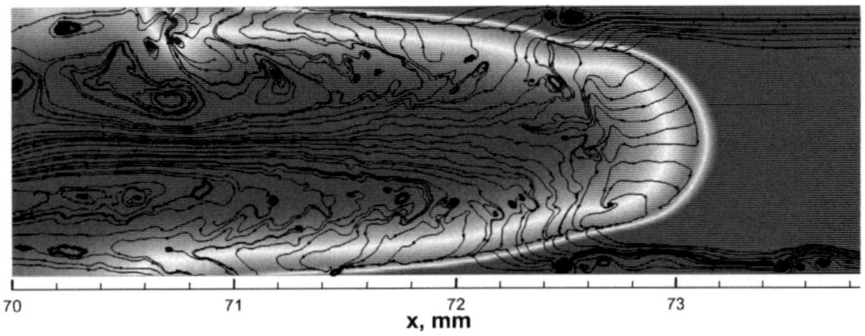

Figure 7. Flame front acceleration process in a smooth tube. Vortices diffusion from boundary layer starts to play significant role on the flame dynamics after a certain time period from ignition.

4. Conclusions

Accurate simulation of gasdynamic processes accompanying flame propagation in gaseous mixtures is crucial for correct interpretation of underlying physical mechanisms therefore low-order dissipation-free methods like CABARET method seem to have high potential for reliable simulation of complex combustion physics phenomena. Obtained results have shown that diminishing artificial viscosity makes it possible to use rougher computational grids without serious loss of precision of reproducing key features of developing flow structures. Solution of non-stationary combustion evolution out from the ignition source in rectangular channel by means of CABARET technique allowed to obtain results similar to that observed in real experiments and to justify the difference occurring while using classical coarse particles method by more accurate modeling of flame surface instabilities end flame induced acoustics provided by CABARET method. Utilization of CABARET method also allowed to analyze possible influence of flame interaction with flow vorticity generated in boundary layer on flame acceleration process that could have strong quantitative effect on flame acceleration in channel.

Acknowledgements

The work was funded by Russian academy of sciences (Grants No. II.4P Fundamental problems of mathematical modeling and I.24P Combustion and explosion). We acknowledge support from the Supercomputing Center of Lomonosov Moscow State University and the Joint Supercomputer Center of the Russian Academy of Sciences.

References

[1] Shu C W 2016 *J. Comput. Physics* **316** 598–613
[2] Lv Y and Ihme M 2014 *J. Comput. Physics* **270** 105–137
[3] Houim R W and Kuo K K 2011 *J. Comput. Physics* **230** 8527–8553
[4] Yu R, Bai X S and Bychkov V 2015 *Phys. Rev. E* **92** 063028
[5] Iserles A 1986 *IMA J. Numer. Anal.* **6** 381–392
[6] Goloviznin V M and Samarskii A A 1998 *Matem. Mod.* **10** 86–100
[7] Goloviznin V M and Samarskii A A 1998 *Matem. Mod.* **10** 101–116
[8] Goloviznin V M 2005 *Dokl. Math.* **72** 619–623
[9] Woodward P and Colella P 1984 *J. Comp. Phys.* **54** 115–173
[10] Ostapenko V V 2009 *Matem. Mod.* **21** 29–42
[11] Ostapenko V V 2012 *Comput. Math. and Math. Phys.* **52** 387–399
[12] Karabasov S A and Goloviznin V M 2009 *J. Comput. Phys.* **228** 7426–7451
[13] Goloviznin V M, Zaitsev M A, Karabasov S A and Korotkin I A 2013 *New Computational Fluid Dynamics Algorithms for Multiprocessor Computers (in Russian)* (MSU Publishing)
[14] Karabasov S A and Goloviznin V M 2007 *AIAA J.* **45** 2861–2871
[15] Karabasov S A, Berloff P S and Goloviznin V M 2009 *Ocean Modelling* **30** 155–168

[16] Kovyrkina O A and Ostapenko V V 2013 *Math. Models Comput. Simul.* **5** 180–189
[17] Kovyrkina O A and Ostapenko V V 2015 *Dokl. Math.* **91** 323–328
[18] Chase M W 1998 *J. Phys. Chem. Ref. Data* **Monograph 9**
[19] Hirschfelder J O, Curtiss C F and Bird R B 1954 *Molecular Theory of Gases and Liquids* (Wiley)
[20] Warnatz J, Maas U and Dibble R 2006 *Combustion: Physical and Chemical Fundamentals, Modeling and Simulation, Experiments, Pollutant Formation* (Springer Berlin Heidelberg)
[21] Belotserkovsky O M and Davydov Y M 1982 *Coarse-particle method in hydrodynamics (in Russian)* (Russian Publ. Inc. Nauka, Mir)
[22] Ivanov M F, Kiverin A D, Yakovenko I S and Liberman M A 2013 *Int. J. Hydrogen Energy* **38** 16427–16440
[23] Efremov V P, Ivanov M F, Kiverin A D and Yakovenko I S 2015 *Results in Physics* **5** 290–296
[24] Xiao H, Houim R W and Oran E S 2015 *Comb. Flame* **162** 4084–4101
[25] Ivanov M F, Kiverin A D and Liberman M A 2011 *Int. J. Hydrogen Energy* **36** 7714–7727
[26] Bychkov V, Akkerman V, Gordon F, Petchenko A and Eriksson L E 2007 *Comb. Flame* **150** 263–276
[27] Kiverin A D and Yakovenko I S 2015 *JPCS* **653** 12061–12067
[28] Vulis L A 1950 *Thermodynaimics of gas flows (in Russian)* (Russian Publ. Gov. Energetic Publisher)
[29] Kuznetsov M, Alexeev V, Matsukov I and Dorofeev S 2005 *Shock Waves* **14**(3) 205–215

Density dependence of dielectric permittivity of water and estimation of the electric field for the breakdown inception

D I Karpov and D A Medvedev

Lavrentyev Institute of hydrodynamics SB RAS, Lavrentyev prosp. 15, 630090, Novosibirsk, Russia

E-mail: karpov@hydro.nsc.ru

Abstract. Density dependence of the dielectric permittivity for water was obtained using the molecular dynamics method for the supercritical region at $T = 670$ K in a broad range of density, and for $T = 300$ K and $T = 350$ K near the coexistence curve. Critical values of the electric field strength which make possible the spinodal decay "liquid-vapor" were calculated.

1. Introduction

The body force acting on a dielectric liquid in electric field is given by the Helmholtz formula [1]

$$\mathbf{F} = -\frac{\varepsilon_0 E^2}{2}\nabla\varepsilon + \frac{\varepsilon_0}{2}\nabla\left[E^2\rho\left(\frac{\partial\varepsilon}{\partial\rho}\right)_T\right]. \tag{1}$$

A phase transition with appearance of liquid and vapor phases is possible in fluids which are initially in a homogeneous liquid or gaseous state if the density dependence of the dielectric permittivity is nonlinear. If a single-phase system was in a stable state near the coexistence curve, it could become unstable to density fluctuation with following decay to liquid and vapor phases when a sufficiently strong electric field (~ 10 MV/cm and higher under usual conditions) is applied. The reason of such instability is the shift of the spinodal curve to higher temperature in electric field.

The influence of electric field on mixtures of dielectric liquids is similar [2]. It was shown for several liquids that the critical temperature of demixing is proportional to the quantity $E^2\left(\frac{\partial^2\varepsilon}{\partial\phi^2}\right)_T$, where ϕ is the concentration of one of the components. If the condition $(\partial^2\varepsilon/\partial\phi^2)_T > 0$ is satisfied, the electric field leads to the decay of a homogeneous mixture into a two-phase system.

The critical value of the electric field E_0 which enables the decay of a dielectric liquid was determined from the equation of the spinodal obtained in the work [3]

$$\left(\frac{\partial\widetilde{P}}{\partial\widetilde{\rho}}\right)_T = \frac{\varepsilon_0 E_0^2\widetilde{\rho}}{2\mathrm{P_{cr}}}\left(\frac{\partial^2\varepsilon}{\partial\widetilde{\rho}^2}\right)_T, \tag{2}$$

Here, $\widetilde{P} = P/P_{cr}$, $\widetilde{\rho} = \rho/\rho_{cr}$, and $\widetilde{T} = T/T_{cr}$ are the reduced variables, where P_{cr}, ρ_{cr}, and T_{cr} are the critical pressure, density, and temperature. Thus, the occurrence of the spinodal decay is defined by the dimensionless parameter $A = \varepsilon_0 E^2/(2\mathrm{P_{cr}})$. In the same work, it was shown that the instability increment is proportional to the second density derivative of the dielectric permittivity. The decay to liquid and vapor phases is anisotropic [3, 4], bubbles in liquid (and droplets in vapor) have elongated form and they are oriented along the electric field. The possibility of the spinodal decay was demonstrated by the molecular dynamics method in the work [5]. Authors of the work [6] showed that the characteristic time of the spinodal decay without the hydrodynamic stage, the formation of a nucleus of the new phase takes ~ 10^{-9} s or less.

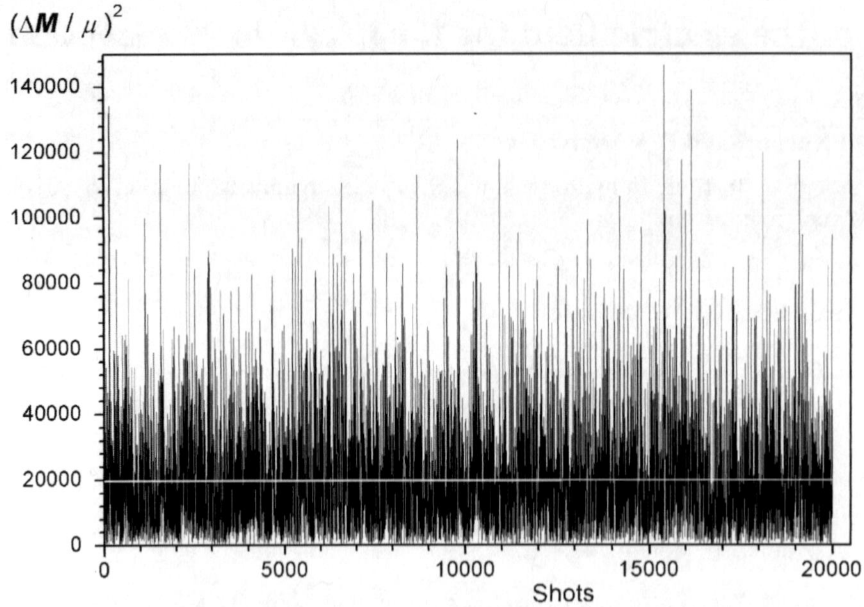

Figure 1. Time sequence of the values of the squared fluctuations of the dipole moments $((\mathbf{M} - \langle \mathbf{M} \rangle)/\mu)^2$ for *N–V–T* ensemble of 5400 water molecules. $T = 350$ K, $\rho = 1.012$ g/cm³. Horizontal line shows the average value of fluctuations squared.

Thus, the anisotropic spinodal decay discovered theoretically in [3] can explain the very fast (time of order of nanoseconds) appearance of vapor channels in a liquid during the electric breakdown of liquids. The substance inside channels has the electric strength which is lower than the one of the liquid itself by an order of magnitude. This leads to the development of discharge in channels and to the formation of streamers. Such mechanism of the inception of streamer channels can explain easily many phenomena observed experimentally at the breakdown of dielectric liquids, namely the ultra-fast propagation of streamer channels (velocity over 100 km/s), nearly cylindrical shape of channel segments, their branching, and the appearance of a sheaf of nonluminous channels with decreased density near a tip with following breakdown in one of the channels [7].

In order to estimate the critical electric field strength necessary for an anisotropic decay, one needs to know the density dependence of permittivity at the constant temperature in a sufficiently broad range of density. Such measurements are extremely rare [8]. Therefore, the development of a method to calculate the dependence $\varepsilon(\rho)$ for a given liquid is important. In this paper, the calculations of the dielectric permittivity of water for the temperatures $T = 670$ K (above the critical one), $T = 350$ K, and $T = 300$ K in a broad range of density were performed. Based on the dependences obtained, the critical electric field for the streamer inception was estimated.

2. Calculation of dielectric permittivity of water by the molecular dynamics method

The algorithm was developed for calculation of the static dielectric permittivity of water in abroad range of temperature and density. The algorithm includes following stages: generation of the ensemble of water molecules with the prescribed density, molecular dynamic simulation of the evolution of the ensemble at the fixed temperature, recording the data on the coordinates of electric charges of oxygen and hydrogen in consequent time intervals, calculation of the dipole moment $\mathbf{M}$, calculation of time averaged fluctuations of the dipole moment squared. Molecular dynamic calculations were performed using the freely distributed package LAMMPS (Large-scale Atomic/Molecular Massively Parallel Simulator) [9].

The SPC/E was used to describe the molecule of water. The symmetric intermolecular interaction is described by the Lennard-Jones potential centered on the oxygen atom (potential parameter is 0.65 kJ/mol, equilibrium distance is 0.3166 nm). For the description of the Coulomb interaction, the positive charges of 0.42e are placed in the centers of hydrogen atoms, and the negative charge of 0.84e is placed in the center of oxygen atom. In our calculations, the bond lengths O-H and H-H were fixed as well as the angle H-O-H. According to [10], the SPC/E model describes the coexistence curve of vapor and liquid for water at the temperatures of 300 K and higher essentially better than several other models (TIP4P, TIP5P, Dang-Chang).

The ensemble of water molecules was placed in a parallelepiped region. The ensemble of 5000 molecules was simulated for the temperatures $T = 670$ K and $T = 300$ K, for the temperature $T = 350$ K the ensembles of 5000 and 5400 molecules were simulated. The periodic boundary conditions were used. At the initial stage, the ensemble was thermalized for 50000 time steps. The calculations were performed for a *N-V-T* ensemble. The calculations of the dielectric permittivity of the ensemble of water molecules converge slowly, therefore calculations were performed for long time intervals of 50 ns ($5 \cdot 10^6$ time steps) and 100 ns (10^7 time steps). Calculations were performed using the HP BL2x220c G7 and HP BL2x220c G6 computer clusters of the Information and Computation Center of the Novosibirsk State University [11].

The values of the dielectric permittivity were calculated similar to the method proposed in [12] using the formula

$$\varepsilon - 1 = \frac{4\pi}{3VkT}\left(<\mathbf{M}^2> - <\mathbf{M}>^2\right). \quad (3)$$

A typical time sequence of 20000 values of the fluctuations of the reduced dipole moment squared $((\mathbf{M} - <\mathbf{M}>)/\mu)^2$ is shown in figure 1. Here μ is the module of the dipole moment of the water molecule. This sequence was calculated for the ensemble of 5400 water molecules in time intervals of 5 fs.

3. Calculation results

The calculations of static dielectric permittivity of water were performed in a supercritical region for the

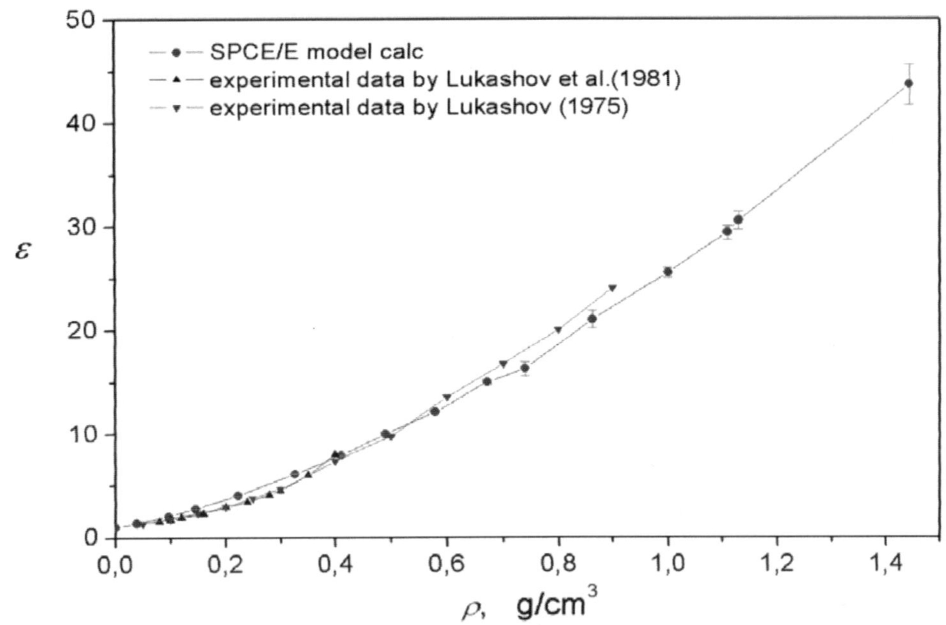

Figure 2. Density dependence of the permittivity of water ε along the isotherm $T = 670$ K. Triangles – the experimental data [8], circles – calculation results.

Journal of Physics: Conference Series **754** (2016) 102004 doi:10.1088/1742-6596/754/10/102004

temperature T = 670 K in the density range from 0.03 to 1.4 g/cm^3 (figure 2). The density dependence of the permittivity $\varepsilon(\rho)$ obtained (figure 2) can be approximated by the formula

$$\varepsilon = 1 + 11.25\rho + 14.05\rho^2 - 0.983\rho^3 . \quad (4)$$

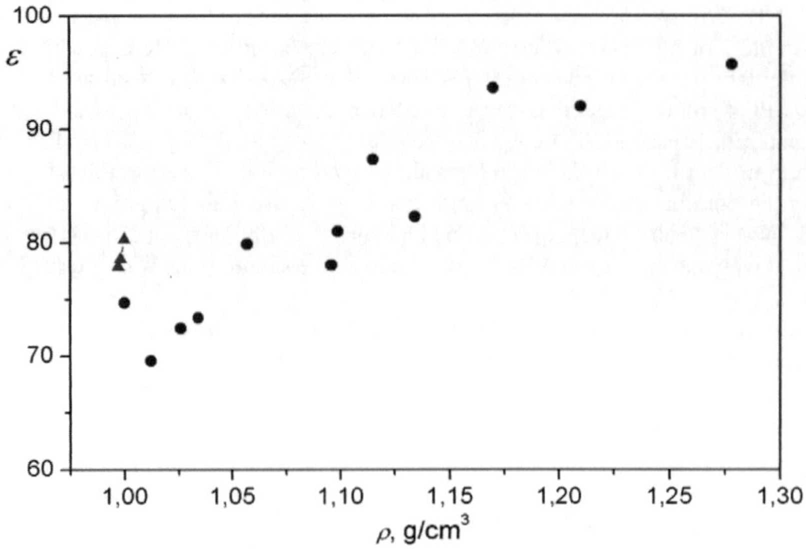

Figure 3. Dielectric permittivity of water at 300 K. Triangles are the data [13]. Circles are the result of calculations for an ensemble of 5000 water molecules (simulation time was 50 ns).

The positive sign of the second derivative $(\partial^2\varepsilon/\partial\rho^2)_T$ in the range of density considered leads to the possibility of an anisotropic spinodal decay under the action of sufficiently high electric field.

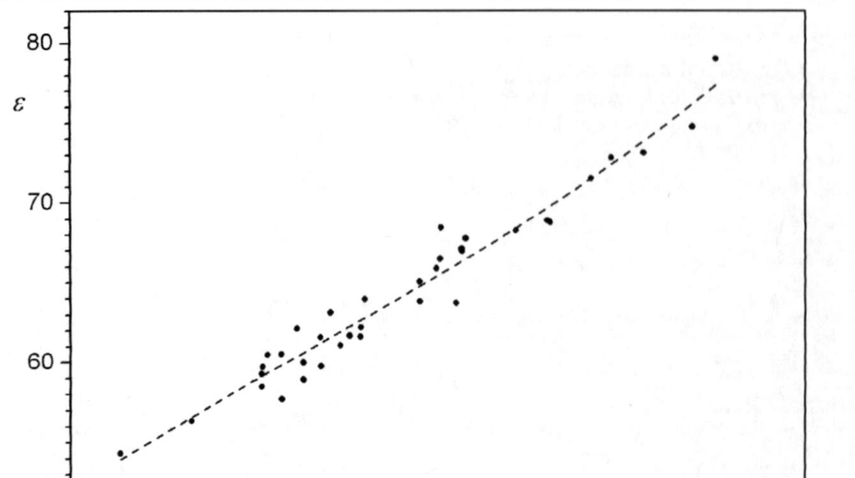

Figure 4. Dielectric permittivity of water at 350 K. The dashed line is an approximation of $\varepsilon(\rho)$ by a fourth-degree polynomial.

At low temperatures, no dependence $\varepsilon(\rho)$ was obtained so far which would allow one to calculate reliably the value of second derivative $(d^2\varepsilon/d\rho^2)$. The dependence obtained looks non-monotonous, the values of $\varepsilon(\rho)$ change irregularly from low to high densities (figure 3). Probably, the potential interaction energy at low temperatures is so high that the correlations of the states of neighbor water molecules appear. In this case, the duration of the time sequence can be insufficient to exclude such "memory" effects.

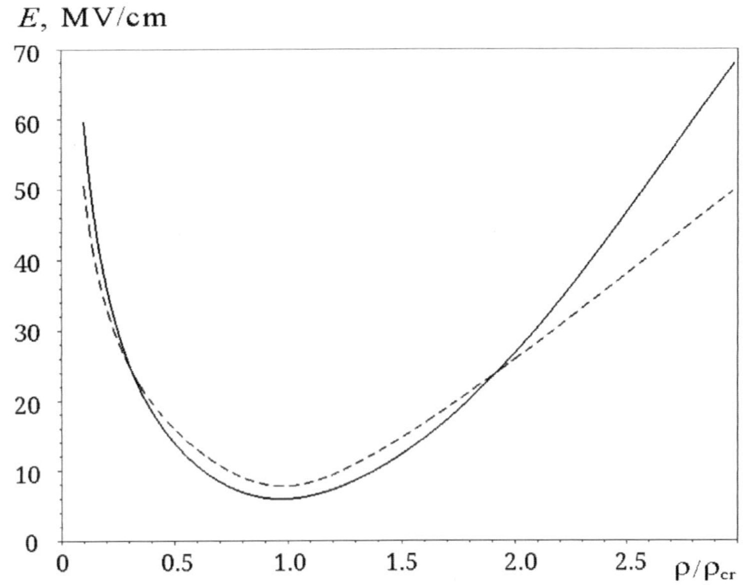

Figure 5. Critical electric field for anisotropic spinodal decay of water for the isotherm $T = 670$ K. The real equation of state from [14] (dashed line) and the Kaplun-Meshalkin-Kupershtokh equation [15] (solid line) were used.

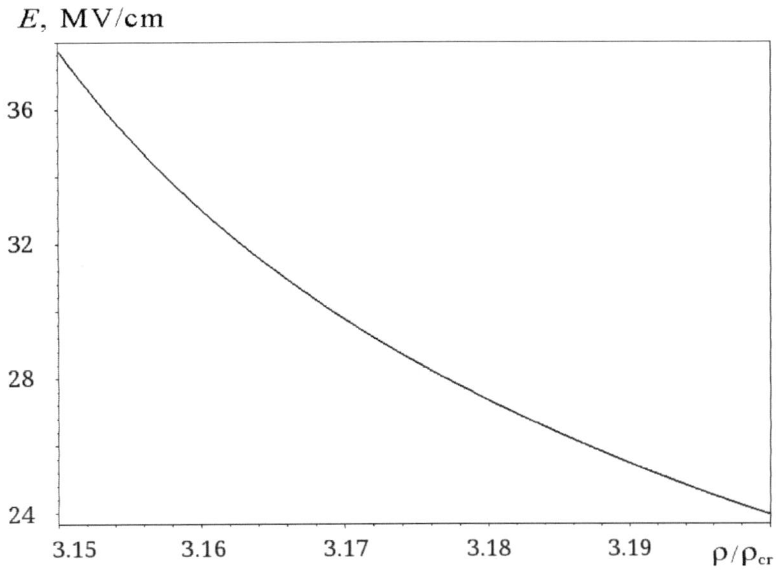

Figure 6. Critical electric field for anisotropic spinodal decay of water for the isotherm $T = 350$ K and real equation of state from [14].

Therefore, simulations at higher temperature of 350 K were carried out for ensembles of 5000 and 5400 water molecules for time intervals of 50 and 100 ns (figure 4). The shape of the simulation region was also changed. The rectangular region with a square base and the aspect ratio from 2.5 to 3.2 was used as well as a nearly cubic region. The results were fitted with the polynomial of the forth order shown with the dashed line in the figure 4.

4. Estimations of the critical electric field strength

Using the data on the density dependence of the dielectric permittivity of water $\varepsilon(\rho)$, the calculation were performed of the critical electric field strength which enables the spinodal decay of the dielectric liquid. The critical fields for the supercritical region at $T = 1.035\,T_{cr}$ (670 K) and the values of the reduced density from 0.1 to 3 are shown in figure 5. For the temperature $T = 0.541\,T_{cr}$ (350 K), the critical fields are shown in figure 6. The real equation of state from [14] was used.

5. Conclusion

The density dependence of the dielectric permittivity for water was obtained using the molecular dynamics method for $T = 1.035\,T_{cr}$ in a broad range of density, and for $T = 0.464\,T_{cr}$ and $T = 0.541\,T_{cr}$ near the coexistence curve. The critical values of the electric field strength were calculated which make possible the spinodal decay "liquid-vapor". The results obtained are in agreement with the estimations of the electric field strength of the electric discharge inception based on the experimental data.

Acknowledgements
The study was supported by the Russian Scientific Foundation (Grant No. 16-19-10229).

References

[1] Landau L D and Lifshitz E M 1960 *Electrodynamics of Continuous Media* (Pergamon Press)
[2] Debye P and Kleboth K 1965 Electrical field effect on critical opalescence *J. Chem. Phys.* **32** 3155–62
[3] Kupershtokh A L and Medvedev D A 2006 Anisotropic instability of a dielectric liquid in a strong uniform electric field: Decay into a two-phase system of vapor filaments in a liquid *Phys. Rev. E* **74** 021505
[4] Kupershtokh A L 2014 Three-dimensional LBE simulations of a decay of liquid dielectrics with a solute gas into the system of gas-vapor channels under the action of strong electric fields *Computers and Mathematics with Applications* **67** 340–9
[5] Karpov D I and Kupershtokh A L 2009 Anisotropic spinodal decay of polar dielectric in strong electric field: method of molecular dynamics *Technical Physics Letters* **35** 479–82
[6] Baidakov V G and Protsenko S P 2004 Computer simulation of nucleation in a liquid under tension *Doklady Physics* **49** 69–72
[7] An W, Baumung K and Bluhm H 2007 Underwater streamer propagation analyzed from detailed measurements of pressure release *J. Appl. Phys.* **101** 053302
[8] Fernandes D P, Mulev Y, Goodwin A R H and Levelt Sengers J M H 1995 A database for the static dielectric constant of water and steam *J. Phys. Chem. Ref. Data* **24** 33–69
[9] LAMMPS Molecular Dynamics Simulator. http://lammps.sandia.gov
[10] Yoo S and Zeng X C 2002 Monte Carlo simulation of vapor-liquid binodal of water *J. Chem. Phys.* **117** 9518–19
[11] Information and computation center (ICC) of Novosibirsk State University. http://www.nusc.ru
[12] Pan D, Spanu L, Harrison B, Sverjensky D A and Galli G 2013 Dielectric properties of water under extreme conditions and transport of carbonates in the deep Earth *PNAS* **110** 6646–50
[13] Kikoin I K 1976 *A handbook of physical constants* (Moscow: Atomizdat), in Russian
[14] The IAPWS industrial formulation 1997 for the thermodynamic properties of water and steam, Water/Steam Thermodynamic Surface on the WEB-site of Moscow Power Engineering Institute, Technical University. http://twt.mpei.ac.ru/MCS/Worksheets/WSP/VTPs.xmcd
[15] Kupershtokh A L, Medvedev D A and Karpov D I 2009 On equations of state in a lattice Boltzmann method *Computers and Mathematics with Applications* **58** 965–74

TPH IOP Publishing
Journal of Physics: Conference Series **754** (2016) 102005 doi:10.1088/1742-6596/754/10/102005

Streamers and partial discharges in water

S M Korobeynikov[1, 2] and A V Melekhov[3]

[1] Lavrentyev Institute of Hydrodynamics SB RAS, 15 Lavrentyev Prosp., 630090, Novosibirsk, Russia
[2] Power Engineering Faculty, Novosibirsk State Technical University, 20 Prosp. K. Marksa, 630073, Novosibirsk, Russia
[3] Institute of Laser Physics SB RAS, 13/3 Lavrentyev Prosp., 630090, Novosibirsk, Russia

E-Mail: korobeynikov@corp.nstu.ru

Abstract. Several types of streamers are considered in the paper, usual streamers, nonelectrode and postbreakdown streamers. Attempts of pulse electrical strength increase with the help of conductive layers and the reasons for unusual streamers appearance are discussed.

1. Introduction

Deionized water is the most suitable dielectric liquid for pulsed power systems for different purposes. This is due to high dielectric permittivity, high pulse electric strength, explosion and fire safety, low decomposition after breakdown. Study of streamers is very important in course of pulse electric strength increase and energy storage increase.

The goal of the paper is an analysis of different forms of streamers in water, using the experimental results obtained in the previous works of the authors [1-4].

2. Streamers

The form of the streamer depends on polarity of the electrode where it originates. Electrical streamer is formation of conductive (may be slightly conductive) area close to electrode and spread of this area in form of branched structure in the direction to the opposite electrode. The form and properties of streamers depends on the polarity of electrode where it was initiated.

2.1. Usual streamers

Streamers are usually initiated on the electrode surface. A cathode streamer (that originates on the cathode and spreads to anode) has form of bush-like structure. These streamers are opaque and have subsonic velocity of propagation to the anode (figure 1). An anode streamers have transparent filament structure which spreads with supersonic velocity (figure 2).

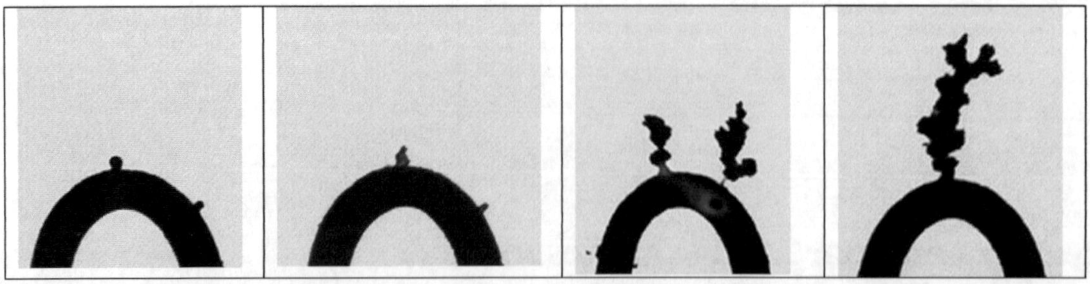

Figure 1. The stages of cathode streamers initiated by the bubble located on the electrode surface [1].

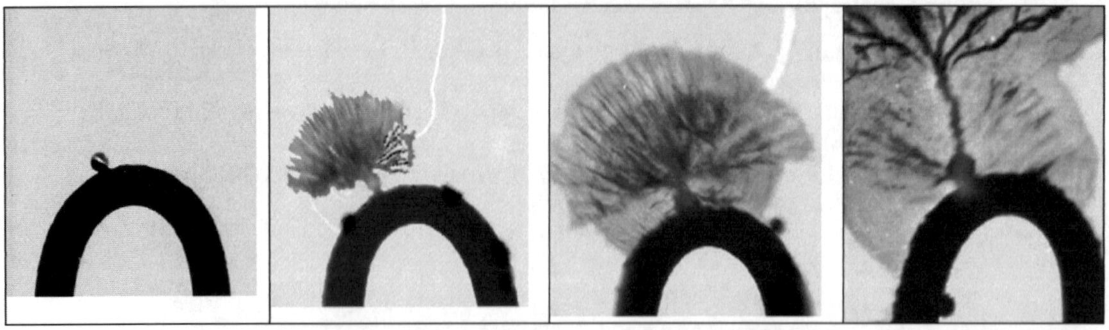

Figure 2. The stages of anode streamer initiated by the bubble located on the electrode surface [1].

2.2. Nonelectrode streamers

These streamers appear in the case of screened electrodes. Screening is electric field decrease near electrodes through formation of conductive layers close to anode or close to both anode and cathode. Nonelectrode streamers are initiated in the region of high field. Their form could be considered as two semi streamers. The anode directed semistreamer is very similar to the usual cathode streamer, and the cathode directed semistreamer is similar to the usual anode directed streame (figure 3). In all registered pictures nonelectrode streamer don't lead to gap breakdown, therefore appearance of this streamer is partial discharge.

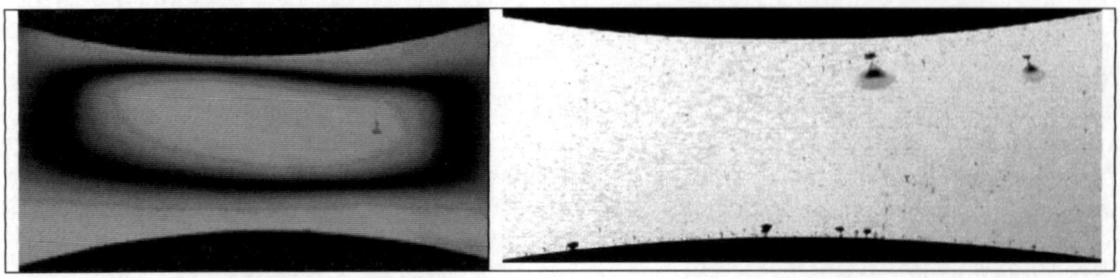

Figure 3. Typical pictures of nonelectrode streamers [2,3].

2.3. Postbreakdown streamers

These kinds of streamers appear in the case of screened electrodes when gap breakdown occurs. It has supersonic filament structure that propagates from main breakdown channel (figure 4).

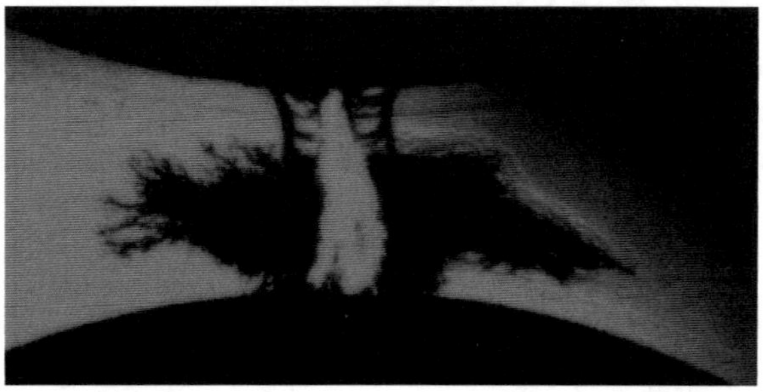

Figure 4. Postbreakdown streamers [4].

3. Discussion

3.1. Attempts to increase pulse electrical strength.

A pulse breakdown strength is controlled by two main factors. First one is the velocity of streamer propagation. Another one is the streamer initiation. In our opinion, the initiation is the key factor of breakdown process. That is why an influence upon initiation is most evident way to increase pulse electrical strength of water. The initiation requires strong electric field, therefore the decrease at the place of streamer initiation should lead to electrical strength increase. There were two successful attempts to increase pulse electrical strength of polar liquids by formation of conductive layers close to electrodes. First one is electrolyte forcing through porous electrode [5]. The pure water is in central region of the gap, near-electrode regions contain electrolyte that lead to high conductivity of near-electrode layers. Under the action of pulse voltage, its conductivity leads to formation of space charges and decrease of electric field at electrode surfaces and increase in the inner part of gap. Experiments [5] had shown that pulse electrical strength was increased 3-4 times in comparison with strength of gap without conductive layers. Figure 5 shows the sketch of interelectrode gap with conductive layers near electrodes and pure dielectric liquid in the central region of the gap. An approximate diagram of electric field through the gap is shown in figure 6.

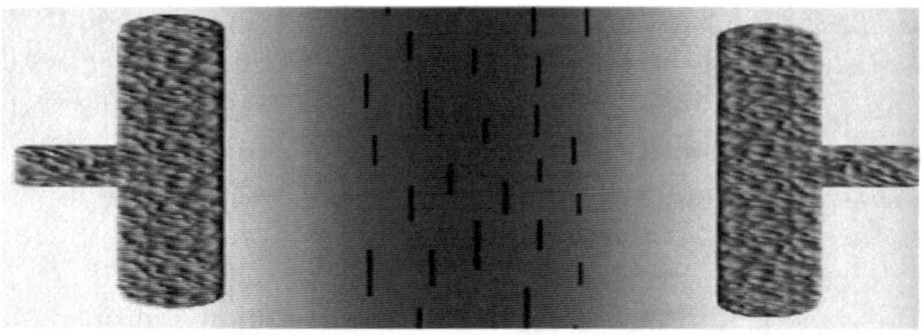

Figure 5. Sketch of the interelectrode gap with the conductive layers.

The same idea concerning near-electrode conductive layers formation was used in [6]. The experiments were performed in glycerol; the conductive layers were formed by pulse heating of electrodes. Heat transfer from electrodes into liquid leads to temperature increase near electrodes that in turn leads to sharp increase of liquid conductivity. In our experiments, the pulse electrical strength of glycerol increased no less than two times in comparison with strength of gap with cold electrodes.

Our experiments [2,3] with screened electrodes exploited the same idea, but conductive layers in water were formed by application of the special modified electrodes. As far as the increase of the pulse electrical strength, the effect was obtained. According to electrooptical measurements, the maximal electric field in the center of the gap was 1 MV/cm, and the electric field near electrodes was 400 kV/cm. The mean electric field strength was 700 kV/cm. If we take into account that the breakdown field strength of the gap without conductive layers was 400 kV/cm, then the pulse breakdown strength in these experiments was increased approximately by 1.7 times.

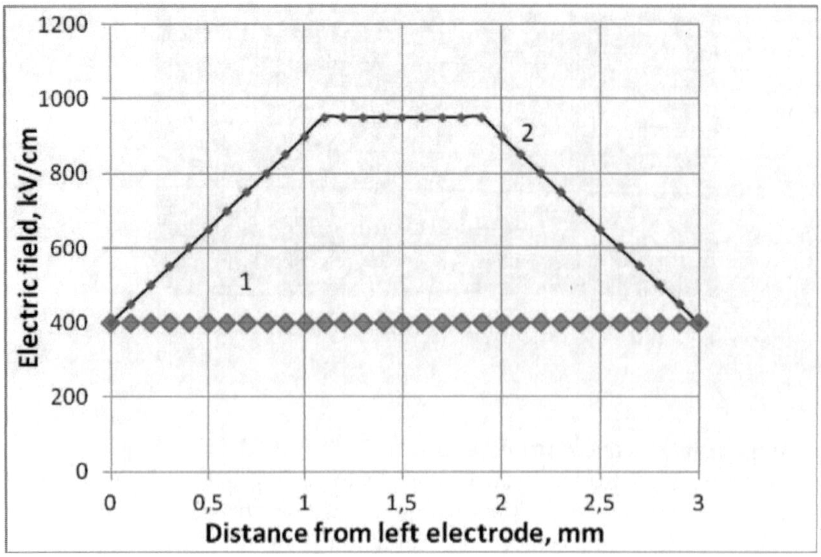

Figure 6. Qualitative diagram of electric field without the conductive layers (1) and with the conductive layers (2). The electric field E near the electrodes is equal to 400 kV/cm.

3.2. Peculiarities of streamers

The main features of positive and negative streamers were discussed earlier [7]. Here one should consider other features that appear in the case of screened electrodes: non-electrode and postbreakdown streamers.

3.2.1. Nonelectrode streamers. As it was discussed in section 3.1 electrodes screening leads to hinder of usual streamer initiation. That is why sometimes another possibility of streamer initiation could be realized. The electric field inside the gap is higher than for usual non modified electrodes. The problem of initiation in this area lies in the absence of initiating electrons. Usually it is supposed that electrode surface is the source of initiating electrons that appears as a result of absorbed negative ions break-up under the action of electric field. In our opinion the initiating electrons in the bulk of liquid could appears due to the same reason in case of microparticles presence in the liquid. If we take into account that particles amount with sizes from micrometer to hundred of micrometer is from 100 to 1000 per cubic centimeter [8], the total specific surface is small, then number of negative ions is small too. Hence, the number of potentially free electrons is small and the probability of breakdown initiation is not so high, so the electric field strength should be greater than in the case of usual electrodes.

Nonelectrode streamers have two other features. First one is form of semistreamers. Anode directed semistreamer looks like usual cathode streamer and cathode directed semistreamer looks like usual anode streamer. It points out on the conclusion that mechanisms of semistreamers propagation is the same as for usual streamers. The additional argument of this conclusion is length of cathode directed and anode directed semistreamers. The cathode directed semistreamer is 4 times longer than the anode directed one. It coincides with the velocity relation of usual streamers. If we take into account that semistreamers are initiated simultaneously then the length of subsonic cathode streamers should be 4 times less than the length of anode supersonic streamers.

Another feature is the mechanism of streamer propagation. It is clear that the field intensity is high at the head of streamer. According to our estimations, its value is E ~10 MV/cm [9]. How could this value be supported during propagation while streamer surface is increased? Evidently that additional charge should be brought to the tip of streamer branches. In the case of the usual streamer, this additional charge is supplied by conductive connection of streamer branches with electrodes. In the case of the nonelectrode streamers the source of additional charge for one semistreamer is propagation of other semistreamer only. As it was shown early [1], a set of usual anode streamers disappear after some growth. In our opinion it is due to impossibility of high field support at the tip of long low conductive streamers. That is why all except one usual anode streamers disappear and only one streamer becomes more conductive and, as a result, the new set of streamers emerges at its tip. In the case of nonelectrode streamers when the cathode directed semistreamer gets in touch with the cathode, this will not lead to breakdown acceleration but to termination of streamer propagation. So, one can conclude that conductivity of cathode directed semistreamer is not high.

3.2.2. Postbreakdown streamers. The reason of this kind of streamers appearance is space charge that appears in conductive layer as a result of voltage rise. When breakdown of the gap occurs this space charge produces high electric field. The mechanism is described below.

The electrical properties of a layer with specific conductivity σ are determined by the Maxwell relaxation time $\tau_m \approx \varepsilon \cdot \varepsilon_0/\sigma$ and the voltage rate, e.g. the characteristic time rise τ_r in case of pulse voltage increase or trailing edge τd in case of pulse voltage decrease. When $\tau_r >> \tau_m$ in near-electrode area, the space charge is formed during the total pulse time rise duration. This space charge leads to electric field redistribution. It is clear from Poisson's equation that gradient conductivity σ leads to electric heterogeneity in the gap and emergence of space charge with density $\rho(x) \approx \varepsilon \cdot \varepsilon_0 \cdot (dE/dx)$. The electric field intensity decreases near the electrodes but increases in central region of the gap. Breakdown is equivalent to sharp decrease of voltage. When $\tau_d << \tau_m$, the current space charge couldn't disappear instantly and according to Poisson's equation should form fundamentally new electric field distribution. The electric field is created here by the space charge only but with the new boundary condition: the potential of both electrodes and of the breakdown channel becomes equal to zero. In this case the strongest field appears at the conductive layers boundary near the breakdown channel. It is interesting that in the cathode layer (negative space charge) electric field is directed from breakdown channel that is why this part of channel becomes transient anode. Analogically, the anode conductive layer (positive space charge) electric field is directed to breakdown channel, therefore this part of the channel becomes transient cathode. One could show that the electric field intensity in both cases is high enough for streamer initiation: subsonic bush-like streamers in the positive space charge zone and supersonic filament streamers in the negative space charge zone. The last kind of streamers can be seen in figure 4. Subsonic streamers are in the afterbreakdown shock wave region and couldn't be displayed due to a change of the refractive index in the shock wave region.

4. Conclusion

As a result of optical and electrooptical investigations of streamers in water, the following conclusions can be made:

1. Conductive layers near electrodes lead to electric field decrease on the electrodes that, in turn, lead to pulse electric strength increase.
2. Nonelectrode streamers develop in form of two semistreamers: the anode directed semistreamer looks like a usual cathode streamer, while the cathode directed semistreamer looks like a usual anode streamer.
3. Registered postbreakdown streamers develop as usual supersonic anode streamers.
4. The reason of both nonelectrode and postbreakdown streamers is the conductivity of thin layers near the electrodes that leads to space charge appearance.
5. Nonelectrode streamers are new type of partial discharges.

Acknowledgments

One of the authors (S.M.K.) is grateful to RNF (grant No. 16-19-10229) in part concerning new form of partial discharge – nonelectrode streamers.

5. References

[1] Ushakov V Ya, Klimkin V F and Korobeynikov S M 2007 *Impulse Breakdown of Liquids,* (Berlin: Springer-Verlag GmbH) p 397

[2] Korobeynikov S M and Melekhov A V 2011 Nonelectrode streamers in deionized water *IEEE Trans. on Plasma Science* **39** 2632–3

[3] Korobeynikov S M, Melekhov A V and Posukh V G 2014 Nonelectrode and postbreakdown ionization processes in water *IEEE Trans. on Plasma Science* **42**(4) 965–8

[4] Korobeynikov S M, Melekhov A V and Vagin D V 2016 Prebreakdown processes and electric fields in water with screened electrodes *Proc. Int. Conf. on Dielectrics* (Montpellier: University Press) 808–12

[5] Vorob'ev V V, Kapitonov V A, Kruglyakov E P and Tsidulko Yu A 1980 Breakdown of water in a system with 'diffusion' electrodes *Soviet Physics Technical Physics* **25**(5) 598–602

[6] Korobeynikov S M, Yanshin E V, Ovchinnikov I T,Yanshin K V, Sarin S G, Kopylov V M and Klepikov A V 1995 Physical processes limiting the pulse energy release in liquid dielectrics. *Conf. Record of Int. Conf. on Pulsed Power* (Albuquerque) p 2

[7] Tobazéon R 1988 Streamers in liquids *The Liquid State and its Electrical Properties* Kunhardt E E, Christophorou L G and Luessen L H eds. (New York: Plenum Press) 465–78

[8] Besov A S, Kedrinskii V K and Pal'chikov E I 1984 Study of the initial stage of cavitation by a diffractive-optics method *Soviet Technical Physics Letters* **10** 100–2

[9] Korobeynikov S M and Melekhov A V 2014 Estimations of the electric field strength of nonelectrode streamers in water *High Temperature* **52**(1) 129–31

Simulation of waves of partial discharges in a chain of gas inclusions located in condensed dielectrics

A L Kupershtokh[1] and D I Karpov[1]

[1] Lavrentyev Institute of Hydrodynamics SB RAS, 15 Lavrentyev Prosp., Novosibirsk, 630090, Russia

E-mail: skn@hydro.nsc.ru

Abstract. A stochastic model of partial discharges inside gas inclusions in condensed dielectrics was developed. The possibility of a "relay-race" wave propagation mechanism of partial discharges in a linear chain of gas inclusions is shown. The lattice Boltzmann method is successfully implemented for three-dimensional computer simulations of flows of dielectric fluid with bubbles. Growth and elongation of bubbles in a liquid dielectric under the action of a strong electric field are simulated. The physical model of propagation of partial discharges along a chain of gas bubbles in a liquid is formulated.

1. Introduction

When the high voltage is applied to an electrode gap filled with a dielectric, partial discharges may occur that do not lead to the loss of the insulating properties of the dielectric as a whole. At this time, the pulses of current are registered in the discharge circuit, the number and intensity of which can predict the probability of a breakdown of the gap. One type of partial discharge in condensed dielectrics are electrical discharges inside the cavities and bubbles, which have dimensions of the order of 10 microns. The works [1-3] are devoted to computer simulations of partial discharges in liquid and solid dielectrics. In this paper, the partial discharges in condensed dielectric between two flat electrodes are studied. Gas inside the small cavities in solid dielectrics or inside the microbubbles in a liquid has a dielectric strength which is much lower than that of the condensed matter. The possibility of the generation of "relay-race" wave propagation of partial discharges in the chain of inclusions is demonstrated. This mechanism differs from hopping spread streamers in the medium with bubbles [4] since for "relay-race" mechanism streamers do not occur in the condensed phase and thus there is no breakdown of the gap. Moreover, the stochastic nature of the phenomenon is taken into account.

2. Partial discharges in a chain of gas cavities located in solid dielectrics

At the same dimensions of cavities in the bulk of the dielectric and at the same gas pressures inside them, the probability of microbreakdowns in inclusions depends on the local electric field within them E_i. The stochastic criterion MESTL [5] was used to describe the occurrence of micro-discharges in cavities. For all nonconducting cavities, the stochastic lag times t_i were calculated in accordance with the density distribution function for the probability of rare events

$$F(t_i) = r(E_i)\exp(-r(E_i)t_i). \tag{1}$$

Here, $r(E)$ is a sharly increasing function of the electric field. For small time step $\Delta t << 1/r(E)$, the probability of a micro-breakdown in a cavity is $f \approx r(E)\Delta t$. It was assumed that the micro-breakdown in a cavity occurs at the current time step if the condition $t_i < \Delta t$ is satisfied.

A chain of several identical gas inclusions placed along the electric field lines was studied (figure 1*a*). The distance between the inclusions is constant Δy. The computational lattice spacing is h. DC voltage V sufficient for the occurrence of partial discharges was applied to the electrodes.

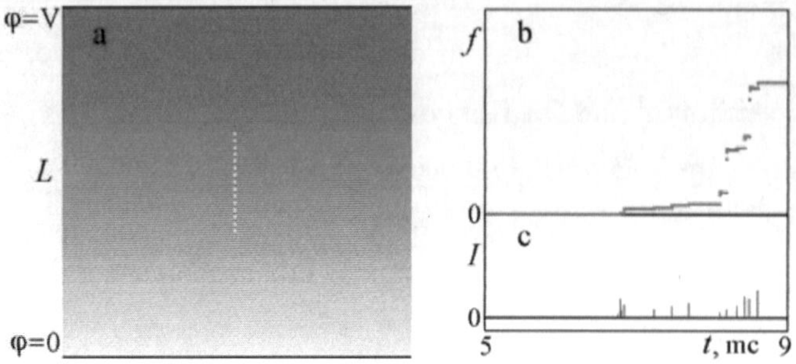

Figure 1. (a) – Chain of gas cavities in the solid dielectric. L = 1 mm. Lattice is 200×200. $\Delta y = 4h$. (b) – Increase of the probability of a microdischarge in one of the cavities with each discharge in other cavities. α = 0.005 1/(V·ms). (c) – Current I in the circuit induced by partial discharges.

The potential of the electric field in the region between the electrodes (figure 1a) is calculated numerically at each time step by solving the Poisson's equation together with the equations of the electric charge transfer

$$\mathrm{div}(\varepsilon\nabla\varphi) = -4\pi q\,, \qquad \mathbf{E} = -\nabla\varphi\,, \tag{2}$$

$$\mathbf{j} = \sigma \cdot \mathbf{E}\,, \qquad \frac{\partial q}{\partial t} = -\,\mathrm{div}\,\mathbf{j}\,. \tag{3}$$

Here, ε is the permittivity of fluid, q is the electric charge density. The periodic boundary conditions are used in x direction. It was assumed that current density $\mathbf{j}$ and conductivity σ are nonzero only within gas inclusions after their breakdown. The permittivity of the condensed dielectric is $\varepsilon = 2$.

Initially, the electric field is somewhat larger inside the inclusions that are at the edges of the chain than inside the inner inclusions. However, it is lower than the field inside single isolated cavity. After the breakdown inside one of the gas inclusions, the electric field and, accordingly, the probability f of a breakdown inside the neighbor inclusions during time step Δt increase significantly (figure 1*b*). The dependences of the function $r(E)$ for the occurrence of partial discharges in the cavity are shown in figure 2*a* that take into account the threshold character of partial discharges [1]

$$r(E) = \begin{cases} 0 & \text{при} \quad E \le E_*\,, \\ \alpha(E - E_*) & \text{при} \quad E > E_*\,. \end{cases} \tag{4}$$

Here, E_* is the minimum electric field at which partial discharges in the cavity are possible. For inclusions $d \sim 10$ microns, the breakdown voltage of air in a cavity is ≈ 350 V [6], which corresponds to the threshold field $E_* \approx 350$ kV/cm.

The sequence of partial discharges in the inclusions has a stochastic character at a relatively weak dependence $r(E)$ or at a weak mutual influence of cavities on each other (figure 2*b*).

If the dependence $r(E)$ is sharper, then, under certain conditions, there is an interesting phenomenon – a "relay-race" mechanism of propagation of partial discharges in the chain of insulation defects. Such waves of partial discharge are shown in figures 3 and 4*a*.

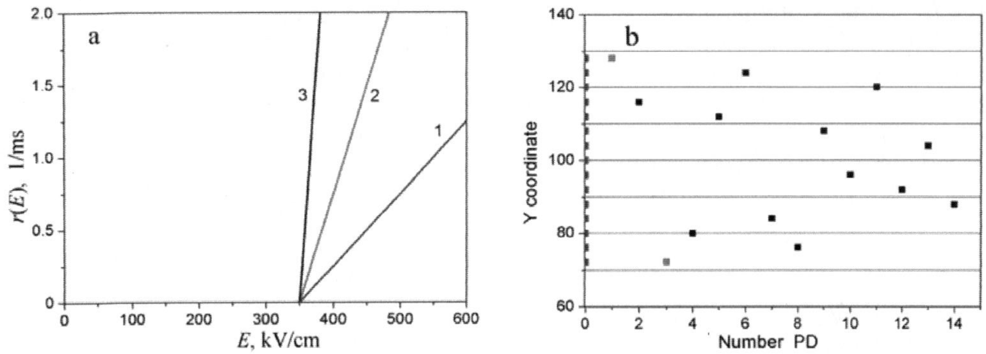

Figure 2. (a) – The probability of the first micro discharge in the cavity $d = 10$ microns in size. α = 0.005 (curve 1), 0.015 (curve 2), 0.065 (curve 3) 1/(V·ms), $E_* =$ 350 kV/cm. (b) – Partial discharges in the chain of cavities at $\Delta y = 4h$ and α = 0.005 1/(V·ms). The initial positions of the inclusions are shown at the left by blue.

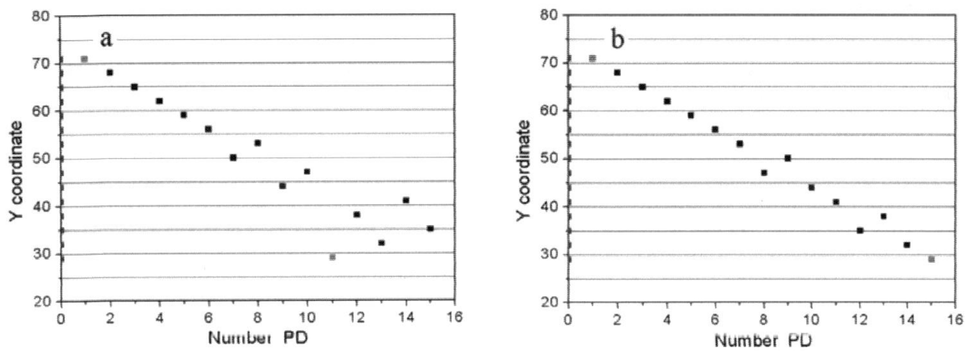

Figure 3. "Relay-race" mechanism of the propagation of a wave of partial discharges along the chain of gas inclusions. α = 0.015 (a), 0.065 (b) 1/(V·ms). Lattice is 100×100. $\Delta y = 3h$.

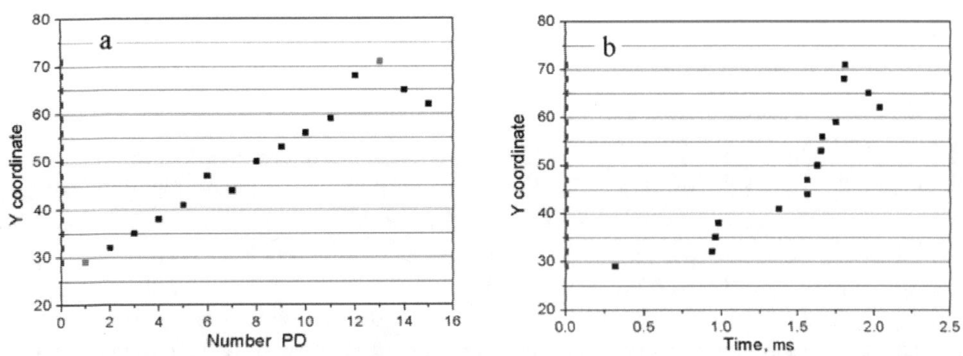

Figure 4. (a) – Wave of partial discharges. α = 0.065 1/(V·ms). (b) – The sequence of partial discharges in time. Lattice is 100×100. $\Delta y = 3h$.

As the partial discharges propagate along the chain of cavities, the electric field increases inside the remaining cavities. At the same time, the statistical time lag of a micro-breakdown decreases, and the average velocity of wave propagation increases (figure 4*b*).

3. The lattice Boltzmann method

The lattice Boltzmann method (LBM) is successfully implemented for three-dimensional computer simulations of liquid dielectric flows. The lattice Boltzmann method describes the viscous flows of fluids with an arbitrary equation of state and simulates the interfaces between vapor and liquid phases with a surface tension.

The three-dimensional version of lattice Boltzmann method D3Q19 [7] on a cubic lattice with nineteen vectors $\mathbf{c}_k$ of pseudo-particle velocities is realized. The evolution equations for the distribution functions N_k have the form

$$N_k(\mathbf{x}+\mathbf{c}_k\Delta t, t+\Delta t) = N_k(\mathbf{x},t) + \Omega_k(N) + \Delta N_k, \quad (5)$$

where Δt is the time step, Ω_k is the collision operator, and ΔN_k is the change of the distribution functions due to the action of the internal and external body forces.

The hydrodynamic variables (the density ρ and the velocity $\mathbf{u}$ of fluid) in a node are calculated as

$$\rho = \sum_{k=0}^{b} N_k \quad \text{and} \quad \rho\mathbf{u} = \sum_{k=1}^{b} \mathbf{c}_k N_k . \quad (6)$$

The collision operator is usually used in BGK form

$$\Omega_k = (N_k^{eq}(\rho,\mathbf{u}) - N_k(\mathbf{x},t))/\tau , \quad (7)$$

where τ is the dimensionless relaxation time.

The Exact Difference Method (EDM) [8] is used for the implementation of the body forces (internal and external forces) in the LBM:

$$\Delta N_k(\mathbf{x},t) = N_k^{eq}(\rho,\mathbf{u}+\Delta\mathbf{u}) - N_k^{eq}(\rho,\mathbf{u}), \quad (8)$$

where the value of the velocity after the action of the total force $\mathbf{F}$ at a node is equal to $\mathbf{u}+\Delta\mathbf{u} = \mathbf{u} + \mathbf{F}\Delta t/\rho$. The corresponding equilibrium distribution functions N_k^{eq} are calculated as

$$N_k^{eq}(\rho,\mathbf{u}) = \rho w_k \left(1 + \frac{\mathbf{c}_k\mathbf{u}}{\theta} + \frac{(\mathbf{c}_k\mathbf{u})^2}{2\theta^2} - \frac{\mathbf{u}^2}{2\theta}\right). \quad (9)$$

To simulate boundaries between liquid and gas, the internal forces acting between neighbour nodes of fluid were introduced by Shan and Chen [9]. For an equation of state of fluid in the form $P(\rho,T)$, the total force acting on a node was introduced by Qian and Chen [10] as a gradient of the pseudopotential $\mathbf{F} = -\nabla U$, where $U(\rho,T) = P(\rho,T) - \rho\theta$. The appropriate isotropic finite difference approximation of gradient operator was proposed in [11]. We use the quite simple well-known "bounce-back" rule to implement the no-slip boundary conditions at the solid electrodes in the LBM simulations. The periodic boundary conditions are used in x and y directions.

The level of wettability of solid electrodes by fluids is simulated by interaction forces acting on a node $\mathbf{x}$ belonging to the fluid from the nearest nodes $\mathbf{x}+\mathbf{e}_k$ representing the solid boundaries [12]

$$\mathbf{F}_k = w_k \psi(\rho(\mathbf{x}))\, B(\mathbf{x}+\mathbf{e}_k)\, \mathbf{e}_k . \quad (10)$$

4. Generation of chain of bubbles in dielectric liquid

The flows arising in a dielectric liquid with bubbles due to the action of an electric field were simulated by the lattice Boltzmann method [13,14]. The dielectric liquid with a dissolved gas is considered. The second set of distribution functions $g_k(\mathbf{x},t)$ is used for gas (the second component). The Graphics Processing Units (GPUs) are exploited for computer simulations.

A chain of small bubbles is placed in liquid between the flat electrodes (figure 5a). The van der Waals equation of state with possible phase transitions is used.

For a complete description of the process, it is necessary to consider the system of equations of electrohydrodynamics [2]. Together with the hydrodynamic description of the fluid flows by LBM, one should take into account the transfer of charge carriers q_i with fluid as well as their effective macroscopic mobility b_i in an electric field. The total electric charge density is equal to $q = \sum_i q_i n_i$, where n_i are the concentrations of carriers of electric charge. In the case of identical and constant diffusion coefficients for all carriers of electric charges $D_i = D$, the following equations are valid

$$\mathbf{j} = q\mathbf{u} - D\nabla q + \sigma\mathbf{E}, \tag{11}$$

$$\frac{\partial q}{\partial t} + \mathrm{div}(q\mathbf{u}) = D\Delta q - \mathrm{div}(\sigma\mathbf{E}), \tag{12}$$

that should be used instead of equations (3). Here, $\mathbf{u}$ is the velocity of fluid. The local conductivity of material $\sigma = \sum_i b_i |q_i| n_i$ depends on the local concentrations of charge carriers and, generally, can vary in space and time. In our simulations, the value of the electrical conductivity σ was assumed constant for simplicity and distinct from zero only in the process of breakdown of the gas inside the bubble. Generation and recombination of charge carriers in the volume of dielectric fluid in this study were not considered. The body force acting on a fluid in electric field has the form

$$\mathbf{F} = q\mathbf{E} - \frac{E^2}{8\pi}\nabla\varepsilon + \frac{1}{8\pi}\nabla\left(E^2 \rho \frac{\partial \varepsilon}{\partial \rho}\right). \tag{13}$$

In this paper, the simulation was limited by the moment of the first partial discharge in one of the bubbles. Up to this moment, no free charges are present in the fluid, and the charges transfer equation (12) may be omitted. In this case, the effect of electric field on the substance is limited by second and third terms on the right side of the equation (13).

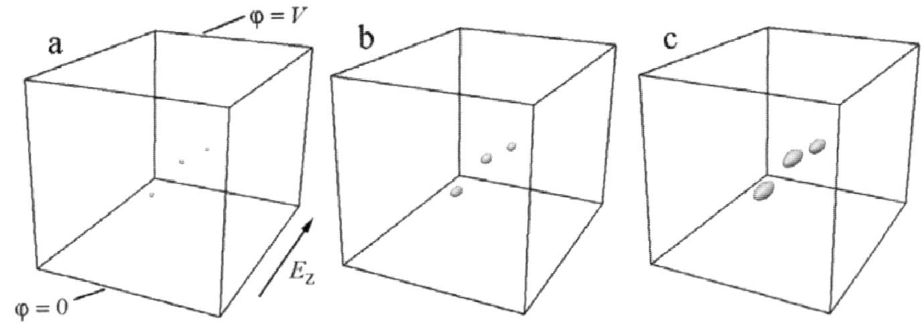

Figure 5. Bubbles growth in strong electric field before generation of the first partial discharge. $t = 700$(a), 800(b), 900(c). Lattice is 192×192×256.

Figure 5 shows the gas phase only, the liquid phase is considered to be transparent. After applying a voltage V to the electrodes, the expansion of existing bubbles occurs (figures 5*b,c*). This is because of the electrostriction forces [13,14] as well as diffusion of the dissolved gas into the bubbles. An inception of micro-breakdown in one of the bubbles is determined by the electric field inside bubble and by the value of parameter Pd (Paschen's law). Later, the elongated bubbles merge and form the approximately cylindrical gas–vapor channel.

5. Conclusion

The possibility of a "relay-race" wave propagation mechanism of partial discharges in a linear chain of gas inclusions is shown. The lattice Boltzmann method is successfully implemented for three-dimensional computer simulations of flows of dielectric fluid with bubbles. Growth and elongation of bubbles in a liquid dielectric under the action of a strong electric field are simulated. The physical model of propagation of partial discharges along a chain of gas bubbles in a liquid is formulated.

Acknowledgments
The study was supported by the Russian Scientific Foundation (grant No. 16-19-10229).

Journal of Physics: Conference Series **754** (2016) 102006 doi:10.1088/1742-6596/754/10/102006

References

[1] Wu K, Suzuoki Y and Dissado L A 2004 The contribution of discharge area variation to partial discharge patterns in disk-voids *J. Phys.* D*: Appl. Phys.* **37**(13) 1815–23

[2] Kupershtokh A L, Karpov D I, Medvedev D A, Stamatelatos C P, Charalambakos V P, Pyrgioti E C and Agoris D P 2007 Stochastic models of partial discharge activity in solid and liquid dielectrics *IET Sci. Meas. Technol.* **1**(6) 303–11

[3] Karpov D I and Kupershtokh A L 1998 Models of streamers growth with "physical" time and fractal characteristics of streamer structures *Conf. Record 1998 IEEE Int. Symp. on Electrical Insulation, IEEE No. 98CH36239* vol 2 (Arlington, USA) pp 607–10

[4] Babaeva N Yu, Tereshonok D V, Naidis G V and Smirnov B V 2016 Initiation of breakdown in strings of bubbles immersed in transformer oil and water: String orientation and proximity of bubbles *J. Phys.* D*: Appl. Phys.* **49**(2) 025202

[5] Kupershtokh A L, Charalambakos V, Agoris D and Karpov D I 2001 Simulation of breakdown in air using cellular automata with streamer to leader transition *J. Phys.* D*: Appl. Phys.* **34**(6) 936–46

[6] Peschot A, Poulain C, Bonifaci N and Lesaint O 2015 Electrical breakdown voltage in micro- and submicrometer contact gaps (100 nm – 10 μm) in air and nitrogen *IEEE Holm Conference on Electrical Contacts* pp 280–6

[7] Qian Y H, d'Humières D and Lallemand P 1992 Lattice BGK models for Navier – Stokes equation *Europhys. Lett.* **17** 479–84

[8] Kupershtokh A L 2010 Criterion of numerical instability of liquid state in LBE simulations *Comput. Math. Appl.* **59** 2236–45

[9] Shan X and Chen H 1993 Lattice Boltzmann model for simulating flows with multiple phases and components *Phys. Rev.* E **47** 1815–9

[10] Qian Y H and Chen S 1997 Finite size effect in lattice-BGK models *Int. J. Mod. Phys.* C **8** 763–71

[11] Kupershtokh A L, Medvedev D A and Karpov D I 2009 On equations of state in a lattice Boltzmann method *Comput. Math. Appl.* **58** 965–74

[12] Kupershtokh A L, Ermanyuk E V and Gavrilov N V 2015 The rupture of thin liquid films placed on solid and liquid substrates in gravity body forces *Comm. Comp. Phys.* **17**(5) 1301–19

[13] Kupershtokh A L and Medvedev D A 2006 Anisotropic instability of a dielectric liquid in a strong uniform electric field: Decay into a two-phase system of vapor filaments in a liquid *Phys. Rev.* E **74**(2) 021505

[14] Kupershtokh A L 2014 Three-dimensional LBE simulations of a decay of liquid dielectrics with a solute gas into the system of gas-vapor channels under the action of strong electric fields *Comput. Math. Appl.* **67**(2) 340–9

Inductive discharge as a method for direct detection of impurities in water

R N Medvedev[1], D S Churkin[2,3] and I A Zarubin[4,5]

[1] Lavrentyev Institute of Hydrodynamics SB RAS, Lavrentyev Prosp. 15, 630090, Novosibirsk, Russia
[2] Institute of Laser Physics SB RAS, Lavrentyev Prosp. 13/3, 630090, Novosibirsk, Russia
[3] Novosibirsk State University, Pirogova St. 2, 630090, Novosibirsk, Russia
[4] Institute of Automation and Electrometry SB RAS, Academician Koptug ave. 1, 630090, Novosibirsk, Russia
[5] Novosibirsk State Technical University, Prosp. K. Marksa 20, 630073, Novosibirsk, Russia

E-mail: ruslan@hydro.nsc.ru

Abstract. This paper reports on the excitation of atomic spectra of tap water vapor using an inductive discharge. It is shown that at a RF source frequency of 27 MHz and a plasma power output from 500 to 600 W, a steady discharge exists to pressures of 10^{-2} atm. Plasma temperature was determined by the Boltzmann plot method and was approximately 5500 K. At this temperature, it was possible to detect sodium ions dissolved in water. This suggests that an inductive discharge in water vapor can be used to determine the presence of impurities.

1. Introduction

There are various methods for studying materials in different states of aggregation. The systems (optical, electrochemical, thermal, magnetic) used in these methods have a complex design based on expensive elements and reagents, each system being typically designed to determine a specific substance. The high cost and complexity of these systems prevents their use for commercial and public services. Until now, control facilities have mostly used chemicals in liquid composition analysis, which limits the number of elements to be defined and require considerable time.

Emission spectral analysis is a well-established technique and is mainly used to determine the composition of metal alloys. The most common methods of emission spectral analysis are those using inductively coupled plasma [1] and flame photometry [2]. Their advantages are the possibility of quantifying the concentration of a large number (up to 50) of elements as well as the low detection limits (a few micrograms per liter). Major drawbacks of these systems are the cost, size, manufacturing complexity, and the need to use byproduct gases (argon, acetylene, oxygen). In addition, flame excitation methods are explosive, which places additional requirements on the design and operation of such plants.

On the other hand, today there is a need for rapid methods of analysis of the composition and quality of water under “field” conditions, which requires minimizing and simplifying the equipment design and reducing the time of sample preparation and analysis. Until now, water analysis in many laboratories has been carried out by chemical methods and only for several parameters: the presence of manganese, iron, pH, and bacteriological analysis. Therefore, the development and introduction of low-cost methods of analysis is an urgent task.

Previous studies [3, 4] have shown that an electrical discharge in water and at the boundary with a liquid can cause dissociation of molecules of both the liquid and the impurities dissolved therein. This makes possible qualitative and in some cases quantitative determinations of impurities, mainly metals. The main disadvantage of direct discharges between liquid electrodes in water is the low energy input into the plasma heating as most of the energy is released in the water column between the electrodes and the plasma. Furthermore, closing the electrode gap is undesirable as it leads to the appearance of the elemental composition of the electrodes in spectrum.

In this regard, we have attempted to initiate a RF inductive discharge directly in water vapor in order to develop new alternative methods of spectral analysis of aqueous media. Unlike the classical methods of spectral analysis using additional plasma-forming gases, the plasma-forming gas in our experiments was water, i.e., hydrogen-oxygen plasma. Furthermore, this type of discharge is electrodeless, which ensures the purity of the plasma.

A major issue is the method of introducing the aqueous medium into the discharge chamber. In this work, the chamber was filled with vapor as the liquid boiled out from the vessel with decreasing pressure. This is due to the need to reduce the chamber pressure because the RF source power is not sufficient for plasma ignition at atmospheric pressure. From the standpoint of detecting impurities, this method is not very good as most of the impurities, particularly heavy molecules, remain in the liquid phase and are not carried away with the vapor. However, even in this case, it was possible to record the spectrum of the light metal sodium. In the future, we plan to increase the RF source power to a level sufficient to maintain the plasma at atmospheric pressure. This will make it possible to introduce the solution directly into the discharge chamber.

2. Experimental

The discharge was excited in a quartz tube 1 of 20 mm diameter with a copper inductor having five turns (figure 1). The inductor was supplied with the 27.12 MHz frequency current from a CESAR 2710 RF source. The load power was from 500 to 600 W. A vacuum pump was placed on one side of the quartz tube, and a tap water container 2 with a metering valve 3 on the other side. The same valve 4 was placed on the side of the vacuum pump.

Before turning on the RF source, air was pumped out of the system to a pressure of ~10^{-3} atm. After that, the valve 4 was closed, and water vapor was fed to the system by means of the metering valve 3. This system made it possible to maintain the inductive discharge and perform measurements to pressures of up to ~10^{-2} atm. The RF generator was then turned to initiate a discharge. The discharge spectrum was recorded from the sidewall of the quartz tube 1 through a fiber optic cable 5 with a Kolibri-2 spectrometer (VMK-Optoelektronika) having a spectral range from 190 to 1100 nm and a resolution of 0.5 nm. The discharge tube pressure was controlled by a vacuum gauge 6 (Testo 552).

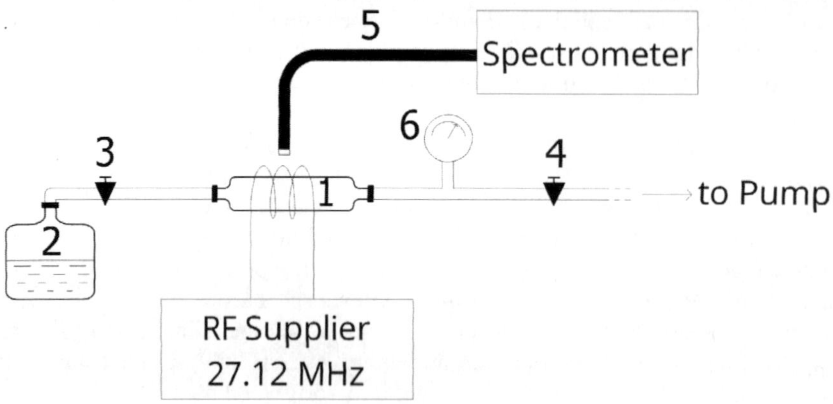

Figure 1. Schematic of the experimental setup.

The discharge time was from 1 to 10 s. The spectrum was recorded with an exposure time of 1 s. The spectrum was then downloaded to a PC and analyzed. The distance from the end of the fiber optic cable 5 to the discharge chamber 1 was varied depending on the brightness of the recorded lines.

3. Results and analysis

In the experiments, the emission maximum was observed on the periphery near the tube wall. This indicates that the plasma temperature is maximal near the wall and, hence, the emission is virtually not absorbed by the colder layers at the wall, i.e. the resulting plasma layer can be considered optically thin.

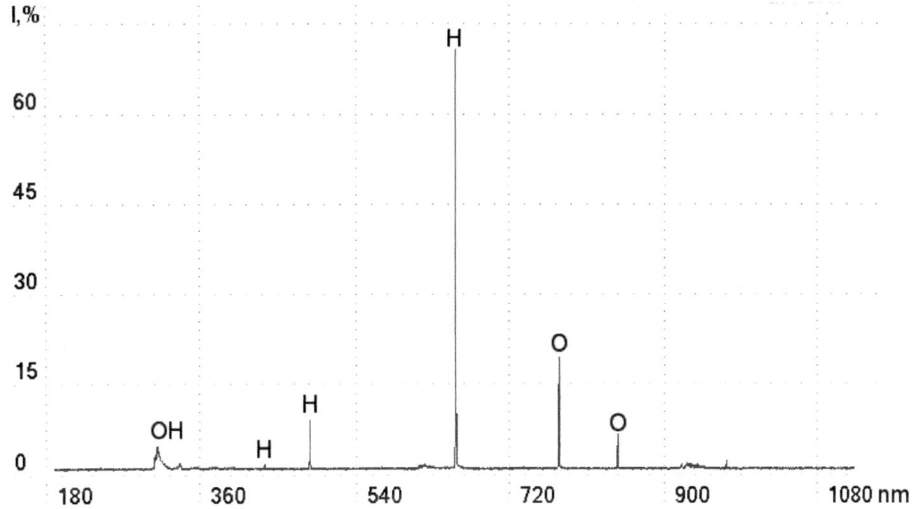

Figure 2. Spectrum of water vapor. The abscissa is the wavelength in nm and the ordinate is in relative units. The distance from the end of the fiber cable to the quartz tube is 300 mm.

Figure 2 shows the spectrum of the inductive discharge in water vapor with a distance between the sidewall of the tube and the end of the fiber optic cable of 300 mm. The Balmer series of hydrogen and the spectra of oxygen and the OH radical are seen. The plasma temperature can be determined from the relative intensities of the Balmer series using the relative-intensity method [1]. Temperature was estimated graphically (Boltzmann plot) using the relation:

$$\ln\left(\frac{I\lambda}{Ag}\right) = -\frac{E}{kT} + C \quad (1)$$

where E is energy of the upper (excited) level, I and λ are the intensity and wavelength of the corresponding spectral line, A and g are the probability of transition (in 10^8/s) and the statistical weight of the excited level, k is Boltzmann's constant, and C is a constant. The temperature can be determined as the slope of the plot in the coordinates $\ln\left(\frac{I\lambda}{Ag}\right)$ and E.

Figure 3 shows the plot for determining the temperature from three spectral lines of the hydrogen Balmer series and two transitions of oxygen atoms: at wavelength λ = 777 nm and λ = 844 nm using formula (1). For the hydrogen lines, the slope of the plot corresponds to a temperature of 5200 K (kT = 1/2.24 eV), and for oxygen lines, to 6100 K (kT = 1/1.88 eV). It is seen that relation (1) for hydrogen is close to linear, suggesting that the state of the plasma is close to local thermal equilibrium.

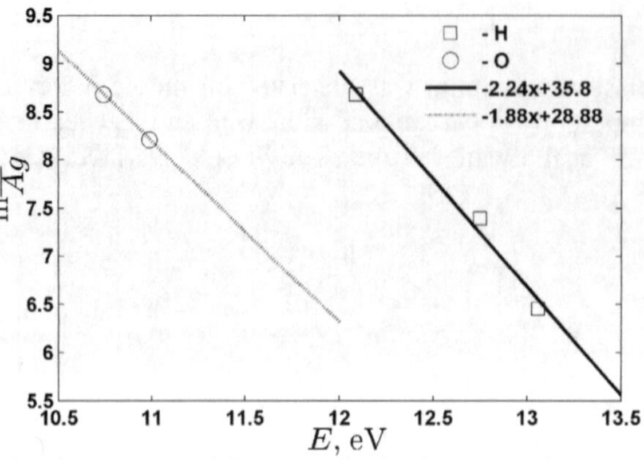

Figure 3. Boltzmann's plot computed from (1) for hydrogen and oxygen spectral lines.

Figure 4 shows the scaled-up lines H_α and H_β. The points correspond to the experimentally obtained spectrum, and the solid line shows the Gaussian function approximation. According to [5], at a Stark width of hydrogen lines from 0.6 to 0.8 nm, the electron density is from 10^{16} to 10^{17} cm^{-3}.

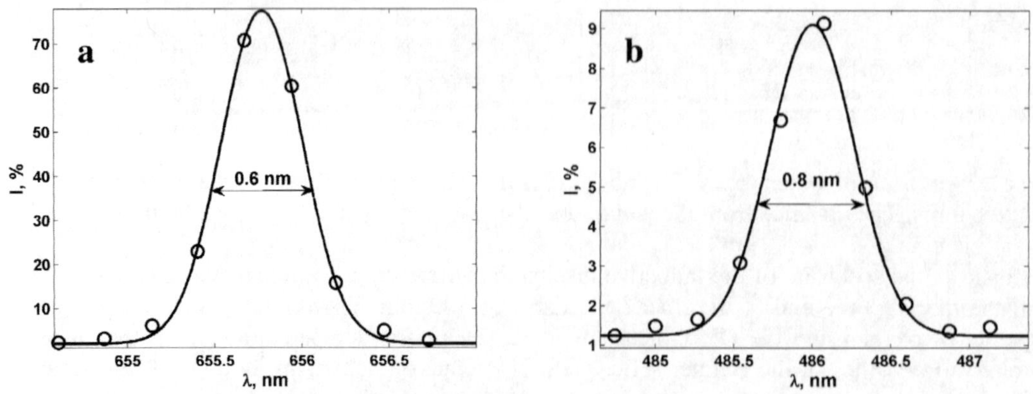

Figure 4. Profile and width of lines. (a) for H_α and (b) for H_β.

Figure 5 shows a portion of the discharge spectrum with a distance between the fiber optic cable and the quartz tube of 10 mm. The Na doublet line at a wavelength of 589 nm is observed. The Na concentration in the test water determined by the procedure specified by environmental regulations F14.1:2:4.167-2000 is 10 ± 1.4 mg/l.

Since water vapor enters the discharge chamber as a result of water boiling, it can be concluded that sodium ions are removed from the solution with water molecules. The sodium concentration in the water vapor was not determined. It is clear that for the entire composition of the solution to enter the plasma zone, it is necessary to optimize the method its supply. Furthermore, this problem is easily solved if the chamber pressure becomes close to the atmospheric pressure, which can be achieved by increasing the power and frequency of the inductive discharge.

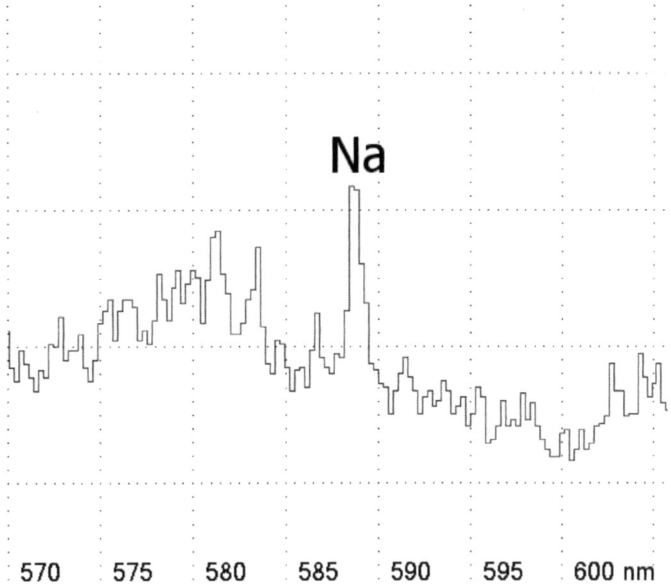

Figure 5. Portion of the water spectrum. The distance between the end of the fiber cable and the quartz tube is 10 mm.

4. Conclusions

The study demonstrated the possibility of initiating an inductive discharge in water vapor at pressures of up to 10^{-2} atm. A plasma discharge temperature from 5000 to 6000 K and a free- electron concentration in the plasma from 10^{16} to 10^{17} cm^{-3} were achieved. Using the Boltzmann plot method, it was shown that the state of the plasma was close to local thermodynamic equilibrium.

The water vapor was found to contain sodium ions, indicating partial evaporation of sodium together with water as the pressure decreased.

After optimization of the method of feeding the aqueous solution into the discharge plasma, inductive discharges in water vapor can be used to determine the content of impurities with low excitation energy, e.g., alkali metals. At the same time, to determine other impurities, it is necessary to increase the plasma temperature. This requires optimizing the characteristics of the discharge chamber and increasing the power input to the inductively coupled plasma, which are the objectives of our future research.

Acknowledgment

This work was supported by the Russian Foundation for Basic Research (Grant No. 16-38-60039).

References

[1] Hill S J 2007 *Inductively Coupled Plasma Spectrometry and its Applications* (Oxford: Blackwell Publishing) p 427

[2] Poluektov N S 1966 *Techniques in Flame Photometric Analysis* (NY: Van Nostrand Reinhold) p 219

[3] Medvedev R N, Zarubin I A and Shokolov A D 2014 Excitation of spectra of metals during discharge in electrolyte *Advances in Nonequilibrium Processes: Plasma, Combustion, and Atmospheric (collected papers)* (Moscow: Torus press) 106–11

[4] Medvedev R N and Sodnomay A B 2015 Spectral determination of the content of dissolved copper ions by electric discharge in a capillary *Sovremennaya Nauka* **6**(1) 115–8 (in Russian)

[5] Namihira T, Sakai S, Yamaguchi T, Yamamoto K, Yamada C, Kiyan T, Sakugawa T, Katsuki S and Akiyama H 2007 Electron temperature and electron density of underwater pulsed discharge plasma produced by solid-state pulsed-power generator *IEEE Trans. on Plasma Science* **35**(3) 614–8

Journal of Physics: Conference Series **754** (2016) 112001 doi:10.1088/1742-6596/754/11/112001

Experimental and computational study and development of the bituminous coal entrained-flow air-blown gasifier for IGCC

N A Abaimov*, P V Osipov and A F Ryzhkov

FSAEI HE "Ural Federal University named after the first President of Russia B. N. Yeltsin", Ekaterinburg, Russia

*nick.sum41@mail.ru

Abstract. In the paper the development of the advanced bituminous coal entrained-flow air-blown gasifier for the high power integrated gasification combined cycle is considered. The computational fluid dynamics technique is used as the basic development tool. The experiment on the pressurized entrained-flow gasifier was performed by "NPO CKTI" JSC for the thermochemical processes submodel verification. The kinetic constants for Kuznetsk bituminous coal (flame coal), obtained by thermal gravimetric analysis method, are used in the model. The calculation results obtained by the CFD model are in satisfactory agreements with experimental data. On the basis of the verified model the advanced gasifier structure was suggested which permits to increase the hydrogen content in the synthesis gas and consequently to improve the gas turbine efficiency. In order to meet the specified requirements vapor is added on the second stage of MHI type gasifier and heat necessary for air gasification is compensated by supplemental heating of the blasting air.

1. Introduction

The coal-fired power industry is widely spread all over the world due to the solid fuels cheapness and availability. The integrated gasification combined cycle units (IGCC) are the most advanced units among the known technologies of solid fuel use [1]. The near zero values of the atmospheric emission of pollutants along with the relatively high electrical efficiency (up to 50-55%) are its basic advantages [2]. As opposed to traditional thermal power plants where coal is burned in steam generation boiler, in IGCC coal is transformed in synthesis gas in the gasifier, the synthesis gas in its turn is a fuel for combustion chamber of the gas turbine unit (GTU). Thus the IGCC operation parameters are generally defined by the solid fuel gasification mode. The basic requirements imposed to the entrained-flow IGCC gasifier are as follows:

1) high efficiency (including cold gas efficiency);
2) elevated carbon conversion rate;
3) minimum pollutant formation rate;
4) operational reliability;
5) generation of the synthesis gas with composition required for GTU high performance operation.

For development of the gasifier which would completely meet the requirements mentioned above it is necessary to carry out the integrated studies, during which the following items should be analyzed:

1) fuel composition, including its kinetic characteristics and ash properties (composition, melting temperature, viscosity etc.);

2) thermochemical processes (TCP) occurring in the gasifier, that is coal drying, pyrolysis, heterogeneous reactions;
3) aerodynamics of the gasifiers interior space;
4) impact of the feeding staged organization on the gasifier operation parameters;
5) features of the unit walls slagging during operation on the project fuel.

The investigation of the fuel properties can be performed by several methods: pressurized drop tube furnace, fluidized-bed process, thermal gravimetric analysis (TGA) etc. The present methods permit to define some TCP regularities but due to the differences in laboratory and industrial units' conditions the actual TCP nature can be understood only by means of studies conducted on the pilot and near-commercial gasifiers. However the measurements' conducting in the entrained-flow gasifiers is rather difficult due to the design features, high pressure (up to 3 MPa) and high temperature (up to 1700°C) in the gasification chambers. Therefore in recent times the physical experimental studies are often combined with the mathematical modeling methods, in particular the Computational Fluid Dynamics (CFD) method which is the most effective one [3]. The entrained-flow gasification models created by means of CFD method consist of several submodels: TCP, turbulence, radiation, convection, dispersed phase motion, slagging etc.

In order to provide the modeling data adequacy all submodels should be verified by means of the experimental data. The TCP submodel is the most important one and at the same time it is the least proven submodel [4]. During the TCP submodels creation and adjustment it is required to use the data obtained at the laboratory and industrial units.

2. Experiment description

The entrained-flow pressurized gasifier designed by "NPO CKTI" JSC was used for the TCP submodel verification and validation through the experimental elaboration of the pulverized coal fuel gasification technology.

2.1. Gasifier design

The vertical down-flow gasifier consists of two chambers: gasification and cooling chambers (figure 1). The gasification chamber with internal diameter of 0.21 m and height of 1.6 m is located in the upper part. The cooling chamber comprising the water quenching section and the slag bath is located below. The double fuel burner is equipped by the swirl-type and uniflow atomizer in case of operation on the air blast and oxygen blast respectively. There is a possibility of the blast supply with necessary air, oxygen and vapor ratio at temperature within the range of 350-550°C.

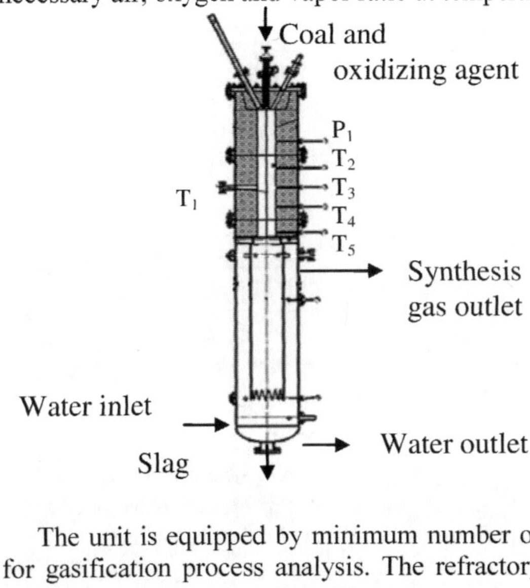

Figure 1. Single stage gasifier designed by CKTI.

The unit is equipped by minimum number of the temperature, pressure and flow sensors required for gasification process analysis. The refractory temperature is controlled throughout the height by means of five platinum platinum-rhodium thermocouples T_2-T_5, the flow temperature T_1 and the chamber pressure P_1 are also controlled. The synthesis gas composition (CO, CH_4, H_2, O_2) is defined at the gas chromatograph.

2.2. Experiment technique

The operation parameters values vary with time in one of the operation modes during experiment. The synthesis gas composition determination method by means of chromatograph assumes that the probe sampling is performed at a given instant therefore it is important to know the basic parameters of the mode in present time. The primary experimental data permits to provide it even in nonstationary conditions of unit operation. The thermal energy necessary for endothermic reactions in the gasifier is supplied to coal as a result of its partial combustion and also from the fire-resistant insulation walls. The high temperature flame core is generally located in the first half of the gasification chamber near the burner; therefore the brickwork temperature possesses the maximum values (figure 2).

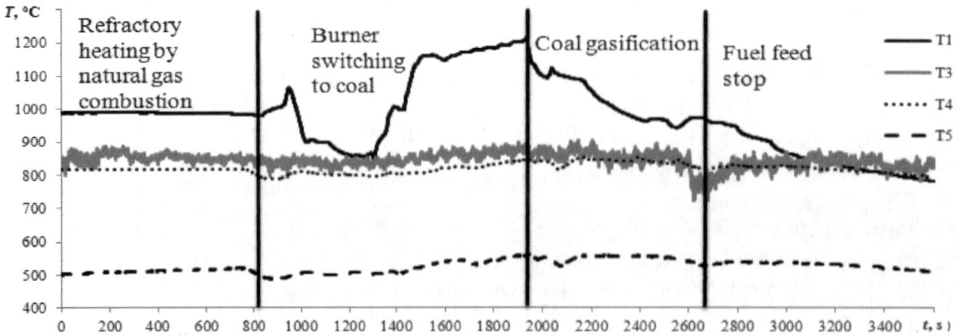

Figure 2. Temperature history, T_1 is the flow temperature, T_3-T_5 are the refractory temperatures.

The set of experiments analyzed in the paper was carried out with the use of Kuznetsk bituminous coal (flame coal). Proximate and ultimate analysis of Kuznetsk bituminous coal (flame coal) and also particle size distribution of pulverized coal are presented in table 1.

Table 1. Proximate, ultimate analysis and particle size distribution of Kuznetsk bituminous coal.

Technical and elemental composition											
Parameter	W^a, %	A^d, %	V^d, %	V^{daf}, %	C^{daf}, %	H^{daf}, %	N^{daf}, %	S^{daf}, %	O^{daf}, %	Q_i^{daf}, MJ/kg	Q_i, MJ/kg
Laboratory values	2.9	23.7	29.9	39.2	78.8	5.97	2.16	0.97	12.1	31.6	21.6
Reference values	4.5	18	31.4	40.5	77.7	5.51	2.62	0.55	13.6	31.9	21.9
Particle size distribution											
Fraction, micron	160		100		80			71		< 71	
Percentage, %	10.25		47.7		26			2		14	

2.3. Experimental results

A set of air experiments (21% of oxygen) with α (stoichiometric factor) = 0.44, 0.53, 0.7 and 1.01 (table 2) is considered in the paper. The coal mass flow was 5-15 kg/h, air flow rate was 20-60 Nm3/h. Pressure in the reaction chamber varied from 0.13 to 0.18 MPa, and blast temperature varied from 420 to 500°C, which corresponds to the compressed blast air temperature in the industrial air gasifiers.

Table 2. Experimental results of the air-blown coal gasification.

Parameter	Case 1	Case 2	Case 3	Case 4
Stoichiometric factor	0.44	0.53	0.71	1.01
Fuel consumption, kg/h	12.5	10.3	15	8.8
Air flow rate, Nm^3/h	31.7	31.5	60.4	51.2
Flow temperature T_1, °C	1095	1126	1291	1022
Brickwork temperature T_2, °C	1155	1151	-	1124
Brickwork temperature T_4, °C	994	987	1123	972
Air temperature, °C	493	502	451	421
Pressure P_1, MPa	0.175	0.131	0.179	0.173
Synthesis gas composition, volume percent				
CO	18.9	16.8	12.9	0.1
H_2	14.2	12.5	8.5	0.3
CH_4	0.16	0.2	0.18	0.1
Lower heating value, MJ/m^3	3.97	3.54	2.54	0.08

The experimental data processing was performed with the use of the rated equilibrium concentrations, defined by the entropy maximization method in the multiple-purpose software system Terra [5]. The rated equilibrium concentrations, presented in figure 3 have weakly nonlinear dependence on α, due to different equilibrium mixture temperatures, which were taken to be equal to the experimental values of the flow temperature 1100-1300°C.

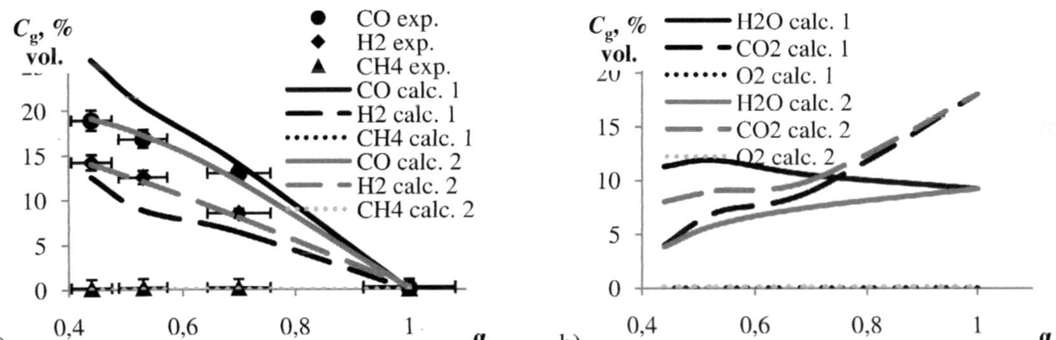

a) b)

Figure 3. Experimental and calculated concentrations of the synthesis gas components in the gasifier: a) CO, H_2, CH_4; b) H_2O, CO_2, O_2.

In the first calculation of the equilibrium concentrations, presented on the diagram, the unburned coal loss and the water gas shift reaction in the water quenching section were not considered. In the second calculation the reagents concentrations necessary for experimental synthesis gas compositions obtaining were defined. The rated unburned coal loss is approximately 10% at $\alpha = 0.44$ and approximately 0% at $\alpha = 1.01$. The quantity of water reacted in the water quenching section exceeded three times the quantity of water in coal at $\alpha = 0.44$. In the complete combustion mode ($\alpha = 1.01$) the water gas shift reaction had insignificant impact on the final synthesis gas composition due to no CO and H_2 in the combustion products.

3. CFD model verification

The CFD model, described in [6] but with consideration of the project fuel kinetic constants (table 3), obtained by the TGA method was used for the analysis of the reverse flows, the pulverized coal fuel motion trajectories and the temperature fields in the chamber analysis. The Shear Stress Transport turbulence model was previously verified with the use of experimental data [7].

Journal of Physics: Conference Series **754** (2016) 112001 doi:10.1088/1742-6596/754/11/112001

Table 3. Kinetic constants of heterogeneous reactions.

Constant	$C + O_2 = CO_2$	$C + H_2O = CO + H_2$	$C + CO_2 = 2\ CO$
A, kg/(m^{2}s)	500	60000	40000
E, kJ/mol	70	160	170

The reverse flows are located in the near-axial zone of the chamber with the peak velocity values near the burner (figure 4a), which makes the process steady. It is seen from the calculated temperature field (figure 4b) that the pulverized coal is retained against the wall by the air twisted by the swirl burner therefore the flame core is located near the reaction chamber wall (figure 4c). The resultant cyclone effect leads to the particles separation, slag tapping and increasing range of the coal burnup.

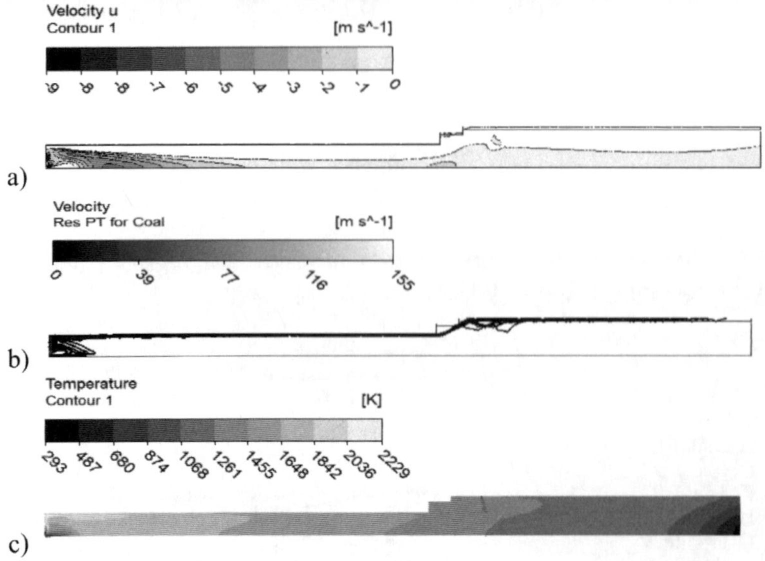

Figure 4. The results of numerical simulation of the entrained-flow air gasification in the gasifier: a) reverse flows velocity (white color means the flow moving to the chamber outlet); b) the pulverized coal motion trajectories on the entrance region; c) temperatures distribution in the segment longitudinal cross section.

In this case CO and H_2 concentrations in the synthesis gas at the outlet from the reaction chamber are the basic verification criteria (figure 5). The calculation results obtained by the CFD model are in satisfactory agreements with experimental data.

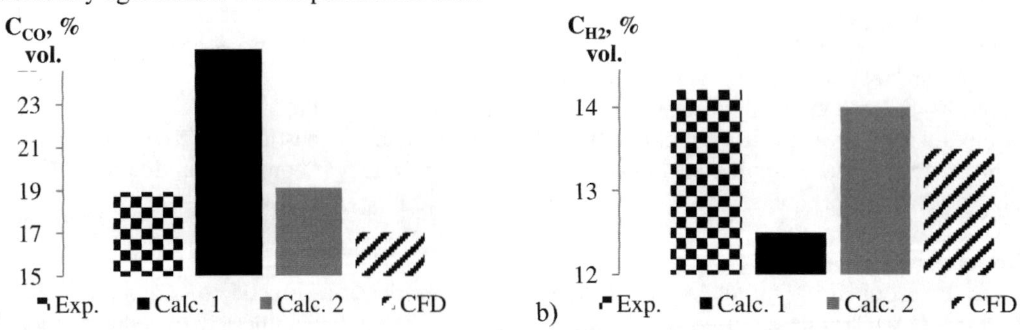

Figure 5. Comparison of CO (a) and H_2 (b) concentrations, obtained during experiment, equilibrium calculations and CFD modeling.

4. Gasifier development

The entrained-flow pressurized air gasifier development was performed with the use of the verified model. The two-stage fuel supply scheme (MHI type) was used due to the low TCP reaction rate in air. With the increase in the water vapor content in the working fluid the GTU efficiency is also increased. For hydrogen content increasing in the synthesis gas the vapor is added on the second gasifier stage and heat, necessary for the vapor gasification is compensated by supplementary heating of the blast air. It allows the gasifier chemical efficiency improvement, the increasing range of the coal burnup and hydrogen content increasing in the synthesis gas up to the required level. The gasifier slagging and pollutant emission impact is planned to be considered in future.

5. Conclusion

1) Complex investigations are necessary to develop state-of-the-art entrained-flow air-blown gasifier for high-efficient IGCC. One of the most difficult issues is the gasifier TCP study. Combination of the experimental and CFD investigations are necessary to solve this problem.
2) Experimental study of gasification technology with initial data acquisition for CFD model verification was performed on the NPO CKTI gasifier. Syngas composition and temperature profiles inside the reactor at different stoichiometric ratio were analyzed. Hydrogen concentration raises and CO concentration decreases due to the water shift reaction in quenching section of gasifier. Unburned carbon level was up to 10%, which leads to combustible gases concentrations decrease.
3) Proposed CFD model consists of calibrate submodels and adjusted kinetic constants for heterogenic reaction that was obtained in TGA experiments. The model was verified with the experimental data and can be used for development of contemporary entrained-flow gasifier for high-efficient IGCC.
4) Two-stage air-blown MHI gasifier is proposed as prototype for further development. Modernization involves high temperature stream steam addition and blast air high-temperature heating.

Acknowledgements

This work was financially supported by the Russian Science Foundation (project no. 14-19-00524). We would like to express gratitude to Shestakov N.S, Shurchalin A.A. and the other staff of «NPO CKTI» JSC for their help in experiments within the framework of the presented project.

References

[1] Clean Coal Research Program, Technology Readiness Assessment - Analysis of Active Research Portfolio (United States Department of Energy, Office of Fossil Energy, December, 2012) / http://www.netl.doe.gov/File%20Library/Research/Coal/energy%20 systems /gasification/TRL-Comprehensive-Report_121112_FINAL_1.pdf
[2] IGCC with higher plant efficiency (Mitsubishi Heavy Industries, LTD., July, 2009) / http://www.jcoal.or.jp/coaldb/shiryo/material/JapanTaiwanCCTWorkshop2009_04.pdf
[3] Lee H-H, Lee J-C, Joo Y-J, Oh M and Lee C-H 2014 Dynamic modeling of Shell entrained flow gasifier in an integrated gasification combined cycle process *Applied Energy* **131** pp 425-40
[4] Vascellari M, Roberts D G, Hla S S, Harris D J and Hasse C 2015 From laboratory-scale experiments to industrial-scale CFD simulations of entrained flow coal gasification *Fuel* **152** pp 58-73
[5] Trusov B G 2002 Software TERRA for computing plasmachemical processes *Materials of 3rd Int. symp. on theoretical and applied plasma chemistry* (Ples: ISUCT) pp 217–8
[6] Abaimov N A and Ryzhkov A F 2015 Development of a model of entrained flow coal gasification and study of aerodynamic mechanisms of action on gasifier operation *Thermal Engineering* **62** pp 767–72
[7] Abaimov N A and Rizhkov A F 2015 Turbulence models verification for numerical Simulation of the cyclone chamber *Proc. of 5th Int. Conf. Heat and mass transfer and hydrodynamics in swirling flows* (Kazan: KAI) pp 429-37

Modeling ohmic heating in the drying zone of the plasma shaft electric furnace, when recycling the technogenic waste

A I Aliferov[2], A S Anshakov[1,2], V A Sinitsyn[1,2], P V Domarov[1,2], A A Danilenko[2]

[1] Kutateladze Institute of Thermophysics SB RAS, Novosibirsk, Russia
[2] Novosibirsk State Technical University, Novosibirsk, Russia

E-mail: Domaroff@yandex.ru, anshakov@itp.nsc.ru

Abstract. Efficient use of ohmic heating in the drying zone of the plasma shaft furnace for gasification of organic and technogenic wastes is studied. It is shown that by using ohmic heating in the drying zone, energy release takes place in the filling along the entire zone.

1. Introduction

Now the problem of disposal and neutralization of annually increasing technogenic wastes, including the organic ones, is urgent all over the world. One of the effective ways to solve this problem is waste gasification in the plasma shaft furnaces. However, they have significant disadvantages: high energy consumption (up to 1 MW·h per 1 ton of waste), and short service life of plasmatron electrodes [1]. Therefore, the development of new approaches to the creation of electrotechnological plasma installations with improved power parameters for the problems of technogenic waste (TW) processing is very popular nowadays. These energy-efficient installations are the electric furnaces with combined plasma-ohmic heating, used for TW processing.

During the processing, the waste are wet (in real conditions, their humidity is up to 60%) in the zone of their supply only within the drying zone. Waste should be fed dehydrated and dried to the pyrolysis zone. Therefore, ohmic heating by the current of commercial frequency will be implemented only in the drying zone. The shape of the drying zone cross-section is selected to be square, to ensure uniform power distribution $q_{\mathrm{Vep}}(x)$ over the waste charge cross-section.

2. Electromagnetic processes in the drying zone of electric furnace

According to analysis of publications, in summer the wet solid domestic waste show such a property as specific electric resistance, whose value is $\rho_{el}=3\div5$ Ohm·m. Moreover, depending on humidity and morphological composition, this value can differ significantly from the value presented above. Therefore, in model calculations it was assumed that $\rho_{el}=1\div10$ Ohm·m.

While modeling, it was taken into account that wastes contain moisture only in the loading and drying zones until it evaporates due to the temperature. Thus, ohmic heating by the current of commercial frequency can be implemented only in the drying zone. Wastes are fed to the pyrolysis zone in the dehydrated and dry form with the specific electric resistance, which is ten times higher than in the beginning of the drying zone.

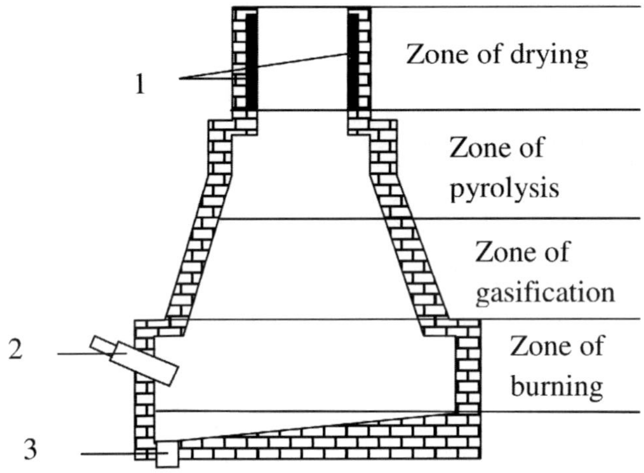

Figure 1 Scheme of plasma-ohmic furnace:
1 – electrodes;
2 – plasmatron; 3 – liquid slag drain

The shape of the drying zone cross-section was chosen square, to ensure uniform distribution of power $q_{Vep}(x)$ over the waste charge cross-section.

2.1 Mathematical model

When organizing a continuous process are formed the established dynamic and thermal regimes in the furnace chamber. Therefore, heat and mass transfer processes in these conditions, the countercurrent solid and gaseous phases can be considered stationary.

While developing the model of electromagnetic processes, some assumptions were made: 1. We neglect the displacement currents. 2. The electromagnetic field is alternating, changing sinusoidal in time. 3. There are no free discharges in the computational domain. 4. The process of mass transfer in the elementary layer of the charge believe quasi-stationary.

The differential equation of are the mathematical description of the processes of electromagnetic field, continuous in space and time, in the drying zone of the furnace.

When forming the electrical circuit "electrode - charge – electrode", the value of current introduced into the system on the face surface of electrodes has been set (calculation was carried out at the stabilized alternating current of the commercial frequency).

The drying zone is produced in different sections of the removal of water vapor into the lower zone of the combustion furnace.

The assigned task is solved using the software complex ANSYS by the method of finite elements.

3. Results of mathematical modeling

The output of integral power in a charge with passing electric current was calculated by the software complex ANSYS. The range of voltage applied to the electrodes was U=50÷250V and current I = 100A. In the filling, specific electric resistance technogenic waste was set as humidity linear function $\rho_{el}=f(W)$, in the range of electrical resistivity ρel = 1 ÷ 10 Ohm·m and waste humidity 60 ÷ 15%.

Dependence of integral power, allocated to the elementary volume release along the height of drying zone at ohmic heating on voltage applied to the electrodes is shown in Figure 2.

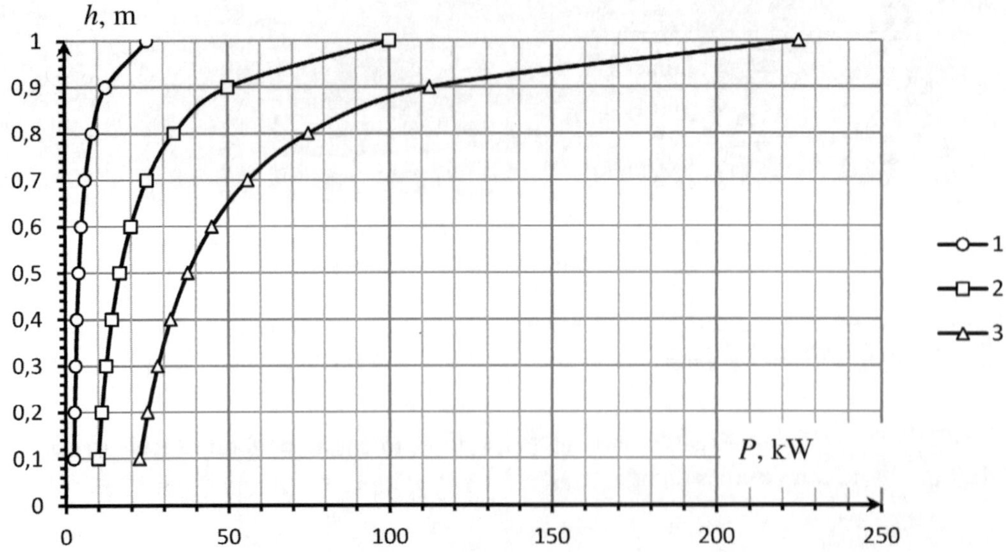

Figure 2. Power output along the drying zone: 1 – U=50V;
2 – U=100V; 3 – U=150V.

It can be seen that when increasing the specific electric resistance of waste, the power output in the filling decreases along the height. This is due to liquid evaporation from technogenic wastes, while they move down the shaft. As it can be seen from the diagram, at the level of 0.5 m from the edge of the loading window, the released power does not change, and this suggests the possibility of reducing the geometric dimensions of electrodes (decreasing their length) from 1m to 0.5m.

As it was shown in [2, 3], specific power inputs for technogenic waste processing depend significantly on their humidity. The calculation-experimental dependence determining a link between the costs required for technogenic waste processing and their humidity is shown in Figure 3.

During the drying process under the action of ohmic heating, moisture evaporates from wastes, decreasing from (60 ÷ 50)% to 20%. Wastes with minimum humidity of 20% contain all chemical elements required for their complete gasification. As it can be seen in Figure 3, while reducing waste humidity from 50 to 30%, the specific power inputs for gasification reduce considerably (almost twice - from 0.75 to 0.4 kW•h/kg). Thus, at combined plasma-resistive heating of technogenic waste charge, we achieve a significant reduction in specific power inputs using heat supply only from the plasmatron for processing 1kg of technogenic waste: less than 0.4 kW • h/kg (see Figure 3.).

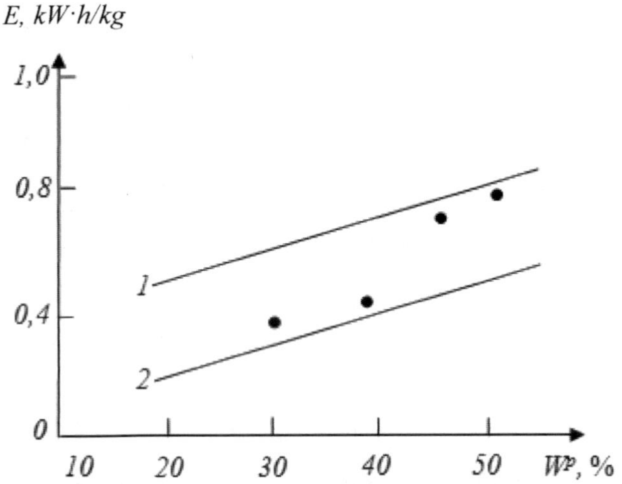

Figure 3. Dependence of specific power inputs on waste humidity

An example of the ohmic heating effect on decreasing the plasmatron loading is as follows. For plasma furnace productivity G_m=100 kg/h, the required plasmatron power is 140kW. Reducing waste humidity to 29% leads to a decrease in plasmatron power to 50 kW. Therefore, additional power input through resistive heating P = 140-50 = 90 kW in the drying zone allows reduction in plasmatron power by 58% for the given furnace efficiency.

4. Conclusions

Distribution of produced power along the height was calculated taking into account the dependence of resistance on humidity. They showed a significant change in the characteristic, and this should obligatory affect heat transfer inside the shaft.

While analyzing the results and summing them up, it should be noted that ohmic heating of wet wastes in the upper part of the shaft electric furnaces is an analogue of preliminary technological process of waste drying before loading them into the furnace, aimed at reduction of specific power inputs for gasification of organic component of the mixed wastes. This also reduces the plasmatron loading, increasing the service life of plasmatron electrodes.

References

[1] Cherednichenko V S, Anshakov A S, Kuzmin M G Plasma electrotechnological devices / - Novosibirsk: NSTU, 2011.- 602p.

[2] Kuzmin M, Cherednichenko V, Anshakov A, Aliferov A, Domarov P, Radko S, Urbakh E and Faleev V 2012 Plasma electric furnace for processing/utilization of carbon-bearing anthropogenic wastes (Proceedings of the XVII Congress) pp 127-131.

[3] Aliferov A I, Anshakov A S and Sinitsyn V A 2009 Numerical modeling of heat transfer in plasma shaft electric furnace at utilization of anthropogenic waste *Thermophysies and Aeromachanics Vol. 16 No. 1* pp 155-161

Journal of Physics: Conference Series **754** (2016) 112003 doi:10.1088/1742-6596/754/11/112003

Analytical calculation of adiabatic processes in real gases

I B Amarskaja, V S Belousov, P S Filippov
Ural Federal University, 19 Mira street, Ekaterinburg, Russia
E-mail: tot@urfu.ru

Abstract. The impact of gases nonideality in the compression and expansion processes on the specific heat ratio and the heat capacity is analyzed. The specific heat ratio variation leads to temperature variation during compression in the compressor and expansion in the turbine and, consequently, the gas turbine cycle efficiency factor variation. It is also essential to consider the gases nonideality in the compression and expansion processes in the compression processes in compressor. Generally it is assumed during calculations that the heat capacities depend only on temperature, in this case the reference data presented by various authors differs markedly. In the real processes the heat capacity and the specific heat ratio depend on temperature and within the particular temperatures and pressures range depend on pressure. Consequently, the operating fluid nonideality in the gas turbine cycle should be considered.

1. Introduction

The processes of gases compression in the noncooled compressor and expansion in the turbine can be assumed as adiabatic to a good approximation. Usually during the thermodynamic analysis of the processes mentioned above it is assumed that the operating fluid is an ideal gas with the thermophysical properties depending only on temperature.

In real processes the heat capacity and, consequently, the specific heat ratio depend heavily on temperature and within the particular temperatures and pressures range depend on pressure, it means that the operating fluid nonideality in the gas turbine cycle should be considered.

The thermodynamic efficiency improvement in the modern gas turbine units is achieved by means of the increase in the compression pressure ratio in the compressor and the combustion products' temperature rise before the turbine, which can result in the necessity of the gas nonideality consideration. For instance, the change of 2% in the specific heat ratio as compared to the k value being equal to 1.4 at the compression pressure ratio $\beta = 25$,at the temperature before the turbine of 1600 K, at the relative efficiency of the compressor and the turbine $\eta_{ri}^{c} = \eta_{ri}^{t} = 0,85$ results in the change of 35 K and 30 K in temperature at the end of the compression process in the compressor and at the end of the expansion process in the turbine respectively, in this case the gas turbine cycle efficiency factor changes in 2%. Therefore the accurate determination of the specific heat ratio values is of high importance.

2. The pressure dependence of the specific heat ratio for the nonideal gas

Usually during the calculations the heat capacity is taken to be not depending on pressure and the temperature dependences at the pressure of 0.1 MPa, presented in the guidelines [1-3] are used. It should be mentioned that sometimes the reference data presented by various authors differ markedly, as it is seen in Fig.1, where the dependences mentioned above are given for air.

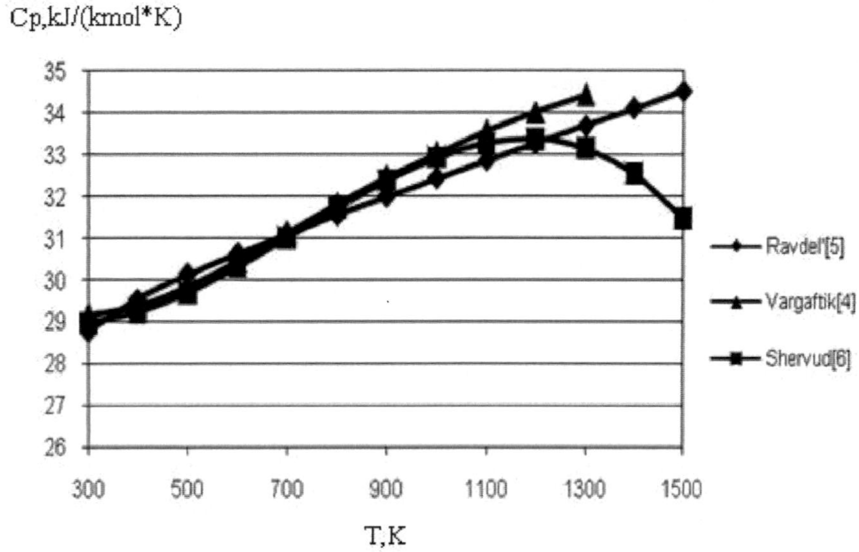

Figure 1. Temperature dependence of the molar heat capacity at constant pressure of 0.1 MPa

The pressure dependence of the air heat capacity was approximated in [4] by the polynomials, obtained by the table data processing, but the heat capacity at constant volume, which is included in the specific heat ratio expression was calculated by the Meyer relation $c_p - c_v = R$ being true only for the ideal gas.

The present expression is not appropriate for the real gas and the relation between the heat capacities is written as

$$c_p = c_v + T\left(\frac{\partial p}{\partial T}\right)_v\left(\frac{\partial v}{\partial T}\right)_p = c_v - T\left(\frac{\partial p}{\partial v}\right)_T\left(\frac{\partial v}{\partial T}\right)_p^2$$

The derivatives in the specific heat expression can be calculated only if the real gas state equation is known.

The pressure dependence at different temperatures was considered with a correction $\varphi(p,T)$ to the temperature dependence of the heat capacity at pressure of 0.1 MP, $c_p^0(T)$,

$$c_p(p,T) = \varphi(p,T) \times c_p^0(T). \tag{1}$$

The correction to the pressure dependence was approximated by data [3] as follows

$$\varphi(p,T) = A(p) \cdot \exp\left[(0{,}1 - p)\Theta \cdot \sum_{i=0}^{4} a_i(p)\Theta^i\right], \tag{2}$$

$$\text{where } \Theta = \frac{T - 250}{100}, \quad A(p) = b_0 + b_1 p, \quad a_i(p) = \sum_{j=0}^{2} \alpha_{ij} p^j . \tag{3}$$

The analytical temperature dependence at pressure of 0.1 MPa was accepted in accordance with [3]:

$$c_p^0(T) = c_0 + c_1 T + c_2 T^2 + \frac{c_3}{T^2}$$

In this case the Redlich-Kwong and Diterichi equations [2] are the most adequate equations [2]. The deviation from ideality can be also taken into consideration by the compressibility factor $z = \frac{pv}{\mathrm{R}T}$, the analytical expression of which can be obtained by the reference data processing [3]. The processing results for air are presented in Fig. 2.

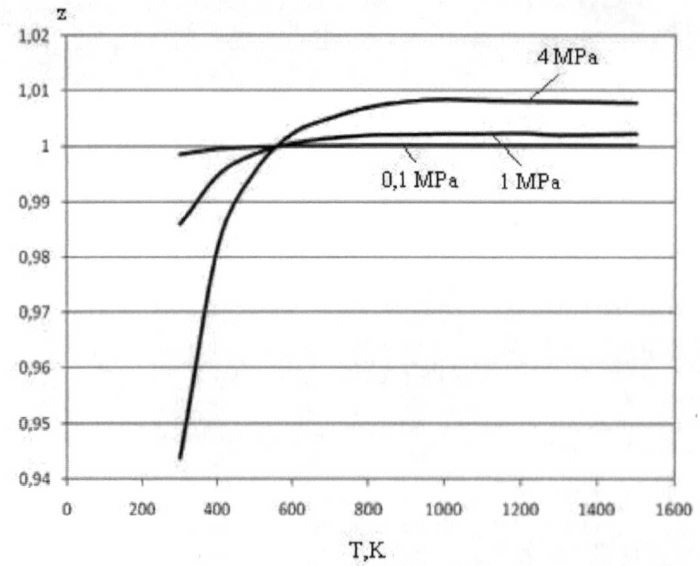

Figure 2. Temperature dependence of the air compressibility at different pressures

In Diterichi state equation $p(v-b)\cdot\exp\left(\frac{a}{\mathrm{R}Tv}\right) = \mathrm{R}T$ the corrections for the interactions between molecules and their volume, defined from the condition that the first and the second order derivatives are equal to zero $\left(\frac{\partial p}{\partial v}\right)_T^{\mathrm{cr}} = \left(\frac{\partial^2 p}{\partial v^2}\right)_T^{\mathrm{cr}} = 0$ in the critical point, are as follows

$$a = 2\mathrm{R}T_{\mathrm{cr}}v_{\mathrm{cr}}, \quad b = \frac{v_{\mathrm{cr}}}{2}.$$

The results of the specific heat ratio calculation by the Diterichi equation are presented in Fig.3.

The nonideality consideration at the pressure of 4MPa results in the specific heat ratio change from k=1.4 to k=1.5 value, which results in the temperature increase from 832 to 932 K at the end of compression at the compression pressure ratio $\beta = 25$ and at the relative efficiency of the compressor $\eta_{ri}^{\mathrm{c}} = \eta_{ri}^{\mathrm{t}} = 0,85$.

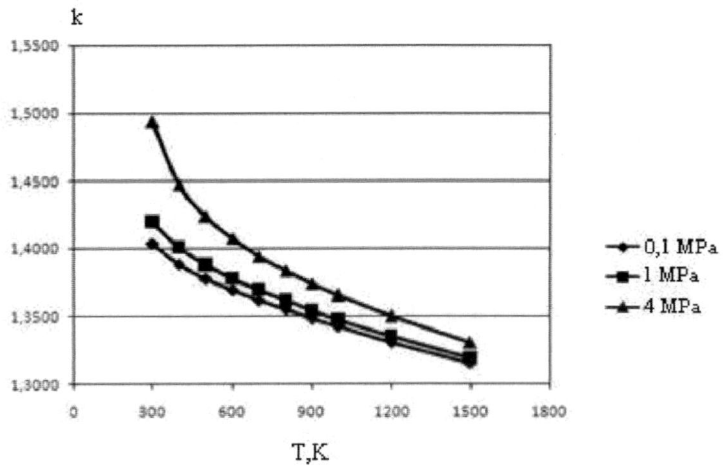

Figure 3. Temperature dependence of the air specific heat ratio at different pressures

The pressure dependence of the heat capacity is sufficient for temperatures less than ~ 700 K, which are typical for the end of compression processes in the compressor.

3. Conclusion

Thus, the accurate determination of the specific heat ratio values with consideration of both temperature and pressure dependences of the heat capacity is of high importance. It is particularly true for the compression processes in the compressor as far as at the temperatures typical for the expansion processes in the turbine the pressure dependence is not so essential and the gas can be assumed to be ideal, and in this case only the temperature dependence of the heat capacities can be considered.

This work was carried out at UrFU and financially supported by the Russian Science Foundation (project number 14-19-00524).

References

[1] Rid R, Prausnots J and Shervud T 1982 *The gases and liquids properties* (L.:Chemistry) p 592

[2] Vargaftic N B 1963 *The guidelines for thermal and physical properties of gases and liquids* (M. Physmathgiz) p 708

[3] Ravdel A 1983 *Quick reference of the physical and chemical values edited by A.A. Ravdel and A.M. Ponomareva* (L. Chemistry) p 232

[4] Hasanov N G and Shigapov A B 2014 The real air properties impact on the stationary gas turbine units performance (*The power industry problems* N 9-10) p 11-20

Journal of Physics: Conference Series **754** (2016) 112004 doi:10.1088/1742-6596/754/11/112004

Simulation of permafrost changes due to technogenic influences of different ingeneering constructions used in nothern oil and gas fields

M Yu Filimonov[1,2] **and N A Vaganova**[1,2]

[1] Ural Federal University, Yekaterinburg, Russia
[2] IMM UB RAS, Yekaterinburg, Russia

E-mail: fmy@imm.uran.ru

Abstract.
Significant amount of oil and gas is producted in Russian Federation on the territories with permafrost soils. Ice-saturated rocks thawing due to global warming or effects of various human activity will be accompanied by termocarst and others dangerous geological processes in permafrost. Design and construction of well pads in permafrost zones have some special features. The main objective is to minimize the influence of different heat sources (engineering objects) inserted into permafrost and accounting long-term forecast of development of permafrost degradation due to different factors in particular generated by human activity. In this work on the basis a mathematical model and numerical algorithms approved on 11 northern oil and gas fields some effects obtained by carrying out numerical simulations for various engineering systems are discussed.

1. Introduction

Regions of permafrost take about 75% of territory of Russia. Preservation of ecological balance is an important problem depending both of climate changes and human activity [1]. These regions are intensively involved in oil and gas production and these efforts have a significant effect on permafrost due to heat from tubes with oil in the wells leads to thawing and possible hazards related with distruction of soils [2, 3]. Not only well and pipelines but also there are a number of buildings, constructions and tanks, which serving the oil production may lead to permafrost degradation. These processes generate significant difficulties in construction and operation of various engineering structures. Most of buldings in permafrost regions are designed such a way to decrease the thermal influence on the soil.

Numerical simulation is an effective method to estimate the results of thermal influence of objects inserted in permafrost and allow to choose more safe insulations and construction. In accordance with [4, 5] simulation of heat distribution in permafrost is a problem of solution of three-dimensional thermal conductivity equations with nonhomogeneous coefficients, including localized specific heats of phase transition — an approach that allows to solve the problem of the Stefan type without explicit separation phase transition boundary. In papers [6–8] are presented a mathematical model of heat distribution with nonlinear boundary condition, numerical algorithm and codes including with parallel computing approach [9]. In [10] this model is also consider filtration of fluid in soil as heat conduction mode. Let consider application of this to solution of a problem of constructing and exploitation of some reservoirs in permafrost area.

2. Object of simulation

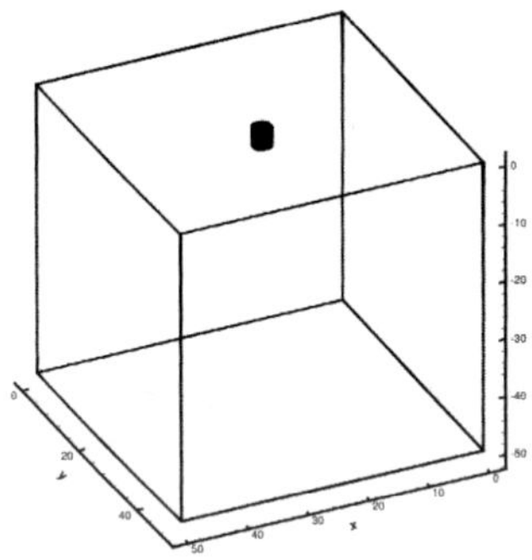

Figure 1. The area with reservoir.

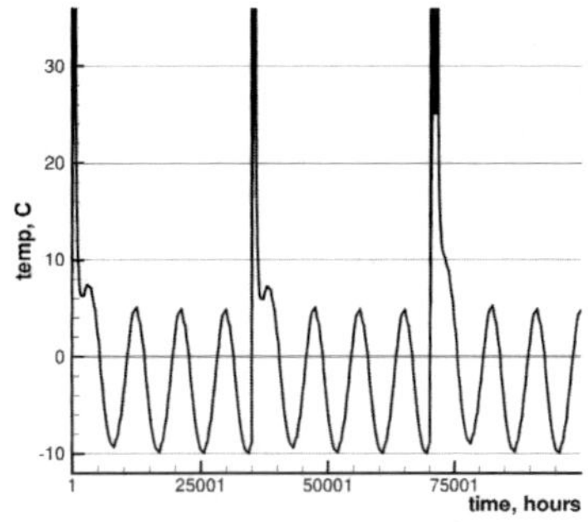

Figure 2. Changes of temperature in reservoir, °C.

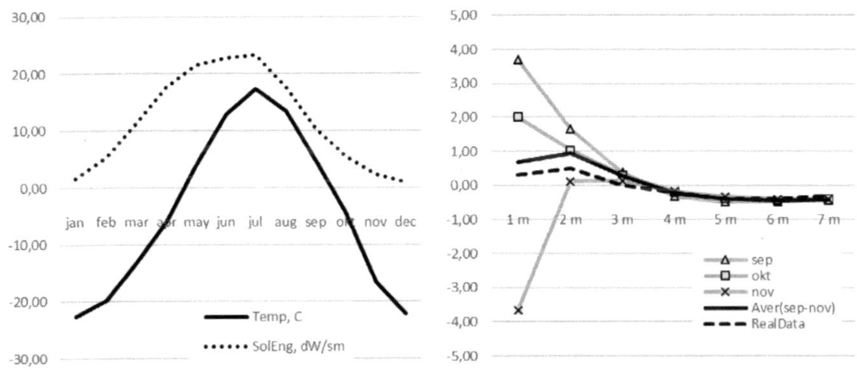

(a) (b)

Figure 3. (a) — montly average air temperature (solid) and solar radiation (dotted). (b) — temperature in soil on autumn in measured (dashed) data and computed data (solid lines).

In a three-dimensional area (fig. 1) a tank which has cylindrical form is considered as a heat source. To simulate long-time influence of this object inserted into permafrost soil we will compute an annual thermal field in the soil taking into account a number of essential climatic factors related with the place under consideration [3, 8]. In Fig 3(a) montly average air temperature and intensity of solar radiation are presented. Temperature of permafrost except for the layer of seasonal thawing (freezing) of soil (ALT), is assumed equal to -0,3°C.

The equation of heat conductivity has the form

$$\rho(c_\nu(T) + k\delta(T - T^*))\frac{\partial T}{\partial t} = \nabla\left(\lambda(T)\nabla T\right), \tag{1}$$

with initial condition

$$T(0, x, y, z) = T_0(x, y, z), \tag{2}$$

where ρ is density $[kg/m^3]$, T^* is temperature of phase transition $[K]$,

$$c_\nu(T) = \begin{cases} c_1(x,y,z), & T < T^*, \\ c_2(x,y,z), & T > T^*, \end{cases} \text{ is specific heat [J/kg K]},$$

$$\lambda(T) = \begin{cases} \lambda_1(x,y,z), & T < T^*, \\ \lambda_2(x,y,z), & T > T^*, \end{cases} \text{ is thermal conductivity coefficient [W/m K]},$$

$k = k(x,y,z)$ is specific heat of phase transition, δ is Dirac delta function.

The boundary condition at the upper boundary are

$$\gamma q + b(T_{air} - T(x,y,0,t)) = \varepsilon\sigma(T^4(x,y,0,t) - T_{air}^4) + \lambda\frac{\partial T(x,y,0,t)}{\partial z}, \tag{3}$$

and includes climatic the parameters such as solar radiation q, air temperature T_{air}, a part of energy γ for the soil heating and coefficient of emissivity of the surface $\varepsilon\sigma$. At the bottom and lateral boundaries the zero flux conditions are assumed.

Numerical methods of solving problems are the most effective and universal method of research for models considered in this paper. A large number of works is devoted to development of difference methods for solving boundary value problems for the heat equation To solve (1)–(3) a finite–difference method is used.

With using these ideas [4, 5], to solve problem (1)–(3) in three-dimensional box a finite difference method is used with splitting by the spatial variables and taking into account the inner boundaries of the reservoir. Solvability of the same difference problems approximating is proved in [6] in the case of thermal traces of of underground pipelines without phase transition in soil [11].

In the condition (3) the values of intensity of solar radiation and seasonal changes in air temperature are given by weather stations. Fig. 3(a) shows the data for the considered field. The soil parameters in (1) are determined as a result of geophysical research of oil and gas field.

Applying the developed iterative algorithm [7, 12] to define some of the parameters in nonlinear boundary condition (3) it is possible to identify them so that the temperature distribution in the soil found as a solution of equation (1)–(3) to be periodically repeated, that allows to implicitly take into account different climate and natural features of the considered geographical location.

To approve the model and algorithm of computations it is necessary to compare the numerical results and measured data. In Fig. 3(b) a real temperature distribution in soil on autumn is presented (dashed black line) in compare with computed data (gray and black solid lines), which are in good agreement and the temperature distribution periodically repeated. Moreover, the thickness of ALT in measures reaches 5,6–6,0m in different incisions, in computations it takes approximately 5,7–5,8m.

3. Numerical results

Consider a computational area having the form of a box (fig 1). As a basic soil we will use a loam with the following parameters. Thermal conductivity: frozen — 2,79 W/(m K), melted — 2,67 W/(m K), volumetric heat: frozen — 2176 kJ/(m^3 K), melted — 2594 kJ/(m^3 K), volumetric heat of phase transition — $1.376 \cdot 10^5$ kJ/(m^3 K). The background temperature of permafrost is -1.5C, except for the layer of seasonal thawing (freezing) of soil (ALT). Soil is a sandy loam with parameters:

We will consider 3 types of tanks:

I-st tank is an open-top reservoir with 10m diameter, deep is 4m. The shell consists of two layers: 0,3m of concrete (thermal conductivity — 1,69 W/(m K) density — 2500,0 kg, specific heat — 840,0 J/(kg K)) and 0,072m of penoplex (thermal conductivity — 0,47 W/(m K) density — 1600,0 kg, specific heat — 840,0 J/(kg K)).

II-nd tank is an underground reservoir with 3m diameter, deep is 4m and height is 2,3m. The shell consists of two layers: 0,1m of concrete and 0,072m of penoplex.

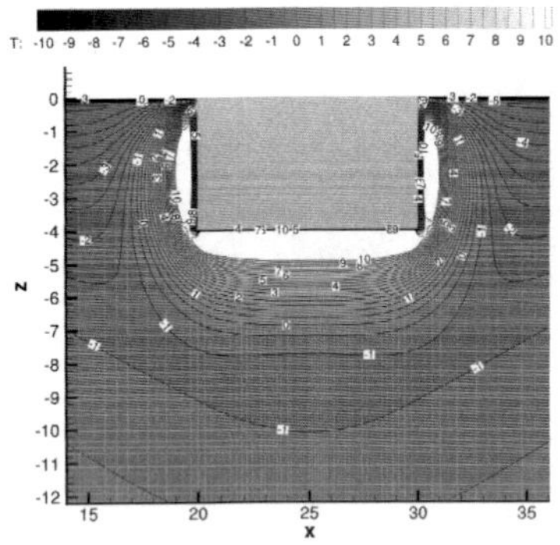

Figure 4. Temperature around I-st tank. May.

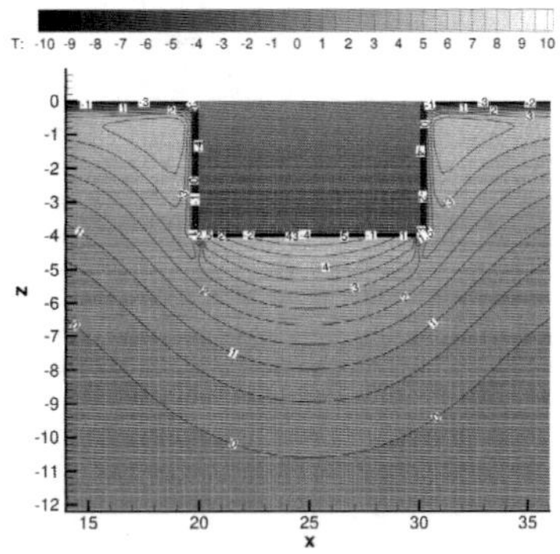

Figure 5. Temperature around I-st tank. October.

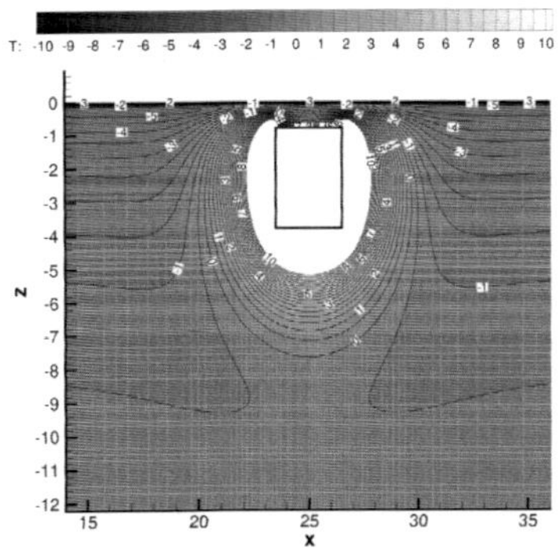

Figure 6. Temperature around II-nd tank. May.

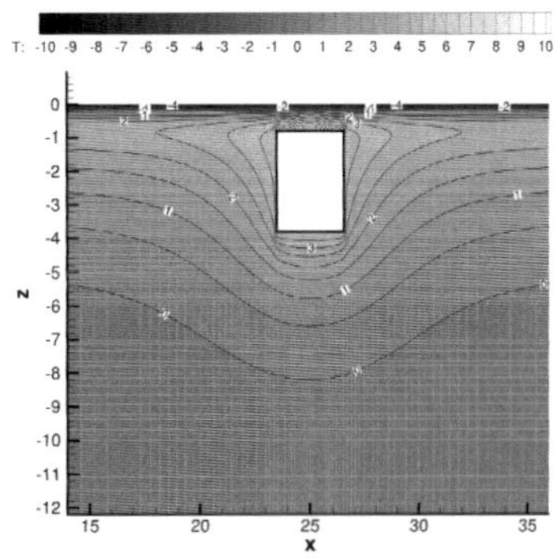

Figure 7. Temperature around II-nd tank. October.

III-rd tank is an underground reservoir with 3m diameter, deep is 4m and height is 2,3m. The shell consists of two layers: 0,1m of concrete and 0,288m of penoplex.

We will assume that temperature in the tank varying as follows: one months (may) in 4 years it is in using (is heated up to $50°C$), and all the rest of the time it takes air (I-st tank) or soil (II-nd and II-rd tanks) temperature. In Fig. 2 the temperature in the tanks are shown during the time.

In Figs. 4–9 the thermal fields in a vertical section of computational domain with tanks are presented for may when the reservoirsare used (figs. 4,6,8) and for october (figs. 5,7,9). Deep of thawing (zero-isotherm) depends not only by insulating of the tanks (II-nd and III-rd tanks), but also and by the size of reservoir (I-st tank).

It is necessary to note that when the insulating layer is larger, it is not always good for conservation of permafrost. In Fig. 10 the profiles of temperature under II-nd (fig. 10a) and III-rd (fig. 10b) tanks are presented. Indeed, for much better insulated III-rd tank the heat

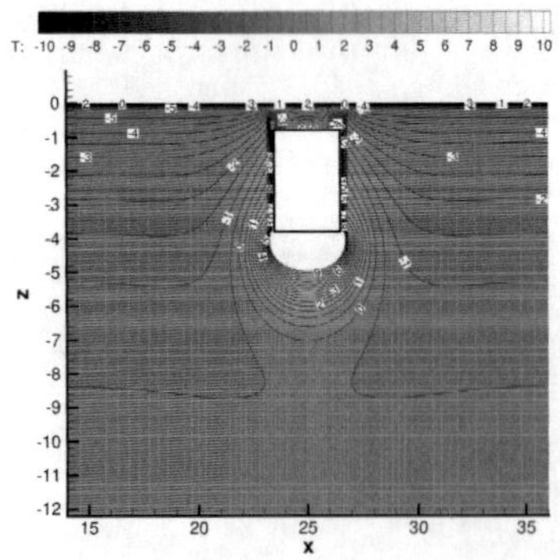

Figure 8. Temperature around III-rd tank. May.

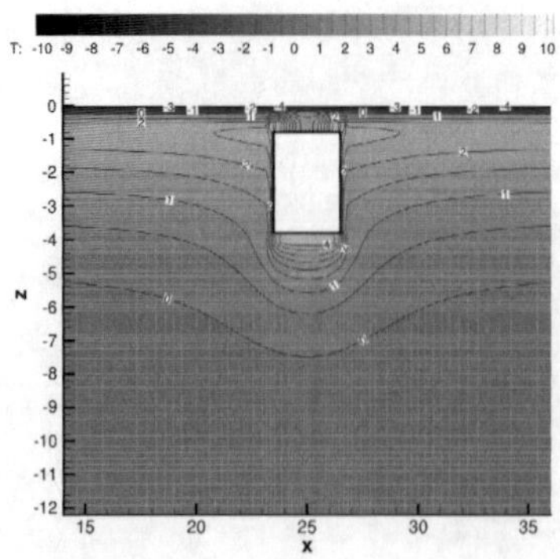

Figure 9. Temperature around III-rd. October.

propagates not so intensively as for II-nd tank, but in long-time perspective, the thick layerof insulation prevent natural heat diffusion and cooling of the soil. So that an overheating of the bottom takes place.

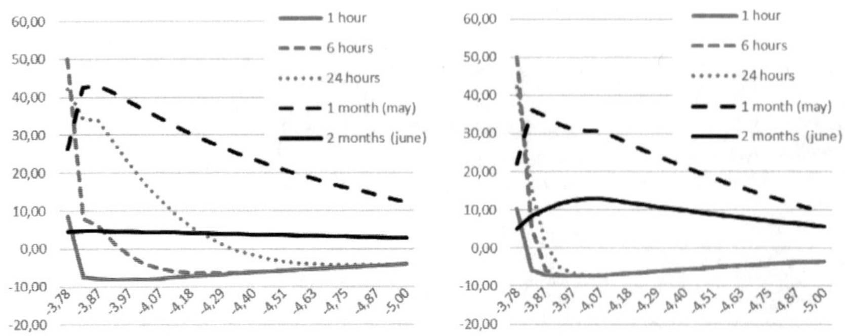

(a) (b)

Figure 10. Temperature under tank.

4. Conclusion

The suggested mathematical model of heat distribution and numerical algorithm of solution and program code allows to carry out a series of numerical experiments to make predictions about long-term dynamics of permafrost thawing related with different regimes of exploitation of engineering constructions. These methods are approved by the comparison with experimental data obtained during a number of exploitations of oil and gas field. The results of numerical simulations allow to evaluate the effectiveness of thermal stabilization of soil.

4.1. Acknowledgments

This work was supported by RFBR projects (16-01-00401, 14-01-00155) by contract 02.A03.21.0006 (reg.N211 of RF Government) and Program of UB RAS, project 15-7-1-13.

References

[1] Nelson F E, Anisimov O A and Shiklomanov N I 2002 *Natural Hazards* **26**(3) 203

[2] Zhang T, Barry R G, Knowles K, Heginbottom J A and Brown J 2008 *Polar Geography* **31**(3) 47
[3] Filimonov M and Vaganova N 2014 *Academic Journal of Science* **3** 121
[4] Samarsky A A and Moiseenko B D 1965 *Computational Mathematics and Mathematical Physics* **5** 816
[5] Samarsky A A and Vabishchevich P N 1995 *Computational Heat Transfer, Volume 2, The Finite Difference Methodology* (Wiley)
[6] Vaganova N A 2008 *Proceedings of the Steklov Institute of Mathematics* **S1** S260
[7] Filimonov M Y and Vaganova N A 2013 *Applied Mathematical Sciences* **7** 7151
[8] Filimonov M Y and Vaganova N A 2012 *9th International Pipeline Conference(IPC 2012)* (*ASME Conference Proceedings* vol 4) p 133
[9] Vaganova N and Filimonov M 2015 (*CEUR Workshop Proceedings* vol 1513) p 42
[10] Vaganova N A and Filimonov M Y 2015 *Lect. Notes Comput. Sci.* **9045** 385
[11] Vaganova N 2014 (*AIP Conference Proceedings* vol 1631) (New York: AIP Publishing)
[12] Vaganova N A and Filimonov M Y 2015 (*AIP Conference Proceedings* vol 020016) (New York: AIP Publishing)

Experimental investigation of heat transfer efficiency on the convex surface of the annular channel of varying geometry intensifiers

A T Komov[1], A V Varava[1], A V Zakharenkov[1], A V Dedov[1], E A Boltenko[2] and B Y Agishev[1]

[1] General Physics and Nuclear Fusion Department, National Research University "Moscow Power Engineering Institute", 14 Krasnokazarmennaya Street, Moscow, Russia
2 Physics laboratory, Elektrogorsk Research Center on the safety of nuclear power plants, 6 St. Constantine Street, Elektrogorsk, Russia

E-mail: ZaharenkovAV@gmail.com

Abstract. The work is a continuation of the experimental studies on the enhancement of heat transfer in the fuel assembly on the experimental stand in National Research University "Moscow Power Engineering Institute". The description of the experimental setup, construction and main geometrical parameters of intensifier are presented. The new experimental data on the pressure loss and heat transfer coefficient using an edge enhancer - twisted wire single-phase convection mode are presented. In the research, the range mode parameters and geometric characteristics of the intensifier were extended. The relation of the coefficients of hydraulic resistance and the Nusselt number of steps twist twisted wire was found, the effect of the ribs on the heat transfer coefficient was shown. It is found that for any twist pitch ranging from 20 to 100 mm corresponds to a maximum heat transfer rib height $\dot{H}$ = 0,35. An increase in the heat transfer coefficient in the convex heating surface was experimentally obtained.

1. Introduction

Experimental studies conducted by a team of authors in [1] described above have shown that the use of intensifiers heat in which a positive effect is achieved through the interaction of swirling and transit (axial) flows in the annular channel is justified and effective. The intensifier, the main element of which was the system, produces formation of swirling and transit flows in the circle gap: edge - twisting wire. The interaction of flows with each other, with the ribs and the walls of the swirl ring wire channel throughout the length of the heated surface occurs within a plurality of additional and powerful vortex flow turbulence.

In [1] presents the results of experiments carried out previously in a wide range of regime parameters at twist wire step $t = 50\ 10^{-3}$ m.

It was found that

- a significant increase in heat transfer (80%);
- depending on the coefficients of hydraulic resistance and heat transfer from the relative height of the ribs are nonmonotonic, are clearly marked peaks;

- the maximum value of the heat transfer coefficient obtained at the value $\dot{H}$ = 0.35, and the coefficient of hydraulic resistance - at $\dot{H}$ = 0.23.

To complete the work in the field of single-phase convection authors believed it was appropriate to conduct new studies, sufficiently detailed thermal-hydraulic characteristics, varying the step intensifier spin at different values of the dimensionless ribs height *H*.

2. The description of experimental installation and test section

The Experimental studies were carried out on the installation whose main systems, equipment, work area design and methodological principles are described in detail in [2].

To determine the temperature field along the length of the heating element 11 thermocouples were installed on the inner surface of test section. Using compensation wires via the switching device thermocouples connected to the Centre for Automation of Thermophysical Research (CATR). To measure the voltage directly at the working area on the inner surface of the heated pipe potentiometric bends installed. The working section is placed in a high-pressure vessel (HPV), the inner wall which with the cylindrical surface of the heating element forms a circular channel. The size of the annular channel is $(d_2 - d_1)/2 = 1.7\ 10^{-3}$ meters, less than the size of the capillary constant. Therefore, this annular channel can be attributed to the type of gap or mini channels. HPV placed vertically in the hydraulic circuit.

Main characteristics of the annular channel (Figure 1):

- the inner diameter of the annular channel (the outer diameter of the heating element)– $d_1 = 12.9 \cdot 10^{-3}$ m;
- the outer diameter of the annular channel (the inner diameter of HPV) – $d_2 = 16.3 \cdot 10^{-3}$ m;
- the inner diameter of the heating element – $d_3 = 10.0 \cdot 10^{-3}$ m;
- area of the flow cross section of the annular channel – $S = 77.9 \cdot 10^{-6}$ m^2;
- hydraulic diameter of the smooth annular channel – $d_g = d_2 - d_1 = 3.4 \cdot 10^{-3}$ m;
- the length of the heating element corresponds to the distance between the selection pressure – $l = 0.7$ m;
- length of the upper current lead – $l_1 = 0.55$ m;
- length of the lower current lead – $l_2 = 0.25$ m.

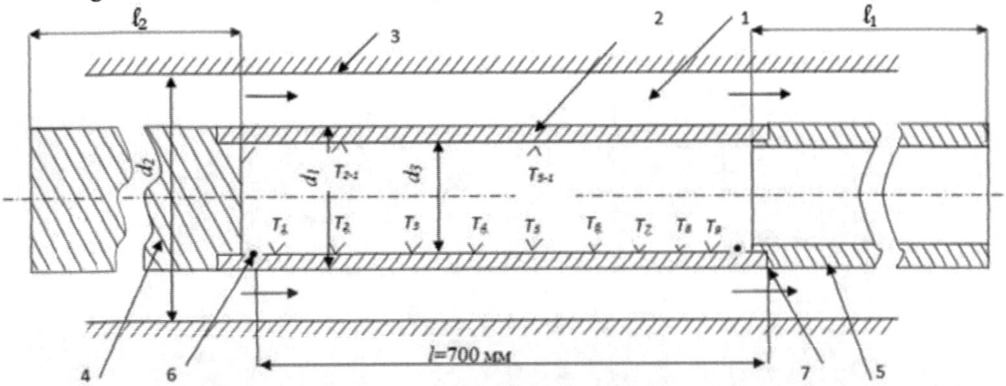

Figure 1. Scheme of annular channel: 1 – annular channel; 2 – heating element; 3 – the inner surface of HPV; 4, 5 – lower and upper current lead, respectively; 6, 7 – potentiometric bend

3. Technical parameters and construction of intensifier

Intensifier is designed to create in an annular channel two interacting flows, the presence of which increases the efficiency of heat transfer. The description of the intensifier presented in [1].

General view of the intensifier mounted on a convex surface of the test section shown in Figure 2.

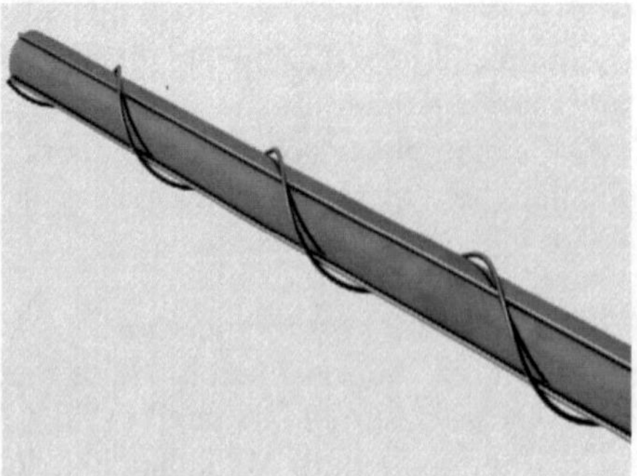

Figure 2. Scheme of installation of ribs and winding wire on the test section

Experiments on heat transfer coefficients and hydraulic resistance were carried out at the steps twist intensifier $t = 20\cdot10^{-3}$ m, $30\cdot10^{-3}$ m, $40\cdot10^{-3}$ m, $50\cdot10^{-3}$ m, $60\cdot10^{-3}$ m, $100\cdot10^{-3}$ m. The series of several experiments to determine heat transfer and hydraulic resistance at various rib heights (obtained over 300 experimental points, including twisting during winding wire directly on a heated work area) was carried out. A comparison of the array of experimental data for the $t = 50\cdot10^{-3}$ m [3] showed good reproducibility of measurements.

In [4], the main geometric characteristics are the intensifier rib height *h* and the diameter of the wire twisting *d*. In processing the experimental data used dimensionless parameter $\dot{H}$ (relative height of the ribs) and the coefficient of swirl flow *k*, associated with a twist pitch *t*:

$$\dot{H} = \frac{h}{(d_2 - d_1)/2}, \qquad k = \pi\bar{d}/t,$$

where $\bar{d}$ - the diameter of the middle line of the swirling flow, and $\dot{H}$ inherently determines the ratio of the swirling flow characteristics and transit (axial) flows.

4. Pressure drops and hydraulic resistance

Use of intensifiers, of course, leads to a substantial increase in pressure loss in the work area. Therefore, one object of this study was to determine the optimum field spin steps in which the increase of the hydraulic resistance, at least not more than the heat transfer coefficient values increase. In the experiments, direct measurements were determined by the pressure loss and mass flow rate of the coolant for various $\dot{H}$. Figure 3 shows the dependence of the mass flow of coolant pressure loss for different twist wire steps with $\dot{H} = 0.44$.

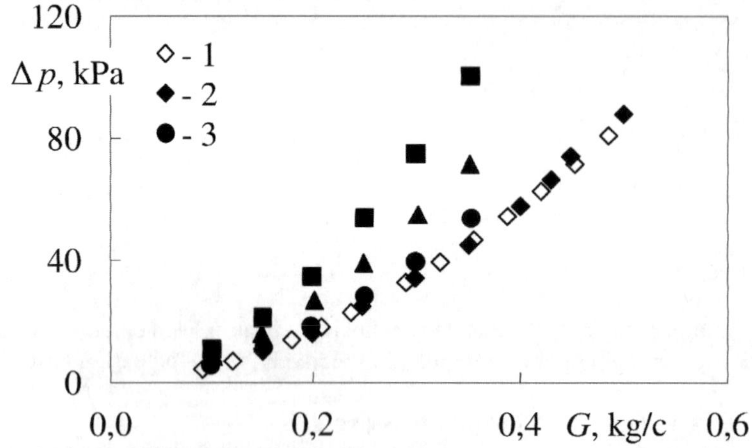

Figure 3. Pressure drops - mass flow rate of the coolant relationship ($p = 3.0\cdot10^6$ Pa, $T_{in} \approx 100$°C, $\dot{H}$= 0.44):1 – smooth annular channel; 2 – annular channel with ribs; 3 - $t = 100\cdot10^{-3}$ m; 4 – $t = 60\cdot10^{-3}$ m; 5 – $t = 40\cdot10^{-3}$ m

According to the data presented in Figure 3 can draw the following conclusions:

- presence of the ribs without wire winding has practically no effect on the hydraulic channel resistance;
- a decrease in step spin (spin rate) pressure loss increases sharply (Figure 4).

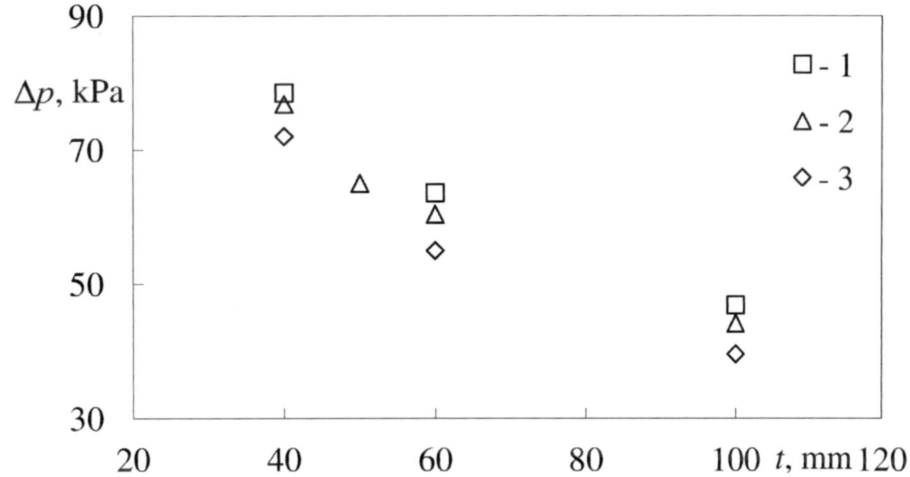

Figure 4. Pressure drops - the intensifier swirl pitch t relationship ($p = 3.0 \cdot 10^6$ Pa, $T_{in} \approx 100$°C, G= 0.30 kg/c): 1 – $\dot{H}$= 0.23, 2 – $\dot{H}$= 0.35, 3 - $\dot{H}$= 0.44

From analysis of Figure 4 shows that there is a bundle of data on the hydraulic losses at the height of the ribs.

On Figure 5 shows the hydraulic drag coefficient on the Reynolds number for different steps twist when $\dot{H} = 0.23$. Hydraulic resistance coefficient ξ is calculated as follows:

$$\xi = \frac{2\Delta p d_g S^2 \rho}{G^2 l},$$

where Δp - pressure loss in the work area; $d_g = 4S/P$ - hydraulic diameter of the annular channel with an intensifier; P - wetted perimeter; S - area of the flow cross section of the annular channel; ρ - density of the coolant; G - mass flow rate of the coolant; l - the length of the working area.

The Reynolds number is calculated by the relation:

$$\mathrm{Re} = \frac{G d_g}{\mu S},$$

where μ - dynamic viscosity coefficient of the coolant.

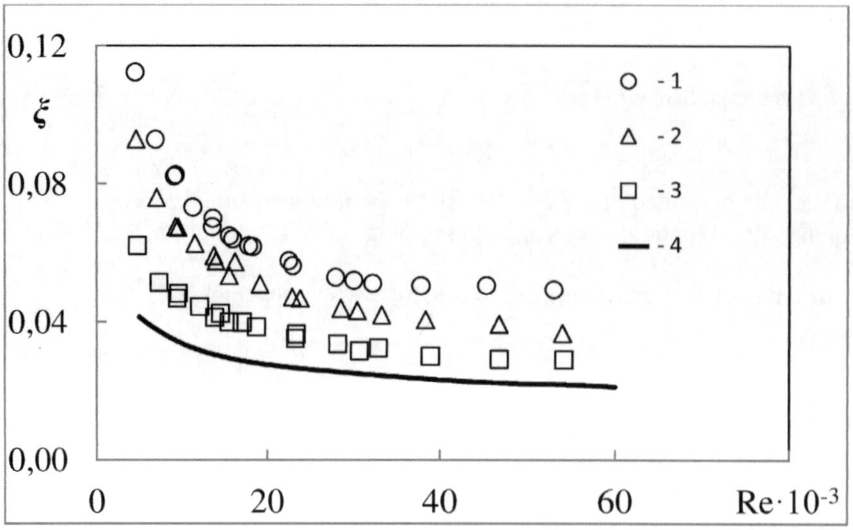

Figure 5. The pressure drop coefficient - Reynolds number relationship ($\dot{H}$ = 0.23): 1 – $t = 40 \cdot 10^{-3}$ m; 2 – $t = 50 \cdot 10^{-3}$ m; 3 – $t = 60 \cdot 10^{-3}$ m; 4 – $t = 100 \cdot 10^{-3}$ m; 5 – smooth annular channel (calculation)

For comparison, the graph shows the calculation of ξ (Re) using a modified Filonenko equation for a smooth annular channel:

$$\xi_0 = \frac{1}{(1{,}82\lg \text{Re} - 1{,}64 - 0{,}19(d_1/d_2)^{0{,}25})^2}.$$

Presented graph suggests a very tangible effect of geometrical parameters intensifier by a factor of hydraulic resistance. This is confirmed by data on the relation of the coefficient of hydraulic resistance in the parameter $\dot{H}$ obtained earlier [1].

The influence of parameters intensifier ($\dot{H}$ and t) on the coefficient of hydraulic resistance is particularly evident in the direct dependence of ξ ($\dot{H}$) for different steps twist t and ξ (t) for different dimensionless ribs heights $\dot{H}$. The dependence of the coefficient of hydraulic resistance by twisting step is shown in Figure 6 (for Re = 60000) at different $\dot{H}$.

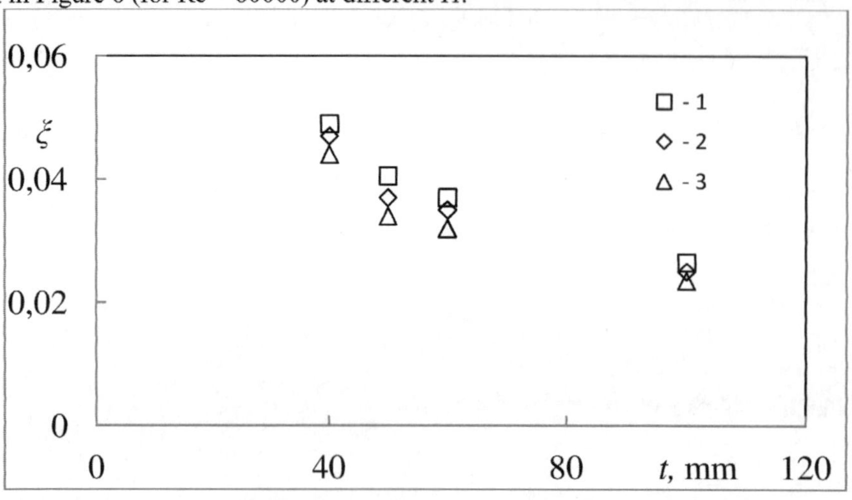

Figure 6. The pressure drop coefficient - the intensifier swirl pitch t relationship (Re = 60000): 1 - $\dot{H}$= 0.23, 2 -$\dot{H}$= 0.35, 3 - $\dot{H}$= 0.44

From Figure 6 that bundle dependence ξ (t) for $\dot{H}$ ranges - (8 - 10)%.

Figure 7 shows the dependence of $\xi / \xi 0 = f(\dot{H})$ at various steps twisting wire, ξ_0 - hydraulic resistance coefficient of fluid flow in a smooth annular channel, is calculated using the equation(1).

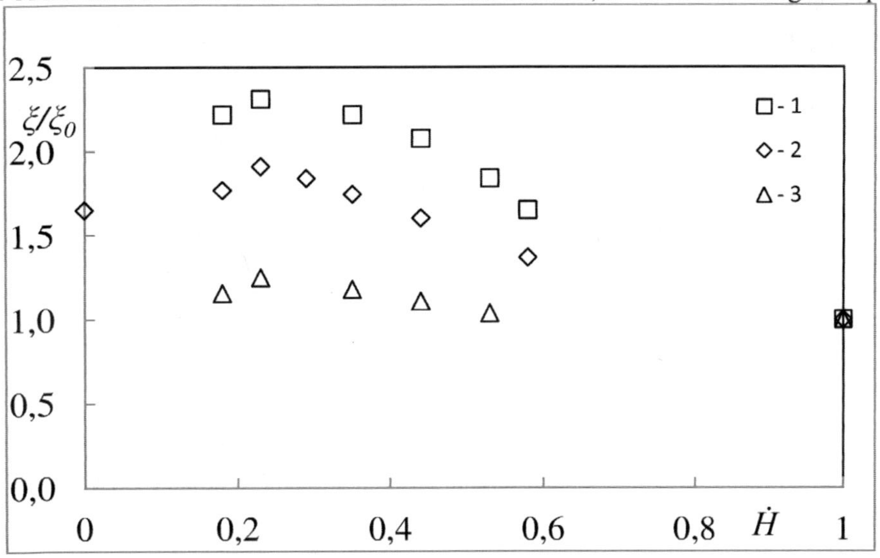

Figure 7. The pressure drop coefficient - parameter $\dot{H}$ relationship (Re = 60000): $1 - t = 40 \cdot 10^{-3}$ m; $2 - t = 50 \cdot 10^{-3}$ m; $3 - t = 100 \cdot 10^{-3}$ m

The experimental data presented in Figure 7 allow the following conclusions:

• the value of the hydraulic resistance coefficient significantly depends on the pitch of twist. With increasing t hydraulic resistance coefficient decreases;

• dependence of the hydraulic drag coefficient on the dimensionless parameter $\dot{H}$ for all steps twisted wire is nonmonotonic and is close to $\dot{H} = 0.23$ there is a maximum hydraulic losses. Such dependence $\xi(\dot{H})$ is retained over the entire range of Reynolds numbers.

• maximum ξ/ξ_0 value for different steps spins at a Reynolds number Re = 60000 in the range from 1.3 to 2.2.

5. Heat transfer coefficient

The aims of the carried out research systematic study are:

• the impact of the step twist t (twist coefficient k) in the heat transfer coefficient at various flow characteristic of the interacting flows determined by the dimensionless parameter $\dot{H}$;

• determining the value $\dot{H}$, corresponding to the maximum heat transfer for different steps t.

Methods of measurement and processing of the experimental data for the study of heat transfer in detail presented in [2].

The modified Isachenko-Galin equation for a smooth annular channel was using for processing the experimental data.

$$Nu = 0.017\,\mathrm{Re}^{0.8}\,\mathrm{Pr}_l^{0.4}\left(\frac{\mathrm{Pr}_l}{\mathrm{Pr}_w}\right)^{0.25}\left(\frac{d_2}{d_1}\right)^{0.18}.$$

The measurement results are presented in the form of Nu/A = f(Re), where $A = \mathrm{Pr}_l^{0,4}\left(\frac{\mathrm{Pr}_l}{\mathrm{Pr}_w}\right)^{0,25}\left(\frac{d_2}{d_1}\right)^{0,18}$.

The parameter A takes into account the change in the coolant properties at different temperatures, as well as the geometric dimensions of the annular channel.

The relationship Nu/A = f(Re) for different swirling steps of intensificator t is illustrated in Figure 8.

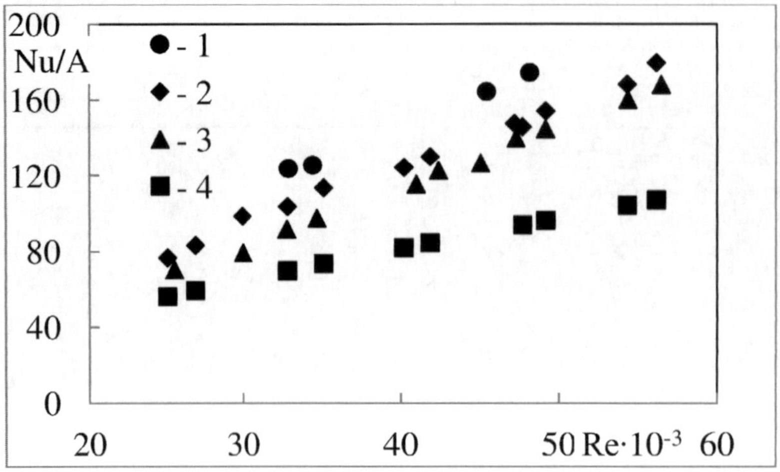

Figure 8. The Nusselt number - Reynolds number relationship (p = 3.0 10^6 Pa, $\dot{H}$ = 0.29): 1 – t = $20 \cdot 10^{-3}$ m, 2 – t = $50 \cdot 10^{-3}$ m, 3 – t = $60 \cdot 10^{-3}$ m, 4 – smooth annular channel (experiment)

As it follows from Figure 8 steps to a decrease spin (spin rate increase) significantly increases heat transfer.

Influence of geometrical parameters on heat transfer intensifier in the form of dependences Nu/A = f ($\dot{H}$) is shown in Figure 9.

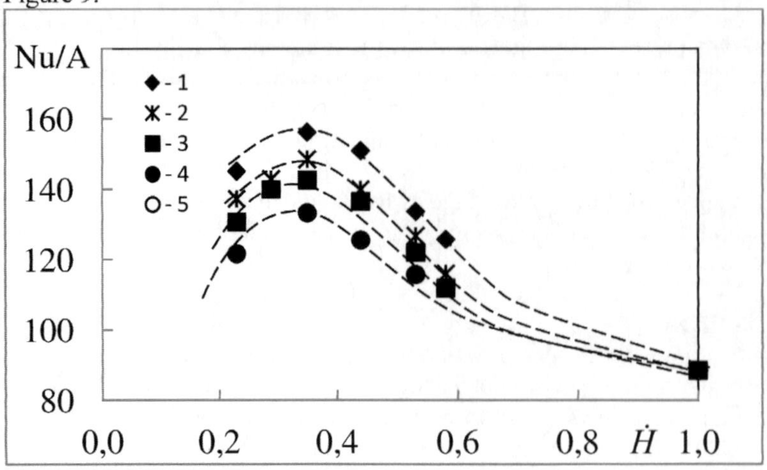

Figure 9. The Nusselt number - $\dot{H}$ parameter relationship (Re = 45000): 1 – t = $40 \cdot 10^{-3}$ m, 2 – t = $50 \cdot 10^{-3}$ m, 3 - t= $60 \cdot 10^{-3}$ m, 4 - t = $100 \cdot 10^{-3}$ m, 5 – smooth annular channel

As appears from analysis Figure 9 dependence Nu / A = f ($\dot{H}$) is nonmonotonic and has a maximum value at $\dot{H}$ = 0.35. The same value was obtained in the previous study, in which the studies were conducted only for the spin step t= $50 \cdot 10^{-3}$ m.

6. Conclusions

The studies of heat transfer and hydrodynamics in the annular channel with heat transfer intensifiers. Operating parameters: mass velocity ρw = 1000 ÷ 4000 kg/(m^2 c), a pressure p = (3.0 ÷ 7.0) 10^6 Pa, the coolant temperature at the entrance of the operating portion T = 100°C, the heat flux q = (0.23 ÷ 0.70) 10^6 W/m^2.

A study of the effectiveness of the method of interacting streams in order to intensify the heat exchange on the convex heating surface. It was found that the coefficient of heat transfer and hydraulic resistance depends essentially on the geometric characteristics of the intensifier: $\dot{H}$ parameter that determines the intensity of the axial and swirling flows and step spin intensifier t. It was established experimentally that for different steps twist intensifier for maximum hydraulic resistance is observed at $\dot{H} = 0.23$, and on heat transfer - with $\dot{H} = 0.35$.

This work was supported by grant 16-19-10457 RNF.

References

[1] Boltenko E, Varava A, Dedov A, Zakharenkov A, Komov A and Malakhavskiy S 2015 Investigation of heat transfer and pressure drop in an annular channel with heat transfer intensifiers *J. Thermal Engineering* **62** 177-182

[2] Zakharenkov A, Komov A, Varava A, Boltenko E, Dedov A, Masnikov V and Ilyin A 2013 Experimental installation for gidrodimaic and heat exchange investigation in model elements of fuel assembly *J. Vestnik Moscow Power Engineering Institute* **2** 39-45

[3] Zakharenkov A, Komov A, Boltenko E, Varava A and Dedov A 2013 Investigation of heat exchange and hydrodinamics parameters of single-phase swirling flows in annular channels with axial ribs *J. Vestnik Moscow Power Engineering Institute* **4** 53-58

[4] Agishev B, Boltenko E, Varava A, Dedov A, Zakharenkov A and Komov A 2016 Investigation of heat transfer efficiency in the fluid flow in an annulus in dependence on the geometric characteristics intensifiers *15 Minsk Int. Heat and Mass Transfer Forum,* vol 3 p. 80-83

Comparative analysis of the influence of turbulence models on the description of the nitrogen oxides formation during the combustion of swirling pulverized coal flow

V Kuznetsov[1,*], N Chernetskaya[2], M Chernetskiy[1,2]

[1]Siberian Federal University, 79 Svobodny, Krasnoyarsk, 660041, Russia
[2]Kutateladze Institute of Thermophysics, Novosibirsk, 630090, Russia
[*]victor_partner@mail.ru

Abstract. The paper presents the results of numerical research on the influence of the two-parametric k-ε, and k-ω SST turbulence models as well as Reynolds stress model (RSM) on the description of the nitrogen oxides formation during the combustion of pulverized coal in swirling flow. For the numerical simulation of turbulent flow of an incompressible liquid, we used the Reynolds equation taking into account the interfacial interactions. To solve the equation of thermal radiation transfer, the P1 approximation of spherical harmonics method was employed. The optical properties of gases were described based on the sum of gray gases model. To describe the motion of coal particles we used the method of Lagrange multipliers. Burning of coke residue was considered based on diffusion – kinetic approximation. Comparative analysis has shown that the choice of turbulence model has a significant impact on the root mean square (RMS) values of the velocity and temperature pulsation components. This leads to significant differences in the calculation of the nitrogen oxides formation process during the combustion of pulverized coal.

1. Introduction

Nitrogen oxides are one of the dangerous pollution from coal combustion at thermal power stations. Influencing the combustion process, i.e. applying technological methods, it is possible to reduce significantly the concentration of nitrogen oxides in exhaust gases [1-3]. One of the effective technological methods to reduce NOx concentration is the use of low NOx swirl burners [4]. The optimization of their operation is largely based on numerical experiment [5-7]. Despite the great strides achieved in the development of the numerical experiment, as well as large variety of various numerical techniques, the complicated structure of coal, yet not fully understood, and the complex chemical processes occurring during coal fuel combustion and gasification, do not allow creating universal models. Therefore, search for mathematical models that would describe more accurately the combustion processes of pulverized coal and the formation of nitrogen oxides in furnace-burner units still remains quite urgent.

Mathematical description of pulverized fuel combustion includes a set of interrelated models describing turbulent gas motion, the transfer of thermal and radiant energy, combustion, gasification, movement of the coal particles, etc. When considering vortex burners with a swirling flow, it is important to select turbulence model, which would accurately describe the averaged fields and the large-scale pulsations of the swirling flows. As shown by existing works, k–ε and k–ω turbulence models poorly describe the swirling flow. To improve the description of swirling flows it is proposed to use modifications of the two-parameter turbulence models, for example, k–ω SST model of Menter,

Reynolds stress transfer models, as well as eddy-resolving methods, such as for example, large-eddy simulation (LES) method.

At that, it can be assumed that the choice of turbulence model will affect also the description of the pulverized coal combustion processes and eventually to predict the formation of nitrogen oxides.

The aim of this work is to conduct comparative analysis of the effect of the two-parametric $k{-}\varepsilon$ turbulence model and Menter's $k{-}\omega$ *SST* model, as well as Reynolds stress transfer models on the results of the simulation of nitrogen oxides formation during the combustion of swirling pulverized coal torch.

2. Problem statement and research methods

For numerical studies and verification of the mathematical model we used experimental data on pulverized coal torch combustion involving flow swirl. The data were obtained from 2.4 MW firing stand [8]. The sketch of the firing stand is presented in Fig. 1. Flow characteristics when conducting experimental studies and calculations were as follows: the primary (dry) air - 0.117 kg/s, the temperature - 343.15 K, the average axial velocity - 23.02 m/s, and the (dry) coal consumption - 0.073 kg/s. The secondary (dry) air flow rate - 0.745 kg/s, the temperature - 573.15 K, the average axial velocity - 43.83 m/s, the average tangential velocity - of 49.42 m/s. Technical content of coal (wt. %, dry): volatile matter - 37.4, fixed carbon - 54.3, Ash - 8.3. Chemical composition of coal (wt. % daf): C - 80.36, H - 5.08, N - 1.45, S - 0.94, O - 12.17. Qv = 32.32 MJ/kg.

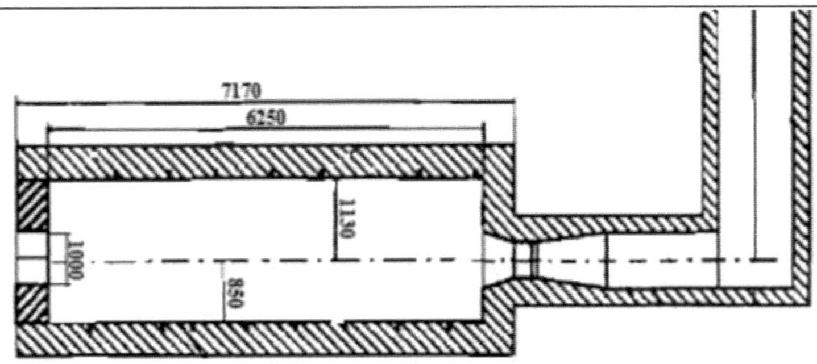

Figure 1. Sketch of the fire stand with 2.4 MW burner [8].

For the numerical simulation of turbulent flow of an incompressible liquid we used the Reynolds equations with due account for the interphase interaction. The Reynolds equations were closed by the following two-parametric turbulence models: standard $k{-}\varepsilon$ turbulence model [9], and $k{-}\omega$ SST model of Menter [10]. In Reynolds stress transfer model components $\overline{u'_i u'_j}$ of the Reynolds stress tensor are found by solving the transport equations:

$$\frac{\partial}{\partial x_l}\left(\rho u_l \overline{u'_i u'_j}\right) = \rho\left(P_{ij} + Diff_{ij} - \varepsilon_{ij} + \phi_{ij}\right),$$

where P_{ij} - is the term describing the Reynolds stress generation, $Diff_{ij}$ - is the transfer due to diffusion, ε_{ij} - is the dissipation rate, ϕ_{ij} - is the pressure cross-term.

To describe the right hand side terms of the equation we take several hypotheses. Diffusion term is modeled by a generalized gradient diffusion hypothesis (GGDH) [11]. Pressure cross-term may be modeled on the basis of [12, 13]. In the present work to describe the motion of particles we used Lagrange multiplier method. The particle motion is described by the dynamics equations for material point inclusive of the drag force and gravity. Accounting for flow turbulence in the particle motion is produced by the introduction of random fluctuations of the gas velocity in the motion equation for the particles. The solution to the equation of radiant energy transport is based on the P1 approximation of spherical harmonics for a two-phase two-temperature gray medium. As shown by previously performed computational studies [6, 14, 15], this integrated model gives good correlation with the experimental data, when burning solid fossil fuel in a pulverized form. The combustion process of coal particle is considered in terms of the following consecutive steps: evaporation of moisture from the fuel, devolatilization and the combustion of the volatile components, and the combustion of the coke residue. Yield of volatiles is considered in the single-component approximation in the form of $C_xH_yO_z$ substance. The rate of devolatilization was calculated using a single-stage kinetic model with constants appropriate to the studied grade of coal [8]. Calculation of the gaseous components combustion was carried out with due account for the reactive power and concentration of fuel and oxidizer, as well as the rate of turbulent mixing of fuel and oxidizer. The combustion rate of coke residue was calculated in accordance with the provisions of the classical diffusion-kinetic theory. Kinetic constants of chemical reaction for the oxidation of coke residue were taken from [8].

A mathematical model of NOx formation during coal combustion involves consideration of three mechanisms taking into account the influence of temperature fluctuations: the thermal NOx calculated based on the known dependence of Zeldovich [16], prompt NOx and fuel NOx calculated using a model proposed in [17] with reactions (1, 2), as well as the supplement of the "reburning" reaction (3), proposed by Chen [18].

$$dx_{HCN} / dt = -3.5*10^{10} \exp(-3370/T) x_{HCN} x_{O_2}^{a}, \quad (1)$$

$$dx_{HCN} / dt = -3*10^{12} \exp(-30200/T) x_{HCN} x_{NO}, \quad (2)$$

$$dx_{NO} / dt = -2.7*10^{6} \exp(-9466/T) x_{NO} x_{C_nH_m} \quad (3)$$

where x – is the molar fraction of the respective component, a – is the order of reaction with respect to oxygen.

The reaction rate of NO formation is strongly temperature dependent due to the large activation energy values. Consequently, temperature fluctuations can greatly increase the concentration of nitric oxide formed in turbulent flows. To account the effect of turbulent fluctuations on the nitric oxide formation process, we produced the solution to the equation for temperature variation, when solving the transfer equations for NO and HCN concentrations. To determine the average rate of NO formation reaction, we carried out integration between the upper and lower temperature limits using the probability density function of the β distribution.

To solve the conservation equations for the gas phase we used the well known control volume method. For calculation of diffusion fluxes on the faces of the control volume, we used the central-difference approximation of the second-order accuracy. When approximating convective terms, we used the second-order accuracy scheme. To solve the resulting system of equations, we used the incomplete factorization method, in which just the diagonal terms were factored out. To link the pressure and velocity fields, we used in the present work the SIMPLE-like algorithm with aligned grids. The proposed model and solution methods were previously tested for solving problems on pulverized coal combustion and gasification [14, 15], and showed a satisfactory agreement with the experimental data in terms of the basic process parameters in the combustion chamber.

3. Results and discussion

Previously conducted computational studies on the influence of computational mash refinement on the calculation results of pulverized coal torch burning in the combustion chamber of fire stand with swirling burner have shown [19] that for the mesh size of more than 500,000 cells, no significant differences were noted in the calculation results. Therefore, in this paper, we have used a computational mesh with 500,000 cells. It was shown previously [19] that for the distribution of axial and tangential velocity components over the sections of the combustion chamber, Reynolds stress model and the two-parameter k-ε and Menter's k- ω SST turbulence models give similar results for pulverized coal combustion. At the same time, there are significant differences in the fluctuating velocity component depending on the selected turbulence model (Fig. 2). From Fig. 2 we can see that the RMS velocity fluctuation, common to axial and tangential velocity, and calculated from the expression $(2k/3)^{0.5}$, has a lower level and better agreement with the experimental data for Reynolds stress model. This in turn affects the calculation of burnout rate of the volatile component. It is obvious from Fig. 3 that higher values of pulsations in the initial section of the torch lead to more rapid burnout of volatile component and, accordingly, to higher temperatures. The temperature differences at the initial section of the torch amount to about 10% when using the two-parameter k-ε turbulence model and Reynolds stress model, respectively.

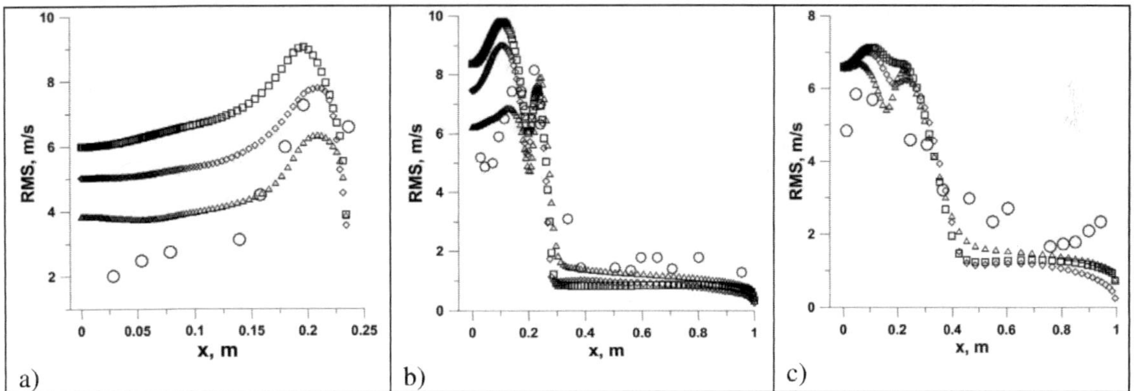

Figure 2. The distribution fluctuating velocity component in cross sections. a) z=0 m; b) z=0.25 m; c) z=0.85 m. Circle - experiment, rhombus – k-w SST, square – k-e, triangle -RSM.

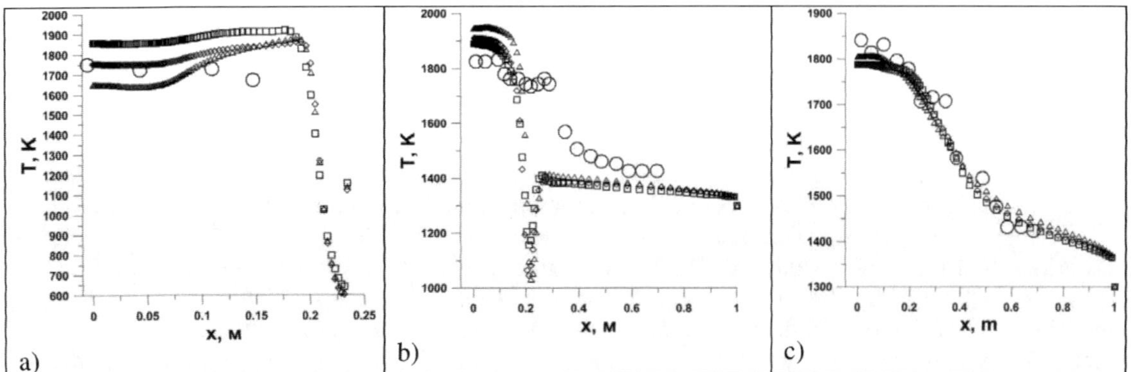

Figure 3. Temperature distribution in the cross sections: a) z=0 m; b) z=0.25 m; c) z=0.85 m. Circle - experiment, rhombus – k-w SST, square – k-e, triangle -RSM.

The calculation of the nitrogen oxides formation was conducted after obtaining the convergent solution for the basic parameters of two-phase reacting flow in the "postprocessing" regime. At that, we solved the additional equation for the temperature variation to take into account the influence of the pulsation components of the flow on the nitrogen oxides formation. Figure 4 presents the nitrogen oxides concentrations in the sections of the combustion chamber, while Fig. 5 shows RMS for temperature.

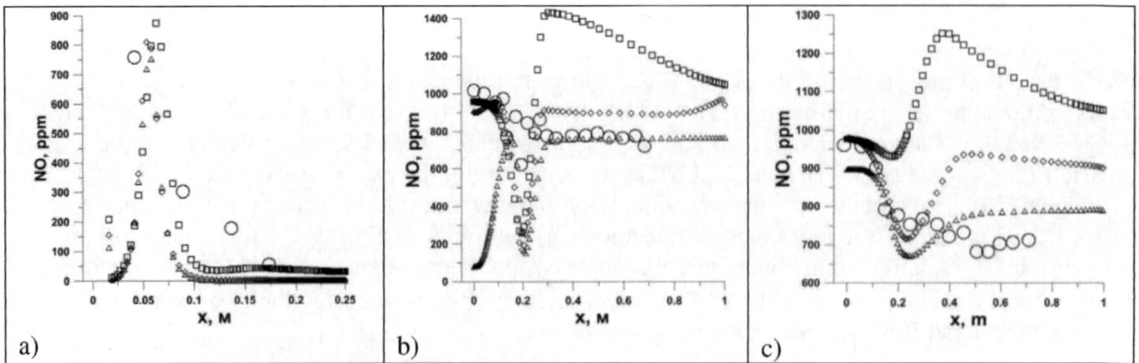

Figure 4. NO of concentration in the cross sections: a) z=0 m; b) z=0.25 m; c) z=0.85 m. Circle - experiment, rhombus – k-w SST, square – k-e, triangle -RSM.

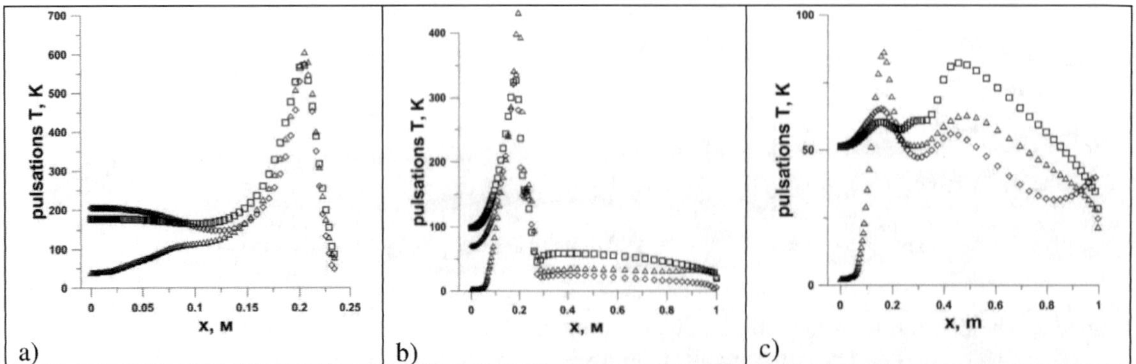

Figure 5. Temperature fluctuations in the cross sections: a) z=0 m; b) z=0.25 m; c) z=0.85 m. Circle - experiment, rhombus – k-w SST, square – k-e, triangle -RSM.

We can see that, when using Reynolds stress model as a turbulence model, nitrogen oxides concentration values are lower than those obtained for the two-parameter turbulence model, and are in satisfactory agreement with the experimental data. For Menter's k-ω SST model and the standard k-ε turbulence model, concentrations of nitrogen oxides exceed the experimental values by more than 25 and 50%, whereas for the Reynolds stress model, deviation of the NO concentration does not exceed 10% in comparison with the experimental data.

This is primarily due to higher values of RMS for both the velocity (Fig. 3) and the temperature (Fig. 5). Deviation of the RMS values for temperature according to the selected model is about 10-30% depending on the considered section of the combustion chamber. Given the exponential dependence of the fuel NO formation reaction rate on the temperature (formulas 1-3), which in this case make the main contribution to the formation of NO, this is one of the major factors explaining the difference in calculation results.

Conclusion

We have performed numerical study on the influence of the two-parametric k-ε, and k-ω SST turbulence models, as well as Reynolds stress transfer model, on the description of the nitrogen oxides formation during the combustion of pulverized coal in the swirling flow. It is shown that the choice of turbulence model when calculating combustion of pulverized fuel in the swirling flow significantly influences the distribution of the RMS values of velocity and temperature fluctuations. This leads to a difference in the description of burning process of the gaseous components, especially at the initial part of the torch, and, accordingly, the difference in temperatures that accounts for more than 10% at the initial part of the torch, depending on the choice of turbulence model.

It is shown that the use of two-parametric turbulence models results in satisfactory agreement of the calculation with the experimental data on the averaged parameters of velocity, temperature and concentration of the gas components (discrepancy does not exceed 5%), though gives a significant deviation (more than 50%) in the nitrogen oxides concentration within the combustion chamber space in comparison with the experimental data. Reynolds stress model gives a deviation in distribution of the nitrogen oxides concentration no more than 10% as compared to experimental data, with underlying slight difference in the distribution of the axial and tangential velocities, temperatures and concentrations of gases as compared to the two-parametric turbulence models.

The research was supported by the Russian Foundation for Basic Research (RFBR), research project No 15-38-20969.

References

[1] Kotler V.R., 1982, Reduction in nitrogen oxides emissions in boilers of thermal power plants during solid fuel combustion. Souzenergo.
[2] Zhou H., Cen K.F., J.R., 2004, Energy, 29(1), pp. 167–183.
[3] Hill S.C., Smoot L.D., 2008, Prog. Energy and Combust. Sci. 26(4/6), pp. 417–458.
[4] Chen Zh. Li Zh., Zhu Qunyi, Jing J., 2011, Gas/particle flow and combustion characteristics and NOx emissions of a new swirl coal burner. Energy 36, pp. 709-723.
[5] Krasinsky D.V., Salomatov V.V., Anufriev I.S., Sharypov O.V., Shadrin E.Y., Anikin Y.A., 2015, Modeling of pulverized coal combustion processes in a vortex furnace of improved design. Part 2: Combustion of brown coal from the Kansk-Achinsk basin in a vortex furnace. Thermal Engineering (English translation of Teploenergetika), 62(3), pp. 208-214.
[6] Chernetskiy M.Yu., Dekterev A.A., Burdukov A.P., Hanjalić K., 2014, Computational modeling of autothermal combustion of mechanically-activated micronized coal. Fuel, 135, pp. 443-458.
[7] Wojciech P. Adamczyk S., Werle A. R., 2014, Application of the computational method for predicting NOx reduction within large scale coal-fired boiler. Applied Thermal Engineering, 73, pp. 343-350.
[8] Peters A.A.F., Weber R., 1997, Mathematical modeling of a 2.4 MW swirling pulverized coal flame. Combustion Science and Technology, 122(1-6), pp. 131-182.
[9] Launder B.E., Spalding D.B., 1972, Lectures in mathematical models of turbulence. London, Academic Press, 236 p.
[10] Smirnov P.E., Menter F.R., 2009, Sensitization of the SST turbulence model to rotation and curvature by applying the Spalart-Shur correction term, Journal of Turbomachinery, 131(4), pp. 1-8.
[11] Daly B.J., Harlow F.H., 1970, Transport equations in turbulence, Physics of Fluids, 13(11), pp. 2634-2649.
[12] Launder B.E., 1989, Second-moment closure and its use in modeling turbulent industrial flows. International Journal for Numerical Methods in Fluids, 9(8), pp. 963–985.
[13] Launder B.E., 1989, Second-moment closure: present... and future? International Journal of Heat and Fluid Flow, 10(4), pp. 282–300.
[14] Maidanik M.N. et al., 2011, Mathematical simulation of the furnace and turning gas conduit of a P-50R boiler during joint combustion of solid and gaseous fuel. Thermal Engineering (English translation of Teploenergetika), 58(6), pp. 483 – 488.
[15] Messerle V.E. et al., 2015, Modeling and full-scale tests of vortex plasma–fuel systems for igniting high-ash power plant coal. Thermal Engineering (English translation of Teploenergetika), 62(6), pp. 442-451.
[16] Zeldovich Y.B, Sadovnikov, P.Y, Frank-Kamenetckiy, D. A., 1947, AS USSR. p. 317.
[17] DeSoete G.G.,1975, Proceedings of the 15th International symposium on Combustion. The Combustion Institute, pp. 1093–1102.
[18] Chen W., 1994, PhD thesis, Brigham Young University.
[19] Chernetskiy M.Y., Kuznetsov V.A., Dekterev A.A., Abaimov N.A., Ryzhkov A.F., 2016, The effect of turbulence models on the pulverized coal combustion in the swirling flow. Thermophysics and Aeromechanics, 23(4) pp.615-626.

Journal of Physics: Conference Series **754** (2016) 112007 doi:10.1088/1742-6596/754/11/112007

Study of the two-stage gasification process of pulverized coal at the hydrodynamic flow separation

V Kuznetsov[1], M Chernetskiy[2,3]* and A Ryzhkov[3]

[1]Siberian Federal University, 79 Svobodny, Krasnoyarsk, 660041, Russia
[2]Kutateladze Institute of Thermophysics, Novosibirsk, 630090, Russia
[3]Ural Federal University, 19 Mir Street, Yekaterinburg 620002, Russia
*micch@yandex.ru

Abstract. The paper presents a numerical study of advanced two-stage gasifier with a combined countercurrent and concurrent flow pattern and dry fuel feed system EAGLE. The Kuznetsk coal was used as a fuel for the gasifier under study. We have conducted studies on the influence of the inclination angle of the upper burners in horizontal and vertical planes, and the amount of steam supplied, on heat and mass transfer processes in the chamber as well as on the composition of coal-derived gases. It is shown that the increase in the inclination angle of the upper burners in the horizontal plane allows intensifying the process of two-stage gasification and makes it possible decreasing the height of the chamber without sacrifice of the composition of the coal-derived gases.

1. Introduction

Gasification of coal is one of the most strategic pathways of coal processing. Market analysis of gasification technologies shows that the most popular are flow gasifiers, whose share is about 80% [1]. Flow technologies can be organized on the basis of both single-stage oxidizer and fuel feeding system, and using two-stage feeding. At that, in terms of the efficiency of conversion process, the lowest performance indicators are typical for single-stage gasifiers operating on coal-water mixture, while the highest indicators can be achieved in two-stage gasifiers with dry fuel feeding system.
The analysis of technical solutions used in new designs of gasifiers in terms of efficiency and economy of their application in solid-fuel combined cycle gas turbine units (CCGT) with integrated gasification has shown that the use of two-stage principle of the fuel conversion process is one of the upcoming trends in technology modernization. According to [2], the transition from single-stage process in the flow gasifier to two-stage process, even without optimization of the synthesis gas composition, can increase the efficiency of the CCGT by almost 1%.
The development of two-stage gasifiers with dry fuel feeding system is carried out according to two conceptual flow patterns: concurrent type with upward flow and combined countercurrent and concurrent flow pattern. The EAGLE-type two-stage gasifier, developed in the frameworks of the Japanese project "Coal Energy Application for Gas, Liquid and Electricity" (EAGLE) [3] is one of the original technical solutions based on a combined counter-current and concurrent flow pattern. In this paper, on the basis of numerical studies, using the Kuznetsk coal as a fuel, we examine the influence of several operating parameters of the EAGLE gasifier on the heat and mass transfer parameters in the gasification chamber as well as the efficiency of the gasification process.

2 Problem statement and research methods

The operational principle of the EAGLE gasifiers is based on a hydrodynamic flow separation providing a combined countercurrent and concurrent flow pattern (Fig.1). One advantage of this solution is reducing the required height of the reaction zone, while increasing the residence time of coal particles in the reactor. One flow, with the oxygen to coal ratio close to stoichiometric value, is delivered to the lower part of the gasification chamber. While combusting, it creates the necessary temperatures for efficient proceeding of the gasification reactions. The second flow, with the lack of air, is delivered into an upper part of the chamber, where it interacts with the high temperature upflow of combustion products. Due to the lack of oxygen for complete combustion, the carbon residue transforms into gasification reaction products through the conversion process. In this work, the Kuznetsk coal was used as a gasifier fuel (Table 1). The estimated coal consumption is 1700 t/day. The dimensions of the gasification chamber are presented in Fig. 2. Gasification air flow was selected based on the assumption that the maximum excess air was 0.32. The pulverized-coal was supplied at a high concentration under pressure. Nitrogen was used as pulverized-coal carrier. Blast burners of the bottom tier were enriched with oxygen (O_2=25 %, vol.) due to fuel transportation in the form of high-density pulverized coal carried by nitrogen. The upper tier of burners is supplied with both air and steam at an amount of 0.137 kg of steam per kg of coal. Coal consumption in the lower tier of burners made up 25% of the total amount of coal fuel supplied to the chamber. The air temperature in the lower tier was 830°C, and the temperature of air-steam mixture in the upper tier was 900°C.

Table 1. Technical and elemental analysis of Kuznetsk coal.

W^a, %	A^d, %	V^d, %	V^{daf}, %	C^{daf}, %	H^{daf}, %	N^{daf}, %	S^{daf}, %	O^{daf}, %	Q_s^{daf}, kc/kg
5.4	22.3	34.7	44.7	75.57	5.66	1.78	0.55	16.44	7086

For the numerical simulation of turbulent flow of an incompressible liquid we used the Reynolds equations taking into account the interfacial interaction. The Reynolds equations were closed using the standard two-parameter k-ε turbulence model. For modeling the wall boundary conditions we used the wall-functions method. The solution to the equation of radiant energy transport is based on the P1 approximation method of spherical harmonics for a gray two-phase two-temperature medium. The gas absorption coefficients are calculated according to the gray gases sum model, particles absorption and scattering coefficients are determined by the approximation of optically large particles. In the present work, to describe the particles motion we used the Lagrange method. Accounting for the flow turbulence in the particle motion is produced by the introduction of random fluctuations of the gas velocity into the particle motion equation. The calculation of chemical kinetics of gaseous fuel combustion is based on the use of global irreversible reactions between the combustibles and the oxidant. The rate of reactants combustion R_{vol}, including volatiles, is determined according to the reactive capacity and concentration of fuel and oxidizer, as well as the rate of turbulent mixing of fuel and oxidant. This model represents the combination of kinetic combustion model of the gas components with the eddy break-up model. The coal particle burning process was represented as the following consecutive steps: evaporation of moisture from the fuel, devolatilization and the combustion of the volatile components, and the combustion of the coke residue. Yield of volatiles is considered in the single-component approximation in the form of a $C_xH_yO_z$ substance. To calculate the rate of devolatilization we used single-stage kinetic model with constants given in Table 2. The combustion rate of coke residue was calculated in accordance with the provisions of the classical diffusion-kinetic theory. The specific reaction rates for the oxidation of the coke residue are given in Table 2. To describe gasification processes, we included into the mathematical model the reactions of steam-air conversion of coal. Reactions and kinetic constants are presented in Table 2.

The conservation equations for the gas phase are written in the form of a generalized conservation law in a control volume. Thus, finite-difference analog of the equation is written for the concerned volume. For calculation of diffusion fluxes on the faces of the control volume we used the central-difference scheme with second-order accuracy. When approximating convective terms, we used the second-order accuracy scheme with the differences against the flow. To correlate the velocity field and pressure we used SIMPLE-C procedure. Source terms that take into account the effect of dispersed phase on the carrier flow, are formed using the Particle-Source-In-Cell (PSI-CELL) method.

The proposed model and solution methods were previously tested for solving problems of pulverized coal combustion and gasification [4-8] and showed a satisfactory agreement with the experimental data on the basic process parameters in the combustion chamber.

Table 2. Specific reaction rates of combustion and gasification.

Reaction	Ks = A exp(-E/RT)		References
	m/s	J/mol	
1. $C + O2 \rightarrow CO + CO2$	$3.3 \cdot 10^4$	135756	[9]
2. $C + H2O \rightarrow CO + H2$	$1.6 \cdot 10^4$	181427	[10]
3. $C + CO2 \rightarrow 2CO$	$3.5 \cdot 10^4$	140365	[11]

3 Results and discussion

When carrying out computational studies, we have examined several options of the two-stage gasification process. Thus, for comparative analysis we considered the influence of the inclination angle of the upper tier burners (Figs. 3, 4) in the horizontal (angle α) and vertical (angle β) planes on heat and mass transfer processes in the chamber, as well as on the gases composition at the outlet. Initial data are presented in Table 3.

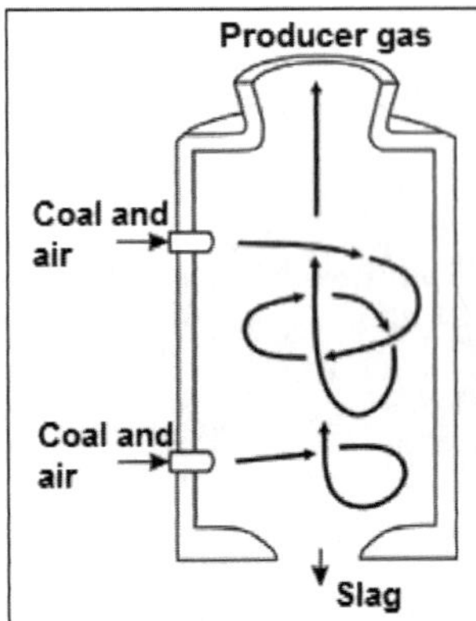

Figure 1. EAGLE gasifier

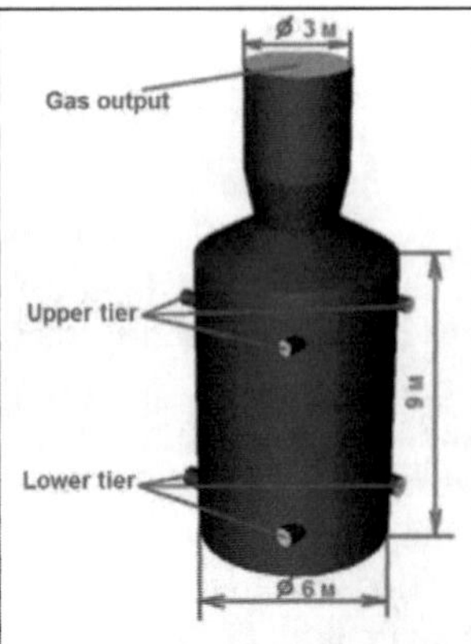

Figure 2. Feeding diagram and gasifier dimensions at a flow rate of 1700 t/h

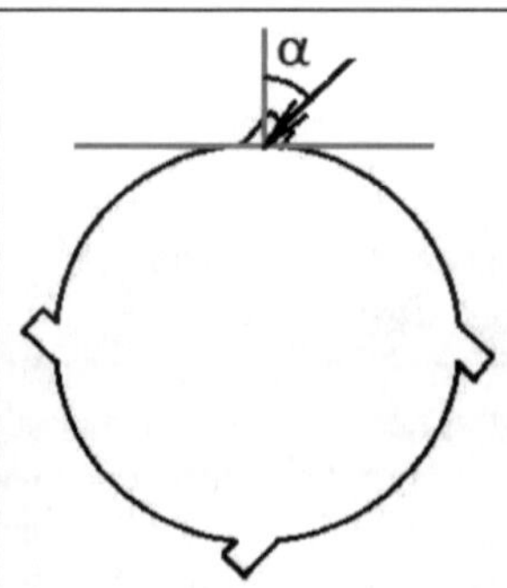

Figure 3. Inclination angle of burners in the horizontal plane of the upper tier

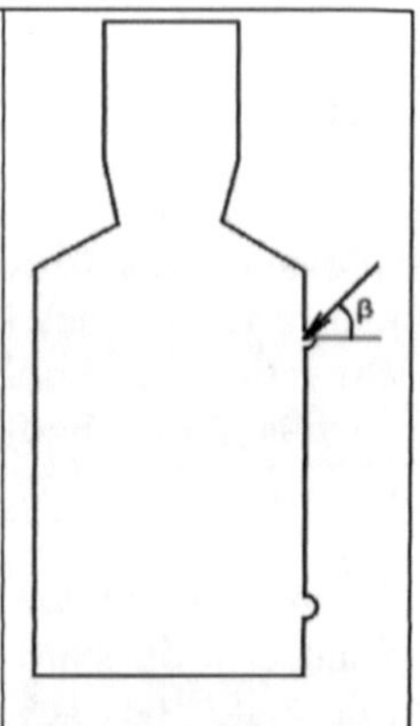

Figure 4. Inclination angle of burners in the vertical plane of the upper tier

Table 3. Gasification process options in the chamber.

Option	1	2	3
Inclination angle of upper burners α, °	30	45	60
Inclination angle of upper burners β, °	15	15	15
Inclination angle of bottom burners α, °	45	45	45
Inclination angle of bottom burners β, °	0	0	0

Figure 5 shows the calculations results presented in the form of the temperature field in the central section of the gasification chamber at different injection angles of pulverized coal and air-steam mixture into the upper tier of the burners. It is obvious that there is a significant influence of coal burners tilt in the horizontal plane on heat and mass transfer processes in the combustion chamber. Thus, the reduction of the inclination angle leads to a reduction in the high temperature region in the combustion chamber. This is due to the difference of the upward flow velocity in the center of the combustion chamber. In Fig. 6 we can see that the upward flow velocity in the center of the chamber at a height of H=2.5 m for the case of α=30° is 12 m/s, while for the rest options the velocity ranges from 5 to 7 m/s. Comparison of axial velocity profiles across the chamber at a height of H=2.5 m

Journal of Physics: Conference Series **754** (2016) 112007 doi:10.1088/1742-6596/754/11/112007

shows that at the angle of α=30° the downward gas flow along the chamber walls is less intense as compared to the angles α equal to 45 and 60°. At heights of H=5 and 6.5 m, we observed the alignment of the axial velocity profiles for all options. The temperature distribution along the height of the chamber is characterized by symmetry with respect to the axis, while temperature in the chamber increases with reducing the flow injection angle α. This is explained by more intense flow of descending gas and fuel from the upper tier of burners at increased injection angle α that contributes to better mixing with the pulverized coal combustion products of lower tier burners, as well as more intense endothermic gasification reactions with the evolution of gasification products in the form of CO and H_2 (Fig. 6). Temperatures and gas compositions at the exit of the chamber are approximately the same for all options (Table 4). The increase in the angle α allows intensifying the two-stage gasification process and makes it possible to reduce the height of the chamber without sacrifice of the coal-derived gas composition.

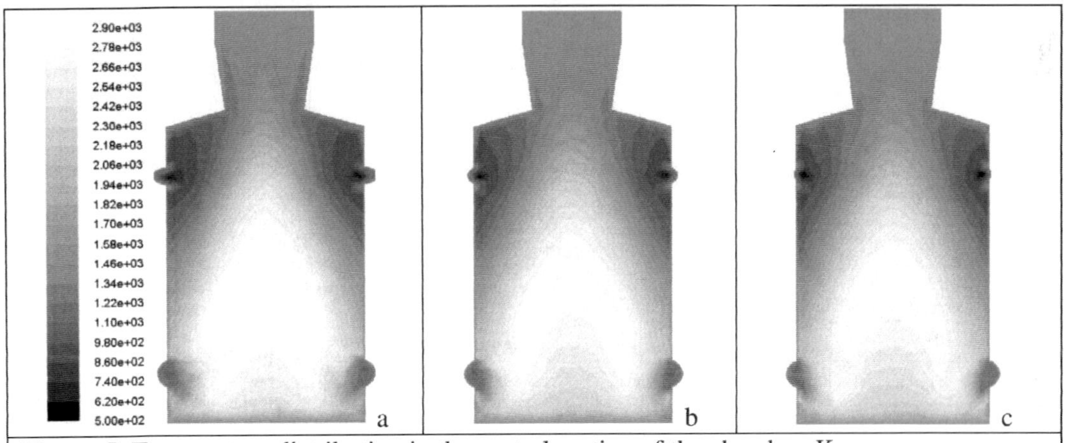

Figure 5. Temperature distribution in the central section of the chamber, K.

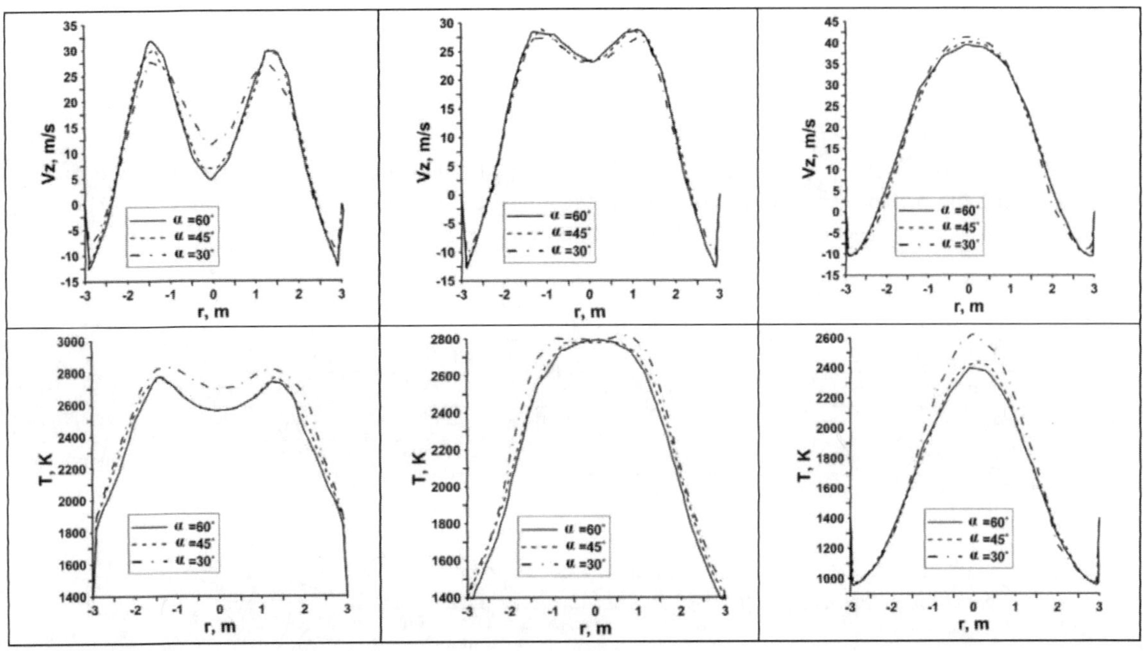

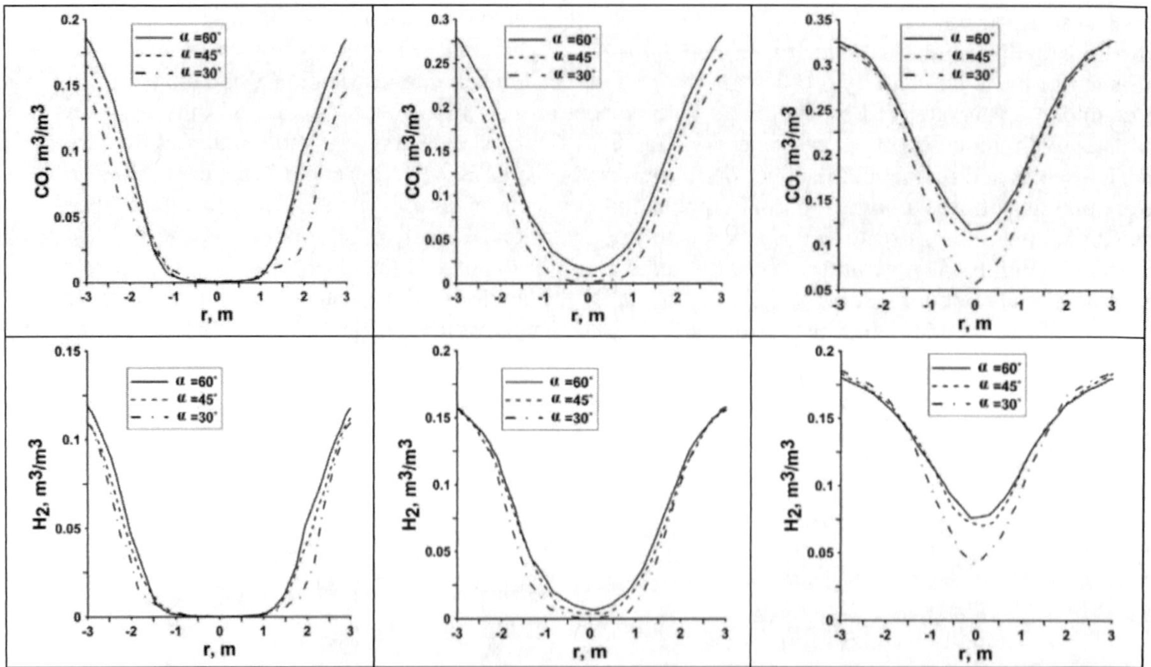

H=2.5 m | H=4.5 m | H=6.5 m

Figure 6. Radial distribution of axial velocity, temperature, and volumetric concentration of carbon monoxide and hydrogen over the chamber height.

The results of computational studies are presented in Table 9 in terms of the coal-derived gas composition at the outlet of the chamber. Comparative analysis shows that increasing of angle α leads to increase of chemical efficiency from 84.7% to 86%.

We have studied also the effect of steam on the pulverized-coal gasification process. As in the above considered simulations, the steam with a temperature of 900°C was supplied to the upper burners. Steam flow rate was ranged from 0 to 0.5 kg of steam per kg of coal. The simulations were performed for option #3 with angles α = 60° and β = 15°. The results of computational studies are presented in Table 5. We can see that the increase in the amount of superheated steam leads to a change in the composition of the coal-derived gas at the outlet of the gasification chamber. Thus, the proportion of CO reduces while the proportion of H_2 increases. Increasing the proportion of H_2 leads to increase in chemical efficiency of the process to 87% at steam flow rate equal to 0.5 kg steam per kg of coal.

Table 4. The composition of the coal-derived gas at the outlet of the chamber.

Option No.	1	2	3
CO, %°	27.87	27.91	28.5
H_2, %°	16.75	16.73	16.6
CO_2, %°	4.7	4.65	4.51
H_2O, %°	2.06	2.1	2.21
CH_4, %°	2.33	2.39	2.3
N_2, %	46.29	46.22	45.88
Gas temperature at the outlet, K	1367	1377	1382
Chemical efficiency	84.7	85	86

Table 5. The composition of the coal-derived gas at the exit of the chamber depending on the amount of injected steam.

Amount of steam, kg of steam / kg of coal	Chemical efficiency, %	$CO+H_2+CH_4$, vol. %	Gas temperature at the outlet, K
0.5	87	20.73+20+0.9	1343
0.268	86	25.7+19.3+1.4	1372
0.137	86	28.5+16.6+2.3	1382
0.089	85	29.8+16+2.3	1386
0.045	85	31.1+14.6+2.65	1398
0	83	31.5+13.8+2.8	1414

Conclusion

Numerical studies of advanced two-stage flow gasifier have been carried out for Kuznetsk coal gasification, based on hydrodynamic flow separation with the implementation of a combined counter-current and concurrent flow pattern.

It is shown that the increase of the inclination angle of the burners of the upper tier in a horizontal plane leads to a decrease in the velocity of the upward central flow, as well as to reduction of the height of the high-temperature combustion zone that enables decreasing the height of the chamber without reducing the chemical efficiency of the gasification process.

Moreover, computational studies on the influence of high-temperature steam on the pulverized-coal gasification process were carried out. It was revealed that the increase of steam supply from 0 to 0.5 kg of steam per kg of coal at a temperature of 1173 K leads to an increase in chemical efficiency of the gasification process from 83 to 87%, as well as lowers the temperature at the outlet of the gasification chamber from 1414 to 1343 K.

References

[1] Higman C., 2013, State of the Gasification Industry – the Updated Worldwide, Proceedings of the International Pittsburgh Coal Conference, Beijing, China, September 16–19.

[2] Higman C., Burgt M., 2008, Gasification. Elsevier Science, 435 p.

[3] Kiso F., Akiyama T., Morihara A., Takahashi K., Kida E., Iritani J., et al., 2000, EAGLE project for IGFC in Japan. Proceedings of the 25th International conference on coal utilization & fuel systems, Clearwater, FL, USA.

[4] Maidanik M.N., et al., 2011, Mathematical simulation of the furnace and turning gas conduit of a P-50R boiler during joint combustion of solid and gaseous fuel, Thermal Engineering (English translation of Teploenergetika), 58(6), pp. 483 – 488.

[5] Messerle V.E. et al., 2015, Modeling and full-scale tests of vortex plasma–fuel systems for igniting high-ash power plant coal. Thermal Engineering (English translation of Teploenergetika), Vol. 62, Issue 6, pp. 442-451.

[6] Chernetskiy M.Yu., Dekterev A.A., Burdukov A.P., and Hanjalić K., 2014, Combustion of mechanically-activated micronized coal, Transport Phenomena, Delft University of Technology, 135, pp. 443-458.

[7] Abaimov N.A., Ryzhkov A.F., 2015, Development of a model of entrained flow coal gasification and study of aerodynamic mechanisms of action on gasifier operation. Thermal Engineering (English translation of Teploenergetika). Vol.62, Issue 11, pp. 767-772.

[8] Krasinsky D.V., Salomatov V.V., Anufriev I.S., Sharypov O.V., Shadrin E.Y., and Anikin Y.A., 2015, Modeling of pulverized coal combustion processes in a vortex furnace of improved design. Part 1: Flow aerodynamics in a vortex furnace. Thermal Engineering (English translation of Teploenergetika), Vol. 62, Issue 2, pp. 117-122

Acknowledgments

Problem statement (section 1 and 2) and results analysis (section 4) were carried out at UrFU and supported by the Russian Science Foundation by Grant 14-19-00524 (M. Chernetskiy and A. Ryzhkov)

Experimental study of fluid dynamics in the pebble bed in a radial coolant flow

Y V Smorchkova, A N Varava, A V Dedov and A T Komov

National Research University "Moscow Power Engineering Institute"
Russia, 111250 Moscow, Krasnokazarmennaya, 14

Abstract. The results of experimental studies of pebble bed hydrodynamics are presented. For the first time experimental data on the pressure loss in a radial flow of fluid through the pebble bed was obtained. Experiments were carried out in the liquid flow rate ranging from 0.09 to 0.4 kg / s, fluid temperature is 20°C.

1. Introduction

In recent years, the problem of improving the efficiency and safety of nuclear power plants becomes urgent. One solution to this problem is to use the microfuel fuel assemblies (MF FA), immediately cooled by single-phase or two-phase coolant [1,2]. The design of MF FA for VVER reactors was proposed in [1]. In such an assembly microfuel is placed between the perforated covers. To minimize pressure losses radial flow of coolant was proposed. Despite the fact that the MF FA geometrical parameters is fully consistent with the traditional assemblies, their implementation on a large scale in nuclear energy is unlikely.

However, now interest in small nuclear power is increasing. In Russia, this interest is primarily related to the need of development of remote regions. Also, low-power nuclear station can solve the problems associated with increasing the energy security of critical facilities, such as industrial facilities with continuous production, urban infrastructure, etc., providing the vital needs of the city.

However, for low-power stations wide spreading it is extremely important to provide a reliable radiation safety in case of accidents. So high temperature resistance, high integrity and good retention of fission products within microfuel up to temperatures of about 1600 ° C are of particular importance. In [3] and other works microfuel tests for strength and corrosion resistance were described.

We regard options of microfuel consideration for the use of low-power reactors to be very interesting and promising.

One of the difficulties, arising in the implementation of microfuel in nuclear power plants, is the lack of knowledge of heat-hydraulic characteristics in the pebble bed with internal heat release. Quite a lot of works are devoted to the study of hydrodynamics and heat transfer in pebble bed, but the experiments in this work were carried out in narrow ranges of operating parameters, which leads to the need for more research in this area. In addition, there are no experimental studies of heat transfer and hydrodynamics in the pebble bed in a radial flow of coolant in the literature.

2. The experimental facility

2.1. The hydraulic circuit
The studies were conducted in an open hydraulic circuit, the circuit is shown in Fig. 1. The distilled water, which is in the tank (1) was used as an operating fluid. The circulation of the coolant is carried out by multistage centrifugal pump Grundfos CRNE 1-4 (2). Water from the tank is pumped through a filter (3) to the working area (6) and flows back to the tank. As the volume of water in the tank is large enough, in the experiments in the circuit water temperature does not change significantly.

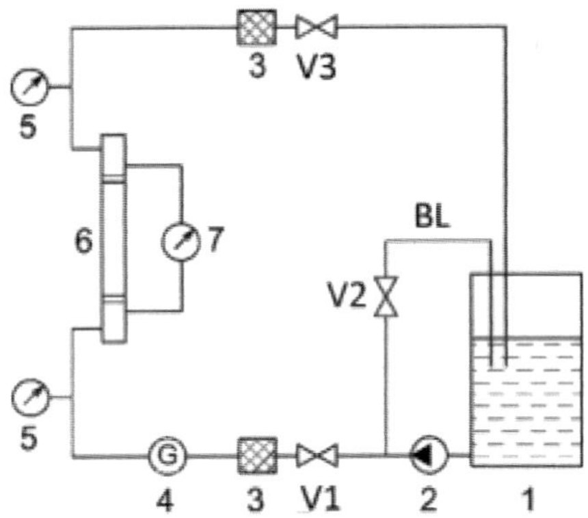

Fig.1. Hydraulic circuit

Flow control was carried out in two ways: step by step by using a frequency regulator that is installed on the pump and smoothly via the bypass line (BL) and the valves (V1, V2). Measurement of flow through the working portion was performed by an electromagnetic flowmeter (4) Vzlet EM Profi-212.

For pressure measurement in the work area exemplary arrow manometers (5) on the inlet and outlet of the working area and the differential pressure sensor (7) Elemer 100 are used.
Experimental set is designed for the following operating parameters: temperature in the circuit is 20 °C to 180 ° C, the pressure in the circuit is up to 1.0 MPa, the coolant flow rate is (0,01-0,50) kg /s.

2.2. Working area
The working area diagram (WA) is shown in Fig. 1. WA is made of material which is transparent to high-frequency radiation, as the further experiments are suggested to be carried out by the heating of the pebble bed via high-frequency generator. Pebble bed (3) is placed between the inner (2) and outer (4) perforated covers. Balls of 2.0 mm diameter are made from steel grade AISI 420. Perforated inner cover has a conical shape, in order to ensure a steady flow rate adjustment of pebble bed. The cover is made of abs plastic using 3D printing. The outer cover is a perforated tube made of polycarbonate. Perforation of the covers was performed by drilling, hole diameter 1.5 mm.

The geometric parameters of the working arrea are the following: inlet diameter (the diameter of the base of the cone) is 11 mm, the wall thickness of the cone (the inner perforated cover) is 2 mm, the diameter of the outer perforated cover is 47 mm, wall thickness is 1,5 mm, the diameter of the outer pipe (6) is 54 mm, wall thickness is 3 mm. The height of the filling is 100 mm.

To minimize pressure losses in such a construction, the flow through the pebble bed should be close to the radial. To do this, we should define inner and outer covers' perforation parameters. Determination of perforation options was performed on the basis of numerical simulation results in the software package ANSYS Fluent. During the simulation, it was found that a steady height perforations of covers shows unsatisfactory results. Therefore, the covers were divided into three sections and for each section we determined the coefficients f living section and number of holes. Covers' perforation parameters are shown in table 1.

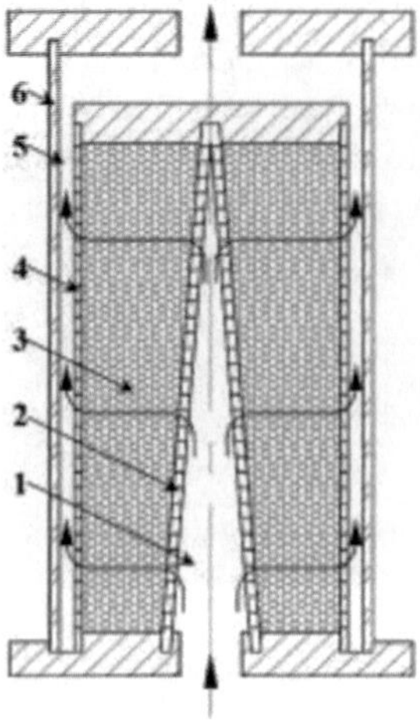

Fig.2. The working area diagram

Table 1. Covers' perforation parameters

	Inner cover		Outer cover	
	f	Number of holes	*f*	Number of holes
Lower section	0,299	350	0,113	400
Middle section	0,395	250	0,148	390
Upper section	0,471	200	0,148	390

3. Experimental data

The porosity of the pebble bed was determined by gravimetric method and is estimated to be 0.385. According to the results of multiple weighing on laboratory scale portions of 100 balls was determined their mass, as all the balls are calibrated, difference with the weighing results is not more than 1%. Further, we know the mass of one ball and a set of balls from the work area, their number and we can calculate the volume. The porosity of the pebble bed was determined using the formula:

$$\varepsilon = 1 - \frac{V_p}{V}$$

V – the volume of the working area section, where the pebble bed placed, V_p– the volume of balls.

Experiments to determine the pressure loss was conducted in the range of flow rate from 0.09 to 0.4 kg /s at 20 ° C fluid. The range of Reynolds numbers (Re = Ud/ν, U is the superficial velocity, d is the diameter of the ball, ν is the kinematic viscosity) was 1800 – 8400. Initial experimental data is shown in Fig. 3 by points. For the development tests of the method of measuring the experimental values were compared with the results of numerical simulations performed in ANSYS Fluent. The simulation results are presented in Fig. 3 by a solid line. The discrepancy between the results does not exceed 12%. Unfortunately, it is not possible to compare the experimental data with known equations for the pebble bed hydraulic resistance because all available in the literature equations are designed to flow through the pebble bed, located in the circular cross-section tube and give overestimated results.

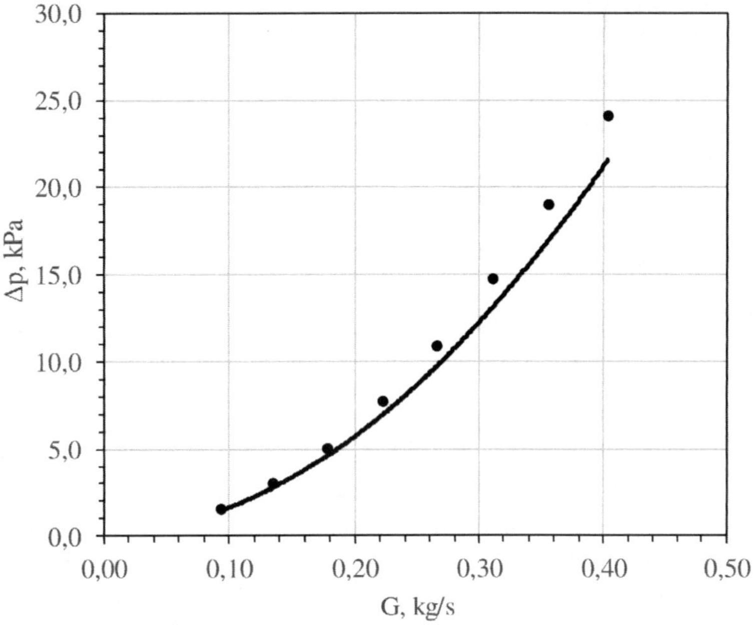

Fig.3. The pressure loss relation on the mass flow rate

For comparison in Fig. 4 shows the results of this work and the dotted line presents the experimental data obtained earlier [4] for the pebble bed, located in the circular tube with axial flow of liquid. The porosity and height of the pebble bed, as well as operating parameters of the experiments in [4] were similar to this article. We can see that the pressure loss by organizing the radial flow through the pebble bed is significantly reduced. This confirms the necessity of radial flow when using ball fillings in nuclear power installation.

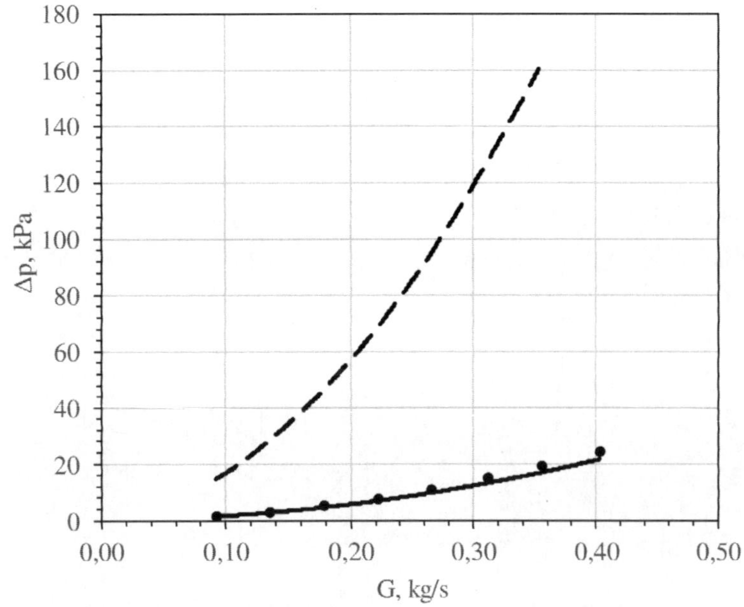

Fig.4. The pressure loss relation on the mass flow rate

Journal of Physics: Conference Series **754** (2016) 112008 doi:10.1088/1742-6596/754/11/112008

4. Conclusion

Experimental data on the pressure losses in the pebble bed in a radial flow of coolant was first obtained, despite the fact that numerical modeling of such geometries was conducted previously and it was applied to a microfuel fuel assemblies for nuclear reactors. Thereby, currently the accuracy of the data is only possible to estimate by comparison with the simulation results, as calculated ratios presented in the literature allow us to determine the hydraulic resistance ball fillings with an axial flow.

Grant RSF № 16-19-10457

References

[1] Ponomarev-Stepnoy N N, Kukharkin N E, Khrulev A A, Degaltsev Y G etc. 1999 Nuclear Energy Prospects of microfuel VVER vol 86 **6** 443-449

[2] Goltsev A O and others 1993 Nuclear Energy The concept of the body safely pressurized water reactor with a heat-generating units based on microfuel HTGR vol 75 **6**

[3] Filippov G A, Grishanin E I, Falkowski L N etc. 2009 Nuclear Energy Evaluation microfuel resistance of protective coatings in the gas-vapor medium in the interaction with structural materials vol 106 **3** 153-158

[4] Borozdin A V, Varava A N, Dedov A V, Komov A T, Malahovskiy S A, Smorchkova Y V 2015 Thermal processes in the engineering Experimental study and numerical simulation of fluid flow and heat transfer in the pebble bed **7** 295-300

Numerical analysis of experiments with gas injection into liquid metal coolant

E V Usov[1], P D Lobanov[1], N A Pribaturin[1], N A Mosunova[1], V I Chuhno[2] and A E Kutlimetov[2]

[1] Nuclear Safety Institute of the Russian Academy of Sciences, Novosibirsk Branch, Novosibirsk, 630090, Lavrentieva Street, 1, Russia
[2] Novosibirsk State University, Novosibirsk, 630090, Pirogova Street, 2, Russia

Abstract. Presented paper contains results of a numerical analysis of experiments with gas injection in water and liquid metal which have been performed at the Institute of Thermophysics Russian Academy of Science (IT RAS). Obtained experimental data are very important to predict processes that take place in the BREST-type reactor during the hypothetical accident with damage of the steam generator tubes, and may be used as a benchmark to validate thermo-hydraulic codes. Detailed description of models to simulate transport of gas phase in a vertical liquid column is presented in a current paper. Two-fluid model with closing relation for wall friction and interface friction coefficients was used to simulate processes which take place in a liquid during injection of gaseous phase. It has being shown that proposed models allow obtaining a good agreement between experimental data and calculation results.

1. Introduction

During the hypothetical accident with damage of a steam generator tubes in the reactor with lead coolant water steam under high pressure may enter the lead coolant. It may cause a pressure spike in coolant with amplitude much more than nominal pressure. A vapour bubble forms as a result of a steam injection and starts to move along a reactor loop. Changing of bubble volume causes the oscillations of coolant pressure and free level of liquid.

To simulate the accidents with steam generator tubes rupture system thermal-hydraulic codes can be used. In the Nuclear Safety Institute of the Russian Academy of Sciences in frame of Federal Target Program "Nuclear Power Technologies of the New Generation for 2010–2015 and until 2020" HYDRA-IBRAE/LM code [1,2] has been developed for best-estimate transient simulation of fast reactors during postulated accidents including SGTR.

To simulate processes that take place in heavy liquid metal coolant during injection of gaseous phase two-fluid [3] model with constitutive relations for wall and interphase friction and heat transfer coefficients is used. Constitutive relations depend on the flow regime map. HYDRA-IBRAE/LM code contains database with thermo physical properties for different type of coolants: air, water, sodium, lead, lead-bismuth and Rose's alloy

To check the correctness of HYDRA-IBRAE/LM code to model SGTR it should be widely validated. For this purpose experiments with gas injection into the Rose's alloy have been conducted. Experiments with water have been carried out to validate models for gas transfer in the water coolant due to fuel rod discharge. These experiments have been performed at the IT RAS with water and Rose's alloy as a coolant and argon and air as a gas.

2. Brief description of experimental setup

The main goal of experiments performed at the Institute of Thermophysics was to obtain data on the pressure evolution in heavy liquid metal and water during gas injection. Dependences of pressure amplitude and its oscillation frequencies on an inlet gas flow rate and duration of injection have been received. A vertical circular tube has been used in the experiments. The tube diameter – 25 mm, its height – 1.2 m. The tube has been partially filled with water or liquid metal. Height of liquid has been equal to 1 m. A gas tank has been placed under the tube. Tank and vertical tube have been connected with valve. The gas has been injected from the gas tank through the valve. The valve has been opened by electrical signal. The time of full valve opening has reached 5 ms and depended on the experiment. The time of injection has been changed from 0.05 to 0.2 s. Experiments with water have been carried out with air. Some experimental parameters are shown in the table 1. Experiments with a Rose's alloy have been carried out with inert gas argon due to high chemical activity of the Rose's alloy. Experimental parameters with Rose's alloy and water are shown in the table 1.

Table 1. Experimental parameters

№	Coolant type	Volumetric flow rate ($l\ s^{-1}$)	Valve opening time (ms)
1	Water	0.88	50
2	Water	0.98	100
3	Rose's alloy	0.66	50
4	Rose's alloy	0.52	100

It is important to tell about accuracy of the experimental data. The error of the valve opening time measurements is 20%, the pressure measurement error is 15% and the volumetric flow rate measurement error is 10% for water and 15% for Rose's alloy. For test with water flow rate was measured with flow meter. For Rose's alloy flow rate was found by the measuring of the liquid free level position due to experimental difficulties in rate measuring with the flow meter at high temperatures. So the flow rate measurement data has less accuracy for Rose's alloy than for water. Also, there are several difficulties in measurement of the flow rate time dependence. To provide calculation we assume that there are fixed flow rate during injection. But this assumption adds errors in the simulations.

3. Mathematical models of the HYDRA-IBRAE/LM code

A basic system of equations for two-phase flow calculation is presented below:

$$\frac{\partial}{\partial t}(\varphi_i \rho_i) + \frac{1}{A}\frac{\partial}{\partial z}(A\varphi_i \rho_i v_i) = \Gamma_i \tag{1}$$

$$\frac{\partial(\varphi_i \rho_i v_i)}{\partial t} + \frac{1}{A}\frac{\partial(A\,\varphi_i \rho_i v_i v_i)}{\partial z} + \varphi_i \frac{\partial}{\partial z}P = \Gamma_i v_i + \tau_{interi} + \tau_{wi} - \varphi_i \rho_i g \cdot \sin\vartheta \tag{2}$$

$$\frac{\partial(\varphi_i \rho_i h_i)}{\partial t} + \frac{1}{A}\frac{\partial}{\partial z}(A\varphi_i \rho_i h_i v_i) - \varphi_i \frac{dP}{dt} = \Gamma_i h_i + \Gamma_w h_i^w + Q_{i,diss} + Q_{interi} + Q_{wi} + Q_i \tag{3}$$

$$\sum \varphi_i = 1, \sum \Gamma_i = 0, \sum \tau_{interi} = 0 \tag{4}$$

The first equation is the mass conservation equation for i-phase. The second one is the momentum conservation equation. And the third is the energy conservation equation. In the equations φ is i-phase volume fraction, ρ is a density of phase ($kg \cdot m^{-3}$), h is a phasic specific enthalpy ($J \cdot kg^{-1}$) and v is a velocity ($m \cdot s^{-1}$) of phase. The values on the right hand of mass equation determine mass exchange rate ($kg \cdot m^{-3} \cdot s^{-1}$). The values on the right hand of momentum equation are the interphase friction ($Pa \cdot m^{-1}$) per unit volume, wall friction ($Pa \cdot m^{-1}$) per unit volume and the gravitational forces. For the energy equation the right hand contains heat flux from the interphase heat transfer ($J\ m^{-3} \cdot s^{-1}$) and wall heat transfer rates per unit volume. The last equations determine the connections between phases.

There are six unknown variable (φ, pressure, enthalpy of gas and fluid and velocity of gas and fluid) and six differential equations. Other variables (density, interphase friction, wall friction, mass and heat transfer) are functions of the basic variables.

Two-phase flow regime map for vertical channel that is used in HYDRA-IBRAE/LM for heavy liquid metal coolant with noncondensible is shown in the table 2. According to works [4] and [5] it is considered three two-phase regimes for liquid metal: bubble, slug and annular-mist. Different constitutive relations for different flow regime is used in the HYDRA-IBRAE/LM code. For example, interphase friction for bubble regime is calculated according to [6]. For slug regime data from [7] are used. Relations for interfacial area calculation are presented in [8]. Some relation for interfacial and wall heat transfer can be found in works [2], [9] and [10].

Table 2. Flow regime map for vertical channel

№	Upper boundary of gas volume fraction	Regime
1	0.3	Bubble
2	0.99	Slug
3	1	Annular-mist

4. HYDRA-IBRAE/LM calculation scheme

HYDRA-IBRAE/LM nodalization scheme has been developed to simulate experiments with gas injection. The scheme consisted of a vertical channel with an upper and bottom boundary condition. A volumetric gas flow rate is defined as the bottom boundary condition. At the beginning of the calculation the flow rate was equal to zero. Then it was increased to nominal level and stayed at this level during time while the valve had been opened. At the end volumetric gas flow rate was decreased to zero level. A typical time dependence of volumetric flow rate is shown in the figure 1. A fixed pressure is defined as an upper boundary condition. A bottom part of the channel was filled with liquid (water or Rose's alloy). An upper part of the channel was filled with argon. The scheme is shown in the figure 2. Proposed calculation scheme contains 200 computational cells in the upper part of the channel and 200 cells in the bottom part of the channel. The value of time step was equal 10^{-4} s.

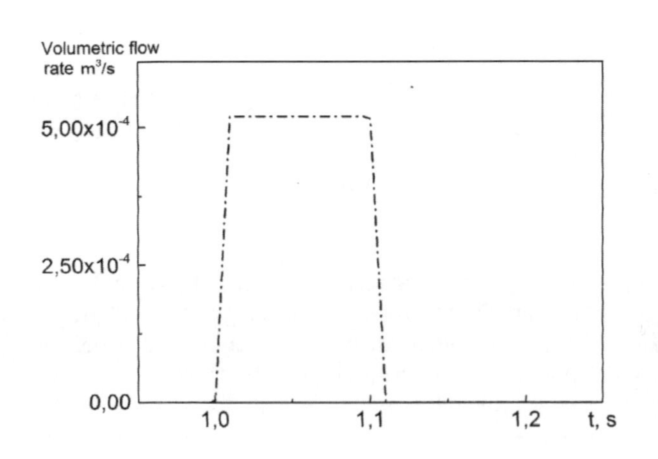

Figure 1. A typical time dependence of volumetric flow rate

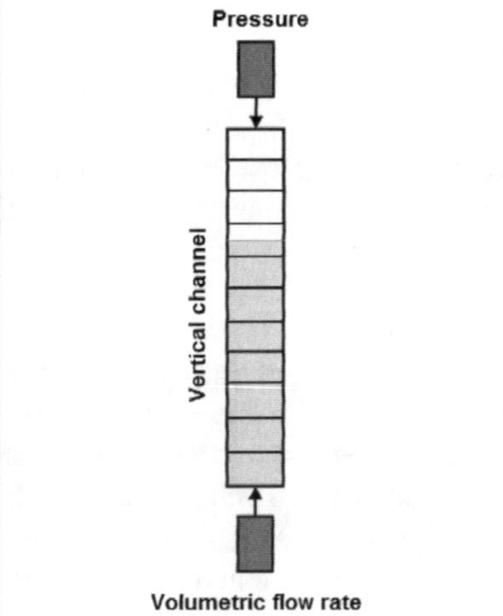

Figure 2. HYDRA-IBRAE/LM calculation scheme.

5. Calculation results

Liquid pressure evolution has been received as a result of the calculation. A pressure transducer was placed at the bottom of the vertical channel as in the experiment. Series of calculations has been carried out. Different valve opening times and different volumetric gas flow rate has been used in calculations.
The results of calculation for water as a coolant are shown in figures 3 and figure 4.

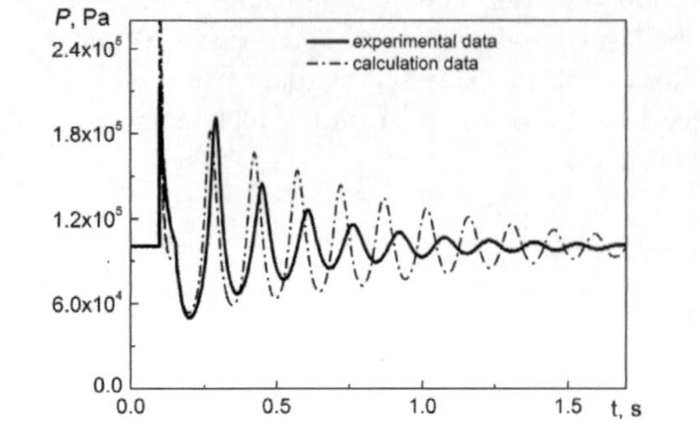

Figure 3. Pressure evolution for water as a coolant. Test № 1.

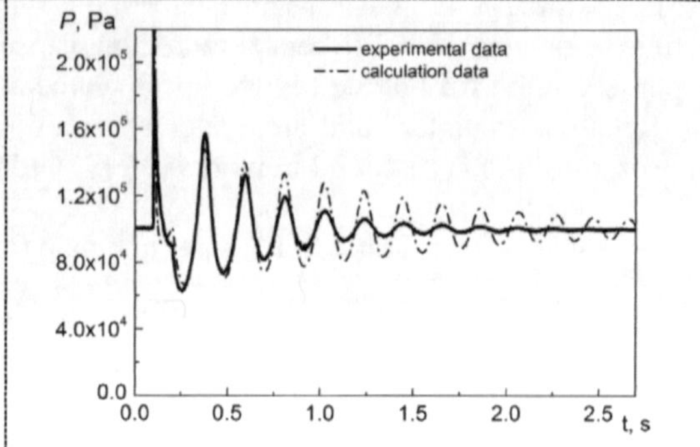

Figure 4. Pressure evolution for water as a coolant. Test № 2.

The results of calculation for Rose's alloy are shown in figure 5 and figure 6.

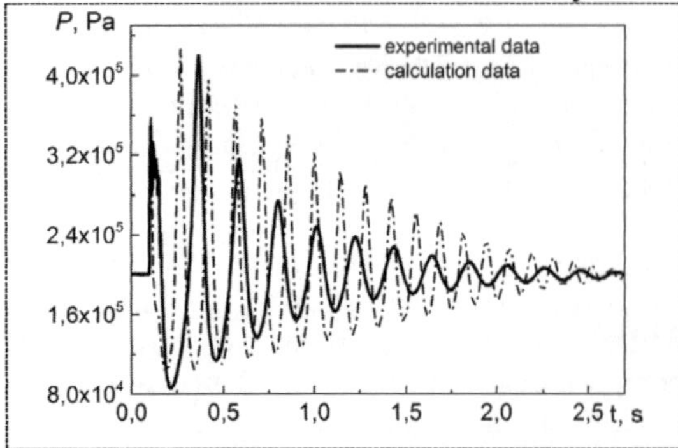

Figure 5. Pressure evolution for Rose's alloy as a coolant. Test № 3.

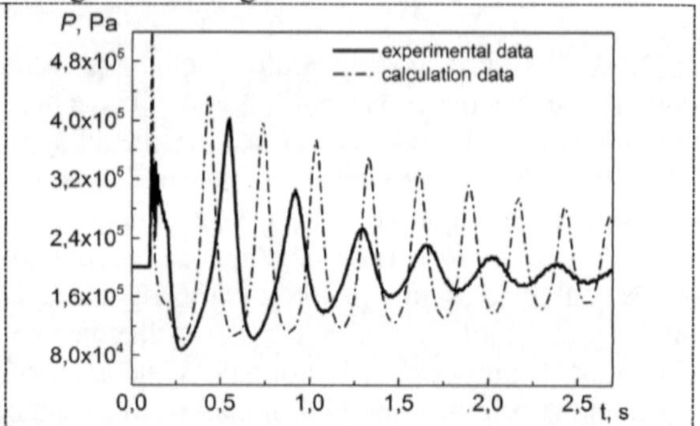

Figure 6. Pressure evolution for Rose's alloy as a coolant. Test № 4

It may be seen that calculation results for Rose's alloy have worse accuracy than the results obtained in water. There are several factors that can influence on results. The first one is the experimental errors. Actually, the flow rate measurement data have a bad accuracy for Rose's alloy. Also, it may be caused by shortcomings of closing relations that have been used for heavy liquid metal due to lack of experimental information on bubble motion in liquid metal coolant.

6. Conclusion

The results of the gas injection experiments have been presented in current paper. The experiments allow receiving information about pressure evolution during gas injection in stagnant heavy liquid metal and water column. Presented data can be used to validate thermal hydraulic code such as developed in Nuclear Safety Institute HYDRA-IBRAE/LM code. With help of the thermal hydraulic code HYDRA-IBRAE/LM simulations of the experiments have been provided. A good agreement between the experimental data and the results of calculation has been obtained. The paper also contains description of the experimental set up, experimental conditions and brief descriptions of the calculation models of the HYDRA-IBRAE/LM code that allow providing independent calculation of presented experiments.

7. References

[1] Alipchenkov V M, Anfimov A M, Afremov D A, *et al* 2016 *Thermal Engineering* **63** (2) 130
[2] Usov E V, Pribaturin N A, Kudashov I G, *et al* 2015 *Atomic Energy* **118** (6) 382
[3] Nigmatulin R I 1978 Fundamentals of Mechanics of Heterogeneous Media (in Russian) (Moscow: Nauka)
[4] Tobita Y 2006 *Exp. Nucl. Tech.* **153** (3) 2006 245
[5] Marten K, Baumgartner E, Khan M, Hoffman H 1982 Impact of Gas in Sodium Flow on the Temperature Variation of an LMFBR Rod Bundle *KFK* (3253) 47
[6] Loth E 2008 *Int. J. of Multiphase Flow* **34** 523
[7] Wallis G B 1969 One-dimensional two-phase flow (New York: McGraw Hill)
[8] Schor A L, Kazimi M S and Todreas N E 1984 *Nuclear Engineering and Design* **82** 127
[9] Barry R E and Balzhiser R E 1966 *Proc. of 3rd Int. Heat Transfer Conf. (Chicago)* vol 2 p 318
[10] Ninokata H and Okano T *Nuclear Engineering and Design* 1990 **120** 349.

Journal of Physics: Conference Series **754** (2016) 112010 doi:10.1088/1742-6596/754/11/112010

Simulation of optimal exploitation of an open geotermal loop

N A Vaganova[1,2] and M Yu Filimonov[1,2]

[1] Krasovskii institute of mathematics and mechanics UrB RAS, S.Kovalevskaya str. 16, 620990, Yekaterinburg, Russia

[2] Ural Federal University, Mira str. 19, 620002, Yekaterinburg, Russia

E-mail: fmy@imm.uran.ru

Abstract. Geothermal aquifers are a renewable resource of heat and energy. To encourage these resources open geothermal systems consisting of injection and production wells are commonly used. As a rule, such systems consist of two wells. Hot water from the production well is used and became cooler, and the injection well returns this cold water into the aquifer. To simulate this open geothermal system a three-dimensional nonstationar mathematical model and numerical algorithms are developed taking into account the most important physical and technical parameters of the wells to describe processes of heat transfer and thermal water filtration in the aquifer. Results of numerical calculations, which, in particular, are used to determine an optimal parameters for a geothermal system in North Caucasus, are presented.

1. Introduction

Geothermal energy is a type of renewable energy that encourages conservation of natural resources. According to [1], the geo-exchange systems save homeowners from 30 to 70% in heating costs, and from 20 to 50% in cooling costs, compared to conventional systems. Geo-exchange systems also save money because they require much less maintenance. In addition to being highly reliable, they are built to last for decades. In Russia, the geotermal units development started in the 1960s in Kamchatka region with Paratunskaya GeoTES, where steam and overheated water used to drive turbines and produce electricity. Geotermal units are in exploatation also in North Caucasus, Stavropol and Krasnodar regions. Up to 30 bln of m^3 of geoternal water is produced with the temperature from 80 to 110°C for these purposes. Water with such temperature can be used directly for industrial processes or for heating buildings.

A geothermal open loop system (figure 1) is a system of two wells which drilled to tap an aquifer with hot water. One well is injection, the second is production one. Temperature in production well is determined by temperature in the aquifer. This water then used for heating and then returns by pumps into the injection well with more cold temperature. Pressure which is generated by pumps forms a flux of water filtrating by an aquifer porous media (figure 2). During the time of exploitation of the system, cold water is distributed from injection to production well. A basic parameter of optimisation is the temperature of water in the production well. The computed temperatures changing during the years of exploitation are shown in figure 3.

Journal of Physics: Conference Series **754** (2016) 112010 doi:10.1088/1742-6596/754/11/112010

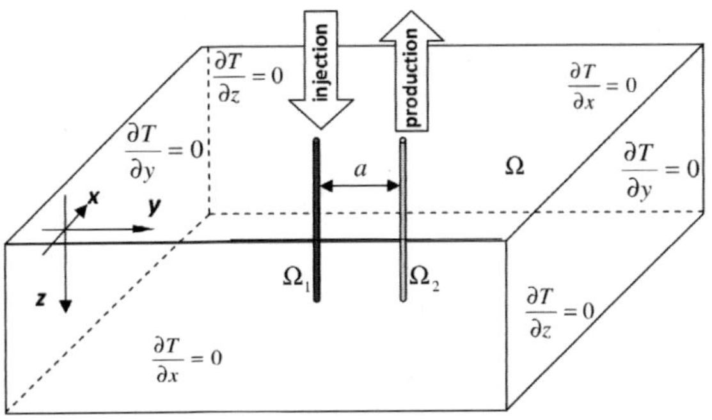

Figure 1. A model of an open geothermal loop.

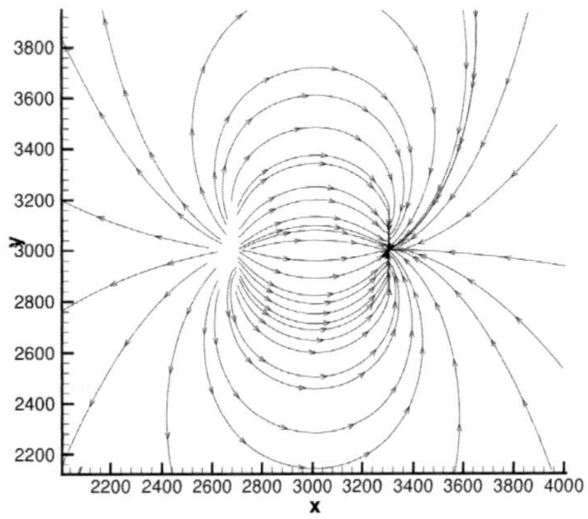

Figure 2. Streamlines of velocity in horizontal plane.

2. Mathematical model

In simulations of underground flow, Darcy's law and law of mass conservation (continuity equation) are used [2]. A convection-diffusion equation with dominant diffusion due to low velocity of filtration is considered. Let $T(t,x,y,z)$ be the temperature in the aquifer, $\mathbf{V}=(u,v,w)$ be a vector of filtration velocity. Thermal exchange is described by equation

$$\frac{\partial T}{\partial t}+b\left(\frac{\partial T}{\partial x}u+\frac{\partial T}{\partial y}v+\frac{\partial T}{\partial z}w\right)=\lambda_0\Delta T. \tag{1}$$

Here, $b=\dfrac{\sigma\rho c_f}{\rho_0c_0(1-\sigma)+\rho c_f\sigma}$, $\lambda_0=\dfrac{\kappa_0}{\rho_0c_0(1-\sigma)+\rho c_f\sigma}$, ρ_0 and ρ_f are density of aquifer soil and of water, c_0 and c_f are specific heats of aquifer soil and of water, κ_0 is thermal conductivity coeffitient of soil, σ is porosity. The aquifer has an initial temperature T_0, and the temperature in injection well is set as "cold water" with temperature T_1, which returns from production well after using. An analytical approach to describe filtration processes meets difficulties and restrictions [3], so it is necessary to use numerical methods [4, 5].

To compute the velocity, we use pseudoviscosity method with splitting by physical processes [6]. We solve an equation for pressure $p=p(t,x,y,z)$

$$\frac{\partial^2p}{\partial x^2}+\frac{\partial^2p}{\partial y^2}+\frac{\partial^2p}{\partial z^2}=0 \tag{2}$$

with the following boundary conditions for the surfaces of injection and producing wells

$$P(t,x,y,z)\Big|_{\Omega_1}=P_1-\rho gz,\quad P(t,x,y,z)\Big|_{\Omega_2}=P_2-\rho gz. \tag{3}$$

At the lateral boundaries of Ω for temperature and pressure, we set the zero flux conditions. The dimensions of the computational domain are large enough as to avoid the influence of boundary conditions. Due to low velocity of filtration, we can use a steady-state flow to describe convective transport terms in equation (1). Following [6] to compute the components of velosity, we may use the following equation for the previously constructed pressure field:

$$\frac{\partial p}{\partial t}+\nabla\mathbf{V}=0. \tag{4}$$

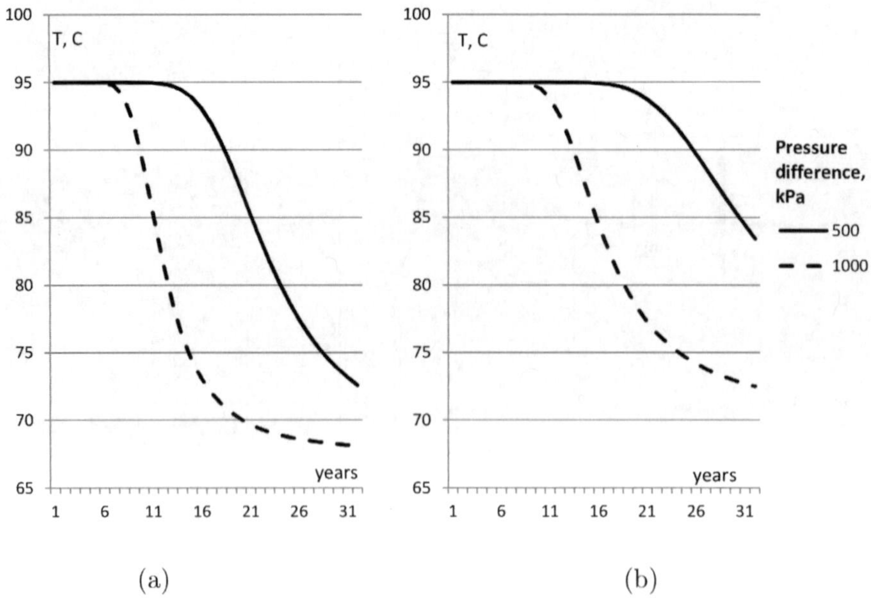

Figure 3. Temperature in productive well during exploitation for two differences of pressure in injection and production wells. The distance between wells is 600(a) and 700(b) m.

The equations (1)–(4) for temperature and pressure in aquifer are solved using a finite difference method based on an approach of works of A.A.Samarskii and P.N.Vabishevich [7]. A finite difference method is used with splitting by the spatial variables in three-dimensional domain to solve the problem. We construct an orthogonal grid, uniform, or condensing near the ground surface or near the surfaces of Ω_1 and Ω_2. The original equation for each spatial direction is approximated by an implicit central-difference scheme and a three-point sweep method to solve a system of linear differential algebraic equations is used. This approach was successively used for the problems of describing a thermal trace from underground pipeline taking into account filtration and evaporation of fluid from soil surface [8, 9]. After finding the pressure field, a vector of velosity of filtered water is determined in the aquifer.

Table 1. Thermal parameters of the aquifer.

	soli	water
Thermal conductivity [W/mK]	2.00	-
Density [kg/m^3]	-	1000.0
Specific heat [J/(kg K)]	-	4.18
Volumetric heat [J/(m^3 K)]	2150	
Filtration velocity [m/s]	$1.7{\cdot}10^{-5}$	
Porousity	0.241	

3. Numerical results

Consider a computational domain be a box 6000 m·6000 m·50 m size (figure 1). Injection and production wells are in the points (2700 m, 3000 m) and (3300 m, 3000 m), respectively. The distance between the wells is 600 m. Initial temperature of water in the aquifer is from 95 to 105°C. Temperature of injected water is 55°C. Soil thermal parameters correspond to Khankal geothermal fields in the North Caucasus and presented in Table 1.

To solve the problems, a finite-difference an implicit central-difference upwind scheme and a three-point sweep method is used with splitting by the spatial variables in three-dimensional domain with orthogonal grid, uniform, or condensing near the surfaces of wells. For computing pressure p, an iterative steady-state method is used with so called "weak" boundary conditions. Because the processes are slow and quasi-stable, we use the sequence pressure (steady state)→velosity field (steady state)→temperature in dependence with time. Computations are carried out with 1 day time step for 31 years.

We present the results of calculations for the differential pressure between production and injection wells for 500 kPa ($P_1 = -P_2$=250 kPa) and 1000 kPa ($P_1 = -P_2$=500 kPa). The thermal fields in a horizontal section of the aquifer for 10 and 20 years of exploitation are preseted in figures from 4 to 7. The injection and production wells are denoted by the black and white circles, respectively.

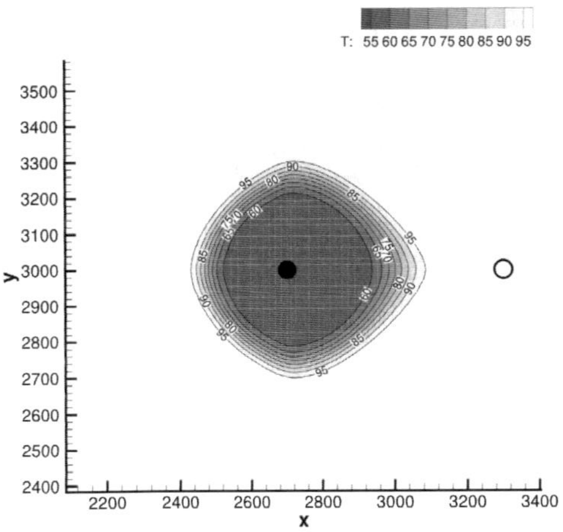

Figure 4. Temperature in aquifer for 10 years of exploitation. Pressure difference is 500 kPa.

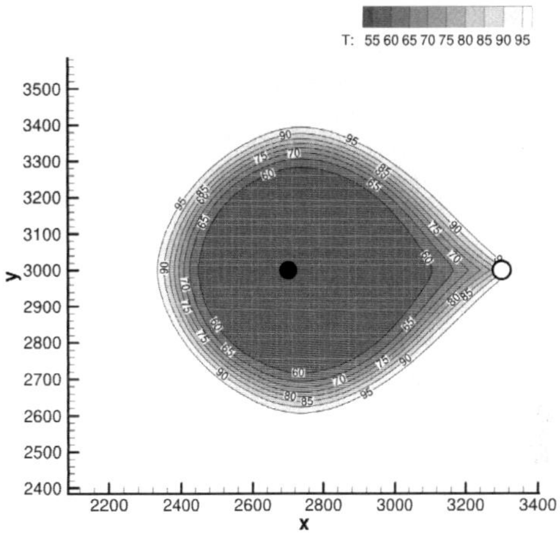

Figure 5. Temperature in aquifer for 20 years of exploitation. Pressure difference is 500 kPa.

An effective life of an open geothermal system depends not only from distance between the wellheads in aquifer (figure 3), but mostly from power of using, i.e. value of pressure difference. The pressure in the wells may be changed due to seasonal needs variation or others restrictions.

4. Conclusion

Computations allow simulate different regimes of exploitation and to estimate parameters of an open geothermal system, in particular to determine an appropriate distance between injection and production wells and pump pressure depending on the operating conditions of the geothermal system. Investigations of different types and forms of injection and productive wells will follow.

Acknowledgments

This work was supported by RFBR (projects No. 16-01-00401 and No. 14-01-00155), by RF Government (contract No. 02.A03.21.0006, reg. No. 211) and by Program of UB RAS (project No. 15-7-1-13).

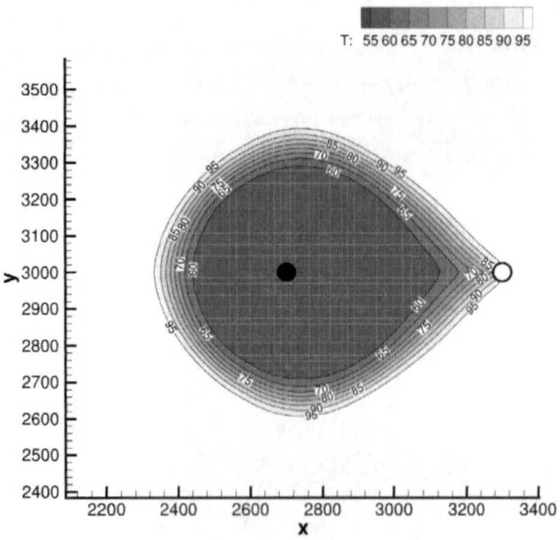

Figure 6. Temperature in aquifer for 10 years of exploitation. Pressure difference is 1000 kPa.

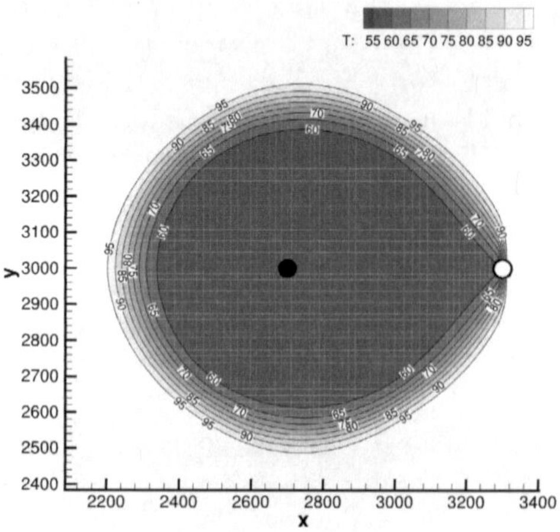

Figure 7. Temperature in aquifer for 20 years of exploitation. Pressure difference is 1000 kPa.

References

[1] Kreith F and Goswami D Y 2008 *Heat Pumps, Energy Management and Conservation Handbook* (Boca Raton: CRC Press)
[2] Polubarinova-Cochina P Y 1977 *Theory of motion of ground water* (Moscow: Nauka)
[3] Vaganova N A 2003 *Proc. Steklov Institute of Mathematics* **S2** S182–93
[4] Vaganova N A and Filimonov M Y 2015 *Lect. Notes Comput. Sci.* **9045** 393–9
[5] Filimonov M Y and Vaganova N A 2014 *Proc. Institute of Mechanics UrB RAS "Problems of Mechanics and Materials Science"* (Izhevsk: Institute of Mechanics UrB RAS) pp 219–22
[6] Chorin A J 1967 *Journal of Computational Physics* **2**(1) 12–26
[7] Samarsky A A and Vabishchevich P N 1995 *Computational Heat Transfer, Volume 2, The Finite Difference Methodology* (Chichester: Wiley)
[8] Bashurov V V, Vaganova N A and Filimonov M Y 2011 *Computational technologies* **16**(4) 3–18
[9] Vaganova N 2014 *AIP Conference Proceedings* **1631**(1) 37–41

AUTHOR INDEX

AUTHOR INDEX